THE PICTURE OF THE TAOIST GENII PRINTED ON THE COVER of this book is part of a painted temple scroll, recent but traditional, given to Mr Brian Harland in Szechuan province (1946). Concerning these four divinities, of respectable rank in the Taoist bureaucracy, the following particulars have been handed down. The title of the first of the four signifies 'Heavenly Prince', that of the other three 'Mysterious Commander'.

At the top, on the left, is Liu *Thien Chün*, Comptroller-General of Crops and Weather. Before his deification (so it was said) he was a rain-making magician and weather forecaster named Liu Chün, born in the Chin dynasty about +340. Among his attributes may be seen the sun and moon, and a measuring-rod or carpenter's square. The two great luminaries imply the making of the calendar, so important for a primarily agricultural society, the efforts, ever renewed, to reconcile celestial periodicities. The carpenter's square is no ordinary tool, but the gnomon for measuring the lengths of the sun's solstitial shadows. The Comptroller-General also carries a bell because in ancient and medieval times there was thought to be a close connection between calendrical calculations and the arithmetical acoustics of bells and pitch-pipes.

At the top, on the right, is Wên *Yuan Shuai*, Intendant of the Spiritual Officials of the Sacred Mountain, Thai Shan. He was taken to be an incarnation of one of the Hour-Presidents (*Chia Shen*), i.e. tutelary deities of the twelve cyclical characters (see Vol. 4, pt. 2, p. 440). During his earthly pilgrimage his name was Huan Tzu-Yü and he was a scholar and astronomer in the Later Han (b. +142). He is seen holding an armillary ring.

Below, on the left, is Kou *Yuan Shuai*, Assistant Secretary of State in the Ministry of Thunder. He is therefore a late emanation of a very ancient god, Lei Kung. Before he became deified he was Hsin Hsing, a poor woodcutter, but no doubt an incarnation of the spirit of the constellation Kou-Chhen (the Angular Arranger), part of the group of stars which we know as Ursa Minor. He is equipped with hammer and chisel.

Below, on the right, is Pi *Yuan Shuai*, Commander of the Lightning, with his flashing sword, a deity with distinct alchemical and cosmological interests. According to tradition, in his early life he was a countryman whose name was Thien Hua. Together with the colleague on his right, he controlled the Spirits of the Five Directions.

Such is the legendary folklore of common men canonised by popular acclamation. An interesting scroll, of no great artistic merit, destined to decorate a temple wall, to be looked upon by humble people, it symbolises something which this book has to say. Chinese art and literature have been so profuse, Chinese mythological imagery so fertile, that the West has often missed other aspects, perhaps more important, of Chinese civilisation. Here the graduated scale of Liu Chün, at first sight unexpected in this setting, reminds us of the ever-present theme of quantitative measurement in Chinese culture; there were rain-gauges already in the Sung (+12th century) and sliding calipers in the Han (+1st). The armillary ring of Huan Tzu-Yü bears witness that Naburiannu and Hipparchus, al-Naqqāsh and Tycho, had worthy counterparts in China. The tools of Hsin Hsing symbolise that great empirical tradition which informed the work of Chinese artisans and technicians all through the ages.

# SCIENCE AND CIVILISATION IN CHINA

The three first men in the world were a gardener, a ploughman, and a grazier; and if any man object that the second of these was a murtherer, I desire he would consider that, as soon as he was so, he quitted our profession and turned builder. It is for this reason, I suppose, that Ecclesiasticus forbids us to hate husbandry; 'because,' says he, 'the Most High has exalted it'. We were all born to this art, and taught by Nature to nourish our bodies by the same earth out of which they were made, and to which they must return, and pay at last for their sustenance. Behold the original and primitive nobility of all these great persons, who are too proud now, not only to till the ground, but almost to tread upon it! We may talk what we please of lilies and lions rampant, and spread eagles in fields *d'or* and *d'argent*, but if heraldry were to be guided by reason, a plough in a field arable would be the most noble and ancient arms.

Abraham Cowley (1618–67),
*The Antiquity of Agriculture*

Therefore the ancient kings made people turn back to agriculture and war. For this reason it is said: 'Where a hundred men farm and one is idle, the state will attain supremacy; where ten men farm and one is idle, the state will be strong; where half farms and half is idle, the state will be in peril.' That is why those, who govern the country well, wish the people to take to agriculture ...

A sage knows what is essential in administering a country, and so he induces the people to devote their attention to agriculture. If their attention is devoted to agriculture, then they will be simple, and being simple, they may be made correct. Being perplexed it will be easy to direct them, being trustworthy they may be used for defence and warfare. Being single-minded, their careers may be made dependent on rewards and penalties; being single-minded, they may be used abroad.

Indeed, the people will love their rulers and obey his commandments even to death, if they are engaged in farming, morning and evening; but they will be of no use, if they see that glib-tongued, itinerant scholars succeed in being honoured in serving the prince, that merchants succeed in enriching their families and that artisans have plenty to live upon. If the people see both the comfort and the advantage of these three walks of life, then they will indubitably shun agriculture; shunning agriculture, they will care little for their homes; caring little for their homes, they will certainly not fight and defend these for the ruler's sake.

*Shang Chün Shu*, tr. Duyvendak (3), p. 191

The plough was invented by the ancient sages, and ever since the first use of cereal grains the people's livelihood has depended upon it. No king or ruler of a state could dispense with it. To eat one's fill and live in peace without having to struggle for survival, is commended as proper conduct, and is what distinguishes us from what Yang Tzu calls 'living like brutes'.

*Lei Ssu Ching*, tr. auct.

李約瑟著

# 中國科學技術史

冀朝鼎

JOSEPH NEEDHAM

# SCIENCE AND CIVILISATION IN CHINA

VOLUME 6

## BIOLOGY AND BIOLOGICAL TECHNOLOGY

PART II: AGRICULTURE

BY

FRANCESCA BRAY

RESEARCH FELLOW

EAST ASIAN HISTORY OF SCIENCE LIBRARY

*The right of the University of Cambridge to print and sell all manner of books was granted by Henry VIII in 1534. The University has printed and published continuously since 1584.*

CAMBRIDGE UNIVERSITY PRESS

CAMBRIDGE

LONDON NEW YORK NEW ROCHELLE

MELBOURNE SYDNEY

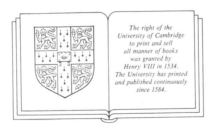

Published by the Press Syndicate of the University of Cambridge
The Pitt Building, Trumpington Street, Cambridge CB2 1RP
32 East 57th Street, New York, NY 10022, USA
296 Beaconsfield Parade, Middle Park, Melbourne 3206, Australia



First published 1984

Printed in Great Britain at the University Press, Cambridge

Library of Congress catalogue card number: 54-4723

*British Library Cataloguing in Publication Data*

Needham, Joseph
Science and civilisation in China.
Vol. 6
Pt. 2: Agriculture
1. Science—China—History 2. Technology—China—History
I. Title II. Bray, Francesca
509′.51 Q127.C5
ISBN 0 521 25076 5

AO

This book is dedicated to the memories of

SHIH SHENG-HAN
of the Northwestern Agricultural College, Wukung,

WANG YÜ-HU
of the Peking Agricultural College,

and

AMANO MOTONOSUKE
of Osaka City University,

without whose pioneering works on the history of agriculture in China this volume would not have been possible.

# CONTENTS

# LIST OF ILLUSTRATIONS

# LIST OF TABLES

# LIST OF ABBREVIATIONS

The following abbreviations are used in the text and footnotes. For abbreviations used for journals and similar publications in the bibliographies, see pp. 618 ff.

| | |
|---|---|
| *CFNS* | *Chhen Fu Nung Shu* |
| *C/HS* | *Chhien Han Shu* |
| *CLPT* | *Cheng Lei Pen Tshao* |
| *CMYS* | *Chhi Min Yao Shu* |
| *FSCS* | *Fan Sheng-Chih Shu* |
| *KCT* | *Keng Chih Thu* |
| *LSCC* | *Lü Shih Chhun Chhiu* |
| *NCCS* | *Nung Cheng Chhüan Shu* |
| *NSCY* | *Nung Sang Chi Yao* |
| *PTKM* | *Pen Tshao Kang Mu* |
| *SC* | *Shih Chi* |
| *SMYL* | *Ssu Min Yüeh Ling* |
| *SSTK* | *Shou Shih Thung Khao* |
| *TKKW* | *Thien Kung Khai Wu* |
| *TSCC* | *Thu Shu Chi Chheng* |
| *WCNS* | *Wang Chen Nung Shu* |
| *YHSF* | *Yü Han Shan Fang Chi I Shu* |

# AUTHOR'S NOTE

This volume differs from its predecessors in that agriculture is not a science but a technology. The mileposts in its history have been, not the deduction of new laws, but the evolution of new implements and the discovery of new crops. The great agronomists of the past, Columella or Chia Ssu-Hsieh, Hsü Kuang-Chhi or Gervase Markham, were concerned not to formulate theories but to transmit a received body of accepted wisdom. General rules had to be so hedged about with qualifications that they were frequently reduced to mere commonplace. Even now, though true sciences such as plant nutrition and genetics have come to play a significant role in dictating advanced farming practice, laboratory results usually prove poor indicators of results in the field. Man proposes, Nature disposes, and theory is still very far removed from practice.

The history of agriculture, then, is not a voyage of intellectual discovery like the history of the sciences. And yet it is no mere catalogue of arbitrary facts. Agriculture is *par excellence* the technological system that mediates between nature and society. On the one hand natural conditions influence the form of agricultural technology adopted and thus affect a society's economic and political relations; conversely social and political constraints will affect patterns of development in agricultural systems. There is thus an internal logic in the history of agricultural development. The interplay between the exigencies of the natural environment and those of state and society lend the study of agricultural history a potent fascination.

Sometimes, for example in the case of classical Rome or early modern Europe, it is possible to reconstruct the agriculture of the period in considerable detail and to evaluate it in its social context. In many other cases, for example the 'Dark Ages' of medieval Europe, or pre-colonial Southeast Asia, our ignorance is almost total. China provides us with an unusually broad historical perspective: the tradition of agronomic writings and economic records stretches back unbroken almost to the earliest historical period, a remarkable opportunity to probe the relations between social and technological change. But a systematic treatment of these problems would be a Herculean task. Here I have restricted my account chiefly to the relations between agricultural systems and natural environment, with only occasional forays into social and economic arguments which will be developed at more length in Volume VII. China is, however, a particularly intriguing subject for theories on the relations between man and nature, superstructure and infrastructure. It possesses, broadly speaking, two highly contrasted natural environments within a single, vast political and cultural unit: the continental zone of the Northern plains and the sub-tropical zone south of the Yangtze. Agricultural technology in the two zones followed quite different patterns of development, and although in the early periods of Chinese history the

North was both economically and politically dominant, by medieval times the balance had shifted to the South. This happy contraposition, together with a consideration of the stimulating role of elements of Chinese technology in Europe's 'Agricultural Revolution', lured me out on to the thin ice of social theory, to hypothesise a fundamental contrast in historical development between systems of dry-grain and wet-rice agriculture, characterised by concomitant trends in the development of the relations of production in general.

These theoretical explorations are confined to the concluding section of this volume, which is otherwise appropriately down to earth. It begins with a brief account of the ecological background to China's agricultural history, together with a discussion of the probable origins of agriculture in East Asia. Since they differ in many respects from those used by Western historians, I go on to consider the nature and scope of the Chinese sources, and the type of agricultural history which they enable us to write. The main body of the work, its centrepiece, is a technological history, divided by topic and arranged along the broad lines of the great Chinese agricultural treatises. It covers three main subjects: field systems, tools and techniques, and crop plants. A separate section on water control and irrigation would have been indispensable, had not many of the technical essentials already been treated in Volume IV, Part 3; further details are given as appropriate in the sections on field systems and rice. Animal husbandry, always an integral part of Western farming, played a very minor role in the Chinese tradition and we have accordingly omitted it here; it will be treated in a later volume. To the Chinese agronomists farming and cloth-making were inseparable occupations—Adam delved and Eve span, and taxes in kind were levied on both. Thus agriculture and sericulture (or textile production in general) both found their place in the traditional Chinese agricultural treatises. The connection is not so obvious to Western eyes, and so here textiles and textile production are treated quite separately from agriculture, in Volume V, Part 1; this volume contains only a brief description of the cultivation of textile plants. Furthermore, crops such as sugar and tea, the production and processing of which played a crucial role in the rural economy of many parts of China, have also been omitted, to be treated in detail in the volume on Agricultural Arts. The substance of this volume is devoted to the production of those crops without which no Chinese could survive: above all cereals, supplemented by legumes, oil crops, tubers, fibre crops, vegetables and fruit. These were the crops produced in every region, the staples of subsistence in times of economic hardship and self-sufficiency, in times of prosperity the basis from which a flourishing commercial agriculture could be developed. I have insisted the more upon these staple crops because, I would argue, even where commercial crops played a significant part in the rural economy, it was always the system by which the main cereal crop was cultivated that determined relations of production and thus patterns of commercialisation.

Many people have contributed to the substance of this volume. My thanks are due first and foremost to the East Asian History of Science Trust, which financed the research; I am also indebted to the British Academy and the Royal Society, which jointly funded a year of field-work in Malaysia in 1976–77, and to Academia Sinica, the Universities' China Committee and the British Council, which contributed to the expenses of a study tour of China and Japan in summer 1980. My work in Malaysia was facilitated by Cik Rohaini Zakaria of the Kemubu Agricultural Development Authority, and by the villagers of Bunut Susu who instructed me most patiently in the art of rice farming. In China and Japan I met with unfailing help and kindness. Particularly fruitful were discussions with Liang Chia-Mien[1] and his colleagues at Huanan Agricultural College, Canton; Ma Yao[2], Wang Ning-Sheng[3] and the staff at the National Minorities Institute, Kunming; Chin Kung-Wang[4] of the Szechwan Agricultural Machinery Laboratory; Ma Tsung-Shen[5] and his colleagues in the History Department, Northwestern Agricultural College, Shansi; Wang Yü-Hu[6], Fan Chhu-Yü[7], Yang Chih-Min[8], Tung Khai-Chhen[9], and the other agricultural historians of the North China Agricultural College and Academia Sinica, Peking; the Agricultural Heritage Group at the Nanking Institute of Agricultural Science and Li Chhang-Nien[10] of Kiangsu Agricultural University; Hu Tao-Ching[11] of Ku-Chi Press, Shanghai; Yu Hsiu-Ling[12] of Chekiang Agricultural University; Amano Motonosuke[13] in Osaka; Nishiyama Buichi[14] in Tokyo; Iinuma Jirō[15] of Kyoto University Humanities Department; Tamaki Akira[16] of the Economics Department of Senshu University; and Katō Yuzo[17] of the Economics Department of Yokohama University.

Those who have kindly read and commented on sections of this volume in draft include:

Amano Motonosuke (Osaka)
Gregory Blue (Cambridge)
Derk Bodde (Philadelphia)
Timothy Brook (Harvard)
T. T. Chang (Los Baños)
Chin Kung-Wang (Chengtu)
G. Dalton (Chicago)
G. E. Fussell (Sudbury)
Clive Gates (Canberra)
Peter Gathercole (Cambridge)
Peter Golas (Denver)
Keith Hart (Montreal)
A. Haudricourt (Paris)
Polly Hill (Cambridge)
Joseph Hutchinson (Cambridge)

[1] 梁家勉 [2] 馬曜 [3] 汪寧生 [4] 金公望 [5] 馬宗申
[6] 王毓瑚 [7] 范楚玉 [8] 楊直民 [9] 董愷忱 [10] 李長年
[11] 胡道靜 [12] 游修齡 [13] 天野元之助 [14] 西山武一 [15] 飯沼二郎
[16] 玉城哲 [17] 加藤祐三

Bill Jenner (Leeds)
E. M. Jope (Belfast)
Dieter Kuhn (Heidelberg)
David Lehmann (Cambridge)
Grith Lerche (Copenhagen)
Georges Métailié (Paris)
Nakaoka Tetsurō (Osaka)
Joseph Needham (Cambridge)
A. F. Robertson (Cambridge)
François Sigaut (Paris)
Nathan Sivin (Philadelphia)
Axel Steensberg (Copenhagen)
Thomas Thilo (East Berlin)
Donald Wagner (Copenhagen)
R. O. Whyte (Hong Kong)

Special thanks are due to Dr C. T. Gates, formerly of CSIRO, Canberra, originally a collaborator on this volume. His insights into the relations between the triad of nature, agriculture and society have been invaluable. Unfortunately, the problems of collaboration at such a distance eventually proved insurmountable, and after much deliberation it was decided that the volume in the *Science and Civilisation in China* series should concentrate on the more technological aspects of Chinese agricultural development, to the exclusion of broader considerations. Dr Gates' work on climatological cycles and their influence on the course of Chinese agricultural and political history is at present being prepared for publication in Australia.

Philippa Hawking provided invaluable help with Japanese translations, in particular with Amano Motonosuke's intricate prose, and much other assistance. Diana Brodie typed the manuscript in its final form, and Liang Lien-Chu wrote in most of the Chinese characters. The index was compiled with speed and efficiency by Christine Outhwaite. Heartfelt thanks are also due to Kusamitsu Toshio, for his last-minute help in correcting errors in the Japanese, and to the staff of the Cambridge University Press for their patience and efficiency.

I should like to thank all those I have named for their kind assistance. My deepest gratitude goes to Joseph Needham, a constant source of encouragement and inspiration, and to my husband A. F. Robertson, without whose stimulating criticism and unfailing moral support the work would never have been done.

F. B.

*East Asian History of Science Library*
*Cambridge*
*March 1982*

# 41. AGRICULTURE

## (*a*) INTRODUCTION

China is, one might say, the agrarian state *par excellence*. The fundamental occupation, the root *pen*[1] and the basis of the nation's wealth and well-being, was agriculture—as all Chinese philosophers and political economists from Confucius on agreed. The 'spirits of the land and grain' (*she chi*[2]) were symbols of national survival, to whom supreme loyalty was due: 'When a Prince endangers the altars of the spirits of the land and grain, he is changed, and another is appointed in his place.'[a] The royal house of Chou claimed their descent from the agricultural deity Hou Chi[3], Prince Millet. Chinese emperors sacrificed at the altar of the spirit of the soil each spring and autumn, and in the spring it was also their duty to drive out to the royal fields near the capital and ceremonially plough a furrow, after which each of the chief ministers would in his turn put his hand to the plough (Fig. 1).[b]

The ritual and symbolic role of agriculture in Chinese political philosophy reflected its true economic importance. The first land tax was supposedly levied by Duke Hsüan of Lu[4] in −594;[c] thereafter and throughout the imperial era the Chinese state drew the chief part of its revenues from the land tax (*shui*[5] or *tsu*[6]), a tax in kind levied directly from the peasants on the produce of their land.

The Chinese have farmed for several thousand years, ever since the neolithic villages of the Wei River valley and the Yangtze delta started growing millet and rice in about −5000. Thus not only do we have several millennia of Chinese farming to consider, we must also bear in mind that the area of China is comparable to that of Europe and stretches from the Mongolian and Manchurian borderlands on the latitude of Paris to the tropical island of Hainan. Agricultural techniques and traditions naturally vary enormously from region to region and from century to century. There are, however, several features characteristic of Chinese agriculture as a whole which differ fundamentally from agricultural traditions in the West. Since one naturally turns to one's own history as a yardstick, it is helpful to define the basic differences between East and West at the outset, justifying what may at first strike some readers of this volume as an over-emphasis on some topics at the expense of others, by identifying the important

[a] *Lun Yü*, *Thai-po*, Book 8, VIII/9; Legge (2), p. 211.

[b] This practice was officially initiated by Emperor Wen of Han[7] (r. −179 to −156), but its roots certainly go back much further. Cf. Keightley (3). A ceremony of royal ploughing was also practised in ancient Ceylon; Paranavitana (1). The European equivalent is, of course, the blessing of the plough by Christian priests; the more pagan origins of this ceremony survive in the festivities of Plough Monday.

[c] *HS* 24A/6*a*; Swann (1), p. 136.

1 本　2 社稷　3 后稷　4 魯宣公　5 稅
6 租　7 漢文帝

Fig. 1. Ceremonial ploughing by the emperor and his chief ministers; *WCNS* 11/2a.

differences in context that have coloured patterns of development in Chinese agriculture.

## (1) General Characteristics of Chinese Agriculture

The modern state of China includes not only the inland mountains and the great floodplains of the Yellow River, the Huai, the Yangtze and the Pearl River, but also the steppes of Mongolia, deserts and oases in Central Asia, Tibet, and the vast rolling plains of Manchuria. But even today the main *agricultural* area, that is to say the area where permanent fields are cultivated, corresponds to the traditional area of Chinese culture, from Kansu and Szechwan in the West to the China Sea, from Peking and Liaoning in the North to the tropical island of Hainan. The boundaries of this area have remained more or less unchanged since the Chhin unification of China, and within it Chinese farmers have lived in their villages, growing millet, wheat or rice in the fields that their families had probably tilled for centuries before them. This area corresponds very closely to the natural boundaries of arable farming—to the north and west it is bounded by arid steppes suitable for pastoralism, to the west and southwest by high, rugged mountain ranges. Only to the south was there no natural obstacle to the extension of agriculture, so that there it was political rather than environmental factors which determined the frontier.

One crucial difference between the farming traditions of China and Europe lies in the importance accorded to livestock. In Europe grain production has always been integrated with animal husbandry in a system of mixed farming. A large proportion of farmland consisted of permanent pastures, downlands, common grazing or meadows, and until the introduction of modern rotations in the 17th century and later, it was usual to pasture animals in the arable fields while they lay fallow, which was one year in every two or three. Folding livestock on the fallow fields was one of the very few ways of maintaining soil fertility in traditional Europe. Oxen and horses were widely used for draught, and the traditional teams, in Northern Europe at any rate, consisted of four, six, or as many as twelve beasts. Most livestock, however, were raised for productive purposes, providing wool, hides, meat and dairy products, and these last formed an important part of the diet. While meat was often the privilege of the rich, milk and especially cheese were a very important source of protein among the poorer classes in Europe.

In China farming has concentrated on grain production throughout the historical period, and may even have done so in neolithic times. In the 1930s only 6 per cent of all farmers owned any pasture, and this constituted a mere 1 per cent of the total farmland.[a] It has sometimes been asserted that such low figures are a comparatively recent phenomenon, the result of steady population growth and pressure on land over the centuries, and it certainly is true that grain production is a far more efficient way to feed a large population than animal husbandry.[b] But

[a] J. L. Buck (2), p. 174.

[b] See Perkins (1), App. F.

a cursory glance at the classical sources shows how small a role animal husbandry has played in the national economy throughout Chinese history: the chapters on stock-raising in the agricultural treatises are short and usually tucked away at the back of the volume,[a] while historical accounts of land reforms or agricultural settlements mention very few provisions for pasture.[b] Animals certainly were kept by Chinese farmers, but in far smaller numbers than in Europe, and they were (and still are) usually grazed on waste land, on uncultivated hillsides or river-banks. Occasionally fodder crops such as lucerne or alfalfa were grown, but usually it fell to children to comb the countryside in search of whatever fodder they could find. Buffaloes, oxen or mules were used for ploughing, but the Chinese plough was much lighter than its European counterpart and could easily be managed by a single animal.[c] Although carts were common in the North, in the mountainous South roads were few and most transport was by boat or by carrying-pole. Apart from draught animals the Chinese have traditionally kept pigs and fowls, which thrive on a diet of household scraps and waste; pigs were particularly valued as a source of manure, a 'one-man fertiliser factory', as Chairman Mao said. A few sheep were kept for their wool, but otherwise few productive animals were reared. Meat plays a very restricted role in the traditional diet of the Chinese, only pork and poultry being really popular, and milk products are totally absent.

The Chinese diet is largely vegetarian.[d] On average well over 90 per cent of the calories are provided by cereal grains,[e] and most proteins come either from cereals, legumes, or protein-rich legume products such as bean-curd and soya sauce,[f] often supplemented in South China by fish, fresh, dried or processed into sauces.[g] Although many vegetarian diets are deficient in certain proteins, notably lysine, in China the additional presence of soya products or fish generally compensates to provide a nutritionally adequate diet.[h]

The Chinese word for 'food' or 'meal' is *fan*[2], which denotes a cereal food such as boiled rice or millet porridge, and *fan* is the essential part of any meal: only *fan* will satisfy hunger. To enliven the bland *fan* the Chinese add *tshai*[3]. This

[a] See *CMYS*, *WCNS*, *NCCS*, etc. If such works truly are a reflection of common practice, then clearly the silkworm was far and away China's most important animal.

[b] Exceptionally the 'equal land allotment regulations' (*chün thien chih*[1]) of the Northern Dynasties included allowances of pasture land for up to four head of livestock (Wan Kuo-Ting (*10*); Han Kuo-Pan (*1*)), but significantly these regimes were of nomadic origin. The −2nd century *Kuan Tzu*, *Pa Kuan Phien*, states that in the average peasant family of the time, 60 % of the total production came from the cultivation of cereals, 20 % from fruit and vegetables, and 20 % from fodder and animal produce together.

[c] A pair of light oxen or donkeys was fairly common in the North, but in the southern rice-fields the water buffaloes always work singly. On the ratio of draught animals to land see Perkins (1) p. 306; Golas (1).

[d] Shinoda Osamu (*6*), (*7*); K. C. Chang (3) provide well-documented historical accounts of the nature and evolution of Chinese dietary patterns.

[e] J. L. Buck (2), p. 414.

[f] *Ibid.* p. 418.

[g] Anderson & Anderson (1).

[h] Buck (2) p. 418. Unprocessed soybeans are not particularly nutritious, for although they contain a high proportion of proteins these remain unavailable unless the beans are hydrolysed.

[1] 均田制 [2] 飯 [3] 菜

character originally meant 'vegetables', and is still used in that sense in combination with other characters: *pai tshai*[1], 'white vegetable', is the Chinese cabbage, and *shu tshai*[2] is now the common term for vegetables in general. *Tshai* used alone soon came to mean any side-dish eaten with *fan*. Today most *tshai* eaten in rural areas consist chiefly of vegetables: fresh greens, pickled cabbage, sweet potatoes or bean-curd, given piquancy with soy sauce, ginger, chili or vinegar; in the Sung dynasty vinegar and soya sauce were considered to be two of the necessities of life essential even to the humblest peasant family.[a] Peasants would grow many vegetables on their own plots, or buy them at rural markets, supplementing them with wild herbs or plants gathered in the countryside.[b] Numerous elaborate meat dishes are mentioned in ancient Chinese texts such as the *Chou Li* and the *Chhu Tzhu*, and some, from the Han dynasty, have even been found desiccated but otherwise intact in royal tombs,[c] but it is unlikely that the common people ate much meat even then—Chou dynasty references to farming and the peasant life all speak of growing cereals, some mention vegetables and beans, but of meat we hear only in descriptions of noble life.

As for milk products, they are so seldom consumed that very few Chinese retain the capacity to digest lactose after they are weaned. This is not a genetic deficiency, for if they are brought up in a country where dairy products are habitually consumed Chinese continue to synthesise the lactase necessary to digest milk throughout their adult life. Indeed, even in China itself, milk products enjoyed a certain shortlived vogue during the period of the Northern Dynasties and early Thang, when upper-class Chinese intermarried extensively with the new rulers, invaders from the steppes with a decided taste for kumiss, cheese, and other products equally distasteful to most true Chinese.[d] The herdsmen who provided these dairy products were probably not Chinese but of nomadic origin, as is the case in China today where Tibetans, Mongols, Uighurs, Kazaks and other nomadic races living in the grasslands of Mongolia and Sinkiang raise the vast herds (mainly sheep and goats, then cattle, horses and camels) that provide the inner provinces of China with meat, wool, and even draught animals (Fig. 2).[e]

The divide between traditional Chinese arable farming and the nomadic pastoralism of the borderlands is very clearly drawn and corresponds, broadly speaking, to the 375 mm. isohyet. The traditional borders of the Chinese empire stopped where the grass began, for even the richest pastures of the steppes do not make good farming land,[f] and the machinery of the Chinese state was designed to

[a] Shinoda (*6*).

[b] Many of the forty-six vegetables mentioned in the *Shih Ching* probably grew wild, for example sow-thistle, cat-tail, smartweeds, wormwood, ferns and bamboo shoots; K. C. Chang (4), p. 28.

[c] Ying-Shih Yü (1).

[d] See Schafer (25), and also *CMYS*, ch. 57.

[e] Tregear (2) pp. 130–139.

[f] According to Mongol tradition, 'the earth is holy and their ancient tribal laws forbid the ploughing of more than the necessary minimum of land, or of any land for two years in succession'; Lattimore (10), p. 202.

1 白菜 2 蔬菜

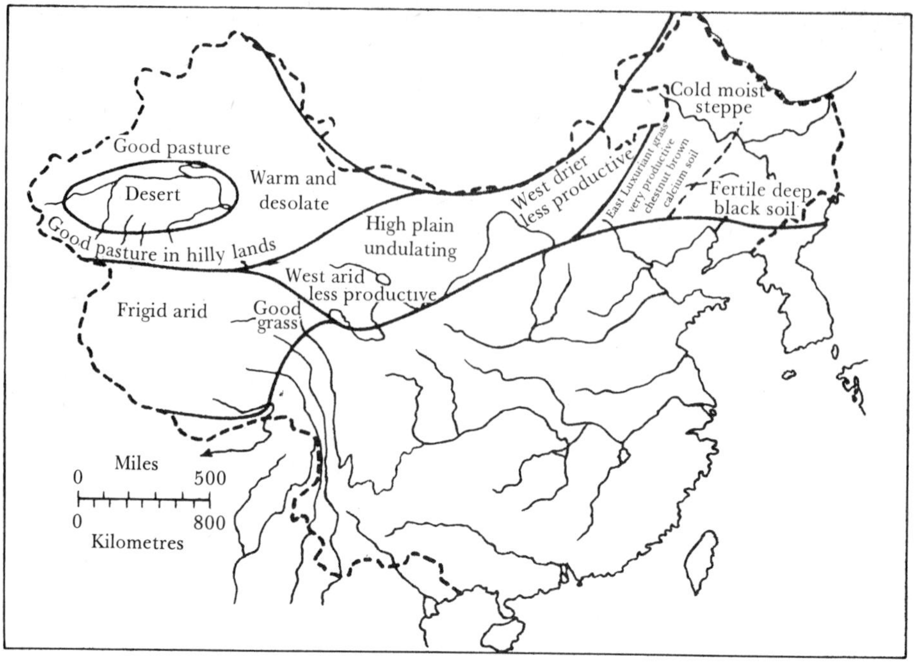


Fig. 2. Map of the northern grasslands; Tregear (2), p. 132.

cope with settled farmers, not with nomadic pastoralists who were difficult to locate, to count and to tax, and whose political loyalties Chinese statesmen presumed (often correctly) to shift with the prairie winds.[a] Any attempt by Chinese settlers to extend agriculture beyond the clearly defined Chinese frontiers was likely to end in disaster, for the rewards of frontier farming were slight and precarious, and the settlers might well be tempted to adopt the better life of the nomads, in which case they would be lost to China.[b] This was very clear in the minds of Chinese statesmen, who felt it necessary to maintain a strict political and physical divide between Chinese farmers and foreign herdsmen—a policy which did not inevitably work to the advantage of the Chinese state:[c]

[In the −3rd century Chhin] created the first unified Chinese Empire. It then put together the sectional walls of the preceding kingdoms and established the Great Wall frontier. It is important to make it clear that this frontier was the voluntarily demarcated limit of the convenient expansion of the Chinese Empire; in other words it was not

[a] Governments have always disapproved of nomads, especially those whose traditional pasture-lands overlap political boundaries, and usually try to find some pretext to make them settle down so that they may be counted, taxed and controlled like everyone else. Lattimore (11) describes Chinese, Russian and Japanese assaults on the Mongol way of life, which surprisingly culminated in the founding of an independent Mongol republic. But few nomadic peoples are so fortunate. In our modern world of sovereign states and sophisticated communications networks it has become increasingly difficult for nomads to retain their independence. The Beduin of Saudi Arabia have been richly rewarded for sacrificing the customs of their ancestors, but most, like the Kirghiz of the High Pamirs, face economic if not physical extinction.

[b] Lattimore (12), p. 483.

[c] Lattimore (12), p. 481.

necessitated by the aggression of the nomads against China. That aggression came later, as a consequence of the demarcation of the frontier by the Chinese, and was largely due to the inequality of the terms of trade: what China wanted of the surplus produced by the nomads (livestock, hides, wool, furs) did not equal in value what the 'barbarians' wanted from China in the way of grain, textiles and ironware. This interpretation is confirmed by the fact that the individual kingdoms of North China, before they were united, dealt with separate tribes out in the steppe, while the creation of a unified empire in China immediately called into being the unified tribal league or nomadic empire of the Hsiungnu.

Unlike in European farming systems, then, animal husbandry played a very small role in China. Cereals were the basis of the Chinese diet, and occupied most of the arable area. Crop rotations did traditionally include soil-enriching crops such as legumes or green manures, and oil seeds and fibre crops such as brassica, sesame and hemp were also important, but most fields were planted at least once a year with cereals. Buck's survey shows that in the 1930s an average of 70 per cent of the total arable area was planted with cereal crops.[a] All the Chinese agricultural treatises, from the Han dynasty on, devote their longest sections to cereal cultivation. Although by modern Western standards traditional Chinese crop yields were not especially high,[b] compared with the yields obtained in Europe before the introduction of modern rotations and chemical fertilisers they were really impressive, commonly giving returns of twenty or thirty times the amount of seed-grain sown, as opposed to the returns of three or four to one usual in medieval Europe (see Crop Systems, p. 423). An important contributing factor was the economy of Chinese sowing techniques such as drill-sowing in the north and transplanting in the south (see Sowing, p. 286). Since hardly any land or grain was required for feeding livestock, a given area of land could support a far greater density of population than its equivalent in Europe, and as the Chinese population grew, farmers exerted their ingenuity to improve their methods and intensify production. It seems that in the crowded Metropolitan Provinces fallowing of land to restore its fertility was regarded as a last resort as early as the Han (see p. 429), whereas in Europe it formed an essential part of all crop rotations until the 17th or 18th century. By that time farmers in the Canton area habitually grew three crops a year, two of rice and one of oilseed, turmeric, indigo or some other commercial crop,[c] and many northern farmers commonly grew three crops in two years. By the 1930s the national cropping index[d] had reached 1.4.[e] The density of population on the land encouraged the spread of labour-intensive techniques such as transplanting and meticulous weeding, which again contributed to raising land productivity. Chinese farmers tended their fields with a care and attention to detail unrivalled in Europe except in some Mediterranean regions, and as in the

[a] (2), p. 209.
[b] *Ibid.* p. 223.
[c] *Kuang Tung Hsin Yü*, ch. 14, p. 371.
[d] The total area of crops sown in one year divided by the total area of arable land.
[e] Myers (2), p. 307.

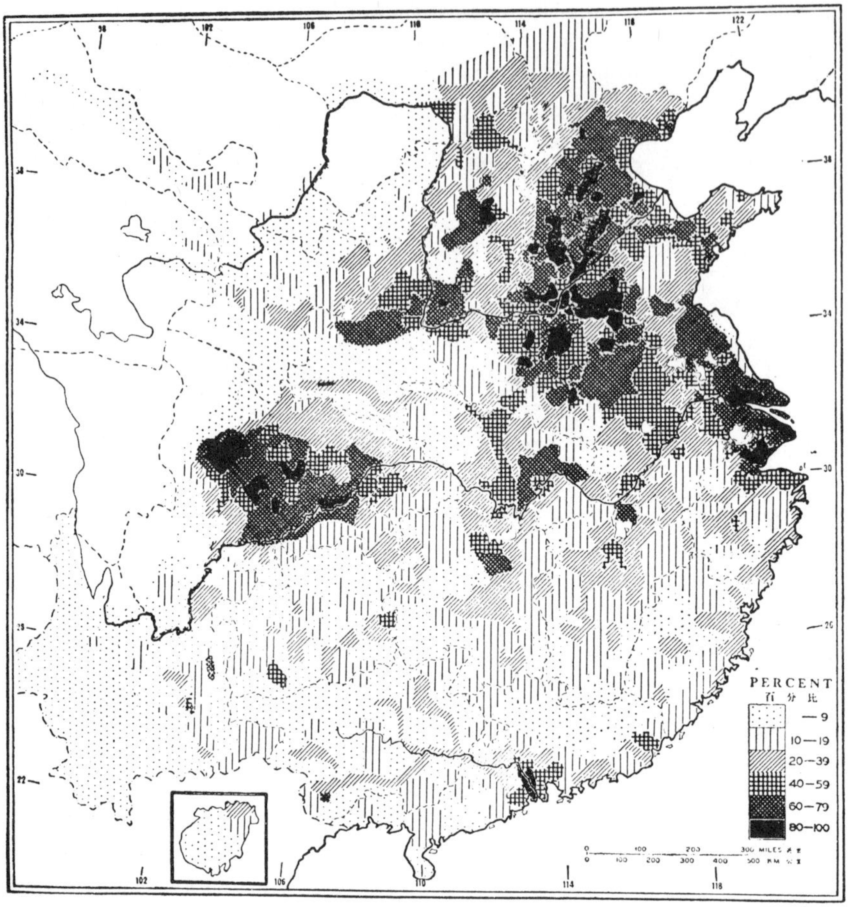


Fig. 3. Map showing the proportion of China's land suitable for cultivation; after Buck (2), p. 168.

Mediterranean regions, farms were very small, especially in the south where rice was grown. Although the ownership of land might be concentrated in a few hands, landholding in China was far more evenly distributed, and farms worked by a single family were the norm for most of China's history (see p. 604).

Despite the comparatively high productivity of Chinese land, as the population grew and grew, new land had to be brought under cultivation. China is a land of many steep hills and rugged mountains, and the proportion of its surface suitable for cultivation is not high (Fig. 3). In many areas of high population density the supply of obviously productive land was soon exhausted, and the landless farmers had two possibilities: to migrate to an under-populated area where land was available, or to find some means of extracting a crop locally from marginal land.

The key grain producing areas of China in the early dynasties were the Metropolitan Provinces (the Wei and Yellow River valleys), and the broad plains along the lower course of the Yellow River. Here were grown the millet and wheat that stocked the Imperial Granaries and supported the Imperial Household, the army, and the bureaucracy. But these regions were overpopulated and many landless farmers migrated south, swelling the ranks of the Yangtze rice-growers. By the Thang the Yangtze provinces rivalled the north in their contribution to the state granaries, and by the Sung the economic centre had unquestionably shifted to the Yangtze. The Delta provinces were already becoming overcrowded, and there was a steady flow of migrants inland to Hupei and Hunan, and south to Kwangtung. This geographical expansion continued uninterrupted until about 1800, by which time even such remote provinces as Yunnan had received an influx of Chinese settlers.

Some land-hungry farmers, unwilling to abandon their ancestral villages, preferred the often bitter toil of reclaiming marginal land, draining and dyking marshy land for rice-fields, hacking narrow terraces from mountain slopes, or patiently flooding and draining fields along the sea-shore until the salt was leached away and rice could be planted (see Field Systems, p. 113). This relentless process of reclamation has not always been wise, particularly among the steep hills of the southern provinces where the result has been to denude the bedrock, leaving nothing but a barren waste. But the culprits rarely had any choice: with so little land and so many people, the alternative to reckless clearing was often starvation. The only parts of Europe where it was necessary to reclaim land with anything approaching similar vigour were the mountainous shores of the Mediterranean, where hillsides were terraced in classical times for vines, olives and wheat, and the densely populated Netherlands, which reclaimed much of their best farmland from the sea. Despite the population growth of the last few centuries, Europe can still afford to leave unspoiled large areas of downland and of forest which in China would long ago have been stripped of their trees and sown with grain.[a]

After noting some of the salient differences between agriculture in China and the West, let us now turn to the variations in farming systems between China's different regions.

## (2) Agricultural Regions

The most thorough classification of China's different agricultural regions was that proposed by the American agronomist J. L. Buck (2) in 1937, on the basis of an intensive survey carried out in the early 1930s. Buck took as his criteria climatic, topographical and pedological conditions, natural flora, cultivated plants and domesticated livestock, local distribution of land use, average crop yields, rural

[a] Imperial China could not supplement its own agricultural production with imports from abroad, as 19th- and 20th-century Europe has done, for it was only comparatively recently that countries such as Thailand and Burma began to produce marketable surpluses of rice; Cheng Siok-Hwa (1); Tanabe (1). As soon as the Thais began selling rice, the Chinese bought all that they could (Tanabe (1) p. 41), but it was a mere drop in the ocean. Europe, on the other hand, was able to provide for a very large proportion of its grain and meat needs by importing food from America and Australia, where European settlers often pioneered advanced farming methods impossible in the more constricted agricultural zones of their old homelands. This meant, of course, that pressure on land in Europe was considerably reduced.

living standards, tenurial relations and so on, on the basis of which data he defined eight principal agricultural areas and two distinct agricultural regions, the northern wheat region and the southern rice region. Buck's division has been accepted in almost every detail by subsequent agronomists and geographers, with the addition of one extra agricultural area in the far Northeast of China, representing the Manchurian provinces that at the time of Buck's survey were under Japanese occupation (Fig. 4).[a]

The precise figures given by Buck for various data in the 1930s are clearly of limited application for a historical study, so we shall not attempt to reproduce them in any detail here. Since general topography and climate have already been discussed at length in Section 4, meteorology in Section 21, geology in Section 23 and pedology in Section 38, we shall confine ourselves here to giving a very brief outline of the more enduring characteristics of the nine agricultural areas, followed by an enumeration of the most important features of the northern and southern agricultural regions and the contrasts between them.[b]

### (i) *The maize-millet-soybean area*

This occupies most of the modern Northeastern provinces of Heilungkiang, Kirin and Liaoning, formerly known as Manchuria. Apart from periods of colonisation in the Han and Ming, this area has only been open to Chinese farmers since the late 19th century. It is, however, of great interest to students of contemporary development in China, since it is the one area where no long-established boundaries, tenurial complexities or customary rights existed to impede the development of a highly productive and uncharacteristically land-extensive farming system. As in Hokkaido in Japan, also colonised very recently, it has been possible to mechanise farming to a degree that would be impractical in any of China's crowded agricultural heartlands. Set down in a vast cornfield on the rolling Manchurian plain, the observer might well be excused for thinking himself in Alberta or the Corn Belt.

The soils of the northeast steppelands are mostly fertile chernozems. The climate is harsh, with long bitter winters when temperatures may fall to −35°C; rivers are frozen from late November to April and only five months are frost-free. But although the growing season is only 120 to 150 days (depending on the latitude), the long summer days of 14 to 16 hours, with temperatures of 15–20°C, compensate for this to some degree. Summer rains are heavy and regular, annual rainfalls averaging 500 mm. of which half falls in July and August, and this regularity usually guarantees good crops.

[a] We shall not consider here the extra areas of Sinkiang, Mongolia and the Tibetan plateau, as they are, for reasons just given, beyond the scope of this study.

[b] Fuller descriptions will be found in Buck (2); Tregear (2); and Cressey (1). Since 1949 communisation and rural industrialisation have brought about many radical changes in the Chinese countryside, but we have taken no account of these in the following descriptions, as they are of little relevance to a historical study which theoretically stops at +1600.

1: Maize-millet-soybean area
2: Spring wheat area
3: Winter wheat-millet area
4: Winter wheat-sorghum area
5: Yangtze rice-wheat area
6: Rice-tea area
7: Szechwan rice area
8: Double-cropping rice area
9: Southwestern rice area

Fig. 4. Map of China's nine agricultural areas, showing the division between the northern, dry grain region and the southern, wet-rice region; after Buck (2), pp. 25 and 27.

Millets are grown extensively in the highlands, while maize has come to replace sorghum in the lower regions, and Manchuria is China's chief producer of soybeans, a crop which may have originated in this area (see p. 512). In the zones adjacent to China proper few animals are kept except for draught purposes (mainly oxen and donkeys), but in the northern zone there are wide pasture lands where horses are still bred today, as they have been for thousands of years.

### (ii) *The spring wheat area*

Buck gave this area its name because it is the only agricultural area of China proper where the climate is too harsh for winter wheat to grow. It is a high-lying area of rugged mountains and badly eroded hills, bordered on the north by the grasslands of Mongolia and the Ordos Desert, which form a natural limit to the extension of settled agriculture. The soils are mainly loessial, fairly rich in nutrients but low in organic content. The chief constraint on productivity is the low rainfall, and agriculture has traditionally been confined to river valleys where irrigation was possible.

The climate is harsh, temperatures varying between −7°C in January and 23°C in July. The frost-free period is five months. The rainfall, which is concentrated in the midsummer months, is not only very low (about 375 mm.) but also very irregular. Droughts are not the only hazard to agricultural production, for violent summer hailstorms or early autumn frosts may also damage crops.

Despite its name, the chief crops in this area are setaria and panicum millets, which tolerate arid conditions better than most other cereals; spring wheat and barley are also important, oats are grown in the far north, and peas, legumes and alfalfa are quite common. Winter cropping and double cropping are more or less non-existent. Livestock are quite numerous: flocks of sheep graze the hilltops, and oxen, donkeys, mules and horses are used for draught and transport. To the north and west, where Mohammedans and pastoralists form a significant proportion of the population, mutton and milk products are consumed.

Agriculture in this area is marginal and fraught with hazards, but Chinese have farmed here since well before the Han dynasty. At times the climate was considerably more favourable, milder and wetter, and at such times Chinese farmers pressed out into marginal areas that eventually reverted to desert, driving out the pioneers and engulfing their settlements.

### (iii) *The winter wheat-millet area*

These are the loess-lands *par excellence* (Fig. 5), where the blanket of wind-blown loess topsoil may reach a thickness of 150 m. Many villages traditionally consisted of cave-dwellings hollowed out of the loess cliffs. The natural fertility of these soils is high, though they are low in organic matter and nitrogen, but again rainfall is a limiting factor in cultivation. The climate is very similar to that of the spring wheat area, but slightly warmer and wetter. In the loess-lands rain is prayed for with fervour but remains something of a necessary evil, for a summer storm can easily wash away half a hillside, taking crops and houses with it. The dangers of erosion can to some extent be averted by judicious contour terracing, a practice which probably dates back in this area to well before the Sung, and by the 20th century over one-third of the cultivated area was terraced. Despite this, over-cultivation and the clearing of hilltops and other land better left under its natural

Fig. 5. Erosion in the Shensi loesslands; S. W. Williams (1), vol. 1, p. 97.

vegetation have reduced many parts of the winter wheat-millet area to a lunar landscape.

The chief crops are millets and winter wheat, sorghum and cotton, which is particularly well suited to the river valleys of the Wei and Fen, where the climate is rather milder. Generally speaking, millets and sorghum are grown on hillsides, wheat in the valleys. Winter cropping and double cropping are both quite common, but it is difficult to maintain soil fertility under such conditions since few livestock are kept: some sheep, but chiefly oxen, mules and donkeys for draught.

This area has been farmed continuously since the early neolithic period, some 7000 years ago, and despite the harsh and unpredictable climate it long remained an area of considerable economic importance, maintaining a surprisingly large population. Nonetheless the severity and frequency of droughts, together with rapid soil erosion, have kept all but a few sheltered areas from achieving prosperity; indeed it seems that farming in this area has become increasingly thankless and precarious over the centuries.

Fig. 6. Hoeing in the broad northern plain; King (1), fig. 115.

### (iv) *The winter wheat-sorghum area*

This is the area of the great Yellow River plains, plus the low hills of the Shantung peninsula. The plains are low-lying and flat (Fig. 6), marked in places by depressions which flood after the summer rains to become lakes or marshes; however, these 'lakelands' usually dry out soon enough for a crop of wheat or sorghum to be sown in the autumn (see p. 451). The rivers in this region bear a heavy load of alluvium from the primary loesslands upstream, and since the plains are so flat the rivers often silt up their channels and flood, or even change their course, with disastrous results.[a] The topsoil provided by these alluvial deposits is generally fertile, though some zones near the coast suffer from salinity so severe that they are completely barren.

[a] The Yellow River has changed its course many times in its history, flooding thousands of square miles and drowning or starving millions of people. Hence its nickname of 'China's Sorrow'; see Vol. 1. As Buck callously remarks: 'Geologically speaking, man has settled these plains thousands of years before they were ready for occupation, and he must therefore take the risks involved' ((2) p. 61).

The climate is warmer and wetter than in the loess-lands. Temperatures vary between just below freezing and 30°C, while the frost-free season is seven months long. Rainfall is again concentrated in the midsummer months, but it averages about 500 mm., significantly higher than in the areas to the west and northwest. Here too, though, it is very unreliable, and farmers often dig wells to provide extra water for their fields, a contributing factor to the problem of salinisation in certain areas.

In recent centuries winter wheat has been the most important cereal crop grown, together with millets, sorghum and, in the last century or so, maize. Sorghum is especially useful as it will withstand either drought or flood; wheat grows well in the low-lying 'lakelands' after the recession of the summer floods. Soybeans and sweet potatoes are common summer crops while cotton, which has been important in the area since Ming times, is planted in the spring. Considerable areas are double-cropped. Very few livestock are kept and those only for draught: a few oxen, donkeys and mules, less than in any other area except the Lower Yangtze. (Pigs are, of course, the exception to this rule, being kept in every region except the northwest in considerable numbers.)

For much of its early history this area was the breadbasket of China, and even though the economic emphasis shifted south in the medieval period it has continued to support a very high density of population, in a state of relative prosperity compared with the northwestern areas. But with the dangers of drought, flood and starvation ever looming, many farmers have preferred to migrate southwards to find a more secure living.

### (v) *The Yangtze rice-wheat area*

This is the floodplain of the Yangtze, an area of flat or gently rolling plains broken by low hills and a few rough mountain ranges. The plains are crisscrossed with rivers and canals which are dyked to prevent flooding, and lakes abound (Fig. 7). The soils vary but are mostly leached pedalfers of medium to low fertility; where rice is grown, however, podzolisation has reduced soil acidity and these soils are fairly productive (see p. 26).

The climate is characterised by warm winters (temperatures are well above freezing in January) and hot, humid summers with July temperatures of 30°C or more. Rainfall is much more evenly distributed than in the north, averaging 1000 mm. a year, and the frost-free period is over nine months long.

By far the most important crop is rice, but it is seldom double-cropped. Instead it is alternated with winter crops of wheat (along the lower Yangtze) or barley (further upstream where its shorter ripening period is appreciated). Green manures are frequently grown to raise the organic content of the severely leached soils, and oilseed and cotton are important commercial crops. Mulberries are grown along most dykes in the Yangtze Delta, where sericulture has been an important industry since the Sung dynasty. Double-cropping is extensive, and in

Fig. 7. Yangtze Delta landscape; *China, Land of Charm and Beauty*, p. 96.

order to maintain the soil's fertility, considerable quantities of river-mud, oilcake and human manure are applied. Few livestock are kept: water buffaloes and oxen are used for draught, and pigs for their meat and as a source of manure. But the fish that abound in the waterways and lakes are an important source of supplementary protein.

This area was for centuries the centre of the Chinese economy, providing large surpluses of rice for the national market. It also developed thriving industries, including for example the production of lacquer-ware and paper goods, though the outstanding product of the lower Yangtze was fine silk cloth. As the population grew, farmers from the Yangtze plains migrated inland or south, taking with them advanced cultivation methods and other techniques, and as these areas were developed the economic predominance of the Yangtze plains was gradually reduced. Nevertheless it remained a key area, drawing some of its wealth from the trade along the Yangtze and up the coast, but also still a highly productive region in its own right. It has long been famous for the wealth, luxurious life-style and beauty of its cities, in particular Hangchou and Suchou, and it remains one of the most densely populated areas of China.

### (vi) *The rice-tea area*

This area includes the inland plains around the Tungthing and Poyang Lakes in Hunan and a few narrow plains along small rivers in coastal deltas, but consists mainly of low mountains and hills where most of China's tea is produced. The soils are all leached and relatively unproductive, and many hills are so badly leached and eroded that cultivation is not practical. Lower slopes are often terraced, but those which are not suffer from rapid and severe erosion.

The climate is subtropical, temperatures varying from 6 to 30°C, and frosts are rare, especially on the coast. Annual rainfalls average 1500 mm., with the heaviest falls in the summer but no months without rain. In the higher hills and mountains mists form which permit the development of the finest teas.

Rice is the chief cereal crop, but double-cropping of rice is not very common. Wheat and barley are often grown in the winter, but wheat gives poor yields as the climate is not suitable. Rapeseed is another common winter crop, and a large proportion of the land is double-cropped. Mulberries, citrus fruits and sugar cane are grown on the coast. On the whole, cereal yields are poor in this area as the soils are unproductive; considerable effort has to go into maintaining soil fertility using lime, human manure, oilcakes, compost and green manures. Oxen and water buffaloes are kept for draught, and pigs are raised for meat.

The coastal zone has been an area of dense population and intense economic activity since Sung times: proximity to the important Fukien ports and the cities of the Yangtze Delta stimulated agricultural advances, the production of commercial crops and handicrafts. The inland zones were underpopulated until the last few centuries and developed more slowly, but immigration and the improvement of rice cultivation techniques steadily raised agricultural production in Hunan until by the Chhing it was contributing a substantial surplus to the national market. Similarly, growing foreign markets for tea stimulated economic growth in the hilly areas. In the 1930s Buck considered this to be the most prosperous agricultural area of China.

Fig. 8. Terraces in Szechwan; *China, Land of Charm and Beauty*, p. 162.

### (vii) *The Szechwan rice area*

This area is delimited by the high mountain ranges surrounding the 'Red Basin' of Szechwan. The mountains rise abruptly from the basin, and only their lower slopes can be cultivated. The Red Basin itself consists mainly of rounded terraced hills (Fig. 8), but in the west it flattens out into the Chengtu alluvial plain, one of the most productive areas of all China. The soils of the Basin are purple rather than red and are all relatively fertile; the irrigated soils of the Chengtu plain are very fertile and high-yielding.

Protected by the Chhinling range from the cold north winds, the Red Basin has a mild climate with winter temperatures of around 5°C and summer temperatures of up to 30°C. Frosts are very rare. The annual rainfall is nearly 1000 mm., spread fairly evenly, but serious droughts do occur.

Rice is the principal cereal crop, though barley, wheat and maize are also quite important. Maize is particularly suited to hillside cultivation. Sweet potatoes, rape and sesame are also widely grown. Commercial crops include sugar cane, cotton, tea and citrus fruits, and Szechwan is also an important silk-producing area. Winter and spring cropping are widely practised and a large proportion of the land is double-cropped, but only a single crop of rice is usually grown in this area. Human manure is extensively used to maintain soil fertility. Farms in

Szechwan are comparatively well stocked with animals; oxen are more numerous than elsewhere in South China, and water buffaloes and pigs are also kept in relatively large numbers.

The high mountains that surround Szechwan on all sides have enabled the province to cut itself off politically from the rest of China on more than one occasion, and the high fertility of the soils and long growing season have meant that it has not suffered economically from its independence. Szechwan was traditionally known as a land of abundance, where rice was plentiful and fruits of all sorts grew in profusion. It was the earliest region in China to produce tea, and was famous for its silks by Han times. But recurrent civil strife has taken a heavy toll of life, keeping population densities and the general level of prosperity lower than might otherwise have been expected.

### (viii) *The double-cropping rice area*

The physical features of this area are similar to those of the rice-tea area, except for the famous 'karst' topography of central Kwangsi and parts of Kwangtung.[a] Most cultivation is found in the flat, irrigable river valleys, and in the broad Canton Delta. The soils are mainly laterites, often badly eroded and leached, and infertile except in the river valleys where centuries of rice cultivation have brought about podzolisation.

The climate is tropical. Mean temperatures vary between 12°C and 30°C; rainfall is abundant, averaging 1750 mm. a year, and well distributed, and humidity is high. The area is practically frostless and has a good growing temperature throughout most of the year.

The principal cereal and predominant crop is rice, much of which is double-cropped, and since the fields are under rice most of the year few winter crops are grown. Sweet potatoes form quite an important component of the local diet. The chief commercial crops are sugar cane, vegetables, and such fruits as citrus, bananas, lichees and longans. Since the soils are very acid, liming is frequently practised, and human manures and oil-cakes are other common fertilisers. Water buffaloes and oxen are kept for draught, and pigs for meat.

This area forms a physiographic unit with Tonkin and the Red River Delta in North Vietnam, and for a long period it also formed an ethnic and cultural unity with them, the whole area being known in China as Ling-nan[1], 'South of the Peaks'. Although the northern Chinese generally dismissed Ling-nan as a region of jungle-covered, fever-ridden mountains inhabited by barbarous savages, the ports of Canton and Haiphong were flourishing on international trade by Han times, and there is evidence that the coastal plains developed sophisticated rice cultivation methods which were later to be copied by provinces further north.

[a] Volume 1, fig. 4.

[1] 嶺南

The early inhabitants were chiefly non-Chinese, and it was only with the medieval Mongol invasions that the area began to fill up with Chinese migrants from the North. The hills and mountains remain sparsely populated by ethnic minorities even today, but the coastal plains maintain very high densities of Chinese population, and emigration to Southeast Asia has been an important factor in the regional economy.

### (ix) *The Southwestern rice area*

This area consists of a high, intersected plateau with deep river valleys and many high mountain ranges, as well as numerous small, deep, oval valleys with terraced sides and bottoms, where most settled cultivation takes place apart from the broad lake plains south of Kunming. Only a very small proportion of the land, most of it terraced and irrigated, is under permanent cultivation (Fig. 9) but shifting cultivation is common in the forested hills. Those soils which have not been podzolised by constant rice cultivation are generally very poor agriculturally, though for purposes of short-term shifting cultivation the mountain soils, enriched by ashes, are reasonably productive.

The climate is very mild and the temperature varies only by 13°C in Kunming, between 9°C and 22°C; the variations are somewhat higher in the mountain areas. Rainfall is quite high, averaging 1500 mm. or so, and well distributed, though concentrated in the summer months. Frosts are rare, and Kunming is known to the Chinese as the land of eternal spring.

The chief cereal grown in the valleys is rice, which is double-cropped in a few southern areas. More common winter crops are wheat, barley, and broad beans, which are the most important food crop in the valleys after rice. Rapeseed, cotton, tobacco, tea and fruits are other important crops in the valleys. The shifting cultivators grow maize and dry rice on their hill farms, together with a wide range of other less important crops like chili.[a] Fertility of permanent agricultural soils is maintained chiefly through the use of human and animal manures. More livestock are kept in this area than in any other area in China. Pigs are kept by valley dwellers and mountain dwellers alike, though the small black mountain pigs lead a much more adventurous existence than their fatter, housebound brothers in the valleys. Oxen and water buffaloes are the chief draught animals, but tiny, fiery ponies, harnessed troika-style to small carts, are a common sight in Kunming.

This area's natural affinities are with Southeast Asia rather than with China proper. The natural channels of communication from Yunnan were along the valleys of the great rivers, east along the Red River to Annam, or south down

[a] Opium was a very important economic crop in this area in the 19th and 20th centuries, as it was also in parts of Szechwan; Fei & Chang (1) give a clear idea of its role in the valley economy, but it was even more important among the hill tribes; see Geddes (1). Now, however, its cultivation is strictly controlled by the government and its use is restricted to medicinal purposes.

Fig. 9. Steep, forested hills of Southwest China; *China, Land of Charm and Beauty*, p. 151.

the Mekong or Salween to Burma, Thailand, Laos, Cambodia or Cochinchina. Kweichow's natural links were with Kwangsi and Hunan, both backwaters of Chinese culture. Only since 1949 have railroads from Chengtu and Changsha, and roads into Szechwan and Hunan, brought these provinces into regular contact with central China.[a] Until recently Chinese inhabitants were rare, and even today they form a minority among a kaleidoscope of different tribes and races.

[a] Buck (2) complains in the 1930s of the severe difficulties encountered by his research officer when attempting to cover this area.

Traditionally the valley dwellers and wet rice cultivators were Thais of one sort or another, as they still are today; Miao and I tribes lived by shifting cultivation in the hills, while many tribesmen were hunters and gatherers and did not farm at all. This is not to say that the area has suffered from cultural backwardness: the −1st millennium kingdom of Tien[1] was subjugated by imperial troops in the Han and their reports, together with archaeological finds, suggest that though somewhat unpolished the culture of −2nd century Kunming was vigorous and very inventive.[a] Rice cultivation was already advanced in the area, and a 9th-century Chinese general on an expedition to the area sent back the earliest description of double-cropping wheat and rice in all China (see p. 464). Going back much further into prehistory, there is reason to suppose that rice may first have been domesticated in this region of highland Southeast Asia (see p. 485). Nevertheless, the difficulty of rapid and regular communications, even with the Southeast Asian states which were of comparatively easy access, ensured that this area remained one where subsistence rather than economic expansion was the keynote.

The characterisations we have just given are based chiefly on data collected in the 20th century, but of course over the centuries and millennia Chinese agriculture has undergone considerable modifications: techniques have changed, new crops have been introduced, populations have grown, migrations have taken place, and we have every reason to suppose that the rural landscape of, say, Hunan today would be quite unrecognisable to a 16th-century inhabitant of the region. Since our historical study ends, in principle, in the 17th century, are Buck's definitions of agricultural areas of real relevance to our work?

Chinese historical documents naturally offer nothing to compare to Buck's work by way of a national agricultural survey.[b] But after a general reading of dynastic histories, agricultural treatises, botanical works and so on, one is left with the feeling that Buck's definitions are in essence valid over a very long period, though of course not in every detail. For example, winter wheat only became popular in the Lower Yangtze area in the Sung, when the introduction of quick-ripening rices from Champa made double-cropping possible (see p. 495); on the other hand in the southern province of Kwangtung double-cropping of rice was already known in the Han dynasty, and winter wheat was first grown on the Yellow River plains at roughly the same time (see p. 466). Perhaps the closest historical equivalent to Buck's agricultural areas is the description of agriculture in the 'Nine Provinces' (*chiu chou*[2]) in the *Chou Li*. The *Chou Li* was compiled in the Western Han, but this section harks back to the much more ancient Tribute of

[a] Von Dewall (3).

[b] One could perhaps draw up rather similar accounts based on the information contained in local gazetteers, at least for the Ming and Chhing, though of course reliable sampling would present problems. But even 20th-century surveys like Buck's are not above reproach in that respect; see Esherick (1) for some of the criticisms that have been raised.

[1] 滇　　[2] 九州

Chi-chou 冀州: Setaria and panicum millets; poultry.
Yen-chou 兗州: Setaria and panicum millets, rice and wheat; horses, oxen, sheep, pigs, dogs and poultry.
Chhing-chou 青州 and Hsü-chou 徐州: rice and wheat; poultry and dogs.
Yang-chou 揚州: rice; poultry, cormorants and peacocks.
Ching-chou 荊州: rice; poultry, cormorants and peacocks.
Yü-chou 豫州: Setaria and panicum millets, beans, wheat and rice; horses, oxen, sheep, pigs, dogs and poultry.
Liang-chou 梁州: no products given.
Yung-chou 雍州: Setaria and panicum millets; oxen and horses.

Fig. 10. The products of the Nine Provinces as given in Han sources.

Yü, *Yü Kung*[1], in the Book of Documents.[a] The *Yü Kung* does not deal with agricultural production as such, but the *Chou Li* does, if rather schematically, and on this basis we can draw a sketchy early Han or pre-Han equivalent of Buck's map, including the most typical cereal crops and domestic animals of each of the nine regions (Fig. 10). This shows us that even then millet was the typical crop of Northwest China and rice predominated south of the Huai, while a more mixed economy prevailed in the eastern plains and Shantung where millet, wheat and rice were all grown.

It may be that the seeming importance of rice in Northeast China in the Warring States and Han period, which contrasts with its rarity in northern

[a] Section 38, Geobotany *in statu nascendi*, discusses the dating of this work as well as its pedological content.

[1] 禹貢

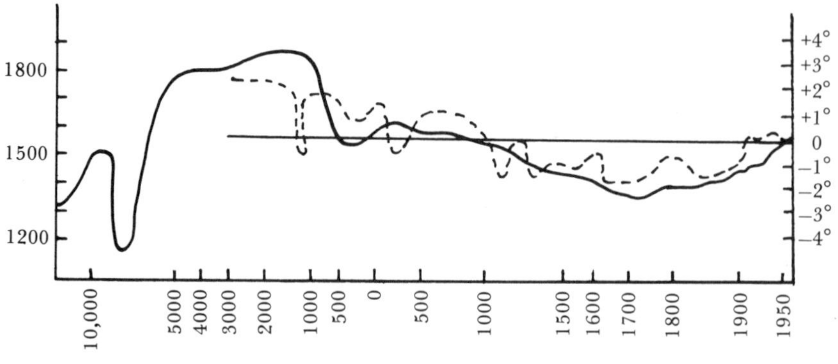


——— : variations in the Norway snow line (in metres above sea level);
- - - - - : variations in average annual temperature in China (in degrees Celcius). Calculations based largely on botanical data.

Fig. 11. Long-term fluctuations in average temperature in Europe and China; Chu Kho-Chen (*8*), p. 495.

provinces at later periods, was due to climatic variations. There have certainly been fluctuations in the Chinese climate over the last few thousand years, as Chu Kho-Chen (9), (*11*) has ably demonstrated. The data contained in the Chinese historical records do not permit any very detailed analysis of these fluctuations; nevertheless, indicators such as the distribution of plum trees or bamboo, or the dates associated in different historical periods with the flowering of certain trees, have permitted Chu Kho-Chen to draw a chart of fluctuations in average temperature in China which he correlates with similar calculations made for Europe (Fig. 11). It is clear that climatic change of any sort, whether in average temperature, rainfall, the frequency of electric storms, or the level of ambient humidity is bound to affect agricultural production. Detailed studies of European local history have shown this to be the case, but they have also shown how the effects of such fluctuations may differ from region to region: for example, a climatic change that produces a fall in average temperature in the Baltic may have the effect in the Mediterranean zone of reducing rainfall. Furthermore, there may be a timelag between the occurrence of similar effects in different regions.[a] Since China covers just as large an area as Europe, one cannot hope to draw any simple conclusions from general historical studies of climate.

Climatic fluctuations can affect agriculture in several ways. First, marked and sustained climatic changes may affect the types of crop that can be grown; if humidity and temperature rose in North China it would be easier to grow wet rice in the marshy depressions there (as was perhaps the case in the Han dynasty), whereas if they fell along the lower Yangtze wheat would do better (which might account for the successful introduction of winter wheat to the Yangtze Delta in the Sung (see p. 465)). Such fluctuations might also alter the boundaries between

[a] Le Roy Ladurie (2); H. H. Lamb (1).

marginal agricultural land and steppe or desert, especially in the far north and northwest, where farming offers a precarious livelihood at the best of times: it is possible that the prolonged occupation of such areas by Chinese farmer-settlers in the Han and Thang corresponds to periods of milder climate. Climatic changes also affect crop yields, but the same change might affect different crops in different ways, and the process is difficult to trace in the absence of any precise figures. In any case, it does not do to overplay the effects of climate, for man is seldom passive in the face of adversity, and who can say how many cunning innovations, involving improved crop varieties or more intensive tilling methods, were devised to overcome a deterioration in natural conditions? Furthermore the effects of climatic change might well be completely overshadowed by other factors, as Chu Kho-Chen himself says:[a]

> It must be stated here that climatic fluctuation is only one of the factors which might affect the agricultural operations; other factors such as the avarice of the landlord class, the demands of the market, the provision of irrigation systems, etc., might take precedence, and their influence might far outweigh the climatic change. We are only dealing with the probabilities that climatic change might have had a share in producing the difference of farm practices at different times.

The precise effects on agriculture of climatic variations, even if we are able to chart these variations accurately, are thus very difficult to evaluate. The same is true of other basic environmental changes, for example in topography or soil composition. Erosion in the loess highlands of Northwest China proceeds very rapidly once the soil is denuded of its natural vegetation. It has been suggested that when the first neolithic farmers settled there the plateau surface was higher, largely unbroken, and with a much higher water-table;[b] on the other hand, most of the village settlements of that period are found on the lower loessic terraces of large and small river banks,[c] so perhaps the landscape has changed less than the fragility of the loess sediments might lead us to suppose. In South China the effects of erosion following deforestation have been more drastic, for in many places the top-soil was so thin that it could only survive a few years of farming before it was all washed away. In some cases, however, the adverse effects of clearing the land were mitigated by such agricultural practices as terracing, or the addition of such fertilisers as river-mud or green manures, which stabilised the soils to some extent. The effects of tillage on Chinese soils have been discussed at some length in Section 38. One of the principal effects of centuries-long cropping is that almost all agricultural regions in China have come to suffer from deficiencies in nitrogen, phosphorus, and/or potassium (Fig. 12). But in the case of certain highly leached, acid soils in south China, the continued cultivation of irrigated rice has actually brought about an improvement in soil structure and productivity:[d]

[a] (9), p. 18.
[b] Andersson (1).
[c] K. C. Chang (1), p. 94.
[d] Buck (2), p. 155.

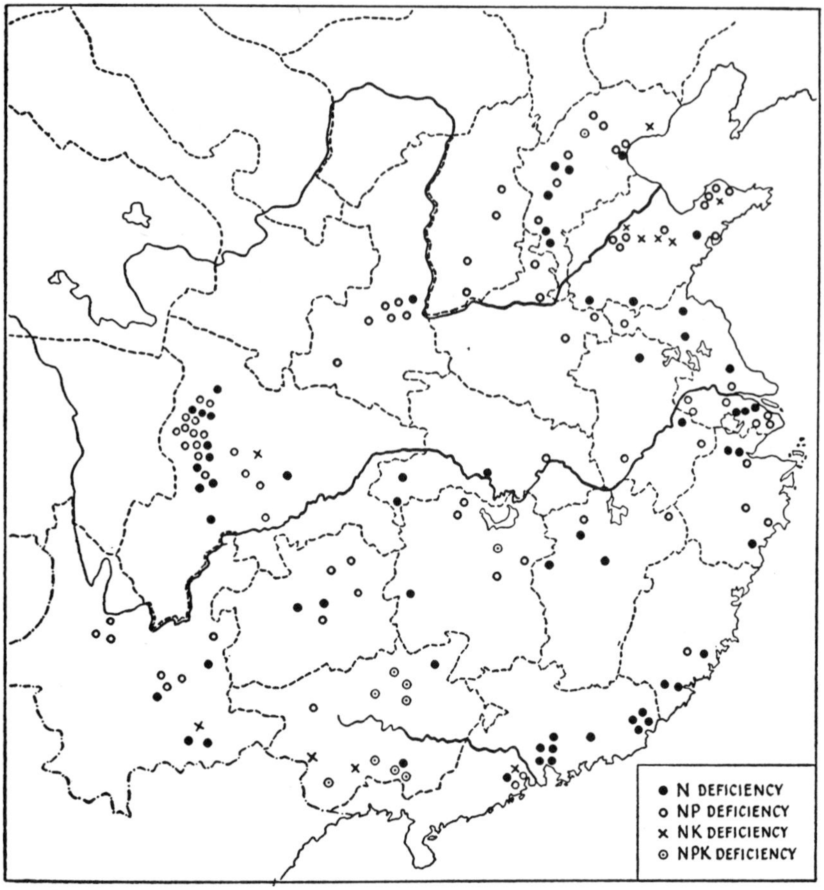


Fig. 12. Map showing the chief mineral deficiencies of Chinese soils; after Shen (1), p. 25.

The podzolisation of soils is caused by the influence of percolating water in conjunction with various organic acids. The latter help to dissolve some of the mineral compounds and to deflocculate the clay particles so that they may be carried down to the subsoil in colloidal suspension. The cultivation of rice requires that the land be kept under water during much of the year; and this irrigation water slowly but continually seeps from the surface horizons to the subsoils, carrying with it more or less colloidal clay and some compounds in solution. The addition of various organic manures, including a considerable amount of night-soil, helps to increase the podzolisation effect. Occasional liming of the soil tends to offset the leaching effect since lime helps to coagulate the clays and prevents their forming colloidal suspensions in the soil water.

Wherever it is possible to obtain water for irrigation, practically all of the soils of the important groups of South China are planted to rice. After formerly well-drained upland soils have been used in this way for a period of several years, their characteristics are changed, much of the red iron compounds are dissolved and removed from the surface to the subsoil or into the river waters, and the soils fade in colour until, after a long time, they are predominantly grey like the rice lands of the alluvial plains ... Practically every bit of land in South China which will hold water and which can be irrigated is planted to rice for at least part of the time. The soils are worked and cultivated and fertilised until they become well adapted to rice culture, regardless of their original characteristics. The 'Rice Region' is indeed an appropriate name for South China.

It is in the distinction between the rice-growing south and the wheat- or millet-growing north, rather than in subtle differences between smaller agricultural areas, that we find the most fundamental and enduring differences in agricultural systems. J. L. Buck (2) draws a line between the northern and southern agricultural regions of China that corresponds roughly to the Chhin-ling range and the Huai River (Fig. 4), and despite climatic fluctuations and technical developments this division seems to have been valid ever since Chinese agriculture first began. It is possible to identify differences not only in climate, crops and agricultural tools and methods, but also in farm management, tenurial relations and patterns of overall economic development, and we shall elaborate on the social and economic dimensions in our concluding section. In the meantime, Table 1 gives a skeletal picture of the most important environmental and practical differences between northern and southern farming systems, to which we shall allude throughout this volume. For a more elaborate treatment we would refer the reader to Buck (2). But Buck was by no means the first to draw this distinction: it was very clearly present in the minds of all Chinese writers, who distinguished systematically between the dryland agriculture of the north, with its associated crops of millet and wheat, and implements such as the seed-drill and bush-harrow which were designed to preserve soil moisture (*pao tse*[1]) and use it to maximum efficiency, and the irrigated farming of the south, where rice was the principal crop, the typical implements were vertical, toothed harrows designed to reduce wet clods of clay to smooth, silky mud, and water control (*shui li*[2]) rather than moisture preservation was the key technique.[a] This division between North and South China is the first fact remarked upon by any Chinese today when engaging in a discussion of agriculture in his country, and it seems to have its roots in the very origins of Chinese agriculture.

## (3) Origins of Chinese Agriculture

Before attacking the problem of Chinese agricultural origins as such, we discuss recent trends in general archaeological theory which have fundamentally affected the interpretation of the Chinese material. These general theoretical issues demonstrate how deeply the choice of model affects, not only archaeological interpretations, but also the collection of the data on which such interpretations are based. Successive models of the origins of Chinese agriculture, and of the evolution of Chinese neolithic cultures, can only be understood and evaluated if we are aware of the wider issues at stake. After discussing general theories of

[a] See *CFNS*, *WCNS*, *NCCS*, *TKKW* passim.

[1] 保澤 [2] 水利

Table 1. *Characteristics of Northern and Southern Chinese agricultural regions*

| | North China | South China |
|---|---|---|
| *Climate* | widely diverging winter and summer temperatures<br>low rainfall (350–750 mm.), unreliable,[a] concentrated in summer months<br>150–250 frost-free days | lower variation between winter and summer temperatures<br>high rainfall (750–2000 mm.), more reliable,[a] more evenly distributed<br>250–365 frost-free days |
| *Soils* | pedocals, relatively productive young sedimentary soils<br>deficient in nitrogen and phosphorus | pedalfers, highly leached, low productivity mitigated by podzolisation in rice-growing areas<br>deficient in nitrogen, phosphorus and potassium |
| *Terrain* | many plains, high percentage of land cultivated[b] | many mountains, high percentage of land uncultivated[b] |
| *Irrigation* | unusual (except in the northwest); water often provided by wells[c] | general, from streams, tanks, canals[c] |
| *Crops* | dryland (fields called *ti*[1])<br>typical cereals: millets, wheat and barley, sorghum | irrigated (fields called *thien*[2])<br>typical cereals: rice (single- and double-cropped), winter wheat and barley |
| *Farm Management* | larger holdings<br>more draught animals (oxen, donkeys, mules)<br>plough, seed-drill, bush- and tined harrows are typical implements<br>more productive animals (pigs, sheep) | very small holdings<br>fewer draught animals (water buffalo, some oxen)<br>plough, vertical harrows, smoothers are typical implements<br>fewer productive animals (pigs), but importance of fish and other aquatic products |

[a] Fig. 13.
[b] Fig. 3.
[c] Fig. 14.

[1] 地 [2] 田

agricultural origins, the following section concludes with a summary of different theories of agricultural beginnings in China itself, and a brief review of the evidence currently available. Many of the general issues raised in the next few pages will recur in the sections on the development of techniques, field systems, and the evolution of individual crop plants.

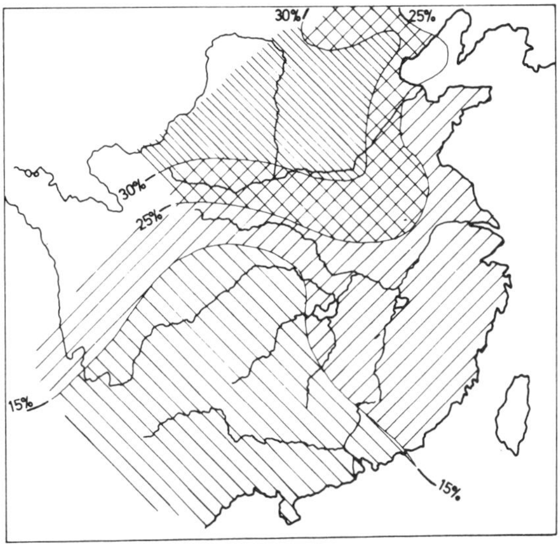


Fig. 13. Map showing the percentage variability of annual rainfall; after Tregear (2), fig. 16.

## (i) *Stimuli to the adoption of agriculture*

A vigorous debate rages as to the origins of agriculture and of crop domestication.[a] Not only does the question arise of when and where a particular crop was first domesticated, but also of how and why. The root causes of the development of an agricultural economy have now become one of the archaeologists' main concerns, though until quite recently such questions were largely ignored. As long as the history of mankind was regarded as a linear progress from simple origins to ever higher levels of culture and comfort, the advantages of agriculture over primitive hunter-gathering appeared indisputable. It seemed unquestionable that farmers enjoyed innumerable blessings denied to the wandering hunter-gatherer: permanent dwellings, a constant supply of food, security, and all the cultural elaborations made possible by the existence of a food surplus: for example, the crafts of pottery, weaving and metallurgy, and a growing complexity of social organisation culminating in the formation of cities and states. Hunter-gatherers, on the other hand, spent long miserable hours scrabbling around in a search for unappetising roots and grubs which left them little leisure or strength for culture.

Since the advantages of agriculture were so evident, then clearly primitive man would not hesitate to avail himself of them as soon as the opportunity presented itself or the knowledge became available. It was suggested that man began by

[a] We shall concentrate on the domestication of plants in the following discussion; the domestication of animals is thought to be closely associated with that of plants, in Chinese as in other neolithic societies (see e.g. Ucko & Dimbleby (1); C. A. Reed (2)), and the arguments rehearsed here apply in essence to both types of domestication.

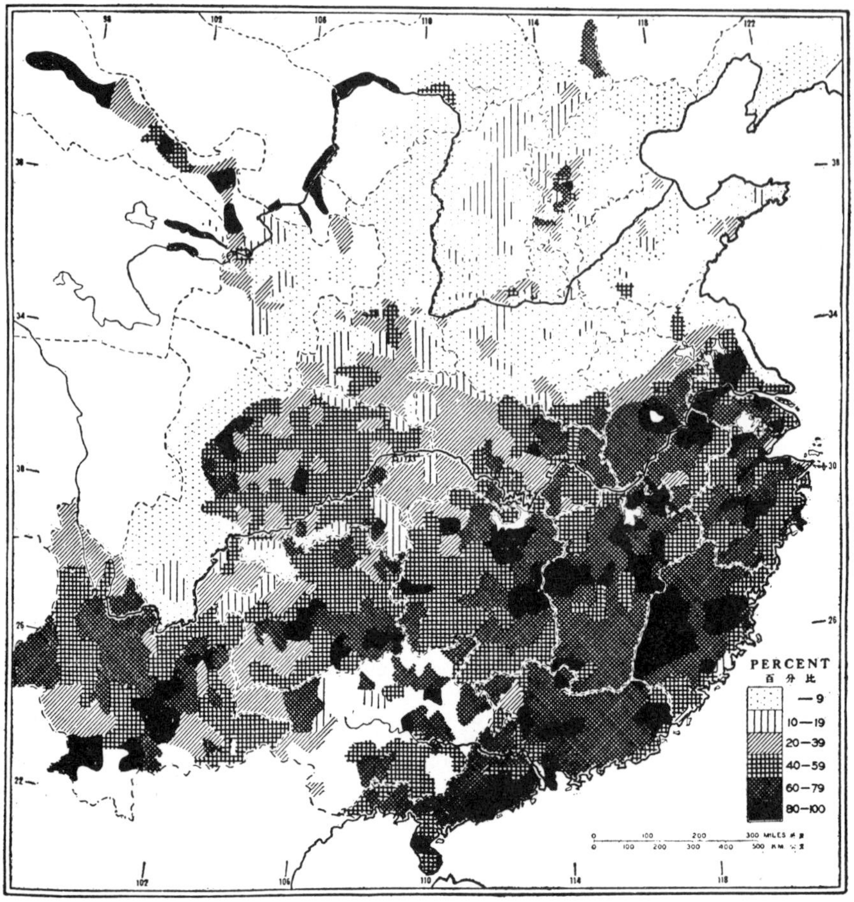


Fig. 14. Map showing the percentage of irrigated land in different regions; after Buck (2), p. 187.

gathering wild cereals to supplement his diet and then observed that the same plants sprouted from the heaps of refuse around his dwelling, put two and two together and discovered the principle of sowing. This hypothesis seems to give early man little credit for intelligent observation of natural processes. Another, more generous, school of thought points out that only the annual forms of cereal are suitable for domestication: perennial cereals do not form seed reliably, but annual forms must do so as it is essential to their survival. Therefore, it is suggested, the adoption of agriculture could not occur until a period when annual forms of cereal became widespread as a result of climatic stress. In about −10,000 or −9000, at the end of the last Ice Age, temperatures in the northern hemisphere generally rose and rainfall levels fell; as a result of progressive desiccation the number of annual plants increased:[a]

[a] R. O. Whyte (3), pp. 214, 218.

Annuality is a mechanism for escaping seasonally unfavourable environments. Annuals produce large amounts of seed to allow for wastage. These seeds are often larger, to provide the seedlings of the next generation with the reserves necessary for their establishment in an inhospitable environment. The evolution from perennials of annual forms of specific or lower taxonomic status favours the dispersal of that genus naturally or by man into new habitats in which the perennial forms could not persist ... Peoples of primitive hunting/collecting/pastoral communities in the desert fringes to the west and southwest of continental Asia who had been accustomed to stripping the ripe heads of perennial grass as one of their sources of plant food were faced with the manifestations of the Neothermal, recurring periods of drought and high temperature. Coincident with the reduction in availability of food from the drought-affected stands of perennial grasses, they began to note the new resource in the form of larger-grained annuals, occurring in ever-increasing numbers and greater diversity and adapted to completing their growing season in the short period of the year available to them. These new annual forms would have become established on the bare ground between the perennial species of a climax grassland that had become impoverished by desiccation and by over-grazing of the flocks and herds of the early pastoralists. The people would note that the annuals had the habit of seed-shattering, and would learn to push the seeds deeply enough into the ground to protect them from marauding flocks of birds and from rodents—so proceeding gradually to the digging stick, the dibbler, and to the beginning of dryland cultivation.

It seems unlikely, however, that the appearance of annual forms was of itself sufficient reason for the adoption of agriculture. Recent studies of modern hunter-gatherer societies, now driven to living in areas just as desiccated as the Neothermal desert fringes mentioned by Whyte (for example, the Australian deserts or the Kalahari), have shown that even under these adverse conditions the search for food occupies a relatively short time, far less than is required for most traditional farming methods, and provides a more varied and often more nutritious diet.[a] In between searching for food, the hunter-gatherers have ample leisure to tell stories, play games, elaborate metaphysical theories, and take long naps—indeed the way of life seems so attractive that Marshal Sahlins has dubbed it 'the original affluent society'. Several people have pointed out that with the spread of settled farming, surviving hunter-gatherers have now found themselves relegated to environmentally unfavourable areas, and suggest that formerly their way of life offered even more attractions. Furthermore, most hunter-gatherers have an expert knowledge of practical botany, many of them tending or sowing plants as well as simply collecting them, but although they have all the knowledge necessary to practise agriculture they choose not to do so:[b]

Australian aborigines are today regarded as hunter/gatherers par excellence, but it is perhaps instructive that we can list a significant number of ecosystem manipulations among them. The use of fire is well known, but they also employ water control techniques to manipulate microenvironments ... Aborigines have been recorded practising a variety

[a] Richard B. Lee (1), (2), (3) describes life among the !Kung bushmen of the Kalahari; Mulvaney and Golson (1) treat the resources of Australian aborigines; more general studies are Sahlins (1); Lee & DeVore (1); Kraybill (1).

[b] Jim Allen (1), p. 184; see also, for example, Binford (1); Flannery (2).

of food preservation techniques including ... elaborate preservation of the nuts of *Cycas media*, one technique for which includes the digging of large grass-lined storage trenches. There are recorded instances of deliberate seed planting, and the habit of planting back portions of harvested yam is widespread ... It is intriguing to consider that some Aboriginal groups may have been for a long time poised a few steps along an agricultural pathway, or indeed may have regressed from further along such a pathway.

It is quite common for non-agriculturalists to grow food-plants in greater or lesser quantity, and so, as Warwick Bray points out:[a]

The critical moment was not the planting of the first seed, or the taming of the first animal, but the achievement of a fully effective system of agriculture and of dependence upon cultivated foods ... Rather than employ arbitary figures,[b] I would prefer to introduce a more flexible definition based on the idea of *dependence*, i.e. that an agricultural society (whatever the proportion of its diet derived from farming) is one which has reached a stage of dependence on cultivated foods ... There is a rather simple criterion for this state of dependence or commitment; it is irreversible. The community, no matter how little cultivated food it consumes, can no longer revert to a hunting or gathering existence without suffering a loss of population through death or migration. In effect, this means that the stage of commitment has been reached as soon as the number of part-time farmers becomes greater than the number of people which the area could support under a purely hunting or gathering regime, or when the landscape has been so modified (for instance, by the removal of the natural vegetation) that a return to the former way of life is no longer feasible.

In a very short space of time, archaeologically speaking, between about −8000 and −2000, agricultural societies appeared, flourished and expanded in West, East and Southeast Asia, Europe, Africa and Central and South America, and the major part of the world's population became farmers. But the 'discovery' that plants could be grown by man could not have been the deciding factor in the development of agricultural societies, nor could the appearance of annual forms suitable for cultivation; both were necessary but not sufficient conditions. Many hunter-gatherer societies grew some plants, more or fewer as they felt inclined, but the progress from this state of knowledge towards an agricultural economy is not inevitable. What, then, were the constraints that obliged neolithic man to abandon his relatively carefree existence as a hunter-gatherer and take up the heavy burdens of the farmer?

Three related hypotheses have been proposed to explain the shift from foraging to farming. The first is that through climatic change, changes in sea-level, increased competition from other species, or similar causes, the naturally available food resources diminished and it became necessary to supplement them by growing food plants and domesticating animals:[c]

[a] (1), pp. 243–4.

[b] It has been suggested, for instance, that an agricultural society is one which obtains, say, 50% of its total calorie intake from farming; Smith & Young (1), p. 58.

[c] D. R. Harris (3), p. 182.

Childe was the foremost proponent of the view that the development of food production in the Near East was the result of 'desiccation' following the retreat of the ice sheets of the last glacial. He envisaged the enforced concentration of human, animal, and plant populations in river valleys and around oases as leading to domestication and the adoption of a food-producing economy (2), (4), but he did not suggest what precise processes of adaptation might have been involved in the establishment of this new mode of subsistence. His vigorous advocacy of postglacial climatic change as the primary cause of the Neolithic Revolution did, however, have a lasting impact on the thinking of many scholars and until recently it was accepted as an explanation of the shift to food production (for example by Clark (3)).

Childe's hypothesis has been challenged on the grounds that no clear archaeological evidence exists for major post-glacial climatic change in the Near East.[a] On the other hand, desiccation on the fringes of the Sahara, particularly where populations depended not only on foraging but also on fishing and so were reluctant to move, could well have had a decisive role in encouraging the shift to agriculture in that area.[b]

Another factor frequently seen as determinant in the adoption of agriculture is that of demographic growth. In hunter-gatherer societies today numbers increase very slowly if at all,[c] but this was perhaps not always the case. M. N. Cohen (1) believes that modern man, who originated in Africa, gradually spread through Africa and the adjacent continents, dispersing in order to maintain population densities at which survival was easy. These densities varied according to local conditions, the preferred habitats being game-bearing savannahs. As population densities inexorably increased, man was gradually forced to occupy less favourable environments and to switch to less palatable diets. Eventually, as his diet came to include less and less game and more and more plant foods, man became a farmer.[d] Yet against such arguments one might object that overall population densities in the early neolithic were not high, and there was no obvious reason why populations should have concentrated in the areas where farming first developed: since hunter-gatherers can live very adequately even in deserts or tundras, why did they not simply wander off 'into the surrounding emptiness'?[e]

If something constrained groups of hunter-gatherers to abandon their nomadic life for a sedentary existence, then population pressures might well build up to the point where farming became a necessity. Various causes have been proposed to explain pre-agricultural sedentarism. Bronson (1) calls these 'locational constraints': a population may find itself confined to an island, or be prevented from occupying adjacent regions for reasons of defence or health, or else there may be an attraction that keeps them in a particular spot, such as a lake well-

[a] Braidwood (1), p. 103; Flannery (1); Van Zeist (1).

[b] Thurstan Shaw (1).

[c] Female fertility is low, for physiological reasons associated with the strenuous life-style; gaps between children are usually at least four years long, and infanticide is fairly common; Orme (1).

[d] Gibson (1) uses a similar model to explain the elaboration of farming techniques and social organisation in early Mesopotamia.

[e] Bronson (1), p. 44.

stocked with fish.[a] Where wild cereals with a short ripening period made an important contribution to the foragers' diet, they might wish to store part of the harvest for future use, perhaps in plaster-lined pits such as are characteristic of Natufian and other pre-agricultural sites in the Middle East, and to settle down near their stores of grain which they would be reluctant to leave both uneaten and unprotected.[b] Other attractions of the sedentary life that have been suggested include the possibility of accumulating possessions and the pleasures of a gregarious existence.[c] Sedentarism is not dependent upon agriculture: non-agricultural village societies are known from the Natufian culture of the East Mediterranean, from the coastal peoples of Northern Peru and the Northwest coast of North America, and elsewhere; in some cases agriculture developed as the settled communities grew, in others the local resources were sufficient to maintain the population without resorting to agriculture.[d] But in most cases sedentarism was probably associated with quite rapid population increases,[e] and as pressure on resources mounted either people would have to leave, or some means would have to be found to increase the local food supplies, as for instance the adoption of agricultural techniques.

We have briefly presented some of the arguments as to why early man might have wished to change his mode of life and practise agriculture.[f] It is unlikely, however, that the same combination of factors operated in every case, and most archaeologists today would agree that there can be no universally valid model for the adoption of agriculture. 'I do not believe', says Flannery, 'that agriculture began the same way, or for the same reasons, in all areas, nor do I believe one model can explain them all.'[g]

### (ii) *General theories of agricultural origins*

Having discussed why a society of hunter-gatherers might settle down and turn to farming, we should now consider the problem of where and when agriculture began. Again, several general models have been proposed, and these must be briefly discussed if we are to understand the various theories about the origins and spread of Chinese agriculture.

Most agricultural societies as we know them today rely heavily on the pro-

[a] See Bronson (1), pp. 34 ff.

[b] Flannery (1).

[c] Orme (1).

[d] Charles A. Reed (4); M. N. Cohen (2).

[e] Women would have more children as under sedentary conditions the age at first pregnancy would be lowered, the period between births would diminish and there would be less infanticide; Reed (4).

[f] For more detailed discussions see C. A. Reed (4); Bender (1); Megaw (1); and Harlan (5).

[g] Flannery (3), p. 272. As increasingly intensive studies of the transition to food production are carried out, for example the minutely detailed studies of the Teotihuacan Valley in Mexico (MacNeish (1), (2), (3)) or sites in the Near East (A. J. Legge (1), (2); Hole, Flannery & Neely (1); Helbaek (1); Noy, Legge & Higgs (1)), the models required to explain particular cases become increasingly complex and therefore increasingly unlikely to be universal in their application. A good example of the possible complexity of such models is to be seen in Harris (3), p. 190.

duction of cereal crops, and it has frequently been assumed that agriculture originated with the domestication of cereals. Although we earlier rejected the notion that the sole presence of wild annual cereals was sufficient to stimulate their domestication by man (and thus the beginning of agriculture), it is clearly impossible to domesticate a plant which is not there, and so, when discussing the early development of agriculture, it is necessary to give careful consideration to the natural distribution of potential domesticates.

The 19th-century botanist Alphonse de Candolle was the first to attempt to locate the region of origin of the various cultivated plants, using any relevant information he could lay his hands on: the distribution of wild relatives, historical references, regional names, linguistic derivatives, and what scanty archaeological evidence was then available. On this basis he published in 1883 a book entitled *The Origin of Cultivated Plants*, in which he considered the regions of origin of domesticated plants, their early cultivation, and the process of diffusion by which they were introduced to other areas. He deduced:[a]

> Agriculture came originally, at least so far as the principal species are concerned, from three great regions, in which certain plants grew, regions which had no communication with each other. These are—China, the southwest of Asia (with Egypt), and inter-tropical America. I do not mean to say that in Europe, in Africa, and elsewhere savage tribes may not have cultivated a few species locally, at an early epoch, as an addition to the resources of hunting and fishing, but great civilisations based upon agriculture began in the three regions I have indicated.

In 1926 Vavilov (3), taking up de Candolle's notion of agricultural centres of origin but applying more advanced botanical techniques, proposed that one could reliably determine a crop's centre of origin by analysing its patterns of variation, the region of greatest genetic diversity being the centre of origin. In his *Origin, Variation, Immunity and Breeding of Cultivated Plants*, Vavilov made use of the world-wide surveys carried out by Soviet botanists in the 1920s and 1930s, and concluded that it was possible to define at least 'eight independent centres of origin of the world's most important cultivated plants',[b] including Chinese, Indian, Central Asian, Near Eastern, Mediterranean, Abyssinian, South Mexican and Central American, and South American centres. It has since been pointed out, however, that although centres of diversity do exist and are very useful in explaining genetic variation, they are *not* necessarily centres of origin;[c] indeed, Vavilov himself had to invent the concept of secondary centres to cope with this discrepancy. It is also possible, of course, that areas of domestication are not coincident with centres of origin.[d]

[a] (1), p. 17. [b] (2), p. 20.
[c] Zohary (3); Harlan (5).
[d] It has been suggested (e.g. Whyte (3)) that a species is in fact so well adapted ecologically to its centre of origin that there is no stimulus there to the development of new (domesticated) forms; only on the very fringes of this region, where wild varieties do not flourish naturally, would conscious tending and selection by man (leading to domestication) become necessary. See also pp. 435 ff. below on setaria millet.

De Candolle and Vavilov, considering the origins of agriculture from a largely botanical point of view, were both prepared to recognise several regions where it might have originated independently, and both of them included China among them.[a] Vavilov believed that: 'the earliest and largest independent centre of the world's agriculture and of the origin of cultivated plants consisted of the mountainous regions of central and western China, together with the adjacent lowlands'.[b] But archaeologists, primarily concerned with cultural evolution, felt that the crucial problem was not where the domestication of various plants took place but where the *idea* of domestication was first conceived and put into practice. They saw the invention of agriculture as part of 'an economic and scientific revolution that made the participants active partners with nature instead of parasites on nature'.[c] At the beginning of the neolithic period, under the pressure of food shortages caused by climatic change, man (or woman) learned to farm and domesticated animals, and closely associated with these innovations were other inventions:[d]

> The practice of farming and life in settled hamlets can be accepted as the first mark of the full Neolithic revolution. Certain particular items of material culture so often accompany it that they deserve mention in any general definition of Neolithic culture. One is the polished axe or adze, made either of igneous rock or flint, the other the straight sickle, made more or less on the Natufian model. The crafts of potting and weaving were soon to become the most important additions to Neolithic culture, but they were ... secondary traits following upon the essential innovations in farming life and equipment.

The domestication of plants was such a revolutionary concept, in the eyes of these archaeologists, that it could only have occurred once in human history. The earliest farming settlements known were those of the Near East, and it was here, they concluded, that wheat and barley were first domesticated and pottery and weaving invented. From this unique centre of origin, the new technology and the sedentary way of life associated with it gradually diffused to other parts of the world:[e]

> It is at once apparent that [the Neolithic Revolution] took place only over a great range of time. Beginning some eight or nine thousand years ago in its cradleland, it took between three and four thousand years to reach Western Europe on the one hand, China on the other ... The Neolithic Age ... represents the period between the end of the hunting way of life and the beginning of a full metal-using economy, when the practice of farming arose and spread through much of Europe, Asia and North Africa like a slow-moving wave.

[a] We should point out, however, that both Vavilov and de Candolle were under the illusion that the legend of Shen Nung[1] corresponded to historical fact. According to a Chinese myth of rather late origin, Shen Nung, the Heavenly Husbandman, was a sage-emperor who lived in the −3rd millennium and taught his people the use of the plough and the cultivation of cereals.

[b] (2), p. 21.

[c] Childe (4), p. 48.

[d] Hawkes & Woolley (1), p. 220.

[e] *Ibid.* p. 219.

[1] 神農

This interpretation of agricultural origins was based on the premise that cereals were the first crops to be domesticated, and that the shift to agriculture was, in the first instance, a response to a growing food shortage precipitated by climatic deterioration. In 1952 Sauer (1) rejected this notion outright. Starving men were not innovative or likely to take risks, he maintained, and we should look for the origin of plant domestication among well-fed, sedentary people living in areas of wide botanical diversity and having the leisure and resources to experiment. Sauer proposed that the first domesticators were not *agriculturalists* but *horticulturalists*, probably fisherfolk living in a mild climate in a forested zone, who tended fibre crops and other useful plants in small gardens, then went on to the vegetative propagation of tropical crops such as bananas and edible tubers before sowing cereals and other seed crops: 'food production was one and perhaps not the most important reason for bringing plants under cultivation'.[a] Sauer postulated three hearths of plant domestication, one in mainland Southeast Asia and two in America, from which the idea of plant cultivation gradually diffused. As the practice of plant cultivation reached different ecological zones, new species were brought under domestication, and in non-tropical areas agriculture became more important than horticulture.

When Sauer advanced his new theory, all the most extensive and detailed archaeological studies of neolithic communities had been carried out in the Near East, and almost no work at all had been done in tropical zones. Given the state of knowledge at the time, Sauer's hypothesis could be neither proved nor disproved, but since it challenged certain long-cherished cultural preconceptions about the location of the cradle of civilisation, it was coldly received. However, the arch-diffusionist concept of a single centre of agricultural origins was no longer proving satisfactory. As more sites were explored in Europe and West Asia, as recovery techniques became more sophisticated and carbon-14 dating clarified cultural sequences, discrepancies appeared in the diffusionist theories and the very concept of 'hearth areas' or 'centres' was eventually called into question. Perhaps the key role assigned to the Near East simply resulted from its being the most thoroughly investigated area:[b]

> Are the factors which we have been accustomed to consider as critically important of more than local significance, and are the areas which we have for so long considered the crucial centres on which subsequent developments depended, part of a much wider phenomenon in time and space?

After all, as Harlan points out, the notion of a 'centre' may reflect long-term success rather than anteriority: diffusion from 'centres' may oust pre-existing but less efficient or successful domesticates in neighbouring regions.

> In the case of the Near East, we seem to have a definable centre in the sense that a number of plants and animals were domesticated within a relatively small region and

[a] (1), p. 27.

[b] Higgs & Jarman (1), p. 3.

were diffused outward from the centre. In Africa, nothing of the sort is apparent. The evidence seems to indicate that activities of plant domestication went on almost everywhere South of the Sahara and North of the Equator from the Atlantic to the Indian Ocean. Such a vast region could hardly be called a 'centre' without distorting the meaning of the word, so I have called it a noncentre.[a]

Partly in response to Sauer's provocative hypothesis, a number of excavations have been carried out in tropical zones in the past two decades, producing evidence which strongly suggests that plant cultivation was widely practised in these regions at a much earlier date than had hitherto been suspected.[b] The vegetable remains discovered at Spirit Cave in North Thailand, and tentatively dated to −8000, have been claimed by some archaeologists as proof of Sauer's theory that Southeast Asia was the earliest Old World hearth of domestication; taken with the results of excavations elsewhere in Thailand, at Non Nok Tha and Ban Chiang, this is held to demonstrate a continuous progression from an early economy based on foraging and plant-tending to the domestication of palustrian species (taro and rice), and eventually to a heavier reliance on rice-farming, the successive changes in economic emphasis permitting the occupation of new ecological zones within the region.[c] Severe doubts have been cast on the stratification and dating of the Thai sites, as well as on the identification of the Spirit Cave remains.[d] However, excavations in neighbouring regions of Southeast Asia have produced further evidence of early cultivation, all of which tends to support the idea of early plant domestication in Southeast Asia.[e] Further afield, recent research in Africa amply supports Harlan's concept of an African noncentre.[f] There is also some archaeological evidence (and a great deal of fervid argument) to support the theory that tropical horticulture in the lowland, marshy plains of Mexico and Honduras, as well as the Amazon and Orinoco basins, preceded the domestication of grains and dryland tubers in the highland zones of Mexico and Peru, the regions once regarded as the earliest centres of American domestication.[g]

On the basis of this widespread evidence for early plant domestication in tropical zones, Harlan has proposed three mutually independent systems of domestication, each with a centre and a non-centre: (i) a Near East centre and an African non-centre; (ii) a North Chinese centre and a Southeast Asian and South

[a] (5), p. 56; Harlan first used the term 'noncentre' in (6) in 1971.

[b] One frequently-raised objection to Sauer's hypothesis was that it would be impossible to clear tropical forest without metal axes. Experiments with stone axes have shown this notion to be false; Steensberg (7).

[c] Solheim (1); Gorman (1); Higham (1), (2).

[d] For a summary of the controversy see Reed (4).

[e] See Davidson (1) on Vietnam; Glover (1) on Indonesia; Allen, Golson & Jones (1) and Smith & Watson (1) for a wide range of studies on the region. Of particular relevance in this context is the discovery of irrigated fields in Highland New Guinea, which may date back to as far as −7000; the excavator Jack Golson describes the digging of large drains across a highland swamp about 6000 years ago, creating 'a web of short channels, so disposed as to define roughly circular clay islands of about a metre diameter', and suggests that water-tolerant taro was grown in the channels and yam and banana on the islands; Golson (1) p. 616; see also Steensberg (7).

[f] Harlan, De Wet & Stemler (1).

[g] D. R. Harris (2); Lathrap (1), (2); Puleston (1).

Pacific non-centre; and (iii) a Mesoamerican centre and a South American non-centre. He visualises 'some stimulation and feedback in terms of ideas, techniques or materials between centre and non-centre in each system'.[a]

As the number of hypothesised zones of domestication climbs from Childe's unique Near Eastern cradle back nearly to the number proposed by Vavilov, the question of diffusion versus independent discovery arises once again. If, as Childe and Hawkes believed, agriculture was only 'invented' once, then clearly some mechanism of diffusion (whether of knowledge or of population) must have been responsible for the appearance of agriculture in other parts of the world. There are still a few proponents of the single-centre diffusionist theory.[b] At the opposite extreme there are those who argue that plant domestication, the manufacture of pottery and so on, are such natural adaptations to widely replicated human needs that they must have developed quite independently, if more or less contemporaneously, in different parts of the world. Regarding the notion of diffusion as tendentious, they would prefer to consider all cultures and technologies as indigenous developments unless powerful proof to the contrary can be adduced.[c] Most archaeologists nowadays incline towards a more flexible approach somewhere between these extremes.

Let us now consider how different theoretical positions have affected interpretations of the origins and spread of agriculture in China.

### (iii) *The origins of agriculture in China*

When sites of the early neolithic Yang-shao[1] culture were first discovered in Honan, Shensi, Kansu, and other northwestern provinces in the 1920s and 1930s, it was assumed that they must represent an influx of culture from the West. J. G. Andersson (4) spoke of the 'striking similarities' in design between Chinese ceramic styles and those of Turkmenia and the Ukraine, and suggested that agriculture based on the cultivation of wheat had been introduced to China together with the painted ceramics, probably in −3000 to −2000. A few years later, by 1934, he had modified his opinion:[d]

> When, in 1925, I wrote my *Preliminary Report on Archaeological Research in Kansu* the idea hovered in my mind that perhaps the higher stage of agriculture, represented by *wheat* as the principal and most suitable crop, had come to China from the west at the beginning of the Yang Shao age ... But the first kind of crop which we were able to identify points in a totally different direction. One day during the early years of our work here in Stockholm we happened to examine more closely a fragment of a jar from Yang Shao Tsun. It was unusually thick in the wall, porous and full of plant imprints. Two Swedish botanists, G. Edman and E. Söderberg, examined this small fragment and their examination led to

[a] (5), p. 56.
[b] E.g. George F. Carter (10).
[c] Meacham (2).
[d] (1), p. 335.

1 仰韶

the most important discovery in this field since the discovery in 1921 of the Yang Shao dwelling-site. It could be shown with certainty that the plant imprints in this fragment were husks of cultivated rice (*Oryza sativa*). The discovery was in a high degree sensational not only because it sets back the history of rice an immense distance in time, but also because it points, not to dry Central Asia, but to rainy Southern Asia, which is the homeland of rice.

Were it not for the current prevalence of diffusion theory, it might seem odd that Andersson and his colleagues were so intent on identifying an extraneous source for Chinese agriculture, since even then it was clear that the crop most typical of the Yang-shao sites was foxtail millet, *Setaria italica*,[a] a crop which both de Candolle and Vavilov considered to be native to China. There were some botanists, however, who assigned an Indian origin to this millet,[b] while on the other hand setaria was known to have been grown by neolithic farmers in Europe as a subsidiary crop to wheat,[c] so that it would be consistent with Andersson's hypothesis to assume either that domesticated millet had been brought to China from India or West Asia, or that once the Chinese had learned how to farm wheat they went on to domesticate the millets native to the area.

Apart from the Yang-shao culture of the Northwest, characterised by its painted ceramics, another neolithic culture had been defined in Shantung. More technically and culturally sophisticated than the Yang-shao culture, the Lung-shan[1] was assumed both on stylistic and stratigraphic grounds to be a later development of the Yang-shao.[d] Sites identified as belonging to Lungshanoid cultures were found over a wide area, not only in the central provinces of northern China but also along the east and southeast coast. In many of these sites remains of rice were found. It was generally supposed that Yang-shao farmers improved their agricultural techniques, progressing from shifting swidden cultivation to settled farming, and that the resulting increase in population resulted in waves of migrations out from the 'Nuclear Area' into East and Southeast China. 'The farmers grew rice and wheat in addition to millet, which presumably continued to be the leading staple in the North ... rice may have been one of the crops successfully cultivated by the Lungshanoid farmers in South China and brought back to the North.'[e]

Although on superficial acquaintance the Yang-shao culture could reasonably be compared with West or Central Asian cultures, as more and more neolithic sites were discovered in China patterns of such distinctive character emerged that the hypothesis of Western origins came to be largely discarded, and it began to be assumed that China must have developed agriculture independently of Western or Indian influence.[f] The most systematic argument for the independence of Chinese neolithic cultures and agricultural origins is that advanced by Ping-Ti

[a] Shih Hsing-Pan (*1*).
[b] See Ping-Ti Ho (5), p. 58.
[c] Hawkes & Woolley (1), p. 255.
[d] K. C. Chang (1), 1963 ed., p. 89.
[e] K. C. Chang (1), 1963 ed., p. 93.
[f] *Ibid.* p. 55.

[1] 龍山

Ho.[a] So far as agricultural origins are concerned, he argues that the earliest Chinese agricultural techniques and crops (namely setaria and panicum millets) were uniquely well adapted to the semi-arid environment of the Northwestern loess-lands where the Yang-shao culture developed. Invoking the great natural fertility of the loess soils, Ho postulates that the earliest Chinese farmers never needed to practise shifting agriculture, but adopted from the outset settled agricultural patterns with short periods of fallow.[b] By 1969 carbon-14 dates were available for Chinese neolithic remains[c] which showed that millet cultivation in the 'nuclear area' dated back as far as −5000, while the earliest Chinese rice finds, from Lungshanoid sites, proved to be several thousand years earlier than anything known from India, *c.* −4000 in the Yangtze Delta compared to −1700 at Harappa.[d] Citing the wide distribution of wild rices in modern China (see p. 483), Ho deduced an indigenous Chinese domestication of rice by Lungshanoid farmers.[e] In sum, then, Ho's widely accepted view was that plant domestication in China was the discovery of the −5th millennium Yang-shao millet farmers of the northern 'Nuclear Area', who brought rice and other plants under cultivation as they migrated east and south during the phase of Lungshanoid expansion. The idea that rice was first domesticated in South China or mainland Southeast Asia by migrant farmers from the North, accustomed to cereal cultivation and unable to grow millet successfully in a tropical climate, had also been proposed by Barrau, just before Ho published his papers.[f]

However, the interpretation of southeastern neolithic cultures as the product of migration from a nuclear area in North China received a considerable setback as more early sites were excavated in the southern provinces. In 1976 domesticated rice was discovered in large quantities at the site of Ho-mu-tu[1] in the Yangtze Delta, and carbon-14 dating put its earliest rice remains at *c.* −5000, that is about the same date as the earliest Yang-shao site at Pan-pho[2] in Shensi.[g] Even before this, however, another school of thought had been gathering strength. On the basis of archaeological research in Southeast Asia, Solheim (2) first proposed in 1967 that the earliest Far Eastern domestication of plants and animals was incipient in mainland Southeast Asia as early as −10,000, among the late

[a] (5), (6), (*1*).
[b] (5), p. 71.
[c] Chinese carbon-14 dates are based on a half-life of 5730 years, not 5570 years as is usually the case in the West, and are sometimes (but not always) calibrated to dendrochronology. The implications of dendrochronological correction are discussed by Renfrew (4), as is the question of half-life. The differences in Chinese convention mean that Chinese dates should be slightly adjusted if they are to be compared with dates calculated elsewhere, but for comparisons of sequences within China the problem does not arise. It should be noted that in any case carbon-14 dates are indicative rather than absolute; with this proviso the Chinese dates are just as accurate as any others.
[d] The Harappan evidence was, in any case, later shown to be based on a misidentification; C. A. Reed (4), p. 918.
[e] (5), p. 71.
[f] Barrau (1a); also K. C. Chang (6), p. 183.
[g] Anon. (*503*), (*504*); Yu Hsiu-Ling (*1*).

[1] 河姆渡 [2] 半坡

Hoabinhian cultures.[a] Solheim and other Southeast Asian archaeologists suggested that in fact the Chinese neolithic cultures, both Yang-shao and Lungshanoid, owed much of their impetus to Hoabinhian influence from the south.[b] The presence of Hoabinhian assemblages from an early site at Hsien-jen-tung[1] in Kiangsi was taken to confirm that the direction of diffusion of agricultural technology was from south to north,[c] not from north to south as had previously been postulated, and Solheim raised the possibility that 'one or more of the cultures of South China was the primary ancestor of Chinese culture [generally].'[d]

While the school of Southeast Asian diffusionists saw North Chinese cultures as a direct product of southern influence, others felt that, although signs of a common palaeolithic heritage could be discerned in the various Chinese neolithic cultures, there was much to be said for regarding them as largely independent developments:[e]

> Scholars used to worry about proving or disproving that the Chinese Neolithic was derived from the Near East. Now the question concerns the detailed process whereby the terminal Palaeolithic people expanded their reliance upon food plants and eventually came to domesticate some of them for principal food sources.
>
> Presently available archaeological evidence points to two regions where the initial switch from the Palaeolithic to the Neolithic way of life occurred, namely the Huang Ho basin of North China, where millets were the centre of attention, and the southeastern coastal areas, where there was probably a greater dependence on roots and tubers. Even though the minute links are not yet completely available, there can be no question now that the Yang-shao and the Ta-phen-kheng[2] [Taiwanese] cultures grew indigenously from their respective Palaeolithic bases. The fact that both the initial phase of the Yang-shao culture and the Ta-phen-kheng culture were characterised by cord-marked pottery (with incised designs) suggests some kind of interrelationship of the two, but in view of the very different material inventories in general, it does not appear that either can be regarded as a derivative of the other. The Yang-shao culture was wholly confined to North China, but the Ta-phen-kheng culture resembles in some respects the Hoabinhian

[a] 'The term Hoabinhian entered the literature of archaeology at the First Congress of Prehistorians of the Far East, meeting in Hanoi in 1932 (Matthews (2), p. 86), where, following spectacular discoveries by French archaeologists in the limestone caves of Tonkin, it was described as a culture composed of implements, flaked with a rather primitive technique, usually split river pebbles worked often on only one face, including discs, short axes, almond-shaped tools and with many hammer and grinding stones and bone tools. Three stages of the Hoabinhian culture were claimed: I, with large and crude flaked tools only; II, smaller core tools, a few used flakes and occasionally pebbles with ground edges; and III, dominated by edge-ground tools and with some pottery, usually with cord impressions' (Glover (1), p. 145).

Recent excavations of Hoabinhian sites have produced some carbon-dates which suggest that the Hoabinhian spans the Pleistocene-Recent boundary and continues in some areas into quite recent times. Glover (1) points out that there are many difficulties attendant on the use of the term Hoabinhian if taken as referring to a culture, since it 'ignores significant differences rather than emphasising precise similarities' (Matthews (1), p. 1). Gorman ((2), p. 300, following D. Clarke (1), p. 357) prefers to define the Hoabinhian as a 'technocomplex', and presents strong arguments for an indigenous development of the techniques of plant and animal domestication in Southeast Asia within the context of this Hoabinhian technocomplex.

[b] Bayard (1).

[c] Aigner (1).

[d] (3), p. 25.

[e] K. C. Chang (1), 1977 ed., p. 141; see also Meacham (2).

[1] 仙人洞 [2] 大坌坑

culture of Vietnam and the rest of Indo-China, so much so that many issues would depend on a consideration of both cultures.

K. C. Chang then identified a third neolithic culture on the Huai and Lower Yangtze plains, the Chhing-lien-kang[1] culture, but felt that given the current state of knowledge it was impossible to draw any firm conclusions as to its origins.[a] At that stage the earliest known Lower Yangtze cultures were tentatively dated to the late −5th millennium, and Chang assumed that the area owed its culture to an influx of neolithic migrants either from the Yang-shao or the Ta-phen-kheng culture.

Since then there have been some very exciting archaeological discoveries in China. Perhaps the most spectacular was the find of the early neolithic village of Ho-mu-tu[2] near Shaohsing in 1976: the lowest cultural stratum, carbon-dated to about −5000, showed considerable cultural sophistication. The wooden houses were built on stilts using relatively advanced carpentry techniques; pottery, though of rather coarse paste, included beautifully decorated incised ware quite unlike anything known elsewhere in China; and large quantities of rice, identified as the domesticate *Oryza sativa*, were found.[b] The early date and relative technological sophistication of the Ho-mu-tu culture seemed to rule out any cultural influx from the contemporary Yang-shao area, as did its heavy reliance on rice. Meanwhile, several neolithic-type sites were excavated along the South China coast and in Kwangtung and Kwangsi, containing cord-marked pottery and polished stone tools but no signs of agriculture;[c] and in the 'Nuclear Area' of North China itself, in Hopei, Honan and Shensi, a number of pre-Yang-shao neolithic sites have recently been excavated,[d] the best known being Tzhu-shan[3] in Hopei and Phei-li-kang[4] in Honan. These last sites are characterised by coarse, cord-marked or comb-marked pottery and what An Chih-Min describes as 'evidence of agriculture': i.e. polished grinding-stones and stone rollers, round-bladed stone 'spades' and stone sickles, as well as carbonised millet, the clay figurine of a pig, and pig and dog bones.[e] Other reports are more cautious, however, interpreting the remains as evidence of a foraging rather than a truly agricultural economy.[f] The millet remains have apparently not been scientifically identified yet.[g]

The archaeological evidence to date for the origins and development of Chinese agriculture is summarised in Table 2. As more and earlier archaeological sites

[a] *Ibid.* p. 142.

[b] Anon. (*503*), (*504*); Yu Hsiu-Ling (*1*).

[c] Anon. (*515*), (*516*). The sites of the carboniferous zones of South China cannot be dated accurately by carbon-14 as the soil conditions affect the results, giving much earlier dates than is reasonable; but although absolute dates cannot be obtained the sequences are accurate; Hsia Nai (*6*).

[d] An Chih-Min (*4*): Chou Pen-Hsiung (*1*); Han-tan *CPAM* (*1*); Hopei *CPAM* (*1*); Khai-feng *CPAM* (*1*), (*2*), (*3*).

[e] (*4*), p. 253.

[f] Han-tan *CPAM* (*1*).

[g] In any case it is difficult to distinguish wild from domesticated setaria under such conditions; see p. 437.

[1] 青蓮崗 [2] 河姆渡 [3] 磁山 [4] 裴李崗

## Table 2. *Cultural sequences in neolithic China*

Time scale printed at both sides of the table: −5,500, −5,000, −4,000, −3,000, −2,000

| KANSU | SHENSI | HONAN | HUPEI | N. KIANGSU/ SHANTUNG | S.KIANGSU/ CHEKIANG | FUKIEN/ TAIWAN | N. VIETNAM |
|---|---|---|---|---|---|---|---|
| | | Proto-Yang-shao Phei-li-kang 裴李崗 (5,500) Tzhu-shan 磁山 (harvesting tools/ ?millet/?pig) | | | | | Băcsón (?incipient horticulture) |
| | Pan-pho 半坡 (4800–4300) (setaria millet, brassica, hemp/ dog, pig, silkworm) | Hou-kang 後崗 (4150) | | Chhing-lien-kang 青蓮崗 (*no* rice, ?? millet/dog, pig) | Ho-mu-tu Phase I 河姆渡 (5000) (rice, aquatic plants/dog, pig, ?water buffalo) | Quemoy (5300–4200) (shell mounds/ ?early horticulture) | Da-bút (?horticulture/?pigs and chickens) |
| Yang-shao 仰韶 (setaria millet, hemp/dog, pig) | Miao-ti-kou I 廟底溝 (3900) | Miao-ti-kou I 廟底溝 (3900) | Yang-shao 仰韶 | Liu-lin 劉林 (4000) | Ma-chia-pang 馬家浜 (4700–4000) (rice, aquatic plants/?water buffalo) | Ta-phen-kheng 大坌坑 (4300) (shell mounds/ ?early horticulture) | Dâu-du'ong (rice/cattle) |
| Ma-chia-yao 馬家窰 (3100–2600) | Ta-ho-tshun 大河村 (3800–3100) | Ta-ho-tshun 大河村 (3800–3100) | Chhü-chia-ling 屈家嶺 (rice, ??peanuts/ dog, pig, sheep, poultry) | Hua-thing 花廳 | Sung-tse 松澤 (4000–3000) (rice) | Feng-pi-thou 鳳鼻頭 (2500–400) (developed rice agriculture) | Red River cultures (lowland rice/ buffalo) |
| Pan-shan 半山 (2400) | Miao-ti-kou II (2800) (chicken, dog, pig) | Miao-ti-kou II (2800) (chicken, dog, pig) | 'Honan' Lung-shan 龍山 | Ta-wen-khou 大汶口 (dog, pig, cattle, sheep) | Liang-chu 良渚 (3300–2300) (both *hsien* 秈 and *keng* 粳 varieties of rice) | | Bronze Age cultures |
| Ma-chhang 馬廠 (2300–2100) | Shensi Lung-shan 龍山 (dog, pig, sheep, cattle, water buffalo) | Honan Lung-shan 龍山 (2350) (pig) | Shang | 'Classic Lung-shan' 龍山 (millet/pig, dog, cattle, sheep) | Hu-shu 湖熟 (sometimes called 'Geometric') | | |
| Chhi-chia 齊家 (2150–1700) (millet/pig/copper) | W. Chou | Erh-li-thou [?Hsia] 二里頭 (2050) | | Shang | | | |
| Shang | | Shang | | | | | |

are discovered in North and South China and neighbouring Southeast Asia, it becomes increasingly difficult to propose any single unifying theory linking the development of individual cultures.[a] Any single-centre diffusionist explanation of China's agricultural origins must be treated with extreme caution, especially those that attempt to account for the origins of southern rice domestication by the arrival of northern millet-growers, or to explain Yang-shao farming as an offshoot of earlier southern rice cultivation. But given the widespread occurrence of pre-agricultural sites with polished stone tools and cord-marked wares, there does seem a case for postulating a widespread East and Southeast Asian Hoabinhian substratum, with a foraging economy supplemented by plant-tending. Increased pressure on food resources, possibly caused by such factors as climatic fluctuation or a rise in sea-level, might have pushed Hoabinhians in some areas to rely increasingly on food production rather than hunting or foraging, domesticating tubers in tropical zones, rice in semi-tropical, marshy areas, and millets in the drier north.

Alternatively one might accept Gorman's and Solheim's view that the principles of plant domestication were in fact discovered in mainland Southeast Asia and gradually spread from there. It can hardly be claimed that the case for a Southeast Asian origin of plant domestication is conclusive; as Reed dryly remarks after a careful review of the literature, 'I find the evidence insufficient to provide the kind of foundation for priority of agriculture for Southeast Asia claimed by some and copied unwittingly by others.'[b] Yet the circumstantial evidence is thought-provoking. Firstly there is the question of the geographical origins of the most important domesticates, in particular setaria and panicum millets, rice and tubers. H. L. Li (15) sees central and southern China as a buffer zone between two separate areas of original plant domestication: one in North China where panicum and setaria millets were domesticated, the other in mainland Southeast Asia where rice, Job's tears (*Coix lacryma-jobi* L.) and various tubers such as taro and yam were first domesticated. Harlan, however, feels that many of the crops listed as originating in North China in fact probably originated in the south; he believes that few domesticates can be assigned with certainty to the northern Nuclear Area, and that the millets are not among these.[c] We discuss the problem of the origins of rice and millets at length later (see pp. 481, 435), and we tend to agree with Harlan. Both millets and rice appear to be of tropical rather than temperate origin. Furthermore, there is ethnographic and linguistic evidence to show that the cultivation of Job's tears, a perennial seed-bearing crop suited to horticultural rather than agricultural techniques,[d] preceded that of millets and rice in much of Southeast Asia, and in its turn was preceded by tubers such as yams and taros.[e]

[a] Chinese archaeologists have always been understandably reluctant to venture into this minefield, preferring to investigate relationships between contemporary neighbouring cultures, or regional cultural sequences; Mou & Wei (*1*); Nanking Museum (*1*); Wu Shan-Chhing (*1*); etc.

[b] (4), p. 909.

[c] (5), p. 214.

[d] A. K. Koul (1).

[e] Kano (1); Golson (1).

On the basis of excavations and palynological studies in Taiwan, K. C. Chang postulates that primitive slash-and-burn cultivation, possibly of tubers, was practised by cord-ware cultures in the forests of the island from about −9000, to be superseded in about −2200 by the cultivation of millet and rice, introduced by migrants from the Chinese mainland.[a] If Chang is correct in his interpretation of the Taiwanese evidence, then there would seem to be a case for assuming that some late Hoabinhians in China practised horticulture, growing mainly tubers but with a few cereals mixed in with them. Millets and tubers are found growing in the same gardens in Southeast Asia today, and the same is true of hill rice. It has been suggested that rice was initially encountered as a weed in taro gardens (taros, like rice, grow well on swampy ground), and that some rice-planting techniques (such as transplanting, and harvesting by cutting single stems with a knife) were perhaps directly borrowed from taro-cultivation.[b] Such horticultural techniques are well adapted to the tropical zones of Southeast Asia; but as they spread to the drier north the Hoabinhian cultivators would have had to abandon the cultivation of moisture-loving tubers to concentrate their skills on plants like millet which, although of tropical origin, could tolerate more arid conditions. In temperate but marshy areas such as the Yangtze Delta both tuber and rice cultivation were practicable, but rice would give higher yields than in the tropics and so might tend to exclude tuber cultivation.[c]

Such explanations are, for the moment, purely hypothetical. Although pre-farming and early farming sites are now emerging in some numbers in several areas of China, archaeological retrieval techniques are still unsophisticated and no really detailed studies of the relationship between economy and environment, of the nature of those carried out recently in Mexico and the Near East, are as yet contemplated in China. Furthermore, to prove or disprove any hypothesis based on the effects of Hoabinhian culture it would be necessary to consider the evidence for the region as a whole, not merely that provided by excavations in South China and Thailand but also evidence from Japan, Indochina, Burma and Assam, and Island Southeast Asia. Even where China is concerned, there are conspicuous regional lacunae: very little is known of early cultures in Szechwan and Yunnan, for instance, both of which are clearly areas of crucial importance. Recent political events in Southeast Asia have probably resulted in the destruction of a great deal of evidence, but Vietnamese archaeologists are now publishing some very interesting material,[d] while in Japan evidence is emerging that

[a] K. C. Chang (6); see also C. A. Reed (4), pp. 909–11 for an evaluation of Chang's case.

[b] Haudricourt (13), (14).

[c] See Grist (1): rice yields are higher in temperate than in tropical zones as they increase in proportion to day length.

[d] Davidson (1) quotes Vietnamese reports which seem to demonstrate a continuous sequence of technological elaboration, with the cultivation of plants beginning *c.* −6000, and the development of rice agriculture in the Red River area in the −4th millennium.

plant cultivation was practised well before the introduction of rice by the Yayoi[1] in the −3rd century.[a]

Another point to be considered is the extent to which the postulated Hoabinhian horticulturalists actually depended on cultivation. It is important not only to identify the origins of plant domestication but also to trace increasing dependence on horticulture or agriculture as a source of food. This clearly will be linked to problems of climatic fluctuation, demographic growth, changes in sea level, etc., but very little attention has been devoted as yet to the constraints which might have led to the development of agriculture in East Asia. All these factors must be taken into account when considering the origins and development of agriculture in China itself, and it will clearly be some time before any more definitive conclusions can be reached. It is quite clear, however, that as early as the −5th millennium (see Table 2) the two distinct agricultural traditions that we associate today with north and south China had already emerged, namely the cultivation of dryland cereals, especially millets, in the northern plateaux and plains, and the cultivation of wet rice in the river valleys and deltas to the south of the Huai. It is the elaboration and refinement of these two traditions that will occupy the rest of this volume.

## (*b*) SOURCES

A unique advantage for the historian of Chinese agriculture is the existence of a tradition of agricultural treatises and monographs stretching back unbroken for over two thousand years. Over five hundred agricultural works are known to have been written in China before the end of the Chhing dynasty,[b] many of which are still wholly or partly extant. The number of titles may seem moderate compared with the torrent of agricultural works that has appeared in Europe since the seventeenth century, and the skimpy remnants of the Han and pre-Han works bear no comparison with the great Roman agricultural treatises which have survived intact. But the Chinese agronomic corpus offers a unique continuity, with exactly the characteristics one would expect of an unbroken literary tradition: as the knowledge and expertise of agronomists grew over the centuries, earlier texts were superseded, neglected, and finally forgotten, replaced by more up-to-date works. The very earliest texts generally survive only as scattered quotations in later works, while the number of titles steadily increases century by century. A total of 78 agricultural works are known to have been written in the pre-Sung period, while 105 were produced during the three centuries of the Sung dynasty, 26 in the century of Yuan rule, and 310 in the five and a half centuries of the Ming and Chhing. True to general Chinese form, later works draw heavily on

[a] Sasaki (*1*). [b] Wang Yü-Hu (*1*).

[1] 彌生

earlier ones for their material, often quoting verbatim and at length; usually such citations are acknowledged, and it has thus been possible to reconstitute substantial sections of several early works. Some Chinese agricultural works carry their zeal for quotation to such extremes that they are almost devoid of original material,[a] while one or two authors are fiercely independent and original. Perhaps the best example is the Sung writer Chhen Fu[1], whose *Nung Shu* of 1149 was written expressly to refute the 'empty words' and 'irrelevances' of the *Chhi Min Yao Shu* and *Ssu Shih Tsuan Yao*,[b] a very unorthodox attitude.

The unbroken tradition of agricultural writing in China is due in part to the importance accorded to agriculture by the Chinese state. Since state revenues were derived in large part directly from the peasantry, statesmen and officials were naturally concerned with maintaining a reasonable level of agricultural production. All the political philosophers of the late Chou insisted on the importance of encouraging the peasants to farm efficiently (and of discouraging them from dissipating their energies in other, less fundamental activities); the Legalists were particularly insistent on the causal relation between agricultural prosperity and political power, and as one might expect, two of the most important agricultural works of the period are contained in Legalist works of political economy, the *Kuan Tzu* and the *Lü Shih Chhun Chhiu*.[c] Almost without exception the authors of Chinese agricultural works served at some time in the bureaucracy. Official service brought them into direct contact with agriculture not only in their administrative capacity, but also (in the earlier dynasties) by providing them with a sizeable estate from which they were supposed to derive their official income. In most cases, of course, such men would anyway have come from the landowning elite. Some writers, like Chhen Fu, lived in retirement and never served as officials, but generally speaking the most renowned agronomists, from Fan Sheng-Chih of the Former Han to Chia Ssu-Hsieh, the Northern Wei author of the *Chhi Min Yao Shu*, Wang Chen, the Yuan author of the *Wang Chen Nung Shu*, and Hsü Kuang-Chhi, the late Ming author of the *Nung Cheng Chhüan Shu*, all received one or several official postings. These authors were mainly concerned with agriculture proper: the clearing and tillage of arable land, irrigation and drainage, and the cultivation of field crops. Another class of agricultural litera-

[a] This is true not only of those works specifically intended as compendia but is a general tendency in the majority of Chinese agricultural works. In this the Chinese agronomic tradition differs fundamentally from the European.

[b] *CFNS* 1/1. Chhen's righteous indignation may be largely attributed to the fact that the Northern Wei and Thang works treated conditions in North China, whereas Chhen's experience was all of the rice-growing areas of the Lower Yangtze; Hu Tao-Ching (*12*), Wan Kuo-Ting (*6*).

[c] The *Kuan Tzu Ti Yuan Phien* is a description of the soils and natural products native to different regions of China; it is a text of fundamental importance in the development of Chinese soil science and has been discussed at length in Section 38; see also Wang Yü-Hu (*1*), p. 3; Hsia Wei-Ying (*2*). The *Lü Shih Chhun Chhiu* contains four sections on agriculture, *Shang Nung*[2] (Putting Agriculture First), *Jen Ti*[3] (The Land's Requirements), *Phien Thu*[4] (Distinguishing Between Soils), and *Shen Shih*[5] (Examining the Seasons). It has been suggested that these chapters draw heavily on an earlier agricultural work, the *Shen Nung Shu*; see Hsia Wei-Ying (*3*).

[1] 陳旉 [2] 上農 [3] 任地 [4] 辯土 [5] 審時

ture seems to have been the domain of retired officials or of those literati who never entered the official world, people who had many peaceful hours to devote to more gentlemanly rural pursuits than pushing the plough, namely a whole series of works on horticulture and the arts of leisure.[a]

The Chinese state apparatus not only contributed significantly to the ranks of agricultural writers, it also actively encouraged the composition and dissemination of agricultural treatises, especially once printing had been developed. The earliest officially commissioned agricultural work to have survived to the present day is the Yuan work *Nung Sang Chi Yao*, but the earliest work in this genre was probably produced in the Thang.[b] In the Sung dynasty both the *Chhi Min Yao Shu* and the *Ssu Shih Tsuan Yao* were printed and distributed by imperial order.[c] The agricultural book in which the Chinese emperors seem to have taken most personal interest is the charming *Keng Chih Thu*, Agriculture and Sericulture Illustrated. In about 1145 Lou Shou[1] presented to the emperor Kao-Tsung[2] a series of twenty pictures on the various stages of rice cultivation and a similar series on sericulture, each with an appropriate poem. Kao-Tsung and his wife, who were both calligraphers, themselves annotated some of the poems. The whole series was engraved on stone by Lou's nephew and grandson in about 1210 and also went through a series of printed editions, but it gained its greatest prestige during the Chhing dynasty when no fewer than three emperors, Khang-Hsi, Yung-Cheng and Chhien-Lung, composed additional sets of poems to accompany the pictures, and the whole work was printed several times by imperial order (Fig. 15).[d]

The development of printing by Sung times enabled the government to disseminate both old and newly commissioned agricultural works, and tempted many private individuals to commit their farming experience to paper (Fig. 16). As we have seen, the number of agricultural works published in China increased exponentially from the Sung on. Woodblock printing also permitted the use of illustrations, yet it is striking that Chinese agronomists made comparatively little use of this facility. The great pioneer in this field was of course Wang Chen, author of the *Wang Chen Nung Shu* of 1313, but sadly for the historian of technology most subsequent authors were content simply to crib their illustrations, if any, from the *Wang Chen Nung Shu* or *Keng Chih Thu*.

The identification and classification of the multifarious Chinese agronomic works is complex and difficult. The Chinese are past masters at the fastidious art of

[a] Hu Tao-Ching (7), p. 52.

[b] Yang Chih-Min (*1*), p. 4.

[c] *Ibid.*

[d] The history of the *Keng Chih Thu* and of the different versions of the illustrations is almost impossible to unravel. O. Franke (11) reproduces two sets of illustrations, one a Japanese edition based on a Ming original, the other the drawings commissioned for the 1739 Chinese edition. P. Pelliot (24) reproduces the Semallé Scroll illustrations which may have been based on a Yuan or even a Sung original. D. Kuhn (2) provides a systematic and critical survey of the different editions and their derivation. See also F. Jäger (4).

[1] 樓璹 [2] 高宗

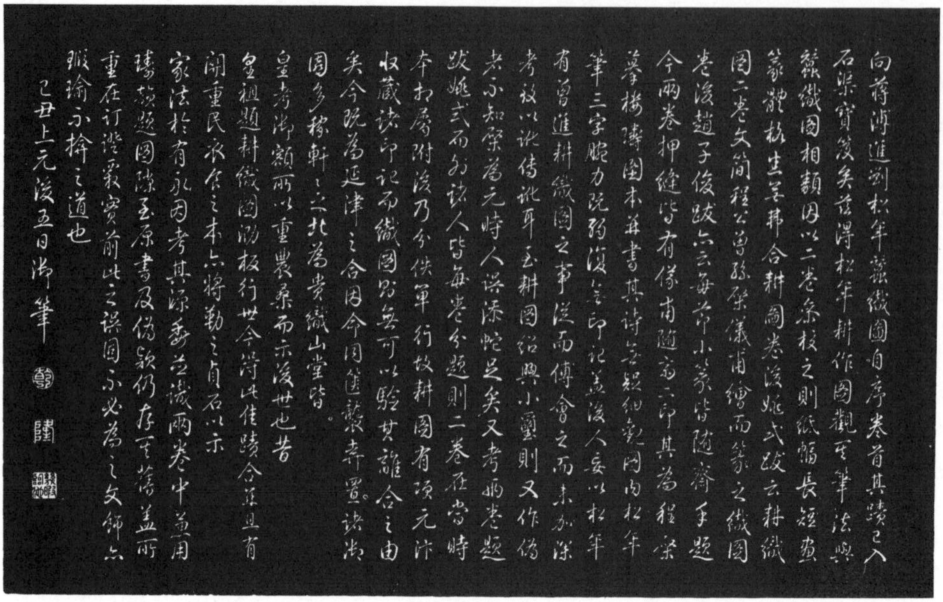

Fig. 15. The Semallé Scroll version of the *Keng Chih Thu*, showing Chhien-Lung's preface of 1769; Pelliot (24), pl. X.

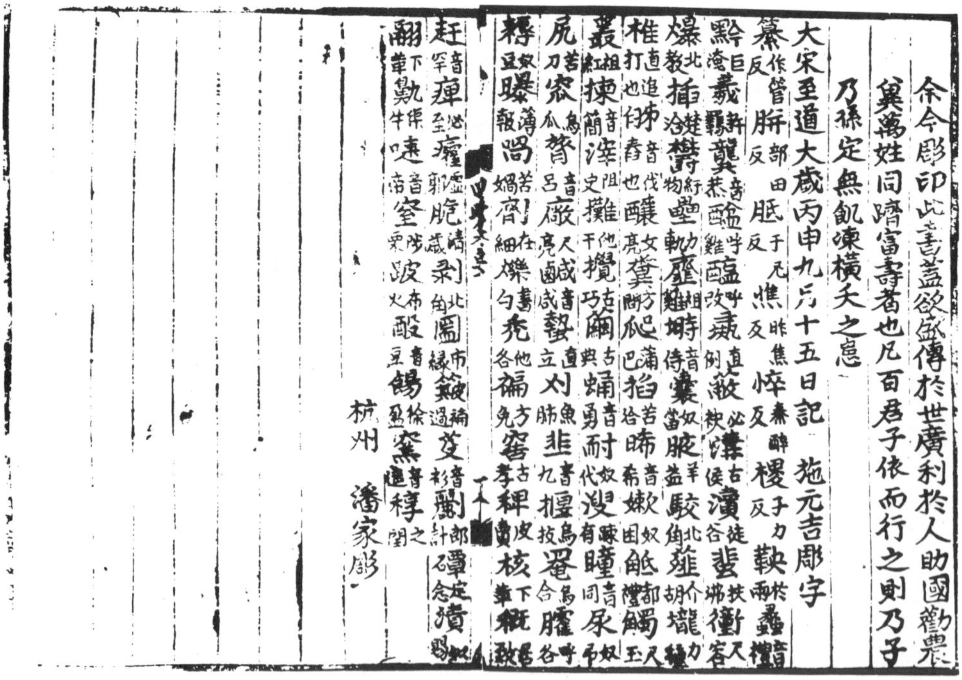

余今彫印此書蓋欲流傳於世廣利於人助國勸農
冀萬姓同躋富壽者也凡百君子依而行之則乃子
乃孫定無飢凍横夭之患
大宋至道大歲丙申九月十五日記 施元吉彫字
杭州 潘家彫

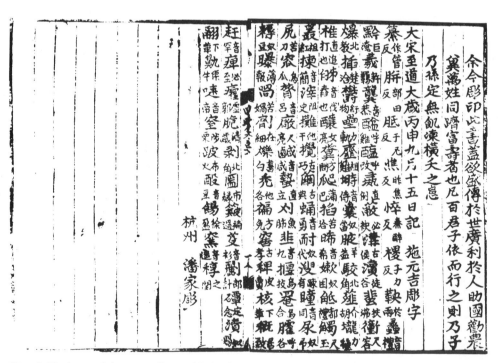

Fig. 16. Final page of the +1590 Korean block-print edition of the *Ssu Min Tsuan Yao*, copied from a popular Hangchou edition of +996 which predated the official printing by at least twenty-five years. No original Sung edition of this, or of any other, agricultural work survives.

bibliography; the earliest bibliography of agricultural works is to be found in the *Chhien Han Shu*,[a] which lists nine agriculturalists (*nung chia*[1]) whose works totalled 114 chapters (*phien*[2]). They include the *Shen Nung Shu* in 20 *phien*, the *Yeh Lao Shu* in 17 *phien*, the *Tsai Shih* in 17 *phien*, the *Tung An-Kuo Shu* in 16 *phien*, the *Yin Tu-Wei Shu* in 14 *phien*, the *Chao Shih Shu* in 5 *phien*, the *Fan Sheng-Chih Shu* in 18 *phien*, the *Wang Shih Shu* in 6 *phien* and the *Tshai Kuei Shu* in 1 *phien*. All these works are now lost except for scattered quotations,[b] though what survives of the *Fan Sheng-Chih Shu*, the best preserved amongst them which is quoted at some length in the *Chhi Min Yao Shu*, is sufficient to show that the Han agronomists were skilled farmers with an eye for practical detail.[c] It is hard now to imagine what form the originals took, but if the *phien* were comparable in length to those of other contemporary literary genres, then the *Shen Nung Shu*, *Fan Sheng-Chih Shu* and their ilk must have been works nearly comparable in size and scope to the Roman agricultural treatises.

Subsequent bibliographies included in the dynastic histories and imperial catalogues are invaluable for tracing the development of Chinese agricultural literature: not only do they list title and length of works many of which have long since disappeared, but they provide the author's name, sometimes with a few biographical details, or even a brief summary of the contents of the work. It has thus been possible to reconstitute a list of Chinese agricultural works from late Chou to modern times which must be very nearly complete, but only in this century has the task been undertaken. A pioneer in this field was Mao Yung (*1*), whose bibliography of Chinese agricultural literature was first published in 1924. But the definitive works in the field are annotated bibliographies by two outstanding scholars, Wang Yü-Hu (*1*) and Amano Motonosuke (*9*). Wang's work was first published in 1964. It contains a complete chronological list of all known Chinese agricultural works up to the twentieth century, with a brief biography of each author, an outline of the contents, and an account of the various editions; it is supplemented with indexes arranged according to title, author's name, and subject.[d] In 1975 the Japanese historian Amano Motonosuke published a Japanese edition of Wang's bibliography (*8*), together with a companion volume (*9*) in which he presents a detailed comparative analysis of the different editions of each work. Together these two books form an indispensable research tool for anyone interested in the history of Chinese agriculture.[e]

In his bibliography Wang distinguishes between fourteen categories of agricultural work, but of course many other classification systems could be and have

[a] *I Wen Chih* 30/20*a*–*b*.

[b] They were all reconstituted by the Chhing scholar Ma Kuo-Han[3] in *YHSF* 69, but this version of the texts has since been severely criticised; Wang Yü-Hu (*1*).

[c] Shih Sheng-Han (2), (*5*); Wan Kuo-Ting (*1*).

[d] Wang distinguishes fourteen categories: general works; meteorology and divination; cultivation techniques and water control; implements; field crops; trees, bamboos and tea; locust control; horticulture; vegetables, cultivated and wild; fruits; flowers; sericulture; animal husbandry and veterinary science; aquaculture.

[e] See also Chhü Chih-Sheng (*1*); Hu Tao-Ching (*6*), (*11*); Huanan Agricultural College (*1*).

1 農家 2 篇 3 馬國翰

been devised.[a] The chief difficulty lies not so much in categorising those texts accepted as agricultural, as in deciding which works to include within the general classification. As we have already pointed out, most educated Chinese were concerned with agriculture in some capacity or other, and a great many Chinese writings reflect this concern. The eminent Sung scientist Shen Kua[1], for example, devoted a great deal of time and energy to developing improved systems of water control. Should his notes on irrigation be included within the agricultural corpus?[b] In the following pages we shall concentrate on those works usually accepted as agronomic treatises, but our categorisation will be much simpler than Wang's. We shall briefly describe four main genres: agricultural calendars or *yüeh ling*[2], agricultural treatises by individual authors, state-sponsored compilations, and specialist monographs; we shall include a brief description and assessment of one or two outstanding works in these genres. We shall then consider in more general terms the other sources which contribute to our overall knowledge of Chinese agriculture.

### (1) The *yüeh ling*[2] or Agricultural Calendars

This literary form dates back to very early times in China. China's harsh and unpredictable climate compelled Chinese farmers to make a careful study of phenology (*wu hou*[3]).[c] Perhaps the earliest text to link natural phenomena, such as the position of constellations or the flowering of certain plants, with agricultural tasks was the *Chhi Yüeh*[4], Seventh Month, in the Odes of Pin in the *Shih Ching*.[d] Chinese calendrical computation of course dates back much earlier than the *Shih Ching* and was a developed science by Yin-Shang times,[e] while Chinese farmers had certainly used natural signals like the blooming of the sweet flag to time their operations ever since farming began. There is a tendency for modern scientists, imbued with the idea of linear progress, to imagine that man's observation of his environment has become more accurate over time. The reverse is rather the case, for as man develops his technical skills and becomes more independent of nature he is bound to pay it less attention. It is among hunter-gatherers and primitive agriculturalists, especially those living in a harsh or unreliable environment, that the most sophisticated correlations of natural phenomena with the tasks essential to human survival are to be found. This inverse development is clearly reflected in the Chinese agricultural calendars: references to plants or stars in the early texts increasingly give way to simple dates reckoned by the lunar month (*yüeh*[5]) or solar term (*chieh chhi*[6]).

[a] Yang Chih-Min (*1*).

[b] Hu Tao-Ching (*7*). Shen Kua did in fact write two works which were unambiguously agronomic in content, the *Meng Chhi Wang Huai Lu* and the *Chha Lun*, but both have unfortunately been lost; Hu Tao-Ching (*8*), (*13*).

[c] Chu Kho-Chen (*10*).

[d] *Shih Ching* 15/3*a*; tr. Legge (8), p. 226.

[e] Tung Tso-Pin (*1*), (*6*); Thang Han-Liang (*1*).

[1] 沈括 [2] 月令 [3] 物候 [4] 七月 [5] 月 [6] 節氣

The first agricultural calendars date from the Chou dynasty. One, the *Hsia Hsiao Cheng* or Lesser Annuary of the Hsia, describes natural conditions in the area between the Huai and the Yellow River: the surviving text of 473 characters is divided into twelve monthly sections, subdivided to include astronomical configurations, phenology, agriculture, sericulture, animal husbandry and hunting.[a] Estimates as to its date differ wildly, one author seeing it on astronomical grounds as a pre-Shang text,[b] another as a late Han compilation.[c] On the grounds that it contains no references to *wu hsing*, *yin yang*, or the sexagenary cycle, we ourselves feel that it is probably pre-Han at the latest. Another early agricultural almanac, perhaps the most influential of all, is the *yüeh ling* section of the *Li Chi*, from which the genre took its name. Since the *Li Chi* was one of the canonical texts of Confucianism, its *yüeh ling* section was required reading for any educated man, and so it was the best known of all the Chinese agricultural calendars. Its contents are far more varied than those of the *Hsia Hsiao Cheng*, including prescriptions on ritual, etiquette and other topics, as well as the basic productive tasks.[d]

Each of these two early works became the model for a long tradition. The *Li Chi Yüeh Ling* was the forerunner of a series of almanacs containing mystical-religious material, matters of etiquette, dietary prescriptions and the like as well as agricultural precepts. Among the important works in this genre we may count the +2nd-century *Ssu Min Yüeh Ling*,[e] the 6th-century *Ching Chhu Sui Shih Chi*,[f] the *Thang Yüeh Ling*, the 8th-century *Ssu Shih Tsuan Yao* (which together with the *Chhi Min Yao Shu* was one of the first agricultural works to be printed by imperial order in the early Sung),[g] the Ming work *Pien Min Thu Tsuan*, and the 18th-century *Nung Phu Pien Lan*.[h] The *Hsia Hsiao Cheng*, with its emphasis on practical agricultural matters and its omission of mysticism or religion, may be seen as the prototype for a whole series of chapters in agricultural treatises, where one or several sections are devoted to a chronological enumeration of operations, subdivided normally into months but sometimes into seasons or solar terms. Among the works that include such sections are the *Nung Sang I Shih Tsho Yao*, published by Lu Ming-Shan[1] in 1314 (which serves in many ways as a companion volume to the *Nung Sang Chi Yao* which we shall discuss below,[i] p. 71), the *Nung Cheng Chhüan Shu*, the late Ming *Shen Shih Nung Shu* and its early Chhing supplement the *Pu Nung Shu*,[j] the Chhing imperial compilation *Shou Shih Thung Khao* and the mid-Chhing *San Nung Chi*. But perhaps the outstanding achievement in this strictly utilitarian genre is the calendrical diagram in Wang Chen's *Nung Shu* of 1313 (Fig. 17), in which 'the Heavenly Stems, the Earthly Branches, the four seasons,

[a] Hsia Wei-Ying (*4*); Grynpas (1) gives a French translation and commentary.
[b] Chhen Chiu-Chin (*1*).
[c] Shih Sheng-Han (*2*).
[d] Tsou Shu-Wen (*1*).
[e] Shih Sheng-Han (*2*).
[f] Tung Khai-Chhen (*1*).
[g] Wan Kuo-Ting (*8*).
[h] Tung Khai-Chhen (*1*); Wang Yü-Hu (*3*).
[i] Wang Yü-Hu (*4*); Tung Khai-Chhen (*1*).
[j] Chheng Heng-Li & Wang Ta-Tshan (*1*).

1 魯明善

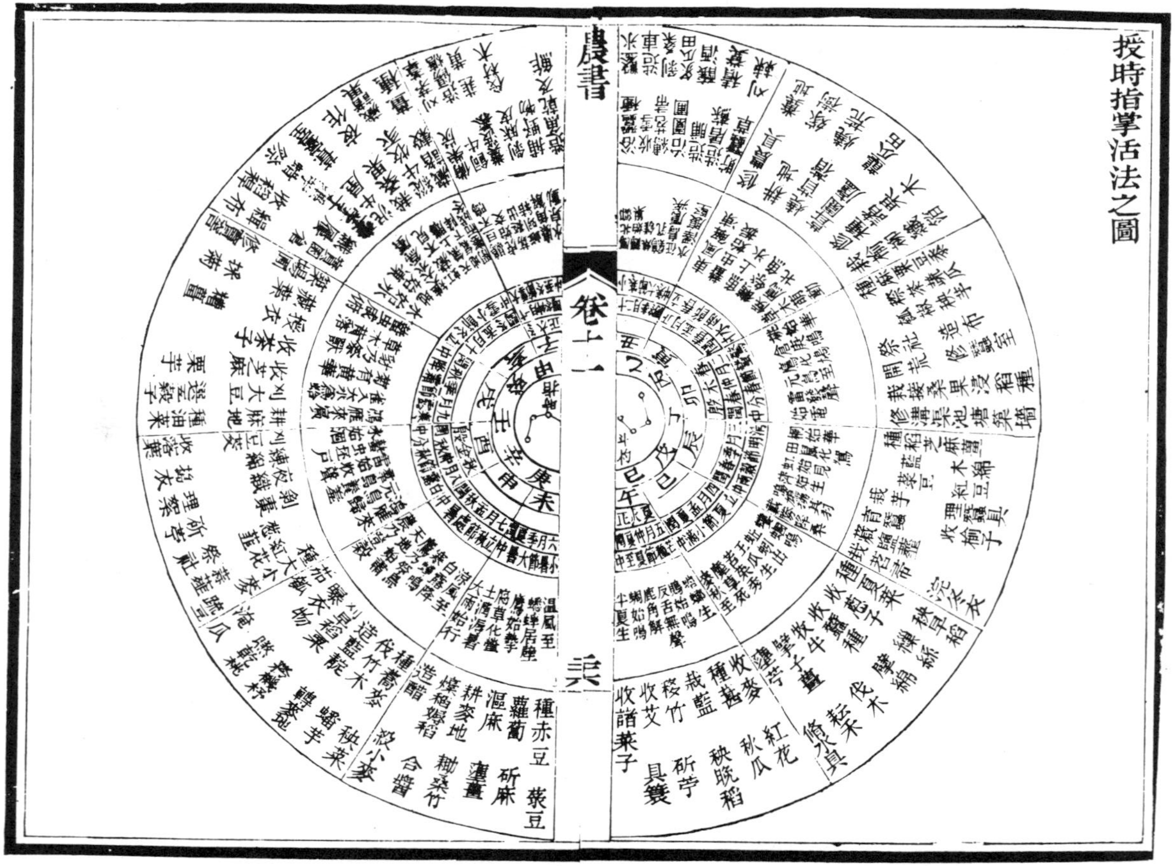


Fig. 17. Calendrical diagram from the +1313 *Wang Chen Nung Shu*; 11/26*a*–*b*.

twelve months, twenty-four solar terms and seventy-two five-day periods, each sequence of agricultural tasks and the natural phenomena which signal their necessity, stellar configurations, seasons, phenology, and the sequences of agricultural production, are cunningly and concisely assembled so that all the essential points of a farming almanac are united in a single small circle'.[a] This marvellous application of the *yüeh ling* principle was to be reproduced extensively in later works.

Wang Yü-Hu (*1*) lists twenty-one works in the category of *yüeh ling*, and although only seven of these still survive, many other Chinese agricultural works, as we have seen, contained sections in *yüeh ling* form. The tradition has survived right up to the present day, but with modifications. Although Chu Kho-Chen (*9*) attempted to create a twentieth-century version of the traditional *yüeh ling* form it met with comparatively little success; far more popular now are the almanacs covering a wide range of topics such as are still put out yearly by some Chinese publishing houses,[b] but the most widely-used farming calendars in China today are those based on an elaboration of Wang Chen's calendrical diagram, detailed chronological tables that permit advanced planning of year-round farm management and include sections on chemical fertilisers, pesticides and modern crop varieties rather than flying elm pollen or the blooming of the sweet flag.[c]

## (2) Agricultural Treatises

By agricultural treatises we mean works that provide a systematic, comprehensive description of agricultural methods and techniques. In China such works fall into two main categories: the works by individual authors considered here, which contain a relatively high proportion of original material, and the state-commissioned compilations considered in the next section, which consist almost entirely of quotations.

### (i) *The* Chhi Min Yao Shu *or Essential Techniques for the Peasantry*

Although earlier works like the *Shen Nung Shu* or *Fan Sheng-Chih Shu*, if we are to judge from the fragments which still survive, may have been agricultural works of considerable importance, the *Chhi Min Yao Shu* of *c.* +535 is the earliest Chinese agricultural treatise to have survived in its entirety. It is a long and impressive work, logical and systematic in its arrangement, comprehensive and detailed in its treatment, and a model for all subsequent Chinese agronomists. Here the author Chia Ssu-Hsieh[1] introduces his masterpiece:[d]

[a] Shih Sheng-Han (*7*).
[b] Tung Khai-Chhen (*1*), (1).
[c] E.g. Anon (*519*), Sinkiang Agricultural Institute (*1*).
[d] *CMYS* 0.12, tr. auct. Our references are to the 1957 edition by Shih Sheng-Han (*3*), which is conveniently divided into numbered chapters and paragraphs.

1 賈思勰

I have gleaned material from traditional texts and from folk-songs, I have enquired for information from old men and learned myself from practical experience. From ploughing to pickles there is no domestic or farming activity that I have not described exhaustively. I call my book 'Essential Techniques for the Peasantry'. In all it comprises 92 chapters divided into ten books. At the head of each book there is a table of contents; although this makes the text more complicated, it means that it is much easier to find things...

My intention in writing this was to instruct the youngsters of my family, and I did not intend it for educated readers. I repeat myself often, trying to drum in instructions on every task, and have not bothered with fine phrases. I hope that no reader will laugh at this.

The conventional literary style of the period was elaborate, flowery and allusive, but Chia's text is resolutely plain and lucid, and considering its many vicissitudes over the next fifteen centuries it is astonishingly comprehensible even today.[a] Chia's logical approach and sobriety of style was a pattern to which later agronomists adhered closely.

The *Chhi Min Yao Shu* is a long book, over one hundred thousand characters, and it quotes from more than 160 other works, sometimes at considerable length.[b] Our present versions of *Fan Sheng-Chih Shu* and *Ssu Min Yüeh Ling* are based almost exclusively on passages cited in *Chhi Min Yao Shu*, and the sections on brewing and culinary technology preserve *in extenso* several pre-Sui treatises of which we should otherwise have no knowledge. The last section, on exotic products, most of which originated from South China or Vietnam, also provides many valuable quotations from early floras. Almost half of the book consists of quotations, but the main body of the text is from Chia Ssu-Hsieh's own hand. Little is known about the author except that he served as a middle-ranking official;[c] however it is generally assumed that his agricultural experience was based on conditions in the Shantung area.[d] The work describes in considerable detail the practical details of running an agricultural estate, cultivating both subsistence and commercial crops and directing a number of household manufactures and culinary preparations. Table 3 gives a list of the contents of *Chhi Min Yao Shu*; the order generally reflects the relative importance accorded to different crops or activities at the time, a principle followed by most later agronomists.

There has been much debate as to whether *Chhi Min Yao Shu* was written with peasant farmers in mind, as the title rather implies, or intended as a handbook for

[a] On the various editions of *Chhi Min Yao Shu*, their derivation, relationship, and fate see Shih Sheng-Han (*3*); Amano (*7*), (*9*) pp. 29–43; Nishiyama and Kumashiro (*1*) pp. 1–23. Shih provides a parallel translation from classical into modern Chinese (see also Tachai Brigade (*1*)), Nishiyama and Kumashiro a modern Japanese version, and while it is true that these scholars disagree on many aspects of interpretation, these are generally points of detail rather than fundamental differences. We ourselves are preparing a translation of *CMYS* into English which we hope will be published in the near future; Bray (5).

[b] Shih Sheng-Han (*4*).

[c] Herzer (2); Chekiang Agricultural College (*1*); Liang Chia-Mien (*4*).

[d] Amano (1), (7); Kumashiro (1), (2); Shih Sheng-Han (1), (*4*); Liang Kuang-Shang (*1*); Li Chhang-Nien (*3*); Wan Kuo-Ting (*4*).

Table 3. *Contents of* Chhi Min Yao Shu

| | | |
|---|---|---|
| Book I | Chapter 1: | clearing and tilling land |
| | 2: | collection of seed grain |
| | 3: | cultivation of setaria millet |
| Book II | 4–16: | cultivation of field crops (cereals, beans, etc.) |
| Book III | 17–29: | cultivation of vegetables |
| | 30: | monthly calendar |
| Book IV | 31: | planting hedges |
| | 32: | transplanting trees (general rules) |
| | 33–34: | fruit-trees and Chinese pepper |
| Book V | 45: | mulberry trees and sericulture |
| | 46–51: | timber trees and bamboo |
| | 52–54: | dye plants |
| | 55: | tree-felling |
| Book VI | 56–61: | animal husbandry (including poultry and fish) |
| Book VII | 62: | commerce and trade |
| | 63–67: | brewing, etc. |
| Book VIII | 68–79: | culinary preparations (soya sauces, vinegars, preserved meats, etc.) |
| Book IX | 80–89: | culinary preparations (meats, cereal dishes, candies, etc.) |
| | 90–91: | glue making, preparation of ink, brushes, etc. |
| Book X | 92: | plants not indigenous to North China |

the owners of large estates.[a] It is, of course, possible that the large estates which are known to have existed at the time were divided up into small farms, leased out by the landowner to individual smallholders to manage as they pleased; but references in *Chhi Min Yao Shu* to the extensive cultivation of commercial crops such as safflower or timber, to the hiring of additional labour or to playing the grain market, all seem to indicate that Northern Wei landlords liked to keep a substantial acreage directly under their own control. Chia himself clearly had considerable practical experience of farming as well as organising domestic labour for the purposes of brewing, pickling, dyeing cloth, etc. Every chapter contains some telling practical detail which demonstrates that his knowledge was far more than theoretical. Though always providing a tried and tested alternative for those more conservative in their outlook, Chia himself was clearly eager to experiment with new techniques, though pointing out possible pitfalls:[b]

I myself once had a flock of two hundred sheep, and since I did not have much fodder or beans I had nothing to feed them on. Within a year over half the flock had starved to death, and all the remaining sheep were scabby, emaciated, weak and ill—as good as dead—with short, thin fleeces, completely dry and oil-less. At first I thought that my farm was not suitable for sheep, and I suspected that it was a bad year for diseases. But in fact the cause of all these ills was simple starvation.

[a] See, for example, Kumashiro (1).
[b] *CMYS* 57.7.4.

Chia's approach to agriculture was, on the whole, strictly pragmatic and unconcerned with mystical and astrological beliefs; although he does quote passages from earlier works that give, for example, planting dates in terms of auspicious days, Chia himself advocates planting dates based on the solar calendar, and the whole work is remarkable for its accurate figures and attention to detail. For each crop Chia gives a series of figures for the amount of seed which should be sown, depending on the date at which sowing takes place and the fertility of the soil (sadly he does not usually say how much one might expect to harvest).[a] He is also the first Chinese writer to describe crop rotations, for example:[b]

Millet fields: ground that has been used (*ti*[1]) for green or ordinary mung beans is best; next comes ground that has been used for hemp, glutinous panicled millet (*shu*[2]) or sesame, and lastly ground used for rape-turnips or soya beans.

It is possible to draw up quite a comprehensive list of crop rotations from *Chhi Min Yao Shu*,[c] with a somewhat wider range of field crops (including hemp, cucurbits, coriander, and so on, as well as the usual cereals and legumes) than the rotation systems of Rome or medieval Europe. In fact the general impression given by *Chhi Min Yao Shu* is that contemporary agriculture in Northern China was at a very high level of productivity, at least comparable to that achieved by Roman farmers in their heyday and far higher than anything Northern Europe had to offer before +1600. The Northern Chinese estate farmer disposed of a wide range of implements including the adjustable mould-board plough, various types of harrow and ridger, seed-drills and rollers. He made judicious use of animal and green manures and had a sophisticated system of crop rotation which enabled him to crop his fields continuously without fallowing. He could choose from a great many varieties of most crops (Chia names nearly a hundred different varieties of setaria millet, for example, and specifies their main characteristics), and as well as cultivating cereals and all sorts of vegetables and fruit for subsistence, he might also grow commercial crops like timber or dye-plants, raise livestock and produce a number of food-stuffs (wines or beers, vinegars, soya sauce and preserves) at home.

The *Chhi Min Yao Shu* is in many ways unique, an outstanding achievement that served as a model for most later writers. The points on which later authors did choose to diverge from Chia's pattern are often significant. For example, *Chhi Min Yao Shu* is the only Chinese agricultural treatise to combine agricultural techniques with the preparation of culinary products, including not only brewing and pickling methods and the preparation of barley-sugar, but recipes for every sort of dish from roast meats and dumplings to scrambled eggs with chives. From

[a] Li Chhang-Nien (*3*) p. 106 gives a table of sowing rates for various crops from the *Chhi Min Yao Shu* and *Fan Sheng-Chih Shu*.

[b] *CMYS* 3.3.1.

[c] Amano (*7*) p. 482 presents them in a table; see also Table 9 below, p. 431.

[1] 底 [2] 黍

Thang times onwards culinary and dietetic material was included not in agricultural books but in more specialised works on food, the *shih phu*[1]. On the other hand Chia devotes relatively little space to sericulture, which at that period did not play an important role in the economy of North China, whereas the Southern-based agricultural books and, after the Sung, even books intended for use in Northern China, pay far greater attention to sericulture and textile production generally. Furthermore, although only one book out of ten of the *Chhi Min Yao Shu* is devoted to animal husbandry, it is clear from the nature of this section, and from the amount of animal-power which Chia assumes his farmers to dispose of, that both draught animals and herds played a far greater role in the economy than they did in later times. There are even references to the consumption of dairy products.

The level of knowledge and technical skill embodied in the *Chhi Min Yao Shu* was very nearly as high as one could expect under pre-modern conditions. Chia describes a system in which land was cropped continuously, a wide variety of crops was grown, the farmer had a wide range of implements at his disposal and knew how to use them to advantage. The system seems to have been both land- and labour-intensive and highly productive. Chia was describing the heyday of Northern Chinese agriculture; subsequently technological advances in the North seem to have been rare, and many of the later authors who quote *Chhi Min Yao Shu* with admiration bewail the fall in farming standards since Chia's day. Works like the Sung *Chung I Pi Yung* and the Chin or early Yuan *Chung Shih Chih Shuo* and *Han Shih Chih Shuo*, or the Chhing works *Chih Pen Thi Kang*, *Nung Yen Chu Shih* and *Ma Shou Nung Yen* do indeed seem to represent a distinct lowering of standards,[a] the result as much as the cause of increasing impoverishment in the northern provinces as the economic centre shifted to the more productive southern regions. *Chhi Min Yao Shu* embodies both the high standards and the adventurous spirit typical of an expanding economy. Thenceforward the great technical and theoretical innovations in Chinese agriculture, and the great treatises which describe them, were all to come from the South.

### (ii) *Wang Chen's* Nung Shu, *or Agricultural Treatise*

Wang Chen was a native of Shangtung province who spent many years working as an official in Anhui and Kiangsi. The preface to the *Wang Chen Nung Shu* is dated +1313, and the work was probably completed a few years earlier.[b] It seems

[a] On the *Chung I Pi Yung* see Hu Tao-Ching (*4*), (*5*). The *Chung Shih Chih Shuo* and *Han Shih Chih Shuo* are preserved only in quotations in *WCNS*, *NSCY* and other texts; see Wang Yü-Hu (*1*), pp. 106–7. The three Chhing works are local handbooks which describe agricultural conditions and practice in Northwest China, where survival was notoriously precarious; they are particularly interesting in that they devote great attention to local climatic and soil conditions and their effect on agricultural practice; Wang Yü-Hu (*2*).

[b] Shih Sheng-Han (*7*), p. 55.

[1] 食譜

that no Yuan editions of the work have survived, but there are several Ming and Chhing editions, the earliest of which is dated Chia-Ching 9 (1530); our references are all to the 22 *chüan* Wu-Ying Tien[1] Palace edition, prefaced 1774 and completed in 1783.[a]

The Mongols united their rule over the whole of China in 1279. The wars between Han Chinese and nomadic invaders had been long and bitter, the Mongols were notoriously savage in their treatment of the vanquished, and by the early fourteenth century life in rural China was anything but Arcadian. Although the Mongols rapidly perceived the futility of their attempt to turn the whole of North China into 'a giant horse-pasture', the devastation wrought by decades of warfare had brought many Chinese peasants to the brink of starvation, and even in more fortunate regions morale and productivity were low. Spurred by pity and a Confucian concern for the people's welfare, Wang Chen wrote a treatise that he hoped would help improve agricultural standards all over China. *Informed* official encouragement and instruction was the only means of bringing about such improvements, Wang believed; left to themselves the peasants could do little, and ignorant officials could easily do more harm than good by imposing misguided policies. The ruling class must have a thorough, practical understanding of the problems and hardships faced by the peasants if they were to achieve positive results. The *Wang Chen Nung Shu* was thus intended in the first instance to instruct local officials in the best agricultural methods currently available, so that they could pass this information on to the peasants under their jurisdiction.

The *Wang Chen Nung Shu* is slightly longer than the *Chhi Min Yao Shu*, about 110,000 characters in all (see Table 4). It is divided into three main sections. The *Nung Sang Thung Chüeh*[2] (Comprehensive Prescriptions for Agriculture and Sericulture) comes first in all editions, and contains short, self-contained pieces on every agricultural task, such as ploughing, sowing, irrigation and the cultivation of mulberries; this section is closely modelled on the *Chhen Fu Nung Shu*. Next in the 22 *chüan* editions (but last in the 36 *chüan* editions) comes the section entitled *Pai Ku Phu*[3] (Treatise on the Hundred Grains), subdivided into sections on cereals, vegetables, fruits, bamboos, and miscellaneous (including tea and fibre plants). It also has a brief concluding section, based mainly on material from the *Chhi Min Yao Shu*, entitled *Pei Huang Lun*[4] (On Preventing Famine), the first of the *nung shu* to devote a separate chapter to the topic. Lastly there is the *Nung Chhi Thu Phu*[5] (Illustrated Treatise on Agricultural Implements), which includes not only agricultural tools, food-processing and irrigation equipment, but also different types of field, ceremonial vessels, various types of grain storage, carts, boats and textile machinery. It is this section which is the unique and outstanding contribution of the *Wang Chen Nung Shu* to the Chinese agronomic tradition.

[a] For a critical comparison of the six Ming and Chhing editions see Amano (*9*), pp. 141 ff; a peculiarity of the *Wang Chen Nung Shu* is that in some editions the text is divided into 22 *chüan*, in others into 36; the order of the three main sections is different in each case, but the contents are the same.

[1] 武英殿 [2] 農桑通訣 [3] 百穀譜 [4] 備荒論 [5] 農器圖譜

Table 4. *Contents of the* Wang Chen Nung Shu

| | |
|---|---|
| Chapters 1–6: | *Nung Sang Thung Chüeh* 農桑通訣 (Comprehensive Prescriptions for Agriculture and Sericulture) |
| 7–10: | *Pai Ku Phu* 百穀譜 (Treatise on the Hundred Grains) |
| | 7: cereals (including legumes, hemp and sesame) |
| | 8: cucurbits and green vegetables |
| | 9: fruits |
| | 10: bamboos and miscellaneous (including ramie, cotton, tea, dye plants) |
| 11–22: | *Nung Chhi Thu Phu* 農器圖譜 (Illustrated Treatise on Agricultural Implements) |
| | 11: field systems |
| | 12–14: agricultural tools |
| | 15: wicker and basket ware |
| | 16: food-processing equipment and grain storage |
| | 17: ceremonial vessels, transport |
| | 18–19: irrigation equipment, water-powered mills, etc. |
| | 19: special implements for wheat |
| | 20–22: sericulture and textile production |

NOTE. This table is based on the 22 *chüan* editions; the arrangement of the 36 *chüan* editions is rather different.

The *Nung Sang Thung Chüeh* and *Pai Ku Phu* section contain very little material that is original. Wang quotes extensively from a number of works, from *Fan Sheng-Chih Shu* to *Nung Sang Chi Yao*; sometimes he fails to atrribute the quotation, or abridges or alters it. Only in a dozen or so places does Wang make contributions of his own. However, there is one distinctive feature of these two sections that is worth noting: throughout Wang makes a systematic and conscientious effort to contrast northern and southern agricultural technology,[a] a contrast which is naturally full of interest for technological historians. Agricultural technology in North China was shaped by the predominance of dryland cultivation, that of South China by irrigated cultivation; not surprisingly northern dryland technology appeared more sophisticated than its southern counterpart, while southern irrigation techniques were much more advanced than those of the North. Wang hoped that by pointing out the comparative advantages of northern dryland techniques and implements he could encourage their adoption by southern farmers, and conversely that he could propagate southern wet-field techniques in the North. More generally he was anxious to disseminate information on any new and advanced technique that could lighten the farmer's burden of work:[b]

The hand-harrow (*yün thang*[1]) [Fig. 18] is a new tool from the Chekiang area, shaped like a wooden patten ... set with rows of a dozen or so short nails [and a long bamboo handle]. With the *yün thang* ... one can weed many more fields in a day. In Chiangtung

[a] See Amano (*9*), pp. 153–4 for a table of contrasts.
[b] *WCNS* 13/28*b*–29*a*.

[1] 耘盪

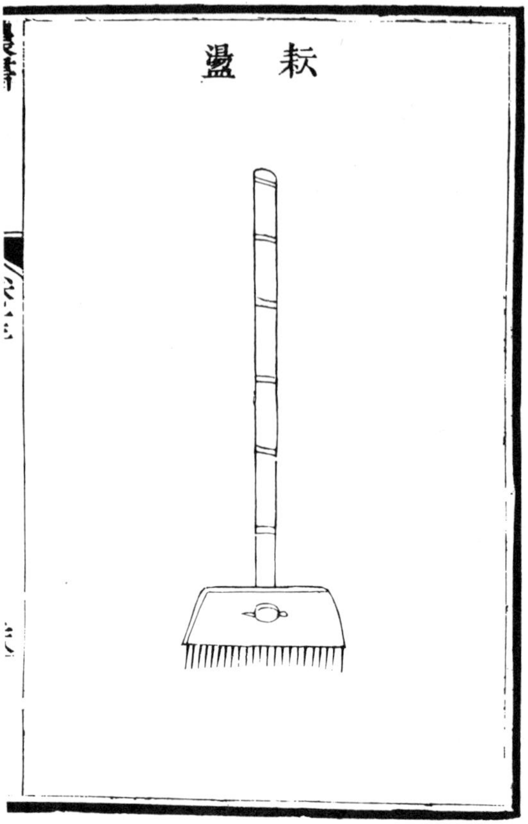


Fig. 18. The hand-harrow, *yün thang*, an innovation of the Yangtze area in the 14th century; *WCNS* 13/28*a*.

and other areas I have seen farmers weeding their fields by hand, crawling between the crops on their hands and knees with the sun roasting their backs and the mud soaking their limbs—a truly pitiable fate . . . It seems to me a great shame that this tool is not more widely known and used, so I have drawn and described it on purpose that philanthropists may disseminate the practice.

The means which Wang Chen hit upon to familiarise his readers with new or strange technology was to provide them with a series of clear and accurate illustrations depicting every conceivable agricultural implement or machine, accompanied by a text describing its construction, use and provenance. Often this was prefaced by a brief lexicographical passage listing its dialectal or ancient names, and followed by a poem. Some of these poems were taken from Mei Yao-Chhen[1] (d. 1060) or other poets, but most were composed by Wang himself. This, then, is the famous *Nung Chhi Thu Phu*.

As we said, the range of objects which Wang describes must be nearly exhaustive, covering as it does everything remotely connected with agriculture from hoes

[1] 梅堯臣

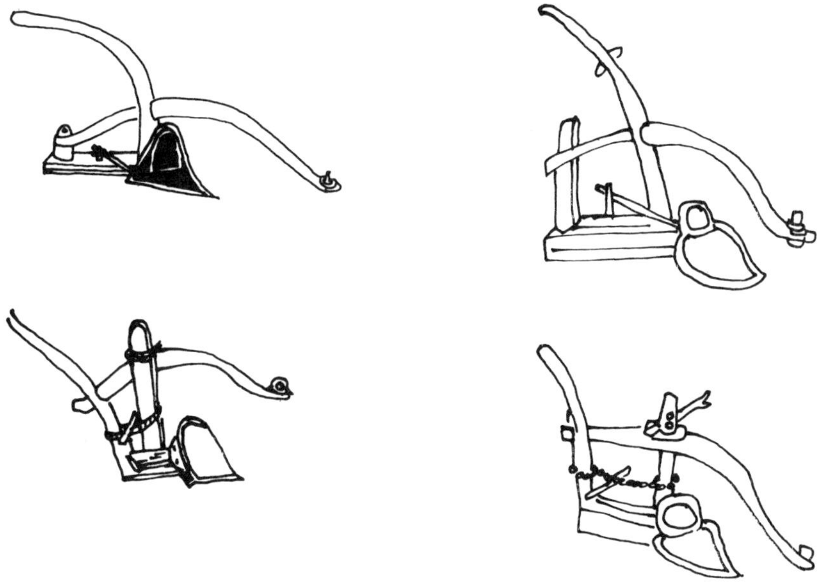

Fig. 19. Variations in the same illustration in different versions of the *Wang Chen Nung Shu*; left, Ming ed., right, Ssu Khu Chhüan Shu ed.

to water-mills to wicker trays for feeding silkworms. It has been said of Wang, by no less an authority than Hsü Kuang-Chhi, that his poetic talents were greater than his agricultural learning,[a] and there certainly are inaccuracies and inconsistencies to be found in both text and illustrations. Yet Hsü himself reproduced the *Nung Chhi Thu Phu* almost intact in his *Nung Cheng Chhüan Shu*, and judged as a whole it is an admirable achievement. The descriptions are mostly accurate, though not always sufficiently detailed to serve as a blue-print for construction (as was presumably the intention). In one or two cases, for example the seed-drill (Fig. 95), it seems that Wang himself was not sufficiently familiar with the implement to provide a really detailed description. As for the illustrations, if they do not always achieve the highest standard it is not fair to assume automatically that Wang is to blame. The original edition was lost long ago, and in later editions many of the pictures seem to have been redrawn rather too hastily; certainly there are often considerable discrepancies between editions (Fig. 19). Many inaccuracies may simply be due to the hazards of transmission, different conventions of perspective or of drawing cross-sections (as in the case of the winnowing-fan shown in Fig. 187). In other cases the artist does seem to have been completely ignorant of the tool he was supposed to depict (for example, the case of the coulter, p. 195). In such cases there is a clear inconsistency between illustration and text, and so presumably Wang was guilty of lazy proof-reading rather than crass ignorance. He does seem to have taken great pains to familiarise himself with all the practical details of agriculture. One feels that Chia Ssu-Hsieh and Hsü

[a] *NCCS* 5/19*a*.

Kuang-Chhi were not ashamed to doff their caps and robes and push a plough or hoe a row of cabbages with their own hands; reading their work one can almost see the mud under their fingernails. Wang, one feels, was more conventionally Confucian: one can imagine him earnestly questioning a farmer as to the dimensions and construction of his plough and harrow, standing in a dry corner of the farmyard at a respectful distance from the water buffalo, but one cannot picture him up to his knees in mud in a rice-field. Perhaps it was this fastidious tinge to the *Wang Chen Nung Shu* that provoked Hsü Kuang-Chhi's scorn.

Nevertheless Wang Chen's *Nung Chhi Thu Phu* was an innovation of genius, a unique contribution to Chinese agronomy. It recorded accurately and in reasonable detail the entire range of agricultural and related equipment, as used in both the dry lands of the North and the irrigated fields along the Yangtze, and even included sections on poldered fields, terraces, and different methods of irrigation. None of these topics had ever been systematically described before. Wang himself lived in troubled times, but the technology he depicted had been perfected in the heyday of the Sung, a period of technical innovation, rapid economic expansion and widespread prosperity. It has often been said that Chinese science and technology reached a peak in the Sung and were never surpassed thereafter.[a] Though we believe this to be in the main a tendentious over-generalisation, as far as agriculture is concerned a comparison of the equipment of medieval China, as represented in the *Wang Chen Nung Shu*, with that still used in the early (and indeed the later) twentieth century, certainly seems to show that few important developments or improvements took place between 1300 and 1950.[b] On the other hand, the enormous popularity of the *Wang Chen Nung Shu* and the enthusiasm with which it was quoted and copied may well have obscured many minor developments which would otherwise have been separately documented. Many of the Ming and Chhing improvements in textile technology, for example, some of which were extremely important, are now very difficult to reconstruct because contemporary books on the subject simply quote the passages from Wang Chen that they suppose to be relevant, instead of taking the trouble to write fresh descriptions (Section 31).

### (iii) *The* Nung Cheng Chhüan Shu, *or Complete Treatise on Agricultural Administration*

The last of the great traditional agricultural treatises was the *Nung Cheng Chhüan Shu* by Hsü Kuang-Chhi[1], written at the end of the Ming. Hsü (1562–1633) was one of the outstanding figures in Chinese history, a principled politician, a first-

[a] See, for example, M. Elvin (2); B. Gille (17).

[b] This is not to say that no developments or improvements took place in other branches of agriculture; on the contrary. We hope to advance some partial explanations for the relative technological stagnation of Chinese agricultural equipment in our concluding section.

[1] 徐光啓

rate scientist, a friend and protector of the Jesuits as well as one of their first converts. We have already outlined Hsü's scientific achievements in Sections 19, 27 and 33;[a] among his other attributes, he was an able administrator with a keen perception of his country's needs, and consequently he applied his scientific talents to agriculture, to irrigation and fertilisers, as well as to astronomy and chemistry.

Ultimately the driving force in Hsü's career was his patriotism. Like Wang Chen, Hsü lived in troubled times. The Ming dynasty was in decline, corruption was everywhere, civil war threatened constantly, while north of the border the Manchus awaited their chance to sweep down and capture the capital. Hsü devoted much of his energies to devising means for protecting and strengthening the Ming state: on the one hand she must perfect her defences (hence his interest in military colonies and artillery), on the other she must build up her economy, for only an increase in general prosperity could restore the Chinese people to strength and regain their loyalty (hence his interest in irrigation, famine relief, and economic crops).

Hsü had considerable practical experience of agricultural administration and technology.[b] Although he spent most of his life in high office, from 1607 to 1610 he was forced to retire from public life and returned to his home in Shanghai, where he experimented with Western irrigation technology that he had learned about from the Jesuits,[c] and with the cultivation of such unusual crops as sweet potatoes, cotton, and the *nü chen*[1] tree.[d] During this period he wrote a short work on the sweet potato, the *Kan Shu Su* (now lost), and probably one or two other short agricultural monographs.[e] He was soon recalled to public office, but continued to gain in agricultural experience. The years 1613 to 1620 he spent on and off in Tientsin, organising self-supporting military settlements (*thun thien*[2]) and attempting to promote improved irrigation and the cultivation of wet rice in the Northeast.[f] For some time Hsü had been contemplating writing a comprehensive agricultural treatise, based on his own experience as well as previous authors' works, which would not only provide a practical guide to farming methods but also tackle the chief economic problems of China as he saw them, namely unequal population distribution, underproductive agriculture, rural poverty and the danger of famine. He had apparently been collecting material and drafting sections of the work, some of which appeared as separate monographs, since before his

[a] See also Anon. (*529*).

[b] Hu Tao-Ching (*10*); L. A. Maverick (2).

[c] Hsü was a close friend of Sabatino de Ursis, who published a short treatise on European irrigation methods, *Thai Hsi Shui Fa*, in Peking in 1612. Hsü included this work as the final section of the chapter on water control in the *Nung Cheng Chhüan Shu* (*NCCS* 19, 20).

[d] *Ligustrum japonicum*, the tree used for producing insect wax.

[e] The *Wu Ching Su* and another work on cotton, both now lost, may well date from this period; Hu Tao-Ching (*9*). Hsü also wrote a series of unpublished notes on fertilisers and other agricultural topics, now in the Shanghai Municipal Library.

[f] During this period Hsü wrote the *Nung I Tsa Su*; Hu Tao-Ching (*9*), (*10*).

[1] 女貞 [2] 屯田

Table 5: *Contents of the* Nung Cheng Chhüan Shu

| | |
|---|---|
| Chapters 1–3: | *Nung Pen* 農本 (Fundamentals of Agriculture): quotations from the classics, etc. on importance of encouraging agriculture |
| 4–5: | *Thien Chih* 田制 (Field Systems): land distribution, field management |
| 6–11: | *Nung Shih* 農事 (Agricultural Tasks): clearing land, tilling, etc., including a detailed exposition of settlement schemes |
| 12–20: | *Shui Li* 水利 (Water Control): various methods of irrigation and types of irrigation equipment, including 19–20: *Thai Hsi Shui Fa* 泰西水法 on Western irrigation equipment |
| 21–24: | *Nung Chhi Thu Phu* 農器圖譜 (Illustrated Treatise on Agricultural Implements): largely based on the same section in *WCNS* |
| 25–30: | *Shu I* 樹藝 (Horticulture): vegetables and fruit |
| 31–34: | *Tshan Sang* 蠶桑 (Sericulture) |
| 35–36: | *Tshan Sang Kuang Lei* 蠶桑廣類 (Further Textile Crops): cotton, hemp |
| 37–40: | *Chung Chih* 種植 (Silviculture) |
| 41: | *Mu Yang* 牧養 (Animal Husbandry) |
| 42: | *Chih Tsao* 製造 (Culinary Preparations) |
| 43–60: | *Huang Cheng* 荒政 (Famine Control): 43–45: administrative measures; 46–60: *Chiu Huang Pen Tshao* 救荒本草 (Famine Flora) |

brief retirement to Shanghai,[a] but his political duties constantly intervened and his *magnum opus*, the *Nung Cheng Chhüan Shu*, was still an unfinished draft at his death in 1633. It was left to the famous Chiangnan scholar Chhen Tzu-Lung[1] to assemble a group of like-minded men with whose aid he edited the draft, publishing it six years later in 1639.[b]

The *Nung Cheng Chhüan Shu* is an enormous work of 700,000 characters, that is to say seven times as long as *Chhi Min Yao Shu* (Table 5). By far the greater part of it consists of quotations, not always entirely accurate, taken from no fewer than 299 sources;[c] for example, despite Hsü's slighting remarks about Wang Chen's agricultural abilities, in the *Nung Cheng Chhüan Shu* he reproduces the *Nung Chhi Thu Phu* almost in its entirety, though in a rather different order. Indeed, in so far as practical peasant farming is concerned, Hsü adds little to Wang's treatment of tools, crops and techniques. The chief innovation in the *Nung Cheng Chhüan Shu* is its stress on agricultural administration. As we have said, Hsü was well aware of the fundamental economic and political problems facing late Ming China. While the Lower Yangtze and other fertile areas were so heavily overpopulated that even a slight drop in the harvest could bring instant famine, huge tracts of land in

[a] Liang Chia-Mien (*3*).
[b] Our references in this volume are to the 1843 edition; a new edition annotated by Shih Sheng-Han (*8*) was published in Shanghai in 1979.
[c] Khang Chheng-I (*1*).

[1] 陳子龍

the northern provinces lay waste and empty, a vacuum into which the Manchu hordes might pour at any time. Rural poverty was causing widespread unrest and peasant rebellions broke out every year in some part of the country. Through mismanagement and corruption, the imperial coffers were nearly empty, and the army was undermanned, underpaid and underfed. To save the situation Hsü proposed a series of closely coordinated agrarian policies.

First, the northern provinces must be strengthened and protected by setting up self-sufficient military and civilian agricultural colonies (*thun thien*[1]). Dryland agriculture in these areas was underproductive, but matters could be improved by constructing large-scale irrigation works and cultivating wet rice, a policy which would have the double advantage of providing an ample economic base for the troops and of attracting rice-farming settlers from the overcrowded provinces of the south. To demonstrate their advantages and explain how agricultural colonies should be organised, Hsü quotes from numerous memorials and reports, and contributes a lengthy chapter containing statistics on manpower requirements, etc., based on his own experience in establishing agricultural colonies near Tientsin; irrigation planning and technology are also dealt with in considerable detail, down to such practical details as the organisation of labour for construction and maintenance (Figs. 20, 21).

Secondly, the ever-recurring threat of local famine must be averted. Even in the late twentieth century, efficient long-distance communications remain one of China's most intractable problems, and in Ming times when famine struck a remote province like Fukien there was little hope of outside relief arriving in time. The public and private granaries established in most districts, which were supposed to cope with such emergencies, in fact were woefully inadequate.[a] Hsü felt that a much more reliable means of famine-relief was the cultivation of special famine foods: plants such as sweet potatoes, maize or tapioca which would grow on marginal land not used for cereals. Few Chinese would eat any of these plants by preference, but in good years they could be fed to livestock, and in bad years no one could afford to be fussy. Hsü devotes 18 *chüan*, a third of the whole book, to *Huang Cheng*[2], Famine Control: 3 *chüan* on administrative measures, and the rest an illustrated flora, *Chiu Huang Pen Tshao*[3], written by Chu Hsiao[4] in 1406, which describes over four hundred wild plants that could be picked and eaten if all else failed; Hsü himself tested many of the plants recommended in this section.[b]

Thirdly, once the danger of starvation was averted, rural incomes must be raised (with the indirect but desirable effects of increasing state revenues and of reducing discontent and the likelihood of rebellion). This was to be achieved by encouraging household industries and the cultivation of economic crops. The *Nung Cheng Chhüan Shu* confirms the importance of sericulture in the Chiangnan economy, but it also advocates the expansion of the comparatively new cotton

[a] See p. ■■ below; P. E. Will (1). [b] Shih Sheng-Han (7), p. 68.

[1] 屯田 [2] 荒政 [3] 救荒本草 [4] 朱橚

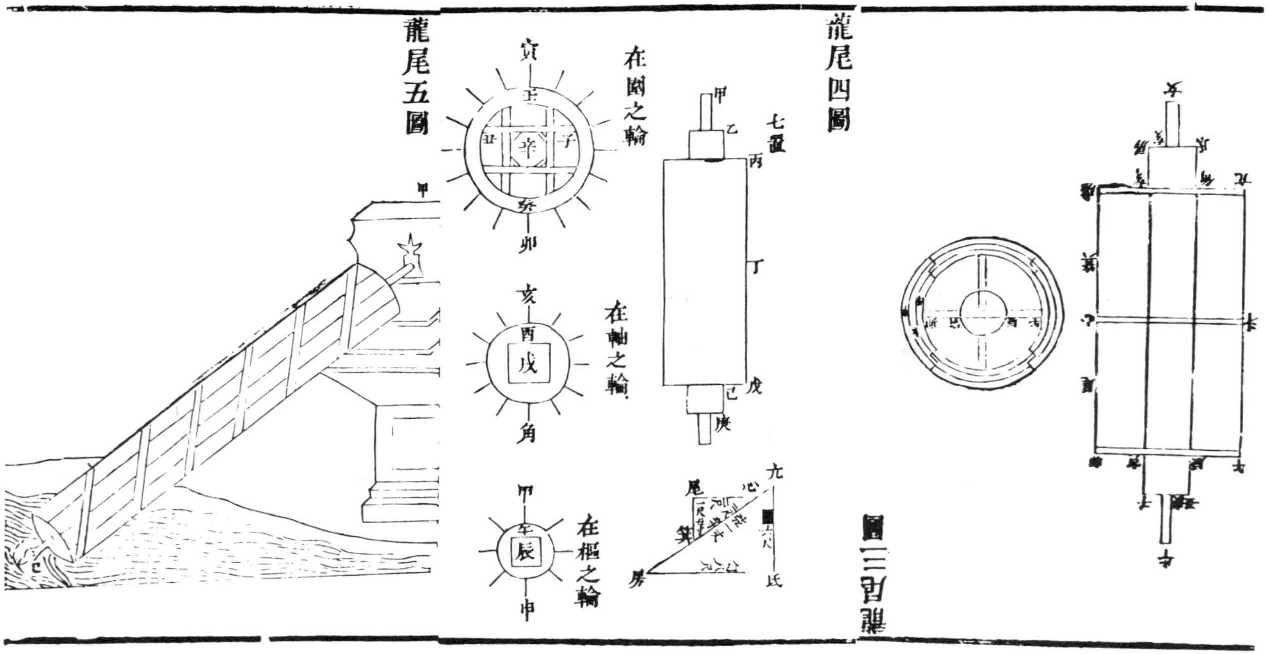


Fig. 20. Archimedean screw from the *Thai Hsi Shui Fa*, a Western device which Hsü Kuang-Chhi hoped would raise standards of irrigation, but which in fact failed to become popular; *NCCS* 19/15*b*–16*b*.

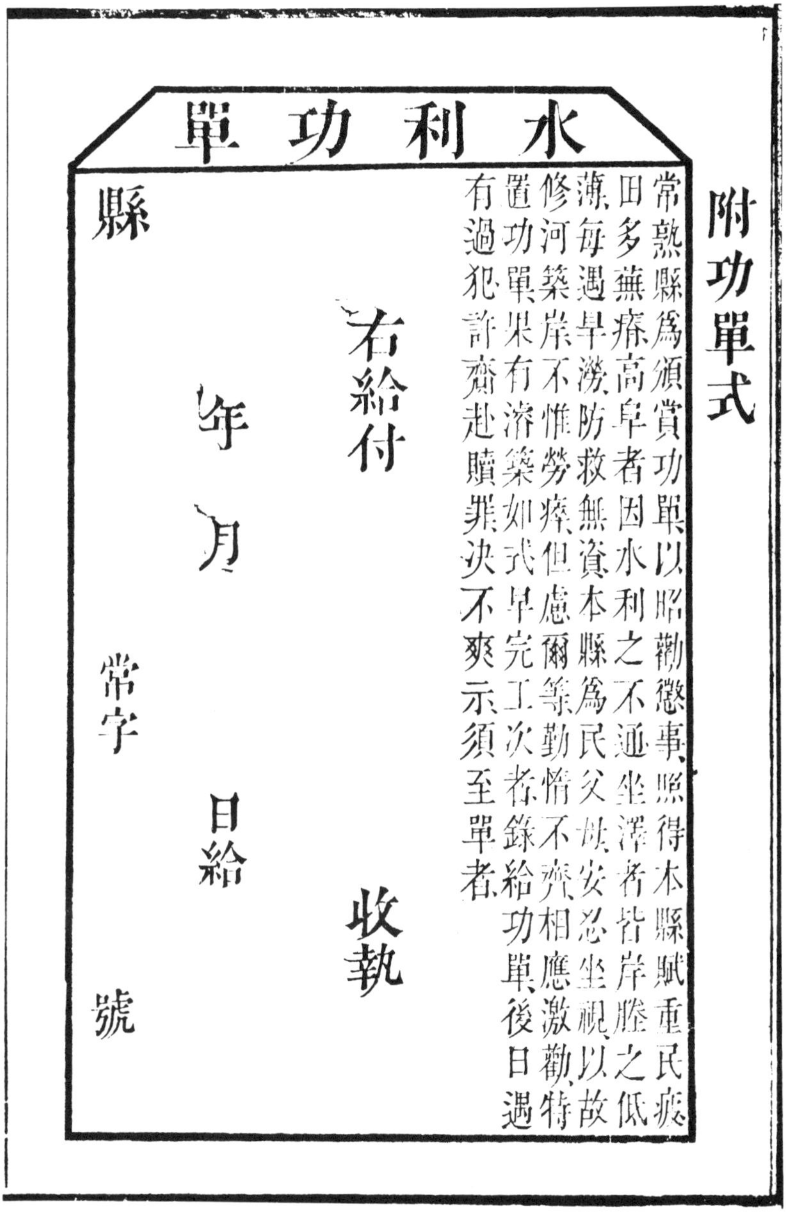


Fig. 21. Hsü's blueprint for a work certificate to be allotted to those who participated in the officially organised maintenance of irrigation works; it was to be surrendered by anyone contravening the regulations on water distribution; *NCCS* 15/13*a*.

and ramie industries. Hsü was particularly anxious to popularise the cultivation of cotton in Hopei, an aim in which he was ultimately very successful. He also described the economic advantages and cultivation methods of a wide range of other economic crops, including tea, oil crops, bananas, sugar-cane, bamboo and various timber trees which could be grown by peasant smallholders.

The great originality and strength of the *Nung Cheng Chhüan Shu* lies in its emphasis on the role of administration in agricultural development. Improved tools, new techniques or better strains of seed might all contribute to agricultural improvement, but only if their introduction and dissemination was properly planned and coordinated could the utmost benefit be extracted from them. The title of Hsü's great work was not chosen idly: it really is a treatise on agricultural administration. Although official encouragement of agriculture had always been recognised as an important responsibility of Chinese bureaucrats at every level, it often tended towards pious platitudes and symbolic gestures. Hsü knew that good intentions were not sufficient. Well-considered, practical and coordinated policies must be devised and administered by experienced men. Plans must be based on accurate figures rather than vague historical models, and details such as the organisation of manpower or the fees to be paid by farmers for irrigation water must be worked out in detail. Sadly Hsü did not live to finish the *Nung Cheng Chhüan Shu*. The Ming dynasty, which he so loyally supported, fell in 1644, and the Manchus, who resented Han patriotism in any form, neglected Hsü's great work in favour of their own compilations like the *Shou Shih Thung Khao*. Sporadic attempts were made to put some of Hsü's proposals into practice,[a] but they lacked the practical expertise, quantitative accuracy and breadth of vision that give the *Nung Cheng Chhüan Shu* such authority. It is particularly ironic that while the Chinese state on more than one occasion had introduced agrarian measures akin to development policies today,[b] none of these has been systematically recorded in the Chinese sources. Yet although Hsü's stratagem for rural development, which under more favourable political circumstances might have proved immensely successful, is recorded in the *Nung Cheng Chhüan Shu* in convincing detail, his grand plan was never put into practice.

### (3) State-commissioned Compilations

These differ from the agricultural treatises in that they consist mainly of citations from earlier works, with almost no original material. Any originality lies rather in the arrangement and balance of the work, in the traditional Confucian style of argument by implication. Such works cannot be omitted from our discussion as they were officially distributed all over the empire and had an influence out of all proportion to their agronomic worth.

[a] See, for example, T. Brook (1) on attempts to popularise wet rice cultivation in the northern provinces under the early Chhing emperors.

[b] See Hsü Cho-Yün (1), F. Bray (3) on Han, James T. C. Liu (2), F. Bray (4) on Sung development policies.

### (i) *The* Nung Sang Chi Yao, *or Fundamentals of Agriculture and Sericulture*

This work dates from 1273, and is the earliest official agricultural compilation to have survived. Two earlier works in the genre are known to have been completed, the Thang *Chao Jen Pen Yeh* and the Northern Sung *Chen-Tsung Shou Shih Yao Lun*, but both have since been lost. According to Wang Phan's[1] preface to the *Nung Sang Chi Yao*, in 1271 the Yuan emperor established a Board of Agriculture, *Ssu Nung Ssu*[2], to be solely responsible for the promotion and teaching of agriculture and sericulture; one of its tasks was to produce a practical handbook that could be used throughout the Yuan kingdom (which until 1279 consisted only of the northern Chinese provinces and parts of Szechwan), and in 1273 the Board of Agriculture presented to the throne the *Nung Sang Chi Yao*, which was immediately printed and distributed all over the country.

None of the Yuan editions of the *Nung Sang Chi Yao* has survived, and the numerous later editions probably contain many additions, perhaps even whole new sections.[a] With this proviso in mind, let us briefly describe the contents. The text is altogether about 60,000 characters long, divided under ten headings (*men*[3]): (i) literary quotations (*tien hsün men*[4]); (ii) tillage (*keng khen men*[5]); (iii) crop plants (*po chung men*[6]) (including cereals, oil and fibre crops); (iv) mulberries (*tsai sang men*[7]); (v) silkworms (*yang tshan men*[8]); (vi) vegetables (*kua tshai men*[9]); (vii) fruits (*kuo shih men*[10]); (viii) bamboos and trees (*chu mu men*[11]); (ix) medicinal herbs (*yao tshao men*[12]); (x) animal husbandry (*tzu chhu men*[13]) (including poultry raising, pisciculture and apiculture). The arrangement thus follows that of *Chhi Min Yao Shu*, minus the sections on brewing and other household arts, with an additional final section entitled *Sui Yung Tsa Shih*[14], A Variety of Annual Tasks, based on the *Ssu Shih Tsuan Yao*. Most of the text consists of quotations, especially from *Fan Sheng-Chih Shu*, *Ssu Min Yüeh Ling*, *Chhi Min Yao Shu* and *Ssu Shih Tsuan Yao*; some Sung and early Yuan works are also quoted, but interestingly no reference is made to the *Chhen Fu Nung Shu*, probably because that described agricultural methods of the rice-growing areas of south China, which at that point were beyond the Yuan borders. The quotations are attributed and arranged in chronological order; not only is *Nung Sang Chi Yao* thus an important source for the text of many earlier works, but also it allows us to see various agricultural phenomena in a historical perspective. The original material in *Nung Sang Chi Yao* deals with new crops whose cultivation the government wished to encourage, for example ramie from South China and cotton from Central Asia, water melons and sugar-cane.

[a] Of which the section on pit cultivation, *ou thien*[15], is probably one. On the various editions of *Nung Sang Chi Yao* see Liu Yü-Chhüan (*1*); Amano (*9*), pp. 130–140; most recent editions are based on the Wu-Ying-Tien[16] Palace edition of 1775.

[1] 王磐 [2] 司農司 [3] 門 [4] 典訓門 [5] 耕墾門
[6] 播種門 [7] 栽桑門 [8] 養蠶門 [9] 瓜菜門 [10] 果實門
[11] 竹木門 [12] 藥草門 [13] 孳畜門 [14] 歲用雜事 [15] 區田
[16] 武英殿

Two of the ten sections of *Nung Sang Chi Yao* are devoted to sericulture, and its predominant position reflects the determination of the Yuan government to increase silk production in North China. Like the Chin government before it, the Yuan government paid for most of its imports in bolts of silk cloth, including the tea which came from South China. Tea-drinking had suddenly become the rage in the North: whereas previously it had been a luxury consumed occasionally by the rich, by the thirteenth century tea was drunk on every possible occasion by the masses as well as their masters. Attempts to limit consumption to citizens of the seventh rank and above failed dismally, as did attempts to cultivate tea in the alkaline soils of the northern provinces. The only solution was to allow tea imports to continue but to increase the amount of silk available to pay for them, first by increasing silk production, and secondly by cutting down national consumption. The *Nung Sang Chi Yao* accordingly describes all the most advanced sericultural techniques in sections (iv) and (v), while in section (ii) advocating the cultivation of alternative fibre crops like cotton or ramie which could substitute for silk within the Yuan empire. The government was largely successful in its campaign: not only did Northern Chinese silk production rise, but ramie and especially cotton soon acquired an important economic role in the North.

The compilers of the *Nung Sang Chi Yao* believed firmly that any new or exotic crop could be cultivated in China provided the soil conditions were right. If one ignores basic climatic requirements, their position was essentially correct. Crops such as cotton and water-melon were suited to both soil and climate in North China, and their introduction was accordingly rapid and permanent. But lacking any scientific knowledge of soil chemistry or plant nutrition, the *Ssu Nung Ssu* occasionally blundered and tried to popularise the cultivation of crops that even today, with the aid of chemical fertilisers and irrigation, cannot be successfully grown in the north: tea, sugar-cane and citrus fruits are cases in point.

On the whole the *Nung Sang Chi Yao* fulfilled its purpose admirably. It was essentially a practical handbook, full of information on crops both old and new, down-to-earth, realistic and accurate. It did embody official policies, like the expansion of textile production and the popularisation of improved tillage methods, but on the paternalistic role of benevolent government in guiding and protecting its myriad subjects, which feature so prominently in other official compilations, the *Nung Sang Chi Yao* is admirably restrained and terse.

### (ii) *The* Shou Shih Thung Khao, *or Compendium of Works and Days*

Exactly the opposite is true of this Chhing compendium, commissioned by the Chhien-Lung emperor in 1737 and presented to the throne by O-Erh-Thai[1] and his team of more than fifty collaborators in 1742. The whole aim of this work was the glorification of the emperor as the benevolent holder of the Mandate of

[1] 鄂爾泰

Heaven, encouraging agriculture in the fine tradition of Yao and Shun; as a result ceremonial aspects are stressed at the expense of the practical, and throughout the work techniques of production occupy second place after ritual practices.

The first imperial edition, the Wu-Ying-Tien Palace edition of 1742, was printed and distributed by royal command in every province of the Chhing empire.[a] It is slightly shorter than the *Nung Cheng Chhüan Shu*, but nevertheless at 78 *chüan* it is one of the longest of all the *nung shu*. It is divided under eight headings: (i) the celestial seasons (*thien shih*[1]) (an agricultural calendar); (ii) the land's requirements (*thu i*[2]) (including chapters on local products, land systems and irrigation); (iii) cereals (*ku chung*[3]) (with an introductory section entitled Excellent Grains: Auspicious Millet and Auspicious Wheat (*chia ho jui ku jui mai*[4]), another delicate compliment to the emperor with whom these adjectives were traditionally associated); (iv) tasks (*kung tso*[5]); (v) exhortations (*chhüan kho*[6]) (including ordinances of the Agricultural Board, spring and autumn sacrifices, etc.); (vi) reserves (*hsü chü*[7]) (on charitable, ever-normal and ordinary granaries, etc.); (vii) supplements to agriculture (*nung yü*[8]) (including horticulture, economic crops and animal husbandry); and (viii) sericulture (*tshan sang*[9]) (with cotton and ramie as a supplement at the end).

The Chhing emperors had already begun collecting rare and ancient books from all over the country to form the Imperial Collection, the *Ssu Khu Chhüan Shu*[10], and although in 1742 the collection was still not complete the compilers of the *Shou Shih Thung Khao* were able to quote from no fewer than 427 sources, among which the *Kuang Chhün Fang Phu*, *Nung Sang Chi Yao* and *Nung Cheng Chhüan Shu* are frequently referred to. A great many observations on the imperial role in fostering agriculture are quoted from classical and historical works, but unfortunately the quotations are quite often inaccurate. The *Shou Shih Thung Khao* contains no original material, though it does add some illustrations to those taken from the *Wang Chen Nung Shu* and *Nung Cheng Chhüan Shu*; in particular it incorporates the complete imperial edition of the *Keng Chih Thu*.

In the interests of promoting the imperial image the *Shou Shih Thung Khao* devotes considerable space to ceremonial, to theories of state agricultural policy, to the idealised 'well-field' (*ching thien*[11]) system of land distribution and so on, with the result that it gives a completely distorted picture of contemporary Chinese agriculture. By the late eighteenth century the Chinese population had recovered from the wars and devastation of the late Ming and early Chhing, and was increasing rapidly. In such conditions the threat of famine is never far removed. Pressure on land was mounting and it was clearly becoming necessary to find

[a] There have been other editions, including the 1956 Chung-Hua edition published in Peking; we refer to the 1847 reprint of the original Palace edition.

[1] 天時 [2] 土宜 [3] 穀種 [4] 嘉禾瑞穀瑞麥
[5] 功作 [6] 勸課 [7] 蓄聚 [8] 農餘
[9] 蠶桑 [10] 四庫全書 [11] 井田

some means of intensifying agricultural production on existing arable land and also to extend the cultivated area. Furthermore, many rural areas were by this time highly commercialised, and the production of economic crops, small-scale rural industry and handicrafts were essential to many peasants' economic survival. One would therefore expect the *Shou Shih Thung Khao* to devote considerable space to such topics as land reclamation, irrigation, famine control and economic crops. That trade should not figure in a Confucian-inspired work is hardly surprising, but it is odd to find land reclamation and famine control omitted entirely from the *Shou Shih Thung Khao*, while economic crops, together with other supplementary income-generating activities such as animal husbandry and horticulture, are squeezed into a single section (vii). Irrigation forms only a minor part of the section on land (ii), apart from a chapter on Western irrigation methods in section (iv) copied directly from the *Nung Cheng Chhüan Shu*. Most of the rest of section (iv), devoted to agricultural techniques, is taken word for word from the *Wang Chen Nung Shu*, written over four centuries earlier.

The *Shou Shih Thung Khao* seems, then, to have little to recommend it. But although one cannot always rely on the accuracy of the quotations, they are drawn from a very wide range of sometimes unexpected sources, literary as well as technical, and so the *Shou Shih Thung Khao* is a very important source book. And perhaps most importantly, it was the last of the great Chinese *nung shu*, distributed and read all over China, and even known to foreigners.

## (4) MONOGRAPHS

A great many agricultural monographs are known to have been written in China, some as early as the Han dynasty, but relatively few have survived up to the present day. The range of topics is quite narrow, and the subjects which captured the Chinese scholar's imagination are on the whole different from those which incited the European writer to take up his pen.

As one might expect in China, administrative topics figure quite prominently on the list; for example, works on famine relief or, from early Chhing times, on locust control. The *Chiu Huang Pen Tshao* and *Fu Huang Khao* are but two examples.[a] Technical subjects were less popular. Quite a number of works on water control exist,[b] as also on the field systems associated with them; for example, the *Chu Wei Shuo* and *Chu Yü Thu Shuo*. The field type most often written about was the method of so-called pit-cultivation (*ou thien*[1]), first mentioned in the *Fan Sheng-Chih Shu*; in the Chhing dynasty nearly a dozen monographs were written on this alone.[c] But farm tools and machines were not considered interest-

[a] Wang Yü-Hu (*1*), pp. 311, 313 lists further titles.
[b] Vol. IV, Pt 3.
[c] Wan Kuo-Ting (*3*); Wang Yü-Hu (*5*).

[1] 區田

ing. It would never have occurred to an educated Chinese to attempt to benefit mankind by improving the design of the plough, and there is none of the technical, experimental literature that we find in the West on such topics. In the ninth century Lu Kuei-Meng[1] wrote a detailed description of a Lower Yangtze plough in the *Lei Ssu Ching*, the Classic of the Plough (see below, p. 181); thereafter this essay was quoted verbatim and without addition by anyone wishing to describe a plough. The *Wang Chen Nung Shu* of course contains a wonderful catalogue of farm equipment, and again this was subsequently quoted word for word whenever the occasion arose. One late seventeenth-century scholar, Chhen Yü-Chi[2], did write a fresh account of agricultural tools, the *Nung Chü Chi*, giving local variants of names and so on, but unfortunately without illustrations. Otherwise, no works on agricultural equipment as such have survived. As for field crops, there are several monographs on cotton and one or two on sweet potatoes,[a] but surprisingly few on cereals. The Sung scholar Tsheng An-Chih[3] wrote a 5 *chüan* monograph on cereals, the *Ho Phu*, but this was lost in the late Ming. All that has survived otherwise is the short list of rice varieties compiled by Huang Sheng-Tsheng[4] in the sixteenth century, the *Tao Phin*, and the historical and philological monograph, *Shih Ku*, On Cereal Grains, by the nineteenth-century scholar Liu Pao-Nan (*1*).

More exotic and delicate plants, on the other hand, were lovingly described by writers in every dynasty. Tea, for example, is first treated in the eighth-century *Chha Lu*,[b] and twenty later monographs on tea are recorded by Wang Yü-Hu,[c] of which twelve survive. Wang also records thirteen monographs on the lichee, the earliest extant, the *Li Chih Phu*, dating from the Sung dynasty; nine others have also survived to the present day.[d]

There are two areas in which the Chinese and European corpus of monographs closely coincide. One is that of veterinary science, the other horticulture. The Chinese were just as fanatical gardeners as the English, and produced countless monographs on chrysanthemums, peonies, orchids and camellias, as well as on gardening generally.[e] This was the urge of inclination; the stream of veterinary books and pamphlets in both East and West was rather an acknowledgement of necessity, for domestic animals, despite their sturdy appearance, are often delicate and susceptible, and prey to dozens of unpleasant ailments. Both in Europe and China, the horse was most frequently the subject of such works, for not only does it fascinate men with its vigour and beauty, but it is also prone to a bewildering variety of painful and disabling diseases. Lengthy treatises have been written on how to judge horseflesh (*hsiang ma*[5]), and on the still more delicate task of keeping

[a] Wang Yü-Hu (*1*), pp. 309–310.
[b] The *Chha Chiu Lun* may possibly be even earlier (Chen Tsu-Lung (1)), but it cannot strictly be counted as an agricultural monograph.
[c] (*1*), pp. 310–11.
[d] *Ibid.* pp. 313–14.
[e] See Section 38.

[1] 陸龜蒙 [2] 陳玉璂 [3] 曾安止 [4] 黃省曾 [5] 相馬

one's steed (or nag, if one had failed to judge wisely) alive and in reasonable health. But although the horse was the most glamorous of the domestic animals, it was the least common in China, and many other authors addressed themselves to the more mundane care of oxen, sheep, goats, poultry, and that valuable beast, the silk-worm.[a]

These, then, are our main sources for the study of Chinese agricultural history. Let us now briefly consider the supplementary material at our disposal, and the picture of Chinese agriculture and rural society that our combined sources enable us to draw.

## (5) Supplementary Sources

First among the supplementary sources there are, of course, the numerous botanical works, like the *Nan Fang Tshao Mu Chuang* and the *Pen Tshao Kang Mu*, which provide all sorts of valuable information on crop plants, their evolution, varietal diversity, geographical distribution, uses, cultivation requirements, and so on. These works have already been dealt with at length in Section 38 and we shall not dwell on them further here, except to point out that historical botanic studies occupy an important place in research on the history of Chinese agriculture today.[b]

Then there are the popular encyclopaedias like the *Thai Phing Yü Lan*, *San Tshai Thu Hui* and *Thu Shu Chi Chheng*, all of which contain sections devoted to agricultural implements and crop plants. These consist of quotations from the classics and from agricultural treatises, and are of little intrinsic interest here except for the illustrations, which sometimes differ slightly from those of the agricultural monographs.[c] Of real interest on its own account, though, is Sung Ying-Hsing's[1] *Thien Kung Khai Wu* of 1637, which is entirely original and provides some very interesting insights into Ming agricultural technology, and particularly regional variations between North and South China.[d]

Next there is the archaeological material. For the prehistoric and early historic periods this is almost our only source of information, while for the early dynasties it provides a valuable supplement to scanty literary sources: grave models, bas-reliefs, paintings and ancient implements help to chart the chronological development of Chinese agricultural equipment and techniques where literary evidence is inadequate—though naturally inconographic evidence must be used with caution.[e] The Chou, Han and Northern and Southern Dynasties, with their

[a] Wang Yü-Hu (*1*) lists over sixty titles of monographs on horses, over fifty on other equines, cattle, ovines and poultry, a dozen on fish, and forty or so on silk-worms.

[b] E.g. Liang Chia-Mien (*1*), (*2*); Li Ming-Chhi (*1*); Pheng Shih-Chiang (*1*); Shih Sheng-Han (*6*); Yu Yü (*1*), (*2*); H. L. Li (11).

[c] E.g. Amano (*4*), p. 777.

[d] Sung Ying-Hsing (1); Amano (*6*); T. Thilo (1).

[e] See p. 151 below; also B. Gille (16), pp. 92 ff.

[1] 宋應星

delightful grave-models and bas-reliefs, are particularly rich in material relating to agriculture (Fig. 22), but from Thang times on little relevant material has been excavated. Nor does the extensive artistic record help us much, for Chinese paintings and drawings, except the illustrations of agricultural treatises, rarely depict agricultural scenes even incidentally: we search in vain for the telling detail in the landscape like the ploughing scene in Brueghel's *Icarus* (Fig. 23).

Most educated Chinese had their roots in a country estate rather than the city, and those who entered the bureaucracy all had charge at some time or other of a prefecture with its attendant rural problems. Some were sufficiently concerned to develop an informed interest in irrigation, methods of fertilisation, or the introduction of new crops, which can be traced in their literary *oeuvre*. Shen Kua[1] was a nationally renowned expert on water control and makes many references to agricultural matters in the *Meng Chhi Pi Than*, but unfortunately the two agricultural books that he wrote, the *Meng Chhi Wang Huai Lu* and the *Chha Lun*, have since been lost.[a] The literary *oeuvre* of another Sung writer, Cheng Yü-Fu[2] (1032–1107), included a *Nung Shu* in no less than 120 *chüan* which sadly was never published and was seen by only a few of his friends.[b] Hsü Kuang-Chhi, the author of the *Nung Cheng Chhüan Shu*, left a number of unpublished manuscripts relating to the introduction of new crops, the use of fertilisers, the development of irrigation and wet-rice cultivation in North China and so on; most of the manuscripts are preserved in the Shanghai Municipal Library. Hsü's published work, quite apart from the *Nung Cheng Chhüan Shu*, also contains many discussions of agricultural topics.[c] Other writers were more light-hearted in their approach to serious matters: an example is the Thang writer Liu Tsung-Yuan's[3] fable of Camel Kuo the Gardener,[d] where horticultural principles are used as a moral allegory. Political essays, the reminiscences of local magistrates, or the lucubrations of legal experts may also contain valuable snippets of information. The systematic sifting of the Chinese literary corpus is a daunting task, but the Chinese Agricultural Heritage Group at Nanking have taken up the challenge and have compiled a truly impressive series of source-books on a wide range of agricultural topics. So far only a few volumes on crop plants have been published,[e] but further volumes on technological history, animal husbandry and water control as well as field crops, fruits and vegetables are expected to appear soon.

Another interesting source is Chinese genealogies. The earliest are those of aristocratic families recorded in the official histories from the Han on. By the Sung less exalted families had also begun to compile genealogies (*tsu phu*[4]) and clan rules. From such documents we can glean information on social mobility

[a] Hu Tao-Ching (*7*), (*13*).
[b] *Ibid.*
[c] See Anon. (*529*); Hu Tao-Ching (*9*), (*10*); Wang Chung-Min (*4*).
[d] *Liu Ho-Tung Chi*, ch. 27; Commercial Press ed. Section 3, pp. 69–70.
[e] Chhen Tsu-Kuei (*1*), (*2*); Hu Hsi-Wen (*2*), (*3*); Li Chhang-Nien (*2*), (*4*); Yeh Ching-Yuan (*2*).

[1] 沈括 [2] 鄭御夫 [3] 柳宗元 [4] 祖譜

Fig. 22. Han pottery model of a winnowing fan and quern; Seattle Art Museum.

Fig. 23. Plough with a straight wooden mould-board, as seen in Brueghel's *Fall of Icarus*; Museum of Fine Arts, Brussels.

over the centuries,[a] as well as clues to farm management and conditions of tenancy on the clan estates. To this extent they are one of the most valuable sources for the history of social relations in rural China.[b] We also possess one or two works relating to private estate management, for example the Later Han *Ssu Min Yüeh Ling*,[c] the Southern Dynasties *Yen Shih Chia Hsün*, and the seventeenth-century *Heng Chhan So Yen*.[d]

After the conventional agricultural treatises, however, the historian of Chinese agriculture finds his two main sources in historical works, in particular the official dynastic histories, and the local gazetteers (*fang chih*[1]). The earliest gazetteers date from the Sung, at which period they are still few and far between, but by the Chhing literally thousands of these works had been compiled, covering every corner of the country. Gazetteers contain sections on local products, with data on agricultural practices, the crops grown in a particular locality and the normal yields; they also contain material on land taxes and rents, prices and trade, as well as any noteworthy local customs. They are an extremely valuable source of economic data of all sorts, often sufficiently detailed to permit quite sophisticated analysis, as in Rawski's (1) study of the relations between agricultural and commercial development in Fukien and Hunan.

The histories contain an even wider range of information. The first place to look in any dynastic history is the Economic Monograph, *Shih Huo Chih*[2], compiled for every dynasty from the Former Han on. As Chinese governments have depended since the −6th century on land taxes for the bulk of their revenue, it is not surprising that the Economic Monographs contain data on land classification and allotment, yields and rents, population figures and government strategies for agricultural development; they also contain, *inter alia*, accounts of famines and disasters, price fluctuations, trade conditions, the construction and maintenance of canals and irrigation systems, and the management of state granaries. Other sections of the histories give accounts of peasant rebellions (and the grievances that triggered them off) or the biographies of statesmen involved in agricultural policies. The Sung reformer Wang An-Shih[3], for example, introduced a set of economic reforms designed to improve the livelihood of the peasantry, including the famous 'green sprouts' agricultural loans (*chhing miao chhien*[4]) and measures to reclassify land and extend the cultivated area. Wang's reforms were hotly opposed by such famous men as Ssu-Ma Kuang[5] and Su Tung-Pho[6], and their memorials of protest, like Wang's policy proposals, are reported at length in the Sung history and in other collections of Sung official documents. It has thus been

[a] See Ebrey (1) for the early medieval period, Beattie (1) for the Ming and Chhing, and Eberhard (28) for a more general approach.

[b] Their value is reflected in the number of studies which have been made of them, including Twitchett (9), Niida Noboru (*3*), Makino Tatsumi (*1*), H. C. W. Liu (1), and Lo Hsiang-Lin (*6*).

[c] See Ebrey (2), Herzer (1).

[d] Tr. Beattie (1), App. III.

[1] 方志 [2] 食貨志 [3] 王安石 [4] 青苗錢 [5] 司馬光 [6] 蘇東坡

possible to reconstruct the course and ultimate failure of Wang's reforms in considerable detail.[a] Other official sources include collections of edicts like the *Thang Liu Tien*, or documents such as the Yellow Registers (*huang tshe*[1]), which were files on labour services based on the enumeration of the entire population, including details of age, sex and occupation together with a summary of the land and property owned,[b] as well as the so-called 'fish-scale maps and registers' (*yü lin thu tshe*[2]), somewhat akin to our own Domesday Book, which recorded each parcel of land, listing the name of its owner.[c]

## (6) The Content of the Chinese Sources and the Implications for Historical Interpretation

Having completed a rapid survey of the main sources at our disposal, it now remains to assess the picture we can draw from them of Chinese rural history. First, and perhaps most importantly in a work such as this, the technical information is copious enough to provide us with a continuous and detailed account of the evolution of Chinese crop plants, cultivation methods and agricultural equipment.[d] Chinese agronomists and technical writers used a wide and precise technical vocabulary, much of which can still be understood today, and the interpretation of difficult or obscure passages is facilitated by comparison with contemporary illustrations or archaeological material. It is possible to reconstruct the technical development of Chinese agriculture from Han times on with a continuity, precision and detail which is quite unmatched in Europe until the 16th or 17th century.

The social and economic history of China is quite another matter. The Chinese government's preoccupation with agriculture as its tax base at least ensured that a great many official documents related to agricultural matters have been preserved, but as the economic historian D. C. Twitchett points out:[e]

> The institutionalised bureaucratic histories of China, to which government finance was a matter of prime importance, pay economic problems as we understand them scant attention. Even within the broad category of state finances the available material has serious shortcomings. The official histories were essentially court-centred works, and thus we are very well-informed on the highest levels of administrative activity, which were

[a] Williamson (1), Meskill (2), and James T. C. Liu (2).

[b] Ho Ping-Ti (4), p. 3.

[c] Although Ho (*ibid*) credits Ming Thai-Tsu with having commissioned the first of these surveys on a national scale, Ray Huang (3), p. 42 points out that while their origins go back at least to the Sung, even in the Ming they could not be described as a national land survey, and there was generally speaking little coordination in the compilation of land data.

[d] For earlier works of this nature, see the Chinese Agricultural Heritage Series on crop plants, and Liu Hsien-Chou (*7*) and (*8*) on the development of agricultural implements and farm machinery; Amano (*4*) treats selected topics, such as the development of the plough or cotton production, chronologically.

[e] (4), p. vi.

[1] 黃册 [2] 魚鱗圖册

conducted at the capital, but given little information on affairs in the provinces, where the policies dictated by central government were actually enforced ... in general it may be said that the material at our disposal stresses theory and general policy pronouncements rather than everyday practice. This of course conforms precisely with the aims of historians writing in the bureaucratic tradition.

These strictures are certainly largely true as far as the official histories are concerned, though for later periods the information they contain can be filled out with details from other sources such as local gazetteers which, operating at a much lower administrative level, were far more specific. But of course generalisation from sources whose scope is so restricted in time and space is fraught with pitfalls.

It is certainly difficult to pass from the level of theory and general policy to administrative practice, but not always impossible, particularly where new developments or changes in traditional practice are concerned. Although there are many gaps in our information, it is feasible, for example, to outline such extremely important agricultural development policies as government distribution of land, the establishment of agricultural colonies at the frontiers (*thun thien*[1]), or the construction of large irrigation schemes and the related attempts to popularise wet-rice cultivation in North China.[a] We even find descriptions of Chinese government agricultural policies that are in many respects reminiscent of present-day development strategies: although we lack the statistical and budgetary data that form such a bulky proportion of modern reports on development projects, it is clear that in many respects Han or Sung policies for agricultural development were similar to those put forward by international development agencies and national government planning offices in Asia today. This is a fascinating area of socio-economic history which deserves further study.[b]

When it comes to more detailed socio-economic analysis of the type which contributes so significantly to our understanding of historical developments in Europe, we are bedevilled less by the lack of adequate information than by difficulties of interpretation. For agricultural historians of China, a particularly teasing problem is the lack of any agreed terminological conventions to describe relations of production. To anyone used to the precise legal definitions of tenurial status that abound in the European literature, with carefully drawn distinctions between, for example, 'copyholder', 'freeholder', and 'tenant in fee simple', the indiscriminate use of a term such as *tien*[2] to describe what may have been either a tenant paying fixed rent, a sharecropper, or even on occasion a serf, depending on (*a*) the period and locality to which the original text refers, and (*b*) the personal predilections of the historian interpreting the text, is both confusing and

[a] Han Kuo-Pan (*1*); Li Chien-Nung (*3*) and Eberhard (26) on land distribution and *thun thien* during the early medieval period; T. Brook (1) on the attempted introduction of rice cultivation to Chihli in the Chhing.

[b] See p. 70, n. 6 above.

[1] 屯田 [2] 佃

frustrating. Occasionally one can refer to the original terms of contract described by terms like *tien* or *tsu*[1] for enlightenment, but such documents are very rare.[a]

Attempts to interpret relations of agricultural production at various periods of Chinese history fall into two categories. First, there is the rigid framework of mainland Marxism whereby the nature of tenurial relations is automatically defined by the historical stage in which it occurs.[b] Secondly, there is the more flexible approach which conceives of a range of tenurial contracts coexisting and evolving at any one time. This approach is to be found among Marxists and non-Marxists alike outside China, but unfortunately the material they have to work on is generally so scanty that few definite conclusions can be drawn. The controversy that still rages, for example, as to whether Sung tenants were mere serfs or legally independent farmers, is unlikely ever to be resolved.[c] Of course, in a country the size of China generalisations on social or economic relations can be misleading, and it is perhaps misguided to place too much importance on identifying what one might call a typical or predominant mode of production. But there is no question that tenurial relations have a fundamental influence on social and technological development, and so it is disappointing to note the complete disregard for accurate definition that characterises even modern studies of rural China. Even where unambiguous evidence must have been available in the 19th and 20th century, most scholars were content when studying tenurial relations to draw only the simplistic distinction between 'owner', 'part-owner' and 'tenant', with no reference to terms or duration of the contract.[d] It is therefore impossible to gain any insight into the degree of control a farmer had over his holding, changes in holding size associated with the domestic cycle, or the likelihood of peasants or their landlords investing in improvements, all of which are clearly crucial in understanding the evolution (or in some cases lack of evolution) of Chinese agriculture.[e]

Closely related to tenure is the question of farm management. Here again we are very poorly informed and have to rely mainly on inference rather than

[a] Only one instance is known of a tenancy contract surviving from the Sung period.

[b] E.g. Chhen Teng-Yuan (*1*); Wan Kuo-Ting (*4*). Most of the historical span with which we are concerned falls, according to Marxist analysis, into the 'feudal' period. In practice the range of tenurial relations which different Chinese historians conceive of as 'feudal' is sufficiently wide to include both serfs and peasants, but no clear conceptual distinctions are drawn that permit a convincing explication of technological development and change in terms of the relations of production.

[c] Golas (1) summarises the different arguments advanced as well as providing a most useful bibliography of Japanese work. As Beattie (1) points out, the main reason for the bitter disagreement between the Kyoto and the Tokyo schools, the chief protagonists in the debate, is that they base their arguments on data taken from different regions of China. One of the most satisfying analyses of Sung tenurial relations in our opinion is to be found in Lewin (1).

[d] Chung-Li Chang (1); J. L. Buck (1), (2).

[e] Scholars working on China are not alone in their failure to grasp the significance of an accurate definition and description of tenurial relations; despite extensive studies by both neo-classical and Marxist economists on the implications of different forms of tenurial contract for patterns of economic development (see Bray & Robertson (1)), modern development literature is still full of statistics based on the distinction between 'owner', 'part-owner' and 'tenant', which is so imprecise as to be nearly useless.

[1] 租

concrete information. Large, gentry-owned estates or 'manors' were certainly fairly common in several regions of China at one period or another, while under different circumstances monastic holdings or lineage estates were important. We do possess certain documents that could be described as works on estate management, for example the *Ssu Min Yüeh Ling* and *Shen Shih Nung Shu*; the *Chhi Min Yao Shu* also contains indirect references to the running of a large estate. However, interpretations of the passages relating to estate management in these works differ substantially.[a] Perhaps the outstanding example of the genre in China is Chang Ying's[1] *Heng Chhan So Yen*, Remarks on Real Estate, probably completed in 1697 and written chiefly for the benefit of his heirs. But even this comparatively explicit and specialised work contains few of the details we would expect from a traditional Western work in the genre. Chang talks in general terms of regulating expenditure, but gives no figures for labour costs, rents, estimates of yields or rates of return on investment. Land in China was usually regarded less as a source of profit than as a safe investment, an alternative to the hazards of industry or commerce, and while Chang does speak of improving his land he gives only the vaguest indications of how this is to be achieved. He also stresses the importance of personally inspecting the estate, and here he is more explicit:[b]

> The young men of the family should in the spring and autumn of every year go in person to the farms and inspect them thoroughly. In the slack season they should also get on their nags and take a trip there. But it is no use going just for the sake of it. The first important thing is that they should know the boundaries of the fields ... The second thing that they should do is see whether the farmers are industrious or lazy in their work, if the ploughing and sowing are early or late, if the stores are abundant or meagre, if men and beasts are many or few, if expenditure is extravagant or moderate and if the land is being well managed so as to improve its quality. The third thing to do is to examine carefully to see whether ponds are shallow or deep and whether dikes are strong or unsound, so that they can be built up. The fourth thing is to inspect the trees on the hillsides to see if they have gone to waste or are flourishing. The fifth thing is to enquire whether the price of grain at the time is high or low. I hope that they will gain accurate knowledge of all this.

These are fine principles indeed, but principles are not sufficient to run an estate efficiently and we are left with little idea of the practical details involved. Still less do we have any idea of how his tenant farmers would have run their farms. Chang Ying does dwell on the necessity of selecting good tenants:[c]

> There is a proverb which says, 'It is better to have good tenants than good land' ... there are three advantages in having good tenants, namely that they are on time with ploughing and sowing, they are energetic in fertilising, and they are resourceful in conserving every drop of water.

[a] On the *Ssu Min Yüeh Ling* see Shih Sheng-Han (*2*); Herzer (1); Ebrey (2). Kumashiro (1) summarises the case for *Chhi Min Yao Shu*. Much more detailed analysis has been possible for the *Shen Shih Nung Shu*; Chhen Heng-Li & Wang Ta-Tshan (*1*).

[b] Tr. Beattie (1), p. 149.

[c] *Ibid.* p. 146.

1 張英

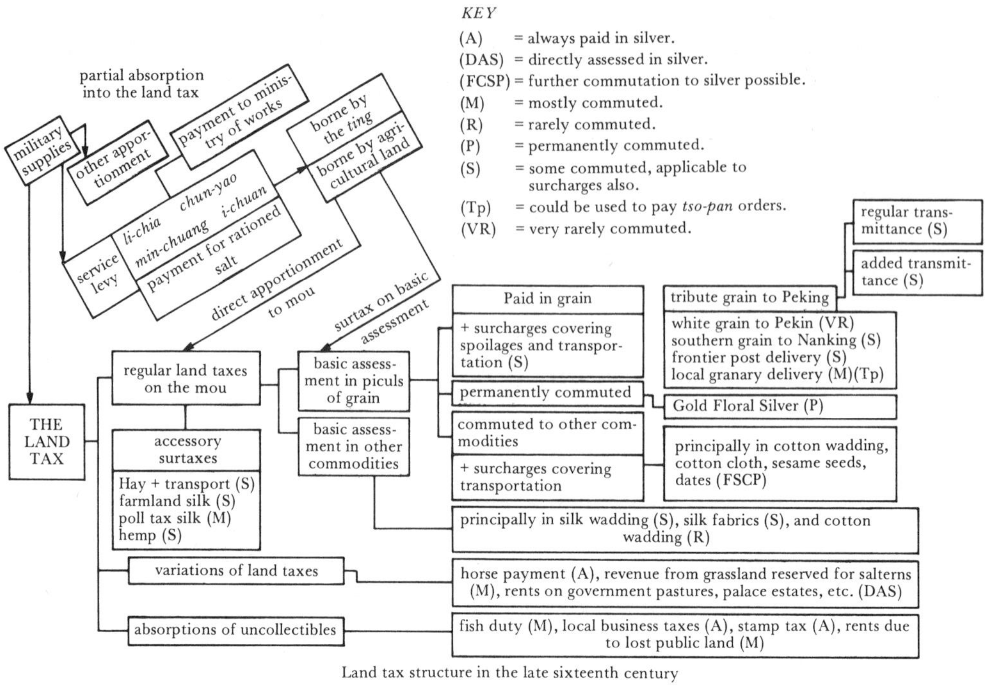


Fig. 24. The Ming land tax system; Ray Huang (3), p. 83.

However, he completely neglects to specify the terms of the tenancy contract.

Although it is true to say that our knowledge of the relations between the rural classes of traditional China is scanty, we do know a little more about the relations between Chinese farmers and administrators. Information about general policies, as Twitchett suggests, is usually schematic, but there are striking exceptions. Certain problems were pressing and recurrent, and were debated at length by Chinese bureaucrats in their private as well as public capacity, or have been minutely described in official handbooks. 'Land taxes in the late Ming period', says Ray Huang,[a] 'were undeniably as complex as personal income tax in the twentieth-century United States, if not more so'; yet by patient and meticulous study Huang has been able to unravel many of the mysteries of the late Ming land tax system to which every peasant was subject (Fig. 24). Another constant preoccupation of Chinese officials was famine relief, the collection, storage and distribution of relief grain. Here again there is sufficient material available on both public and private relief systems to analyse them in considerable detail.[b]

In fine, then, our Chinese sources provide us with detailed technical information on the evolution of Chinese agriculture, as well as some telling insights into

[a] (3), p. 82.

[b] A very general treatment is given below, p. 519. An impressive analysis of famine relief in 18th-century China has recently been published by P.-E. Will (1).

the relations between central government and the farmers upon whom it depended for its revenues. At the level of relations between the various rural classes, however, we are extremely ill-informed, and given the nature of our sources there is little hope that subsequent research can throw much additional light on the subject.

## (7) A Comparison with the European Tradition

We have dwelt at some length on the nature and content of the Chinese sources because we hoped that this would help explain why Chinese agricultural history is necessarily so different from what is written about Europe. European historians may be disappointed to find so little in this volume on problems they regard as fundamental to the understanding of agrarian history; they may regard detailed accounts of the evolution of farm tools as a poor substitute for an analysis of the evolution of land tenure and its consequences. We can only reply that the nature of our sources has dictated our choice of subjects, and hope that the very different picture of Chinese agriculture that we present may contain something novel to stimulate or intrigue them. For those readers who are not familiar with Western agricultural literature, we shall now draw a brief comparison to highlight how the European differs from the Chinese tradition in content and form.

As we have seen, the Chinese tradition of agricultural works stretches back unbroken to before the Han dynasty. True, many of the earliest works have been lost or survive only in fragmentary form, but the titles recorded in the bibliographies show that there was no hiatus in the Chinese tradition, and as the centuries advanced agricultural works were produced in increasing numbers. The popularisation of printing in the Sung ensured more accurate transmission of the texts and of course allowed them a far greater dissemination. There was not, however, a quantum leap in the number of agricultural works produced comparable to that which followed the introduction of printing in Europe.

The Western tradition of agricultural literature begins with the Greeks and Phoenicians, but apart from Hesiod's *Works and Days* and Xenophon's *Oeconomicus*, little has survived from those early times even in quotation.[a] It was the great Latin works that were to stand as models for European farmers right up to the 17th century and beyond, and with good reason. K. D. White draws a significant distinction between the Roman writers and their predecessors[b]: 'Whereas the Greek contribution ... comes mainly from the works of philosophers and men of science, Roman agricultural writing was based from its inception on practical farming experience. The earliest and the latest surviving writers are linked by a common fund of maxims and proverbs ... embodying the experience of generations of farmers.' Here we are reminded of the Chinese tradition, yet the classical Roman works reflected very different concerns from the Chinese.

[a] See K. D. White (2) for a much fuller account.
[b] (2), p. 18.

The first of the great Roman writers was Cato the Censor, whose *De Agri Cultura* was written in the middle of the −2nd century, slightly earlier than *Fan Sheng-Chih Shu*.[a] Since it is a pioneer work, its structure is less than perfect; some scholars have questioned its authenticity on the grounds of its rambling style and structural incoherence, but the majority accept the whole text as genuine. The most important sections deal with the organisation and management of estates which produced mainly wine and olive oil.[b] Later Roman writers were quick to improve on Cato's model: Varro's *De Re Rustica* (in three books) was published in −37 and is considerably more sophisticated and complex, while Columella's *De Re Rustica* (in twelve books) and *De Arboribus* (in one) is an original and comprehensive work which treats the cultivation of field crops, vines and olives with a fine grasp of technical detail.[c]

While Varro and Columella represent the heights of Roman agronomic knowledge, they were not to exert the greatest influence in later times. This honour was reserved for Virgil and Palladius. Virgil's *Georgics* is unquestionably a literary masterpiece, but Virgil himself had little agricultural experience and had to draw heavily on such sources as his contemporary, Varro, for practical information. Even then many of his notions are unreliable. 'Examination of a few passages dealing with such basic topics as ploughing, harrowing and soil testing has shown the poet to be correct in half the operations under discussion.'[d] But lack of technical expertise did not deprive him of a following; of over 750 printed editions of the classical agricultural authors that are known to have appeared in the 15th and 16th century, no less than 412 were of Virgil's *Georgics*.[e]

Palladius' *De Re Rustica*, probably written in the +4th century, was a textbook in calendar form. It contained almost no original material, consisting chiefly of the reasoned explanations of Columella and other authorities condensed into the form of precepts. This simplicity of form and content guaranteed Palladius' popularity right up to the Renaissance; his work was even translated into English verse in the early 15th century.[f]

The Roman writers described the management of large, slave-worked estates producing for the market. Vines and olives were the most important crops for Cato and Varro; with Columella and Palladius wheat production gained in importance, reflecting changes in the political economy of the Roman Empire, but it never ousted viticulture as the most important occupation. Stock-raising was an important side-line, providing traction power and manure for the farm as well as being a source of income. During the early years of the Republic no agricultural works were produced, and it seems that it was the decline of the independent peasantry during the later Republic and the concomitant growth of

[a] For full descriptions, analyses and assessments of the works of the great Latin agronomists see K. D. White (1), (3), and especially (2).

[b] K. D. White (2), pp. 19ff.; in Appendix A he sets out in detail the contents of *De Agri Cultura*.

[c] K. D. White (2), pp. 22–28.

[d] *Ibid.* p. 40.

[e] Beutler (1), p. 1297.

[f] Ernle (1), pp. 33, 419.

large, market-oriented slave-run estates that stimulated the development of the new literary form.[a]

Between Palladius and the first of the European agronomists, Pietro de Crescenzi, there was a hiatus of nine centuries. During this time the great Latin treatises continued to be consulted on many monastic estates,[b] but only in Moorish Spain were any new agronomic works produced in medieval times. Outstanding among these was the *Kitab al Filāḥah* or Book of Agriculture, an encyclopaedic work written in the 12th century by Ibn al-'Awwām of Seville.[c] Although in many ways the *Kitab al Filāḥah* may be considered a successor to the great Roman treatises, it must also be seen in the context of the political and economic expansion of the Muslim world, for at about the same date valuable treatises were produced not only in its Mediterranean dominions of Spain, North Africa and Egypt, but also in Mesopotamia and Persia.[d] The Mediterranean works had much in common with the Latin treatises, yet they clearly belonged to a distinctive Arab tradition, the chief particularity of which was its extensive use of irrigation. The wide sweep of the Arabs' trading contacts had familiarised them with such exotic crops as sugar-cane, oranges and rice, and their skills in irrigation allowed them to transform even the arid confines of Persia and Spain into sub-tropical gardens where these valuable species might flourish. The agricultural treatises of Moorish Spain owed much to their Roman predecessors, but must be seen as a highly specialised side-line dependent on certain forms of political organisation for their survival. Unlike the works of Varro and Columella their application was severely limited—where in feudal Europe was irrigation to be found?—and even in Spain itself, after the fall of the Moorish state and the consequent deterioration of irrigation facilities, the great Arab encyclopaedias soon ceased to be read.

This was not, however, the case with the great Latin agronomists, whose work continued to be copied and consulted well into the 17th century, even though their contents must often have been quite inappropriate to local conditions.[e] These works were written with the dry climate and light soils of the Mediterranean in mind, and the techniques they contained were not really suitable for the wet climate and heavy soils of the Atlantic coast. But it may have been the managerial rather than the agronomic content of the Latin treatises which most appealed to medieval readers. Certainly the only new agricultural works of any

[a] 'One may say that from Cato the Elder until Augustus enterprises based on slave labour grew steadily larger and in fact became the dominant form of economic activity. Roman writers on agriculture take it for granted that slaves would do most of the work, free labourers being used only for the harvest...' (M. Weber (4), p. 318). The nature and organisation of the Italian *latifundia* is discussed at some length in K. D. White (2), pp. 384 ff., that of the provincial villa in Percival (1).

[b] See, for example, Duby (1), Fussell (5).

[c] Tr. Clément-Mullet (1); see also Imamuddin (1); Vallicrosa & Azīmān (1); Bolens (2).

[d] Lewis, Pellat & Schacht (1) under *Filāḥah*; Bonebakker (1); Ḥusām (1); Lambton (1).

[e] See Duby (1) pp. 22 ff. on the application, or mis-application, of Mediterranean agricultural technology on estates in Northern Europe.

merit to appear during this period were treatises on estate management written for the benefit of either monastic foundations or noble landowners.[a]

These were remarkable less for their agricultural than their managerial content. The first really original agronomic work to appear in Europe, after a break of some nine centuries, was Pietro de Crescenzi's *Opus ruralium commodorum* of 1304. Crescenzi relied heavily on the Latin authors, but his approach to them was critical and he supplemented quotations from the classics with numerous personal observations based on his extensive travels in Lombardy and his close acquaintance with the farming practice of the Dominican monastery at Bologna. He was particularly interested in soil erosion, and suggested several alternative or complementary strategies for its control. Crescenzi's work marked the renaissance of European agronomy: the *Opus ruralium commodorum* was most enthusiastically received and was soon translated from Latin into Italian; in 1373 Charles V of France ordered its translation into French, and in 1471 it was published by a press in Augsburg, thus receiving 'the distinction of being the first book on agriculture ever printed'[b]—the first Western book, that is to say, for, as we know, Chinese agricultural treatises were being printed as early as the 10th century.

The revival of European agricultural writing coincided with the development of printing, and has often been regarded as one of its consequences. But although printing certainly facilitated the dissemination of such works, and probably encouraged many novices to try their hand—as Corinne Beutler points out (1), there is nothing like reading someone else's work for tempting one to write something better on the same subject—it seems to us that more basic reasons should be sought for this new flowering. Just as the growth of the *latifundia* had provided a readership for the Latin agronomists, so the breakdown of feudal tenure and the growth of what one might call proto-capitalist farming provided a readership for the new European agronomists. Though these authors naturally came from the educated elite, they wrote not in Latin for scholars, but in the vernacular, for the benefit of their peers: 'Their instruction was directed towards ... the other landowners of the area, so as to teach them how to increase their yields, instruct their peasants, acclimatise new species and profit fully from the sale of their surplus.'[c] These landowners aimed not at self-sufficiency, as had the feudal lords,[d] but at commercial production. It is no coincidence that Italy and Germany were the two countries which first experienced the agronomic revival:[e] there the new free cities provided an ever-expanding market for agricultural produce, encouraging local landowners to improve their farming practice and so

[a] For example, the 13th century English works, which include Walter of Henley's *Husbandry* (*c.* 1280), an outline of husbandry and management practices in the Midlands; Bishop Robert Grosseteste's *Rules* of estate management, written in 1240, in French, for the use of the Countess of Lincoln, and the anonymous *Seneschaucy* which includes many legal details of the various manorial offices. Oschinsky (1) gives a full analysis as well as reproducing the texts.

[b] Olson (1), p. 40.

[c] Beutler (1), p. 1291, tr. auct.

[d] Kula (1).

[e] Meuvret (1).

increase their profits. The profit motive was certainly important: one of the most popular of the new English works had the best-selling title *A Way to Get Wealth.*[a]

The classical Latin treatises, rather like the Chinese, had aspired to universality. Naturally such experienced farmers as Varro and Columella recognised that soil conditions or labour requirements would vary from one region to another, but on the whole they tried to provide advice of general application. But the agronomists of a 16th-century Europe divided into independent nations and city states were more strongly conscious of regional differences. Most agricultural works of the period are explicitly restricted to local application: 'Here I am on German soil, amongst the good, honest Germans ... what others have written is often not suited to us here in Germany for their skies, air, water and earth are quite different from ours ... So I, being merely a German writer, speak in these volumes only of the German countryside', says Coler (1). Some writers confined themselves to national boundaries, others felt obliged to restrict themselves still further: the works of the prolific English writer, Gervase Markham, include one entitled *On the Inrichment of the Weald of Kent* (4). But a heightened consciousness of the particularity of local conditions did not prevent frequent translations of agricultural works from one European language to another, although substantial amendments and additions were often necessary if the foreign works were to be useful.[b]

The rapid development of printing and the ready market for any works pertaining to agriculture or horticulture encouraged writers to capitalise on their specialist knowledge. Whereas before 1600 most agricultural writing was encyclopaedic in scope, throughout the 17th century monographs on surveying, horsemanship, hop-growing and bee-keeping proliferated.[c] By 1700 the list of European agricultural works must have equalled, if not surpassed, the Chinese.

This seems a suitable point at which to take stock and compare the agronomic traditions of Europe and China, for until the 18th century they remained in almost total isolation one from the other.[d] The great Latin treatises had shared the encyclopaedic approach of such Chinese masterpieces as *Chhi Min Yao Shu* and *Wang Chen Nung Shu*, stressing general principles rather than specific application. But with the formation of the independent European states and the development of printing, European agronomists became highly conscious of variation and change, and many of the most important works of that period, Olivier de Serres'

[a] Markham (3); first published in 1623, by 1631 it was already in its fifth edition.

[b] For example, Markham's additions to Estienne & Liebault's *Maison Rustique* (1) form a good proportion of the 1616 edition.

[c] See, for example, the bibliographies in Bourde (1), Ernle (1), App. I. The multiplicity of titles gave ample scope to plagiarists: Poynter (1) p. 33 describes the indignities suffered by Gervase Markham in this respect (*Markhams Maister-peece* (5), a handbook on the care of horses first published in 1610, was his most plagiarised work). Of course the authors of unpublished works were even more vulnerable. The main substance of Samuel Hartlib's *Legacie* (1) of 1651, for example, was a verbatim piratical printing of a treatise by Sir Richard Weston (1), bequeathed to his sons in manuscript form a mere six years previously.

[d] With the one exception of Hsü Kuang-Chhi's interest in European hydraulics.

*Théâtre d'Agriculture* (1), Fitzherbert's *Boke of Husbandrye* (1), Markham's *English Husbandman* (2) or Heresbach's *Rei Rusticae* (1), were explicitly written with a limited geographical region in mind, whereas, though the Chinese agricultural literature did show a clear distinction between northern and southern agriculture, only minor monographs were written for regional use.

One might have expected the interplay between general and specialised works in 16th- and 17th-century Europe to herald the development of a scientific, experimental approach to agriculture such as was developing in other fields at that period. This was not generally the case, however. European agricultural writers seem to have seen their task as the crystallisation of traditional experience, an attitude they held in common with most of their Chinese counterparts. Here we must make a distinction between introduction and innovation. Neither in China nor in Europe were the concepts of improvement and change alien or repugnant to agronomists; on the contrary, they were often remarkably quick to spot the potential for their own country of some alien crop or technique. One thinks of Wang Chen's (admittedly fruitless) attempts to popularise northern wheat-harvesting technology in southern China, or perhaps of François I of France '[qui] par certains salaires et sommes de deniers proposés et donnés à plusieurs pérégrinateurs, a fait que notre France a été enrichie de plusieurs plantes, herbes et arbres exquis, desquels non seulement la figure et culture nous étaient de tout inconnues, mais aussi les noms d'iceux'.[a] Both Chinese and European agronomists enjoyed introducing new ideas from other areas, but it rarely occurred to them to develop new ideas on their own account. Almost without exception, in both East and West, innovations in agriculture before the 18th century were achieved not through systematic experiment by educated men but by the painful efforts of the peasants tilling the soil. The classic agronomic works do record changes and innovations, but these resulted from gradual modifications or sudden inspiration by totally anonymous husbandmen: technological change was thus imposed from below rather than above. Only when the scientific spirit of the Enlightenment was extended to the ancient craft of husbandry did agronomists start consciously to rationalise, to define systems, to *invent* new devices and techniques, and generally to impose innovation from above.[b]

Thus in many respects the agronomic traditions of Europe and China were similar. Do our sources then allow us to draw similarly detailed pictures of agriculture in the two great civilisations? The answer, of course, is no. One area in which the Western sources far surpass the Chinese is the legal aspects of agriculture. We possess a wealth of legal documents even for Rome, but especially for the European nation states, defining and elaborating the rights and duties of landlord and tenant, changes in tenurial form, and so on. Account books and contracts inform us as to conditions of labour hire, while documents such as parish records,

[a] Charles Estienne (1).

[b] On the development of the scientific spirit among European agriculturalists, and the possible role of China in stimulating such innovation, see pp. 571 ff. below.

inventories and wills allow us to reconstruct in some detail patterns of inheritance. As far as the status, property, rights and duties of different rural classes or even individuals are concerned, we are faced in European history with an *embarras de richesses*. Historians naturally vary in their interpretations of European rural society, but it is the very wealth of detail that intensifies the debate: for every general rule proposed, dozens of particular exceptions can be cited. The depth of detail in which it is possible to reconstruct the history of small European communities or even the lives of individuals is quite without parallel in China. One of the most accomplished attempts to flesh out the bare bones of Chinese rural history is Spence's description of life in a Shantung county in the late 17th century (1). Fascinating as his essay is, Spence can provide nothing on the local economy or social relations to compare with Le Roy Ladurie's account (1) of the southern French village of Montaillou in 1300. The methodology first adopted by the French *Annales* school[a] has been both technically and theoretically refined and elaborated,[b] but clearly an enormous range of raw material is the prerequisite for such subtlety of method. Japanese and American sinologists are nowadays resorting to computers in their exhaustive studies of local gazetteers, yet it is unlikely that they will ever be able to produce regional studies or monographs of a depth and quality to match those written on Europe.[c] In such circumstances works of synthesis like those by Marc Bloch, B. Slicher van Bath, Georges Duby or M. M. Postan carry a conviction that one rarely feels after reading historical interpretations of Chinese rural society, however skilled the historian: there is simply too much that remains hypothetical.

The Chinese data allow us to infer nothing like the European wealth of detail on individuals, their personal property and place in rural society. In some respects, though, the sinologist does have an advantage over his European colleagues. The Chinese preoccupation with the workings of bureaucratic government, together with records of a tax system based on revenues from land, provide us with much information on government agricultural policies, taxation methods, fiscal reforms and development plans which affected even the humblest peasant, as well as providing a series of figures for population density, distribution of land-holding and occupational type which are illuminating if not wholly reliable. Furthermore, the best Chinese agricultural treatises, in our opinion, surpass anything produced before the 18th century in Europe in their systematic presentation of technical detail. Only in the treatment of such specialist topics as viticulture

[a] Its founder was Marc Bloch, whose *Caractères originaux de l'histoire rurale française* (7) represented a milestone in historical method. Perhaps the archetypal work of the *Annales* school is Braudel's (1) *La Méditerranée*, a monumental attempt at 'total history'. (For a useful discussion of the *Annales* 'paradigm', its aims and methods, see McLennan (1), pp. 129–44.) A new history of rural France has recently been completed under the general editorship of Duby & Wallon (1).

[b] See, for example, Macfarlane, Harrison & Jardine (1), Dahlman (1).

[c] Of note among the regional studies is the French series *La Vie Quotidienne*, e.g. Soulet (1); one thinks also of more specialised works like P. Léon (1). Still more specialised topics such as regional field systems (Baker & Butlin (1)), the evolution of local tenurial forms (Merle (1)), local patterns of inheritance (Cicely Howell (1)) or social mobility (Raftis (1)) can also be treated in depth.

can the great works of Varro and Columella compare with the *Chhi Min Yao Shu* and its successors for pragmatic details. The Chinese works cover a much wider range of crop plants than Western works, including not only numerous varieties of cereals and fibre crops but also vegetables, fruits, citrus, sugar-cane and tea. The number of species, and also of varieties, recorded and described in the Chinese agronomic literature is incomparably greater than anything to be found in the pre-modern West.

Where technology is concerned too, we feel that up till the 18th century it is possible to draw a far more elaborate picture of Chinese agricultural equipment and its use than can be done for the West. True, individual items may sometimes receive more detailed treatment in the European literature, but Europe has nothing to compare with Wang Chen's systematic illustrated treatment of farming equipment, covering everything from straw sandals to water-mills, and we shall see in section (*d*) below that Wang's information can generally be supplemented from other sources to provide a full historical survey of the most important items of Chinese farm equipment.

This, then, was the situation up to about 1700. Although contacts between Europe and China were multiplying throughout the 17th century, neither agronomic tradition was influenced by the other at this stage, and both could be regarded as having attained a similar level of sophistication, though with rather different emphases. During the 18th century all this was to change. Chinese agricultural literature continued to progress gradually, along the same lines it had been following for centuries; little if any Western influence was felt until the end of the 19th century, and it is only in the last fifty years or so that such concepts as science and modernisation have come to play an important role in Chinese agronomy. Since 1975 the Chinese government has been particularly anxious to transform its agriculture to something closely resembling the Western model, which paradoxically enough seems to have been heavily influenced at its inception by Chinese models (see pp. 558 below). In 18th-century Europe, on the other hand, a quantum leap occurred: agriculture was transformed from a traditional skill to an experimental science. As a consequence of social, economic and political developments which we cannot begin to elaborate here,[a] landowners became actively involved in agricultural improvement. They not only experimented with new methods such as crop rotation or selective stock-breeding but were willing to invest in new, more efficient and labour-saving technology. Engineers, soil scientists, surveyors, botanists, professional and amateur, all rushed to publish their latest theories and discoveries, the merits and demerits of which were subsequently debated in print, and from the 18th century we are faced with an avalanche of agricultural publications, books, pamphlets, patent applications and journals. Europeans visiting China in the 17th century were impressed by the sophistication of Chinese agriculture and brought Chinese agri-

[a] See, for example, Bourde (1), Brandenburg (1), Mingay (1), Winch (1).

cultural treatises back to Europe to learn from them what they could. But from the 18th century both the knowledge and practice of agriculture in Europe took a new turning, and soon European agricultural science far surpassed anything the Orient had to offer.

## (c) FIELD SYSTEMS

In this section we shall describe the main field types found in China, their function, distribution and history. At first glance it may seem strange that we bother to categorise field types so minutely. In Europe we are familiar with different types of field: in England we can trace the historical evolution from square 'Celtic' fields to the vast medieval open-fields, resembling a patchwork of individual strips, through the fenced or hedged fields (*bocages* in French) of the enclosures, to the huge unbounded fields of modern farms in East Anglia. But in our temperate climate with its gentle relief, there are few physical constraints on field shape or size, and the reasons for such changes as we have mentioned are mainly social or technological.[a] In parts of China, as we shall see, the general pattern of land division does seem to have been influenced by political or even ideological factors, but generally speaking it has been determined by the demanding nature of the terrain and, in rice-growing areas, by the peculiar requirements of irrigated cultivation. We shall not dwell here on such political factors as the continuous conflict between the state and the landowning classes, the many systems of egalitarian land distribution initiated by the government and the inevitable resurgence of large private estates. The theories and realities behind the 'well-field' (*ching thien*[1]), the 'equal-field' (*chün thien*[2]) and related tenurial systems will be examined in detail in Volume VII. Here we shall concentrate on the physical and technological factors in the evolution of the Chinese rural landscape.

### (1) Land Clearance and Reclamation

Before land can be cultivated it must first be cleared and turned into fields. The extent to which clearance is necessary depends on the intensity with which land is

[a] The change from square 'Celtic' fields to the medieval strips, for example, has been attributed to a variety of technological and social causes. Some scholars see the change as due purely to technological factors: they believe that the square fields were that shape because they were ploughed with a light ard, or scratch-plough, that was only efficient when cross-ploughing was practised; when the heavy turn-plough was introduced it became more convenient to distribute the land in the long, narrow strips classically associated with the medieval common field system (H. C. Bowen (1); C. S. & C. S. Orwin (1); Lynn White (7)). Others believe that 'it is fallacious to argue that field systems were determined by the existing plough technology, although of course this was one of the many inputs' (Baker & Butlin (2), p. 634), and see the development of the open-fields and commons as a response to largely social factors, whether the egalitarian ideology of the feudal village (G. C. Homans (1); Vinogradoff (2); Ernle (1)), the effects of partible inheritance (Joan Thirsk (1)), cooperative land clearing (T. A. M. Bishop (1)), or the optimal transaction costs of maintaining a mixed-farming economy (Dahlman (1)).

[1] 井田 [2] 均田

cultivated and on the density of population. It used to be generally assumed that agriculture progressed through a series of increasingly intensive methods of cultivation: first shifting cultivation, where land was cleared from forest or savannah and cropped for only two or three years before the farmers moved on; next short-fallow rotations, where fields were cropped for one or two years and then allowed to lie fallow for a similar period; and finally continuous cropping, where the land was never allowed to lie fallow but would be cropped once, twice or even three times a year.[a] The farmers of Northern China, for example, were believed by many scholars to have practised shifting cultivation until well into the Chou period, when they progressed to short-term fallowing, continuous cropping becoming general practice only in medieval times.[b] Other sinologists would argue that both short-term fallowing and continuous cropping developed very much earlier.[c] There is also an increasing tendency to regard cropping systems of different intensity not as a developmental sequence but rather as appropriate responses to different natural environments: thus in tropical forests where natural soil fertility is low and erosion is rapid there are clear advantages to shifting cultivation, whereas in flat, well-watered regions continuous cropping may have been possible from earliest times.[d] Furthermore, under certain conditions farmers may well practise more than one type of cropping simultaneously: it is quite common for wet-rice farmers, whose main fields are continuously cropped, to increase their incomes by clearing small patches of jungle for shifting cultivation.[e] Less well known, perhaps, is the fact that many European and North American farmers living in areas of forest or moorland continued to practise shifting as well as permanent cultivation until the nineteenth or even the early twentieth century.[f]

Where populations are fairly small, it is possible to increase agricultural production simply by clearing more of the same sort of land. But once all the available land suitable for arable has been occupied, it becomes necessary either to intensify production on existing arable or else to reclaim land hitherto rejected for agricultural purposes. This is not always as desperate a measure as it might sound. Much of the most fertile land in Northwestern Europe was left uncultivated throughout Roman times, for the rich clays that are now some of the best wheat land were too heavy for the Roman ard or scratch-plough to work; only with the introduction of the mouldboard plough in the early centuries A.D. did it

[a] Boserup (1) correlates this progression with distinct stages of agricultural technological development; see p. 131 below.

[b] Yang Khuan (*11*) on changes in Chou agriculture, Li Chien-Nung (*4*) on the early development of crop rotations in the Han.

[c] Yu Yü (*3*) argues that continuous and even multiple cropping were common in Northern China by the Han.

[d] D. Freeman (1) lists the advantages of shifting cultivation as practised by the Iban in the tropical rain forests of Sarawak; Rawski (2), p. 24 postulates that irrigated rice cultivation in China may never have passed through the phase of intermittent cropping.

[e] Leach (2) gives a classic description of this practice in Ceylon.

[f] Sigaut (3).

become possible to reclaim these fertile soils. Similarly the Fens of East Anglia, though the initial work of draining was immense, are now one of the most fruitful regions of Britain. In China the pressure of population on land has always been greater than in Europe. Intensive production methods were developed at an early stage (see section (*d*) below), but in some regions it soon became necessary to extend the arable area from the plains into the hills and even up mountain sides, in dizzying cascades of terraces. Lakes were drained, salt-marshes turned to rice-fields, and floating fields constructed on rattan rafts. Nowhere else can such sustained ingenuity have been applied to solving the problem of land shortage, and we see the Chinese becoming ever more inventive as population pressure increases over the centuries.

The Chinese terms for land clearance were many and various. Since medieval times *khai huang*[1] and *khen thien*[2] have been most common, but before the Han terms such as *lai thien*[3], *tso thien*[4], *phou thien*[5], *tien*[6], or simply *thien*[7] were used. Most scholars believe that in Shang and Chou times land could only be cultivated for a few years at a time before losing its fertility, and consequently new land had constantly to be opened up; most land clearance was organised by the state and carried out by corvée labour.[a] One poem in the *Shih Ching* describes the legendary emperor Yü personally directing land clearance:[b] 'Extended is the Southern mountain, it was Yü who put it in order [for cultivation].'

Whatever the case in Shang and Chou times, we know for certain that the government took an active role in land clearance and reclamation throughout the imperial era. Settlement schemes in border areas, whether military (*thun thien*[8]) or civilian (*ying thien*[9]), had an important dual function: first, the influx of Chinese settlers into areas whose political allegiance was often uncertain had a stabilising and 'civilising' effect; secondly, the new agricultural colonies served as very useful overspill areas for refugees and landless peasants from areas of dense population. During the Former Han literally hundreds of thousands of displaced persons (*liu min*[10]) were resettled by the government in sparsely populated areas such as Kiangsu or the Shuo-fang[11] region in the northern bend of the Yellow River,[c] and a large number of military colonies was established throughout the Northwest. Land was cleared, irrigation was provided in many places, and colonists were provided with food, seed-grain, animals and implements either as outright gifts or on credit, while for several years they were not required to pay taxes.[d]

[a] The corvée labourers involved in this task seem to have been known as *chung jen*[12] (Keightley (3); Chang Cheng-Lang (*2*)), and were conscripted only temporarily. Some scholars would say, however, that most of the work-force involved in land-clearance, and in agriculture generally, in Shang times were not corvée labourers but slaves (Shen Wen-Cho (*1*); Yü Hsing-Wu (*2*)).

[b] *Shih Ching*, tr. Karlgren (14), no. 210, 1: *Hsin pi nan shan, wei Yü tien chih*[13].

[c] C/*HS* 23B315*b*; *Hsi Han Hui Yao* 56.

[d] Bray (3).

| | | | | |
|---|---|---|---|---|
| 1 開荒 | 2 墾田 | 3 萊田 | 4 作田 | 5 裒田 |
| 6 甸 | 7 田 | 8 屯田 | 9 營田 | 10 流民 |
| 11 朔方 | 12 衆人 | 13 信彼南山維禹甸之 | | |

Agricultural colonies filled a very useful function and the institution was revived at various stages throughout Chinese history right up to the Ming and Chhing, indeed agricultural colonies and state farms still play an important role in the People's Republic today.[a] We must not over-estimate their overall effect, however, for the area occupied by *thun thien* was only a minute proportion of the total cultivated area at any time.[b]

Some see the early clearance of land as a long and onerous task. Chang Cheng-Lang (*2*) believes that it took three years to prepare the land for cultivation: in the first year the trees were felled and the grasses weeded out (*tzu*[1]), in the second the soil was loosened (*yü*[2]), and in the third the land was levelled, furrows and ditches were made and irrigation ditches dug (*hsin thien*[3]). Shang farmers must certainly have worked very hard for a living if it took them three years to clear land that could then only be cropped for two or three years. Other sinologists would rather interpret *tzu*[1], *yü*[2] and *hsin thien*[3], all terms found in the *Shih Ching*, as the three successive years during which the field could be cultivated, after which it was abandoned.[c] Another school of thought holds that by Shang times agriculture was already settled, while by the Chou continuous cropping was practised in many areas;[d] for reasons given in section (*d*) below, we tend to support this last interpretation, particularly as many later texts use the *Shih Ching* terms to refer to spring and autumn ploughing, not to successive phases of cultivation.

The earliest use of the term *khen thien*[4], still commonly used today, is in the *Kuo Yü*,[e] but the earliest surviving description of the technique of land clearance is in the *Chhi Min Yao Shu*. It is plain that in the following passage Chia Ssu-Hsieh is describing not the clearing of land that has lain fallow but the reclamation of woodland or scrub. After the many years of warfare that had devastated North China following the fall of the Han, large tracts of land had lain waste for many years and though much of it was cleared for settlement schemes or military colonies under government direction there seems to have been plenty available for private enterprise. The methods in both cases were certainly the same:[f]

> When clearing land (*khai huang*[5]) in mountains or marshes for new fields, always cut down the weeds in the seventh month; the weeds should be set on fire once they have dried out. Cultivation should begin only in the spring. (Once the roots [of the weeds] have rotted the work becomes much easier.)

[a] Han Kuo-Pan (*1*), (*2*); Li Chien-Nung (*3*), (*4*); Twitchett (10); Aoyama Sadao (*12*); Ping-Ti Ho (4); *NCCS* 8, 9.

[b] Perhaps 10,000 km² *in toto* in the Yuan dynasty, 30,000 km² in the Ming (much of it in Hopei), and 18,000 km² in the Chhing (Leeming (1), p. 161).

[c] Maspéro & Balasz (1). The various authorities conflict in their opinions as to which term refers to which year of cultivation.

[d] Hu Hou-Hsüan (*3*), (*4*); Yu Yü (*3*).

[e] *Chou Yü* section.

[f] *CMYS* 1.2.1–1.2.3, tr. auct.

1 菑 2 畬 3 新田 4 墾田 5 開荒

Larger trees and shrubs should be killed by ring-barking (*weng*[1]). Once the leaves are dead and no longer cast any shade ploughing and sowing may begin, and after three years the roots will have withered and the trunks decayed enough to be burned out. (The fire burns down below the surface of the soil and kills the roots.)

Once the ploughing of the waste land is completed, draw an iron-toothed harrow across it twice. Broadcast millet and then run the bush-harrow over the field twice. By the next year it will be fit for grain land.

Wang Chen goes into far greater detail in his description, and subsequent *nung shu* merely quote Wang or Chia:[a]

Clearing land in the spring is called burning the waste (*liao huang*[2]) (for instance flat land thickly overgrown is fired in the spring when the soil has thawed and the young shoots are about to appear; their roots are [still] so soft that it is easy to clear them by ploughing). Clearing land in summer is called covering the green (*yen chhing*[3]) (the young shoots are ploughed in as green manure), and in autumn it is called cutting the weeds (*shan i*[4]) (the undergrowth is at its thickest and has to be cut with a sickle before firing; it is then ploughed in in the spring when the roots have rotted) ...

On land that is overgrown with reeds and below the water-table (*pho hsia*[5]) one must first use a coulter on a separate frame before following with the plough, thus turning up the soil will be particularly easy and the oxen will be spared effort.

On wet hills or areas that have been uncultivated for some time there will be many roots and stumps which must be got rid of with mattocks and hoes. If the wrought-iron share-point (encasing the base of the old ploughshare) should happen to hit a stump it may not actually split, but it will impede one's progress. Of course if [the area to be cleared] is very extensive, it cannot all be hoed and one must then break off the stems and branches of the vegetation, lay them on top of the stumps, and fire it as soon as it has dried. The roots will be killed and easily rot away, and after the summer rains one can use an ox-drawn roller (*thu nien*[6]) or threshing-roller (*kun tzu*[7]) wherever one notices broken-off stumps, and roll them to ground level; when they dry they can be prised out. After one or two years the whole field can be ploughed and sown.

If there are large trees or bushes they should be killed by ring-barking ... [here Wang quotes the *Chhi Min Yao Shu*] ...

Whenever one is clearing land, one should do so after the rains. One should also adjust the depth and breadth of the furrow carefully. If it is too shallow then it will not get out all the roots, if it is too deep then it will not turn up proper ridges; if it is too wide then for all one's efforts the soil will have a poor tilth, and if it is too narrow the tilth will be fine but the area covered small. One must take care to choose exactly the right adjustment.

When you have finished ploughing the new land, run over it with an iron-toothed harrow and broadcast panicum millet (*shu*[8]), or setaria millet (*chi*[9]), or linseed (*chih ma*[10]) or green mung beans (*lü tou*[11]) [as green manure]. Harrow once again. The next year the field can be used for cropping.

[a] *WCNS* 2/1*a*–3*a*, tr. auct. But for the organisation of land clearance see also *NCCS* 8, 9.

[1] 劗 [2] 燎荒 [3] 掩青 [4] 芟夷 [5] 泊下
[6] 碡碾 [7] 輥子 [8] 黍 [9] 稷 [10] 脂麻
[11] 綠豆

Nowadays on the Han[1], Mien[2] [a tributary of the Han], Huai[3] and Ying[4] [a tributary of the Huai] rivers, they usually begin clearance by sowing linseed and other seeds the very first year, and sometimes they manage to fill their barns and chests with the rich harvest.

Once you have finished clearing the land between old padi bunds, broadcast padi rice and leave it till it ripens—there is no need to weed or transplant.

In newly cleared land, once the roots of the weeds have dried out so no more come up, whatever type of crop you sow you must keep it quite clean [from weeds] every year to avoid any echinocloa (*pai*[5]) or tares from growing. Then after some years there will be no tares at all, and the crop will always be double that on an old field, for the longer the ground lies fallow, the more fertile it will be. There is a saying that 'even setting up in trade is not as profitable as clearing new land'.

Wang was quite right to recommend newly-cleared land as a gold-mine. Newly-cleared forest is well known to be more productive than most old fields, although its initial abundance may be short-lived.[a] In Southern China it was lake margins that were considered the prize land, and wealthy men invested heavily in their drainage and dyking, but most reclamation work required heavy initial investment and was beyond the means of all but the rich:[b] 'Only households with many workers and much land can afford to open up new land.' But although labour and capital were needed for the complex work of draining marshes or building polders around lakes, the clearing of patches of forest up in the mountains where no landlords or officials went required no capital and comparatively little labour, and was sometimes the last resort of destitute Chinese peasants. Aboriginal tribes had of course practised shifting cultivation in the mountains of China since time immemorial and continue to do so to this day.

## (2) Shifting Cultivation

We do not agree with those scholars who believe that shifting cultivation was the prevalent mode of cultivation throughout China until late Chou times: we feel that in certain areas, for example the loesslands or the river valleys of the south, shifting cultivation was practised for only a short period (if at all) before permanent cultivation began. On the other hand in many parts of China, generally in mountainous areas, several non-Han tribes have practised shifting cultivation from antiquity to the present day, while in newly colonised areas Han Chinese themselves sometimes took to shifting cultivation, which must therefore be considered to have contributed to the Chinese landscape as we know it now.

[a] See Slicher van Bath (1) on the eventually catastrophic results of assartage in medieval Europe: the reckless 12th- and 13th-century expansion into marginal lands which rapidly deteriorated is seen as one of the principal causes of the 14th-century agricultural depression.

[b] *Chih Pen Thi Kang*, Wang Yü-Hu (*2*) p. 17.

1 漢 2 沔 3 淮 4 穎 5 稗

Shifting cultivation goes under a great many names and is or has been practised all over the world. Some of its most common names are swidden (an Old English term), *culture à brûlis* (French), *chena* (in Sri Lanka), *huma* or *ladang* (Malay), *rây* (Indochina) and *milpa* (parts of South America). The Chinese term is *she*[1].[a]

Shifting cultivation is not, as is often thought, inherently inefficient or destructive of the environment. On the contrary, it is a highly adaptable system which under optimal conditions produces higher yields for considerably less labour than most permanent cultivation systems.[b] Only when land becomes scarce so that the swidden fields are cultivated for too long do erosion, low yields and the growth of perennial weeds become serious problems, though of course amateurs at this skilled art can ruin huge areas of forest very quickly.[c]

Shifting cultivation in China is generally a southern rather than a northern phenomenon. The *Shih Chi* refers to Shen Nung[2], the Heavenly Husbandman, apparently practising shifting cultivation:[d] 'He whipped the plants and trees with a scarlet whip (*i che pien pien tshao mu*[3]).' Shen Nung appeared rather late in the Chinese literature, and Mencius emphasises that his followers came from the south and were 'foreigners'.[e] Eberhard in his study of local cultures identifies the Yao[4] tribes as the only shifting cultivators of China;[f] Yao groups were scattered all over China south of the Yangtze, and some of them are still known today by the name of She[1]. In a much-quoted phrase Ssu-Ma Chhien referred to the inhabitants of Southern China as 'ploughing with fire and weeding with water (*huo keng shui nou*[5])', which has often been understood to refer to shifting cultivation, but since shifting cultivators do not irrigate their fields, we think it is more likely to refer to the cultivation of wet rice in fields where the stubble is burned off before ploughing.[g] The Yao in their mountain fields cultivated mainly taro (*yü*[6]), which they grew without irrigation, using spades reminiscent of the *taclla* used in highland Peru for the cultivation of tubers, but shifting cultivators in China also grew beans, millet, Job's tears, upland rice, and many other crops,

[a] This character is pronounced *yü* (see p. 96 above) when it refers to clearing fallow land, *she* when it refers to shifting cultivation.

[b] Some idea of the range and variety of the environments inhabited by shifting cultivators, the crops they grow and the techniques they use can be gained from general works like Sigaut (3) (Europe), J. E. Spencer (4) (Southeast Asia), and D. B. Grigg (1), as well as from numerous ethnographic works among which we may cite W. Allan (1), N. A. Chagnon (1), Condominas (1), Freeman (1), Geddes (1) and E. R. Leach (2).

[c] This appears to be happening in South Yunnan today, where Northern migrants, with no comprehension of the delicate balance between terrain, climate and vegetation that obtains in the tropics, have burned whole mountainsides bare and allowed the soil to wash away in the monsoon rains.

[d] *SC* 1/2b.

[e] *Meng Tzu* 3A/4.

[f] (2), p. 92 ff. The Tai[7] and Yüeh[8], Eberhard says, were wet-rice cultivators, while most other groups did not practise agriculture at all.

[g] *SC* 129/12b; see Amano (7), Yoneda (*1*), Bray (3).

[1] 畬 [2] 神農 [3] 以赭鞭鞭草木 [4] 瑤
[5] 火耕水耨 [6] 芋 [7] 傣 [8] 越

among which maize has played an increasingly important role since its introduction in the 16th century.[a] The Thang poet Liu Yü-Hsi[1] gives a vivid description of a Vietnamese tribe burning the forest to make their fields:[b]

Wherever it may be, they like to burn off the fields,
Round and round, creeping over the mountain's belly.
When they bore the tortoise and get the 'rain' trigram,
Up the mountain they go and set fire to the prostrate trees.
Startled muntjacs run, and then stare back;
Flocks of pheasants make *i-auk* sounds.
The red blaze forms sunset clouds far off,
Light coals fly into the city walls.
The wind draws it up to the high peaks,
It licks and laps across the blue forest.
The blue forest, seen afar, dissolves in a flurry,
The red light sinks—then rises again.
A radiant tarn brings forth an old *kau*-dragon;
Exploding bamboos frighten the forest ghosts.
In the colour of night we see no mountain,
Just an orphan glow by the Starry Han:
It is like a star, then like the moon,
Each after the other, until at daybreak the wind dies away.
Then first comes a light which beats on the stones,
Then follows a heat which glows up to heaven.
They drop their seeds among the warm ashes;
These, borne by the 'essential heat' (*yang*), burst into buds and shoots.
Verdant and vivid, after a single rain,
Spikes of trumpet vine come out like a cloud.
The snake men chant with folded hands;
Neither ploughing nor hoeing involve their hearts.
From the first they have found the temper of this land,
Whose every inch holds an excess of the 'essential cold' (*yin*).

Shifting cultivation was not confined entirely to tribal peoples, however. The preparation of land for permanent cultivation can be onerous, and Chinese migrants to frontier regions sometimes abandoned the familiar techniques in favour of shifting cultivation. In the highlands of the Upper Han River, that is to say the mountainous area on the borders of Shensi, Kansu and Honan, 18th-century settlers from the Central Plains cleared swidden fields on marginal land to grow

[a] Fogg (1) gives a very interesting description of shifting millet cultivation amongst modern aboriginals in Taiwan.

[b] *Liu Meng-Te Wen Chi* 9/4b–5a, tr. Schafer (16) p. 54. Schafer appends the following comments to his translation: 'To "bore the tortoise" is to perform an act of divination with a tortoise's carapace. The Starry Han is the Milky Way. The snake men appear to be shamans or ritual chanters. The doctrine expressed here is that the folk wisdom of the highlanders has led them to compensate for the natural coolness and humidity (*yin*) of the floor of the southern forests by applying natural heat (*yang*), bringing about perfect fertility.' Further references to shifting cultivation (*she*) in Thang and Sung literature are given in Li Chien-Nung (5) pp. 20 ff.

[1] 劉禹錫

maize, while at the same period settlers in Manchuria were able to grow millet, sorghum and buckwheat on newly cleared land for six or seven years in succession before the field had to be abandoned.[a] But the Chinese were not very good at shifting cultivation. During the Chhing dynasty the expansion of maize cultivation in the Upper Han region denuded the mountains and the consequent erosion caused considerable problems.[b]

## (3) Permanent Fields

### (i) *Northern China*

Although there is enormous variety in the Chinese agricultural landscape, it is perhaps helpful to think of the north of China as a region essentially of dryland cultivation, and of southern agriculture as principally irrigated. The fundamental distinction between wet and dry fields is marked linguistically in Chinese: wet fields are called *thien*[1] and dry fields *ti*[2]. Although wet fields may be drained and planted with dryland crops, the differences in layout, construction and tillage methods between *thien* and *ti* are fundamental.

Perhaps the archetypal *ti* landscape is that of the North China Plain. Here fields are generally thin, rectangular plots averaging 3000 to 4000 m.$^2$ in size,[c] laid out in highly regular blocks or huge strips several hundred metres wide. According to tradition the land in North China had from earliest times been distributed by the state to the peasants in equal units or in units proportional to the size of each family. According to Mencius[d] the first of these systems was known as the well-field (*ching thien*[3]) system, whereby the land was divided into units of nine squares: the eight outer squares in each unit were allotted to individual peasant families, who tilled the ninth, central square in common, supposedly for the benefit of the feudal lord. Later the Northern Dynasties introduced a similar system of equal land allotments (*chün thien*[4]) whereby each adult was allotted a fixed proportion of land. Serious doubts have been cast on whether the *ching thien* system ever really existed (see Vol. VII), but even those who believe that some system of land allotments was practised under the Chou rarely consider that these allotments actually consisted of square, regularly laid-out plots such as Mencius describes. Similarly students of the *chün thien* regulations usually believe that the size and layout of individual plots was far more haphazard than the written documents imply. But in a recent study of large-scale topographical maps of North China, Frank Leeming (1) has shown that rectilinear planned layouts are in fact one of the principal features of the landscapes of North China, and that the dimensions and forms of these layouts can be shown to be consistent with

[a] Rawski (2), p. 11.
[b] Ping-Ti Ho (4), p. 150.
[c] J. L. Buck (2), Statistics Volume, p. 47.
[d] *Meng Tzu* 3A/3, tr. Legge (3), p. 116.

[1] 田 [2] 地 [3] 井田 [4] 均田

dimensions and forms proposed for various phases of the equal allotments system. In fact Leeming even goes so far as to suggest that the *chün thien* layout was a conversion of a much earlier rectilinear layout, namely the fabled well-fields:[a]

> The fundamental means of conversion appears to have been the opening up of the chequerboard boundaries [*chhien mo*[1]] of the *ching thien* in one direction, but only one, in each landscape, converting the original system of squares into the systems of strips which feature so prominently in the landscapes as mapped. These strips must have been the framework upon which the *chün thien* surveyors built, but in the centuries between Chhin and Northern Wei, they must have been the foundation of all holdings and transactions in land. All working landscapes need land boundaries for purposes of inheritance, purchase and tenancy, as well as for taxation and official allocation systems where these exist. Whatever the land system, the physical strips of ancient times (if that was indeed their origin) would be the means of definition of the land units. During the same centuries, these strips must have been no less the foundation of whatever official control over land allocation or community organisation was maintained ...
>
> It is important to recognise that even if state allocation of land to households was the exception rather than the rule under Chhin and Han, as the traditional sources indicate, the strip systems would not be likely to disappear altogether, though they would no doubt be corrupted in detail. The preservation of formal layout over vast areas through the centuries since Thang, and since large-scale allocation ceased to be practised in China, is itself a significant indicator of the permanence of land boundaries in these landscapes.

Many no doubt will consider Leeming's conclusions over-bold, yet he does provide an attractively simple explanation for the regularity of the northern Chinese landscape, so different from anything to be found in the South (Fig. 25).

Within the large, regular blocks that Leeming describes, the land was divided into plots consisting of narrow strips. The unit for measuring land, the *mu*[2], was traditionally a strip one pace (6 Chinese feet) across and 240 paces long, and although most family plots presumably consisted originally of a number of contiguous strips, the effect of partible inheritance over the centuries was to reduce many plots to strips only a few feet across, giving an overall effect reminiscent of the open-fields of medieval Europe. In fact, of course, the Chinese system had nothing more in common with the European than the shape of the fields, for the Chinese strips within a single block of land were not managed by common consent, nor were they ever used as communal pasture land. Wang Chen describes the ploughing techniques used by the farmers of the Northern Plains:[b]

> There is apparently a custom in North China that in spring one should plough early or late in the day, in summer one should plough at night, and in the autumn when the sun is high. The regions of the Central Plain are flat and vast, and in the dry fields a single plough requires two, three or even four oxen, and is driven by a single man; the amount of land ploughed is commensurate with the strength of the oxen. They always plough in a particular way:

[a] Leeming (1), pp. 202–3.

[b] *WCNS* 2/5*a*–*b*.

1 阡陌 2 畝

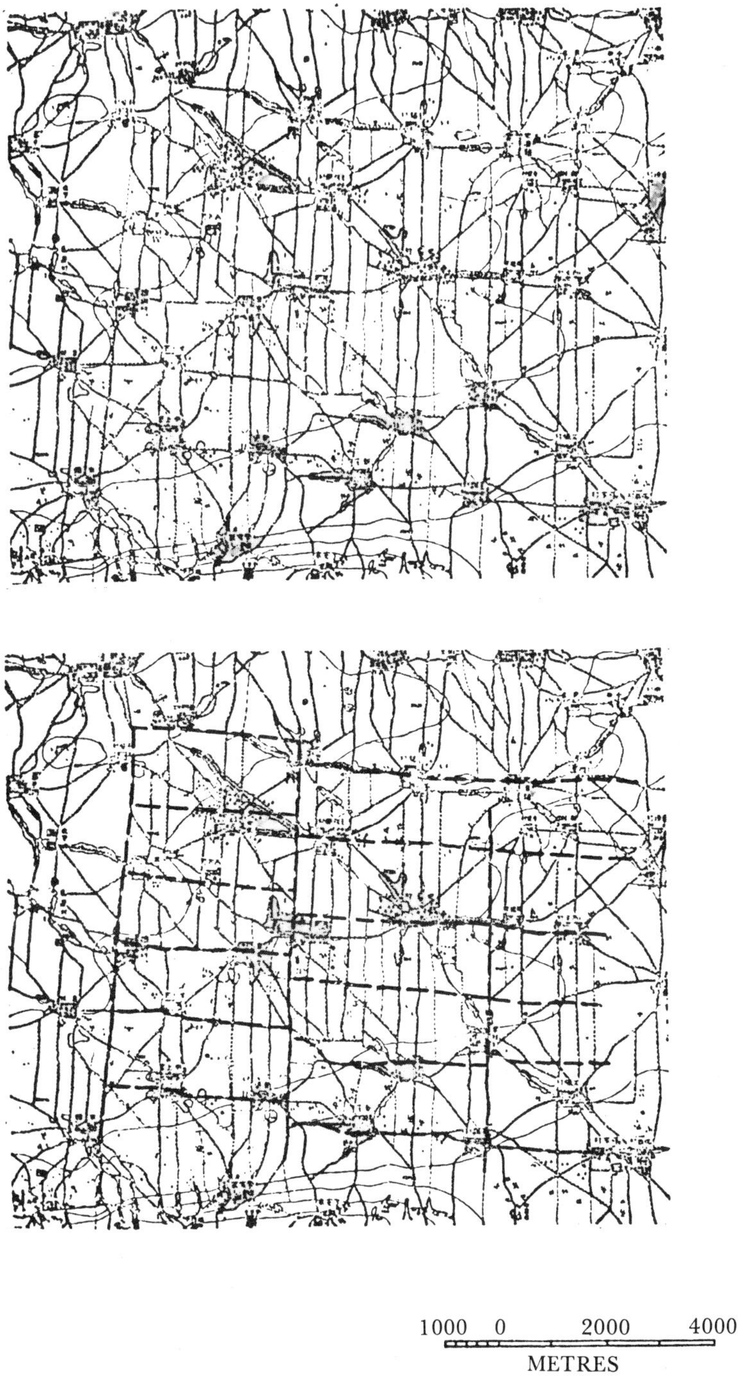


Fig. 25. Layout of fields in modern North China, showing traces of the ancient strips; Leeming (1), pl. 5.

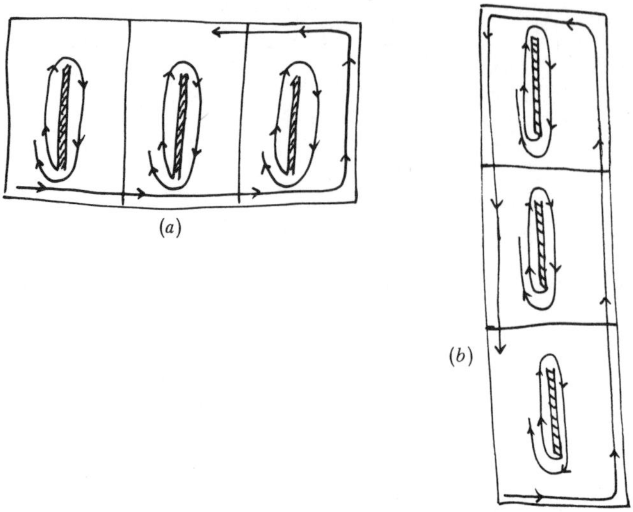


Fig. 26. Alternative reconstructions of the 'skein' ploughing technique.

In the centre of the field they first of all throw up two parallel rows, turning the plough-slice in to the centre so as to form a ridge (*lung*[1]) called a floating lynchet (*fu lin*[2]). Then, taking this lynchet as their point of departure, they wind (*chiao*[3]) outwards. Once this section is finished it is called a skein (*chiao*[3]). They then make another skein next to the first. They make three skeins and then stop. Then they wind back in the opposite direction, from the outside towards the centre, to make a selion (*chhang*[4]), three skeins going to make up one selion. Most flat land is ploughed according to this method.

This method (see Fig. 26) presumably permitted a very thorough working of the soil, breaking the long strip-fields up into squarish units. The effect of the final counterploughing, since the Chinese plough had a fixed mouldboard that always threw the furrow-slice to the same side, would be to level the surface and to prevent the build-up of balks of soil at the edge of the field.

In the hills of North China it was not possible to lay out the fields in rectangular patterns, nor could the farmers plough according to such a regular pattern as those in the plains. In order to prevent the soil washing away down the hillsides, strip lynchets had to be made:[a]

When tilling hillsides, use a plough without a mouldboard (*chiang*[5]) to plough horizontally [i.e. along the contours] and dibble the seed in [instead of using a seed-drill] (*heng chiang tan yen*[6]). [Commentary] ... On hillside fields it is not possible to plough in circles or to turn around; all one can do is use a plough without a mouldboard along the contours and dibble the seed in on the lower side of the slope. If this is done year after year without change, then gradually the field will become level and it will be easier to till.

[a] *Chih Pen Thi Kang*, Wang Yü-Hu (*2*), p. 9.

[1] 壠 [2] 浮疄 [3] 繳 [4] 畼 [5] 耩
[6] 橫耩單掩

This passage describes conditions in the loesslands of Shensi, where erosion was a severe problem. Attempts were made to control the ravages of wind and rain by highly specialised methods of tillage, using light ploughs and harrows to create a dust mulch that retained the soil moisture, and by terracing the hillsides. Lower down the Yellow River, the provision of adequate drainage was just as important as the preservation of soil moisture.

The North Chinese farmers had to cope with short periods of very heavy rain, not with the constant and depressing drizzles with which we Europeans are sadly familiar, and they never evolved anything akin to the drainage and trenching systems so typical of British farming,[a] into which rainwater gradually seeped and slowly oozed away. If the summer rains in China were really heavy, the ensuing floods were beyond the farmers' control, but just as often they failed. Measures were required, therefore, which would guarantee against either flood or drought as the occasion arose, and Chinese farmers found the solution in a method of tilling which produced pronounced ridges and furrows that could either drain off excess water or protect the crops' roots against drought depending on climatic conditions. This system is first referred to in the −3rd-century *Lü Shih Chhun Chhiu* as ridges and furrows (*mu chhüan*[1]): the ridges should be broad and level, the furrows narrow and deep so that the roots of the plants will be shaded while the stems get the sun, ensuring vigorous growth.[b] We have suggested elsewhere that *mu chhüan* was closely related to the 'changing fields' (*tai thien*[2]) system which Chao Kuo[3] attempted to introduce into Kuanchung (Shensi) early in the −1st century.[c] There has been much argument as to the precise nature of the *tai thien* system,[d] but clearly it consisted in essence of ploughing the fields in pronounced ridges and furrows which were inverted the next year, so that a furrow lay where the ridge had been the year before. Similar practices are widespread: ridging was a common ploughing method in Britain,[e] and in Ireland lazy-beds for potatoes are constructed on exactly the same principle. But the Irish lazy-beds are dug with a spade,[f] and the *tai thien* were closely associated with the mouldboard plough.[g]

Today the practice of ridging is widespread in both North and South China, and of course it can be adapted for irrigation: during dry spells, or when cultivating thirsty crops like ginger or indigo, water can be run into the furrows, which in periods of heavy rain act as drains instead. F. H. King describes ridged fields in 20th-century Manchuria:[h]

In a field beyond, a small donkey was drawing a stone roller 3 feet long and 1 foot in diameter, firming the crests of narrow, sharp, recently formed ridges, two at a time.

[a] Fussell (2), chapter 1.
[b] Hsia Wei-Ying (*3*), p. 72.
[c] Bray (3), p. 5.
[d] One of the most recent discussions is in Hara Motoko (*1*).
[e] Sigaut (4).
[f] Gailey (2).
[g] Bray (3), p. 4.
[h] (1), pp. 310–312.

[1] 畮甽 [2] 代田 [3] 趙過

Millet, maize and kaoliang were here the chief crops ... Fields in Manchuria are larger than in China and some rows were a full quarter of a mile long ... In fields where the close, deep furrowing and ridging was being done the team often consisted of a heavy ox and two small donkeys driven abreast, the three walking in adjacent rows ...

It appears probable that the strong ridging and the close level rows so largely adopted here must have marked advantages in using the rainfall, especially that portion of it which comes early, and that which comes later also if it should come in heavy showers. With steep narrow ridging, heavy rains would be shed at once to the bottom of the deep furrows without over-saturating the ridges, while the wet soil in the bottom of the furrows would favour deep percolation with lateral capillary flow under the ridges from the furrows, carrying both moisture and soluble plant food where they will be most completely and quickly available. When the rain comes in heavy showers each furrow may serve as a long reservoir which will prevent washing and at the same time permit quick penetration. The ridges never becoming flooded or puddled, permit the soil air to escape readily as the water from the furrow sinks. This it cannot easily do in flat fields when the rains fall rapidly and fill all the soil pores, because when this happens the soil pores are closed to the escape of air from below, which must take place before the water can enter.

When rows are only 24 to 28 inches apart, ridging is not sufficiently wasteful of soil moisture—because of the greater evaporation due to increased surface—to compensate for the other advantages gained, and hence the practice described in the preceding paragraph, for these conditions, appears sound.

The ridged fields described by King can surely be little different from the broad, level ridges and narrow, deep furrows of the *mu chhüan* system first described in the *Lü Shih Chhun Chhiu* in the −3rd century. The antiquity of field forms in North China suggested by the ridged fields and the rectilinear layouts discussed by Leeming really do seem to bear out those interpretations of Asian society which see it as 'timeless' and 'unchanging'. Turning to Southern China, however, we are aware of a continuous process of evolution, expansion and change.

### (ii) *Southern China*

In some modern irrigation schemes, for example the Muda scheme in West Malaysia, the Murumbidgee Irrigation Area in Australia, or the rationalised communes of the Chengtu Plain, wet rice fields are laid out with a regularity and neatness comparable to dry land fields. This is possible where modern equipment is used to level the terrain to a very high degree of accuracy, but minute differences in level can have disastrous effects on the efficient distribution of water, and in most traditional wet rice landscapes the boundaries of the fields closely follow contours that would be imperceptible to a dryland farmer. Irrigated fields are thus irregular in shape and generally small, partly because it is easier to level small areas effectively,[a] partly because in a small field water flow, depth and

[a] *CMYS* 11.6.3: 'There is no fixed size of plot: you must just decide what suits the soil and make sure the depth of the water is uniform.'

temperature can be controlled more accurately. Even under modern conditions, with access to sophisticated levelling equipment and pumping machinery, a Chinese expert still reckons the optimal size of rice field to be about one sixth of an acre.[a] To a large extent, then, the shape and size of a wet-rice field is imposed by natural conditions and is impervious to social pressures: for example, rice fields are rarely physically divided on inheritance. Although one field may be left to two or more heirs, no new dykes or bunds will be constructed to separate their property. Instead they will mark the division by planting trees or boundary stones on the existing bunds, and the field will continue to be cultivated as if it were a single entity.[b] The same is often true of division between tenants.

It is the water supply that is the crucial factor in growing wet rice. It is important not only that the water be of the right quality[c], but also that it be available *in the right amount at the right time*:[d]

> Successful paddy cultivation depends on adequately inundating the fields during the greater part of the growth period of the plant. This sounds simple enough, but in practice this desideratum is reached only after solving a number of problems that vary in importance in each particular case. In many areas it becomes not so much a matter of supplying sufficient water as of controlling the water; not always the supply of water but its drainage; or again, it may be the supply of water at one time of the year and its drainage at a later date.

Furthermore, rice is highly sensitive to water temperature, and suffers especially if the water becomes too hot in the summer months.[e]

In some parts of the world, Malaya for example,[f] wet rice is grown using only the rain that falls directly into the bunded fields, but given the sensitivity of the rice plant it is clearly preferable to make use of a separate water supply so that the amount and temperature of the water in the field can be properly regulated. This is what we mean by irrigation, and most Chinese *thien*[1] were irrigated so that, under normal conditions, water could be run in or drained off whenever necessary.

Rice cultivation in China reached a fair level of sophistication at an early date. The Western Han writer Fan Sheng-Chih, though a Northerner, recognised the rice plant's sensitivity to temperature:[g]

> When the rice is first sown it needs to be kept warm. To do so, breach the bunds so that the flow of water runs straight [across the field]. After the summer solstice it gets too hot, so the flow of water should run indirectly [across the field].

[a] Ting Ying (*1*), pp. 294, 436.
[b] Fei Hsiao-Thung (2), p. 195.
[c] Grist (1) p. 38: 'Yields of paddy depend in no small measure on the quality of water used for irrigation Water may have a considerable fertilising value because of its mineral nutrients, or may cause damage to the crop by poisonous or indirectly harmful substances.'
[d] *Ibid.* p. 37.
[e] *Ibid.* p. 46.
[f] R. D. Hill (1).
[g] *FSCS*, quoted *CMYS* 11.15.3.

1 田

Fig. 27. Han model of a rice field with water-flow control; Canton Museum.

Wan Kuo-Ting[a] gives the following explanation:

> The water in the field, being shallow, is much heated by the sun, while that in the mountain streams, in the pools or irrigation canals is usually at a lower temperature. Thus when the young rice first begins growth, by causing a direct flow of water from inlet to outlet the main body of water in the field is little disturbed and the temperature maintained. In summer the temperature is somewhat too high, and by causing flow right through the field the general temperature is lowered.

Han grave models of rice fields also show how carefully the water-flow was regulated from a very early period (Fig. 27).

Grandiose irrigation schemes were established in Northern China in Han times and even earlier, supplying water to farmers in a series of channels, flumes and ditches described in idealised form in the *Chou Kuan*.[b] The water was used by northern farmers for their wheat and millet crops, but it did not actually stand in the fields, and little or no wet rice was grown in the North. These centralised projects of North China have been discussed in detail in Vol. IV, pt 3. The irrigation systems of the south were very different. They were not dependent on the centralised control of rivers and canals, for in the south water was generally abundant. Individuals or small groups of farmers could tap the sources by leading a small channel off a stream, digging a tank, or setting up a swape or a square-pallet chain-pump where their fields bordered the river. Ever since Wittfogel (9) propounded his theory of 'hydraulic civilisations' the presence of irrigation works in China and other Asiatic societies has been assumed by many people to imply elaborate bureaucratic control, but in fact even comparatively complex irrigation

[a] (*1*), p. 123, tr. Philippa Hawking.

[b] See Vol. IV, pt 3, p. 257.

works, given favourable natural conditions, can be constructed and maintained by small autonomous groups. D. & J. Oates (1) suggest that even the great hydraulic networks of Mesopotamia originated in local networks constructed independently by farming communities, while the famous Balinese *subaks* were all initiated and maintained by small groups of peasant farmers.[a] Twitchett (6) points out that in the Thang dynasty much of the routine maintenance of irrigation canals was carried out locally by small groups; only in emergencies was labour organised on a large scale by the county authorities.[b]

Nishiyama (*1*) has identified three types of irrigation in China. First, he says, there was the Yellow River system, characterised by contour canals (*chhü*[1]) which distributed water from the Yellow River's tributaries to agricultural land further downstream; contour canals were constructed in this region from the Warring States onwards. Secondly, in the Huai[2] and Ssu[3] area in Central China, river water was dammed (*pho*[4]); reservoirs were thus formed from which the water was released through sluices (*shui cha*[5]) into canals (*chhü*[1]) and led to the fields under the force of gravity. Thirdly there was the Yangtze system, characterised by tanks (*thang*[6]); the land here was flat and swampy and the water drained off into artificial pools. In an interesting study of Japanese irrigation, Tamaki Akira (*1*) suggests that in Japan the earliest irrigation systems used ponds belonging to very small units of perhaps ten or a dozen households, a hamlet. As the political structure of Japan evolved, the tank system was largely superseded by gravity-fed irrigation systems drawn from mountain streams and rivers, each catchment area corresponding to the domain of a 'feudal lord'. Later on much larger and more powerful polities were able to organise the construction of canal networks linking a number of rivers.

Attractive as this correlation of technological and social form is, it seems to have little or no relevance to irrigation in South China, where the form of irrigation seems to have been dictated principally by ecology and population density. It is difficult to discern a chronological evolution in China. While centralised irrigation systems were being constructed in North China in the Han, in the South, in Szechwan, the Yangtze area and Kwangtung, individual farmers relied

[a] Liefrinck (1).

[b] It is not easy to see where the dividing line lies between irrigation systems that can be managed individually or cooperatively and those that must be centrally organised. Generally speaking if the water supply is sparse or irregular, conflicts will arise more often, and the area that can be managed by cooperation will be smaller, than in irrigation systems where water is freely available: 'There is a threshold of complexity in irrigation systems at which cooperation must give way to coordination; at which those served by the systems relinquish their decision-making power and their direct role in settling disputes. Authority and responsibility for these vital functions are then transferred to managerial structures of one sort or another. This is not to say that cooperation is then absent, but rather that it is no longer the dominant pattern of operation. The transfer to managerial coordination is not simply dependent on the size of the irrigated area. It is also—and more directly—dependent on the number of farmers drawing water *from a single source*' (B. Pasternak (3), p. 194).

1 渠 2 淮 3 泗 4 陂 5 水閘
6 塘

on tanks and ponds that they dug themselves. In these tanks they grew lotuses and water-chestnuts and raised fish and turtles, while on the banks they planted trees and tethered their water-buffalo; the rice plots were small, following the contours of the land, and surrounded by sturdy bunds[a] (Fig. 27). Although these Han grave models are our earliest proof of irrigation in South China, bunded, irrigated fields must have been known for some time previously otherwise it is unlikely that so many grave-models would have been found over such a wide area. Furthermore, Ssu-Ma Chhien's description of the Southern Chinese 'ploughing with fire and weeding with water' (see p. 99) corresponds perfectly to rice cultivation in bunded fields. The Sung writer Chhen Fu distinguishes two types of tank irrigation. First there is the gravity-fed system of the hillsides:[b]

On high land identify the places where water accumulates and dig out tanks (*pho thang*[1]). Out of ten *mu* you must be prepared to waste two or three for water storage. At the end of the spring when the rainy season begins, heighten the banks (*thi*[2]) and deepen and widen the interior [of the tank] until it has a fair capacity; strengthen the banks with mulberries and *che*[3] trees[c] to which buffalo may be tethered in the shade as their nature requires; meanwhile the buffalo trampling the banks will strengthen them, the mulberries being well-watered will grow into fine trees, and even in the dry season there will be sufficient water for irrigation, yet in heavy rains the tank will not overflow and harm the crops.

Such tanks were in use in South China as early as the Han, and it seems that several official campaigns were mounted at that time to construct irrigation tanks in the Huai area to the North.[d] This tank system is a slightly smaller-scale version of the Indian and Ceylonese village tanks which fed the fields of whole villages.[e] Larger tanks must also have existed in China, for we read of 'dammed lakes' or reservoirs (*pho hu*[4]) which were presumably larger than the modest ponds described by Chhen Fu. The Sung writer Ma Tuan-Lin[5] tells us:[f]

In the Ming-Yüeh[6] area[g] there are reservoirs (*pho hu*[4]) everywhere. Generally the lake lies higher than the fields, which in turn are higher than the river or sea. When there is a drought they release the lake water to irrigate the fields, and when it is too wet they drain the water off from the fields into the sea, so there is never any natural disaster.

In the case of these hillside tanks or reservoirs, it was sufficient simply to provide channels for the water to run down to the fields. On valley floors, since the water naturally accumulated in the lowest areas, the procedure was somewhat different:[h]

[a] Amano (*4*), p. 185; Chhin Chung-Hsing (*1*); Kweichow *CPAM* (*1*); Liu Chih-Yuan (*4*).
[b] *CFNS* 1/2.
[c] *Cudrania tricuspidata*, also used for feeding silkworms.
[d] Chhin Chung-Hsing (*1*).
[e] E. R. Leach (2).
[f] *Ma Shih Thung Khao*, ch. 6.
[g] Presumably in the far south, though we have been unable to identify the area precisely.
[h] *CFNS* 1/2.

1 陂塘 2 隄 3 柘 4 陂湖 5 馬端臨 6 明越

Low-lying land is easily flooded, so you must observe the spots to which the water runs fastest. Surround them with high, broad dykes (*yü*[1]). On the sloping edges you should plant pulses, hemp, barley, millet and soya, and along the dykes you can also plant mulberries and tether cattle which can conveniently eat the water-weeds.

These were the poldered fields (*yü thien*[2]) of which we shall hear more later. In order to control the water supply, that is to say generally to drain the excess water from the fields, pumping equipment of some sort was necessary. This has been described at length in Vol. IV, pt 3, and so we shall not discuss it further here.

There were few rules as to how irrigated fields should be ploughed:[a]

In the wet fields of the south the soil is ploughed to mud. Whether the field is high or low, broad or narrow, the method does not vary. Each plough is drawn by a single buffalo, and whether it goes forward, halts, turns back or goes in circles is purely at the ploughman's convenience.[b]

The idea is to produce a liquid layer of puddled mud at the surface of the field, leaving an impervious layer of hardpan beneath to prevent leaching. The hardpan may take several years to form, and it is curious to reflect that quite unlike dry fields the fertility and productivity of paddy fields tends to increase as the years go by.[c]

Irrigated rice fields are often planted with other crops in the off-season. Sometimes high ridges are thrown up so that vegetables or ginger can be grown under irrigation (Fig. 28), sometimes the field is drained so that a dry crop, wheat or beans for example, can be planted:[d]

High fields are tilled early. In the eighth month they are ploughed dry [that is, without waiting for rain or irrigation as is usual before ploughing] to parch the soil, and then sown with wheat or barley. The method of ploughing is as follows: they throw up a ridge to make lynchets (*lin*[3]), and the area between two lynchets forms a drain (*chhüan*[4]). Once the section has all been tilled they split the lynchets crosswise and let the water drain from the ditches. This is known as a 'waist drain' (*yao kou*[5]). Once the wheat or barley has been harvested they level the drains and ditches to accumulate the water in the field which they then plough deeply. This is vulgarly called a twice-ripe field.

Irrigated fields are highly productive. They may produce two or even three crops of grain a year, or a mixture of grain and beans, green vegetables, tobacco or some other crop. But their dependence on an assured water supply and the necessity for perfect levelling of the field surface limits ordinary irrigated fields to deltaic plains or valley floors. Often wet-rice farmers will cultivate dry crops in fields on the lower slopes of their river valley, or even take advantage, Inca-style, of a whole range of micro-environments corresponding to different altitudes and

[a] *WCNS* 2/5*b*.
[b] In the case of a young animal it was more likely to be at the convenience of the buffalo.
[c] See for example Liefrinck (1); C. Geertz (1).
[d] *WCNS* 2/5*b*.

[1] 圩 [2] 圩田 [3] 疄 [4] 甽 [5] 腰溝

Fig. 28. Irrigated fields ridged for ginger; King (1), p. 91.

soil types from the river-bed to the hill-tops. Fei & Chang[a] describe a village in a steep river valley in Yunnan where the land fell into four distinct categories: 'dry land' at the top of the mountain, where small amounts of peanuts, tobacco, cotton and other dry crops could be grown; next 'dry fields' where one crop of gravity-irrigated rice could be planted on the yellow clay; thirdly the 'irrigated' or 'wheel fields', irrigated by huge water-wheels, where one crop of broad beans and one of rice were grown each year; and fourthly the 'sandy land' lying by the river, fertile but too loose for rice, where maize, beans, wheat and vegetables were grown.

The irrigated fields were, however, the most productive of all, the most valuable and the most desirable, especially in the eyes of subsistence farmers. As the Chinese population increased and the valley floors and river plains were filled to

[a] (1), pp. 136–40.

overflowing, Chinese farmers applied their ingenuity to constructing irrigated fields in places where hitherto it had been impossible: along the seashores, in river bends and along lake margins, and even on mountain sides.

### (iii) *Special field types*

Different methods of land reclamation have been known in China from a very early period, but it was in the Sung dynasty that reclamation became an important economic factor. The devastation of North China by nomadic invaders led to a rapid influx of refugees to the Yangtze area, while the population of the South had been growing steadily under the peace and prosperity of the early Sung. The resulting pressure on land drove Chinese at every level of society, from the government down through rich landowners to poor peasant families, to develop old and perfect new ways of converting waste land to arable. Although poldered fields, terraces and silt fields had in fact been known for some centuries previously, many former reclamation works had by the early Sung fallen into disrepair, but now they were revived and expanded,[a] and from the Sung and Yuan onwards the process of land reclamation was continuous. Only in the 19th century were the limits of viable expansion reached.[b] However, the reorganisation of land, funds and labour under the People's Communes, together with advances in science and technology, have permitted a renewal of reclamation work in China over the past thirty years, particularly in North China and Central Asia.[c]

*Poldered fields*

Poldered fields, known in Chinese as *yü thien*[1] or *wei thien*[2], are fields enclosed by a high earth dyke which protects them from flooding. They are a feature of low, marshy regions and Europeans are familiar with the poldered fields of Holland. They seem to have appeared rather early in China. The *Yüeh Chüeh Shu*,[d] written in the −1st century, refers to cultivated fields created amidst the flood near the Thu-Men[3], the southern gate of ancient Suchou, and these fields were presumably dyked. The same text also refers to former kings of Wu[4] making dyked fields, and Miao Chhi-Yü[e] quotes other texts which seem to indicate that poldered fields were being constructed in the Thai-Hu[5] area as early as the Spring and Autumn

[a] Li Chien-Nung (5), pp. 12 ff.

[b] Ping-Ti Ho (4).

[c] The most famous example of land reclamation in the People's Republic is, of course, the commune of Tachai[6] in the loesslands of Shansi. Doubts have since been cast on the wisdom of such drastic reclamation as was practised at Tachai, but directly or indirectly Tachai set an example that was to transform large areas of Northwest China. In southern China little land remained worth reclaiming, and flood control and improvements in irrigation have made a more important contribution to increased agricultural production than land reclamation as such. K. Buchanan (1), (2); F. Leeming (2); N. Maxwell (1).

[d] Ch. 2, *Wu Ti Chuan*[7].

[e] (*1*), p. 140.

[1] 圩田 [2] 圍田 [3] 蛇門 [4] 吳 [5] 太湖
[6] 大寨 [7] 吳地傳

period. Miao believes that poldered fields were quite common in this region of the Lower Yangtze from the Han until the mid-Thang, when the development of large estates led to their destruction. Wang Chen says[a] that, although their origin was very ancient, poldered fields only began to flourish in the Sung. On the other hand there are Thang references to powerful families in the Nanking area reclaiming lake shores.[b] The Sung writer Fan Chheng-Ta[1] describes *yü thien* with dykes several tens of *li* long, like great city walls, with rivers and canals inside and gates and sluices outside, which took extra water from the Yangtze in time of drought and were able to supply grain to neighbouring regions during famines.[c]

Wang Chen makes a technical distinction between *wei thien* and *yü thien* (Fig. 29):[d]

*Wei thien* [lit. 'encircled field'] means a field with an earth wall built round it. In between the Yangtze and the Huai there are many marshy areas or river banks which are not seasonally flooded, which are separated off for crops. Powerful families examine the lie of the land, then build an earth dyke (*thi*[2]) with no breaks in it, encircling perhaps hundreds of *chhing*[3] [a *chhing* is one hundred *mu*], perhaps a few dozen *mu*, all of which is planted with crops. In the past it sometimes happened that generals in charge of military colonies, in their efforts to become self-supporting, also followed this method, and so it has been followed in different circumstances both by officials and by the people.

There are also dyked fields (*yü thien*). This means building up layers (*tieh*[4]) [of soil], making dykes to protect the land from the water outside, a similar procedure [to the *wei thien*].

Both these types will withstand either flood or drought, and the surplus from a single harvest will not only be sufficient to feed the local population but can also be exported to neighbouring provinces.

Another very similar type of reclamation was the 'counter field' (*kuei thien*[5]) (figs. 30, 31):[e]

In 'counter fields' (*kuei thien*) the soil is built up [into a dyke] to protect the fields, as for the *wei thien* but smaller. On all four sides are placed flumes (*chien hsüeh*[6]), and the shape is similar to a shop counter [i.e. rectangular] and is convenient to cultivate. If the area is very marshy then the fields should be of smaller size, built with firm dykes on higher ground so that the water outside cannot easily get in but the water inside can easily be pumped out with a chain pump (*chhe*[7]). The parts that are only shallowly flooded should be sown with yellow quick-ripening rice (*huang lü tao*[8]) (... which only takes 60 days from sowing to harvest, and thus is not endangered by floods). If the water is excessive and aquatic plants start growing of their own accord, then *hsien pai*[9][f] can be harvested. In

[a] *WCNS* 11/15*a*–*b*.
[b] *Chhüan Thang Wen*, ch. 314; see also the references given in Twitchett (6).
[c] Quoted Li Chien-Nung (5), p. 15.
[d] *WCNS* 11/15*b*.
[e] *WCNS* 11/17*b*.
[f] Probably a wild form of *Echinocloa crus-galli* L. Beauv., or barnyard millet, a plant which grows happily in swamps and is a well-known weed of rice; in some parts of the world it is used for food, and in Japan a cultivated form *E. frumentaceum* is common.

1 范成大 2 堤 3 頃 4 疊 5 櫃田
6 濺穴 7 車 8 黃穋稻 9 穇稗

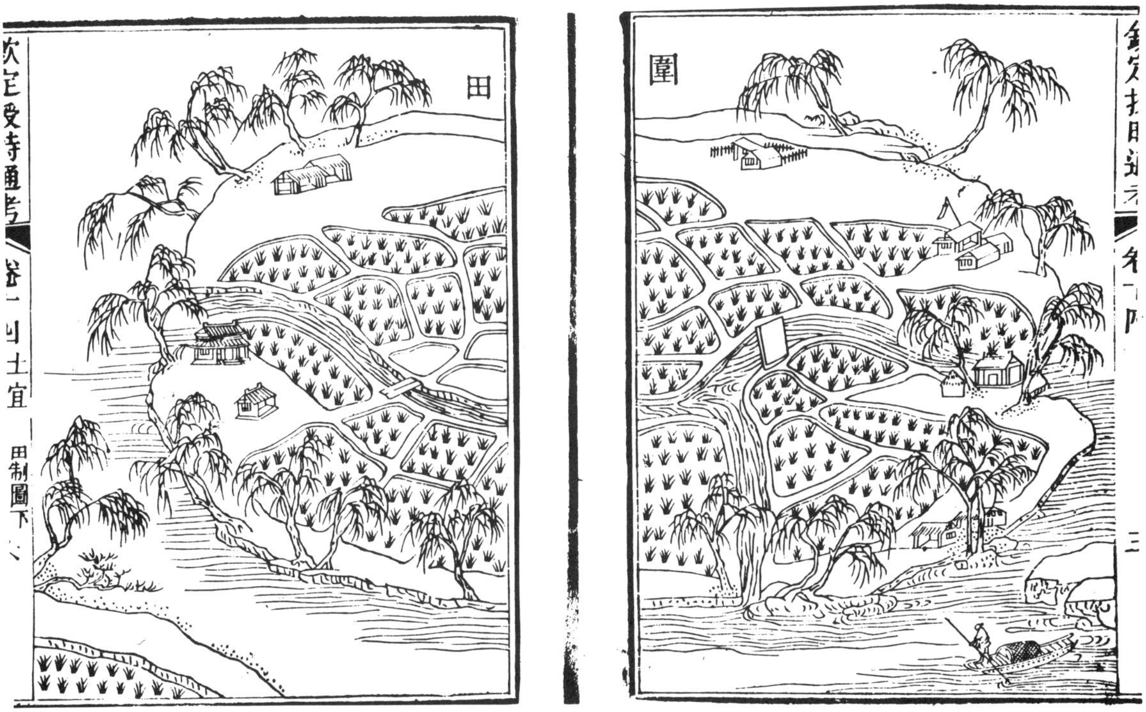

Fig. 29. Dyked fields (*wei thien*). This is a self-contained unit, subdivided into separate fields, with its own drainage channels and with houses and trees on the high dykes; *SSTK* 14/5*b*.

the high, well-drained areas all types of dryland crops can also be grown. These crops will all stave off famine. This is the best method of reclaiming marshy land.

Sandy banks and eyots might also be dyked to make 'sand fields' (*sha thien*[1]) (fig. 32):[a]

'Sand fields' are the fields made on sandbanks found in the South between the Yangtze and the Huai, or on the banks of the Yangtze, or sometimes on islands up in the hills, surrounded on all sides by thick-growing reeds. The fields are protected by dykes. Their soil is always rich and fertile and will guarantee good harvests. They are generally bunded and drained, and are suitable for ordinary or glutinous rice. In between the fields stand the hamlets, and there mulberries or hemp can be grown. Sometimes canals and pools thread between the fields and during droughts their banks can be irrigated. Sometimes a large creek winds along the side of the fields and drains away the water during floods. In this freedom from drought and flood lies their advantage over other fields. The former so-called 'fields collapsing into the Yangtze' (*than Chiang chih thien*[2]), which would appear and disappear at random so that their acreage was never constant and the land tax could not be fixed, were in fact 'sand fields'.

Most dyked fields were extremely fertile and they soon spread from the Thai-Hu area to many other parts of South China, including Hupei, Hunan,

[a] *WCNS* 11/24*b*.

[1] 沙田 [2] 坍江之田

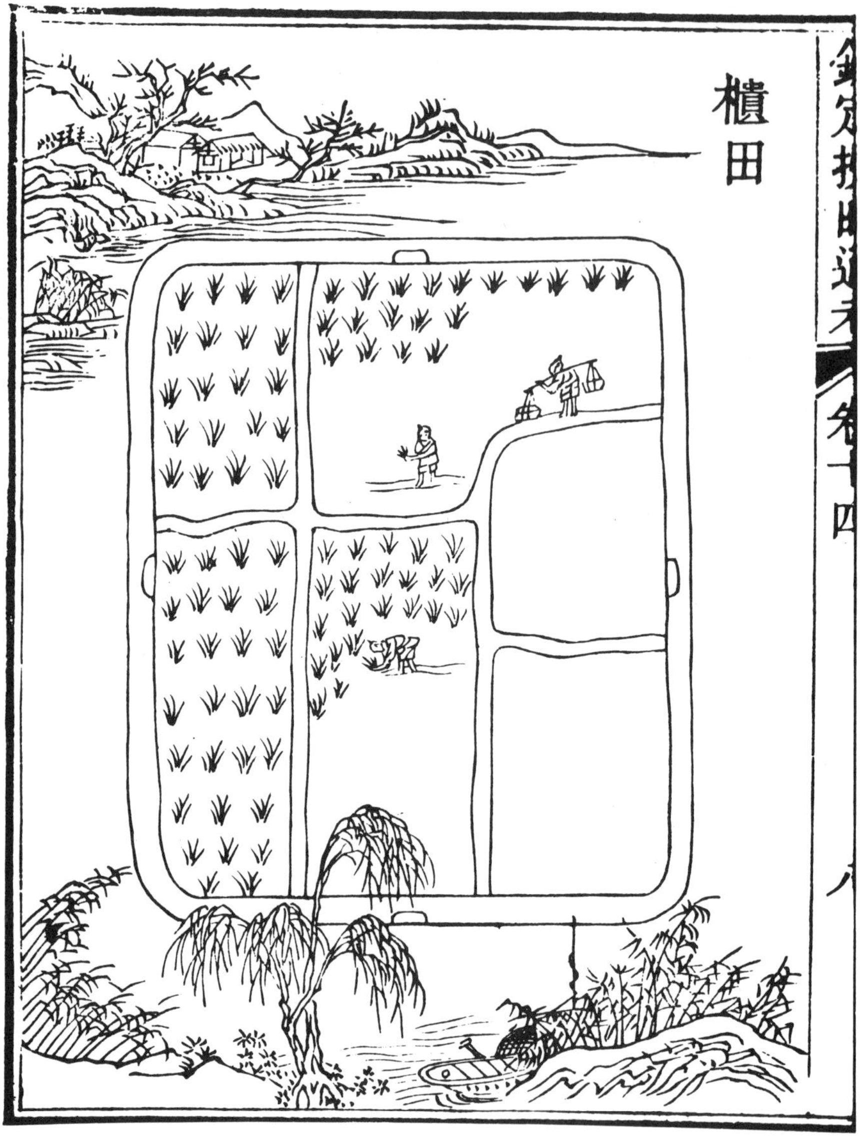


Fig. 30. 'Counter fields' (*kuei thien*): note the flumes in each of the outer dykes; *SSTK* 14/8*b*.

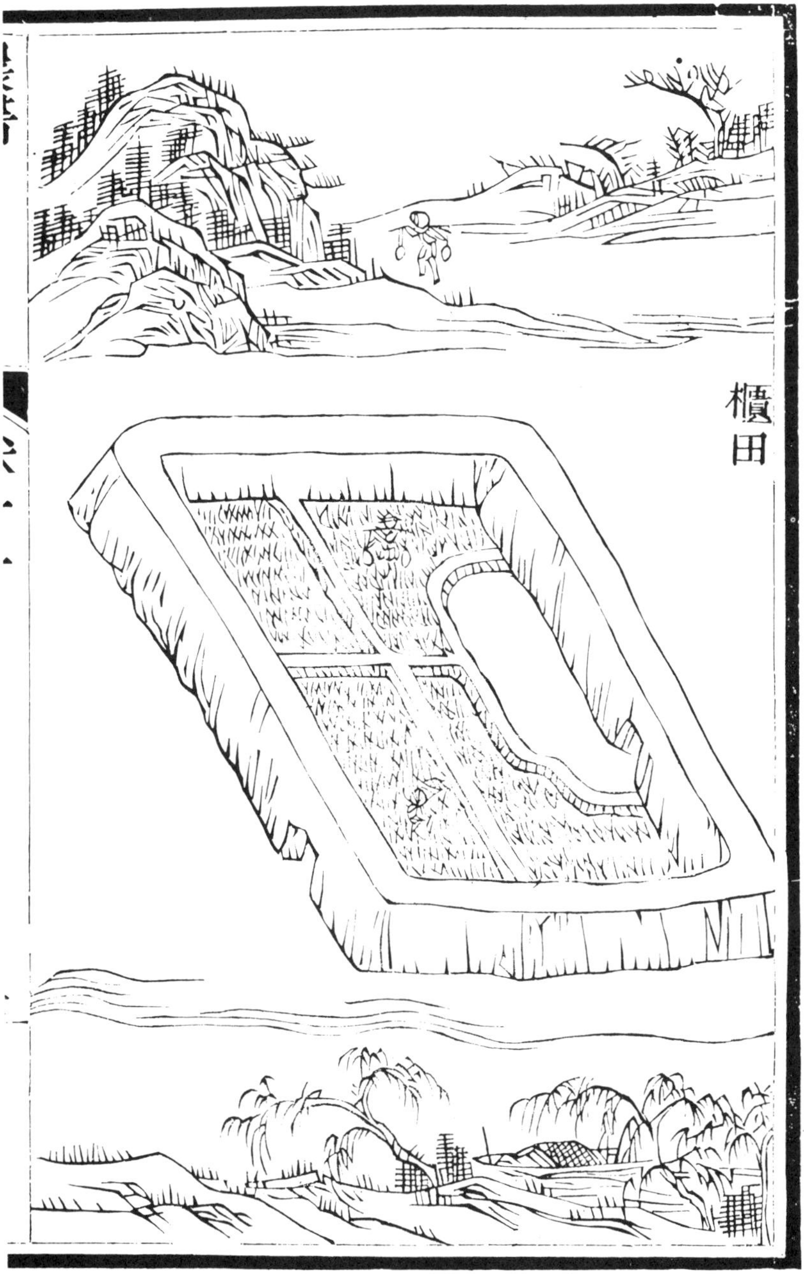


Fig. 31. 'Counter fields': here part of the land is used as a pond, through which water can be let in or out of the rice fields in a controlled flow. This is comparable to the grave models of rice fields; *WCNS* 11/17*a*.

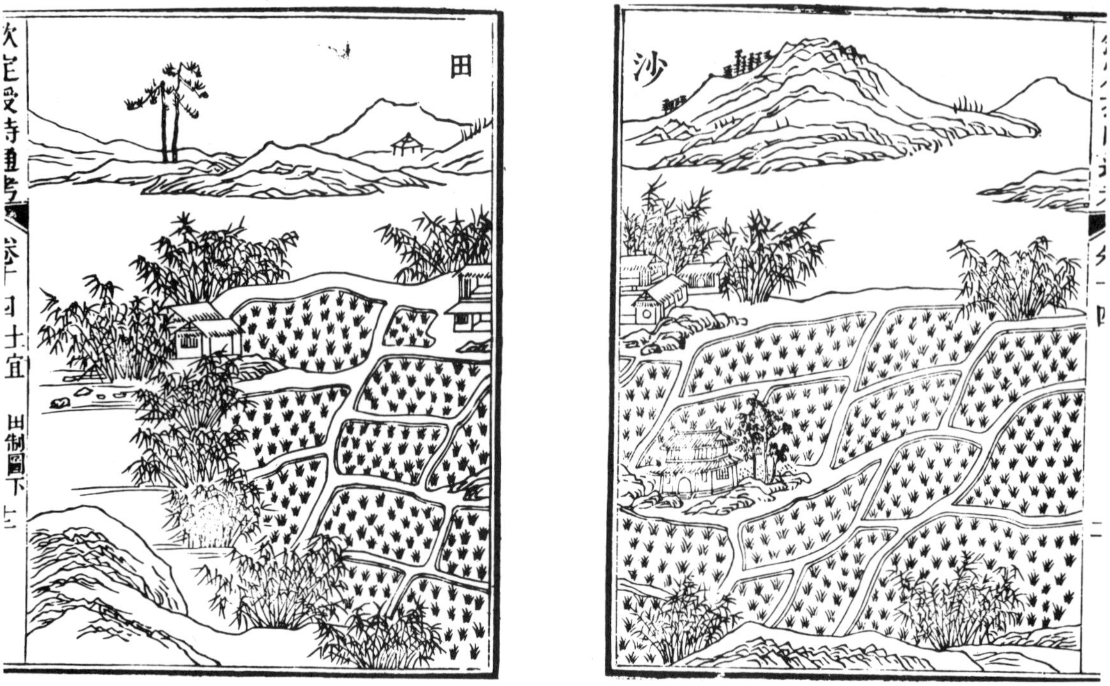


Fig. 32. 'Sand fields' (*sha thien*): note the thick clumps of reeds growing around the fields and beside the creek in the foreground; *SSTK* 14/11*b*.

Kwangtung and Kwangsi.[a] By the end of the Yuan so much of Tung-thing[1] Lake in Hunan, which acted as an overspill when the Yangtze was in spate, had been converted to dry land that there was severe danger of flooding and the government forbade any further reclamation in the area.[b] Most of the swampy lakes in the Yangtze Delta had already been turned into poldered fields by the +12th century,[c] and there *yü thien* have survived to the present day. Fei Hsiao-Thung, who is a native of the region, describes a village of about 1500 people on the shores of the Thai-Hu, whose land consists of 11 *yü thien* totalling 450 acres.[d] Each *yü thien* is divided up into dozens of small plots (Fig. 33), and since the water comes from the stream at the edge of the *yü*, the nearer a plot is to the centre of the *yü* the more difficult it is to supply water. The levels of the plots must therefore be graded like a dish, and to prevent the formation of a pool in the middle, the bunds between the plots have to be constructed parallel to the margin. In order to raise water from the stream outside into the outer plots, square-pallet chain-pumps are

[a] Today the practice continues to spread: large areas of Lake Tien[2] outside Kunming in Yunnan have been reclaimed by the process in the last few years, but unfortunately the new fields are mostly of poor quality, and the beautiful views over the lake from the Lung-Men[3] Temple have been sadly spoiled.

[b] Li Chien-Nung (*5*), p. 17.

[c] Ma Tuan-Lin, quoted Li Chien-Nung (*5*), p. 16.

[d] (2), pp. 17, 156.

[1] 洞庭 [2] 滇池 [3] 龍門

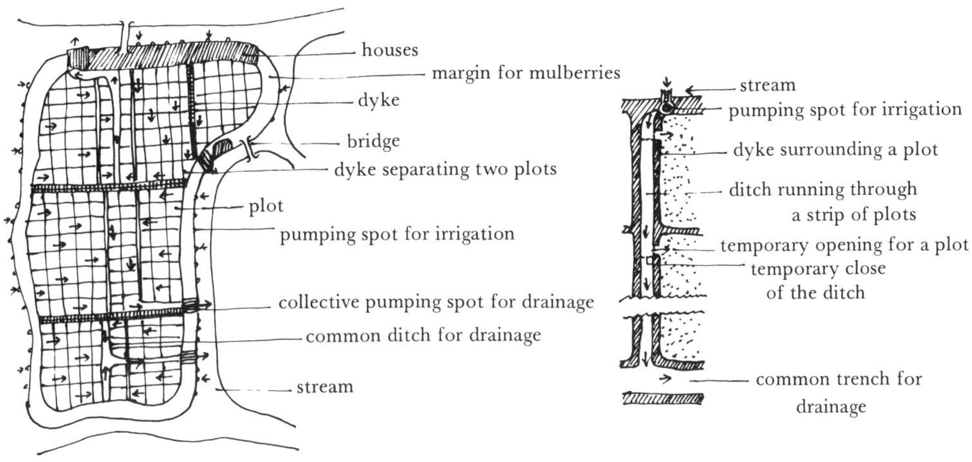


Fig. 33. Poldered fields (*yü thien*); after Fei (2).

fixed at selected spots on the bank and the water is pumped into small irrigation channels that also serve as drainage ditches, conducting the water from plot to plot and eventually into a deep trench dug in the lowest part of the *yü thien*, from which the used irrigation water is pumped back into the stream outside.

*Floating fields*

Sometimes lake margins were so marshy that the construction of dykes was impossible, and in this case the Chinese sometimes built floating fields known as 'frame fields' (*chia thien*[1]) or 'zizania fields' (*feng thien*[2]). Chhen Fu describes them as follows (Fig. 34):[a]

> If the water is so deep that the field turns boggy you must have a *feng thien*, binding wood together to form the surface of the field. This will float like a raft on the surface. Cover the top of the wooden frame with mud and water-weed (*feng*[3]), and then sow seed on it. The wooden frame forms the surface of the field and rises and falls with the level of the water, never flooding.

Wang Chen elaborates by explaining that the frame is like a bamboo raft (*fa*[4]), and that *feng*[3] means the roots of the aquatic plant *Zizania latifolia* (*ku*[5]). He says that floating fields are found more or less everywhere in Southeast China, and quotes a poem by Su Tung-Pho[6] which describes floating fields on the West Lake at Hangchou: 'The water drains away, the wild grass sprouts, and gradually a *feng thien* appears.'[b]

The earliest mention of floating fields is in a Liang Dynasty text, Kuo Phu's[7] poem on the Yangtze:[c]

[a] *CFNS* 1/2. [b] *WCNS* 11/19*a*.
[c] *Chiang Fu*[8], written *c.* +300; *Wen Hsüan* 12, quoted Amano (*4*), p. 175.

[1] 架田 [2] 葑田 [3] 葑 [4] 筏 [5] 菰
[6] 蘇東坡 [7] 郭璞 [8] 江賦

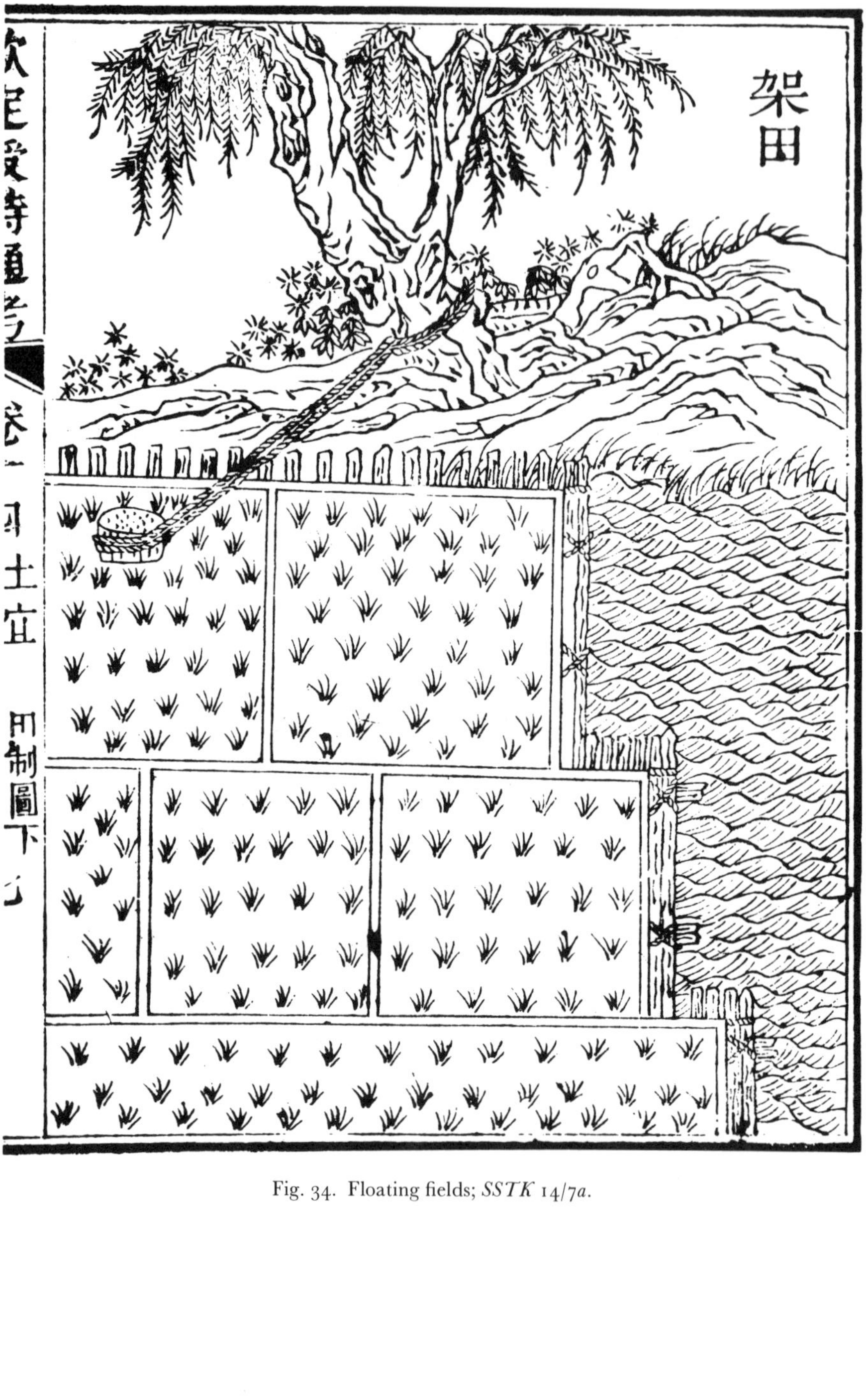


Fig. 34. Floating fields; *SSTK* 14/7*a*.

Covered with an emerald screen,
They drift buoyed up by floating zizania.
Artlessly the seeds of awned cereals are scattered,
And fine rice-plants thrust up of their own accord.

The Chhing work *Chou Kuan I Su*[a] says that floating fields are to be found not only in the Huai and Yangtze area but also on Lake Tien in Yunnan. Floating fields are also found in Kasumigaura in Northeast Japan,[b] and in Lake Inlé in Burma and Lake Dal in Kashmir, as well as in Mexico.[c]

*Reclamation of saline land*

It seems that sea walls were built along the Kiangsu coast to convert the salt-marshes to rice fields long before the Sung dynasty, for in 1026 Fan Chheng-Ta organised the reconstruction of several hundred *li* of sea-wall (*hai yen*[1]) along the coasts of Thai-Chou[2] and Chhu-Chou[3] in order to reclaim the ravaged, salty fields. The project was so successful that many grateful peasants named their children Fan[4] after their benefactor.[d] These fields were known as 'silt fields' (*thu thien*[5]) (Fig. 35):[e]

In the deltas of the Yellow River,[f] the Huai and the Yangtze firm soil (*thu*[6]) is lacking; there is only silt (*thu*[7]) and mud (*ni*[8]), and it is usual for all irrigated crops to be grown on silt or mud. Then along the seashore they practise a system of stepped fields (*teng thien*[9]). The mud and silt brought in by the tide pile up to form little islands which are sometimes sucked away by the whirling currents so that the exact area of land is difficult to calculate. On top [of the silt] grow thick clumps of salt grasses. They wait until there is a high tide, and then they gradually tickle the mud into life. First they sow barnyard millet (*shui pai*[10]),[g] and when this has eradicated the salinity of the soil (*chhih lu*[11]) the land can be used for crops. 'Wash away the salt so rice and millet grow', says the poem.

Where river banks overflow, and along the sea shore, walls are built or standing posts are planted to prevent the tide encroaching. At the side of the fields drains are dug to draw off surplus rainwater and to irrigate the field in times of drought; these are called sweetwater drains (*thien shui kou*[12]).

The yields of silt fields may be as much as ten times those of normal fields, and the common people often take them as heritable property (*yung yeh*[13]).

[a] Quoted *ibid.* p. 174. [b] *Ibid.*

[c] In these last regions they are generally used for growing vegetables. Ojea (1), p. 3 refers to 'unos huertos mobiles de 20 y 30 pies de largo y del ancho . . . en los quales siembran los almácigos de sus legumbres'. Far more widespread than floating gardens in Latin America, however, were the reclaimed fields known in Mexico as *chinampas*: islands formed by filling wattle walls with alternate layers of decaying vegetation and lake-mud; this was an efficient way of draining swampy land that converted a marsh into a grid of canals and fields or gardens. See Nigel Davies (1), p. 38; R. A. Donkin (1) passim; A. Palerin (1), pp. 19, 31, 88, 101; West & Armillas (1); on the floating gardens of Kashmir, see Ames (1), p. 77.

[d] Li Chien-Nung (5), p. 14. [e] *WCNS* 11/22*a*.

[f] Then following its southern course (1194–1853).

[g] In modern Egypt barnyard millet is frequently used for reclaiming saline swamps.

1 海堰 2 泰州 3 楚州 4 范 5 塗田
6 土 7 塗 8 泥 9 等田 10 水稗
11 斥鹵 12 甜水溝 13 永業

Fig. 35. 'Silt fields' (*thu thien*): the foreground shows high waves breaking against a sea wall, while in the background the reclamation process is still incomplete, for we see reeds and salt-grass growing in the swampy soil; *WCNS* 11/21*b*.

In Central China too, beside the Yellow River, in the meanders of the Huai or in places where two rivers converge, as well as in marshy river-bends, lakes, ponds or backwaters and anywhere else where silt accumulates and the water withdraws leaving mud-banks (*yü than*[1]), crops can also be grown. When autumn is over the mud will dry out and the soil will crack, and wheat or barley is broadcast on top, giving double the usual yield. This is how profitable these mud fields (*yü thien*[2]) are.

Both *thu thien* and *yü thien* are created by flood tides, and although they may be different from the point of view of their soils, their high yields are identical.

The reclamation of brackish land was not confined to the Yangtze area, nor was it always officially sponsored as in the case of Fan's sea-walls. J. L. Watson (1) describes how, when the founders of the Man clan arrived in the New Territories of Hong Kong in the 13th century, four other major lineage groups had already possessed themselves of all the best rice land. Since there was nowhere further to go the Mans had to occupy the only land left, a marshy, brackish area in the river delta draining into Deep Bay. The Mans built dykes with locks, enclosing the existing land and reclaiming extra land from the bay: the dykes retained the rain water and excluded salt water floods, and the locks were manipulated so as to minimise the salt content of the irrigation water. The process of reclamation was aided by the gradual geological uplift of Deep Bay and the consequent silting up of the delta. In their 'new fields' (*hsin thien*[3]) the Mans grew one crop a year of special red rice highly resistant to salt. Although this variety gave only medium yields and was not considered of very high quality, the *hsin thien* had the advantage of requiring no fertilisers and the red rice took very little labour. Nowadays the Mans have mostly given up farming for more lucrative pursuits, but the reclaimed salt marshes did provide them with an adequate if hardly bountiful living for over six centuries.

*Terraces*

If asked to describe a typical Chinese landscape most Westerners would probably mention terraced fields, yet terraces are apparently a rather recent development in China proper. Agricultural terracing is a world-wide phenomenon. The most famous examples are probably the rice terraces of Bali, the stone-walled terraces of Peru in which the Incas grew irrigated maize, the serried rows of the Ifugao terraces in Northern Luzon in the Philippines, and the terraced rice fields of Southern China. But agricultural terracing is also found throughout Latin America, around the Mediterranean, in the Middle East, the Himalayan kingdoms, hilly regions of South and Southeast Asia, Japan, and many parts of Africa. There are irrigated and dry terraces, terraces for growing rice, maize, millet, potatoes and beans, shallow terraces and steep terraces, stone-buttressed and earth-walled terraces, in fact terraces of every kind. In China itself we must distinguish the irrigated terraces typical of the South from the dry terraces of the loess region.

[1] 淤灘 [2] 淤田 [3] 新田

Terracing is an ancient art, and its origins are obscure. There is a diffusionist school which holds that agricultural terracing originated in the Middle East in the form of dry field terracing which diffused through the rest of Eurasia, undergoing technical modifications as appropriate. This school is represented by Spencer and Hale (1),[a] who tentatively postulate that in South China in about −2000 the tradition of dry field terracing merged with the technology of East Asian hydraulics to produce the wet-field terrace, a form which subsequently diffused from South China to the rest of Asia. Spencer and Hale do not believe that the wet terrace could have evolved *in situ* from flat irrigated fields:[b]

> There does seem to be an hiatus in [the East Asian] sequence: natural wet marsh; stream-bed, wet-field, mud terrace; dry-wall, stone-bunded, wet-field terrace; sod-covered, mud-bunded, wet-field terrace, if one presumes the wet-field terrace to be an eastern Asian independent invention. It is our present judgement that the natural wet marsh did not constitute the starting point for wet-field terracing, and that the 'disconformity' in evolutionary sequence is present because the whole concept of terracing came into eastern Asia from outside, affording early rice growers a markedly new approach to the whole problem of growing wet field rice.

Wheatley (4), on the other hand, sees no reason why terracing should not have been developed independently in mainland Southeast Asia in response to the peculiar local ecology, and spread gradually northwards through China during the historic period. He cites Colani's (7) discovery, for example, of a highly idiosyncratic local terracing tradition, combining both agricultural and ritual functions, that may date back to −2000 in the Gio-Linh uplands in Quang-Tri province, Vietnam, and points out very pertinently (p. 132) that

> the virtual inseparability of rudimentary terracing from wet-padi cultivation is a point to be borne in mind in any discussion of wet-field terracing in East Asia. Although the act of bunding may produce initially only insignificant inequalities in the levels of fields, the potentiality is always present for increasing vertical differentiation as cultivation laps against a zone of foothills or is pushed towards the head of a valley.

We entirely concur in Wheatley's opinion. In his classification of Latin American terrace forms, Donkin (1) includes, as well as 'cross-channel terraces' and 'contour terraces', a category of 'valley-floor terraces' in which the walls or excavated faces lie roughly at right angles to the direction of drainage. Although these are a comparatively rare form in Latin America (where one of the chief purposes of terracing was to take advantage of the higher slopes so as to avoid frost damage to the crops), valley-floor terraces are to be found in the Middle East[c] and in Southeast Asia,[d] and in Yunnan we have observed valley-floor

[a] See also J. E. Spencer (5).

[b] P. 26.

[c] Zohary (1), p. 22 and Mayerson (1), p. 31, writing of the Negev.

[d] Wheatley (4), p. 132 mentions their presence in Negri Sembilan, West Malaysia, where we have also seen them. Negri Sembilan is technically the most advanced area of traditional rice cultivation in Malaya, thanks to a large influx of Minangkabau migrants from Sumatra during the last two hundred years.

terraces on the banks of a tributary of the Mekong, where the difference in level between the fields at the side of the stream and those at the edge of the valley must easily have been thirty feet or more. And if irrigated terraces can be convincingly explained as a natural evolution from bunded wet fields, it is equally plausible to explain dry field terraces as an extension of the strip-lynchets which are a common method of ploughing hillsides in Europe as well as China.[a]

Given the enormous labour involved in constructing terraced fields, it seems more useful to consider agricultural terraces in general as evolving locally in response to natural conditions and population pressure, rather than seeking diffusionist explanations.

The purposes of terracing may be manifold. In some cases ritual or religious considerations led people to construct terraces, while in others people simply preferred to cultivate hillsides rather than the valley floors which were considered unhealthy. Agricultural terraces in the Cuzco area of Mexico were used for defence purposes too, while the Shans of North Burma also used their terraces as observation posts from which they could swoop down to levy tolls on passing travellers.[b] But most terraces had more prosaic functions: in China terraced fields are known as 'three-fold conservers' (*san pao*[1]), that is to say they prevent erosion and conserve soil moisture and nutrients, and in many cases crops grown on terraced fields give yields several times higher than those grown under ordinary conditions.[c] In the semi-arid loesslands of Northwest China, grain yields on non-irrigated terraces are usually 200 % to 400 % higher than those on unterraced hillsides, while in times of drought they may be even ten times as high.[d] Liefrinck (1) describes how the productivity of irrigated terraces increases over the years as the irrigation water brings fertile sediments down from the mountainside which accumulate as a layer of rich topsoil. But the foremost advantage of wet-field terracing is that it enables irrigated crops to be grown on steep slopes which would otherwise be planted with dry crops, or even left bare.

It is difficult to say exactly when terracing began in China proper, or whether the dry terraces of the Northwest represent a separate tradition from the mainly irrigated terraces of the South. Wheatley (4) postulates that irrigated terracing may already have been practised by the Lac[2] tribes of Tonkin when they were absorbed into the Han empire in the −2nd century, while the Thang general Fan Chho[3] alludes to irrigated mountain fields (*shan thien*[4]), which presumably must have been terraced, among the Man[5] tribes of +9th-century Yunnan. There are also descriptions of landscapes in Szechwan and Kiangsi by the Thang poets Tu Fu[6] and Chang Chiu-Ling[7] that apparently refer to terracing. The first writer to

[a] Curwen (1) and p. 104 above.
[b] Donkin (1); E. R. Leach (3).
[c] Raikes (1) has described agricultural terracing as a form of proto-engineering for mitigating climatic stress.
[d] Fang Cheng-San (*1*); Leeming (2).

1 三保　2 雒,駱,洛,鵅　3 樊綽　4 山田　5 蠻
6 杜甫　7 張九齡

use the modern Chinese term for terrace, 'ladder field' (*thi thien*[1]), was the Sung author Fan Chheng-Ta[2] in his account of a journey from the capital to Kueilin in +1172.[a] Chhen Fu in his *Nung Shu* of +1149 briefly describes the construction of what he calls 'upland fields' (*kao thien*[3]),[b] but Wang Chen's description is fuller:[c]

> Terracing means cutting steps in the mountain to make fields. In mountainous areas where there are few level places, apart from stretches of rock, precipices or similarly barren areas, all the rest, wherever there is soil, from the valley bottom right up to the dizzying peaks, can be split to make ledges (*teng*[4]) where crops can be grown. If stones and soil are in equal proportion then you must pile up the stones in rows, encircling the soil to form a field. There are also mountains where the slope is extremely steep, without even a foothold, at the very limits of cultivation where men creep upwards bent close to the ground. There they pile up the soil like ants, prepare the ground for sowing with hoes [because the fields are too narrow to use ploughs], stepping carefully while they weed for fear of the chasm at their side; such fields are not stepped but mount like the rungs of a ladder, hence their name of 'ladder fields' (*thi thien*[1]).
>
> If there is a source of water above the field, then all kinds of rice may be grown, but if it is only suitable for dry crops then setaria millet and barley are the best.

Wang distinguishes between stone-faced terraces, hewn like ledges into the mountainside, and fields on even steeper slopes formed of piled-up soil. Li Chien-Nung (*5*) suggests that the first could be irrigated while the latter were only suitable for dry crops, but Wang himself does not seem to make this distinction, and similar earth-faced terraces are known to be irrigated today in the loesslands. There great care is taken to ensure that the old topsoil remains on the surface of the new fields (Fig. 36). Whether the terrace walls are made of stone or simply of earth generally depends on the slope: where hill-sides are steep terrace walls will have to be higher than on gentler slopes, and will often be faced with stone for strength. In the Ifugao region of Northern Luzon and the Kachin region of Upper Burma, both tropical regions, terrace walls may reach enormous heights,[d] but in arid zones like Shensi and Kansu terracing of slopes over 25° is not recommended.[e]

Although terracing has contributed to the prevention of erosion in Northern China the balance is very delicate, and fierce winds or streams in summer spate could quickly wear away huge fissures and gullies in the loess (fig. 5). The problem of erosion has been less severe in the South, and nowadays many of the loess terraces have been remodelled and provided with irrigation, which has greatly contributed to the reduction of erosion.

[a] *Tshan Luan Lu* 14*b*.
[b] *CFNS* 1/2.
[c] *WCNS* 11/20*b*.
[d] Some Ifugao terrace walls are as much as fifty feet high; Keesing (1), p. 312.
[e] Leeming (2).

[1] 梯田 [2] 范成大 [3] 高田 [4] 磴

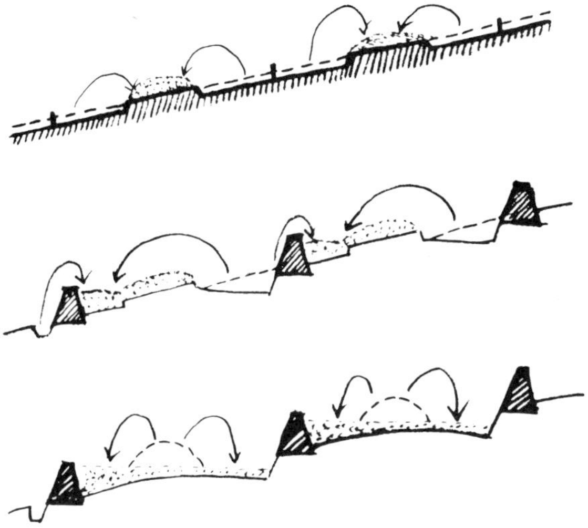


Fig. 36. Preserving the topsoil in loess terracing; after Leeming (2), fig. 5.

*Pit cultivation*

During the reign of the Emperor Thang[1] [legendary founder of the Shang dynasty] there was a long and severe drought. His Prime Minister I Yin[2] therefore developed the method of pit cultivation *ou thien*[3] and taught the people to treat seeds and carry water to irrigate their crops. Pit cultivation depends on the power of manure to fertilise the soil, so good land is not necessary. Hillsides, high land and steep slopes near towns, and even the inside slopes of city ramparts can all be used for pit cultivation ...

On the best soil the pits should be 6 inches square and 9 inches apart. 1 *mu* will hold 3700 pits. In one day [one man can] make a thousand pits. In each pit sow 20 grains of millet. Take 1 pint (*sheng*[4]) of good manure [per pit] and mix it with the soil. You will need 2 *sheng* of seed per *mu*. At the autumn harvest each pit will yield 3 *sheng* of millet, so 1 *mu* will yield altogether 100 bushels (*hu*[5]). An adult couple can manage 10 *mu*, which will give a harvest of 1000 *hu*. Since the yearly requirement of grain is 36 piculs (*shih*[6]), this is enough for 26 years.[a]

The −1st-century agronomist Fan Sheng-Chih wrote of pit cultivation in such glowing terms that it is hard to believe him, yet Chia Ssu-Hsieh, who quoted Fan's essay on pit cultivation at length in the *Chhi Min Yao Shu*, added:[b]

Liu Jen-Chih[7], the Governor of Western Yen-Chou[8] [in Hopei], an experienced and truthful man, told me that when he was in Lo-yang he took a piece of land of 70 paces on his property and experimented with pit cultivation, and he harvested 36 piculs of millet. If this was true then 1 *mu* would have yielded over 100 piculs. Such a method would be most appropriate for households with little land.

[a] *FSCS* quoted *CMYS* 3.19.1–12.

[b] *CMYS* 3.19.12.

1 湯 2 伊尹 3 區田 4 升 5 斛
6 石 7 劉仁之 8 西兗州

Since the average yield of grain during the Han was probably not more than 2 piculs per *mu*, pit cultivation certainly did seem to offer distinct advantages over more conventional methods of cultivation, especially since it used both water and fertiliser in a very concentrated and efficient manner. Chinese agronomists through the ages have all found the idea of pit cultivation attractive for its productivity (and also, we suspect, because it afforded the opportunity of drawing lots of pretty geometric diagrams (Fig. 37), a pastime to which Chinese intellectuals seem to be addicted). Wang Chen and Hsü Kuang-Chhi both recommend pit cultivation as a suitable method for poor peasants, particularly in areas where water is scarce; Hsü suggests that pit cultivation could with advantage be substituted for the system of irrigation from wells that was common in Northeast China.[a] The idea was also taken up with enthusiasm by several Chhing gentleman-farmers, who again recommended pit cultivation as a suitable occupation for poor peasants, extolling the high yields.[b] Modern experiments carried out in Honan and Hupei in 1958 have indeed shown that the yields obtained from pit cultivation, though not quite as astronomically high as suggested by Fan Sheng-Chih, were much higher than those obtained by conventional methods,[c] yet the method seems never to have been put into practice except by a few well-to-do eccentrics. Why did poor peasants not find any attraction in pit cultivation when it was apparently so admirably suited to their needs? It must have been because of the very high requirements of fertiliser and labour, both of which would be in short supply in a poor peasant family.[d]

In conclusion, then, the Chinese have over the centuries developed a whole range of field types that have enabled them to bring all but the steepest mountains and most barren sands under cultivation. They have turned hillsides, marshes, lakes and sea-shores into fertile land with an energy and ingenuity unknown in the West, where natural conditions were more propitious and populations less numerous. The social status of the reclaimers of new land was inversely proportional to the capital required: the construction of sea-dykes and large canals could be undertaken only by the government, while wealthy landlords with numerous tenants had the capital and labour to reclaim the highly profitable areas around lakes and next to rivers.[e] But even the poor and landless had some avenues open to them: they could hack small terraces from mountain-sides that might not seem very desirable compared with the fertile lowlands, but at

[a] *WCNS* 11/11*a*; *NCCS* 5/6*a*.

[b] Wang Yü-Hu (*5*).

[c] Hsü Cho-Yün (1), p. 119.

[d] Many poor peasants could never afford to marry, and only rich families could afford to have several children. Furthermore, a poor peasant would own no animals that could supply manure, and its purchase was costly. Interestingly though, methods very similar to the *ou thien* do seem to have been practised fairly extensively in medieval Iraq (Husam (1), p. 72) and in Wales (Sir Joseph Hutchinson, pers. comm.).

[e] Landlords would require their tenants to supply labour for reclamation work. Lake fields, being extremely fertile, were subject to very high taxes; Lewin (1); Li Chien-Nung (*5*).

Fig. 37. A diagram showing the layout of pits in the *ou chung* system; *NCCS* 5/2*a*.

least the land was their own, and it would probably be many years before the government got around to registering it as taxable land. Given such incentives to rich and poor alike to bring new land under cultivation, it is not surprising that by the end of the 18th century hardly any land that was not sheer precipice was without some form of cultivation.

## (*d*) AGRICULTURAL IMPLEMENTS AND TECHNIQUES

### (1) Tillage Implements

The most obvious feature of a farming system, apart from its crops, is the system of tillage by which they are grown. This is at once the most fundamental and the most variable characteristic of a farming system, and indeed it is usual to classify different types of agriculture according to the chief implement used for tillage: thus Highland New Guinea is known as a 'digging-stick culture', pre-Columbian America and sub-Saharan Africa as 'hoe cultures', and Europe, North Africa and most of Asia as 'plough cultures'. Indeed, the very word 'farming' in English immediately conjures up an image of sturdy Shire horses driven by a muffled ploughman under an autumn sky or, to the more up-to-date and less romantic, of a four-wheel-drive tractor and a multi-furrow reversible plough opening great black swathes through the golden stubble. Since the implement used to till the fields is often typical of particular cropping or field patterns,[a] the use of the term 'plough culture' or 'hoe culture' implies far more than the use of a particular implement. We must also bear in mind that, although the plough or the hoe is probably the most conspicuous implement used for tillage in these cases, it is by no means the only one.[b]

In this context the term 'complex', defined by American anthropologists as a 'functionally integrated cluster of traits'[c] has been aptly applied by Leser[d] to agriculture, to define and describe mutually dependent combinations of agricultural implements. The traditional 'plough complex' of Northwestern Europe, for example, he defines as the combination of plough, harrow and roller: the plough turns the soil, the harrow breaks up the clods, and the roller pulverises them and smooths the soil; when all three have been used, the soil is ready for sowing. The equivalent complex in the jungle-clad hills of Borneo would be the steel axe and wooden digging-stick,[e] while in the English kitchen garden it would be the spade, fork and rake. The concept of a 'tool complex' is useful not only because it defines more precisely the nature of the task performed, but also because the recurrence in two separate places of a complete complex is a more reliable index of cross-cultural contacts or technological diffusion than the recurrence of just one or two of its individual elements. Conversely, variations in the same complex in neighbouring areas might alert one to look for environmental or socio-economic factors to account for these differences. The concept of a 'complex' in agricultural terms is most usefully applied to tillage, the preparation of the soil for growing crops, which has the most extensive and variable of the agricultural tool complexes, and

[a] Boserup (1); H. C. Bowen (1); Baker & Butlin (1).

[b] In medieval Europe, where the use of the heavy plough drawn by a large team of oxen was commonplace, the peasant was still referred to as 'the man with the hoe' (Ernle (1)).

[c] Jacobs & Stern (1).

[d] Leser (5).

[e] D. Freeman (1).

moreover dictates the form of several subsequent operations such as sowing, weeding, and even harvesting.

In different cultivation systems 'tillage' may mean anything from felling and burning a patch of virgin jungle and making holes in the cleared ground with a digging stick, to creating a fine tilth and deep straight furrows in a field that has been continuously cultivated for centuries. In her analysis of cropping patterns, Boserup[a] convincingly links different tools to different cropping systems: a digging stick is adequate to cultivate the soft ashes and leaf mould of land cleared from virgin jungle, but where land has been cleared from bush and thick undergrowth, even after burning off the vegetation, the remaining roots and the stiffer soils necessitate the use of a more powerful tool, namely the hoe; on frequently or permanently cropped land where grasses and other weeds must be cleared the most effective method is cross-ploughing with an ard,[b] or turning the weeds in with a turn-plough.[c]

Boserup reverses the Malthusian theory to explain agricultural development as a response to, rather than the cause of, population growth. Her hypothesis is that agricultural development generally takes the form of increasing cropping frequency: long-term forest fallow where patches of primary forest are cleared, used for one or two years, then left fallow for twenty years or so (a highly productive system but covering only small areas), is succeeded by shorter-term bush fallow where the land is used for anything from three to eight years in succession, then left fallow for a similar period before re-use. The bush fallow system is in turn succeeded by permanent cultivation, where the land is never fallow for more than one or two years. Boserup considers that each successive system represents a decline in the returns to labour[d] (that is, in a plough system one must work harder to produce the same amount that one would in a jungle garden), and feels that only a heavy increase in population density could warrant such a change.

Here we must disagree. Although some studies of different cultivation systems have implied that jungle swidden farms require less labour than permanent cultivation, in fact these studies do not take into account the arduous and often dangerous task of clearing the land.[e] Whereas land under permanent cultivation only has to be cleared of natural vegetation once, in forest swidden or bush fallow farms this may be a yearly occurrence.[f] Furthermore, the different systems seem to us to be more closely confined to particular ecological niches than Boserup implies, so that one cannot regard her succeeding stages of agricultural development as a world-wide model. The long-term forest fallow, for example, is most practicable in areas which have a marked dry season, and less so in a constantly

[a] (1), pp. 24 ff.

[b] Or 'scratch-plough', a symmetrical-shared plough which makes a shallow furrow.

[c] A heavier plough, often with an asymmetric share, and with a mould-board that inverts the soil.

[d] (1), p. 41; see also Sahlins (1).

[e] For a telling account of the difficulties of clearing jungle without the benefit of steel axes, see Chagnon's account of Yanamamö gardens in the Orinoco basin; (1), p. 33.

[f] See Freeman (1) and Geddes (1) on jungle swidden, W. Allan (1) on bush gardens.

humid climate such as we have in Britain.[a] On the other hand, forest fallowing is only really necessary in tropical and equatorial areas where soils are leached, infertile, and rapidly eroded when exposed. Here the system of long-term fallowing provides fertility (in the form of nitrogen-rich leaf-mould and potassium-rich ashes) for short periods of cultivation, but this fertility is soon exhausted and if cultivation is prolonged beyond one or two years, yields quickly diminish, the soil is eroded by tropical storms, and the garden becomes unreclaimable waste-land.[b] In the absence of large quantities of manure to maintain fertility, and of soil stabilisation techniques such as terracing, provided population density is not high forest swidden cultivation is a most appropriate and effective method of exploiting the tropical jungle;[c] it is far less suited, however, to temperate regions.

In temperate regions, on the other hand, soils are generally more fertile and less vulnerable to erosion, and so cultivation periods can be longer and fallowing periods shorter than in the tropics; in fact on particularly fertile loams and sediments permanent cultivation may be possible from the start. Furthermore, since growing seasons are shorter and crop yields generally lower than in the tropics, a larger area has to be cultivated in a temperate climate than in the tropics to maintain the same population.[d]

Generalisations are dangerous, but it is largely true to say that except under some form of coercion, whether economic, political or religious, man will not work harder than is necessary to live comfortably; he will prefer labour-saving to labour-intensive methods of work. Since it is quicker and easier (for the man if not for the animal) to cultivate one acre of land with an ox-plough than it is to dig it with a spade or work it with a hoe, all other things being equal, a farmer will prefer to till with a plough rather than a hoe.[e] Crop yields on ploughed, continuously cropped land are generally lower than those on forest- and bush-fallow, partly because the latter are temporarily extremely fertile and are usually planted with several crops simultaneously,[f] partly because the plants are better cared for.[g] Yet given the choice, cultivators who wish to increase their production will generally extend their area of cultivation rather than intensify their methods. Indeed, if extra land becomes accessible, farmers will quickly revert to less labour-intensive cultivation methods, as for example the early colonists in North America whose large and ill-run farms shocked British visitors.[h] As long as there is extra

[a] See Freeman (1) on the difficulties of successful firing in wet weather.

[b] Freeman (1).

[c] In tropical valleys, the permanent cultivation of wet rice avoids the problems of low fertility and erosion, and calls for quite different tools and techniques.

[d] Harris (1). This statement only refers to pre-modern systems of agriculture.

[e] In very boggy rice fields in Malaysia the land has to be hoed because buffaloes would founder in the mud; that the local farmers consider the extra labour a heavy burden can be seen from the conditions of rent for such fields: only one third of the crop goes to the landlord as opposed to one half or two-thirds on ordinary land; Bray & Robertson (1).

[f] The squash-maize-beans complex of Mexico, for example, or the combination of tubers, cereals and vines common in Southeast Asia and Oceania.

[g] The patches being small, individual care of each plant is possible.

[h] F. G. Payne (1).

land available for cultivation, farmers will extend rather than intensify their operations; but once the limit of cultivable land has been reached,[a] the only possibility in the face of further population increase is to intensify production. This can be achieved by a further increase in cropping frequency (growing one crop each year, three in two years, or even several crops each year) and a consequent increase in the use of fertilisers; by more thorough tillage, improving soil structure and fertility; by careful regulation of the water supply; and by better care of the plants, weeding frequently, hoeing and watering to ensure that each single stem bears fruit. The farmer's aim is no longer to procure his living as easily as possible but to procure it at any cost; returns to land have become more important than returns to labour, and as much produce must be wrung from each inch of land as possible. As population pressure mounts and land hunger grows acute, the fields are divided and subdivided until they are scarcely larger than pocket handkerchiefs, and each plant is individually transplanted, manured, watered and weeded. Land is too precious to spare for grazing oxen, and in any case the fields are too small to use a plough. In the 'garden farms' of Japan, Southeast China and Java, we see spade and hoe displace the plough and harrow.[b]

It has generally been considered that the digging-stick and hoe, being the most simple, are the most ancient of the agricultural implements; and many authorities think that the more complex implements are later developments of these basic simple tools:[c]

> A digging stick was a strong, straight, pointed stick, possibly weighted with a stone ... and from it have developed the spade, the garden fork, and perhaps the series of angular digging-sticks of which the Hebridean caschrom is the best known example. The hoe was an instrument consisting of a blade, made of hard wood, stone, or metal, set at an acute angle to its handle; it was the ancestor of the mattock and the pick.

Curwen[d] goes on to adopt Bishop's thesis[e] that there are basically two different types of plough, one developed from a traction spade (the 'spade plough'), and one developed from a primitive hoe or a simple forked branch (the 'crook plough'). Some scholars have maintained that all ploughs are ultimately derived from the hoe,[f] some claim it is an elaboration of the digging-stick,[g] while still others believe it derived from the spade.[h] (We shall see later that among the scholars working on the development of the plough in China there have been

[a] Sometimes large tracts of land remain unoccupied but cannot be farmed until the local population devise special techniques to convert them to agricultural use. This was the case of the fertile but stiff clays of Western Europe before the introduction of the turn-plough, and of hills in South China and Southeast Asia that were too steep for cultivation until terracing was invented.

[b] This process of intensification has been splendidly described by Geertz (1) in colonial Java; he calls the ever-increasing input of labour in an effort to increase total production 'agricultural involution'.

[c] Curwen and Hatt (1), p. 63.

[d] *Ibid.* p. 73; see also Nopsca (1), Montandon (1).

[e] C. W. Bishop (13).

[f] E.g. Hahn (1), Chevalier (3).

[g] E.g. Wissler (1).

[h] E.g. Leser (1), Steensberg (2).

adherents of all these theories.) Haudricourt and Delamarre[a] have demonstrated convincingly, however, that each of these implements must have developed independently: not only are their functions and the motor patterns involved in their use completely different,[b] but the existence in early cultures of such 'tillage complexes' as spade and plough in Mesopotamia, or hoe and plough in early Egypt, shows clearly that the functions of the different tools have always been complementary rather than identical.

The question of the origins and diffusion of the different tillage implements is extremely complex. When examining the technologies of past societies it is important to remember that a particular type of implement rarely exists in isolation but is usually used as part of a complex of complementary tools. Thus 'primitive' or 'simple' tools are frequently used along with 'developed' or 'complex' implements within a single society at any given time. Wheat-growing societies traditionally use ploughs to till their fields, hoes for weeding, and spades for digging ditches and for cultivating their vegetable gardens; areas dependent on tubers may use a spade to dig lazy-beds, a dibbler to sow potatoes, and a quite different type of spade to dig the potatoes up once they are ripe. To complicate the picture, local or regional differences may influence the choice of tools within a single culture: turn-ploughs are used in the heavy soils of Northern France, for example, while ards are common in the dry regions of the South. Furthermore, the choice of implement may reflect the economic status of the user within his society. In a country where landowners use ploughs with large teams of oxen to till their fields, an impoverished share-cropper may only be able to afford a hoe and a spade for the whole range of his agricultural labours.[c] Unequal distribution of wealth will affect not only what agricultural tools a man can afford to use but also the materials of which they are made. Metal remained a luxury in many societies long after the discovery of smelting ores and working the metal, and stone and bone workshops are often found on archaeological sites that also contain smithies. The former produced knives and sickles for the common herd, while the latter made swords and spears for the nobility. So even in a society well acquainted with metallurgy one might find no metal agricultural tools, not because agriculture was not practised, but because all the peasants could afford was wooden ards and bone hoes. This was the case not only for poor peasants in many parts of China throughout its history, but also for such wealthy foundations as the French monasterial manors of the +8th and 9th centuries.[d] So we must carefully resist the tendency to think of societies, whether ancient or modern, as culturally and technologically homogeneous.

[a] (1), pp. 31 ff.

[b] The digging-stick is used to stir the soil or make holes in it, the hoe chops and breaks the soil, the spade cuts and inverts it, and the scratch-plough makes a shallow furrow along the surface.

[c] B. Moore (1); W. Hinton (1).

[d] G. Duby (1), p. 22, though in fact iron tools were probably more common in medieval Europe than Duby allows.

The digging-stick, on the basis of modern ethnographic evidence at least, seems to be used for agricultural purposes proper only in jungle swidden gardens,[a] though it is used over a wider ecological range by hunter-gatherers. We should perhaps bear this in mind when considering the distribution of the implement in antiquity: the usual interpretation of stone discs from Neolithic sites as 'digging-stick weights' may be a mis-identification. The hoe, on the other hand, seems to be a truly universal tool, found wherever there is agriculture. It was essential to the earliest farmers, as is clear from the number found in early Neolithic sites all over the world, and it is still essential to most farmers today. While in certain places it is the chief tool used for cultivation, in most areas it is used principally for weeding, earthing-up, or cultivating small, awkward patches of ground. It occurs in an enormous variety of sizes and forms,[b] rounded, pointed and square, but nowadays the square type is the most common. It seems impossible to postulate any single date or point of origin for the hoe, which must have been as universal as the knife and chopper ever since Neolithic times.

The spade proper seems rather more restricted both in chronological and geographical distribution. It is particularly useful in heavy soils where drainage is necessary, and its prime function in several cultures seems to be the digging of drainage ditches.[c] In humid countries where tubers are grown, spades are commonly used to make well-drained ridges or 'lazy-beds'.[d] A significant function of many spades is their use for breaking up turf; in Europe at least the use of the spade is most common where grassland has to be cleared for cultivation, and its importance has grown with the spread of agriculture to such marginal areas as the peatlands of Northwest Scotland and Ireland, and the steep grassy slopes of the Basque country. The spade sometimes came to replace the plough in these areas from the 18th century on, as the peasantry became increasingly impoverished.[e] The Scottish caschrom or 'crook-spade' (Fig. 38), which has sometimes been thought of as the ancestor of the plough, in fact evolved in the Highlands and Islands only in the 17th or 18th century, where it was used to clear hilly pastures of grass and stones.[f] The spade varies widely in size and form. The handle may be long or short, straight or crooked, the blade flat or convex, with one or two cutting edges, straight-edged, round-edged, or even double-pronged.[g] Spades are usually manipulated by a single person, but traction spades used by a

[a] See Leroi-Gourhan (1), Hirschberg & Janata (1). It seems justifiable to classify the varieties with broad, spatulate ends as spades rather than digging-sticks, especially when they are used to construct raised beds for tuber cultivation; see Lerche & Steensberg (1).

[b] E.g. Leroi-Gourhan (1), vol. I, p. 91 and vol. II, p. 127; Hirschberg & Janata (1).

[c] Was this why the spade, known as *marr*, was so important in the agriculture of Sumer and Assyria (Haudricourt & Delamarre (1), p. 64)? On the use of spades for irrigation and drainage in China, see below, p. 215.

[d] Gailey (1); Gailey & Fenton (1) on Western Europe, Lerche & Steensberg (1) on Highland New Guinea.

[e] Estyn Evans (1); Fenton (1) and (2); Caro Baroja (1), pp. 146 ff.

[f] Fenton (2).

[g] This form is usually associated with turf-paring tools such as the *taclla* of Peru (Estyn Evans (1)), the *laya* of the Basques (Caro Baroja (1)), and the *gabhal* or *gob* of County Mayo (Fenton (2)).

Fig. 38. Different types of caschrom; J. Macdonald (1), p. 57.

team of two or more people are not uncommon.[a] The long history, wide distribution and enormous typological variation of the spade in both the Old World and the New indicate that it would be difficult to trace any single centre of origin, but as an agricultural implement it has traditionally been less important than the hoe or plough: it is hard work to use, and such deep cultivation is rarely necessary for cereal crops;[b] for growing tubers, on the other hand, the spade is the ideal implement, and with the recent spread of potato cultivation to the outlying grasslands of Europe and elsewhere, the agricultural use of the spade has grown considerably over the past three or four hundred years. The spade has always been essential for drainage purposes, for digging trenches and ditches, and this was probably its chief use in most early societies.

The functions of the digging-stick, hoe and spade, and the ways in which these tools are used, are so different that it seems foolish to try and explain one as a development of the other, still more so as in most societies at least two of these implements are to be found in simultaneous use. This functional argument applies equally well to the plough, and we agree with Haudricourt and Delamarre that the plough should be considered as an independent invention rather than as a development of a 'simpler' tillage tool. (Although the idea of an Assyrian peasant suddenly inspired to harness an ox to his hoe is superficially attractive, imagine the difficulty of pulling the square blade at an angle through the soil.) The origins

[a] E.g. Scotland (Fenton (2)) and Inca Peru (Prescott (1), p. 82). On drawspades in China and Korea, see p. 215 below.

[b] It is, however, used for maize, which unlike most cereals requires really deep cultivation; the spread of maize through Europe had a significant effect on the development of the spade (François Sigaut, pers. comm.).

of the plough are far from clear, but it was certainly in widespread use by the mid-Neolithic, and on the assumption that man prefers to work no harder than is necessary, we ourselves would expect its development to be closely associated with the domestication of oxen or buffalo.[a] The ard or 'scratch-plough' certainly pre-dates the heavier turn-plough.

To summarise, the most ancient agricultural implements are the hoe, spade and digging-stick, all of which have probably been in use since earliest Neolithic times. Since these implements are functionally quite distinct they must have been developed independently of each other, rather than the hoe and/or spade having evolved from the digging-stick as has often been suggested. All were invented in at least two independent centres, for they are found in the pre-Columbian New World as well as the Old. The hoe was the most universal of the first agricultural hand tools; the digging-stick was probably confined to forest environments, and the spade used principally for drainage or for clearing turf. The ard or scratch-plough first appears in the mid-Neolithic period. It is a more complex implement than the hoe and spade, and is drawn by animal traction; its appearance seems closely associated with the domestication of cattle. Its complexity indicates that the ard probably had a single centre of origin and diffused gradually through the rest of the Neolithic Old World along with the idea of animal traction. Its precise place of origin is not known, but it seems to have spread rapidly throughout West Asia, South Asia and Europe in the Neolithic and early Chalcolithic period. Since the plough spread so rapidly through the cereal-growing cultures of the rest of the Old World, we would normally expect to find evidence for the plough at a similar period in China too, especially as such typically West Asian crops as wheat and barley were familiar to the Chinese culture-area towards the end of the Neolithic period. It therefore comes as rather a shock to find that in most sinologists' opinions ox-drawn ploughs were unknown in China till about −500. The evidence with which they support this interpretation is not, however, wholly convincing, and on the basis of somewhat different evidence—though again far from conclusive—we shall make what we hope is a reasonably persuasive case for its much earlier appearance.

When tracing the evolution of agricultural implements in China we must bear in mind the various complexes of which each implement is a part, for a development in one element of the tillage complex will usually affect the other elements of the complex. In traditional China tillage is called *keng*[1]. Untilled land is 'raw' or 'unripe', *sheng*[2], but after being tilled it is called 'cooked' or 'ripe', *shu*[3].[b] In tradi-

[a] See p. 141 below.

[b] This surprisingly Lévi-Straussian opposition is fundamental to Chinese classification and is applied in a variety of contexts; it indicates the undergoing of some process or treatment, thus: *sheng fen*[4], fresh manure, ages to give *shu fen*[5], rotted manure; *sheng thieh*[6], cast iron, is the raw material for *shu thieh*[7], wrought iron (see Needham (32), p. 10); *sheng jen*[8], an unknown person or stranger, after a period of initial wariness becomes *shu jen*[9], familiar. As far as we know, no proper study, comparative or otherwise, has been made of this Chinese usage.

[1] 耕 [2] 生 [3] 熟 [4] 生糞 [5] 熟糞
[6] 生鐵 [7] 熟鐵 [8] 生人 [9] 熟人

tional China, an elastic term which we shall use here to cover roughly the period from the Han dynasty to the early +20th century, there were, broadly speaking, three different tillage complexes corresponding to the three most widespread types of agriculture. In the light dryland areas of North China, particularly in the primary loesslands of the Northwest, it was desirable to create a fine dust mulch to prevent evaporation in the arid climate, and also to firm the soil to prevent fierce winds from carrying it all away; the tillage complex consisted of a light plough, often an ard (symmetrical or scratch-plough), a tined harrow, a bush-harrow and a roller. In areas of heavier clay soils such as the lower valley of the Yellow River, drainage was more important than moisture conservation, deep furrowing replaced cross-ploughing, and the heavy turn-plough with a well-developed mould-board became essential; with it went stout metal-tined harrows, but the roller and brush-harrow were less important. In the wet rice fields of Central and Southern China a light turn-plough, drawn by a single buffalo, was used together with a variety of long-tined harrows and spiked rollers to puddle the soil to a smooth mud; in particularly wet areas where a buffalo might flounder, the field was tilled by hand using a heavy iron drag-hoe.

For convenience we shall now trace the evolution of the chief tillage implements of traditional China, plough, hoe, harrow and roller, under separate headings, but taking care to point out wherever possible how a particular development or variation in one element of the tillage complex affected the other elements.

### (i) *The plough*

The ard and the turnplough both exist in a great variety of forms[a] (Figs. 39 and 40). An ard is a symmetrical implement with a symmetrical share that traces a shallow furrow perhaps 2 in. to 5 in. deep but does not invert the soil. It is a light implement usually made of wood with a metal share. Although it is not well suited to clearing new land (where ards are the common tillage implement, hoes and mattocks are usually used to clear grass and undergrowth), an ard can clear annual weeds effectively by cross-ploughing, that is ploughing the field both lengthwise and across. The ard opens up a shallow furrow ideal for most cereals. If the seed is sown broadcast the ard can be used to cover the seed in rows. Indeed, it has been suggested that in Southwest Asia the ard was developed for the purpose of covering the seed rather than tilling the land,[b] and this theory would certainly explain the early development of the seed drill in Mesopotamia. The ard is most useful on light soils such as loams or sands, and in mountain fields where the soil is thin; it can safely be used in areas where deep ploughing would turn up hardpan or cause salination or erosion.[c]

[a] See Leser (1) on German ploughs for an idea of the enormous variety to be found even within a relatively small area; Leser (1), Haudricourt & Delamarre (1) and Šach (1) give terminology and classifications of plough types.

[b] Haudricourt & Delamarre (1), p. 62.

[c] Arnon (1); Lambton (1).

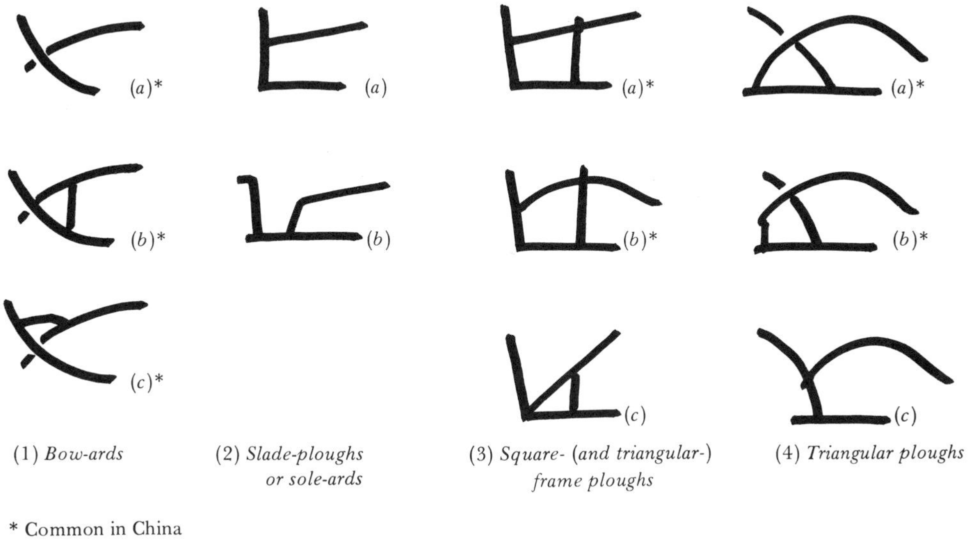


Fig. 39. Common plough types.

The turn-plough is an asymmetrical implement with a mould-board that inverts a ridge of soil; the furrow is usually 4 in. to 8 in. deep. It is a heavier, stronger implement than the ard and may be drawn by anything up to twelve beasts. It can be used in stiff clay soils where an ard would make no impression: the development of the turn-plough in Northwestern Europe in the early centuries of the +1st millennium permitted an enormous expansion of the cultivated area.[a] The turn-plough was a relatively late development, and one can see quite easily how it might have evolved, for if an ard is tilted to one side it can also be used to invert the soil, especially if (as is often the case) it is equipped with an extra device to direct the soil thrown up and deepen the furrow, such as symmetrical wooden boards or bundles of straw.[b]

The earliest form of plough was probably a simple ard consisting of a pointed share fixed to the end of a straight or curved stilt and pulled by ropes; examples made of wood and stone have been found respectively in a −5th millennium site in Satrup Moor, Schleswig and in a −3rd to −2nd millennium site in Hama, Syria.[c] The earliest pictorial representations of the ard are from Uruk, *c.* −3000, and show a more highly developed implement with a proper beam and sole, drawn by two or more oxen,[d] as do the −2nd millennium cave paintings from Bohuslän in Sweden.[e] Archaeological evidence shows that various forms of ard

[a] Slicher van Bath (1), p. 56; Fussell (1), pp. 47 ff., J. Percival (1), pp. 106 ff.
[b] E.g. the *aures* or *tabellae* of the Roman authors (K. D. White (1), p. 139; Leser (1), fig. 25); more recent examples of such devices are to be found in France, Spain and North Africa (Leser (1), pp. 317 ff.) and also in North China (Amano (*4*), pp. 747–8).
[c] Steensberg (2).
[d] Haudricourt & Delamarre (1), pl. 1, fig. 1.
[e] Leser (1), pl. 7.

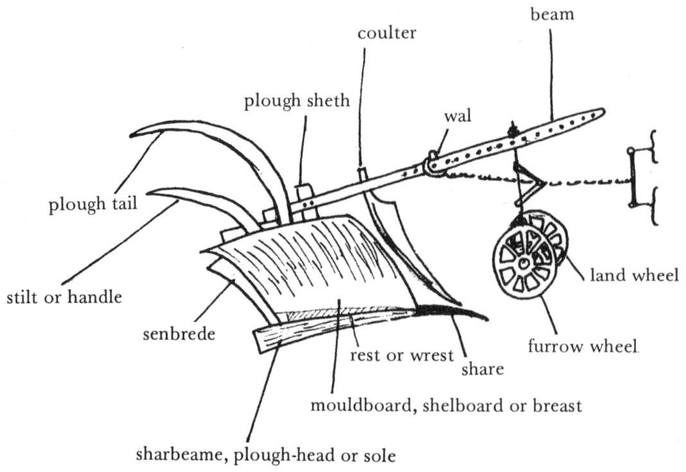


*Main features of an English wooden plough*

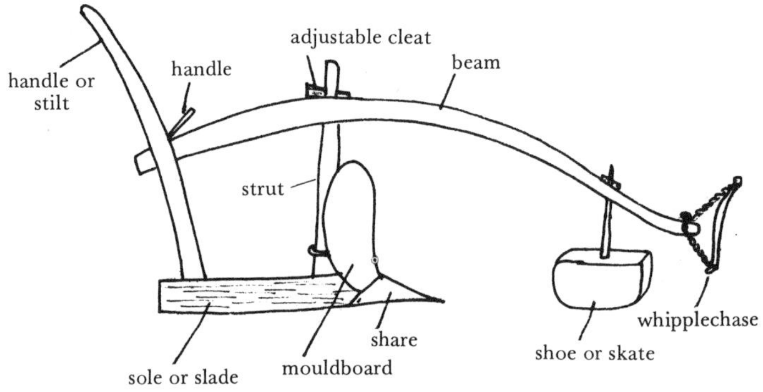


*Main features of a Chinese wooden plough*

Fig. 40. Typical wooden turn-ploughs: (*a*) English: after Fitzherbert (1), Blith (2) and Fenton (3); (*b*) Chinese: after *WCNS* 12/13*b*, Hommel (1), p. 41; Alley & Bojesen (1).

were in use not only in Assyria and Egypt but also in much of West Asia, in Harappan India, and over large areas of Europe (including Britain) by the −4th or −3rd millenium.[a]

It is still a matter of debate whether human or animal traction was first applied to the plough. Historians of China generally favour the theory that human traction was earlier there; in a wider context they are supported by Rau, Leser, Bishop and Steensberg among others.[b] Haudricourt and Delamarre[c] believe that, on the contrary, as far as the plough was concerned, animal traction preceded human; as human traction of the plough is known to be extremely arduous and represents no saving of labour over hoeing, we concur in this view, more espe-

[a] G. Clark (3); Hutchinson (1); Renfrew (3).
[b] Rau (1); Leser (1); C. W. Bishop (15); Steensberg (2) and (3).
[c] Haudricourt & Delamarre (1), p. 62.

cially as the dates for the domestication of cattle and the first appearance of the plough in West Asia and Europe are very close.[a] Certainly all the earliest pictures of the plough show it drawn by oxen,[b] while pictures of human traction only appear much later. It is true that several instances of human traction are to be found, both in ancient and recent times,[c] but all the evidence indicates that it was only resorted to *in extremis* when draught animals were in short supply, either temporarily as the result of an epidemic or war, or more permanently as a consequence of population increase and an expansion of the arable area at the expense of pasture land. Such a lack of oxen seems to have been quite frequent in Egypt during the Second Empire,[d] as it was in the early −1st century Han China:[e]

> Some of the people suffered from lack of oxen, and had no way to avail themselves of the moisture [of the rains]. In consequence the chief magistrate (*ling*[1]) of Phing-tu[2],[f] [known as] Kuang[3], taught [Chao] Kuo[4][g] to use men to pull ploughs. [Chao] Kuo memorialised to the throne that Kuang be appointed his associate in office in order to teach the people to help each other in pulling ploughs.[h]

Despite the fact that animal-drawn ploughs were commonplace in most of the rest of Asia by the −4th and −3rd millennia, most scholars of Chinese history believe that the ox-drawn plough was not known in China until the −5th century.

### *The Chinese plough in antiquity*

The great Yuan agriculturalist Wang Chen[5] seems to have been the first to assert that the ox-drawn plough was not known in China until the Spring and Autumn period,[i] thus contradicting the legend that Shu-Chün[6], grandson of the Emperor Shun's minister Prince Millet, Hou Chi[7], had invented the ox-plough in

[a] C. A. Reed (1) believes that in the absence of more positive archaeological proof we might take evidence of population increase linked with the domestication of cattle as an indication of the adoption of the plough.

[b] For example Egyptian temple paintings and the cave paintings of Southern France (Haudricourt & Delamarre (1), pl. 2, fig. 5; Leser (1), pl. 12).

[c] Human traction was quite common in recent times in North China, where poverty frequently prevented the farmers from keeping oxen (Hommel (1), p. 44); it was also found in some parts of Finland where a very light *socha* plough was used to turn ashes into the soil of swidden plots (Leser (1), p. 176). More anciently it occurred in 9th Dynasty Egypt (Haudricourt and Delamarre (1), pl. 2, fig. 4) and in −1st millennium Assyria (Leser (1), p. 249).

[d] Haudricourt & Delamarre (1), p. 72.

[e] Han Shu 24A/16*b*, tr. Swann (1), p. 189.

[f] In modern Szechwan; perhaps it was the mountainous terrain of the area which made animal traction impractical and so familiarised Kuang with human traction.

[g] The government post of which Chao was the incumbent was called *sou su tu wei*[8], and involved procuring and producing supplies for the army. The emperor Wu's prolonged and expensive campaigns in Central Asia had necessitated the creation of this post in −57, and Chao was the first incumbent.

[h] This passage has sometimes been quoted to prove that even in the −1st century many Chinese peasants were still not familiar with the ox-drawn plough (Hsia Wei-Ying (*3*)), but in fact the text states explicitly that the people were so accustomed to ox-drawn ploughs that they could not cope without them.

[i] *WCNS* 12/6*a*.

[1] 令 [2] 平都 [3] 光 [4] 趙過 [5] 王禎
[6] 叔均 [7] 后稷 [8] 搜粟都尉

the dimmest past.[a] The idea was not entirely new, however, for long before Wang Chen, the great Han lexicographers and commentators on the classics had put forward the notion that in days of old people had tilled their fields by hand, using a strange, rather impractical-looking implement called the *lei ssu*[1] (Fig. 41). Forced in the absence of less ambiguous evidence to rely chiefly on philological analysis, most subsequent scholars have agreed with Wang Chen and have argued that both the Shang and the early Chou dynasties were supported by very primitive cultivation techniques.[b]

In recent years, however, archaeologists in China have excavated numerous early sites throughout the country, and we now have a far clearer picture of the earliest historical dynasties and of the neolithic cultures from which they sprang. We also have more detailed archaeological evidence of the comparatively advanced state of agriculture in the Han dynasty. In the light of this new information it is certainly worth re-examining the problem of the early development of the plough in China.

One of the reasons why students of ancient China have found it difficult to reconstruct the agriculture of that period is that there is so little evidence of implements or techniques: a few stone and bone hoes are almost all that survives from the period prior to −500, and these have usually been taken at face value as the principal tools of ancient China. But it is impossible to draw an accurate picture of an ancient society just from the artefacts that happen to have survived, for such articles are unlikely to be representative. Apart from the obvious factors of loss or damage, anything made of wood, cloth, reeds or any other perishable materials will disintegrate rapidly in most climates, yet may have played a predominant role in the culture. It has been remarked of several present-day societies that, although their material culture is actually extremely rich, a future archaeologist would deduce that they had lived in abject poverty, so few are their durable possessions. Even today agricultural implements in many parts of the world are almost entirely of wood, and this was certainly the case in China for much of its history.[c]

Students of ancient China, faced with a dearth of material directly related to agriculture, have relied instead upon the extensive use of textual evidence, the more eagerly since in China the written word has always been deeply respected and treated as more reliable evidence than any other sort. Unfortunately, the early written material is more concerned with political, social and religious matters than with such lower-class occupations as farming, and the few references

[a] *Shan Hai Ching*.

[b] See, for example, Hsü Chung-Shu (*10*), Wan Kuo-Ting (*5*), Amano (*4*), Yang Khuan (*11*) and Hsü Cho-Yün (*1*).

[c] Where wooden remains have been preserved in China, as in the −5th millennium site at Ho-mu-tu near Shanghai, the variety and sophistication revealed is amazing. See Anon. (*503*) and (*504*).

[1] 耒耜

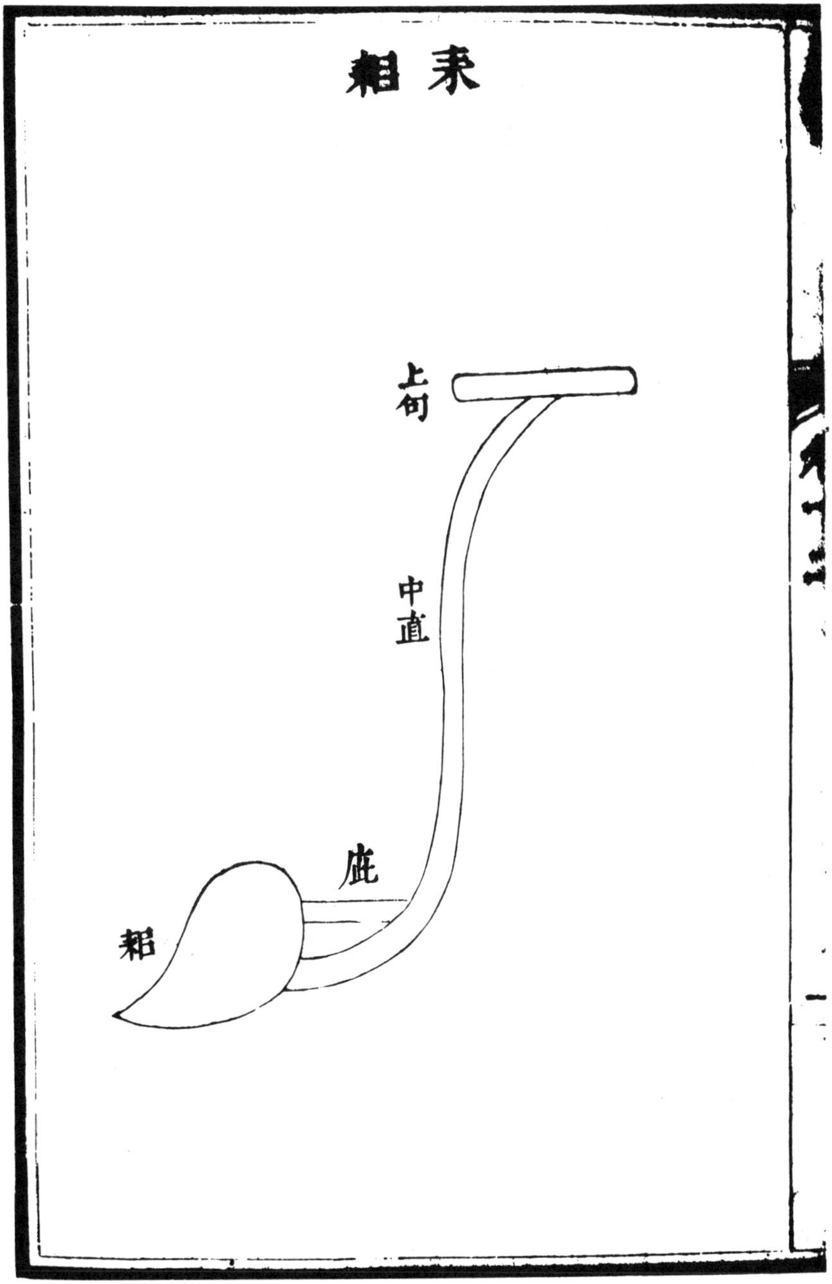


Fig. 41. Yüan reconstruction of the *lei ssu*; *WCNS* 12/2*b*.

to matters agricultural are usually short and cryptic. Furthermore, there has been a tendency to study these snippets in isolation rather than in the context of early Chinese society as a whole. Although quite a lot is known of such features of Chinese society as settlement patterns, political organisation and general technological level even in the neolithic period, historians of Chinese agriculture have generally made little use of studies by archaeologists, historians, economists and ethnographers linking social features with patterns of technological and economic development;[a] nor have they seriously compared early Chinese agriculture with its contemporary counterparts in the rest of the Old World. If we do take a more general and comparative view, we would certainly expect that a civilisation such as the Shang, which had achieved such a high level of technology and organisation and whose agriculture was not based on irrigation or high-yielding crops, would have been familiar with the ox-drawn plough.[b]

The traditional view of the early plough in China was that it had been invented by the Heavenly Husbandman Shen Nung[1] during the golden age that preceded the legendary Hsia[2] dynasty:[c]

> In the time of the Heavenly Husbandman millet fell as rain from the skies. The Heavenly Husbandman then tilled the land and planted the millet ... he fashioned ploughs (*lei ssu*[3]) and hoes with which he opened up the waste-lands.

Another legend previously alluded to attributed the invention of the ox-drawn plough to Lord Millet's grandson long before dynastic times.

Although myths can hardly be accepted as historical fact, they are usually indicative of some symbolic truth, and this pair of legends could easily be interpreted as a folk-memory of ploughs dating back to neolithic times.[d] But Han scholars and their successors were wary of legends except where the Duke of Chou was concerned, and what captured their imagination was not the implications of these myths but the precise nature of the implement which the Heavenly Husbandman is credited with inventing, namely the *lei ssu*.

The binomial expression *lei ssu* first appears in late Chou and early Han texts; for example, the *Chou Shu* and the *Shih Pen*. The former attributes its invention to Shen Nung, the latter to Kua[4], one of his ministers.[e] The *Erh Ya*, though it did not achieve its final form until the Chhin or early Han, is composed mainly of Chou material, and it is certainly significant that this work does not contain any reference to the *lei ssu*. It seems that the concept of the *lei ssu* as one single

[a] E.g. Goody & Tambiah (1); Haudricourt (14); Ucko, Tringham & Dimbleby (1).

[b] Bray (1).

[c] This passage is quoted in *CMYS* 1.1.1, where it is attributed to the *Chou Shu*, now lost.

[d] These two myths, together with many others, were first recorded in fairly late Chou texts, and this has led many authorities to assume that they are really of little antiquity. However, as J. S. Major (3) points out, it may well be that the more ancient and sacred the myth the later it will be recorded, for the act of writing it down (and thus exposing it to profane eyes) would be regarded as sacrilegious.

[e] Both these texts are now lost, but these passages are quoted in *CMYS* 1.1.1.

1 神農　2 夏　3 耒耜　4 倕

implement was not formalised until Han times, when we find the *Shuo Wen Chieh Tzu* giving the following definitions: '*Lei*[1]: a piece of curved wood used for tilling by hand ... *Ssu*[2]: the piece of wood at the end of the *lei*.'

Thereafter it was accepted by Chinese agriculturalists and lexicographers that *lei* and *ssu* were the two component parts of a single ancient tillage implement. In +880 Lu Kuei-Meng[3] entitled his monograph on the contemporary plough of the Yangtze Delta *Classic of the Lei Ssu*,[a] Wang Chen included a section on the *lei ssu* in the chapter of his agricultural encyclopaedia on farming implements,[b] and all the later encyclopaedists followed his example.[c] The classic illustration of the *lei ssu* as a skeletal ard with a strut but no visible means of traction (Fig. 41) was based on Han interpretations of a passage in the 'Craftsmen' chapter of the *Ritual of Chou*, a text which probably dates from the Spring and Autumn period:[d]

> The cartwright (*chhe jen*[4]) makes the *tse*[5] of the *lei*, which is 1.1 feet long; the central straight part is 3.3 feet long, and the upper curved part is 2.2 feet long. The dimensions from the outer part of the *tse* to the head [of the *lei*] measure 6.6 feet. On hard soils a straight *tse* is necessary, on soft soils a curved *tse*; a straight *tse* will push strongly [through the soil, whereas] a curved *tse* will turn [the soil] up well. If it is long on one side and short on the other like a stone chime (*chü kou chhing che*[6]) [the *tse*] is good for any soil.

Cheng Hsuan[7], who wrote the first commentary on this text in the Han dynasty, interprets the *tse* as being a piece at the end of the *lei* to which the *ssu*, the metal share, is attached; since the *ssu* is made of a different material, Cheng says, it is not mentioned in this section. In fact the *Khao Kung Chi* mentions the *ssu* only in the chapter on hydraulic workers, *chiang jen*[8],[e] which describes a complicated hierarchy of drains, ditches and canals to be dug with a *ssu* having a blade 6 inches wide. This *ssu* was clearly a type of spade and had no connection with the *lei* or tillage implement mentioned in the 'cartwright' section, but this did not deter the Thang commentator Chia Kung-Yen[9] from saying that *lei* and *ssu* went together like pestle and mortar.[f] It is unfortunate that the chapter of the *Khao Kung Chi* devoted to the makers of agricultural implements, *tuan jen*[10], is missing in its entirety.

Later scholars attempted to reconstruct the original *lei ssu* down to the exact angles between the different parts[g] (see Fig. 42). Interpretations included the beamless ard of Wang Chen's *Nung Shu*, where the *tse* is interpreted as a strut

[a] *Lei Ssu Ching*.
[b] *WCNS* 12/1*b*–3*a*.
[c] E.g. *NCCS* 21/1*b*–2*b*; *SSTK* 32/2*a*–*b*.
[d] *Chou Kuan, Khao Kung Chi*; *Khao Kung Chi Thu* 12*c*, tr. auct.
[e] *Khao Kung Chi Thu*, ch. 2, p. 120.
[f] It has been argued that rather than *lei* and *ssu* being different parts of the same implement, the binome *lei ssu* was in fact an attempt to reproduce a single word containing a complicated consonant cluster.
[g] E.g. the Sung work *Khao Kung Chi Chieh*, the Chhing works *Khao Kung Chi Thu* and *Khao Kung Chhuang Wu Hsiao Chi*.

1 耒 2 耜 3 陸龜蒙 4 車人 5 庛
6 倨句磬折 7 鄭玄 8 匠人 9 賈公彥 10 段人

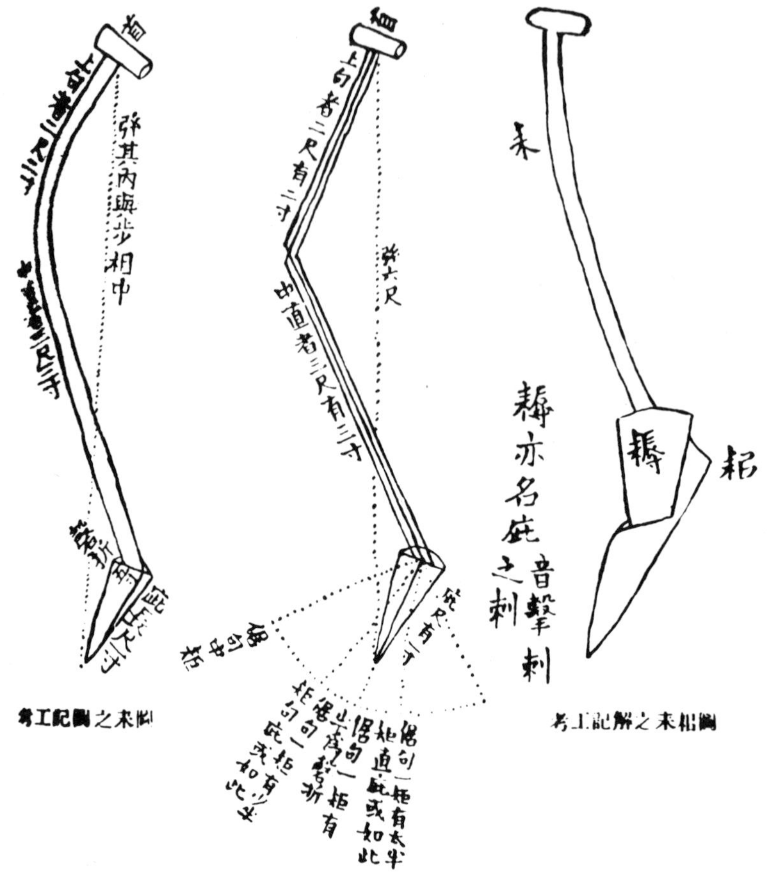


Fig. 42. Reconstructions of the *lei ssu*: the left-hand drawing is from the *Khao Kung Chi Thu* of 1746; the right-hand drawing is from the Sung *Khao Kung Chi Chieh* by Lin Hsi-I, who believed that the *lei ssu* was a primitive turn-plough; the central drawing is from Chheng Yao-Thien (*2*).

between the wooden body and the metal share; an identification with the *nenohite karasuki*[1], the sacred plough used by the emperor of Japan in the +8th century for his ritual spring tilling of the fields;[a] and a suggestion that the *lei ssu* was provided with a wooden foot-rest enabling it to be used in a fashion similar to the Hebridean caschrom.[b]

But as more and more early Chinese bronze vessels and bone inscriptions were discovered and examined, the suspicion began to grow that the Han scholars had been mistaken in their interpretation. The Chhing scholars Hsü Hao[2] and Tsou Han-Hsün[3] were the first to state that *lei* and *ssu* were in fact two separate tools.[c] The enormous hoards of Shang divination records discovered at An-yang[4] in the

[a] See Iinuma (1) for a detailed account of this implement, which is still preserved in the Shōsō-in collection. Hsü Chung-Shu (*10*) and Amano (*4*), p. 720 both consider this interpretation seriously.

[b] Sun Chhang-Hsü (*1*), p. 7.

[c] See Yang Khuan (*11*), p. 2.

[1] 子日手辛鋤 [2] 徐灝 [3] 鄒漢勛 [4] 安陽

Fig. 43. Han reliefs of Shen Nung and Yü the Great wielding two-pronged digging implements; Nagahiro (*1*), p. 65; Hayashi (*4*), fig. 6–4.

1920s and 1930s opened up a whole new perspective on early Chinese society, and a comparative study of bone and bronze inscriptions enabled scholars to establish that the characters *lei* and *ssu* did not occur together in these inscriptions. *Lei* was more common in the Shang texts, and *ssu* in the texts of the early Chou. Studies were also made of the various Shang and Chou 'pictograms' that had obvious agricultural connotations, in an attempt to discern what early Chinese agricultural implements had actually looked like.

As well as the bronze and bone inscriptions, they took into account various Han bas-reliefs depicting such legendary characters as Shen Nung and the Emperor Yü digging with what appear to be double-bladed spades (Fig. 43), together with Cheng Hsüan's commentary on the hydraulic workers section of the *Khao Kung Chi*:[a] 'In ancient times the *ssu* had only one blade but two men dug side by side. Nowadays the *ssu* is forked and has two blades . . .' Hsü Chung-Shu was

[a] *Khao Kung Chi* 12/18*a*. Since the implement described here, like the one depicted in the Han bas-relief of the Emperor Yü, was specifically intended for the construction of irrigation ditches, there is no obvious justification for assuming that the two-pronged *ssu* was ever used as a tillage implement. For a further discussion of these two-pronged implements, see p. 216 below.

Fig. 44. Archaic graphs of *lei*, based on Hsu Chung-Shu (*10*).

Fig. 45. Reconstruction of the *ssu* according to Sun Chhang-Hsü (*1*), p. 32.

the first to write a monograph on the *lei* and *ssu* as separate implements in 1930.[a] After examining various pictograms with agricultural meanings (Fig. 44) he concluded that the *lei*, which he assumed to be the chief agricultural tool of the Shang, was a two-pronged wooden spade such as we see in the Han bas-reliefs, and was used to make a double furrow in the soft soils of Northeast China. The *ssu* he thought was a flat, oval piece of wood attached to the end of a digging-stick (Fig. 45); it was used as a spade and was the main tool in Northwest China where the Chou dynasty was based. Hsü thought the spade-like *ssu* gradually developed into the ox-drawn plough, achieving this form by −500. His theory of the *lei* as a two-pronged spade was based partly on the Han bas-reliefs and partly on the fact that the pronged, spade-shaped *pu* coins common in the later Chou are found chiefly in the areas where the Shang dynasty had formerly held sway.[b] Several distinguished scholars support Hsü's basic theory of two separate implements regionally distributed.[c] The Japanese scholar Sekino[d] believes that it was the *lei* and not the *ssu* which developed into the ox-drawn plough. He sees the *lei* as a pointed implement with either one or two prongs, used like a spade in the heavy soils of Shansi (where pointed *pu* coins are found in the Chou). The *ssu* he sees as a straight-edged hoe, used in the lighter soils of Honan (where straight-edged *pu* coins were common (Fig. 46). Yang Khuan[e] believes that both *lei* and *ssu* were

[a] Hsü Chung-Shu (*10*).

[b] Bronze coinage was in common use during the Chou dynasty. Some of the coins were knife-shaped, some circular, and some spade-shaped. The last type, known as *pu*, fell into two categories according to Hsü and others: those with straight edges and those with points. Several sinologists seem to assume automatically that the *pu* coins must represent the most important agricultural implements of the period during which they were in use. But the form of a coin is obviously severely restricted by practical considerations, so any three-dimensional object, such as a chariot, ship or plough, however important it may have been in the economy, was unlikely to be chosen as a model for coinage. It therefore seems unlikely that any significant deductions about agriculture can be made from Chou coinage. Even the regional variations in coin shape could be attributed to purely stylistic values.

[c] E.g. Amano (*4*), Sun Chhang-Hsü (*1*), Hsü Cho-Yun (*1*).

[d] Sekino (*1*), (1).

[e] Yang Khuan (*11*), p. 29.

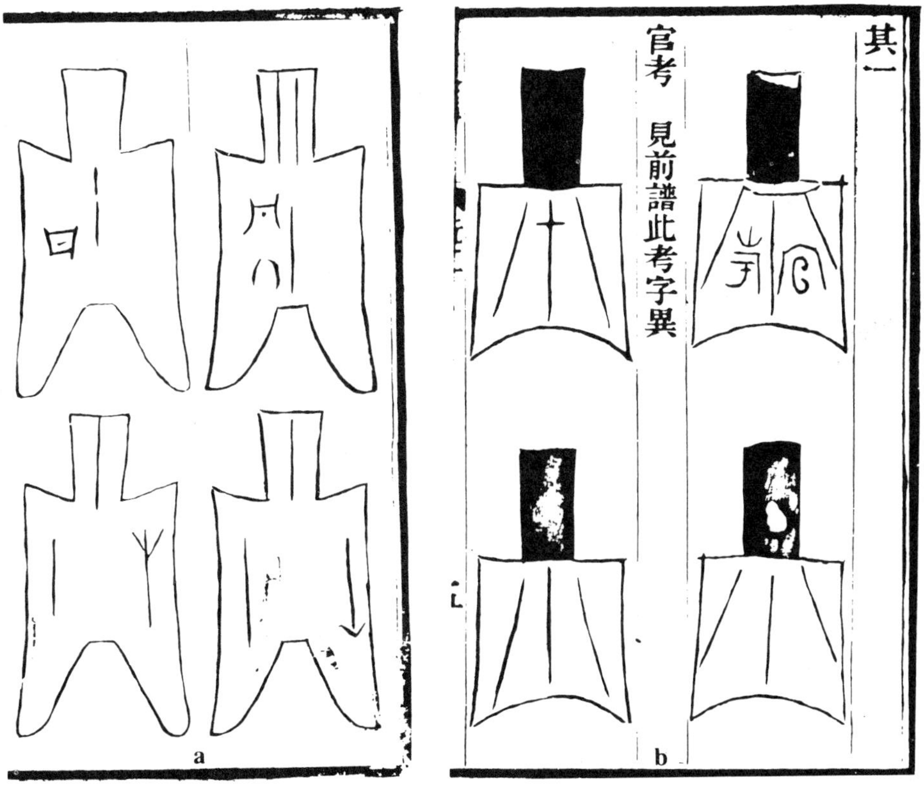


Fig. 46. Various types of *pu* coin; Li Tso-Hsien (*1*).

used spade-fashion as digging implements, the chief difference between them being that the former was pointed and the latter straight-edged; whether either was forked or single-bladed made no difference, Yang believes, inclining to the view that the pointed *lei* was forked in antiquity but not in Han times, whereas the straight-edged *ssu* was certainly single-bladed in ancient times but by the Han was sometimes forked.

These then are the main modern interpretations of early Chinese agricultural technology: the chief tools in use were spades and/or hoes, made of wood, bone or stone in the early period and then acquiring metal tips or blades as bronze and iron became more common. The spades were sometimes used by two or more men: one held the shaft and the others, harnessed to the blade by a rope, pulled the spade forward.[a] It was this application of human traction that gradually

[a] Modern examples of such traction spades have been found in Shansi in the 1930s, where an implement called *chhiang li*[1] was worked by two men (Amano (*4*), p. 275), and in Korea in the 1950s, where four men pulled a V-shaped spade with a beam 12 feet long while one man guided the spade handle (Clayton Bredt, personal communication). Both these implements were in fact used spade-fashion, digging in deeply, then being raised to turn the sod before digging again; this discontinuous application of power is very different to the steady tension exerted on the beam of a plough.

[1] 鏹,搶犁

transformed the spade into a plough, say Hsü Chung-Shu and his adherents. At first only human traction was used, but in the −6th century iron became common enough to be used for plough-shares, at which point the plough became too heavy for men to pull, so they harnessed up oxen instead. Proof of this theory, says Hsü, is to be found in the fact that prior to −500 the semantic element for ox, *niu*[1], does not occur in any of the characters associated with tillage, whereas in the −5th century two of Confucius' disciples have names containing the word *keng*[2], 'to till', and are nicknamed *niu*.[a]

On the assumption that the early Chinese used only hoes or spades to till their fields, various scholars have described Shang and early Chou agriculture as slash-and-burn cultivation. Some interpret the character *thien*[3], which from late Chou times at any rate always means 'field' or 'to cultivate', as originally meaning 'hunting grounds' and deduce that agriculture was rarely mentioned in Shang texts.[b] Others interpret the characters *shao thien*[4] or *fen thien*[5], which occur several times in the oracle bones, as meaning 'to fire the fields [in preparation for cultivation]', that is to say, 'swidden cultivation'.[c] References in the *Odes* to 'first-year fields' (*tzu thien*[6]), 'second-year fields' (*hsin thien*[7]), and 'third-year fields' (*yü thien*[8])[d] have convinced one or two scholars that not only did the Chou peasants practise swidden cultivation but they could only cultivate a plot for at most three years at a time before abandoning it.[e]

Various reasons have been advanced by the spade-and-hoe school for the sudden appearance of the ox-drawn plough (and consequently the transformation of Chinese agriculture) in −500: this was a period of rapid population growth; a number of contending states now existed in China, requiring large agricultural surpluses to support their standing armies; and iron was by this time a fairly common commodity. The Chinese ox-drawn plough is generally regarded as indigenous, a natural development of the traction-spade, and it is often suggested that the use of oxen for ploughing did not become really common till the later Han, and that hoe and spade cultivation were still usual in many areas for some time longer.[f]

To regard Shang and Chou agriculture as based on short-term swidden cultivation seems unsatisfactory for a variety of reasons. Several recent articles have

[a] Hsü also quotes various Han commentators to the effect that *li*[9], which from the Warring States period to the present day has been the common word for ox-plough, in Confucius' time was still used to mean 'spotted' or 'dappled' ((*10*), p. 57) and had no connection with tillage.

[b] E.g. Wu Chhi-Chhang (*4*).

[c] This is refuted by Hu Hou-Hsüan (*4*).

[d] *Shih Ching*, Odes 178 and 276; 17/29*b* and 27/1*b*; tr. Legge (8), pp. 284, 582.

[e] E.g. Chang Cheng-Lang (*2*); Yang Khuan (*11*) says this was the case in the Western Chou, while by the Eastern Chou the terms had come to refer to two- and three-course crop rotations similar to those of medieval Europe.

[f] E.g. Hsia Wei-Ying (*3*); Amano (*4*), p. 723.

1 牛 2 耕 3 田 4 燒田 5 焚田
6 菑田 7 新田 8 畬田 9 犂, 犁, 犁

suggested that maybe early Chinese agriculture was more sophisticated than has generally been thought, and there is a growing body of archaeological evidence to suggest that ploughs were quite widely known in prehistoric China (see p. 155). Moreover, the evidence adduced to demonstrate the primitive nature of agriculture in the late Shang and Chou is far from satisfactory.

The first objection that one might raise is to the interpretation of Shang and Chou inscriptions as 'pictographic'. The earliest Chinese script now known consists of inscriptions on pottery from the neolithic settlement of Pan-pho[1] and dates back to about −4000.[a] Although many of the very early inscriptions are rudimentary and really only resemble tallies, some of them are in fact strikingly similar to characters still in use today. By Shang times, some two thousand years later, the Chinese script already contained a large proportion of phonetic compounds, and obviously many of the surviving pictograms and ideograms were highly stylised versions of characters invented long previously. Since in any case scribes are specialists in writing rather than in agriculture, carpentry or any other craft, however early the pictogram, it will always be a general impression rather than reliable pictorial evidence.[b] Moreover, any inscriptions on the Shang oracle bones that deal with agriculture are concerned only with generalities: will the harvest be good? will the rains come at the right time? how much wheat did a particular province send as tribute this year? They do not mention any of the practical details of farming, and so the vocabulary included in these documents is not likely to be as extensive as that in vulgar use at the time; the bronze inscriptions are even more limited in scope. Both will give a misleading impression of the complexity of contemporary agricultural practice.

With these caveats in mind let us now take another look at some of the early Chinese graphs connected with tillage. As we have seen, the element *lei*[2] that is common to many of them is usually interpreted as a two-pronged digging-stick with a foot-rest. Yet it is also remarkably similar to a Sixth Dynasty Egyptian hieroglyph, also incorporated in many agricultural terms (Fig. 47), which in fact denoted the Egyptian ard, the principal difference being that the Egyptian ard had a double handle. The horizontal stroke in the Chinese graph could well represent the cross-bar that is found in so many later Chinese ploughs, rather than a foot-rest. It is also similar to the type of sole-handle ard still in use in many parts of China in the 1920s and 1930s (Fig. 48). On the basis of a pictorial analysis alone it would be quite as justifiable to assume that some sort of plough was in use at the time as to assert categorically that it was not.

Equally it is not valid to assert that animal traction was not used for ploughing in China before −500 simply because the early agricultural graphs do not con-

[a] Kuo Mo-Jo (*12*), Cheng Te-Khun (17), Ho Ping-Ti (5).

[b] For a striking example of the unreliability of pictorial representation see Haudricourt & Delamarre (1), p. 41.

[1] 半坡　　[2] 耒

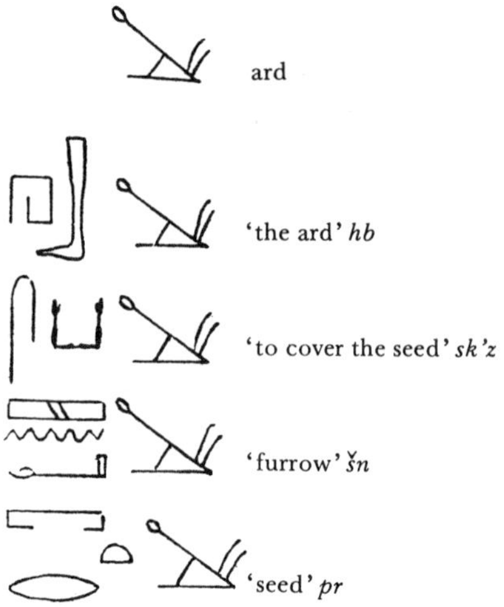


Fig. 47. Egyptian glyphs for 'ard' and related words.

tain the semantic element for 'ox'. The Shang harnessed horses to their chariots at least as early as −1350,[a] and as horses are notoriously more difficult to use as draught animals than bovines, it seems likely that the ox-cart, if not the ox-plough, was known in China some time previously. None of the various forms, either Shang or Chou, of the character *chhe*[1] meaning 'chariot' or 'cart' shows any trace of a horse (Fig. 49), yet it would be absurd to claim that Shang war chariots were drawn by manpower. In the latter case we have incontrovertible archaeological evidence to support the judgement of common sense, but it does show clearly the drawbacks of relying too heavily on unsupported philological evidence. In fact, one modern authority on the interpretation of archaic script[b] believes that several Shang characters do actually depict ploughing with oxen, in particular the archaic form of *jang*[2] (Fig. 50), which now means 'to assist', but which the Han lexicographers also defined as meaning 'to plough'.[c]

The various references to agriculture in pre-Han literature are generally just as unhelpful as the characters on the oracle bones and bronzes. They are few and far between, and very inexplicit. None of the surviving texts of the period, even those dealing with political economy, provides us with an accurate picture of how a farm was run or what equipment was used.[d] But it should come as no surprise that the early Chinese literature does not provide us with accurate technical details.

[a] Vol. IV, pt. 2.

[b] Hsü Chin-Hsiung[3], personal communication in 1972.

[c] *Shuo Wen Chieh Tzu*.

[d] E.g. *Lü Shih Chhun Chhiu*, *Shang Chün Shu*, *Kuan Tzu*, *Meng Tzu*; the first of these does describe in some detail a particular method of cultivation, *mou chhüan*[4], which it advocates as highly productive, but even this passage is difficult to interpret precisely; see Hsia Wei-Ying (*3*) and p. 168.

[1] 車 [2] 襄 [3] 許進雄 [4] 畝畎 or 甽

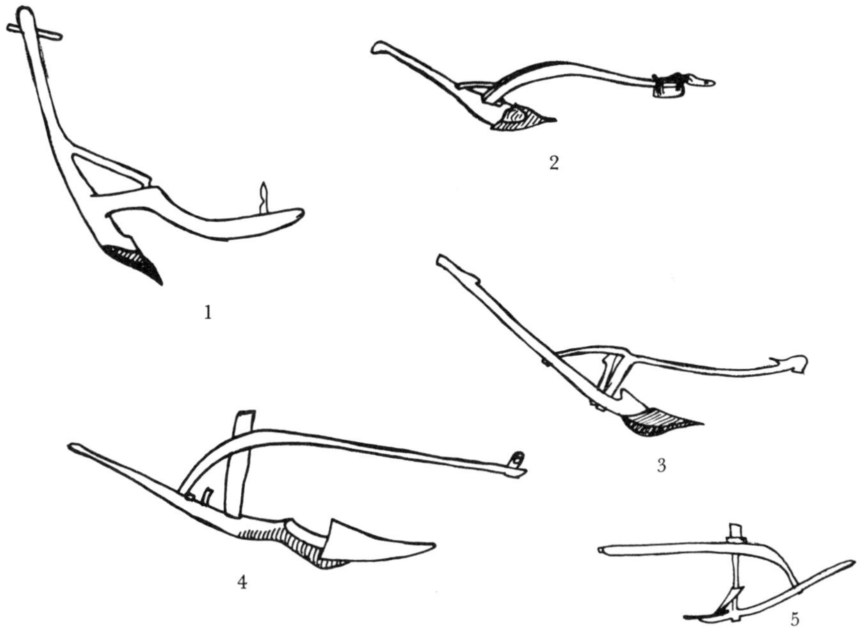


(*a*) *Chinese sole-handle ards*

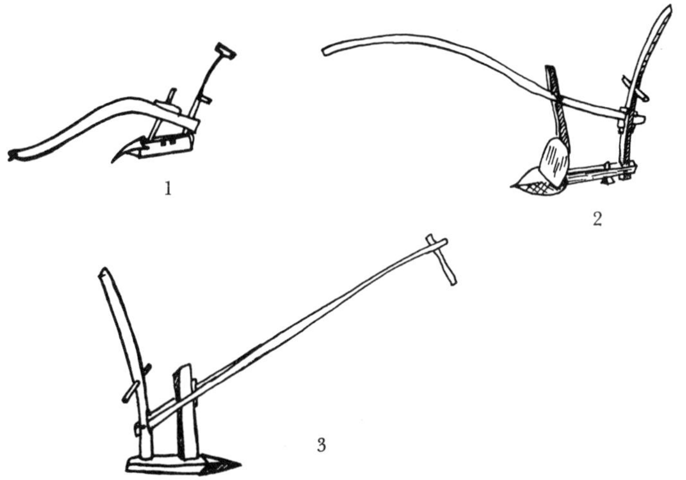


(*b*) *Chinese ploughs with cross-handles on stilt*

Fig. 48. Modern Chinese ards: (*a*) sole-handle ards: (1) from Kansu; Anon (*502*), p. 4; (2) from Shensi; *ibid.* p. 5; (3) from Shensi; *ibid.* p. 8; (4) from Kwangsi; *ibid.* p. 52; (5) from Hainan Island; Amano (*4*), p. 799; (*b*) Chinese ploughs with a cross-handle on the stilt: (1) from Manchuria, Amano (*4*), p. 798; (2) from North East China; Wagner (1), p. 200; (3) from Shantung; Hommel (1), p. 44.

Fig. 49. Archaic graphs for *chhe*, cart or chariot.

Fig. 50. Archaic graphs which may depict ox-ploughing.

Even close study of such a well-informed work as the *Georgics* has failed to reveal the construction of a Virgilian plough,[a] though Virgil himself must have been perfectly familiar with the details of the implement, and we may be sure that most of the pre-Han philosophers upon whom we depend so heavily for technical information were much more ignorant than the Latin poet and could hardly tell one end of a hoe from the other.

One thing is certain: if the ox-plough had first been invented in the −6th or −5th century, as so many sinologists claim, then its invention would have been accredited to somebody contemporary, and it would have been mentioned repeatedly in the treatises of political economy. The seed-drill first appeared in China some time in the Han, and its invention is specifically attributed to an individual official, Huang-Fu Lung[1] [b] (though it may be that the 'triple plough' which Chao Kuo[2] is supposed to have invented in the −1st century was its prototype).[c] During the Warring States period, when every state was striving to increase production[d] and when political philosophers were discussing soil science and cultivation methods,[e] it is extremely unlikely that the replacement of the hoe by an instrument as revolutionary as the plough would go unnoticed.

Let us turn to the archaeological evidence, which is now much more extensive than when Hsü wrote his study in 1930. The absence of an object from a site is not conclusive evidence that it did not exist. 'A curious feature [of Jericho, *c.* −6000] is the almost complete lack of heavier tools, such as axes or adzes for heavy woodworking ... [yet] we know from the presence of sockets in the walls that the inhabitants were able to fell and dress substantial timbers.'[f] Any implement or artefact made of wood, as primitive ploughs usually are, survives only in exceptional climatic conditions. The hot deserts of Upper Egypt and the peat

[a] K. D. White (1), pp. 128 ff.
[b] *CMYS* 0.7.
[c] See p. 262 below.
[d] 'A rich state is a strong state' was the constant theme of the Legalist philosophers.
[e] In particular the chapters on soil science in the *Kuan Tzu* and on agriculture in the *Lü Shih Chhun Chhiu*.
[f] K. Kenyon (1), p. 57.

[1] 黃甫隆 [2] 趙過

bogs of Jutland are both ideal for preserving wood and organic materials, and both these regions have yielded, among other remarkable finds, complete wooden ards.[a] In other parts of the world the ploughs themselves have not survived, but traces of furrows have been preserved under burial mounds or other monuments, showing that ploughs were in use much earlier than had previously been thought.[b] The only early Chinese site where wooden remains of any importance have been found so far is the Ho-mu-tu[1] site in the Yangtze Delta near Ningpo,[c] situated in a peaty, acid marsh which has preserved in near perfect condition large numbers of organic remains, including wooden tools and what has been identified as cultivated wet rice. The Ho-mu-tu site was occupied very early, from about −5000 to −3500,[d] so it is not surprising to learn that no traces of any sort of plough were found there, the more so as the site is in a very marshy area where even today draft animals cannot always be used as they flounder in the deep mud. Unfortunately, in drier areas where one might be more likely to find indications of early ploughing, the conditions of preservation are far less favourable; in particular the extremes of temperature and humidity in North China have ensured that very few wooden objects except those coated with lacquer or paint have survived more than a few centuries.

Although we have no evidence of wooden ploughs in early China, at a slightly later date than Ho-mu-tu there are signs of neolithic plough cultivation in the form of triangular stone shares[e] (Fig. 51a); at least three have been found near Hangchou, on the Great Lake in the Yangtze Delta, one of them in a stratum dated to approximately −3000, another to −2000,[f] while others have been found in a stratum dated to −2000 in Shanghai.[g] A foliate stone share of much later date (probably contemporary with the Western Chou), has been found in Inner Mongolia (Fig. 51b),[h] while in the Mongolian People's Republic archaeologists have reported a 3000-year-old carving depicting ox-ploughing.[i] And although no Shang bronze shares have been found, Vietnamese archaeologists have discovered bronze ploughshares dated to −1500 in the Red River Delta.[j]

Another neolithic stone type that could well be construed as a ploughshare comes from the Liang-chu[2] culture of Chekiang, which may date back as far as −3300 but which was certainly well established throughout the −3rd millennium.[k] These V-shaped stone pieces have hitherto been identified simply as 'winged implements of cultivation',[l] but they are strikingly similar to the bronze and iron 'cap-shares' (*kuan*[3]) of late Chou times (Fig. 52a); moreover, the single perforation would not be suitable for fastening a percussion implement (such as a

[a] Haudricourt & Delamarre (1), p. 39.
[b] C. Renfrew (3), p. 547; Steensberg (3).
[c] Anon. (*503*) and (*504*).
[d] Hsia Nai (*6*).
[e] Liu Hsien-Chou (*8*), p. 10.
[f] Anon. (*514*), fig. 15.
[g] Chang Ping-Chhüan (*2*).
[h] Liu Hsien-Chou (*8*), fig. 13.
[i] *Far Eastern Economic Review*, 21 July 1978, p. 35.
[j] J. Davidson (1), p. 90.
[k] Hsia Nai (*6*).
[l] K. C. Chang (1), p. 181.

[1] 河姆渡 [2] 良渚 [3] 冠

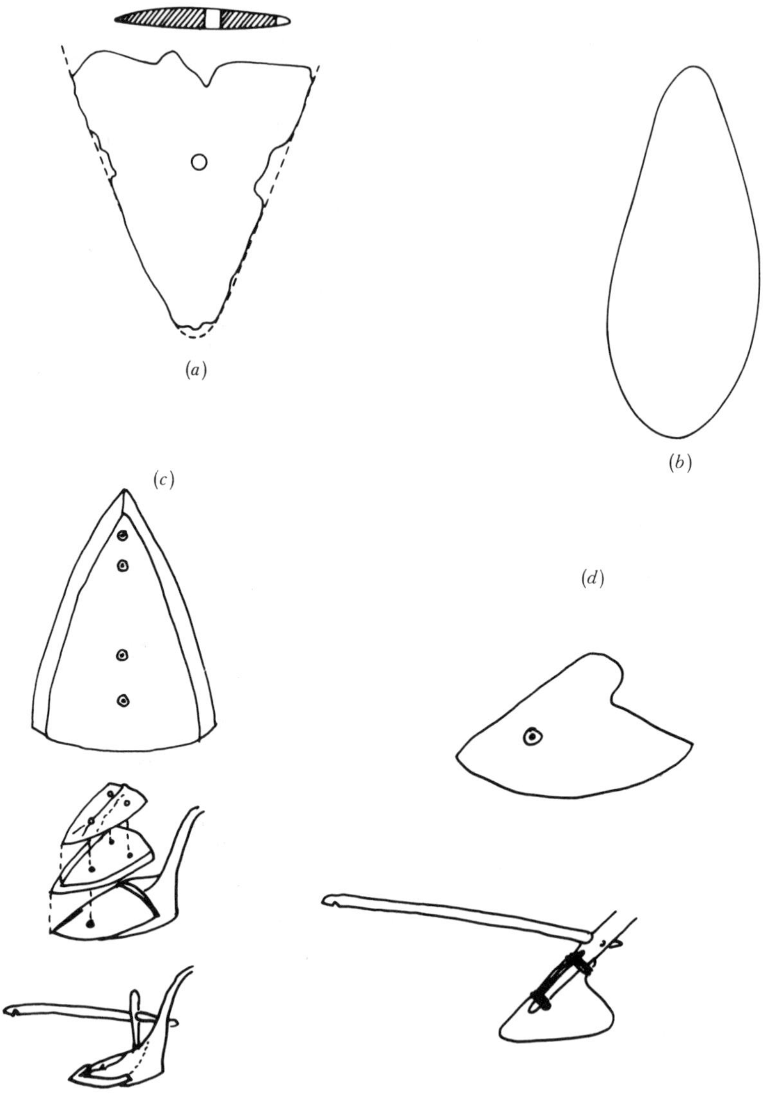


Fig. 51. Neolithic stone shares; (*a*) from Hangchou, (*b*) from Inner Mongolia; Li Hsien-Chou (*8*); (*c*), (*d*) from the Yangtze Delta; Mou & Sung (*1*).

hoe) to a vertical or angled shaft, but would be sufficient to pin a share on to a sloping wooden plough-slade, where the shearing force against the pin would be much less. (Interestingly, no traces have been found in China so far of the hafted stone or wooden blades that were used as ploughs in West Asia and Europe.)[a] Additional weight is lent to this interpretation by the recent discovery in several Yangtze Delta sites of stone 'plough-shares' and 'soil-breaking implements', at sites dating from as early as the Ma-chia-pang cultures and as late as the Shang

[a] Steensberg (2) and (4).

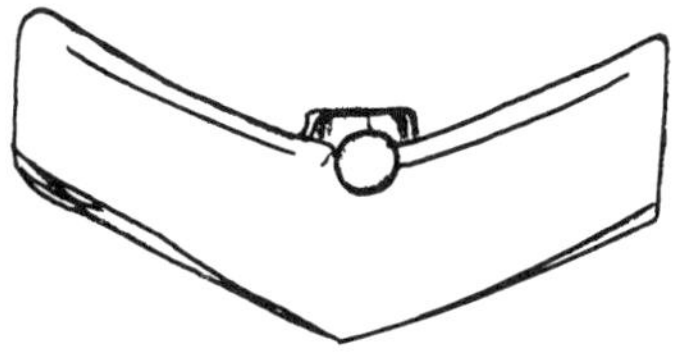

(*a*)

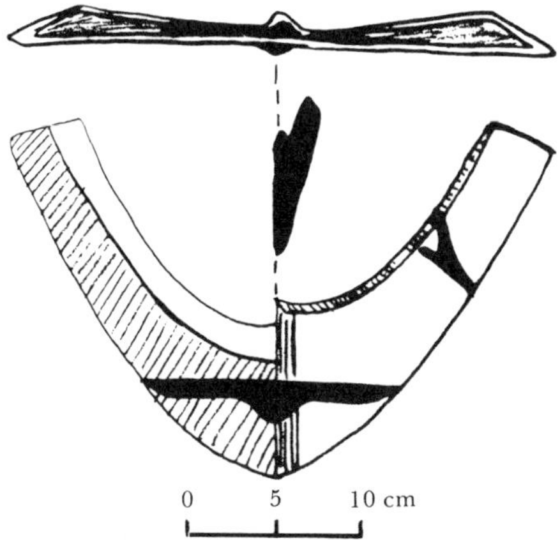


(*b*)

Fig. 52. V-shaped shares: (*a*) stone, from the neolithic Liangchu culture; after Anon. (*43*), p. 29; (*b*) cast iron, from Honan; after Amano (*4*), p. 736.

and Chou[a]. The stone 'shares' vary in shape and size, but are essentially large, thin triangles of shale, pierced at several points in the centre, and with sharply honed edges. The shape and wear-patterns strongly suggest that these large, somewhat fragile objects were ploughshares that were attached to a broad wooden slade, often protected by an additional triangular wooden cover that left only the stone blade protruding (Fig. 51*c*). The 'soil-breaking implements', thin, irregularly shaped blades designed to cut the soil vertically, were a type of coulter, Mou and Sung (*1*) suggest (Fig. 51*d*), used to facilitate the clearing of marshy rice-land and the construction of drainage ditches. The stone ploughs, they think, were used after the blades, both for trenching and for the cross-ploughing of rice-fields.

The earliest of these stone shares—if ploughshares they are, for the unambiguous identification of archaeological artefacts is notoriously difficult—date from the period when the Lungshanoid cultures of China were fast developing. The earliest neolithic cultures of China fall into four distinct groups. The Yang-shao culture of the Yellow River basin was concentrated in closely grouped villages on

[a] Mou Yung-Khang & Sung Chao-Lin (*1*).

loess terraces above streams and rivers; the chief crop was setaria millet. The Ta-phen-kheng cultures of Taiwan and the South China coast lived chiefly from fishing, and though they may well have cultivated tubers such as taro and yam, and possibly even rice, there is no concrete evidence of plant domestication yet. The cultures of the Huai and the Yangtze coastal plains have often been grouped as a single culture,[a] but in fact it seems more sensible to distinguish the Chhing-lien-kang[1] culture north of the Yangtze in Kiangsu and Southern Shantung from the Ma-chia-pang[2] culture to the south in Chekiang.[b] The former area has a basically northern climate and even today grows mainly dryland cereals; in early neolithic times it was probably, like the Yang-shao culture with which it has many affinities, a millet-based culture. The Ma-chia-pang culture, on the other hand, grew large quantities of rice from very early times, and seems to have domesticated the water buffalo at an early date.[c] These four different cultures were all well established by the −5th millenium, and continued to develop more or less independently for over a thousand years.

In the late −3rd millennium these different cultures start to show signs of mutual influence. A series of broadly similar cultures known as Lungshanoid grew up and spread rapidly over most of North and Central China and the Southeast coast:[d]

> The earlier Lungshanoid sites are large, thickly and densely spread over a wide and continuous area of much of China, which points to a period of considerable population size and density ... The earlier Lungshanoid period, from about −3200 to −2500, was ... a period of rapid expansion of all neolithic cultures, during which they came into contact and interacted with each other.
>
> Everywhere one recognises two major stages of cultural development: an early Lungshanoid phase, seemingly transitional, and a late Lungshan phase, in which the earlier peaceful and largely egalitarian village life had been transformed into a warlike and ranked society preparatory for the formation of civilisation and the state.

Lungshanoid sites were from the beginning large, well-populated and closely set. Now once a settlement reaches a certain size it is impossible for it to be self-supporting unless the land within walking distance is sufficiently productive. In tropical cultures, based on the cultivation of such highly productive cultigens as taros or yams, swidden cultivation can sometimes support a considerable population, especially as the climate permits year-round cultivation. But this is unusual, and even in the tropical zone a swidden settlement rarely contains more than one or two hundred people.[e] In temperate zones where cereals are the main crops even

[a] Wu Shan-Chhing (*1*).

[b] K. C. Chang (1), pp. 134 ff. elaborates the arguments for and against this distinction, which in fact is now usually made by Chinese archaeologists, e.g. Anon. (*505*).

[c] Anon. (*503*), (*504*), (*505*).

[d] K. C. Chang (1), p. 172; p. 144.

[e] If larger settlements do occur then they are very widely spaced; D. R. Harris (1).

[1] 青蓮崗 [2] 馬家浜

small settlements would be impermanent under a system of swidden cultivation,[a] and in general the transition from a pattern of small, dispersed, autonomous communities to that of a dependent peasantry under centralised control is 'a critically difficult one which populations wholly dependent on swidden cultivation appear unable, or at least most unlikely, to make'.[b]

Speaking principally of Europe and West Asia, Fenton says:[c]

In recent decades, the growing evidence from finds, cultivation traces, and comparative studies has effectively shown that cultivating tools were already sophisticated at an astonishingly early date. The old text-book concept of digging-sticks and 'scratch' ploughs is now scarcely tenable. Additionally, the range of types of cultivating implements is remarkable.

Although the origins of agriculture in China are probably not as ancient as in the Fertile Crescent, nevertheless the Chinese development from mesolithic to neolithic culture and to its bronze age closely parallels events in Mediterranean area, and it is difficult to conceive that the eastern culture rivalled its western counterparts in all technological fields except the fundamental one of food production. Archaeological research in China has not made use so far of the intensive recovery techniques and the spatial studies that have thrown so much light on early societies elsewhere,[d] yet there is already a respectable amount of evidence for the use of the plough in China during the Lungshanoid period: stone 'ploughshares' have been found; domesticated cattle and water buffalo, from a few rare instances before the −4th millennium, became common throughout China; the population (dependent chiefly on the cultivation of cereals) grew and spread rapidly; and society became increasingly stratified and occupationally differentiated.[e] Furthermore, we must remember the myth of the Heavenly Husbandman, purported to have invented a primitive plough not only before the foundation of the Shang dynasty but well before the legendary Hsia dynasty too,[f] which according to the traditional dates goes well back into the neolithic period. Taken all together, these factors strongly indicate that the Lungshanoid farmers had knowledge of the ox-drawn plough, more especially as this implement had by now (the late −4th millennium) become common throughout most of the rest of Asia, from which the Chinese culture-area was not completely isolated.[g]

When we consider that the Lungshanoid culture was to develop into the civilisation of the Shang period, it seems even more probable that the ox-plough must

[a] Harris cites evidence of fission in cereal-growing settlements in Neolithic Czechoslovakia, *ibid.* p. 255.

[b] *Ibid.* p. 256.

[c] A. Fenton (4).

[d] See E. S. Higgs (1); also Byers (1), MacNeish (1), (2), (3) on Mexico; Vita-Finzi & Higgs (1) and A. J. Legge (1) on Palestine.

[e] K. C. Chang (1), ch. 4.

[f] Perhaps 'undiscovered' would be a better term, for the Shang was considered legendary by most sinologists before the excavation of An-yang, and recently Chinese archaeologists have hinted of one or two highly developed late neolithic cultures that they might actually have been the Hsia state, e.g. Thung Chu-Chhen (*1*).

[g] W. Watson (6).

have been known. Most subsistence level agriculturalists can produce a surplus sufficient to provide for their own needs in a poor year, but even if they have progressed beyond swidden cultivation to settled agriculture, 'this is as much as the hoe-cultivator can do on fallow-rotation soils under rain-fed conditions. The surpluses he produces may allow of tenuous trade or support a chiefly hierarchy, but they are far too unreliable to maintain complex societies with a high degree of occupational specialisation.'[a] The Shang cities were large and were surrounded by huge tamped-earth walls that would have taken thousands of workers years to build. Society was highly stratified. The nobility fulfilled religious and ceremonial roles as well as governing their fiefs and leading their troops, while luxury goods were provided for their use by what must have been a fairly large class of skilled artisans and craftsmen living within the confines of the city. The farmers who lived in the suburbs of the capital and other towns would have to support a considerable number of agriculturally non-productive people, some of whom lived in great luxury. Since transport of heavy goods such as grain must have been even slower and more costly than it was in imperial China,[b] most of the necessary foodstuffs must have been produced in the immediate vicinity of the cities, which means that the land must have been quite intensively farmed. Certainly shifting cultivation could not have supported the Shang city state, let alone the Chou states with their huge standing armies.[c]

If we look at other early civilisations, we see that there are several ways in which small areas can be made sufficiently productive to support a city. First, very high-yielding crops can be grown, as in Mexico and Peru.[d] Secondly, natural or artificial irrigation can be used as in Egypt or Mesopotamia.[e] Thirdly, the actual area under cultivation at any one time can be increased, and labour requirements reduced, if animal power is used. Large fields can be cultivated permanently under a rotation system, instead of small plots being cleared for short-term fallowing, and although the ploughed fields may be slightly less productive than hand-tended plots in terms of absolute production, overall production will be greatly increased, especially if the draft animals also contribute

[a] W. Allan (2). One must be careful when implying that the production of a surplus is a causal mechanism in the rise of civilisation (G. Dalton (1)), but we can at least say that once an agriculturally non-productive sector has appeared, the production of a surplus is necessary to maintain it.

[b] See Hoshi (1), Ray Huang (3).

[c] It has been suggested (Bill Jenner, personal communication) that animal foods formed a large proportion of the diet in Shang and Chou times, so small crop yields could be adequately supplemented by hunting and fishing; but the *Odes* show clearly that the peasant diet, by early Chou times at least, was almost exclusively vegetarian, and K. C. Chang (1), p. 290 holds that hunting was probably of marginal economic importance in the Shang: 'There is no question that agriculture was the basis of subsistence, or that the techniques were highly developed and yields considerable.'

[d] The Mayan civilisation of Mexico seems to have been based on the extremely productive ramon tree (*Brosemium alicastrum*), which yields 2000–3500 kg/ha of carbohydrate seed per year (Puleston & Puleston (1)); the chief Inca crops were maize and potatoes, both of which produce large yields and require relatively little attention.

[e] In Egypt the yearly flooding of the Nile covers the fields with a thick layer of fertile alluvium; in Mesopotamia large and elaborate irrigation works only evolved once civilisation was well established, though small-scale channels and reservoirs go back to pre-imperial times (C. Gabel (1), p. 49).

manure to the fields. Plough agriculture was soon used to supplement irrigation in Egypt,[a] and probably antedated irrigation in Mesopotamia.[b] In dry-farming systems the plough is generally accepted as a prerequisite to successful expansion.[c]

Now the Shang civilisation, though it ruled, or claimed suzerainty over, large areas of West and Central China, was centred on the Yellow River plains, in an area of dry climate and porous loessial soils where millets were the chief crop.[d] Compared with other cereals millets do not give particularly high yields even under modern cultivation methods,[e] though they are extremely resistant to drought, an important factor in the semi-arid climate of North China. Irrigation, though it may perhaps have been known in North China in Shang times, was not practised there on any significant scale until the Warring States Period.[f] Without high-yielding crops or even irrigation (which in any case, as we have seen, is usually accompanied by the use of the plough) to boost their agricultural production, the Shang farmers must surely have used ox-ploughs to till their fields. There was an economic necessity to do so, they were familiar with the principles of animal traction, and the plough was probably known to their Lungshanoid predecessors and certainly to their Bronze Age contemporaries in Tonkin.[g] We may hope that future archaeologists excavating Shang village sites will devote attention to finding concrete corroboration for this largely circumstantial evidence. – 16th-century bronze shares have been found in Tonkin, and might be found in North and Central China, though probably most northern Chinese ploughs were made entirely of wood at that period.[h] One type of evidence that archaeologists should look for is traces of ancient furrows, such as have been found under grave mounds and other monuments in India and parts of Europe.[i]

*Warring States ploughs*

The arguments that we have applied to the Shang dynasty are even more compelling in the case of the Chou. Cities were larger and more numerous than in

[a] F. Hartmann (1); W. L. Westermann (1).

[b] '... hoe or digging-stick agriculture ... is unsuited to irrigation agriculture with its expansion potential and the need not only to till the fields but to excavate irrigation ditches and run-off channels as well' (J. Oates (1), p. 303).

[c] H. Aschmann (1).

[d] Though the West Asian crops of wheat and barley had been introduced by this time (whether directly from Mesopotamia or Iran or indirectly via India it is impossible to say), they played a relatively minor role in the economy until Sung times.

[e] Under traditional cultivation methods yields of maize are generally over twice as high as those of setaria or panicum millet, while wheat and barley yields are slightly higher (see Arnon (1), vol. 2, Purseglove (2)); however, since millets have small grains, they do at least have the advantage of lower sowing rates.

[f] Vol. IV, pt 3, pp. 260 ff.; Ho Ping-Ti (5), p. 46; Chi Chhao-Ting (1), pp. 63 ff.

[g] The Shang state had commercial or tributary links with Tonkin, importing turtle plastrons for divination as well as pearls and other luxury goods.

[h] It is significant that while neolithic shares have been found in the heavy soils of Central China and the steppes of Mongolia, none have been found in the loess lands. Wooden ploughshares are often used where soils are friable or easily worked. In Babylonia and early Egypt, where the soils were reduced to soft mud by irrigation, wooden ploughs were common (Leser (1), pp. 245–6, 251, 257). Cato says that in Campagna the ploughs had wooden shares (K. D. White (1), p. 132), and ards made entirely of wood are still to be found in the Middle East and India (Leser (1), pp. 369, 374, 375).

[i] D. J. Breeze (1); J. Hutchinson (1); V. Nielsen (1); C. Renfrew (3).

the Shang,[a] and were an increasing drain on the agricultural economy, as were the large standing armies that feuding states maintained from the −8th century on. The *Odes* show that in the Western Chou new land was constantly being cleared and brought under cultivation; at this time most fields were fallowed at least one year and often two years out of three (see p. 429). By the Warring States period several states appear to have been so thickly populated that cultivation was continuous and fallowing was no longer practised, while double-cropping was even found in some areas.[b] This was the time when the use of green manures and other fertilisers first became common,[c] large-scale irrigation works were undertaken, and a free market in land grew up.[d]

It is also from this period onwards that iron ploughshares are found in China. The earliest iron implements found in China so far come from the southern state of Chhu and date from the late Spring and Autumn period, but as yet no ploughshares have been found among them.[e] By the Warring States period iron ploughshares, though less common than iron hoes and spades, are found in sites all over China.[f] These shares are of the type known as 'cap-shares' (*kuan*[1]) (Fig. 52*b*), and although they have been described by some authors as small, insecurely mounted and incapable of turning the soil to any considerable depth, they do in fact represent a relatively sophisticated form of share. Early iron shares in the West were usually very light and simple in comparison, and often far more precariously attached. In Greece and Rome the share of the ard was often just tied to the bottom of the sole with rope.[g] The earliest iron shares in Europe were no earlier than those of China, and they were of two simple types.[h] Stangle shares were arrow-like, the shaft of the arrow was fitted through the sole or the brace of the plough, and the arrow-head usually entered the ground at an angle (Fig. 53*a*). This type of metal share evolved from hardened wooden shares of similar form.[i] The second type, the sleeved share, consisted of a flat triangle or pentagon of metal that was wrapped round the tip of the wooden slade to form the point of the plough (Fig. 53*b*). Both types might be considered flimsy and insecure compared with the early Chinese shares, but both types are still in use today, indeed the stangle share has spread right through Europe, West Asia and India as far as

[a] P. Wheatley (2).

[b] *Hsün Tzu, Fu Kuo Phien*. Yu Yü (*3*) cites a range of textual evidence to show the increase in agricultural intensity throughout the Chou period, and although several scholars (e.g. Yang Khuan (*11*)), might dispute his interpretation, it is undeniable that by −100 even temporary fallowing was regarded as a last resort in the Metropolitan area (*FSCS* 1.7).

[c] See p. 293.

[d] This is usually an indication of land hunger. Duyvendak (3), p. 41 gives details.

[e] Perhaps because most of the Chhu sites excavated so far have been royal tombs; K. C. Chang (1), p. 430.

[f] Their comparative rarity is probably due to old and worn implements being recast.

[g] See Leser (1), figs. 92 and 96.

[h] Leser (1), fig. 161; Marinov (1); Balassa (1).

[i] Hansen (1).

1 冠

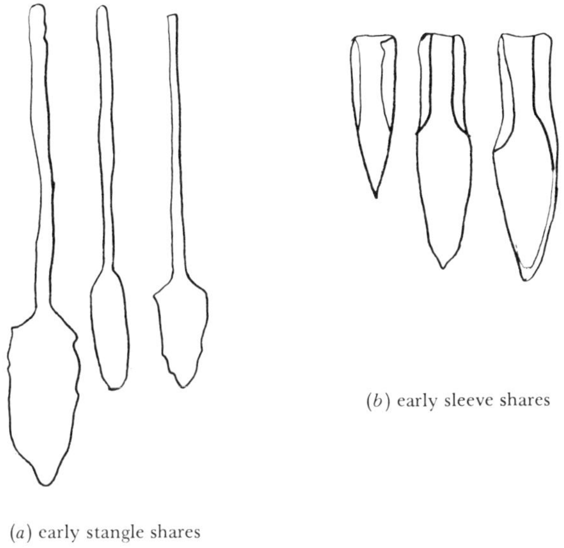


Fig. 53. Early European iron shares: (*a*) stangle shares; (*b*) sleeve shares; after Balassa (1), figs. 1 & 4.

Southeast Asia.[a] That the cap-share has only survived in a few isolated instances[b] (Fig. 54) is presumably due to the improvement in iron supply and casting techniques that enabled the *kuan* share to be developed into a more complete, slipper-like share, the *hua*[1], during the course of the Han dynasty. Indeed it seems probable that the *kuan* itself is a refinement of a simple metal strip bound round the front edge of the plough-slade, and it has been suggested that some early iron objects identified as spade-tips may in fact be small plough-shares.[c] Even the early iron *kuan* show a considerable refinement of design, with a central ridge terminating in a sharp point to cut the soil efficiently, and wings sloping gently up towards the centre so as to throw the soil off the plough and reduce friction. The *kuan* are usually completely flat underneath, which leads us to the conclusion that they were designed for a plough with a flat slade rather than a simple bow-ard. Since the type of plough known as 'sole-ard' (Fig. 39.2)[d] is extremely uncommon in the Chinese culture-area, these *kuan* must have been designed for the other type of flat-sladed plough, namely the frame-plough (Fig. 39.3) so common in China by Han times.

[a] Leser (1), fig. 196 shows a sleeve-share on a modern ard from the Rhône; stangle shares from all over Europe and Asia can be seen in Leser (1) and Haudricourt & Delamarre (1), while Leser (3) traces their path from the Mediterranean to Southeast Asia.

[b] Modern examples are to be found in Shantung (Amano (*4*), p. 800), Kansu (Anon. (*502*), p. 4), Yokohama in Japan (Leser (1), fig. 254), and also, surprisingly, in the Baltic states (Leser (1), figs. 79, 80).

[c] Anon. (*506*), p. 23; in Han times a number of regional terms for 'spade' and 'share' were interchangeable, as can be seen in the *Fang Yen*.

[d] 'Araire dentale', Haudricourt & Delamarre (1), p. 78.

1 鏵

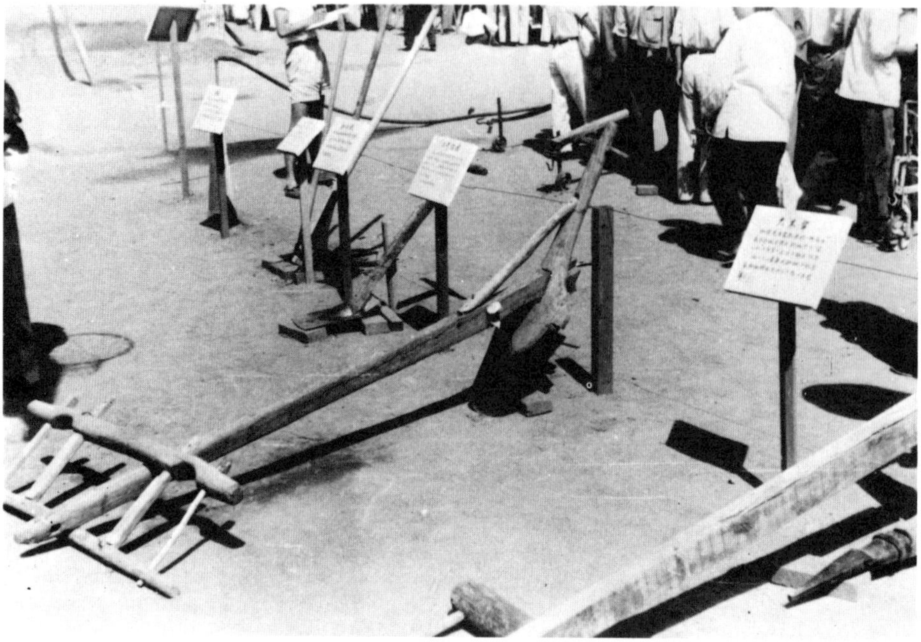

Fig. 54. Modern ploughs from Kansu with cap shares; JN orig. photos.

This is an important point, because it could very well provide an explanation for the appearance of a new term, *li*[1], at the same period as *kuan* first came into use. Let us here try to summarise the early stages of development of the plough in China. Throughout Chinese history, with one or two very rare exceptions, all Chinese ploughs have been either of the type known as 'bow-ard' or else a form of frame-plough. The bow-ards (which are sometimes not true ards as they are asymmetrical and turn the sod) are nowadays usually confined to minority tribes and/or to mountainous regions,[a] which indicates that they are an ancient type that has gradually been ousted by a more advanced form of plough better suited to lowland cultivation. The frame-plough has been common in all the chief farming areas of China at least since the Han dynasty and perhaps earlier. Now taking into account the various pieces of archaeological and literary evidence, we have suggested that the Chinese first started to use an ox-plough during the Lungshanoid period, perhaps as early as −3000. In the light loess lands of North China the early ploughs were made entirely of wood, but in areas where wet fields were cultivated (the rice fields of the Yangtze Delta) or where grass-land had to be dealt with (Mongolia), small stone shares were occasionally pinned to the point of the plough. On the evidence of surviving plough types it seems probable that the earliest Chinese ploughs were bow-shaped; indeed this very simple type seems to have preceded more complex forms throughout the Old World. The rock drawings of Bohuslän in Sweden[b] and Fontanalba in France[c] all depict bow-ards with straight beams drawn by a pair of oxen. The earliest Egyptian and Babylonian ards were also bow-shaped, although they were double-handled, the beam being wedged between the two handles.[d] In West Asia and the Mediterranean area the bow-ard seems to have been superseded quite early by the sole-ard. The earliest examples are found in Mesopotamia, and sole-ards are the earliest plough-type found so far in Greece and Etruria.[e] In Northern Europe the bow-ard seems to have maintained its popularity until Roman times,[f] but then was gradually ousted by the heavy square-framed plough with coulter, asymmetric share and mould-board,[g] a rather similar development to that which took place in China.

It is possible that the Chinese bow-ard evolved independently of outside influence, but it seems more likely that it was introduced from the West, where the earliest evidence for ploughs so far has been found, perhaps accompanying such

[a] E.g. Yunnan (Hu Hou-Hsüan (*3*), p. 80*a*), Kansu and Shensi (Anon. (*502*), pp. 3–8); Kwangsi (*ibid.* pp. 52–3).
[b] Leser (1), pl. 7.
[c] *Ibid.* pl. 12.
[d] *Ibid.* figs. 102, 103, 106, 107.
[e] *Ibid.* figs. 104, 93–8; Haudricourt & Delamarre (1), pls. 10–13.
[f] Hansen (1) describes a −350 bow-ard from Døstrup in Denmark; Leser (1), figs. 25, 26 shows late Roman bow-ards from Cologne.
[g] Pliny, 18.172; K. D. White (1), pp. 141 ff.; F. G. Payne (1), p. 77.

[1] 犂, 犁 or 黎

crops as wheat, barley and hemp which were probably introduced to China from West Asia in the late neolithic. It is impossible to say what this early bow-ard was called, for undoubtedly it had different names in different regions,[a] but some version of *lei*[1], *ssu*[2], or *lei ssu*[3] was certainly among them.

A light bow-ard would be drawn by two beasts at that period, for the combination of curved shaft and whipple-tree that permits the harnessing of a single animal did not appear in China till Han times or later (see p. 180 below). Now oxen are expensive animals to keep, especially over the winter. A small peasant might be able to maintain one animal, but it is unlikely that any but the well-to-do could afford to keep more. In medieval Europe many peasants could not even afford one beast,[b] and so it was common for peasants to pool their resources to form plough-teams.[c] On the medieval manors of Europe the plough-team sometimes consisted of as many as twelve or fourteen oxen.[d] In early China two animals were the norm, and herein lies, we feel, the explanation of the term *ou keng*[4], 'paired tillage', that is found in the *Odes* and other Chou and Han texts.

Most interpretations of the term *ou*[5] have been based on two texts: first, a passage from the Craftsmen section of the *Chou Li:*[e] 'The smith makes the irrigation implements; the *ssu*[6] is five inches wide, and two *ssu* make up a pair (*ou*)'; and secondly, a reference in the *Han Shu* to Chao Kuo[7], who popularised a new system of ploughing in the early −1st century:[f] 'He used a "paired plough" (*ou li*[8]), with two oxen and three men.' The first gloss of *ou keng* was given by the great Han exegesist Cheng Hsüan[9], and the explanation is still debated today.[g] Interpretations include: two men standing facing each other digging a trench with two spades;[h] two men using one spade, one digging while the other pulls the spade with a rope;[i] one man driving and one pulling a simple plough;[j] one man ploughing while another covers the seed with a rake or harrow;[k] and one man driving a plough while one leads the oxen and another presses down the plough-beam.[l] Various texts refer not only to tilling together (*ou keng*) but also to

[a] The *Fang Yen* gives us a fair idea of how many local terms were in use for such basic tools as the hoe and spade even in Han times.

[b] On some manors of the Bishop of Winchester in the +13th century half the tenants had no working stock at all (Postan (2), p. 555).

[c] Baker & Butlin (2), p. 635; see also H. G. Richardson (1).

[d] Though four was the most common numbers; H. G. Richardson (1); G. E. Fussell (1), p. 68.

[e] 12/18*b*.

[f] *C/HS* 24A/16*b*.

[g] For a detailed list of the different interpretations and their supporters see Wang Ning-Sheng (*1*), p. 74.

[h] Yang Khuan (*11*), pp. 9, 41.

[i] Sun Chhang-Hsü (*1*).

[j] Amano (*4*), p. 721.

[k] Based on a passage in the *Lun Yü*, *Hui Tzu*; Wan Kuo-Ting (*5*).

[l] Sung Chao-Lin (*1*).

1 耒　2 耜　3 耒耜　4 耦耕　5 耦
6 耜　7 趙過　8 耦犂　9 鄭玄

weeding together (*ou yün*[1]) or to tilling and raking together (*ou keng yu*[2]); *ou* is also used in conjunction with *yung*[3], a term meaning exchange labour, in the *Tso Chuan* and other pre-Chhin texts.[a] These many examples make it plain that, as Wang Ning-Sheng suggests, *ou* refers to partners or groups who regularly work in the fields together; such practices were common until very recently amongst the minority peoples of Yunnan, where all agricultural tasks were performed in pairs.[b] Now ploughing is a task which readily lends itself to the formation of cooperative pairs or groups. If two peasants each own an ox, together they can make up a team; furthermore while one man drives the plough, the other can lead the oxen. This was very common in the West from earliest times on; it is illustrated in the Fontanalba cave-drawings[c] and in Egyptian murals,[d] and Payne tells us:[e] 'In the Celtic countries, in order to ensure a slow, steady pace and cooperation with the ploughman, the driver walked backwards in front of the team ... He kept his team moving steadily by singing to them.'

Such ploughing partnerships survived in Wales and Scotland until the 18th century.[f] One can easily imagine that the team might consist of more than two partners, and this would have several advantages. First, if all the partners owned oxen it would allow the beasts to be used in turn, thus easing their burden. Secondly, it would allow a man with no animals to plough his field, in exchange presumably for helping his partners with other tasks such as weeding or reaping, or perhaps providing grazing and fodder for the animals. Thirdly, it would mean that extra hands were available to man the plough: it is often necessary to weigh down the plough by leaning on it or even riding on it, so as to deepen the furrow and steady the plough;[g] alternatively, extra hands would be available to broadcast the seed behind the plough or to rake soil over the seed. The partnership could be prolonged to include other tasks, such as weeding or digging ditches, not only for convenience but also because communal work is more agreeable.

This, then, seems to be the most plausible explanation for the term *ou keng*: a ploughing partnership that was carried over to other agricultural tasks. By Han times the term seems to be dying out except as a conscious archaism, perhaps because the growth of large estates had brought about a re-organisation of labour.

The bow-ard used in early China has the advantages of being light and extremely manoeuvrable, but it makes a very shallow furrow and is not efficient on

[a] Wang Ning-Sheng (*1*), p. 75.
[b] *Ibid.* p. 76; Sung Chao-Lin (*1*), p. 6. In Japan, Thailand, Malaya and Indonesia cooperative groups were intimately connected with wet-rice cultivation; they were generally larger and confined to the tasks of transplanting and harvesting; see J. F. Embree (1), M. Moerman (1), Bray & Robertson (1), etc.
[c] Leser (1), pl. 12.
[d] *Ibid.* fig. 112.
[e] F. G. Payne (1).
[f] Baker & Butlin (2).
[g] Payne (1), p. 80; Sung Chao-Lin (*1*).

[1] 耦耘 [2] 耦耕耰 [3] 庸

heavy soils. It is normally used several times in criss-cross fashion to pulverise the soil thoroughly. We would suggest that the rapid expansion of the Chinese population and economy that began at the end of the Spring and Autumn period was facilitated by the development of a square-framed plough with an iron share, heavier and more efficient than the bow-ard and permitting the expansion of the cultivated area into heavy and waterlogged soils that previously had not been farmed. This expanded agricultural potential would account for the preoccupation of political economists of the period with maintaining a sensible rate of agricultural growth where manpower and arable area were properly balanced.[a] It would also account for the passages in the *Lü Shih Chhun Chiu* which refer to a system of ridges and furrows (*mu chhüan*[1]) which, it seems to us, could only be produced with a heavy, metal-shared, square-framed plough:[b] 'These are the ridges made by a six-foot *ssu*; its width of eight inches is what makes the furrows (*shih i liu chhih chih ssu so i chheng mu yeh; chhi po pa tshun so i chheng chhüan yeh*[2]).' Now it has been argued[c] that this refers to the use of a spade-like implement (similar to the Irish loy or ridging-spade) to make raised beds, but there is no reason in that case why the width of the ditch should correspond exactly to the width of the spade-blade: a spade can dig a trench of any size. On the other hand if Lü were talking about a frame-plough, the width of the slade would exactly determine the breadth of the furrow (unless a mould-board or similar device were attached). The measurements given by Lü, a length of six (Chou) feet (approximately 120 cm) and a width of eight (Chou) inches (approximately 16 cm.) correspond very closely to those given by Lu Kuei-Meng for the slade of his frame-plough in +880, namely 125 cm. by 12.5 cm.,[d] as well as to those of 20th-century Chinese frame-ploughs.[e] A frame-plough with a broad convex share would make exactly the ploughing patterns described in the *Lü Shih Chhun Chhiu*, namely wide ridges separated by deep narrow furrows.[f]

The agricultural chapters of the *Lü Shih Chhun Chhiu* were probably taken from an earlier work, the *Hou Chi Shu*[3], and written during the −4th century.[g] They repeatedly stress the advantages of ridging and furrowing (to control soil moisture) and of planting cereals in properly spaced rows, and deplore the fact that so many farmers are ignorant of this system. This indicates that by the −4th century the frame-plough was just common enough to have caught the attention of the literati, and that the more productive ridge-and-furrow method of cultivation

[a] E.g. *Shang Chün Shu* II §6; tr. Duyvendak (3), p. 214.
[b] *LSCC*, *Jen Ti*; Hsia Wei-Ying (*3*), p. 40.
[c] *Ibid.* pp. 41–4, 133–5.
[d] *Lei Ssu Ching*, p. 1; these figures are based on the dynastic conversion tables in Wu Chheng-Lo (*2*).
[e] E.g. Hommel (1), p. 40; Anon. (*502*), pp. 8, 11, 55, 56.
[f] *LSCC Pien Ti*, Hsia Wei-Ying (*3*), p. 80.
[g] *Ibid.* p. 2.

[1] 畝畎 or 甽 [2] 是以六尺之耜所以成畝也；其博八寸所以成甽也
[3] 后稷書

was then officially promoted at the expense of the shallow cultivation produced by the bow-ard.[a] Although the *Jen Ti* chapter of the *Lü Shih Chhun Chhiu* speaks of a *ssu*[1], it may well be that among farmers the frame-plough was more generally known as *li*[2], and that this term gradually spread until by Han times it was, like the square-frame plough to which it was applied, ubiquitous.[b]

*Han ploughs*

It may seem to many people that we have been unwarrantedly confident in ascribing such a full development to the plough in early China. However, a glance at the level of perfection that the Chinese plough had already reached by Han times—a perfection and sophistication that had been achieved nowhere else in the world—lends considerable weight to our contentions.

Chinese ploughs were almost all square-framed (Fig. 55*a*, *b*, *c*, *e*) even in the Western Han; bow-shaped ploughs are still found in some parts of China, for example in Kiangsu,[c] but even these are not simple bow-ards but a more sophisticated type with a heavy share, a mouldboard and adjustable strut (Fig. 55*f*). The adjustable strut is a fundamental development in the Chinese plough, for by altering the distance between the slade and beam it permits a precise regulation of ploughing depth. The depth of furrow made by a single plough can thus be adjusted to suit different crops, soil types, seasons or weather conditions; such flexibility was taken for granted by the agricultural authors:[d] 'Autumn ploughing should be deep, but spring and summer ploughing should be shallow.' 'When ploughing in the autumn ... the first ploughing (*chhu keng*[3]) must be deep, the second ploughing (*chuan ti*[4]) must be shallow.'

In Roman Europe the only means of adjusting the depth of furrow was for the ploughman to alter the pressure on the stilt.[e] In Northern and Central Europe, where the heavy wheeled frame-plough may already have been in use in late Roman times,[f] it may have been possible to alter the depth of the furrow by adjusting the height at which the beam was attached to the wheels, though there is no firm evidence of this till medieval times or later,[g] but even in medieval times the regulation of ploughing depth depended chiefly on the skill of the ploughman, or even on the weight of an extra man riding on the ploughbeam.[h] Such devices

[a] In this context a geographical correlation of finds of metal shares and of historical references to the replacement of the well-field (*ching thien*[5]) system of square fields by the *mu*[6] system of strip fields (see, for example, Yang Khuan (*11*), p. 112) would be very useful.

[b] *Li* may also have been used originally to designate a unit of ploughland, similarly to the oxhide or *charruée* of medieval Europe; see, for example, the passage in *Kuan Tzu* (1.20.5): 'In three days a grown man can manage two *li*, and a five foot lad a single *li*.'

[c] Chang Chen-Hsin (*1*), fig. 4.

[d] *CMYS* 1.3.3. and 1.3.5; trs. auct.

[e] K. D. White (2), p. 176.

[f] Pliny 18.172; K. D. White (1), p. 141; F. G. Payne (1), p. 77.

[g] Haudricourt & Delamarre (1), pp. 340, 381; Gille (15).

[h] F. G. Payne (1), p. 80.

1 耜 2 犂, 犁 or 棃 3 初耕 4 轉地 5 井田 6 畝

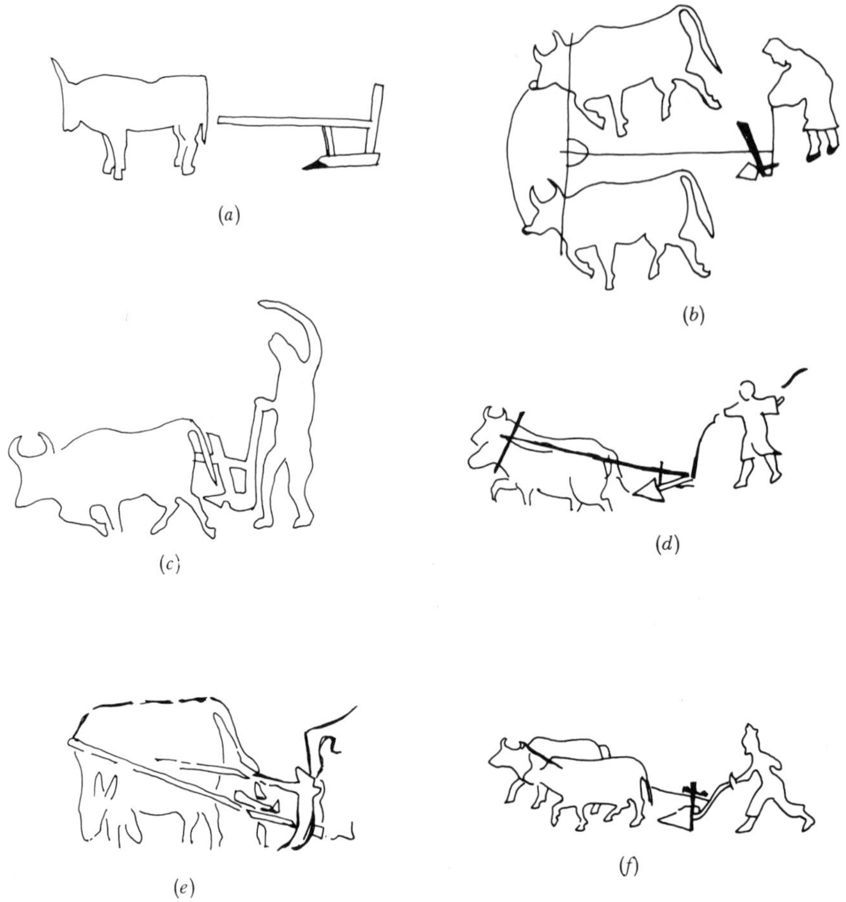


Fig. 55. Han ploughs: (*a*) Late Han wooden model from Wu-wei, Kansu; (*b*) E. Han stone relief from Mi-chih, Shensi; (*c*) E. Han stone relief from Wei-te, Shensi; (*d*) Wang-Mang mural from Phing-lu, Shansi; (*e*) E. Han stone relief from Theng-hsien, Shantung; (*f*) E. Han stone relief from Sui-ning, Kansu.

were resorted to in the remoter parts of China even during this century,[a] but almost all the paintings and bas-reliefs from the Wang-Mang period on show ploughs with adjustable struts (Fig. 55*e*, *f*). The adjustment mechanism[b] is exactly described by Lu Kuei-Meng over six centuries later (see p. 182 below), namely a stepped wedge sitting on top of the beam, inserted through a slot in the strut just where it passes through the beam, and this device was still in use in the present century.[c]

As for ploughshares and other metal parts, the Han saw an enormous improvement over the Warring States period not only in number but also in quality,

[a] E.g. among the minority peoples of Yunnan; Chang Chen-Hsin (*1*), p. 57; Wang Ning-Sheng (*1*), p. 76; Sung Chao-Lin (*1*), p. 5.

[b] See Chang Chen-Hsin (*1*), figs. 2 and 3 from Shantung and Kiangsu.

[c] Hommel (1), p. 40.

variety and distribution. Iron became much more freely available; small private foundries sprang up all over the empire in the first decades of the Han, and by −100 large state foundries had been established in most provinces and commanderies. Iron was in universal use for anything from cooking-pots to swords in an area that stretched from Liaoning to Yunnan and from Shantung to Sinkiang,[a] and the quality of the goods had improved greatly: the cast iron of the Han had a lower sulphur content and a higher carbon content than Warring States iron, which made it less brittle and easier to sharpen.[b]

Han ploughshares show an enormous range both in type and size (Fig. 56*a*). They include slipper-shaped shares, squared-off share-caps, flanged shares, stangle shares, foliate bronze shares, and small, pointed or waisted shares; in weight they range from 0·3 kg. for a simple cap-share[c] to over 12·5 kg. for some of the largest 'tongue-shaped' shares,[d] and in size from a few centimetres in length to giant shares nearly 40 cm. wide and long.[e] Despite the great variety in Han share types, by far the most common was still the V-shaped *kuan*[1]. During the Warring States this was apparently fixed directly on to the end of the wooden slade, but in the Han it appears that it was usual to fit a heavy iron 'tongue' on to a wooden projection at the end of the slade (Lu Kuei-Meng refers to this as the 'turtle flesh' (*pieh jou*[2]) (see p. 182 below). The *kuan* was pushed tightly over the end of this tongue (Fig. 56*b*), which appears to have been known as *feng*[3].[f] *Feng* were usually flat underneath and convex above, with a marked ridge running down the centre which fitted a corresponding ridge in the centre of the *kuan*; indeed the fit was so tight that sometimes archaeologists have had great difficulty in separating them.[g] The advantage of a metal tongue compared to a wooden one is that, despite the extra weight, friction is considerably reduced. The soil is neatly sheared by the point and flanges of the *kuan* and turns easily back over the smooth sloping sides of the *feng*, making a shallow furrow. The combination of *feng* and *kuan* was so successful that by the Eastern Han they were often cast in one piece.[h] The flanges could be hammered out or resharpened if they were damaged,[i] and the amount of metal required was less than if the two pieces were cast separately. This flanged share continued in use through Thang times and indeed right up to the present day[j] (Fig. 57), but has generally been superseded by lighter and less complex forms.

The inversion effect achieved by the combination of *feng* and *kuan* is greatly enhanced if the plough is equipped with a mould-board. It was previously thought that the mould-board, most commonly known as *pi*[4], *li erh*[5] or *ching*[6], was not

[a] Anon. (*506*), n. 7; Anon. (*510*).
[b] Anon. (*507*), p. 52 and (*509*), p. 71.
[c] Anon. (*509*), p. 69.
[d] Anon. (*510*), p. 22.
[e] *Ibid.*
[f] Hayashi (*4*), p. 269.
[g] Anon. (*510*), p. 23.
[h] Liu Hsien-Chou (*8*), fig. 19.
[i] Anon. (*510*), p. 24.
[j] Amano (*4*), p. 772; Anon. (*502*), p. 5.

[1] 冠 [2] 鼈肉 [3] 鋒 [4] 鐴 or 壁 [5] 犁耳 [6] 鏡

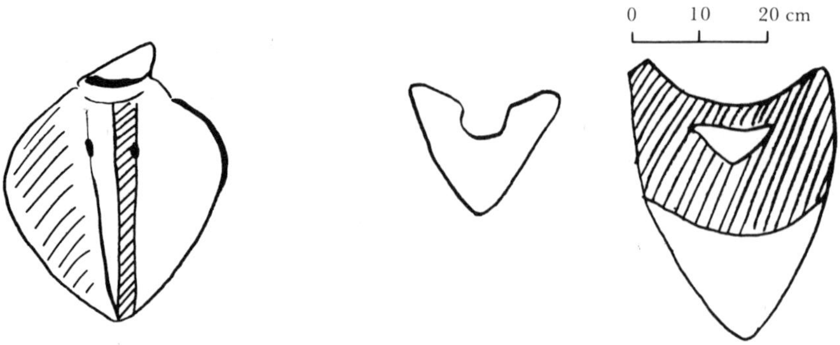


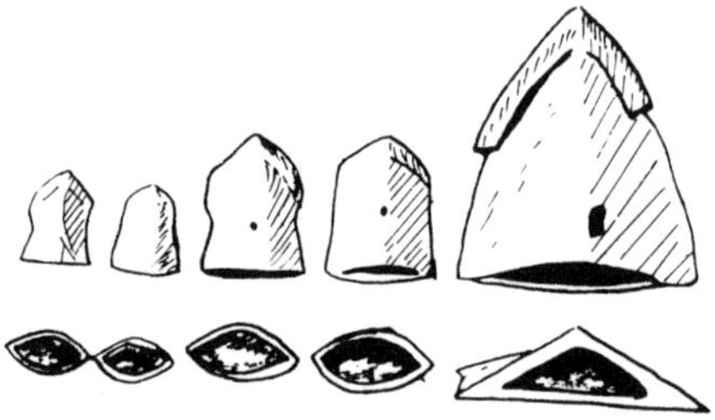

(*a*)

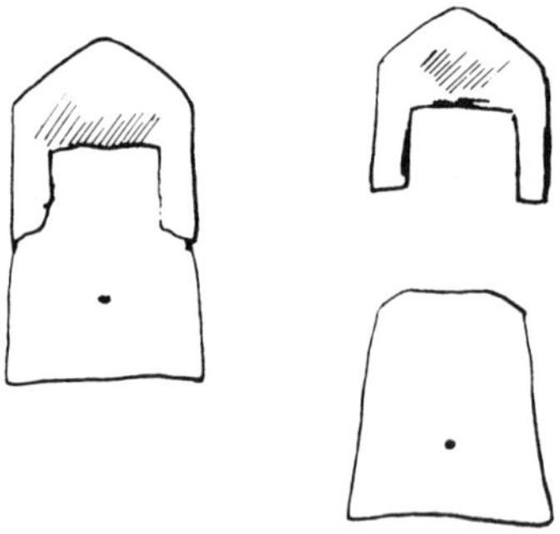

(*b*)

Fig. 56. Chinese ploughshares: (*a*) Han shares; after Hayashi (*4*), figs. 6–16, 6–17, Liu Hsien-Chou (*8*), figs. 18, 20; (*b*) *feng* and *kuan*, after Hayashi (*4*), 6–15; (*c*) Archaeological finds of shares: Thang iron and bronze shares after Liu Hsen-Chou (*8*), figs. 21, 22; Chin/Yuan iron share from Liaoning, after Amano (*4*), p. 781; (*d*) Ming shares; *WCNS* 13/10*a* and 13/11*a*.

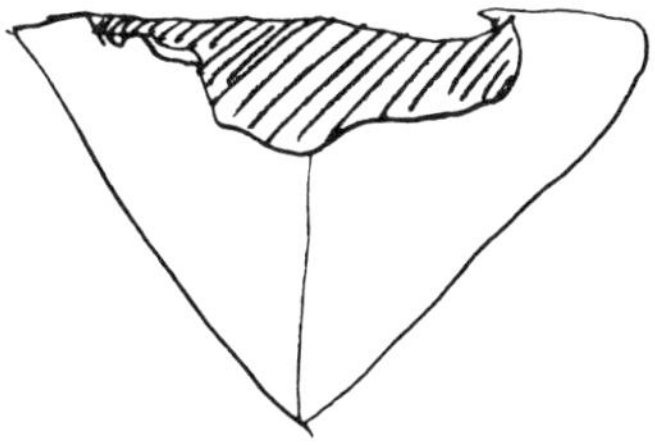

Thang share from Honan

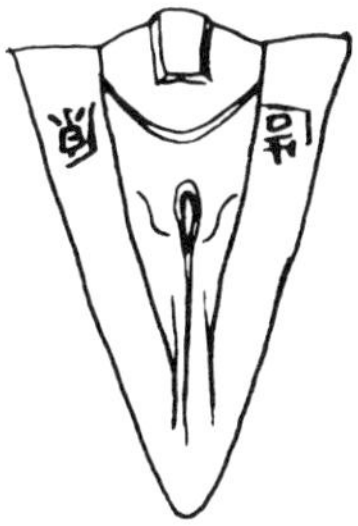

Thang bronze share

Chin/Yuan share
from Liaoning

(*c*)

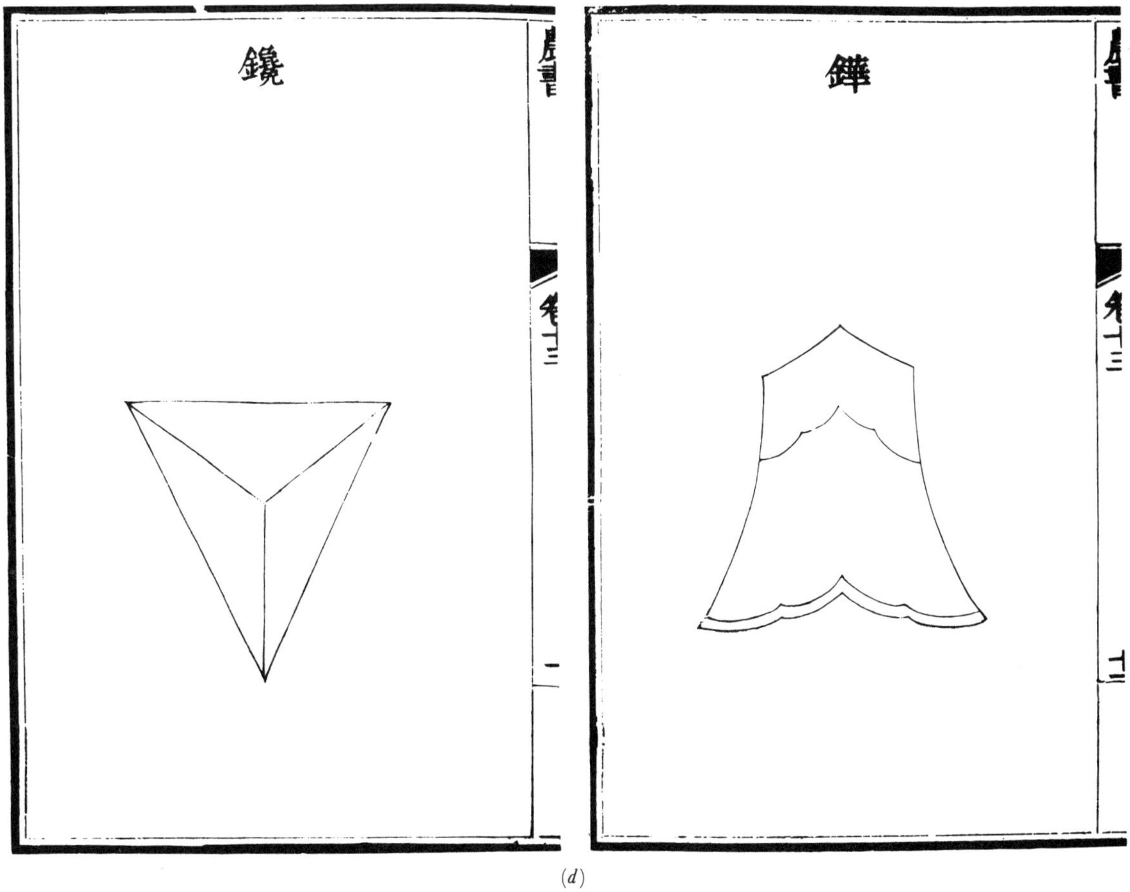


(*d*)

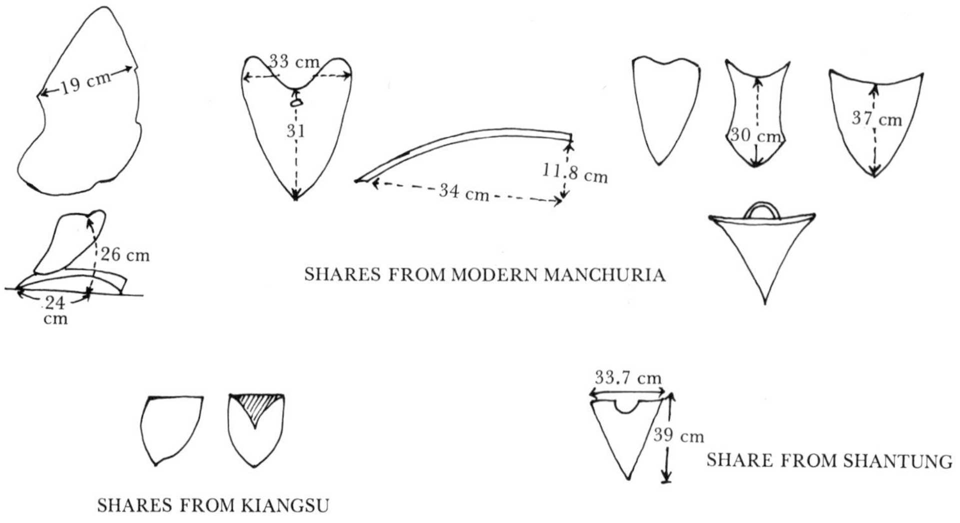


Fig. 57. Modern Chinese ploughshares; after Amano (*4*), p. 800.

known in China until the +4th century[a] (though some Han illustrations such as the engraving from Kiangsu (Fig. 55*f*) seem to show a mould-board fairly clearly). But recent archaeological discoveries have shown that iron mould-boards were in use in China even during the Western Han.[b] Moreover, Han ploughmen had the choice between four different types of mould-board: 'saddle-shaped', 'leaf-shaped', 'gourd-shaped', and square with one corner missing (Fig. 58*a*). The first type, sometimes cast in two separate pieces,[c] threw up a ridge on either side, while the last three types turned the furrow to one side only. The saddle-shaped mould-boards fitted comfortably into a nick in the central ridge of the *feng*, while the upper part rested, presumably, against the front edge of the strut. The asymmetrical mouldboards required more careful anchoring and were tied to the slade and the strut by cords passed through three or four iron lugs on their obverse. The most remarkable thing about Han mould-boards is that most of them are dished and so designed that they marry smoothly to the surface of the share.

A curved metal mould-board is an extremely efficient device, for it reduces friction considerably.[d] In 1784 James Small, the Scottish pioneer of scientific plough design, enunciated the following principles soon after the introduction of the curved mould-board into Europe:[e] 'The back of the sock [share] and mould-

[a] See Amano (*4*), p. 758 for references and details.

[b] Anon. (*510*); Hayashi (*4*), pp. 268 ff.

[c] There is a reference in the *Shuo Wen* giving a specific word, *fei*[1], for a plough with two mould-boards; see Hayashi (*4*), p. 270.

[d] Leser (1), pp. 454 ff.

[e] James Small (1), quoted in G. E. Fussell (2), p. 49.

1 犇

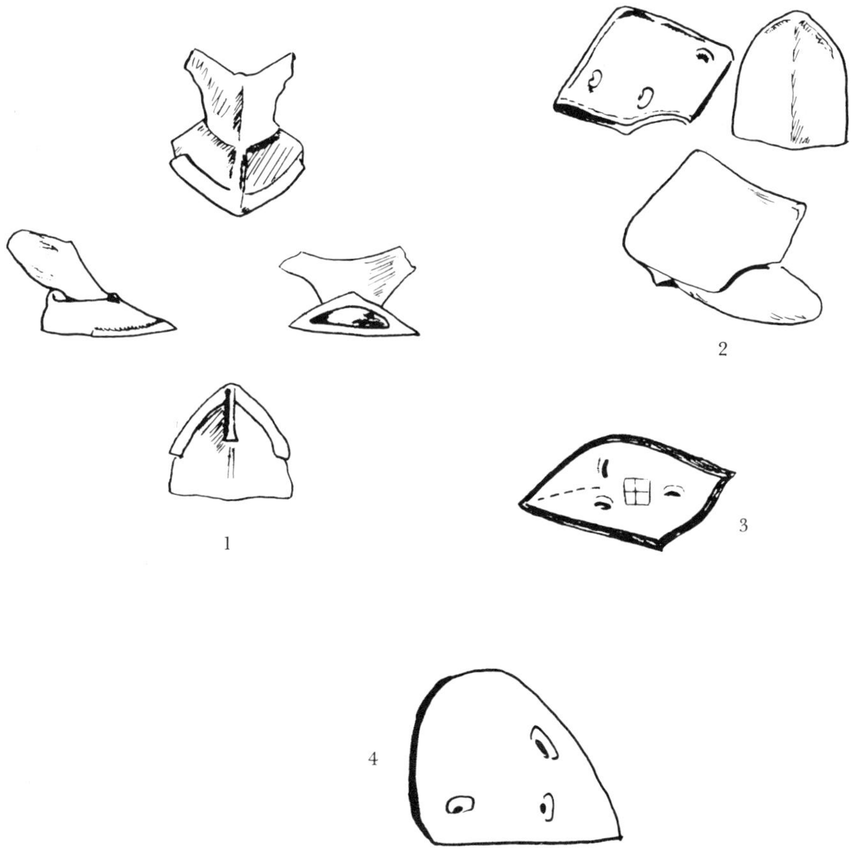


Fig. 58 (*a*)

board shall make one continued fair surface without any interruption or sudden change. The twist, therefor [sic], must begin from nothing at the point of the sock, and the sock and the mouldboard must be formed by the very same rule.' The profile of the saddle-shaped mould-board in fig. 58*a* shows that Small's principles were realised in at least some of the Han mould-boards. The soil would slide over the plough without sticking, which is presumably why we find no references in the Chinese literature to any form of the ploughstaff so commonly used in the West for cleaning the mud off the share.[a] The extra weight of the iron share and mould-board would be amply compensated for by the reduction in friction, for as Arthur Young said in 1797:[b] 'It appears that the weight of the plough is of little consequence very contrary to common ideas ... The weight of the plough is the least part of the horse's labour; the great object is the resistance

[a] In Rome it was called *rallum* (Pliny 18.179; K. D. White (1), p. 140); in medieval England the plough-staff or pattle (Fenton (3), p. 186).

[b] Judging a competition in plough design; quoted in G. E. Fussell (2), p. 54.

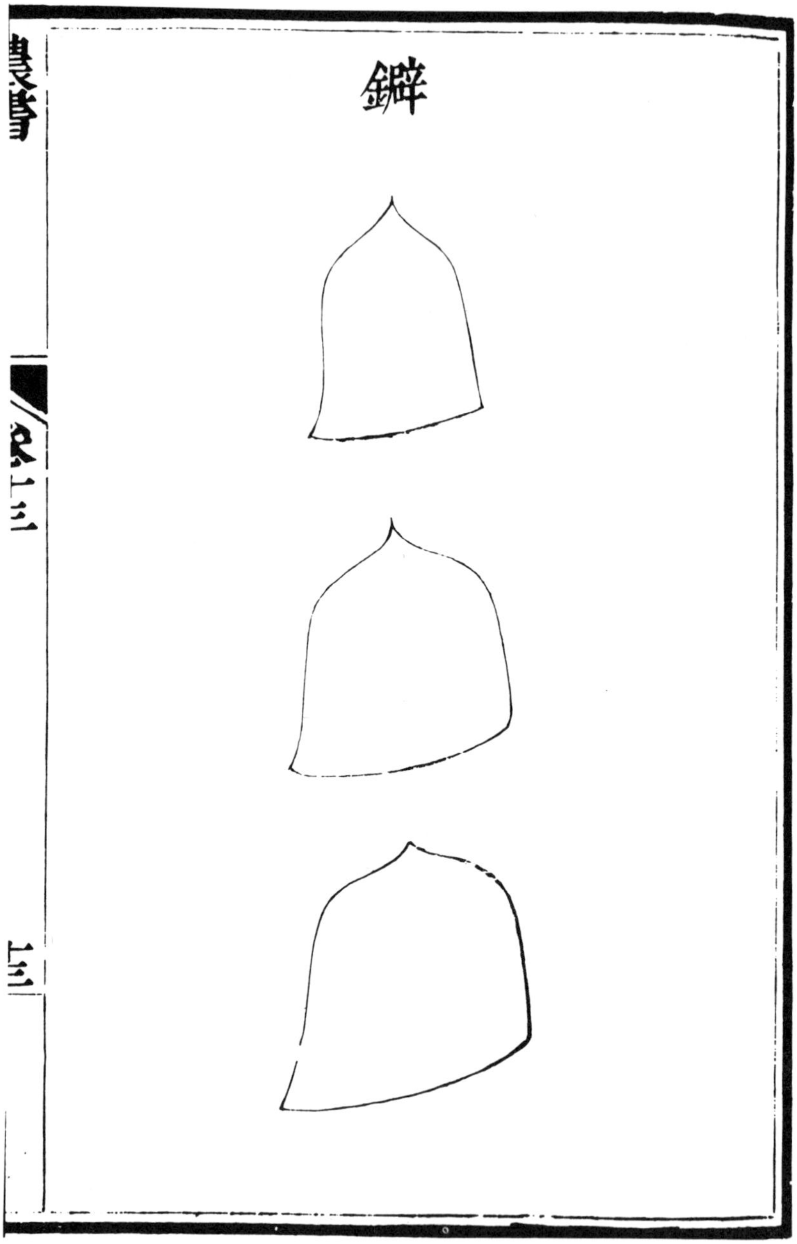


Fig. 58 (*b*)

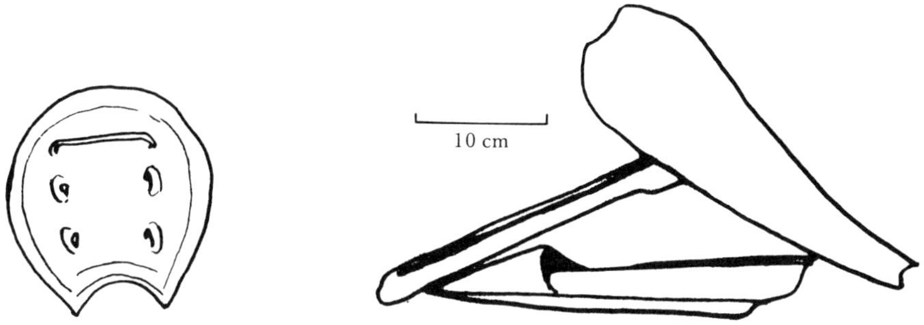


1) Sung mouldboard
from Szechwan

2) Modern share and mouldboard
from Chekiang

Fig. 58 (*c*)

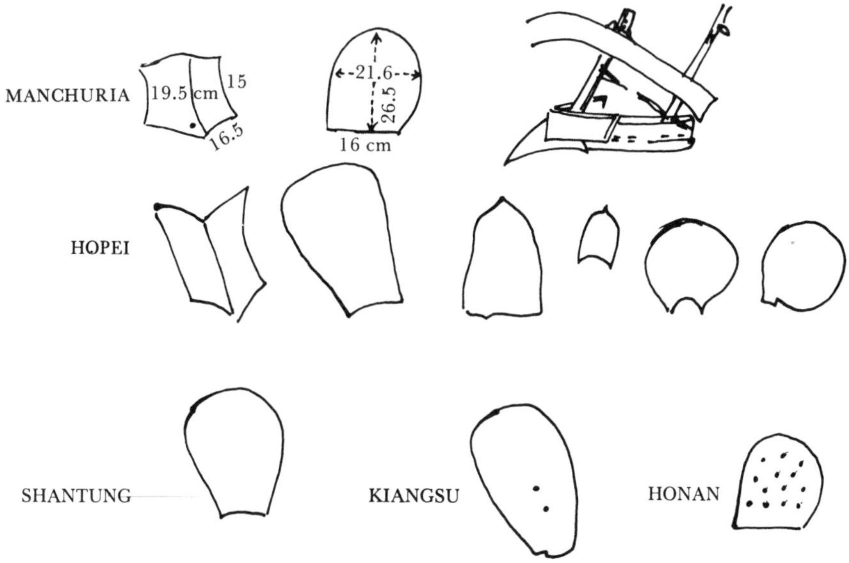


Fig. 58 (*d*)

Fig. 58. Chinese mould-boards: (*a*) Han, symmetrical or 'saddle-shaped'; square with one corner missing, leaf-shaped and gourd-shaped; note the lugs on the asymmetrical forms; after Anon. (*510*) and Liu-Hsien-Chou (*8*), fig. 28; (*b*) Ming; *WCNS* 13/3*a*; (*c*) secured to the frame by metal loops: Sung; after Liu (*8*), fig. 30; modern, from Chekiang, after Hommel (1), fig. 62; (*d*) modern types; after Amano (*4*), p. 800.

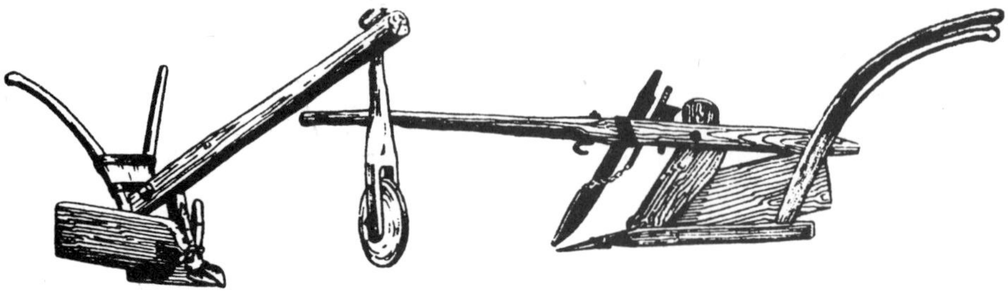

Fig. 59. Straight wooden European mould-board; Leser (1), fig. 148.

met with in the cohesion of the earth; lightness does nothing to overcome this; it is effected by just proportions only.' In fact the greater the weight of the plough, the easier it was to drive a straight and even furrow. Young goes on to say: 'It also appears very decidedly that the share should be nearly, if not quite, as broad in the fin, as the plough is wide in the heel,[a] in order that all the furrow may be cut, and not turned up by force.'

Chinese ploughs from Han times on fulfil all these conditions of efficiency nicely, which is presumably why the standard Han plough team consisted of two animals only, and later teams usually of a single animal, rather than the four, six or eight draught animals common in Europe before the introduction of the curved mould-board and other new principles of design in the +18th century. Though the mould-board plough first appeared in Europe in early medieval, if not in late Roman, times, pre-eighteenth century mould-boards were usually wooden and straight[b] (Fig. 59). The enormous labour involved in pulling such a clumsy construction necessitated large plough-teams, and this meant that large areas of land had to be reserved as pasture. In China, where much less animal power was required, it was not necessary to maintain the mixed arable-pasture economy typical of Europe: fallows could be reduced and the arable area expanded, and a considerably larger population could be supported than on the same amount of land in Europe.

If the Chinese mould-board had already reached an advanced stage of development in Han times, it is reasonable to assume that its origins go back to the Warring States, but no evidence from that period has survived. This could be because the earliest mould-boards were made of wood, not metal; and here we must mention an interesting suggestion made by the Sung scholar Li Hsin-I,[c] namely that the mysterious *tse*[1], mentioned in the cartwright chapter of the *Khao*

[a] Many European ploughs, unlike their Chinese contemporaries, had slades that widened or even forked at the back; this was supposed to increase the breadth of the furrow.

[b] Haudricourt & Delamarre (1), ch. 16. There is some evidence that curved wooden mould-boards were being developed as early as the +14th century in some parts of Europe (F. Sigaut, pers. comm.), but these would still have been considerably less efficient than metal ones.

[c] *Khao Kung Chi Chieh* 2/59*b*; his theory is approved by Liu Hsien-Chou (*8*), pp. 15 ff.

[1] 庇

*Kung Chi* in the passage on the construction of the *lei*[1], was actually a wooden mould-board (Fig. 42):[a] 'The *tse* is 1·1 (Chou) feet long ... For hard soils a straight *tse* is required, and for soft soil a curved *tse*; if it is bent at the same angle as a musical stone then it is good for medium soils.' Lin points out correctly that for opening hard ground the mould-board should be nearly straight and form an acute angle with the length of the plough, while in soft soils a widely curved mould-board turns a better furrow. This is an interesting hypothesis, but the text of the *Khao Kung Chi* is obviously corrupt and incomplete, and it would not do to draw any positive conclusions from it.

It is clear, then, that by the Han, Chinese ploughs were quite highly developed: there were ploughs with sharply pointed shares for opening up new land, heavy ploughs with giant shares for trenching, and, most commonly, broad-shared ploughs with mould-boards for tilling arable fields.[b] The large clods thrown up by turnploughs required breaking up, and thus we see the drag-harrow developing in China towards the end of the Han (see p. 223); the ridge-and-furrow system is better suited to sowing in rows than to simple broadcasting, and the seed-drill also appears in China during the Han (see p. 262). Different regions showed their own course of development. In Kansu and Shensi, where soils were light and winds fierce, the use of the turnplough was neither necessary nor wise, and light, small-shared ards continued in use.[c] In some regions, for example parts of Vietnam and Tunhuang, the plough was supposed to have been first introduced by Chinese officials of the late Han or Chin.[d] But these were very remote parts of the Chinese world, and in most of Han China ploughs were common. Although a phrase from the *Shih Chi*,[e] '[South of the Yangtze] they till with fire and weed with water (*huo keng shui nou*[2]),' has often been interpreted as meaning that the Southern barbarians grew rice by primitive slash-and-burn methods, in fact this phrase certainly refers to wet-rice cultivation in permanent fields (for in swidden cultivation of rice water cannot stand in the 'fields' and so could not be used to control the weeds), and it sounds remarkably similar to methods still prevalent in South China and Southeast Asia today, where the rice stubble is burned just before ploughing to fertilise the soil, and several inches of water standing in the bunded fields prevent weeds from growing amongst the young rice shoots. Considering the high level of agricultural technology reached by the Yangtze cultures in neolithic times it is not surprising to find that ploughs were just as developed in Chiang-nan as in the rest of China, as has been made clear by finds of large hoards of iron ploughshares in Kiangsi and elsewhere.[f]

a See p. 145 above.
b On trenching ploughs see Anon. (*10*), Hayashi (*4*), p. 271, Sung Chao-Lin (*1*), p. 8.
c Chang Chen-Hsin (*1*), figs. 1 and 7.
d *CMYS* 0.6; see also Wang Ning-Sheng (*2*).
e *SC* 30/15*a*; 129/12*a*.
f Anon. (*511*).

1 耒　　2 火耕水耨

*Post-Han developments*

We have just described the diversity and high degree of specialisation shown by Han ploughs. One thing they did have in common, however, was a long straight shaft joined to a wide wooden yoke, so that they were drawn by a team of two or more oxen. But shortages of draught animals seem to have occurred frequently during the Han so that sometimes the peasants had to resort to pulling the plough themselves (see p. 141); probably they used small ploughs equipped with very light shares.[a] It may have been these recurrent crises that stimulated the development of a single-ox plough, with a short curved beam ending in a whipple-tree, joined by traces to the yoke *o*[1], which was individually fitted and made fast by a throat-band *yang pan*[2]. Such a harness neither slipped nor chafed, and was a distinct improvement on the straight yokes of the Han (Fig. 60). The advantage of the curved beam and whipple-tree is that it can be used to harness almost any number of animals:[b]

> In the old days the whipple-tree *keng phan*[3] was slightly shorter, to harness just one or two oxen, and so it formed part of the plough itself. But now ploughing methods vary in every district and sometimes three or four oxen are used. In this case the whipple-tree is made of a straight timber which may be as much as 5 feet (i.e. 150 cm.) across; in the centre is fixed a hook or ring, and during ploughing this pivots at the point where it is fixed to the plough-head.

The first evidence for the single-ox plough is Chin at the earliest. A mural from Chia-yü-kuan[4] in Kansu, dated between +220 and +310, shows a ploughing scene where a single ox is harnessed by traces attached to a cross-beam near the base of the plough,[c] while a pottery grave model from Kwangtung, dated to about +300, shows both a single-ox plough and a single-ox harrow.[d] How rapidly this harness-type spread we do not know, but in +880 the *Lei Ssu Ching* describes the curved plough-beam and whipple-tree as a standard feature (see p. 182). In most of China from Thang times on ploughs were usually drawn by a single ox or buffalo, but in North and Northwest China the long beam and two-animal team are still to be found today, often drawn not by two oxen but by a donkey and a mule or even a camel.[e]

By +880 when the *Lei Ssu Ching*, or Classic of the Plough, was written the Chinese plough had acquired all the essential features that still characterise it

[a] Anon. (*510*), fig. 2.

[b] *WCNS* 12/22*a*; tr. auct.

[c] Anon. (*512*), fig. 52.

[d] Hsü Heng-Pin (*1*), fig. 1. Unfortunately the model, which we have seen, is not clear enough to distinguish further details. Hayashi (*4*), fig. 6–21 and Anon. (*512*), fig. 6, show a +1st century painting from Phing-lu[5] in Shansi depicting a seed-drill drawn by a single ox, but whether it is in traces or shafts is impossible to make out. Shafts were of course commonplace on Han chariots and carts; Vol. IV, pt 2, p. 308.

[e] The modern long-beam plough is called *huo tzu*[6]; see Chang Chen-Hsin (*1*). Amano (*4*), p. 738 shows a Sung example in a painting from Tunhuang.

1 軛 2 鞅板 3 耕槃 4 嘉峪關 5 平陸
6 耠子

Fig. 60. Chinese whipple-tree; *Keng Chih Thu*, Franke (11), pl. XIV.

today, and when 20th-century agronomists remark disparagingly that the Chinese plough has not changed for centuries they are unwittingly paying tribute to its fitness for survival.[a] The Sung was a period of extensive agricultural improvement and innovation, yet we read of no significant modifications to the plough in the agricultural treatises of the Sung and Yuan; indeed the standard description of the plough, quoted verbatim in every subsequent treatise and encyclopaedia, is Lu Kuei-Meng's *Lei Ssu Ching* of +880:[b]

[The plough] has eleven parts altogether, counting both wood and metal: The soil thrown up by the plough is called the ridge (*fa*[1]) or clods (*khuai*[2]), and the part which throws up the ridge is called the ploughshare (*chhan*[3]); the part which inverts the ridge is the mould-board (*pi*[4]). Weeds always spring up on the ridge unless their roots are cut by inverting the soil.

The share lies flat below and the mouldboard slopes back above; the share is sharp on top and the mouldboard curved at the base [to fit tightly against the share]. The part which carries the share is called the slade (*ti*[5]): the very front of the slade fits into[c] the

[a] F. Bray (8).

[b] *Lei Ssu Ching*, pp. 1–2; tr. auct., helped by discussions with Professor Thomas Thilo of the Academy of Sciences, East Berlin; cf. Thilo (3).

[c] Reading *kuan*[6] for *shih*[7]

1 墢 2 塊 3 鑱 4 鐴 5 底
6 貫 7 實

share, and carpenters call this part the 'turtle flesh' (*pieh jou*[1]).[a] The part adjoining the slade is called the 'share press' (*ya chhan*[2]). [The mouldboard] has two lugs on its back joined to either side of the share-press.[b] The part adjoining the share is called the *tshe o*[3], and it protects the mouldboard. All these parts are connected.

The part which comes down from the *tshe o* to the slade, into which it is morticed at right angles, is called the strut or 'arrow' (*chien*[4]). The part at the front of the plough which is curved like a carriage-shaft is called the beam (*yuan*[5]), and the part at the back rising up like a handle is called the stilt or 'rudder' (*shao*[6]). Above the beam projects an extension of the strut which can be tightened or loosened: along the top of the beam there is a groove corresponding to [a slit in] the strut, [into which fits] a piece that is cut into steps, high in front and lower behind, which can be pulled back or forward [as necessary]; this piece is called the wedge or 'adjustor' (*phing*[7]). When it is pushed forward the strut is loosened and so the plough bites deeper into the soil; when it is pulled back the strut is raised and the depth of ploughing is shallower. The strut is called 'arrow' because its height is adjusted like [an arrow in a cross-] bow; the wedge is called 'adjustor' because it is adjusted to exactly the right position. The piece that transfixes the top of the wedge is called the bolt (*chien*[8]); it holds beam and wedge together and without it the two pieces would spring apart and the strut would not stay in place.

The piece across the end of the beam is called the whipple-tree (*phan*[9]); it can pivot, and it is attached on either side by traces (*hsien*[10]) to the yoke (*o*[11]). The very back of the beam is called the 'mid-stilt' (*shao chung*[12]), which is where the plough is actually held.

Lu then gives the dimensions of the various parts, on the basis of which we can draw a reconstruction of the implement (Fig. 61). The only parts which are not entirely clear are the *yen chhan* and the *tshe o*. The *tshe o* seems to have been a wooden brace projecting forward from the strut which supported the mould-board,[c] while the *yen chhan*, to which the mould-board was fastened with cords passing through the lugs, could perhaps have been a wooden peg or pegs fastened in the slade, similar to the arrangement on some modern Chinese ploughs.[d] The measurements that Lu gives do not tally exactly with this interpretation, but Lu himself confesses that he was completely ignorant of how a plough was constructed until one day he asked a farmer to tell him what the different parts were called. 'I was as amazed as if I had stepped into the cot of the Heavenly Husbandman to receive instruction in farming', he says, so perhaps the details of his account were not entirely accurate.[e]

Lu was describing the plough used in the wet-rice fields of the Yangtze Delta. It had a square frame, a curved beam and a whipple-tree, and was constructed

[a] Lu is describing a plough with a flanged share.

[b] This must mean that the mould-board has four lugs altogether.

[c] See fig. 64d, the plough from Hopei.

[d] See Leser (1), fig. 243; Ctarikov (1), fig. 3; also the *WCNS* illustration, Fig. 62 (a).

[e] An alternative reconstruction of Lu's plough by Yen Wen-Ju[13] in the National History Museum is illustrated, not very clearly, in Liu Hsien-Chou (*8*), fig. 33.

1 鼈肉 2 壓鑱 3 策額 4 箭 5 轅
6 梢 7 評 8 建，鞬 9 槃 10 摼
11 軛 12 梢中 13 閻文儒

1. *shao* 梢—stilt
2. *shao chung* 梢中—mid-stilt
3. *ti* 底—sole or slade
4. *chien* 箭—'arrow' or strut
5. *ya chhan* 壓鑱—'press-share'
6. *tshe o* 策額—mouldboard brace
7. *pieh jou* 鼈肉—'turtle flesh'
8. *pi* 鐴—mouldboard
9. *chhan* 鑱—share
10. *phing* 評—'adjustor' or wedge
11. *chien* 建—bolt
12. *yuan* 轅—beam
13. *phan* 槃—whipple-tree

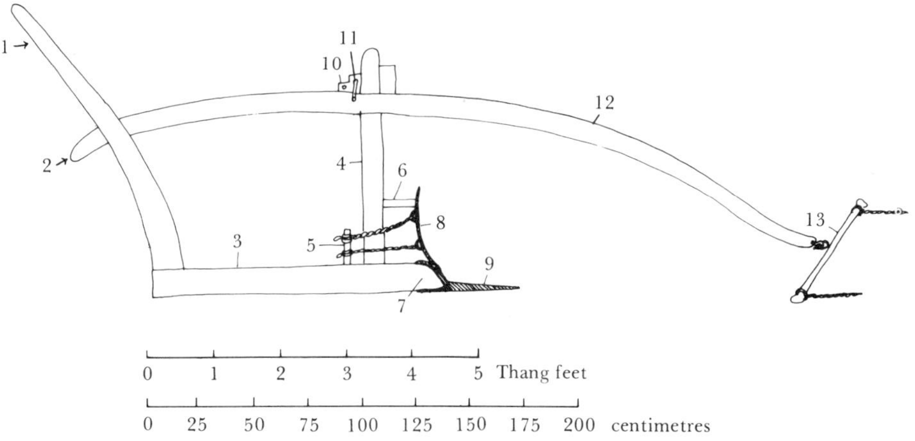


Fig. 61. Reconstruction of the plough described in the *Lei Ssu Ching*.

entirely of wood except for a sharply pointed triangular share (Lu gives its dimension as 1·4 feet long by 6 inches wide, that is approximately 45 cm. by 19 cm.) and an oval mould-board 1 foot (31 cm.) in diameter. As far as we can tell from early illustrations and modern examples, the only important post-Thang modification in the plough probably first appeared some time in the Sung or Yuan dynasties, for the earliest surviving pictures are in the Ming edition of Wang Chen's *Nung Shu*.[a] This is the development of a curved triangular frame, where the strut is prolonged and curves back to serve as a handle. The *Nung Shu* illustrations (Fig. 62) show the beam fixed to a rudimentary stilt, a form which is still found in parts of China today.[b] Sometimes the beam was fixed directly to the back of the slade, forming a very sharp curve, or even just to the front of the strut (Figs. 39·4*c*, 63*b*).[c] Amano[d] declares that this form is contingent upon the use of an iron beam, but though several modern Chinese examples do have metal beams, all those of Southeast Asia are made entirely of wood. Hommel[e] describes farmers training tree-branches specially for such purposes.

[a] Probably based on the original Yuan illustrations. See Amano (*4*), p. 777 for a critical account of the derivation and authenticity of the illustrations in the different editions of the *Nung Shu*.

[b] E.g. Hunan; Amano (*4*), p. 799.

[c] These forms are now widespread both in China and in Southeast Asia: Amano (*4*), pp. 798–9; Anon. (*502*), pp. 7, 9, 10, 58, 60, etc.; Raffles (1), p. 113, shows the 18th century Javanese form as do Mayer (1), Leser (1), fig. 269; *ibid.* fig. 285 shows the Siamese version.

[d] (*4*), p. 789.

[e] (1), p. 64; this practice was also known in Rome (Virgil, *Georgics* 1.169–70).

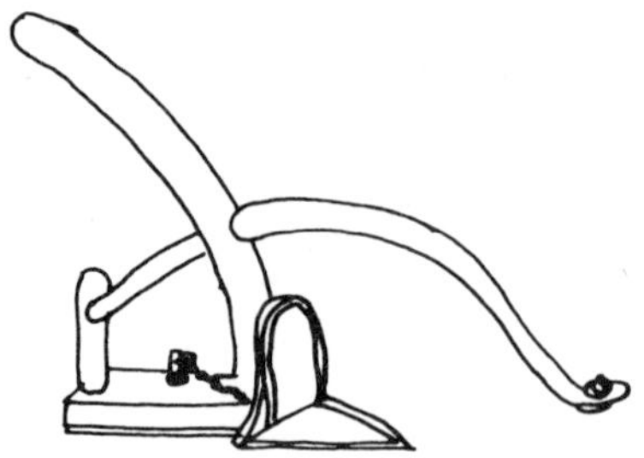

(*a*)

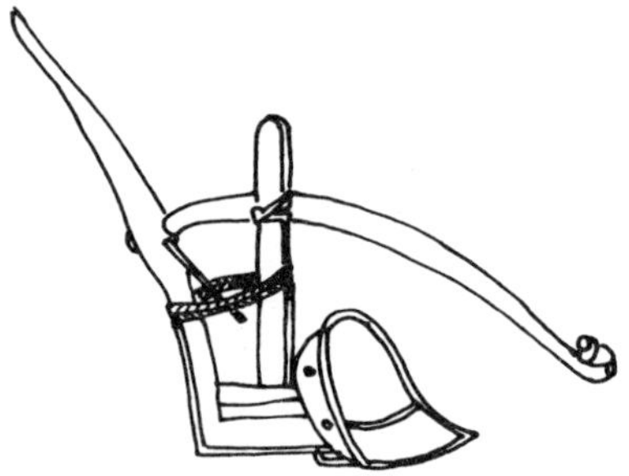

(*b*)

Fig. 62. Variations in the *Wang Chen Nung Shu* illustrations of the plough.

Lu's description of the plough is quoted as it stands, without comment or elaboration, in all subsequent agricultural works. Perhaps since many later writers of agricultural treatises also came from the Yangtze Delta, they felt that Lu's account was sufficiently accurate for their purposes; perhaps they were overwhelmed by the prospect of describing the hundreds of variations and modifications of which they were aware. The only useful written addition to the *Lei Ssu Ching* is a separate section in Wang Chen's *Nung Shu* of +1313 on shares and mould-boards. Wang distinguishes between the small sharp share called *chhan*[1], and the broader, more convex *hua*[2] (Fig. 56*d*):[a]

> The *chhan* and the *hua* are completely different. The *chhan* is narrow and thick and can only be used for straight ploughing (*wei kho cheng yung*[3]), while the *hua* is broad and thin and will turn the soil right over. There is an old farming proverb: 'to clear virgin soil use a *chhan*; to turn well-tilled soil use a *hua*' ... But nowadays in North China they usually use a *hua* while in South China everyone uses a *chhan*.

This indiscriminate use is a mistaken prejudice, says Wang, and the appropriate share for the task should always be chosen. But perhaps the Yuan farmers were less careless than Wang implies, for the regional use of the terms *hua* and *chhan* still

[a] *WCNS* 13/12*a–b*.

[1] 鑱 [2] 鏵 [3] 惟可正用

persists today, yet both broad and pointed shares are to be found in North and in South China.

The Ming illustrations to Wang Chen's *Nung Shu* (Fig. 62)[a] show one plough with a large pointed share and a comparatively small mould-board, the other with what appear to be a share and mould-board cast in one piece; most other encyclopaedia illustrations, such as those in the *Thien Kung Khai Wu*,[b] are equally improbable, except for those which wisely show only the upper part of the plough, the rest being hidden under mud and water.[c] However, the *Nung Shu* illustrations of *chhan* and *hua* (Fig. 56*d*)[d] clearly show a sharp pyramidal share (*chhan*) and a broader *hua*, convex and apparently downward-curved like so many modern examples (Fig. 48b; 64).[e] Chinese ploughshares have continued to be made of cast iron up to the present day; steel, being sharper, would be more efficient but also more expensive to produce, and expense was always a primary consideration for Chinese farmers. Wang Chen says:[f] 'If [the share] has cut through a great deal of soil then its sharp point will inevitably turn blunt, but it can be melted down and recast. However, poor people just resharpen it.' There is no indication that the Chinese were aware of the differential cooling process whereby self-sharpening cast-iron shares could be made, but in Europe this process was only patented by Robert Ransome of Norwich in 1755.[g]

Of mould-boards Wang Chen says:[h]

> The mouldboard (*pi*[1]) is the ear (*erh*[2]) of the plough ... There are different types of mouldboard. The type for ploughing wet fields is called the 'tile twist' (*wa chiao*[3]) or 'high leg' (*kao chiao*[4]), while the type for ploughing dry fields is called the 'mirror face' (*ching mien*[5]) or 'bowl mouth' (*wan khou*[6]). The type used depends on the nature of the soil.

Wang here makes a basic distinction between round and elongated mould-boards, but interestingly does not distinguish between asymmetrical and symmetrical.[i] Archaeological finds have produced mould-boards from all dynasties and of all the types described earlier for the Han dynasty. The only significant change is that from Sung times on some of them are secured to the plough not by lugs but by one or more metal loops or sleeves through which the strut is passed (Fig. 48*c*). If the height at which the loop passes round the strut is regulated by

[a] Copied by the illustrators of *Nung Cheng Chhüan Shu* and *Shou Shih Thung Khao*.
[b] *TKKW* 1/6*b*.
[c] E.g. the various illustrations of *Keng Chih Thu*, *Pien Min Thu Tshuan*, etc.
[d] *WCNS* 13/10*b*, 11*b*.
[e] Amano (*4*), p. 800; Hommel (1), fig. 61; examples seen and photographed by Joseph Needham in Heilingkiang.
[f] *WCNS* 13/11*a*.
[g] G. E. Fussell (2), p. 61.
[h] *WCNS* 13/13*b*.
[i] The latter, cylindrical type Amano (*4*), p. 783 calls *chhang thou*[7].

[1] 鐴 [2] 耳 [3] 瓦繳 [4] 高脚 [5] 鏡面
[6] 碗口 [7] 踃頭

blocks of wood of different thicknesses,[a] the angle of the mould-board can be adjusted. On a lugged mould-board a similar adjustment can be effected by altering either the lengths of the securing cords or the position of the pegs to which they are fastened (see the *Nung Shu* illustration, Fig. 62), and the earliest reference to this, it seems, must be a passage in the *Chhi Min Yao Shu*[b] which advises the farmer, when he wants to plough between rows of stubble without uprooting or covering them, literally to 'tie back the ears of his plough (*mi fu li erh*[1])', in other words to tie the mould-board cords so tight that the board lies almost flush with the strut and so does not turn the furrow.[c] Although such adjustments of the mould-board must always have been made, they are not referred to in the classic texts. Indeed, Wang Chen says:[d] 'To adjust the depth of furrow we have the strut ... to adjust the width of furrow we have the stilt', implying that the mould-board could not be adjusted to change the width of furrow. Both left-handed and right-handed mould-boards have been found in China, but surprisingly, although the share was invariably symmetrical (so that theoretically the furrow could be turned to either side of the plough provided the mould-board was changed), we have no evidence of any reversible ploughs such as the European turnwrest, of which Fitzherbert says:[e] 'In Kente they haue other maner of plowes ... some wyll tourne the sheldbredth at euery landes ende, and plowe all one waye.' This is doubtless a reflection of the very different ploughing patterns of China (see p. 104).[f]

We now come to the question of patterns of intercultural diffusion and influence. The only generally acknowledged exchange of plough technology between China and the West is the introduction from China to Europe in the 18th century of the curved iron mould-board, together with such associated implements as the spiked roller and the winnowing-fan (see p. 553).[g] Further east, full-fledged square-framed plough types are found not only in Korea, Japan and Indochina, but also in the Philippines and Indonesia, where they are associated with typically Chinese forms of harrow and roller,[h] and this implies another wave of technological diffusion. It is worth while briefly reconsidering the development of the Chinese plough since earliest times.

The earliest plough in China, a simple bow-ard, may well have come from West Asia or Europe at the same time as the domestication of cattle (see p. 159). On the other hand, it is just possible that it developed indigenously. The bow-ard

[a] Hommel (1), fig. 62.
[b] *CMYS* 14.6.1.
[c] On the various interpretations of this passage see Amano (*4*), p. 762.
[d] *WCNS* 2/3*b*.
[e] (1), section 2.
[f] See Leser (1), pp. 393–4.
[g] Leser (1), pp. 442 ff.; Leser (2); Slicher van Bath (2).
[h] Leser (1), figs. 264, 270, 296–8.

[1] 弭縛犁耳

is still found all over East and Southeast Asia, particularly in mountainous terrain or in areas occupied by minority peoples. Bow-ards, often strengthened with a strut, are still found in Japan, Korea, Vietnam and Java as well as in Kansu, Shensi, Kwangsi, Kwantung and Yunnan,[a] and this distribution rather indicates that the form spread from China to the other areas; this occurred at a date unknown, but perhaps at a time of migration into new areas.[b]

It is a remarkable coincidence that the first evidence of the square-frame mould-board turnplough appears in China in the early centuries before our era and in Northwestern Europe in the first centuries of our era. At first glance this seems a clear case of technical diffusion: either the turnplough was developed first in China and spread thence to Northwest Asia and Europe during the phase of Chinese expansion and foreign contacts in the early Han, or it was invented by a Slavic farming tribe while they were still living in Asia, the Chinese acquired this knowledge by indirect contact, and the plough travelled with the Slavs to Europe. But we must also consider that the European and the Chinese turnploughs were very different, not only in the full flower of their development (see Table 6) but also in their very early stages. The Han frame-plough, like later Chinese ploughs, had a curved metal mould-board, a symmetrical share, no coulter, no wheels, and a single stilt. But the earliest evidence for turnploughs in Europe is the remains of asymmetrical shares and large coulters; the Roman texts speak of heavy wheeled ploughs used in Gaul and Germany; until the 18th century European mould-boards were wooden and straight; European frame-ploughs (with a few modern exceptions) had two handles or stilts. These are not trifling differences,[c] they are so fundamental to the function and performance of the plough that we must regretfully conclude that the turnplough evolved independently in East and West. However, in both cases it probably developed as a response to the same stimulus, namely population growth leading to an expansion of agriculture into areas of heavy soils.

We now come to the complicated question of relations with Southeast Asia. It has usually been assumed that the exchanges were all one way, from China southwards, and this is certainly true in the case of the square-frame plough, for in early 19th century Java it was actually called *luku China*;[d] that it was still known in 1817 as the 'Chinese plough' indicates that it was a fairly recent introduction; in fact it was probably brought by the early Chinese settlers in Java in the +17th

[a] Leser (1), pp. 403 ff.; Raffles (1), p. 113; Huard & Durand (1), fig. 28; Hu Hou-Hsüan (*3*), p. 80*a*; Anon. (*502*) passim.

[b] There is as yet no proof that ploughing was known in Japan prior to the Yayoi period (−4th century), and the early history of Southeast Asia is still shrouded in obscurity, but archaeological work is progressing in these areas and showing them to have been more advanced than was thought hitherto. Some proponents of mainland Southeast Asia as a very early centre of plant domestication, metallurgy, etc., believe that cultural development spread northwards from Southeast Asia into China (rather than the other way round) during the neolithic period (e.g. Meacham (2)), and they might feel that we should look for the origins of the Chinese bow-ard in Vietnam or Thailand.

[c] See Bratanic (1) on the significance of such dissimilarities.

[d] Raffles (1), p. 113.

Table 6. *Comparison between Chinese and European turnploughs c. + 1600*

| | China | Europe |
|---|---|---|
| Frame | Square or triangular<br>Stilt always single<br>Slade narrow<br>Beam usually curved + whipple-tree | Square<br>Stilt usually double<br>Slade-boards usually wide<br>Beam straight + wheels or whipple-tree |
| Share | Symmetrical<br>Always single | Asymmetrical<br>Sometimes double[a] |
| Mould-board | Usually cast iron<br>Symmetrical and asymmetrical<br>Convex curve<br>Placed vertically above share<br>Often adjustable | Wooden<br>Asymmetrical<br>Flat<br>Placed longitudinally behind share<br>Sometimes reversible (turnwrest) |
| Coulter | None | Essential |
| Ploughing depth | Varied by adjustment of strut | Varied by adjustment of coulter, plough ear, beam height |
| Furrow width | Varied by angle of stilt, mould-board | Varied by adjustment of coulter, share, plough foot |
| Reduction of friction | No wheels<br>Narrow slade<br>Curved iron mould-board<br>Float or slipper quite common[b]<br>Stone or wooden heel under slade quite common[d] | Wheels<br><br><br>Float sometimes replaces wheels[c]<br>Stones sometimes embedded in slade[e] |
| Weight | Very light, carried by one man | Heavy, often carried on cart or sled[f] |
| Team size | Usually 1 buffalo<br>Sometimes 3 or even 4 oxen in Northern China<br>1 man to drive plough | Usually 4 oxen or horses<br>Up to 14 animals for heavy plough<br>Often 2 or even 3 men, one driving, one leading animals, one pressing down beam |

[a] Blith (1).

[b] Leser (1), figs. 234–7; Amano (*4*), p. 785 shows a Ming example from *Pien Min Thu Tshuan*; modern examples are shown in Anon. (*502*), pp. 6, 8, 10. The float is used only in dry fields, and nowadays is often replaced by a single small wheel; Anon. (*502*), pp. 16 ff.

[c] Leser (1), fig. 34, p. 501; Haudricourt & Delamarre (1), pp. 343, 363; this practice became more common in Northwest Europe in the 17th and 18th centuries as the swing plough, more versatile than the wheeled plough, became popular.

[d] Leser (1), figs. 234, 236 from Shantung and a photograph by Joseph Needham from Heilungkiang are the only examples we have found.

[e] A common practice in Northwest Europe; see Lerche (1), (2); D. V. Clarke (1).

[f] Leser (1), pp. 538–40.

century.[a] This is borne out by the fact that the *luku China* was used mainly in market gardens, many of which were run by Chinese, and not in the rice fields. The square-frame plough was probably introduced to the Philippines at a slightly earlier date, for Chinese migration to the Philippines was extensive at the time of the Spanish conquest.[b]

The triangular plough presents a different problem altogether. The pure triangular form (Fig. 63*a*) is found in China, in Shansi and Hopei, as well as in Siam;[c] the almost quadrilateral form (Fig. 64) is found as far as we know in China only.[d] What one might call the Z-shaped plough, where the beam no longer curves right down to join the slade but terminates where it joins the backward-curved stilt (Fig. 39.4*c*), is found in China; for example in Shensi, Shantung, Kiangsu and Chekiang,[e] and also in Annam, Siam, Sumatra and Java, as well as Japan.[f] Now the triangular form and its derivations are certainly later in China than the square-frame plough, yet these forms precede the square-frame plough in Indonesia; moreover the square-frame plough does not seem to be known in mainland Southeast Asia, and is known in insular Southeast Asia only as a late introduction from China. Could it be that the triangular plough came from Southeast Asia and travelled thence to China, perhaps during the period of vigorous Chinese trading and exploring during the Yuan and early Ming? Certainly large areas of Southeast Asia owed less to China than to India in pre-colonial times,[g] so that one can well conceive of a northward rather than a southward flow of influence. But no triangular ploughs are to be found in India itself, so it seems that this particular form must have been indigenous to Southeast Asia (unlike many other Southeast Asian forms of plough which show clear Indian influence;[h] the influence of India is to be found in some plough forms within the Chinese empire too, but only in such far-flung regions as Tibet and Turkestan[i]).

In the case of 18th-century Europe, contacts are easier to demonstrate. Europe had become familiar with China through the reports of the Jesuits and other travellers. China was idolised, and all things Chinese considered the acme of excellence. It may seem strange that farmers as well as artists and philosophers were infected with this enthusiasm, but this was a period when agriculture, in Northwest Europe at least, had become a gentleman's profession: it was developing rapidly and had acquired almost the status of an experimental science, and there

a C. Geertz (1).

b V. Purcell (1).

c Amano (*4*), pp. 800–1; Leser (1), fig. 285.

d Fig. 62; Amano (*4*), p. 801 shows a modern example from Hunan.

e Anon. (*502*), pp. 10 ff.; a modified modern form with straight lines and supporting strut is found in Hunan and Kwangsi, *ibid.* pp. 61 ff.

f Leser (1), figs. 268–9, 274, 277 and 261.

g Coedès (5).

h E.g. some Siamese, Malay, Khmer and Annamese ploughs; Leser (1), figs. 280–1; A. H. Hill (1), fig. 1; Hickey (1), p. 136.

i Anon. (*502*), p. 4, R. A. Stein (6), p. 126; Leser (1), fig. 217.

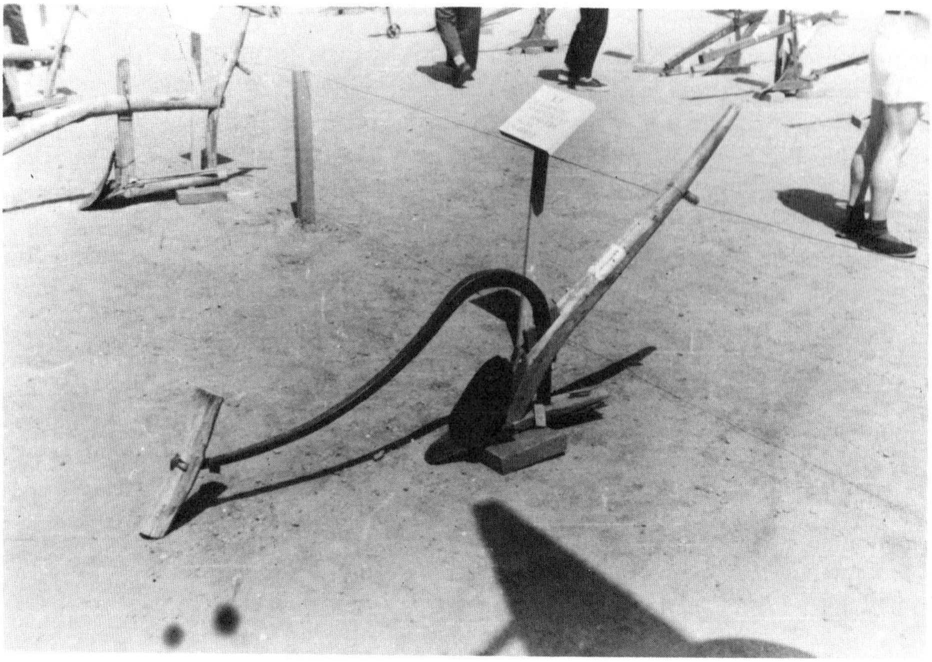

Fig. 63 (*a*)

was a wide market for useful or profitable innovations. As early as 1651, Samuel Hartlib had deplored the inefficiency of most contemporary ploughs and stressed the necessity for scientific investigation:[a]

I wonder that so many excellent *Mechanicks* who have beaten their brains about the perpetual motion and other curiosities, that they might find the best way to ease all Motions, should never so much as honour the *Plough* (which is the most necessary Instrument in the world) by their labour and studies ... Surely he would deserve well of this *Nation* and be much honoured by all, that would set down exact rules for the making of this most necessary but contemned Instrument, and so for every part thereof; for without question there are exact rules to be laid down for this, as for Shipping and other things.

Now the inefficiency of the European plough in Hartlib's time was largely due to the enormous friction engendered by the flat wooden mould-board, which lay along a different axis from the share so that the soil did not turn smoothly over it, and by the wide heavy slade which was thought necessary for a wide furrow. This problem was not solved until the introduction in the late 17th century of curved metal mould-boards, to the Netherlands and East Anglia. In the first section of our Conclusions (pp. 553 ff.) we show that this innovation almost certainly derived from Chinese influence, either directly or through Southeast Asia.[b] It spread

[a] (1), p. 5.

[b] In many places the adoption of the curved iron mould-board coincided with the introduction of such typically Chinese implements as the spiked roller and the winnowing-fan; p. 377 below, Leser (1) p. 442.

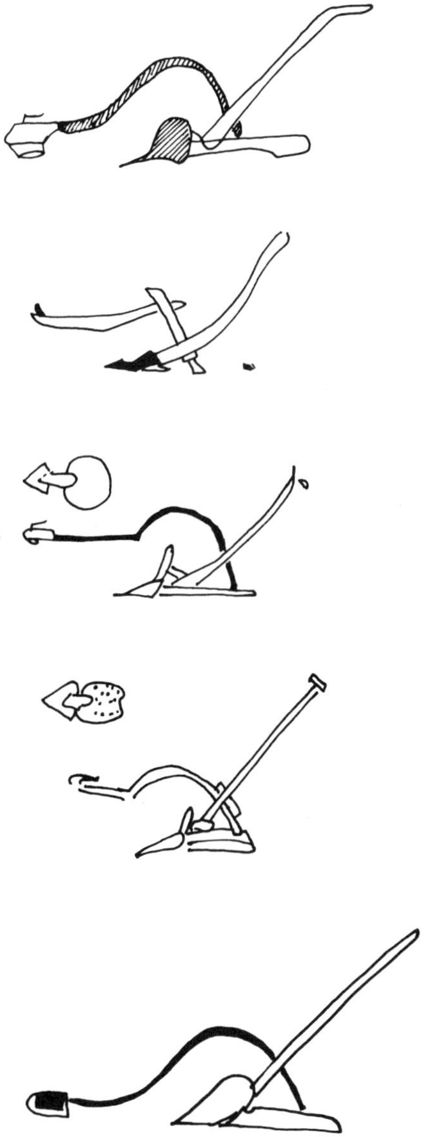

Fig. 63 (*b*)

Fig. 63. Triangular Chinese ploughs from Shantung (JN orig. photo) and from Shansi and Peking, after Amano (*4*), pp. 798–9.

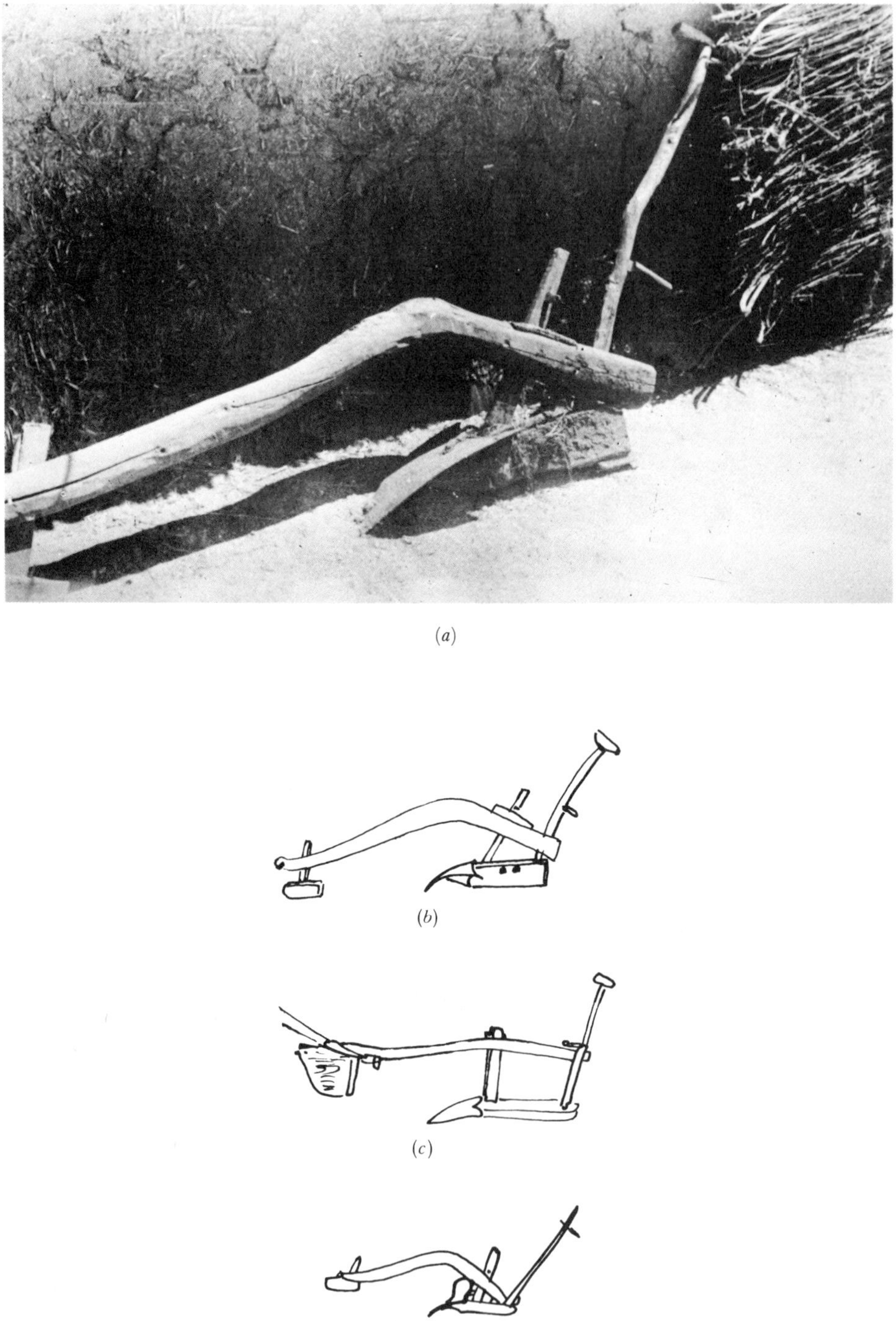

(a)

(b)

(c)

(d)

Fig. 64. Ploughs with downward-curving shares: (*a*) from Heilungkiang; JN orig. photo; (*b*) from Hopei; (*c*) from Shantung; (*d*) from Hopei; after Amano (*4*), p. 798.

rapidly through the areas of the West where agriculture was comparatively advanced, that is to say Holland and Flanders, England, Scotland and parts of North America, but was slower to gain popularity in France and Germany.[a]

The Scot James Small was the first to enunciate, in 1784, the principles on which a curved mould-board should be designed (see p. 174), yet the search for perfection in design continued for many years more.[b] Nevertheless, the combination of curved iron mould-board and a sladeless plough of the 'Rotherham' type[c] was so strong, light and easily managed that it made a revolution in European farming,[d] greatly reducing the number of draught animals and the labour required.

The great difference between Chinese and European mould-boards was that the latter lay along the plough rather than between the share and the strut like the Chinese mould-boards. This extra space gave the European designers far greater scope and freedom, and the great innovation of the +19th century European ploughwrights was to give their mould-boards a convex as well as a concave twist.[e] This, together with the light yet strong sladeless frame, made the new European plough so efficient that when it was introduced to China in the late 19th and early 20th centuries it made the traditional Chinese plough appear clumsy and inefficient by comparison:[f]

> [The Chinese plough] is not suited to deep ploughing. Moreover at the point where the share and mould-board meet it does not curve flexibly and therefore it cannot completely overturn the clods of earth ... [In the Western plough] the surfaces of share and mould-board link in a [smooth] curved line, which causes the clods to be completely inverted.

Devotees of progress attempted to introduce the new style of plough throughout China[g] (with remarkable lack of success) together with such other Western marvels as winch-ploughing and steam-ploughing,[h] and very soon agriculturalists were agreed that the traditional plough of China was a very primitive implement indeed. Yet less than two centuries previously the Chinese plough had been taken as a model of perfection in the West, and for nearly two millennia beforehand the Chinese plough had outstripped its western counterparts in versatility and efficiency.

[a] Curved mould-boards were still a rarity in Germany in the early 19th century; Leser (1), p. 443.

[b] G. E. Fussell (2), ch. 2 gives an excellent account.

[c] *Ibid.* pl. 20. It has been suggested (see Haudricourt & Delamarre (1), pp. 381 ff.) that the wheel-less swing-plough that became popular in England in the 18th century, and hence the sladeless plough, were attributable to the influence of Chinese wheel-less ploughs.

[d] F. G. Payne (1).

[e] Leser (1), p. 448.

[f] *Mien Yeh Thu Shuo* vol. I, quoted Amano (*4*), p. 790, tr. Philippa Hawking.

[g] E.g. *Nung Hsüeh Tshuan Yao* 1/28*a*.

[h] See Amano (*4*), pp. 728 ff.; since these developments are all post-1600 we shall not discuss them here. On the subject of ploughs powered by non-animal traction, we should mention that Samuel Hartlib, inspired by hearing of Chinese wind-powered carriages, attempted to develop a wind-powered plough in the mid-17th century, but it did not catch on (G. E. Fussell (2), p. 40).

*Plough accessories*

Where fallowing is a regular feature of crop rotation, as it was in most of Northern and Central Europe until very recently, the coulter is an essential part of the plough. The coulter is a sharp knife firmly attached to the ploughbeam, which cuts through the grassy sod before the share turns it, thus reducing friction considerably:[a]

> The culture is a bende pece of yron set in a mortes in the myddes of the ploughe beame fastened with wedges on euery syde, and the back therof is halfe an ynche thycke and more then thre ynches brode, made kene before to cut the erthe cleane and it must be well steled, and that shall cause the easyer draughte, and the yrons, to laste much lenger.

The first European coulters date from Roman times and are closely associated with the spread of the turn-plough.[b] In their original form they were probably not attached to the frame of the plough but instead a knife was bound to a separate wooden handle and used independently of the plough as a simple 'ground-opener'; this form is still encountered in mountainous districts of Europe.[c] But it cannot have been long before it was found convenient to attach the knife directly to the ploughbeam; such an arrangement can be seen in the Bayeux tapestry and other early illustrations.[d]

Though coulters were an essential component of most European and Russian turnploughs, they were little known outside these zones of grass and steppe. Ploughs in the Middle East, Central, South and Southeast Asia, China and Japan do not usually have coulters even today. Yet coulters were known in China as early as the Sung dynasty, for the *Chi Yün*[e] refers to a blade, the *li tao*[1], which was used for cutting through fallow soils. Wang Chen includes it in his section on harvesting tools, presumably because of its shape:[f]

> In form it is like a short sickle, but the back is thicker, and I have seen it used for opening up land overgrown with reeds, artemisia scrub and so on, where the roots grow thick and close, and even with strong oxen a plough will not suffice. Therefore before using the tilling-plough a single ox is harnessed to a small plough to which is attached a single blade that slices the soil, opening a single trace. Next the proper ploughshare follows this trace, throwing up a ridge to either side. The labour is more than halved. One can also attach the blade [directly] to the end of the ploughbeam, on the near side, and since this requires no extra men or beasts it is even more economical and convenient.

The illustration provided to Wang Chen's text (Fig. 65) is clearly quite wrong, for Wang describes the *li tao* as a sickle-like blade similar to a European coulter. It is

[a] Fitzherbert (1), fol. iii.
[b] F. G. Payne (1), p. 79.
[c] Leser (1), p. 302; Haudricourt & Delamarre (1), fig. 48, etc.
[d] Anglo-Saxon drawings of ploughs with coulters dating from the +10th century are discussed by Haudricourt & Delamarre (1), pp. 357 ff.
[e] Quoted *WCNS* 14/9*b*.
[f] *Ibid.* 14/9*a–b*.

[1] 劚刀

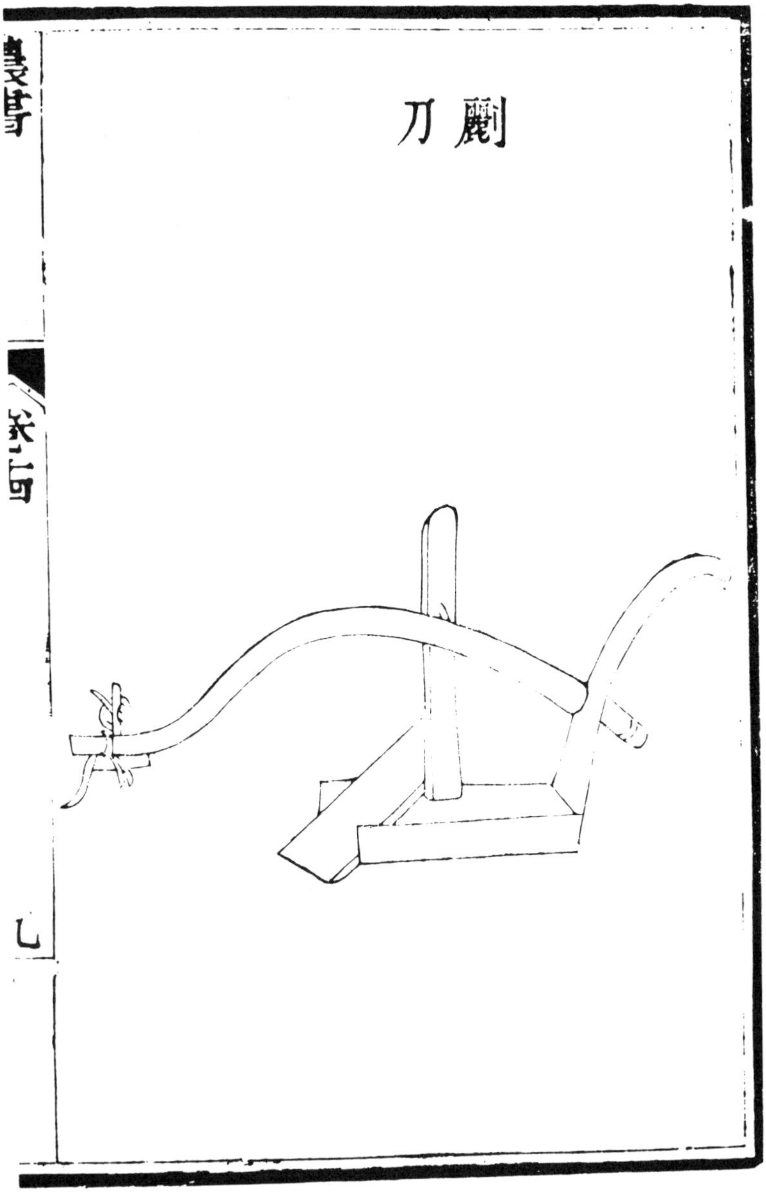


Fig. 65. Coulter (*li hua* or *li tao*): it is clear that the artist had no idea of what he was supposed to be drawing; *WCNS* 14/9*a*.

interesting that he refers to two distinct types, one separate from and one attached to the frame of the turnplough, but unfortunately he gives no indication of how frequently it was used, or of the regions where it was common. Later compendia merely quote Wang's description, though the *Shou Shih Thung Khao* does relocate it in the section on tillage,[a] and none of the many descriptions in *nung shu* of clearing land refers again to the *li tao*. Coulters are occasionally seen on Chinese ploughs today,[b] chiefly those from the far north (nowadays they are called *li tao*[1]), but these are more likely a recent modification than a traditional feature. The *Nung Hsüeh Tsuan Yao* of 1902 refers to the coulter only in the section on 'modern ploughs', as an innovation from Europe and America.[c] Perhaps the coulters mentioned in the *Chi Yün* also came from the West, but from the plains of Russia which, at that time, like the plains of North China, were falling under the sway of the same Tartar horsemen.

Wang Chen refers to one more accessory to the plough, the 'scraper' (*chhan*[2] or *pang*[3]) (Fig. 66):[d]

> Its blade is like a weeding-hoe but wider; it has a deep socket which is wedged onto the sole of the plough at the place where the share is usually fitted.[e] This plough is small and light and only requires a single ox, or it may be pulled by a man. It is used in North China, in dark and gloomy spots where the lowlying soils become waterlogged during the winter. When the thaw begins and the soil quickens in the spring, they scrape the *chhan* along the ground, ploughing up the roots of the weeds and blocking the veins of the soil.[f] The *chhan* is suitable for spring-grown barley and wheat, and should be used wherever the soil is marshy and weedy, for the soil will be rich and fertile even without deep cultivation. If one piles up the [cut] weeds and fires them before sowing, the crop will be doubled.

### (ii) *Hand tillage: hoes,[g] mattocks, spades*

The preponderance of hand tools in 19th- and 20th-century China and Japan led many Western travellers to describe oriental agriculture as a 'horticultural' system, in which draught animals and the implements associated with them were comparatively unimportant. This is a fairly recent state of affairs in China, but although ploughs were traditionally the basis of field agriculture, hand tools such as mattocks, hoes and spades have always been indispensable for opening new land, tilling awkward corners, preparing vegetable gardens, digging ditches, building bunds, and innumerable other tasks. Wang Chen in the *Nung Shu*[h] describes

[a] *SSTK* 32/23*a–b*.

[b] Anon. (*502*), pp. 15, 20, 22, etc.

[c] Ch. 1/38*a*.

[d] *WCNS* 13/14*b*.

[e] I.e. the 'turtle's flesh'; see fig. 61.

[f] Preventing the moisture from rising to the surface.

[g] For convenience we shall refer to heavy hoes used for breaking the soil simply as 'hoes'; the lighter hoes used for weeding, earthing up and so on (see section 4) we shall call 'weeding hoes'.

[h] *WCNS* 13/1*a*.

[1] 犁刀 [2] 剗 [3] 鎊

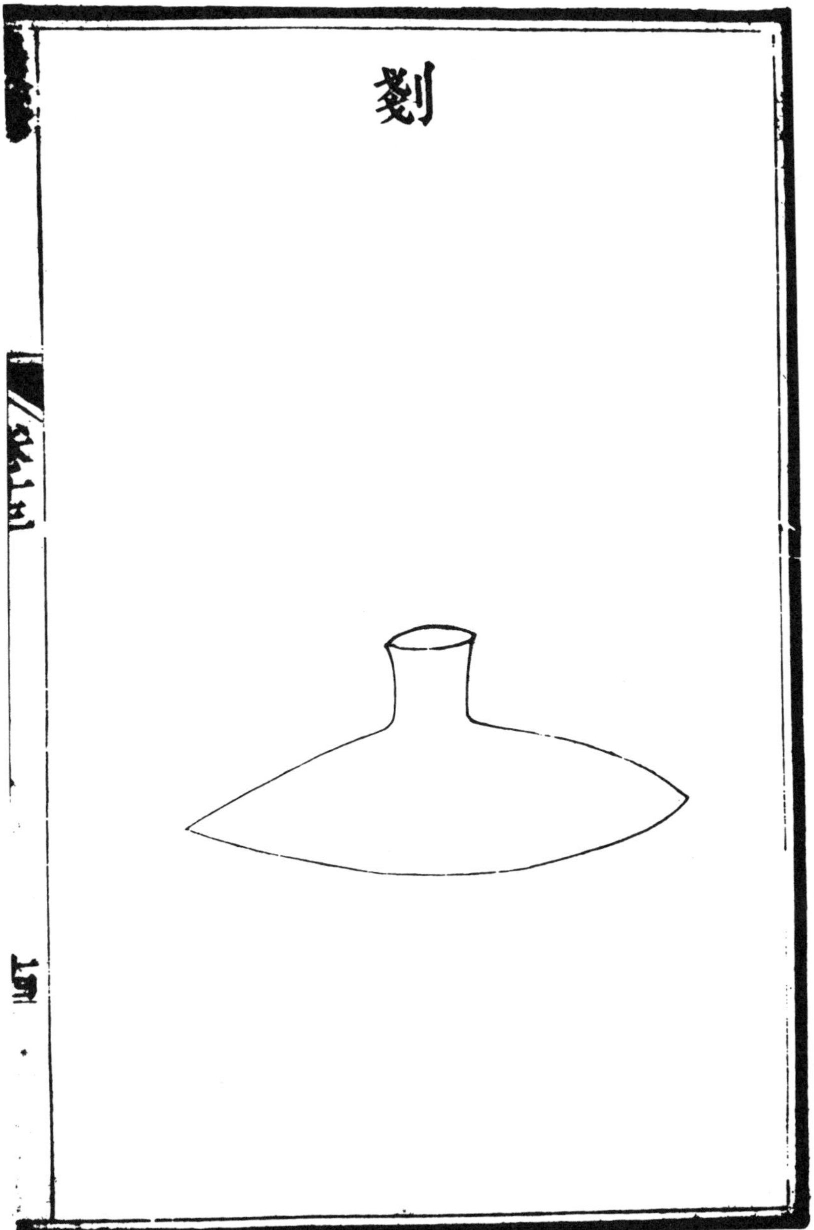


Fig. 66. Scraper (*chhan*), an attachment to the plough designed to pare off weeds; *WCNS* 13/14*a*.

them as 'indispensable to the farmer for all tasks of clearing land or constructing drainage channels'; of hoes and mattocks he says:[a] 'they are used in established fields and gardens as well as for clearing new land, and they come in all shapes and sizes'. The importance of these hand tools grew as the rural population increased and grazing land was converted to arable; by the Edo period in Japan in the 17th century 'the hoe was the most important tool in the whole farming system',[b] and in 20th-century Kiangsu, Alley and Bojesen found the use of the pronged drag-hoe (*thieh tha*[1]) so common that they dubbed it 'the Universal Tool'.[c]

In Britain true digging tools, that is the fork and spade, are most often used for horticulture and tillage.[d] In China the digging fork is unknown, and though spades are used for irrigation and construction work they do not play a very important role in tillage. The reason is not far to seek:[e] 'Spades to be foot-driven require strong foot-gear. The Chinese who work mostly bare-footed in their fields cannot, of course, push the spade with the foot for any length of time.'[f] Percussion implements like hoes and mattocks are the most common hand tillers not only in China but throughout Asia and Africa, and have been since neolithic times.

Many stone implements from Chinese neolithic sites have been identified as 'spades', implying presumably that they were bound to vertical shafts and used for digging.[g] We believe that this is a misinterpretation: a heavy stone blade offers no advantages over a lighter wood or bone blade in a digging implement. For maximum efficiency a digging blade requires to be flexible, and the haft must be very strong and very firmly attached to the blade (digging tools such as forks and spades usually break where haft and blade join). Until the introduction of metal blades with projecting sockets into which the haft could be firmly attached, almost all spades were made of a single piece of wood (perhaps fitted with an iron rim).[h] Stone spades seem an unlikely proposition, therefore, but for a percussion implement such as a hoe a stone head would be very advantageous: the extra weight of the head would add momentum to the swing and so increase efficiency, and since the head would strike the ground at an angle to the haft the strain on the bindings would be far less than for a digging tool. We suggest therefore that many of the neolithic implements classified as 'spades' should in fact be considered as hoes. Such spades as existed in neolithic times were almost certainly made entirely of wood.

[a] *WCNS* 13/2*a*.
[b] Horio (1), p. 174.
[c] Alley & Bojesen (1), p. 89.
[d] See Gailey & Fenton (1).
[e] Hommel (1), p. 61.
[f] This should be borne in mind when assessing the various arguments as to the nature of early Chinese tillage implements; see p. 148 above.
[g] See the reconstruction in Liu Hsien-Chou (*8*), fig. 6.
[h] Examples in Hayashi (*4*), K. D. White (1).

[1] 鐵搭

A possible exception is the type of bone implement found at the site of Ho-mu-tu in Chekiang (Fig. 67). Those found in the lowest stratum date back to about −5000, but the tools continued to be manufactured well into the −4th millennium. Chinese archaeologists have simply labelled these tools *ssu*[1], a rather ambiguous term as we saw earlier (p. 148), and from their structure it is not entirely clear whether they were attached to a straight wooden haft to be used as a spade, or to an angled haft to be used as a hoe. Since angled wooden hafts have been found even in the lowest stratum at Ho-mu-tu (Fig. 67*a*), and since many of the so-called *ssu*[1] are double-pronged rather than single-edged, we incline to the view that these early bone implements were frequently mounted on angled hafts and used as percussion rather than digging tools, similar to the mattocks and double-pronged hoes still used in China today. They are also, however, strong, flexible, and easily enough bound to a straight haft to serve as spades,[a] and may well have been used to dig dykes and ditches round the wet fields in which the people of Ho-mu-tu grew wet rice.

A wide variety of broad-bladed stone 'hoes' and narrower-bladed stone 'mattocks' have been found in Chinese neolithic sites (Fig. 68). Any very precise ascription of function is impossible when dealing with cultures so far removed from our own,[b] but presumably such implements were used to clear overgrown land as well as to break up the soil, and it is even possible that the smaller, narrower blades were used to clear weeds from between crop plants. Hoes and mattocks made of heavy stone would certainly have been supplemented with tools made of wood, bone and shell, especially for lighter tasks such as breaking up clods, smoothing soil and earthing up young plants.[c] The earliest stone hoes were probably not hafted but simply held in the hand, though the smaller or narrower ones could well have been bound to an angled wooden haft. Perforated stones, which could serve either as hoes or axes, appear early in the Chinese neolithic; it may well be that the incised motif on a pot from Hua-thing[2] in Shantung (a site belonging to the late Chhing-lien-kang[3] culture, dated to the −4th millennium) is the earliest extant drawing of a hoe (Fig. 69).

Among the most curious of neolithic stone implements are the type recently discovered in Kwangtung and Kwangsi.[d] They are of uncertain date, since carbon-14 dating is unreliable in that region,[e] but they come from clearly neolithic sites. The simpler versions (Figs. 70*a*, *b*) seem to be heavy hoes, with broad blades narrowing into a short shaft which would fit tightly into a wooden handle.

[a] Hua Chhüan (*1*) believes that straight hafts were more common than angled; here an examination of wear patterns would be useful.

[b] K. C. Chang (1), p. 64.

[c] A rare example of a bone hoe is illustrated in Liu Hsien-Chou (*8*), fig. 5.

[d] Anon. (*515*), (*516*).

[e] Because of the prevalence of carboniferous rocks.

[1] 耜 [2] 花廳 [3] 青蓮崗

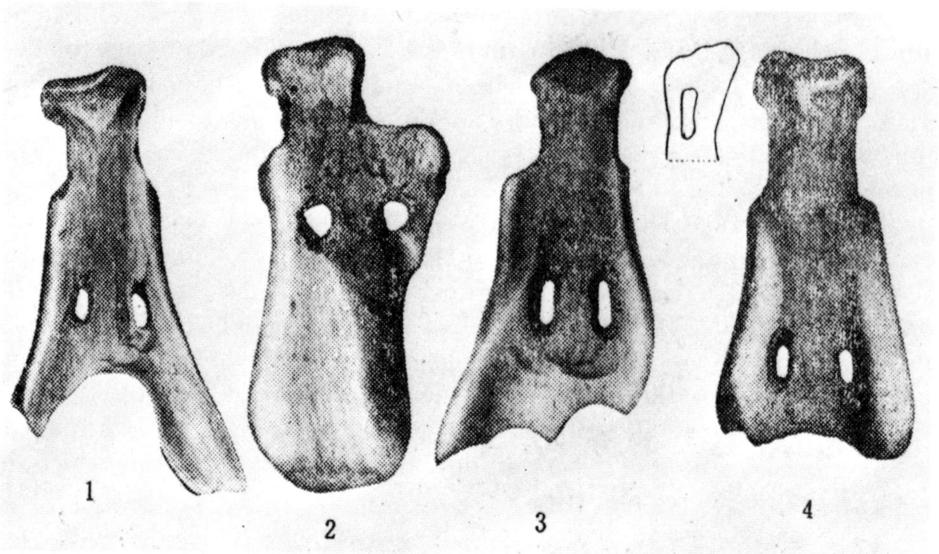


(a)

(b)

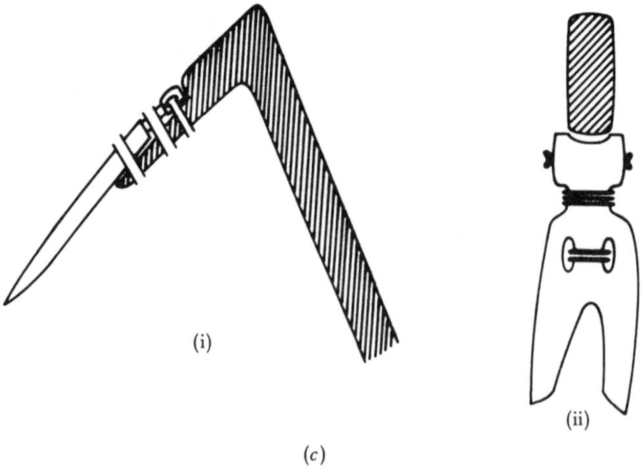


(c)

Fig. 67. Bone digging implements found at the early neolithic site of Ho-mu-tu: (*a*) the bone blades; Anon. (*503*), fig. 7; (*b*) a wooden haft; after Anon. (*503*), fig. 5; (*c*) author's reconstruction of the attachment of blade to haft.

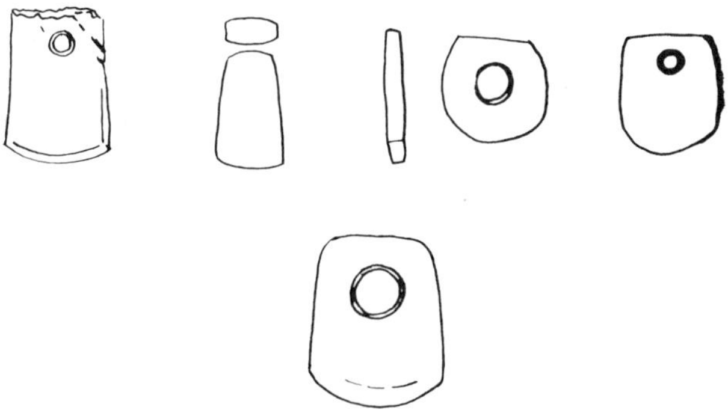

Fig. 68 (*a*)

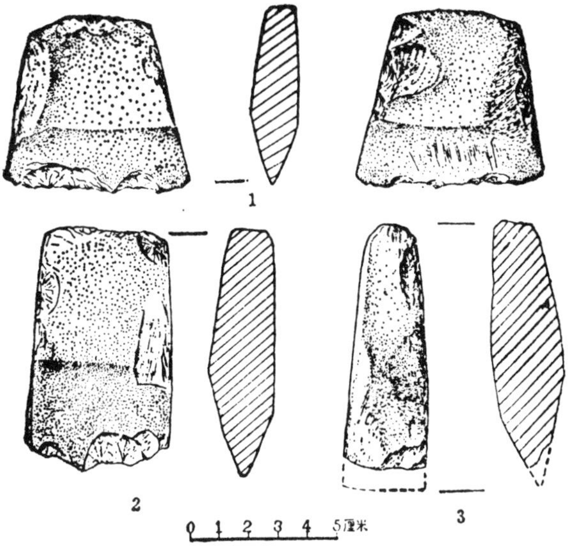


Fig. 68 (*b*)

But the more elaborate versions are so elaborate and apparently impractical that they may perhaps be cult objects or symbols of status.[a]

Throughout the Shang and well into the Chou dynasty most agricultural tools, including mattocks and hoes, continued to be made of shell, stone, bone and wood.[b] They may have been mass-produced, for shell and bone workshops have been found in several Shang sites.[c] Unfortunately, descriptions or illustrations of these humble tools are hardly ever published, for they naturally attract far less attention than the more glamorous products of the age, such as the magnificent bronzes. Bronze was reserved for sacrificial vessels, swords, bells and other high-status articles, and tools made of bronze are very rare.[d]

[a] Officially they are all classified as spades.

[b] K. C. Chang (1), p. 227.

[c] *Ibid.* p. 236.

[d] See Amano (*4*), pp. 687 ff.; K. C. Chang (1), p. 351; Anon. (*517*), p. 7. 259, d: Bronze hoes, sickles, and possibly ploughshares did become quite common in South China during the Warring States period, as can be seen from museum collections in Hangchou, Canton, Kunming and elsewhere.

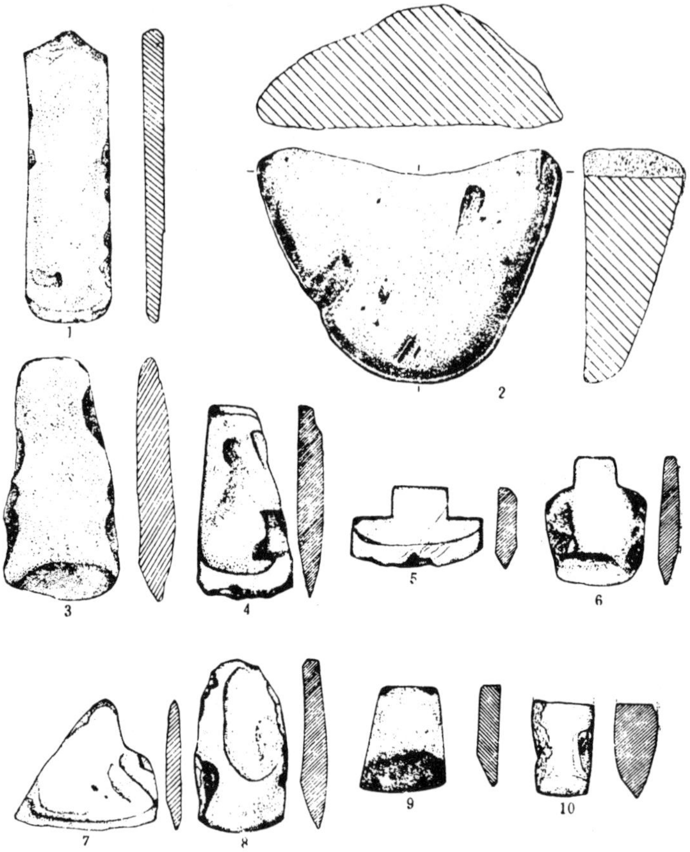


Fig. 68 (*c*)

Fig. 68. Chinese neolithic stone hoes; (*a*) after Anon. (*43*), p. 29; (*b*) from Anon. (*503*), fig. 6; (*c*) from Anon. (*515*), fig. 7.

During the late Spring and Autumn period the use of iron began to spread through China. Unlike bronze, iron was used not only for weapons but also for tools, and by the Warring States period iron axes, ploughshares, hoe-heads and sickles were quite common. Fig. 71 shows cast-iron mattocks and hoes from Warring States sites; wooden hoes and mattocks in late Chou and Han were provided with narrow metal casings around the tip, also a common device for spades.[a] The blades of the early metal hoes were roughly 10 cm. wide by 15 cm. long,[b] smaller than many modern Chinese hoes which may be up to 15 cm. wide by 50 cm. long.[c]

[a] On hoes see Hayashi (*4*), fig. 6–27; Liu Hsien-Chou (*8*), figs. 83–4; on spades, Anon. (*43*), fig. 29 (i), (iv); Hayashi (*4*), figs. 6–2, 6–6, 7, 8, 9, 10.

[b] Hayashi (*4*), p. 273.

[c] Wagner (1), p. 205.

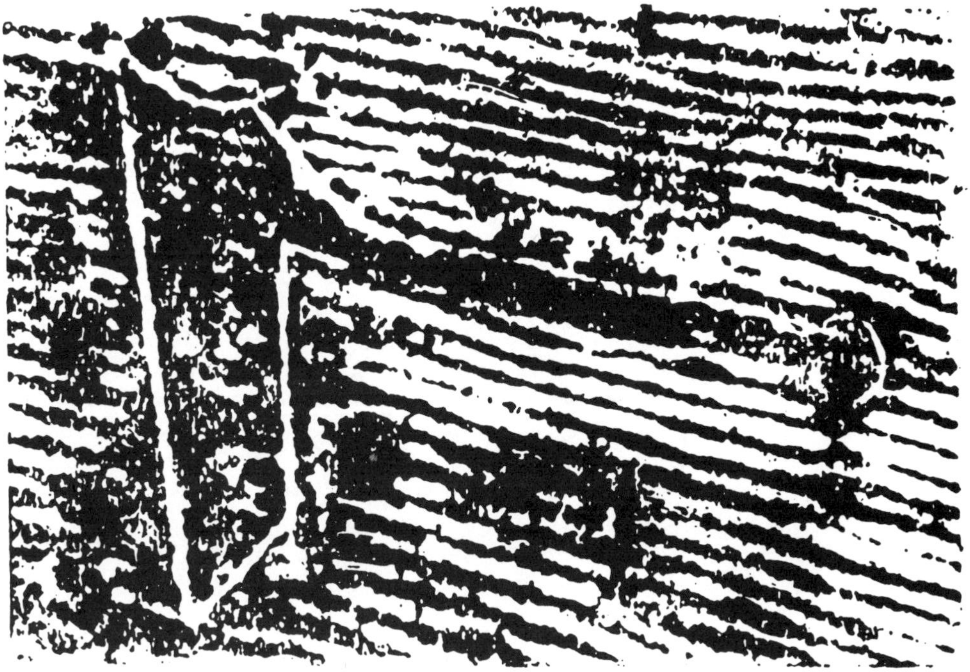

Fig. 69. Incised sign on a pot from a Hua-thing site, depicting what seems to be a heavy hoe; K. C. Chang (1), p. 163.

The early dictionaries give a number of terms for these heavy hoes or mattocks: *khuo*[1], *cho*[2], *chu*[3], *chüeh*[4]. A far greater number of terms existed for tools defined as weeding-hoes (see section 4), but it may well be that the functions were interchangeable as in Edo Japan:[a] 'A newly-made hoe, being heavy, should be used for breaking the soil; a heavy hoe, well swung, can pierce the earth deeply. A worn hoe, being light and thin in the blade, should be used for weeding; it is useful too for heaping the soil up round the roots of the crops.' Nowadays in China heavy rectangular hoes are often called *chhu thou*[5], though in earlier times the term *chhu*[6] was theoretically reserved for (round-headed) weeding-hoes; Hommel says of the *chhu thou*:[b] 'After the first ploughing it is used to break up big clods all over the field. Then it is used to make ridges for planting wheat or vegetables, for weeding, and for making holes beside the young plants to put in manure, etc.'

It seems that hoes are less specialised now than they were in the days of the Han lexicographers and the Yuan and Ming encyclopaedists. Yet although such tools must have been extremely common, we were not able to find more than one

[a] *Hyakushō Denki* 44–45, tr. Horio (1), p. 174, adj. auct.
[b] (1), p. 62.

[1] 钁 [2] 斫 [3] 斸, 欘 [4] 櫡, 鐯 [5] 鋤頭
[6] 鉏, 鋤

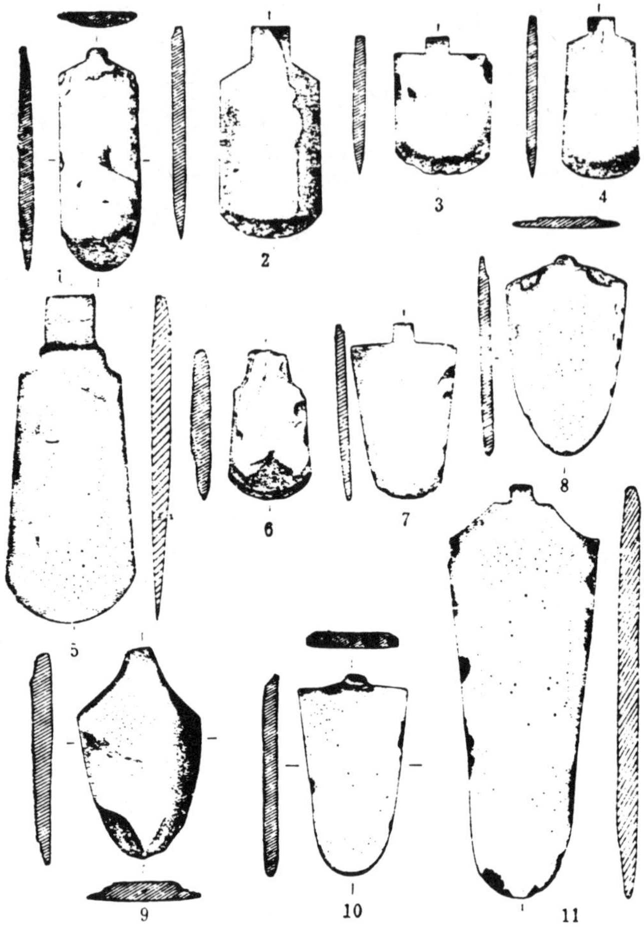


(*a*)

Fig. 70. Stone 'spades' from neolithic sites in Kwangsi; (*a*) simple forms; Anon. (*515*), figs. 4, 5; (*b*) elaborated forms, *ibid.* fig. 6.

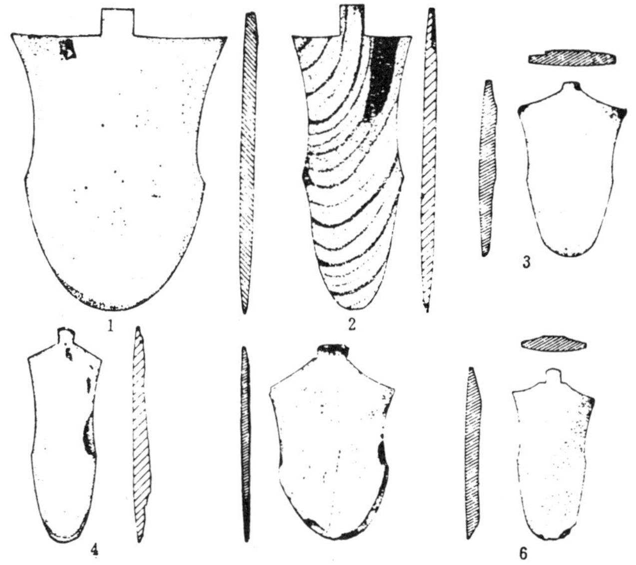


(*a*)

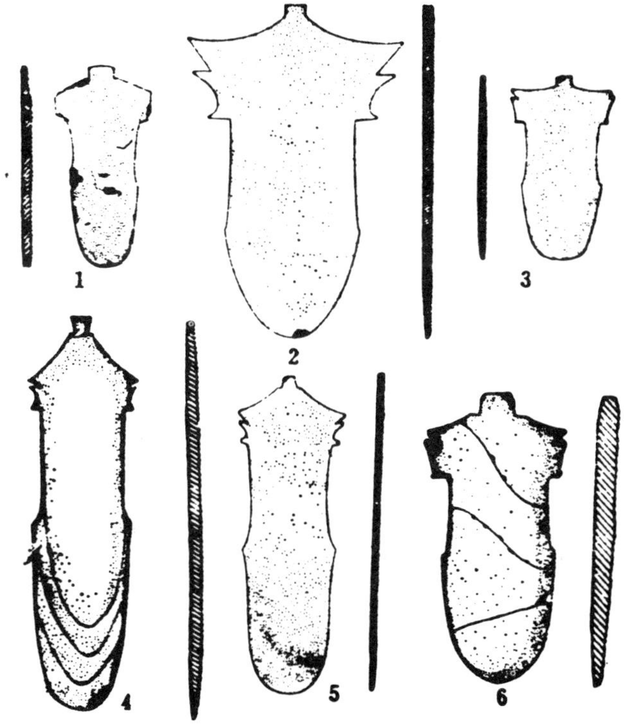


(*b*)

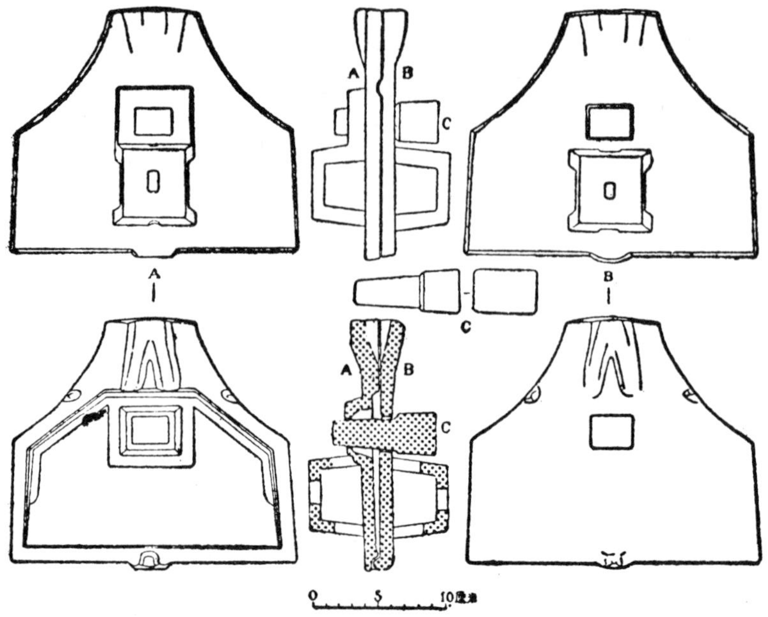


Fig. 71. Warring States mould for cast-iron hoe; Anon. (*43*), p. 65.

illustration among the many bas-reliefs and wall-paintings of the Han and subsequent dynasties (Fig. 72*b*), nor do the early agricultural treatises such as the *Fan Sheng-Chih Shu*, *Ssu Min Yüeh Ling* or *Chhi Min Yao Shu* refer explicitly to their use—though this is probably attributable to their very universality: no farmer or gardener then or now would need to be told how to use a hoe. The earliest illustrations and descriptions are to be found in Wang Chen's *Nung Shu*,[a] and his illustrations are reproduced with some modifications in the *Nung Cheng Chhüan Shu* (Fig. 73*a*, *b*).[b] Fig. 73*b* rather resembles the modern hoes illustrated by Hommel,[c] and Fig. 74*a* is a classic hoe such as can still be found in many parts of the world;[d] both would break up the soil efficiently, whereas Fig. 73*a* shows a mattock more suitable for paring new or fallow ground. Heavy hoes are still in use throughout the Far East today (Fig. 74), though in several countries where they were traditionally manufactured by local blacksmiths the blades are now imported or mass-produced nationally. Their use in China is still universal; in Yunnan in the 1930s 'the ordinary household, which does not possess buffaloes and so hires others to do its ploughing, usually has three hoes and three sickles, each costing approximately 50 cents and lasting more than ten years'; the hoes were men's tools (sickles were women's) and were used for everything from irrigating rice fields to making

[a] *WCNS* 13/1b–2*a*.
[b] *NCCS* 21/20*a*.
[c] (1), figs. 93, 94.
[d] These days they are often manufactured in bulk in Europe. Crocodile hoes from Birmingham have been found as far afield as Malaysia and Uganda.

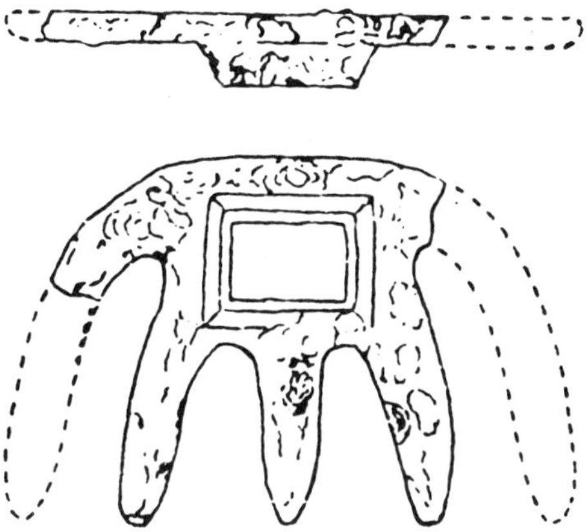

Fig. 72 (*a*)

Fig. 72 (*b*)

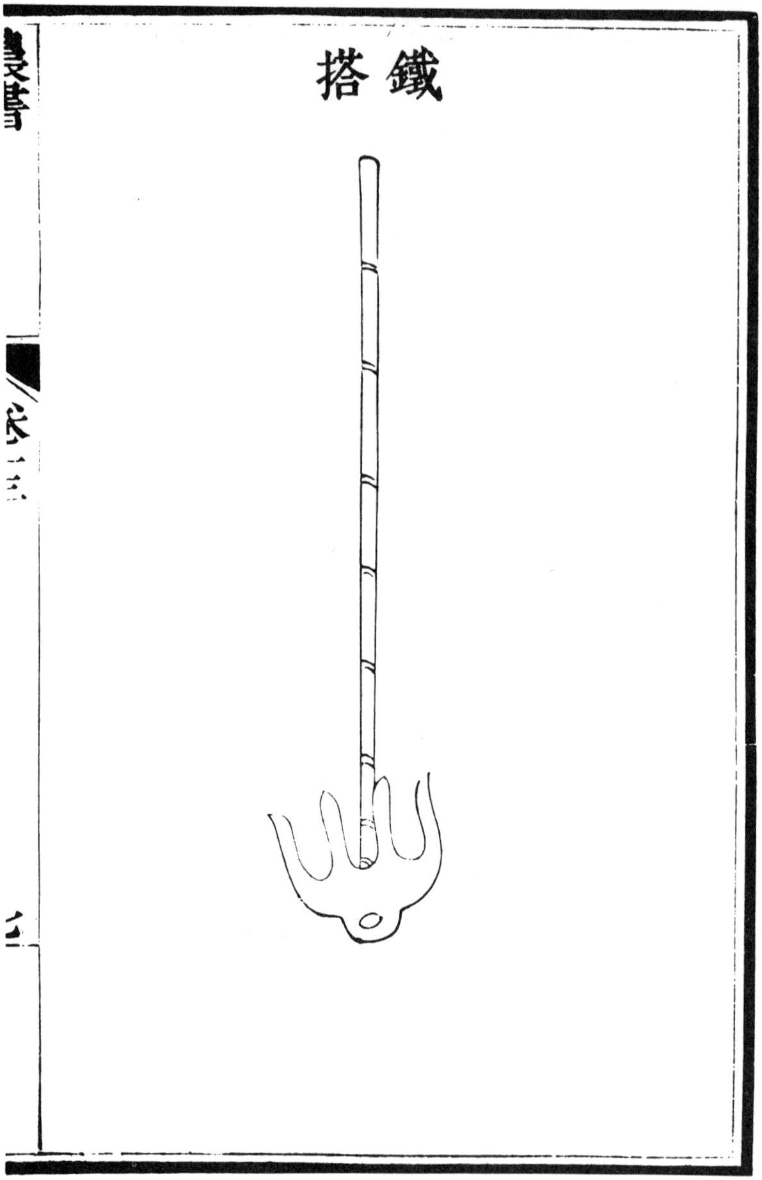


Fig. 72 (*c*)

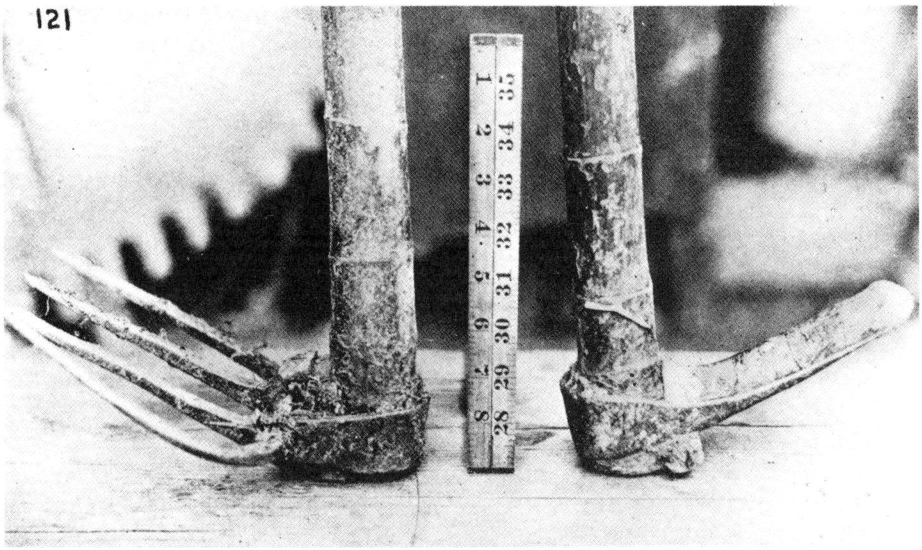

Fig. 72 (*d*)

Fig. 72. Iron drag hoes: (*a*) Warring States, from Hopei; Hayashi (*4*), 6–30; (*b*) E. Han mural from Holingol, Mongolia; *ibid.* 6–31; (*c*) Ming; *WCNS* 13/7*a*; (*d*) modern; Hommel (1), fig. 93.

trenches for broad beans.[a] Wagner points out[b] that while in North China hoe blades are usually of solid iron, in the South they are often made of wood shod with an iron casing round the blade, a similar practice to that of Han times.

Another percussion tool with a long history is the iron-pronged drag-hoe (*thieh tha*[1]) (Fig. 72). Amano[c] believes that this tool probably developed at the same time as paddy cultivation, but quotes the Ming scholar Chu Kuo-Chen[2] [d] who thought that it first came to China after the Thang, probably from Hainan Island. Since the earliest example of the *thieh tha* found so far comes from a Warring States tomb excavated in Hopei in 1965[e] it would seem that Chu was mistaken. The *thieh tha* is similar in form and function to the Roman *rastrum*.[f] The Warring States example is about 15cm. broad and has five prongs, and a Han *thieh tha* with three prongs has been found in Shantung,[g] while a Han mural from Holingol, Inner Mongolia (Fig. 72*b*) shows two men tilling a garden, one using a heavy hoe and the other a *thieh tha*. But we do not know what this tool was called in Han times; the first writer to call it *thieh tha* is Wang Chen:[h]

The *thieh tha* has four or six teeth, sharp and slightly hooked (like a rake yet not a rake) that chop the soil ... the head has a round socket for a straight wooden handle four feet long. Some southern families have no ox-ploughs and use this implement to chop the soil

[a] Fei & Chang (1), pp. 73, 22.
[b] (1), p. 205.
[c] (*4*), p. 812.
[d] *Yung Chhuang Hsiao Phin*, ch. 2.
[e] Hayashi (*4*), p. 275.
[f] K. D. White (1), pp. 52 ff.
[g] Hayashi (*4*), p. 276.
[h] *WCNS* 13/7*b*.

1 鐵搭 2 朱國禎

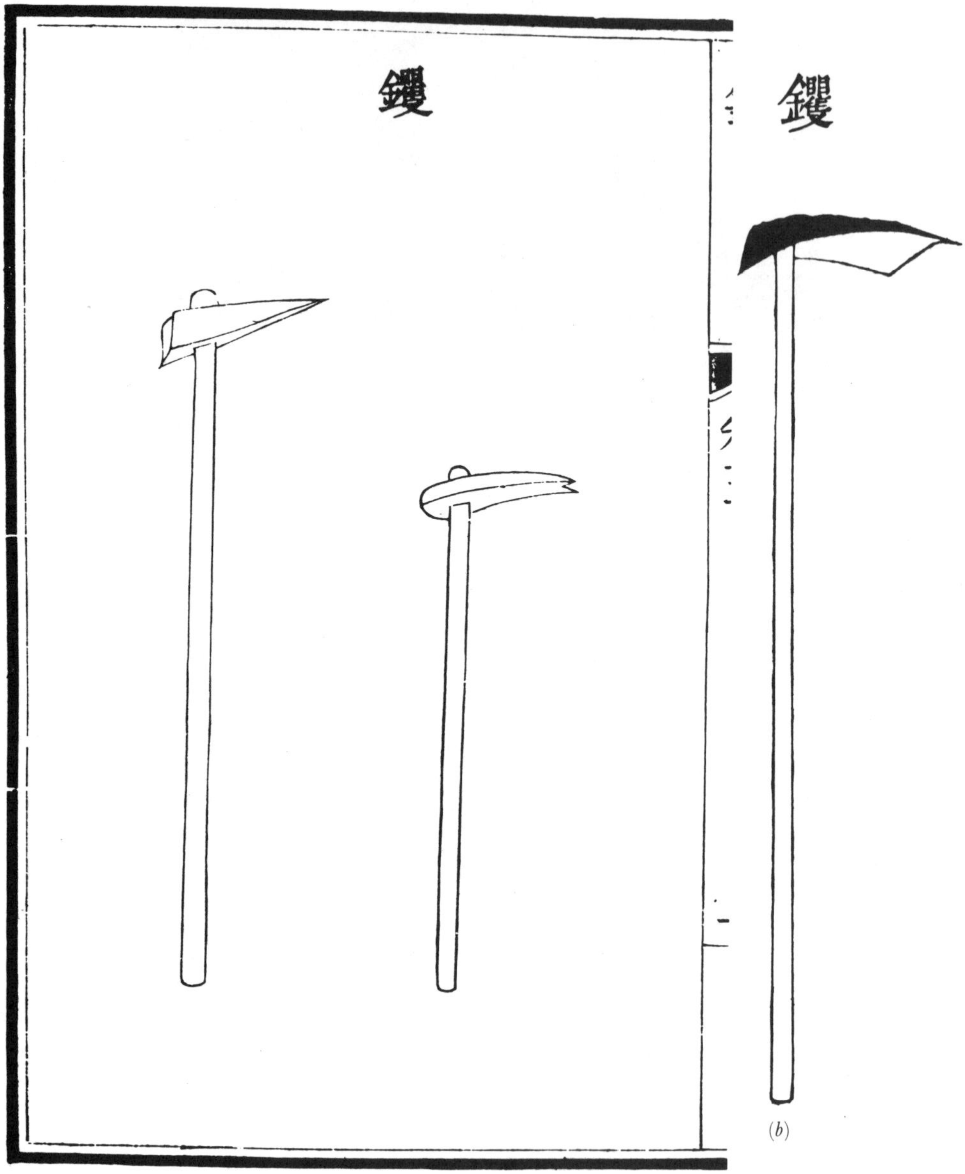


(*b*)

(*a*)

Fig. 73. Chinese mattocks: (*a*) *WCNS* 13/1*b*; (*b*) *NCCS* 21/20*a*.

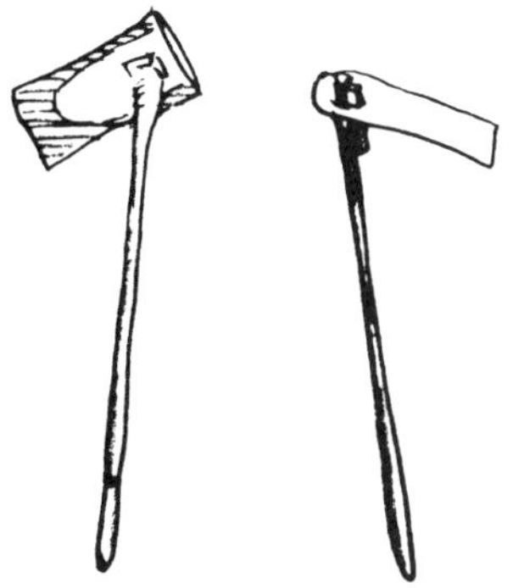

(*a*)

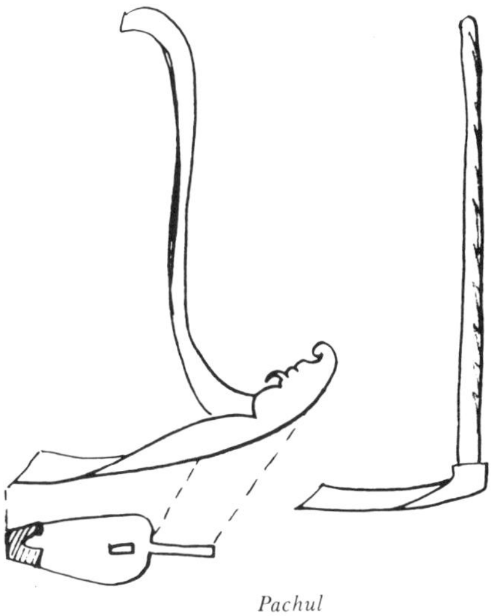

*Pachul*

(*b*)

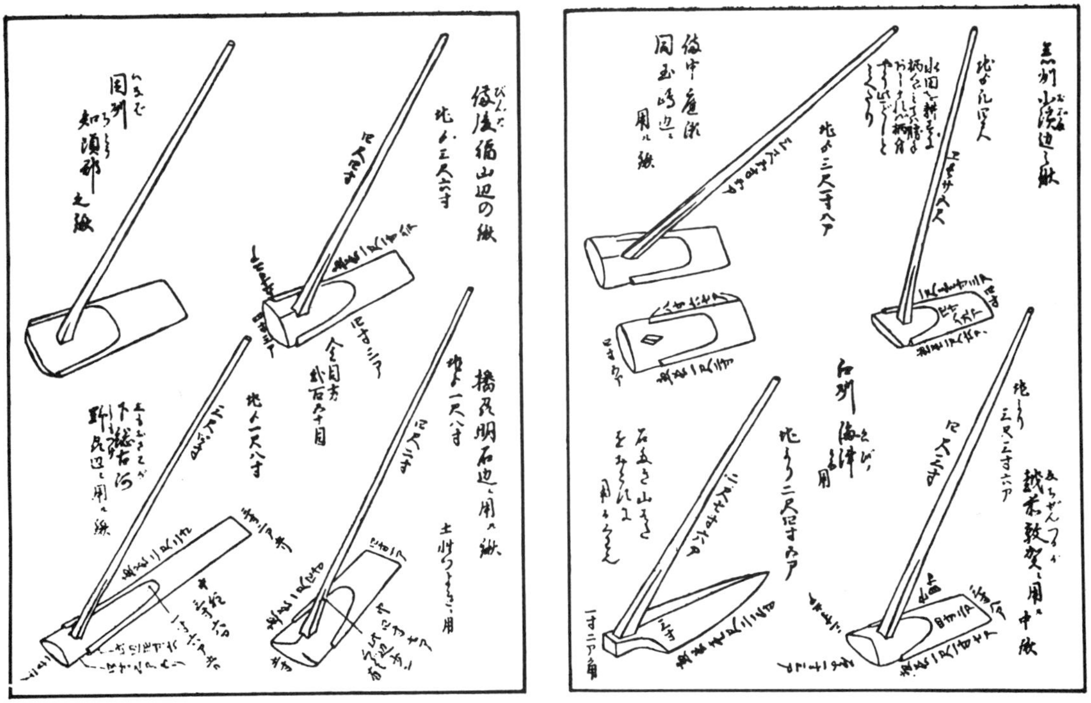

(*c*)

Fig. 74. Asian hoes: (*a*) Chinese field hoes; after Wagner (1), fig. 60; (*b*) Javanese *pachul* of the late 18th century, after Raffles (1), p. 114; (*c*) Japanese hoes, *Nōgu Benri Ron* 1/6*b*–7*a*. Cf. also Fig. 119.

instead of ploughing. Because the cutting edges are separate the effect is similar to harrowing the clods, so it is a cross between a harrow and a hoe. One often sees several friendly families working [with *thieh tha*] in a group, and in this way they can till several *mu* in one day.

In the 1930s the *thieh tha*, or *tid'a* as it was called in local dialect, was almost the only tool used for tilling the padi fields in the Thai-hu area of Kiangsu, since fields were too small and holdings too fragmented to allow the use of draught animals.[a] Groups of farmers still banded together to use them,[b] but the amount of ground covered averaged only 1/4 *mu* per person per day, even in soft soils.[c] The size, number and length of prongs and the angle of the head to the handle depended on the type of land to be worked, whether flat rice-fields or stony mountain-sides.[d]

Wang Chen's description of the *thieh tha* puts one in mind of an implement often referred to in the *Chhi Min Yao Shu*, the *feng*[1]. Although Chia Ssu-Hsieh frequently recommends its use he does not describe the tool, and the earliest description is given by Wang Chen who includes the *feng* in his section on tillage implements. But by Yuan times the term had long been obsolete:[e]

> The *feng* was an ancient agricultural tool with a blade smaller and more pointed than a ploughshare (*chhan*[2]), and with a handle similar to a *lei*[3] [f] ... If the soil was very hard and compacted one would use the *feng* before ploughing so as to save the oxen's strength ... The old farming adage goes: 'In untilled soil *feng* deeply, between young shoots *feng* shallowly' ... The farmers of today do not know what this tool is, or even recognise the name.

Wang Chen does not tell us where he got his information as to the *feng*'s construction. In the *Nung Shu* illustration[g] the *feng* is depicted as a cross between a plough and a spade; Wang places it immediately next to the paragraph on another spade-like tool, the *chhang chhan*[4]. Such distinguished historians as Liu Hsien-Chou[h] and Shih Sheng-Han[i] agree with Wang Chen that the *feng* was a type of spade, yet this hardly tallies with the functions ascribed to it in *Chhi Min Yao Shu*. The literal meaning of *feng* is 'sharp point', as the point of a sword or spear, and Chia Ssu-Hsieh says the *feng* should be used for making ridges, earthing up young plants without compacting the soil and grubbing up dead crowns of millet, as well as for opening up fallow land.[j] It seems that a *thieh tha* would fulfil all these functions admirably, and we therefore suggest that *feng* was an earlier name for that implement, which we know had long been current in North China, even though it was known by a different name in South China when Wang Chen was composing his *Nung Shu*.

[a] Fei Hsiao-Thung (2), pp. 159–60.
[b] Alley & Bojesen (1), p. 89.
[c] Fei (2), p. 160.
[d] Alley & Bojesen (1), p. 89.
[e] *WCNS* 13/4*b*–5*a*.
[f] I.e. curved.
[g] *WCNS* 13/4*b*–5*a*.
[h] (*8*), p. 22.
[i] (*3*), vol. I, p. 7.
[j] *CMYS* 1.5.1; 3.11.2; 6.2.4; 7.2.1; 13.2.1.

[1] 鋒 [2] 鑱 [3] 耒 [4] 長鑱

One tool that really was used for digging rather than hoeing fields was the *chhang chan*:[a]

The *chhang chhan* is an implement for digging [literally 'treading', *tha*[1]] fields. It is considerably narrower than a ploughshare *chhan*[2] and is fixed to a long shaft ... over three feet long and sloping back in a curve; at the top there is a cross-bar which is grasped in both hands. One presses the foot down on the 'heel' protruding at the back of the shaft so that the sharp point bites into the soil, then one jerks back the shaft to lift the clod. This tool can be used instead of a plough in vegetable gardens and pit fields *ou thien*[3]; it requires less strength than a hoe and covers a larger area of ground. In the old days it was known as a 'tread-share' (*chi hua*[4]), and now it is called a digging-plough (*tha li*[5]).

Wang Chen's illustration[b] shows a peasant digging with an implement of which all but the shaft is hidden under mud and water. But later works such as the *Nung Cheng Chhüan Shu*[c] and the *Shou Shih Thung Khao*[d] show a strange arabesque-like implement (Fig. 75) which appears most impractical both to construct and to use. An ordinary spade with a projecting foot-rest, of the type drawn by Juan Yuan[6] in Shantung during the early Chhing (Fig. 76)[e] or referred to by Hommel[f], seems a more likely explanation of Wang Chen's *chhang chhan* or *tha li*, especially if the blade were pointed rather than straight-edged.

Such a tool is doubtless very ancient, and it is probably what *Huai Nan Tzu* means by a 'digging mattock' (*chi khuo*[7])[g]: 'When I Yin[8] was carrying out construction work, he set strong-legged men to the *chi khuo*.'[h] The name *chhang chhan* must already have been current during the Thang, for Tu Fu refers to it in one of his poems.[i]

Amano[j] lists seven references to the *tha li* or 'digging plough' in Chinese literature. Although in Yuan times a *tha li* was, according to Wang Chen, used by a single man, the term later came to be applied to traction spades. Amano quotes a description by Professor Nishiyama Buichi[9] of a 20th-century traction spade:[k]

With the foot-plough (*tha li*) one cultivates by driving it into the earth, throwing the sod forward and retreating. If a rod is fixed to this foot-plough in a V-shape, it is operated by two men facing each other: A digging in the blade with his foot, B raising it with his hand; A retreating, B advancing. Such a plough is still used in Shansi and is called a *chhiang li*[10]; it is about three times as efficient as an ordinary *tha li*.

[a] *WCNS* 13/6*a*–*b*.
[b] *WCNS* 13/5*b*.
[c] *NCCS* 21/21*b*.
[d] *SSTK* 32/9*a*; 32/20*a*.
[e] Amano (*4*), p. 719.
[f] (1), p. 61.
[g] *Huai Nan Tzu* 11/11*b*.
[h] Wang Chen quotes this passage but substitutes *chi hua*[11] for *chi khuo*[7].
[i] *Chiu Chia Chi Chu Tu Shih*, ch. 6, *Ku shih* no. 16.
[j] (*4*), p. 727.
[k] (*4*), p. 725, tr. Philippa Hawking.

[1] 踏 [2] 鑱 [3] 區田 [4] 蹠鏵 [5] 踏犁
[6] 阮元 [7] 蹠钁 [8] 伊尹 [9] 西山武一 [10] 鏹犁
[11] 蹠鏵

Fig. 75. Cultivating with drag-hoe and *chhang chhan*; *SSTK* 32/9*a*.

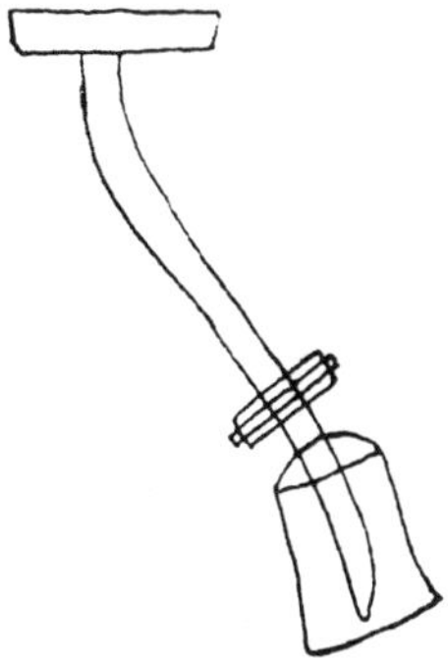

Fig. 76. Spade sketched by Juan Yuan in Shantung in the early Chhing.

A sketch by Yamagata of the *chhiang li*[a] shows a straightforward spade with a lightly curved handle and a pointed blade, nothing like the extraordinary object in Fig. 75.

Traction spades have been reported in modern Korea:[b] one man digging in the blade while a team of four others drag it forward by means of a shaft some 4 m. long. This large spade is used for the first spring ploughing in an area where there is no shortage of draught animals, but where the heavy clay soils are often crusted on top and waterlogged underneath by the end of the winter. The deep cultivation possible with such a spade exposes the waterlogged soil so that it can dry out in the spring winds. The Korean and Shansi traction-spades are the only examples we know in the Far East of true traction spades; other so-called traction-spades reported in Southwest China were actually man-pulled ards.[c]

While spade-like tools such as the *chhang chhan* or *chhiang li* were used for European-style digging of fields, the chief agricultural use of the spade proper, or *chha*[1], was the construction and maintenance of irrigation canals,[d] and this is still true today:[e]

> The spade is not used much on the field itself except for keeping in trim irrigation ditches or digging new ones, and for digging over the field along the edges which cannot be reached with the plough. Otherwise it is used for digging deep where required, as for instance, around the roots of trees in transplanting or getting out stumps.

Spades and shovels were also much used in construction work, and are certainly very ancient in origin. We argued earlier that the problem of secure hafting would require that the earliest forms be constructed of wood rather than stone, so we would not expect to find any archaeological remains before the introduction of

[a] Amano (*4*), p. 725.
[b] Clayton Bredt, personal communication 1977.
[c] Steensberg (3), p. 50.
[d] See the definition in *WCNS* 13/3*a*.
[e] Hommel (1), p. 61.

1 臿

metal blades.[a] Bronze spades have been found at Shang sites,[b] but they look more like weeding-spades (see p. 301) than digging-spades. In the Warring States and Han periods, when iron was still comparatively scarce, it was common for wooden spades to be shod with a thin band of iron;[c] as well as square blades, rounded and pointed blades were also common (Fig. 77). The pre-Han texts refer frequently to spades, often in connection with Yü the Great and his hydraulic achievements, and the spade was known by a variety of names including *chha*[1], *chhiao*[2], *yao*[3], *wei*[4], *ssu*[5] and *hua*[6]. The *Shuo Wen* defines *hua*[7] as a spade (*chha*[1]) with two blades, and this has been variously interpreted, either as a spade with a pointed blade, or as a double-pronged digging-tool, the famous *ssu*[8] (Fig. 43).[d] Several Han representations of such an implement survive, mostly connected with the legend of Yü the Great, so perhaps these iron-shod forks were quite common in early China, but they do not appear in any post-Han iconography. The Han implements were straight and probably used for ditching. In outlying regions of Korea an angled two-pronged implement, the *tabi*, has long been used for tillage (Fig. 78). An engraving on a recently discovered −2nd or −3rd century bronze depicts one man digging a narrow bed of soil with a *tabi* while another breaks up the clods with a hoe. As Korean and Japanese scholars point out[e], similar techniques can be observed in the Andean highlands today. The modern *tabi*, used occasionally in rice fields but chiefly in gardens, exists in both a single-bladed and a double-bladed form; it is shod with iron, and its construction is indeed very similar to that of the Andean *taclla*.

The blades of the Han spades came in a variety of shapes and sizes but were generally flat, though the handle was sometimes slightly curved. The Ming illustration of Wang Chen's *Nung Shu* (Fig. 79) shows us a spade with a cross-bar and a slightly dished square blade, but most modern Chinese spades have flat blades, and the cross-bar is not a standard feature. Finds of cast-iron spades or spade-moulds are conspicuously lacking for all periods, though hoes, weeding-tools, axes and ploughshares are common, so it is possible that spade blades were forged rather than cast as far back as Han times. This is certainly the case today,[f] for while cast-iron sharpens well it is brittle, and a spade should be resilient and flexible for real efficiency.

[a] For the opposite point of view, see Liu Hsien-Chou (*8*), p. 23; Anon. (*515*), (*516*).

[b] Amano (*4*), p. 688; Liu Hsien-Chou (*8*), p. 24.

[c] This practice was known in Rome, but complete blades of iron seem to have been commoner than in Han China; see K. D. White (1), pp. 17 ff. In Yunnan and the Yangtze Delta, bronze was extensively used for a wide range of tools until well into the Han, as can be seen from the collections in the Kunming and Hangchou museums.

[d] Hayashi (*4*), p. 262.

[e] Amano (*4*), 1979 ed., p. 1018.

[f] Hommel (1), p. 61.

1 鍤　2 鍬，喿　3 㧑，銚　4 鍏　5 梩
6 鏵，芣　7 芣　8 耜

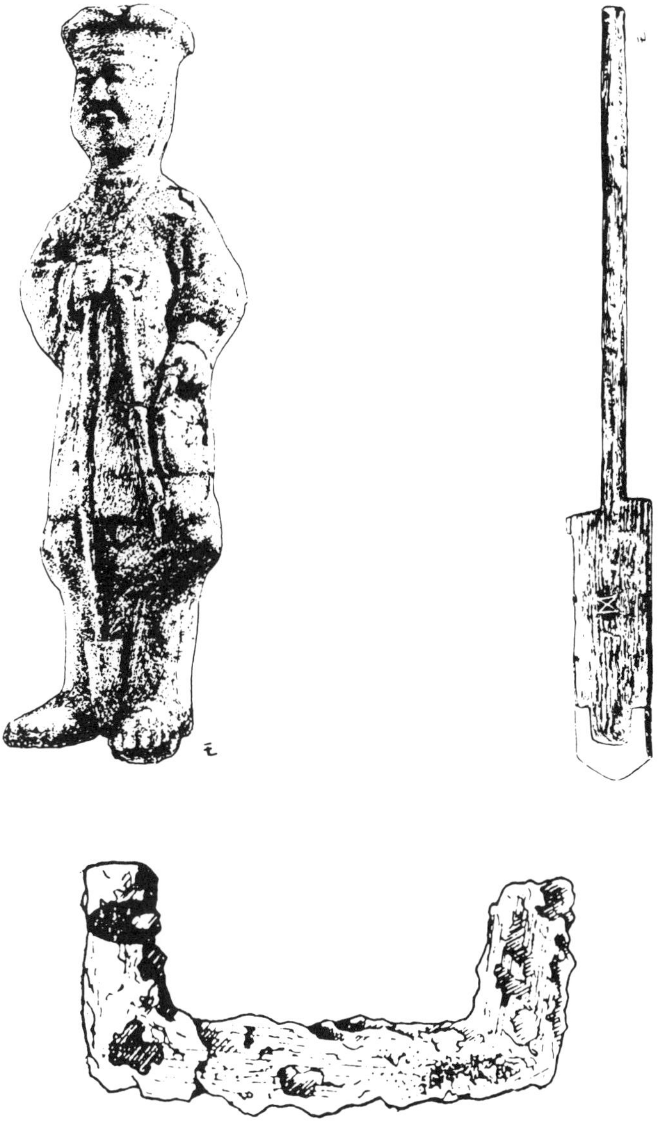

Fig. 77. Han spades, with blades or tips of iron; Hayashi (*4*), 6–1, 6 and 10.

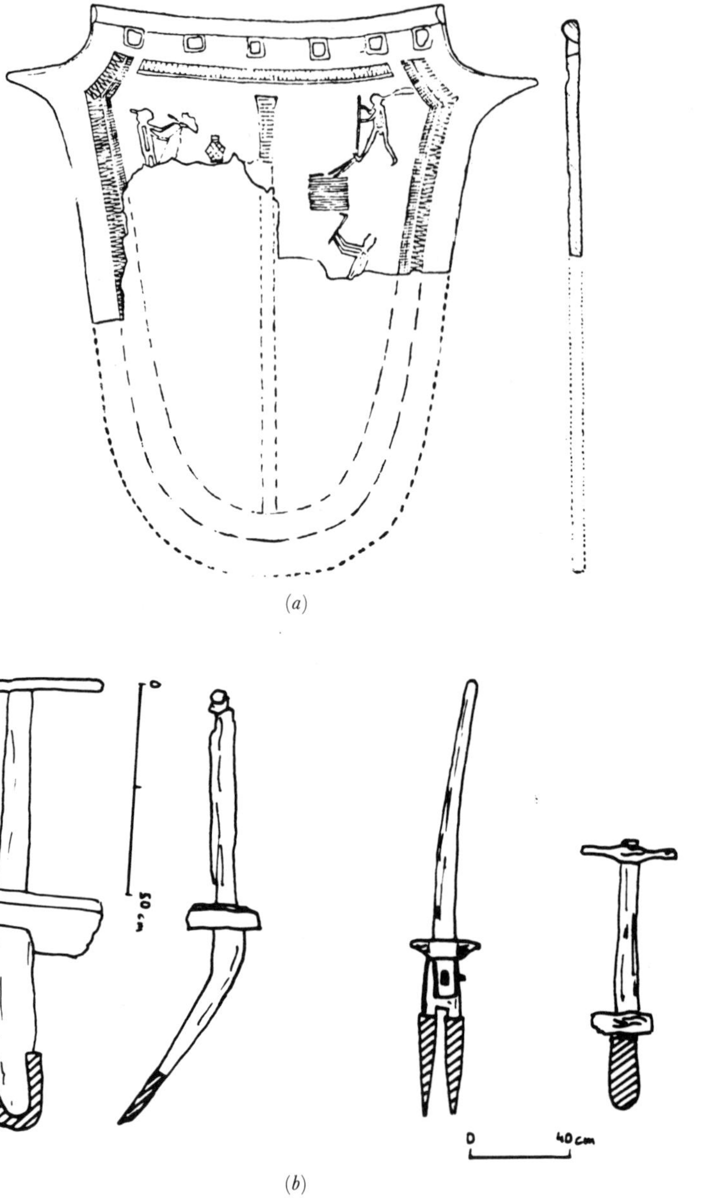


Fig. 78. Korean *tabi:* (*a*) ancient engraving on bronze; Amano (*4*), p. 1019; (*b*) modern forms; Pauer (1), fig. 27.

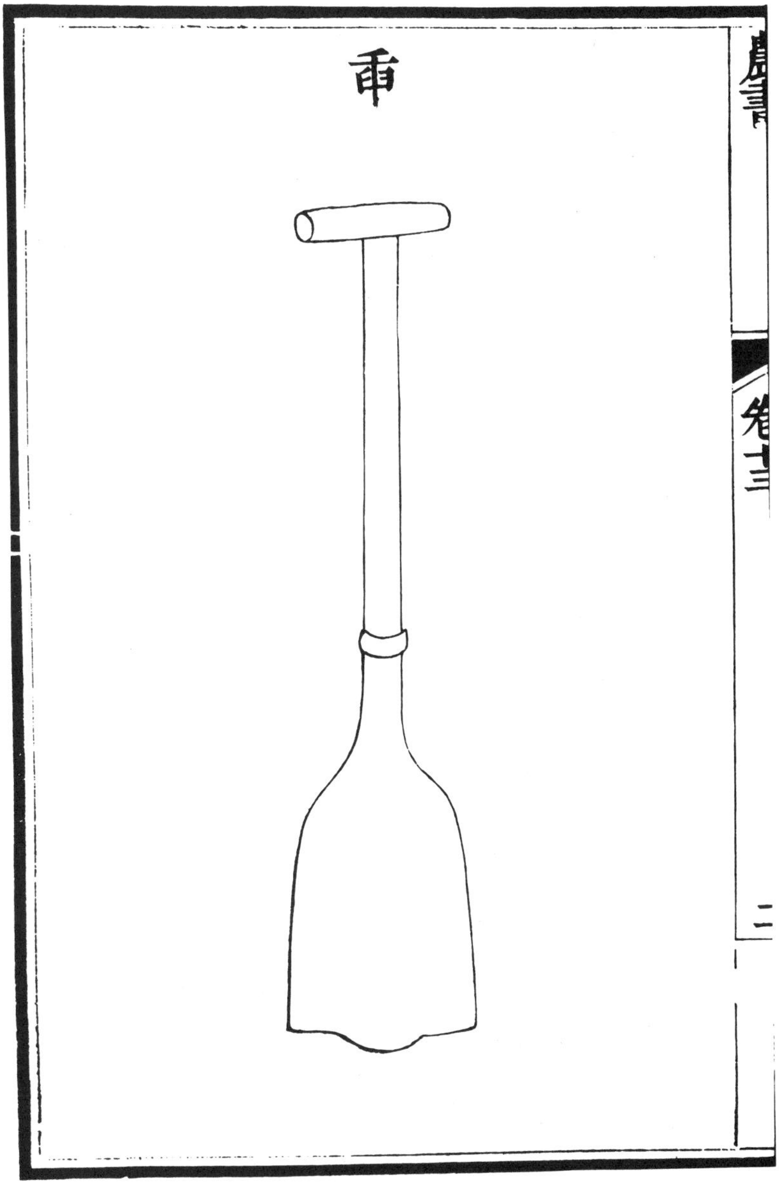


Fig. 79. Ming spade; *WCNS* 13/2*b*.

### (iii) *Smoothing and levelling: beetles, rakes, harrows and rollers*

Many early agricultural and botanical theorists believed that plants drew their nourishment from minute particles of soil which they assimilated: the smaller the particles of soil were, the more easily the plants could absorb them.[a] This hypothesis was certainly based on empirical experience, for every farmer knows that 'among great clods no sturdy plants grow'.[b] Ploughing alone is not sufficient to produce the firm, moist yet crumbly seedbed that will get the plants off to a good start, but the judicious use of harrows, rollers and other devices can not only guarantee a good seedbed on loamy, fertile soils but can also improve considerably light sandy soils or heavy, waterlogged clays that otherwise might yield poor crops. Chinese agriculturalists were aware of this from earliest times, as is apparent from the following passage from the *Lü Shih Chhun Chhiu*, which purports to quote an even earlier work, the *Hou Chi Shu*[1]:[c]

Are you able to make low, wet land fruitful? Are you able to conserve dry soils and temper them with damp? Can you purify your soils, making ditches to wash them clean? ... Can you ensure that your millet heads will be rounded and the husks thin, that the grains will be numerous and plump so that food is plentiful? How may you do all this? By these fundamental rules of tillage:

The strong [soils] must be weakened, the weak strengthened (*li che yü jou, jou che yü li*[2]). The rested must be set to work, the hard-worked rested. The lean must be fattened, the fat made leaner. The compact must be loosened, the loose compacted. The damp must be dried out, and the dry dampened ... Till five times and use the weeding hoe five times (*wu keng wu nou*[3]). Observe all these rules closely.

Lü's terminology would have seemed familiar to European farmers. In his *English Husbandman* of 1616 Gervase Markham gives a fine description of the characteristics of moisture retention, texture and consistency associated with soils which he defines as either 'loose' or 'fast' (according to the *Lü Shih Chhun Chhiu* classification, 'strong' and 'weak'):[d]

Now to give you my meaning of these two words, *loose* and *fast*, it is, that every soyle which upon parching and dry weather, even when the Sunne beames scorch, and as it were bake the earth, if then the ground upon such exceeding drought doe moulder and fall to dust, so that whereas before when it did retaine moysture it was heavie, tough, and not to be separated, now having lost that glewinesse is light, loose, and even with a mans foote to bee spurnd to ashes, all such grounds, are tearmed loose and open grounds,

[a] This theory of plant nutrition was first proposed in Europe by the classical writers: its germ is found in Aristotle and Theophrastus. But the writer who developed the argument at greatest length, and who evolved from it a whole new approach to farming, was the 18th century agriculturalist Jethro Tull (1). For a full account of early Western theories of plant nutrition, see Fussell (3).

[b] *Yen Thieh Lun*, quoted *CMYS* 1.3.2.

[c] *LSCC*, Hsia Wei-Ying (3), pp. 27 ff.

[d] Markham (2), pp. 96–7.

[1] 后稷書 [2] 力者欲柔，柔者欲力 [3] 五耕五耨

because at no time they doe binde in, or imprison the seede (the frost time only excepted, which is by accidence, and not from the nature of the soyle:) and all such grounds as in their moisture, or after the fall of any sodaine raine, are soft, plyable, light, and easie to bee wrought, but after, when they come to lose that moistnesse, and that the powerfulnesse of the Sunne hath as it were, dryde up their veines, if then such earthes become hard, firm, and not to be separated, then are those soiles tearmed fast and binding soiles, for if their Ardors be not taken in their due times, and their seede cast into them in perfect and due seasons, neither is it possible for the Plow-man to plow them, nor for the seede to sprout through, the earth being so fastned and as it were, stone-like fixt together.

In the distinction between 'loose' and 'fast' soils and the appropriate choice of tools and methods 'consisteth the whole Arte of Husbandry', says Markham,[a] and the whole first volume of the *English Husbandman* is devoted to the techniques of ploughing, harrowing and rolling suitable for soil types of differing degrees of fastness or looseness.

The *Lü Shih Chhun Chhiu* does not elaborate on the manner in which strong soils are to be made weaker or damp soils dried out, beyond its exhortation to till *keng*[1] and hoe repeatedly. Jethro Tull would have approved of this insistence on working the soil constantly even after sowing, for he believed that 'a plant is almost as imperfectly nourished by tillage without hoeing as an animal would be without gall and pancreatic juice'.[b] The reference to hoeing in *Lü Shih Chhun Chhiu* is clear enough, but what does *keng*[1] imply? In slightly later texts[c] it was usually taken to include not only ploughing but all the allied processes, such as harrowing and rolling, that were necessary to produce a really fine tilth. As the country saying had it: 'the art of tilling lies in harrowing (*keng thien mo lao yeh*[2])'.[d] The early Han writer Fan Sheng-Chih elaborates on one (slightly modified) phrase from the *Lü Shih Chhun Chhiu* passage:[e]

When the apricots start to bloom, till light soils and weak soils immediately; plough them again when the apricot blossoms fall. After ploughing roll down (*lin*[3])[f] the surface at once ... If the soil is very light it should be trampled (*chien*[4]) by cattle or sheep. This is what is meant by 'making weak soils stronger (*jo thu erh chhiang chih*[5])'.

Elsewhere Han agriculturalists like Fan Sheng-Chih and Tshui Shih tend to stress the importance of timing ploughing correctly and make few specific references to other processes of soil comminution. The *Chhi Min Yao Shu* is the first work to elaborate the practices of harrowing which were to become such a fundamental element of traditional Chinese farming practice, and this fits well with the archaeological evidence which suggests that harrows only started to appear during the Han dynasty. Up until the end of the Han the Chinese must have used much simpler and more laborious methods to pulverise and smooth the

[a] Markham (2), p. 96.
[b] Tull (1), p. 89.
[c] Notably *CMYS*, ch. 1.
[d] Quoted *CMYS* 1.3.2.
[e] *FSCS* 1.4, quoted *CMYS* 1.12.1.
[f] See p. 240 below.

[1] 耕 [2] 耕田摩勞也 [3] 藺 [4] 踐 [5] 弱土而強之

soil. We have seen that Fan Sheng-Chih mentioned driving animals to trample and compact light soils. Fan was describing conditions in Kuan-chung in North-west China, where the soils were loessial and the chief problems faced by the farmer were aridity and erosion, which necessitated 'retaining the moisture throughout the soil' and 'making the weak strong'. Lower down the Yellow River and in the regions between the Yellow River and the Huai, where heavier alluvial soils predominated, the problem was rather to pulverise the heavy clods left by the plough and to release moisture so that the seeds did not become waterlogged, in other words to 'dry the damp' and 'loosen the compact'. However carefully the ploughing was timed, however severe the winter frosts were that weathered the soil, some clods were bound to remain, and the farmer would have to break them up either with a hoe or with a clodding maul or beetle, *yu*[1] or *chhui*[2]. This large wooden mallet is first mentioned in the late Chou texts such as the *Lun Yü*, *Lü Shih Chhun Chhiu* and *Kuan Tzu*;[a] it was apparently used not only for breaking clods but also for covering broadcast seed.[b] The oldest surviving example is a Western Han maul from Niya in Sinkiang,[c] and the implement is still in common use in North China today, though nowadays it is usually called *lang thou*[3] or *mu lang thou*[4].[d] Chia Ssu-Hsieh and Wang Chen both refer to the beetle as a 'wooden hammer' (*mu cho*[5]).[e] It may seem odd that a comparatively primitive instrument such as the maul should have continued to coexist in China for fifteen centuries together with such sophisticated and efficient implements as the iron-tined harrow, but it is presumably more useful in restricted areas such as kitchen gardens, and also when the lumps of soil are so severely compacted that even a heavy iron harrow can hardly break them. Certainly mauls and beetles were a common item in the European farmer's equipment as late as the 17th century,[f] and in Britain were not superseded by the roller until about 1800.[g]

Another tool used to break up clods in Han times and possibly earlier[h] was the hand rake *pa*[6] (which according to the *Fang Yen* was known in Central China as *chhü nu*[7] or *chhü su*[8] [i]). Rakes were used for a variety of purposes, including spreading grain to dry on the threshing-floor or weeding rice or millet, but the strong iron-headed variety was also used in Han times, like the iron drag-hoe *thieh tha*[9], to break and smooth clods in the fields. From the rake developed,

[a] Hayashi (*4*), pp. 276–8; Liu Hsien-Chou (*8*), p. 25.
[b] Cheng Hsuan's gloss on the *Lun Yü* 38.18.6 gives *yu* as 'to cover seed' (*yu, fu chung yeh*[10]).
[c] Liu Hsien-Chou (*8*), fig. 45.
[d] Anon. (*502*), p. 99.
[e] *CMYS* 11.6.1; *WCNS* 12/14*a*.
[f] G. Markham (1), pp. 12–13.
[g] Fussell (2), p. 67.
[h] Rakes (*pa pa*[11]) are mentioned in the *Shu Ching* and *Tso Chuan*.
[i] Quoted *WCNS* 14/15*b*.

[1] 耰 [2] 椎 [3] 榔頭 [4] 木榔頭 [5] 木斫
[6] 杷 [7] 渠挐 [8] 渠疏 [9] 鐵搭 [10] 耰，覆種也
[11] 杷朳

probably in late Han, the animal-drawn tined harrows, and as the harrow spread the use of the hand rake became restricted to gardens. It continued to be a popular tool, and Wang Chen lists a great number of varieties,[a] but by the Yuan only one is used for 'breaking up clods and getting rid of stones and gravel', and its use is restricted to the vegetable patch. In the 1920s Hommel could only find grain rakes and weeding rakes.[b]

The decline in the rake's popularity was a result of the rapid spread of the animal-drawn tined harrow. There are two types of tined harrow in China, the flat harrow (Fig. 80) known as *pa*[1], and the vertical harrow (Fig. 81) known as *chhiao*[2]. Both are presumably a refinement of the hand-rake,[c] but while the former is used throughout China on both wet and dry fields the latter is restricted to the wet rice fields of the south. Both seem to have been developed at the end of the Han, for while neither is represented in Han art, the *pa*[1] is depicted three times in the murals from Chia-yü-kuan[3] in Kansu, painted in the Wei-Chin period (+220 to 316) (Fig. 82),[d] while the earliest known representation of the *chhiao* is a pottery grave model from Kwangtung which is dated between +310 and +312.[e]

K. D. White has speculated that the development of the tined harrow or *irpex* in late Roman times was a local response to labour shortages which made it inconvenient to pulverise the soil, as had been done previously, merely by repeated ploughing; in support he points to several references in both Columella and Pliny to labour shortages and labour-saving devices.[f] While it is possible that rural China also suffered from labour shortages in the period of civil unrest and foreign invasion that heralded and followed the fall of the Han, the growth of large estates, run by powerful local families who offered their tenants and retainers protection in return for labour services, must have precluded labour shortages on the estates themselves.[g] It is unlikely that independent small-holders, however short-handed, developed the ox-drawn harrow, because in those uncertain times they would have found it difficult to keep and protect draught animals. On the other hand the large estates certainly had at their disposal a far larger concentration of animal power than individual peasants could ever hope to muster, even in peace time. The number of domestic animals, particularly draught oxen, depicted on the estate in the mural at Chia-yü-kuan is very striking, so it may be that the number of available animals on such large estates encouraged the development of animal-powered implements such as the harrow

[a] *WCNS* 14/15*b*–17*b*.
[b] (1), pp. 66–7.
[c] In the *Lei Ssu Ching*, Lu Kuei-Meng glosses *pa*[1] by the term *chhü su*[4], used in Han times to denote the rake, *pa*[5], and Wang Chen (*WCNS* 2/7*a*) remarks that *chhü su* was still a common term for harrow in Yuan times.
[d] Hayashi (*4*), figs. 6–32, 33.
[e] Hsü Heng-Pin (*1*).
[f] (1), p. 149.
[g] See P. B. Ebrey (1), pp. 17, 42–5 on great family estates in the late Han and Northern Dynasties.

[1] 耙，耀，爬 [2] 耖 [3] 嘉峪關 [4] 渠疏 [5] 杷

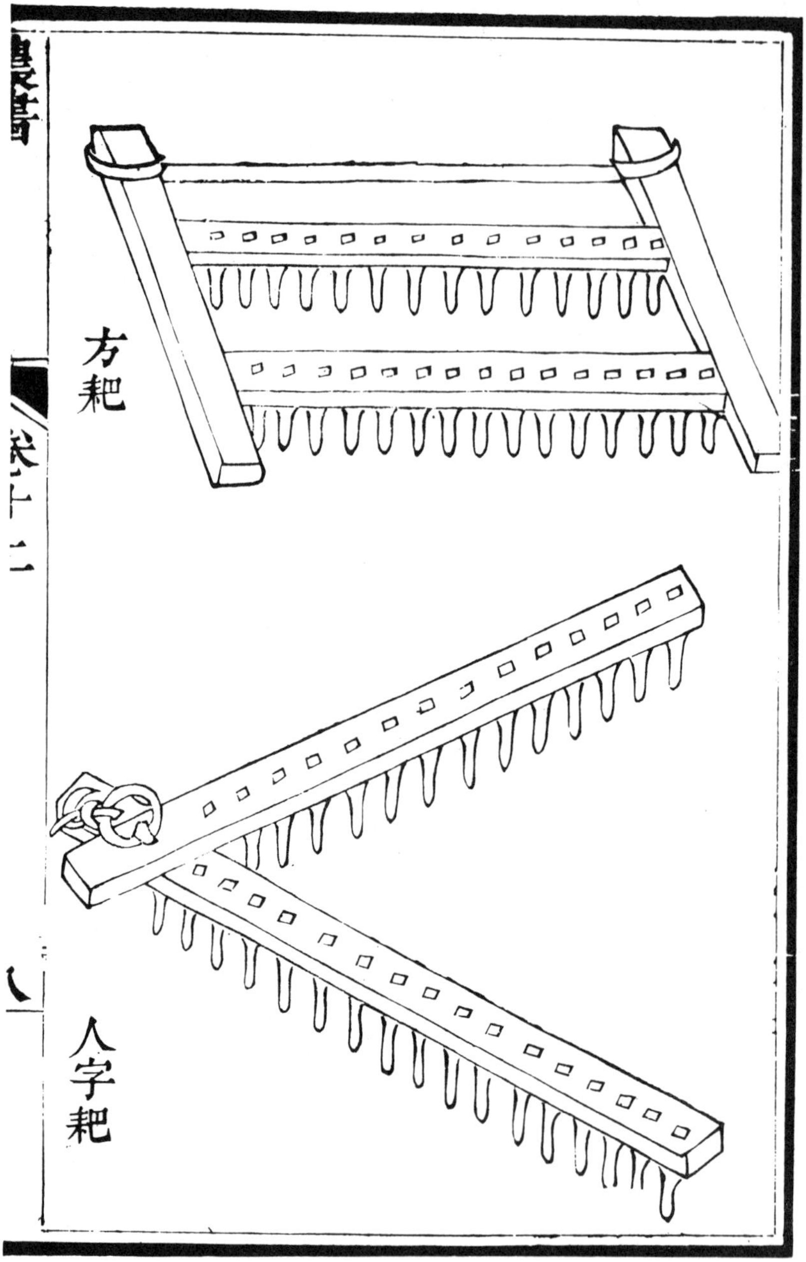


Fig. 80. Flat harrows (*pa*); *WCNS* 12/8*a*.

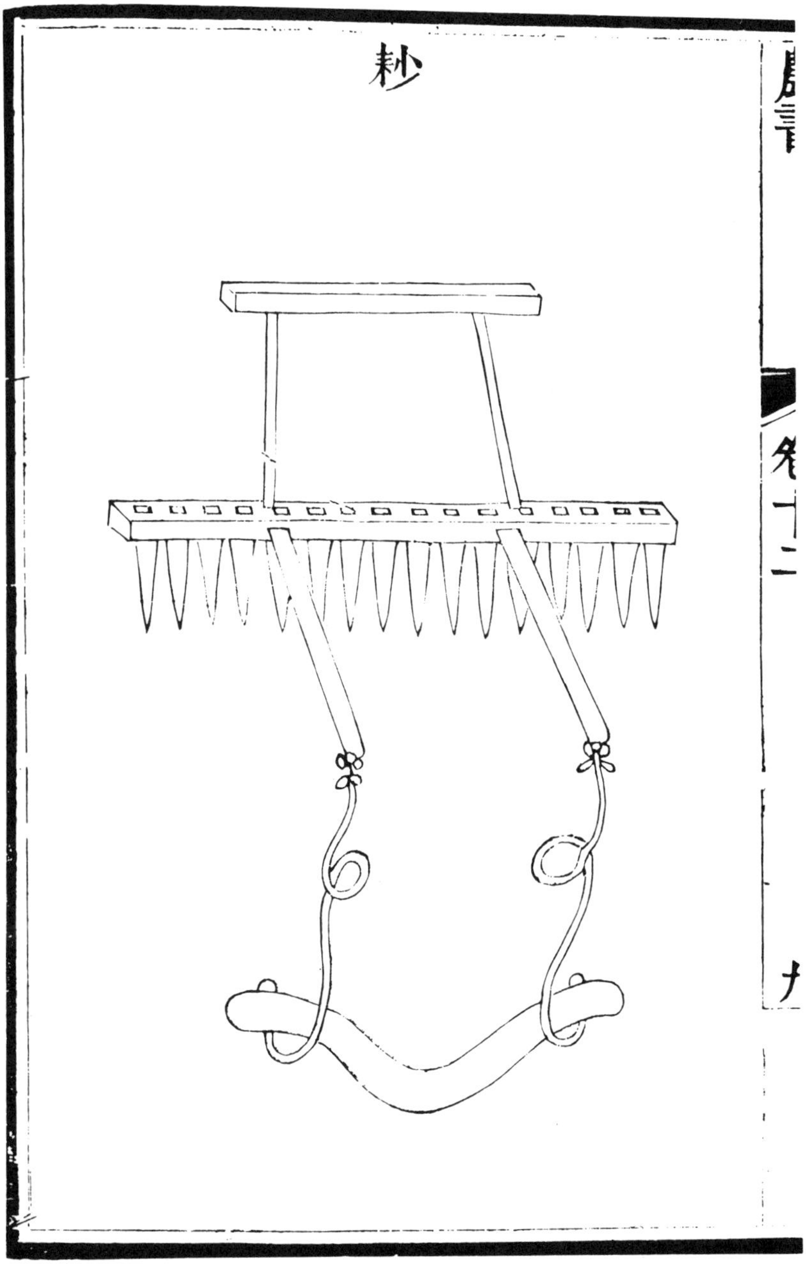


Fig. 81. Vertical harrow, *chhiao*, as used in Southern rice fields; *WCNS* 12/9*b*.

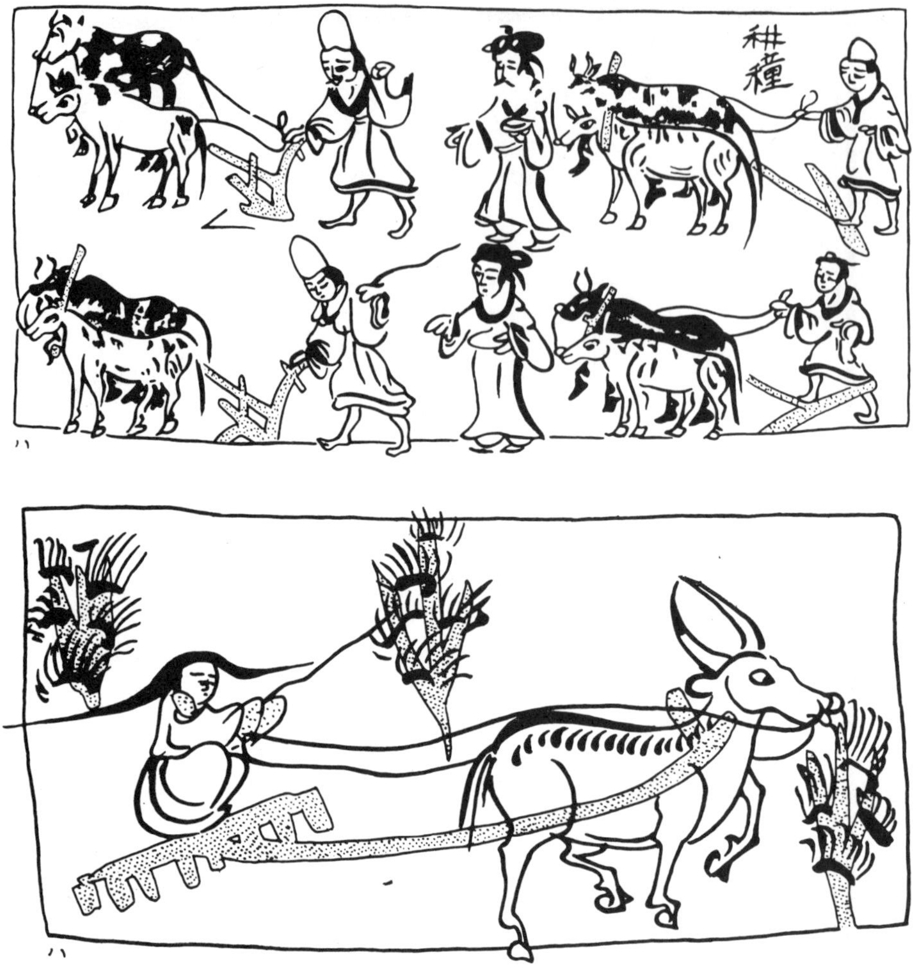


Fig. 82. Wei/Chin murals from Chia-yü-kuan, Kansu, showing ploughing, sowing and harrowing. Note how the driver of the harrow weighs it down with one foot (above), or with his whole body (below). Hayashi (*4*), 6–32, 33.

and different types of mill[a], in order to free skilled human labour for the many profitable but demanding and labour-intensive household industries such as brewing, pickling and weaving which are mentioned in the late Han handbook *Ssu Min Yüeh Ling*[b] as part of efficient estate management.

Although tined harrows apparently developed both in China and in Rome at roughly the same period, it is unlikely that any contact or transmission is involved, for the Western harrows are all constructed on the cross-bar principle (Fig. 83), while Chinese harrows usually consist of one, two or three parallel bars mounted across a pair of beams (Fig. 84).

[a] On the evolution of different types of animal-powered mill see Vol. IV, pt 2, pp. 192 ff.
[b] *SMYL* 3.2; 6.3.3; 6.6; 12.7.4; etc.

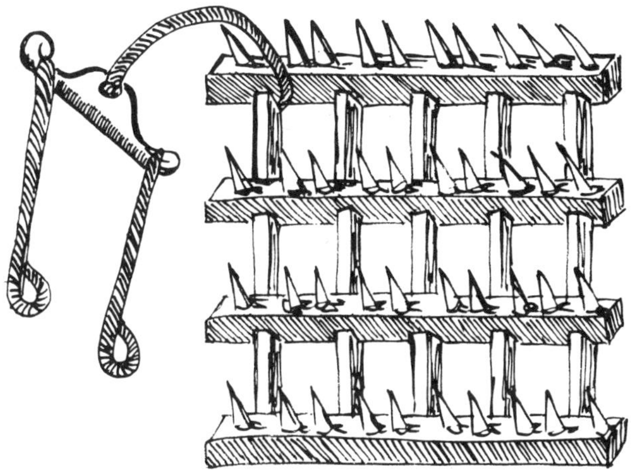

Fig. 83. Elizabethan English harrow; note the square construction of the frame; Markham (2), p. 64.

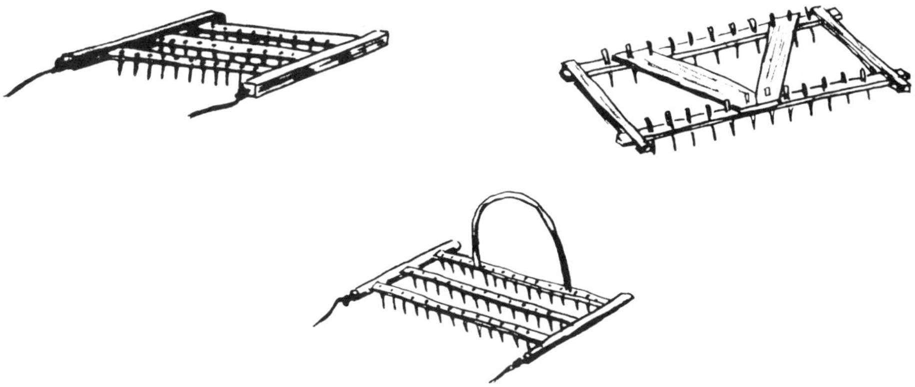

Fig. 84. Modern Chinese harrows, after Wagner (1), p. 203.

The earliest textual references to tined harrows occur in the +6th century *Chhi Min Yao Shu*, where Chia Ssu-Hsieh refers several times to an iron-tined implement, *thieh chhih tou tshou*[1],[a] which Wang Chen[b] explains as an iron-tined flat harrow (*pa*[2]). The *Chhi Min Yao Shu* advocates its use after turning over fallow land,[c] for breaking up impacted clods formed by ploughing wet soil,[d] and for weeding young millet plants:[e]

Once the [setaria] shoots have emerged from the ridges, every time the surface of the soil has turned pale after rain [i.e. the soil has begun to dry out] you should immediately

[a] *CMYS* 1.2.1; 1.3.1; 3.10.4.
[b] *WCNS* 12/8*b*.
[c] *CMYS* 1.2.3.
[d] *CMYS* 1.3.1.
[e] *CMYS* 3.10.4.

[1] 鐵齒鍋榛 [2] 杷

rake and smooth the field criss-cross fashion (*tsung heng pa erh lao chih*[1]) with an iron-tined *tou tshou*[2]. (Method: let a man sit on it and continually pull away the weeds, for if the teeth become choked with them the young plants will be damaged ...).

This quotation shows clearly that the *tou tshou* was a tined harrow of the *pa* variety, for if it had merely been a rake then it would have been impossible for a man to sit on it during operations. Furthermore, the use of the *tou tshou* appears to have been standard in millet farming at that period. This confirms the impression given by the Chia-yü-kuan murals (Fig. 82), which show ploughmen closely followed by men broadcasting seed, which is being covered over by single-barred harrows each drawn by a pair of oxen; the driver of the harrow is himself perched on the bar to weigh it down.

We have no idea whether the *tou tshou* of the *Chhi Min Yao Shu* had only a single bar, like the harrows shown in the Chia-yü-kuan murals, or whether the V-shaped and parallel-barred harrows common in later times had already evolved. The specific use of the term 'iron-tined' in *Chhi Min Yao Shu* does, however, imply that wooden-tined harrows also existed, as in later times. The next literary reference, in the *Lei Ssu Ching* of +880, is hardly more illuminating, for it tells us only that the harrow was always used after ploughing to break up clods and eliminate weeds, and that it had teeth. An interesting point, though, is that Lu Kuei-Meng is the first author to refer to the harrow as *pa*[3]. He is describing conditions in the rice-growing area of the Yangtze Delta, so we know that flat harrows were commonly used then in South China as well as in North. Lu does not mention the vertical harrow, *chhiao*, later so typical of Southern rice-fields, but we know that they must have been used in some parts of South China if a pottery model was made in Kwangtung five hundred years earlier. It may simply be that no clear terminological distinction was made between *pa* and *chhiao* in Thang times.[a] But another possibility is that the *chhiao* was of Southeast Asian origin and had not yet penetrated China north of the Cantonese area (Ling-nan[4] was not considered part of China proper until very late in Chinese history, and its cultural affinities lie rather with Tonkin and the Vietnamese traditions). The vertical harrow is a typical feature of Southeast Asian wet-rice cultivation, and is found in Vietnam (where it is called *cái bwà*),[b] in Burma (*htun*),[c] in Malaya (*gĕrap*),[d] in Java (*garu*),[e] and in the Philippines.[f] It is noteworthy that the flat tined harrow (*pa*) is unknown in these countries, which might be considered

[a] The distinction is certainly not clear in modern China, where the differentiation in nomenclature appears to be confined to the Southern provinces; see Anon. (*502*), pp. 127–9.
[b] Huard & Durand (1), p. 124; G. C. Hickey (1), p. 140; Leser (1), fig. 298 shows an example from Tonkin.
[c] Cheng Siok-Hwa (1), p. 32.
[d] A. H. Hill (1), fig. 3; Bray & Robertson (1).
[e] Raffles (1), p. 114; L. Th. Mayer (1); Leser (1), fig. 303.
[f] Leser (1), fig. 296.

[1] 縱橫杷而勞之 [2] 鐵齒楱 [3] 爬 [4] 嶺南

additional evidence that the vertical harrow is the native Southeast Asian form, and that it spread into China from the South at a comparatively late date.

The art of harrowing was brought to such a high pitch in China in the +1st millennium that by Sung and Yuan times writers were already bewailing its decline. The *pa* or flat harrow now had two standard forms:[a]

The length of the cross bars (*thing*[1]) should be five feet, the width four inches; the two beams should be five inches or more apart ... Each is pierced with square holes in which are set wooden teeth six inches or more long. At each end of the cross-bar is a wooden beam about three feet long, curved slightly upwards at the front end and pierced by a wooden peg to which the ox's traces are attached. This is the square harrow (*fang pa*[2]).

There is also the V-shaped harrow (*jen tzu pa*[3]), which has cast-iron tines ... When using this harrow a man stands on top so that it works the soil more deeply. He must also frequently set foot to earth to clear away the weeds that get caught [between the tines]. This implement should always be used in wet fields.

Although Wang implies a clear distinction between the wooden-tined square *pa* and the iron-tined V-shaped *pa*, iron tines (often curved) are now more common than wood in the square *pa* (they are usually staggered to increase efficiency).[b] The V-shaped *pa* has disappeared completely.[c]

Of the vertical *chhiao* Wang says:[d]

The *chhiao* is an implement for dividing and mixing the mud in [wet] fields. It is about three feet high and four feet wide, with a horizontal handrail at the top and a row of tines at the bottom. The tines are twice as long as those of the *pa* and more closely spaced. One must hold the *chhiao* with both hands, and it is drawn by a beast harnessed between the shafts in front. One *chhiao* requires one man and one beast, but sometimes two *chhiao* are linked together using a pair of men and beasts: this is a special arrangement for large fields so that the work can be finished more quickly. The *chhiao* is used after the plough and the *pa* to make sure the mud is really well worked.

Although it is clear from the attention that Wang devotes to these tined harrows that their use was widespread, a century or so before, the author of the *Chung Shih Chen Shuo* was complaining that no-one paid sufficient attention to harrowing properly:[e]

The ancient farming practice was to use the tined harrow (*pa*[4]) *six times* [italics ours] after each ploughing, but nowadays people are only concerned with the benefits of deep ploughing and do not realise that actually really fine harrowing is much more effective. If harrowing is not done properly then the soil will be coarse and unfruitful, and although young shoots will appear after sowing, because their roots are growing among coarse

[a] *WCNS* 12/8*b*.
[b] Hommel (1), p. 56; Anon. (*502*), pp. 105 ff.
[c] Wagner (1), p. 204 says that none of his Chinese students had ever seen such a thing.
[d] *WCNS* 12/10*a*.
[e] Quoted *WCNS* 2/7*a*; *SSTK* 33/2*a*.

[1] 桯 [2] 方耙 [3] 人字耙 [4] 䎱

lumps of earth they will not marry properly with the soil and so they will be unable to withstand drought. Some will droop and die, others will be attacked by insects and die, some will shrivel up and die, or succumb to other diseases. But if harrowing is properly done, then the soil will be finely worked and fruitful, and because the roots are set in fine and fruitful soil . . . the plants will be drought-resistant and will not suffer from disease.

So harrowing not only produced a fine tilth, but also helped to conserve soil moisture and to build fine, healthy, resistant plants. Wang Chen[a] says that repeated harrowing should produce a layer four inches thick (deep enough to cover a hen's egg) of fertile, pinguous soil (*yu thu*[1]) on the surface: one is reminded of Tull's asseveration that thorough comminution of the soil plays just as important a role as manuring in fertilising the soil.[b]

Despite the complaints and forebodings of most agricultural writers from Yuan times on, tined harrows have continued to play an important part in soil preparation throughout China right up to the present day, as witness the number and variety exhibited at the National Agricultural Exhibition held in Peking in 1957.[c] As we mentioned earlier, this type of harrow was also used after the crops had emerged to weed and stir the soil, but their main role was to break up the soil *before* the seed was sown. This offers an interesting contrast with the way that harrows were used in the West:[d] 'There are some kinds of soil which are so luxuriant that it becomes necessary to comb the crop [of young wheat] while in leaf—the comb is another kind of harrow fitted with pointed iron teeth (*cratis est hoc genus dentatae stilis ferreis*).' In general the Romans regarded it as poor husbandry to be obliged to harrow after the seed had been ploughed in,[e] but this view was not held by later European farmers. Markham, like many of his near contemporaries, recommends pulverising the soil after rather than before the crop has been sown:[f]

After the land is sowen, you shal then harrow it as small as may be, first with a paire of woodden Harrows, and after with a paire of yron Harrowes, or else with a double Oxe harrow; for this earth being somewhat hard and much binding, will aske great care and diligence in breaking.

After your Barley is sowne, you shall about the latter end of *April* beginne to smooth and sleight your land, both with the backe Harrowes and with the rowler, and looke what clots they faile to breake, you shall with clotting beetles beate them asunder, making your mould as fine, and laying your Land as smooth as is possible.

It is difficult to explain why Western farmers should have preferred to harrow after the seed was already sown, when the advantages of a really fine seedbed are so obvious. The cause may lie in the fact that many Western cereals, unlike the Chinese millets and wheats, are winter crops, sown in some haste before the onset

[a] *WCNS* 12/8*b*.
[b] See Fussell (3), ch. 6.
[c] Anon. (*18*), pt 2.
[d] Pliny 18.186; see K. D. White (1), p. 146.
[e] Columella II.iv.2.
[f] G. Markham (2), vol. I, p. 86.

[1] 油土

of frost and snow. Of all the English field crops sown in the +16th century only rye was deemed to require a really fine seedbed:[a]

> For barley and pease, harrow after thou sowe:
> for rye harrow first, seldome after I trowe.
> Let wheat haue a clodde, for to couer the hedde:
> that after a frost, it may out and go spredde.
>
> Threshe sede and goe fanne, for the plough may not lye:
> September doth bid, to be sowing of rye.
> The redges well harrowde, or euer thou strike:
> is one poynt of husbandry, rye land do like.
> Geue [give] winter corne leaue, for to haue full his lust:
> sowe wheate as thou mayst, but sowe rye in the dust.

Harrowing was, as common English usage implies, a gruelling experience:[b]

> It is a greate labour and payne to the oxen, to goo to harowe: for they were better to goo to the plowe two dayes, thanne to harowe one daye. It is an olde saying, 'the ox is neuer wo, tyll he to the harowe goo'. And it is bycause it goeth by twytches, and not alwaye after one draughte.

This being so, perhaps it is not surprising that the European farmer liked to allow his oxen a rest between ploughing and harrowing.

Though a tined harrow is ideal for breaking up recalcitrant clods, the best implement for creating a smooth layer of fine crumbs of soil, ideal for sowing and retaining moisture, is the bush-harrow (Fig. 85):[c]

> The bush-harrow (*lao*[1]) is a flat harrow (*pa*[2]) without tines; instead thin sticks are woven between the bars of the harrow to smooth the field. The ploughman ploughs and smooths alternately, with an eye to the moistness of the soil. The *lao* is used to make sure the field is smooth and the soil fertile (*jun*[3]) ... After the plough and tined harrow have been used, it is essential to use the bush-harrow before the seed is sown. The proverb goes: 'To plough without smoothing is to court disaster (*keng erh pu lao, pu ju tso phao*[4])'.

The bush-harrow seems to have evolved slightly earlier than the tined harrow in both East and West. In the Mediterranean the bush-harrow was apparently not known in pre-Roman times, but it is mentioned by a number of Roman writers, including Virgil and Varro.[d] In China the earliest evidence for its use is in the *Fan Sheng-Chih Shu* of the early −1st century:[e]

> In spring the *pneuma* (*chhi*[5]) of the earth flows freely and one may plough stiff, strong land and heavy black soils. Level the clods immediately with a bush-harrow (*mo*[6]) ...

[a] T. Tusser (1), §§72, 10–11.
[b] Fitzherbert (1), fol. 15.
[c] *WCNS* 12/11*a*.
[d] K. D. White (1), p. 146.
[e] *FSCS* 1.3.1, quoted *CMYS* 1.11.1.

[1] 勞 [2] 耙 [3] 潤 [4] 耕而不勞，不如作暴
[5] 氣 [6] 摩

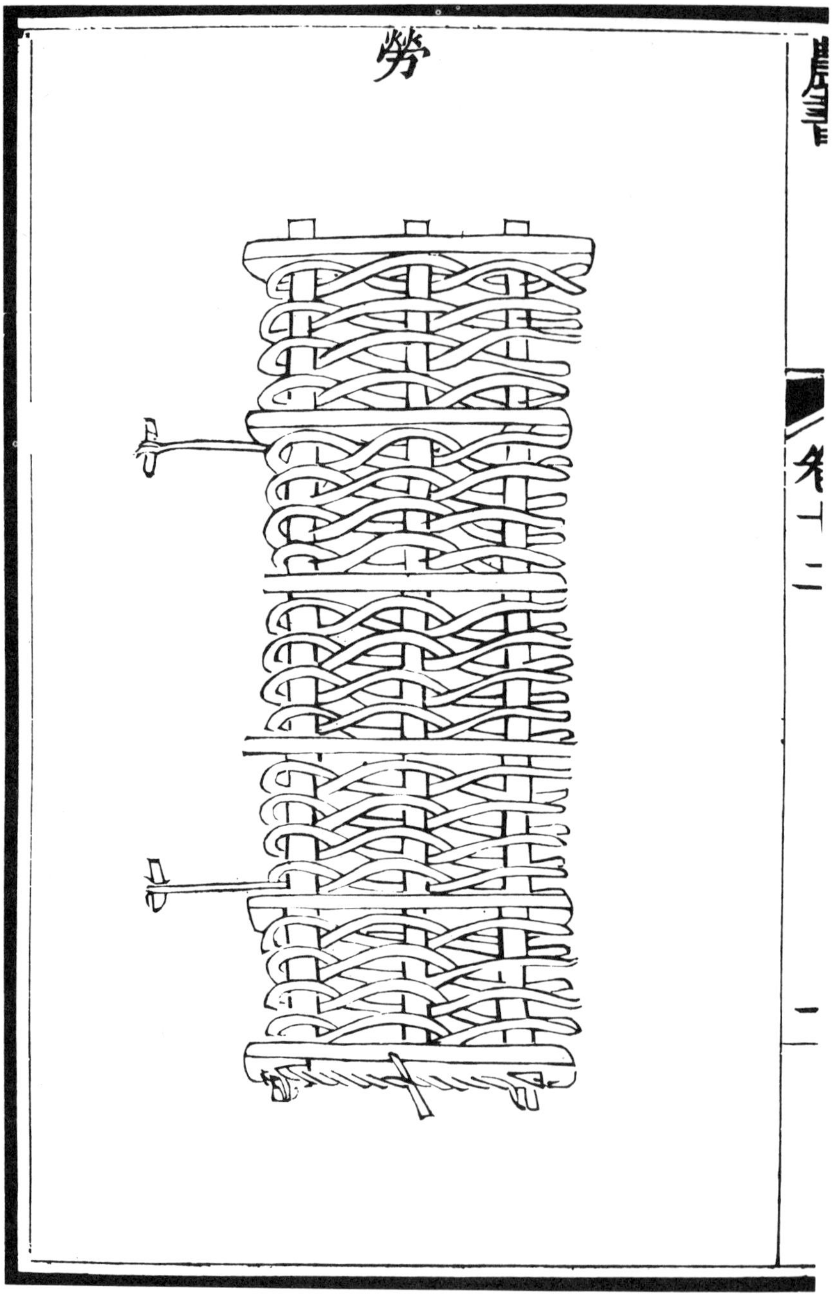


Fig. 85. Chinese bush harrow (*lao*); *WCNS* 12/10*b*.

Make the soil even-textured and free from clods ... This is what is meant by making strong soils weaker.

Chia Ssu-Hsieh[a] explains that *mo*, which literally means 'to rub smooth', is an alternative form for *lao*[1], the bush-harrow, and this is borne out by the fact that in parts of Honan and Shensi *mo*[2] is still used for bush-harrows even today.[b] So the bush-harrow was already in use in the Western Han, and the proliferation of soil-pulverising tools at this period should hardly surprise us as it corresponds to the development and spread of the turn-plough. By the time that the *Chhi Min Yao Shu* was written (*c.* +535), the bush-harrow was almost as important as the plough itself: both spring and autumn ploughing had to be followed by at least two turns with the *lao*:[c]

> Immediately after spring ploughing the *lao* should be used ... After the autumn ploughing, wait until the surface of the soil turns pale, then smooth it with the *lao*. Since spring is very windy, the soil will certainly become dry and alveolar (*hsü*[3]) if it is not harrowed immediately. In the autumn the soil is full of water, so if you harrow while the surface is still wet it will dry out hard ... Spring and summer ploughing should be shallow, the furrows should be narrow, and the *lao* should be used twice (... if the *lao* is used twice the soil will have a fine tilth and will retain its moisture even in a drought).

In effect the bush-harrow is used to create the dust-mulch so beloved of agronomists in dry regions today.[d]

The references to the *lao* in the *Chhi Min Yao Shu* are too numerous to cite. The term *lao* has survived through Yuan times to the present day, but the bush-harrow is known by several other names as well. In the preface and commentary to *Chhi Min Yao Shu*[e] the term *kai*[4] occurs frequently (*kai* literally means 'to cover', and this usage is presumably derived from the harrow's being used to cover over broadcast seed); these sections were probably written in the early Thang,[f] and the term *kai* is still in use in modern Hopei; *lao*[5] is common nowadays in Shantung and Northern Honan, while *mo*[2] is used in Southern Honan and Shensi.[g]

Although the bush-harrow is occasionally found in areas of heavy soils in Europe,[h] it is much less important than the tined harrow in such regions.[i] As for China, in Wang Chen's time while the *lao* was not used in wet fields it was, he claimed,[j] used even by Southerners to work dry fields to a really fine tilth. But today its use is confined to North China, and it is often used in association with the ard. The form is also widely spread throughout the Mediterranean, the Near

[a] *CMYS* 1.3.2.
[b] Liu Hsien-Chou (*8*), p. 27.
[c] *CMYS* 1.3.2–3.
[d] Arnon (1), vol. I, ch. 13.
[e] E.g. *CMYS* 00.8.
[f] Kumashiro (1), p. 431 gives a résumé of the evidence for and against this dating.
[g] Liu Hsien-Chou (*8*), p. 27.
[h] It has been reported in parts of Germany and Bohemia; Leser (1), p. 489.
[i] See Fitzherbert (1); G. Markham (1), (2); G. E. Fussell (2), etc.
[j] *WCNS* 2/8*a*–*b*.

1 劳 2 糖 3 虚 4 蓋 5 劳，耮

East, Transcaucasia, Iran and India,[a] also areas where the ard is used. Leser suggests that bush-harrows are closely linked to the *Pflüge mit Krümel* (or *araires chambiges*, ards with curved beams), a form which he believes to have originated in the Mediterranean area and spread gradually eastwards, probably in post-Roman times.[b] In view of the indigenous and idiosyncratic development of plough forms in China this seems an unlikely explanation of the bush-harrow's appearance there in early Han times. The bush-harrow may have come to China independently of any plough form: the idea could have been brought back by the armies sent by Han Wu-ti into Central Asia and beyond. But since so very little is known of its history, either in Occident or Orient, it would be imprudent to make any categorical assertions as to its ultimate origin or subsequent dispersal.

Whereas in North China the bush-harrow was used to perfect the field's tilth before sowing, in the wet rice fields of South China rollers were used to put the finishing touches to the puddled mud. *Chhi Min Yao Shu* refers to a roller called *lu chu*[1] [c] used for this purpose; another early reference to the roller occurs in the Thang work *Lei Ssu Ching*: 'After the *pa*[2] the *ko chih*[3] or the *lu thu*[4] are used. The *pa* and the *ko chih* are both spiked, but the *lu thu* is just ovoid and is made completely of wood; hard, heavy wood is best.' The illustration of the *lu thu* in the Ming edition of Wang Chen's *Nung Shu* shows an ovoid, ribbed roller (Fig. 86), and Wang Chen explains:[d]

> I myself believe that since the characters *lu thu* contain the radical for stone (*shih*[5]) this roller must originally have been made of stone, and indeed in the North such stone rollers are often found. But in the South they use wood, for wet and dry fields have different requirements ... The *lu thu* is about three feet wide ... with a wooden frame and a wooden axle through the middle [of the roller] so that it turns easily.

The 1769 illustrations of *Keng Chih Thu*, which are almost certainly based on Sung originals of 1237 (Fig. 87),[e] as also the Japanese illustrations of 1676, based on Ming illustrations of 1462,[f] show the *lu thu* being used in the wet fields as the final process before sowing the rice. The Sung-based picture shows a large-roller, apparently ribbed but cylindrical rather than ovoid, while the Japanese drawing depicts a simple flat-surfaced cylindrical roller. It is significant that the Chhing edition of 1739,[g] though the picture is clearly labelled *lu thu*, in fact shows a tined harrow: in modern China smooth rollers are no longer used in wet fields.[h]

On the other hand the spiked roller (*ko chih*)[i] was improved and elaborated, and became a most important part of the rice-farmer's equipment:[j]

[a] Leser (1), pp. 488–90.
[b] Leser (1), pp. 492 ff.
[c] *CMYS* 11.2.3.
[d] *WCNS* 12/15*a*.
[e] Pelliot (24), pl. XV.
[f] O. Franke (11), pl. XIX.
[g] *Ibid.* pl. XX.
[h] Alley & Bojesen (1), p. 89; Anon. (*502*) and Hommel (1) offer no examples of flat rollers used in wet fields.
[i] *Ko* is the pronunciation given in the *Lei Ssu Ching* and *WCNS* but the more orthodox pronunciation would be *li*.
[j] *WCNS* 12/17*a*.

[1] 陸軸 [2] 爬 [3] 礰礋 [4] 磟碡 [5] 石

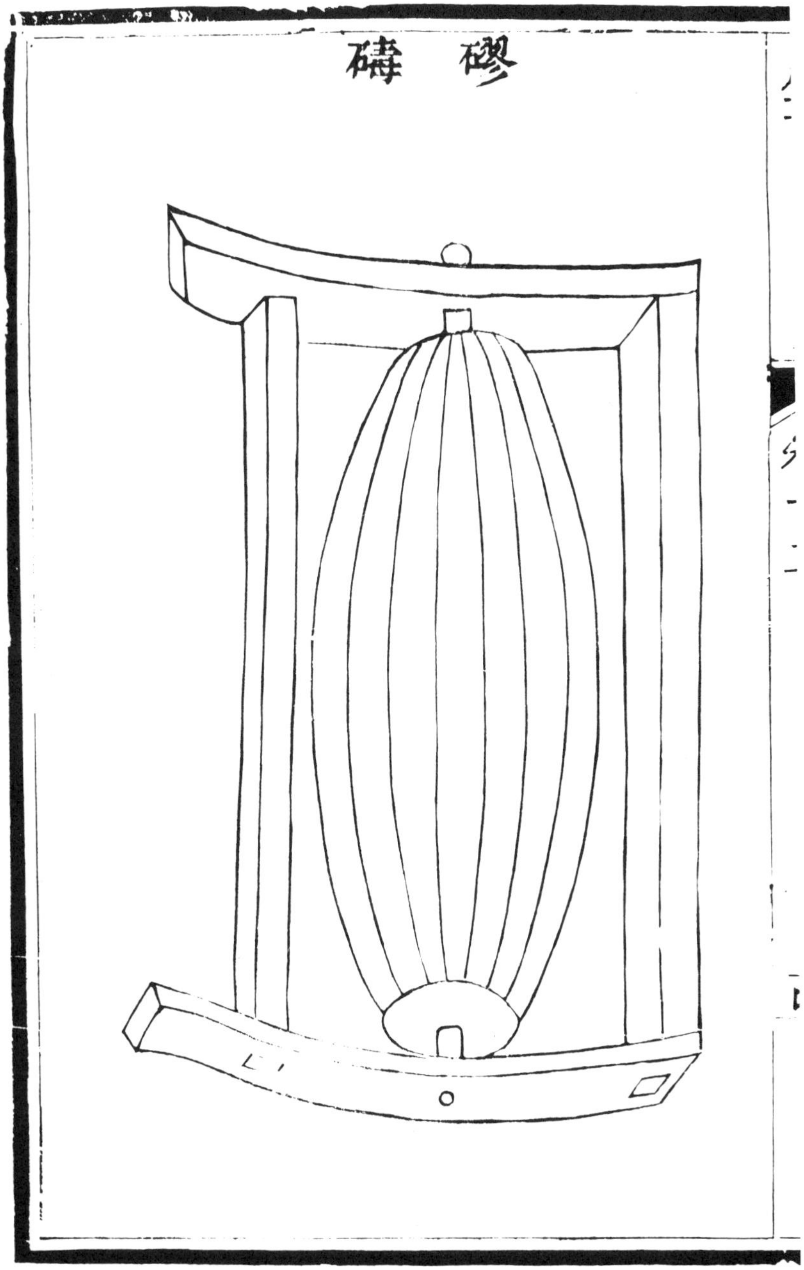


Fig. 86. Ovoid roller (*lu thu*); *WCNS* 12/14*b*.

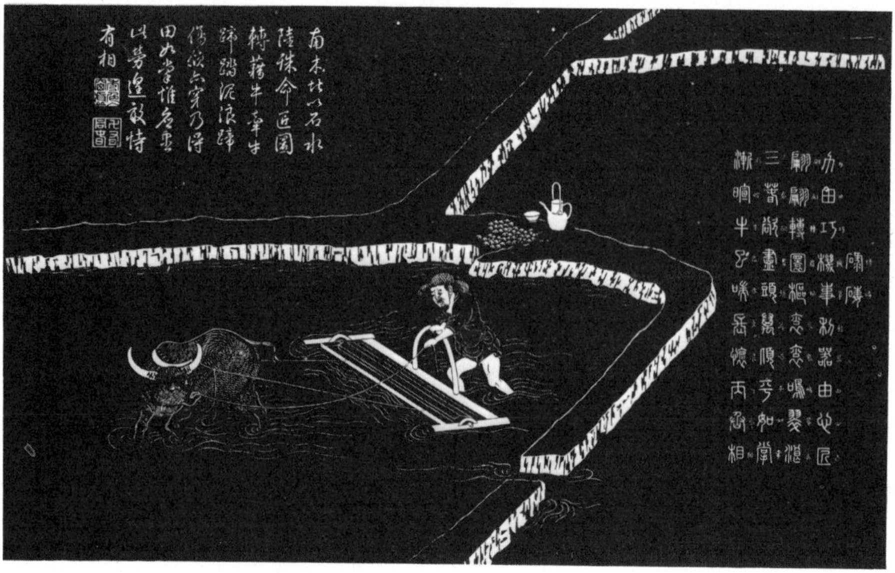

Fig. 87. Cylindrical ribbed roller, from the Sung-based illustration to the *Keng Chih Thu*; Pelliot (24), pl. XV.

> ... With serried teeth, so sharp and hard, it cuts through clods of soil,
> First turns the ground to fertile earth, then churns it to mud in spring;
> Roll the field and roll it again, till the texture's as smooth as can be ...

Although the textual entry in the *Nung Shu* is comparatively short, Wang Chen illustrated two types of *ko chih*, one with spikes and the other with blades (Fig. 88). Both are very effective for stirring and puddling mud, and are still found in great numbers in China today, though nowadays the roller is much smaller in diameter than those shown by Wang Chen. Hommel[a] describes in considerable detail a roller equipped with blades and a wicker mud-guard to protect the driver from splashes (Fig. 89), and a great many variations on the roller theme, some with a single roller, some with several, armed with a variety of spines, teeth, paddles or blades, some with a seat on top to provide extra weight, are shown in Anon. (*502*); the usual term for such rollers nowadays is 'roller harrow', *kun pa*[1] or *phu kun*[2].

In Europe the spiked roller was used chiefly for aerating pasture land or leys, and although several enterprising mid-18th-century farmers laid claim to its invention[b] it was certainly in use, though not commonly, as early as +1700, for Randall describes a 'rowl fitted with Oaken pins' that was in use in Essex at that date.[c] But there are no known examples of spiked rollers in Northern Europe

[a] (1), pp. 57–8; figs. 89–90.
[b] Notably Cuthbert Clarke (1); see Fussell (2), pp. 67–8.
[c] John Randall (1); a spiked roller is in fact mentioned by Olivier de Serres (1) in +1600, but de Serres was writing of Mediterranean agriculture.

[1] 滾耙 [2] 蒲滾

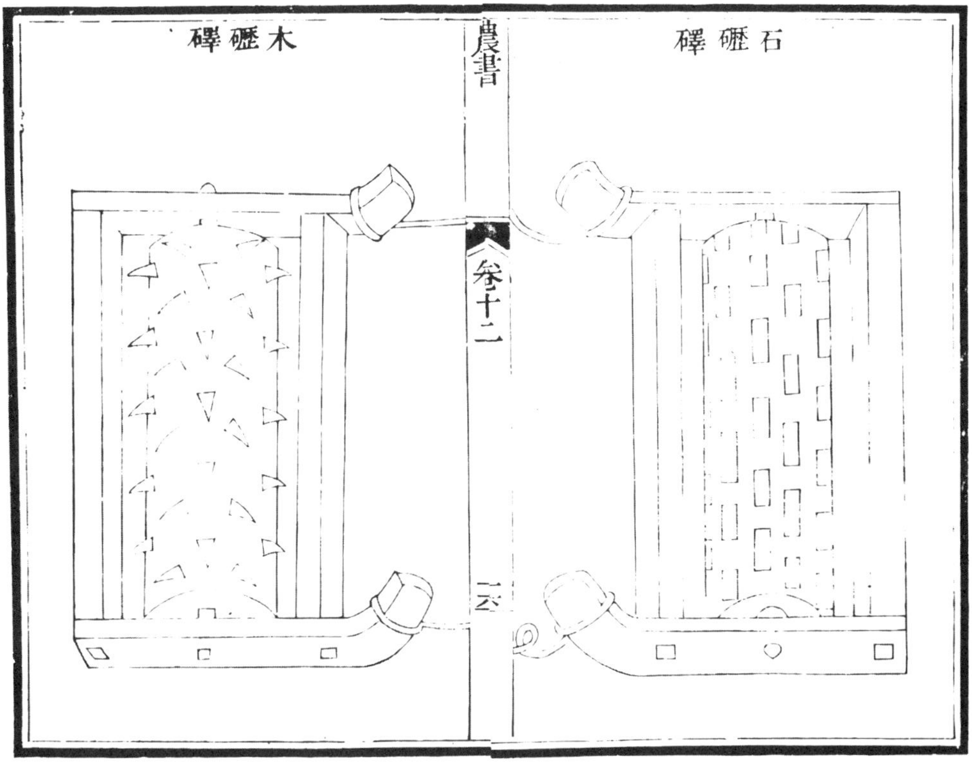


Fig. 88. Rollers with spikes (left) and blades (right) (*ko chih*); *WCNS* 12/16.

Fig. 89. Bladed roller of the *ko chih* type from modern Chekiang; note the wicker cover which protects the driver from splashes of mud; Hommel (1), fig. 89.

previous to the +18th century, so perhaps they were, as Leser suggests,[a] an introduction from East Asia that reached Europe in association with the curved iron mould-board.

When the mud had been stirred or puddled to a really even consistency it was often smoothed, particularly in seedbeds, where an even surface is especially important:[b] 'The field must be smoothed absolutely flat for the seeds so that when the water is let in the depth is equal everywhere and the seedlings grow evenly.' Several tools were used for this purpose. The 'flattening board' (*phing pan*[1]) was 'a smooth-surfaced square wooden plank with two wooden lugs on top, roped to the buffalo's yoke or pulled by a man'.[c] The 'field stirrer' (*thien thang*[2]) had 'a six foot long forked stick forming the handle, with a plank five feet or so wide mortised across the fork ... it is pushed back and forth along the surface to mix soil and water and even out bumps and dips so that the seed can easily be sown'.[d] The 'scraping board' (*kua pan*[3]) was:[e]

> an implement for paring the soil, made out of a wooden board two feet or so wide and twice as long; some have a tongue of wrought iron [as a cutting edge]. To the back of the board are nailed two straight wooden battens that project above the board and are joined by a cross-bar which serves as the handle. On either edge of the board an iron ring is fixed to which the traces are tied [Fig. 90]. The *kua pan* is held steady with both hands, and can be drawn by a man or by a buffalo. It pares the soil as it is drawn along broadside on.

This paring board was used to smooth seedbeds as well as for 'repairing embankments, throwing up polders, excavating drains, piling up soil, making bunds, raising threshing floors, sweeping up grain, collecting chaff, and freeing soil of stones and rubble'. Wang Chen says it was a very common farmer's tool. A number of similar levelling devices are still used in China today, including a wheeled 'scraping board' that is now more prosaically called a 'soil smoothing implement' (*phing ti chhi*[4]).[f]

In the dry fields of North China a smooth roller is used to put the finishing touches to the field, ironing out bumps and compacting loose soils so as to prevent erosion and retain soil moisture. It may be that such rollers were already in use as early as the Western Han. *Fan Sheng-Chih Shu* contains two passages which might refer to rollers:[g]

> In the winter as soon as it stops snowing immediately press down (*lin*[5]) with a [character missing] the snow covering the ground to prevent it being blown away in the wind; repeat every time it snows, and so in the spring the soil will retain its moisture (*pao tse*[6]).

[a] (1), p. 451.
[b] *WCNS* 14/18*a*.
[c] *WCNS* 14/18*a*.
[d] *WCNS* 14/19*a*–*b*.
[e] *WCNS* 14/30*b*.
[f] Anon. (*502*), p. 172.
[g] *FSCS* 1.8.1, quoted *CMYS* 1.16.2; *FSCS* 4.3.3, quoted *CMYS* 10.11.6.

[1] 平板 [2] 田盪 [3] 刮板 [4] 平地器 [5] 藺
[6] 保澤

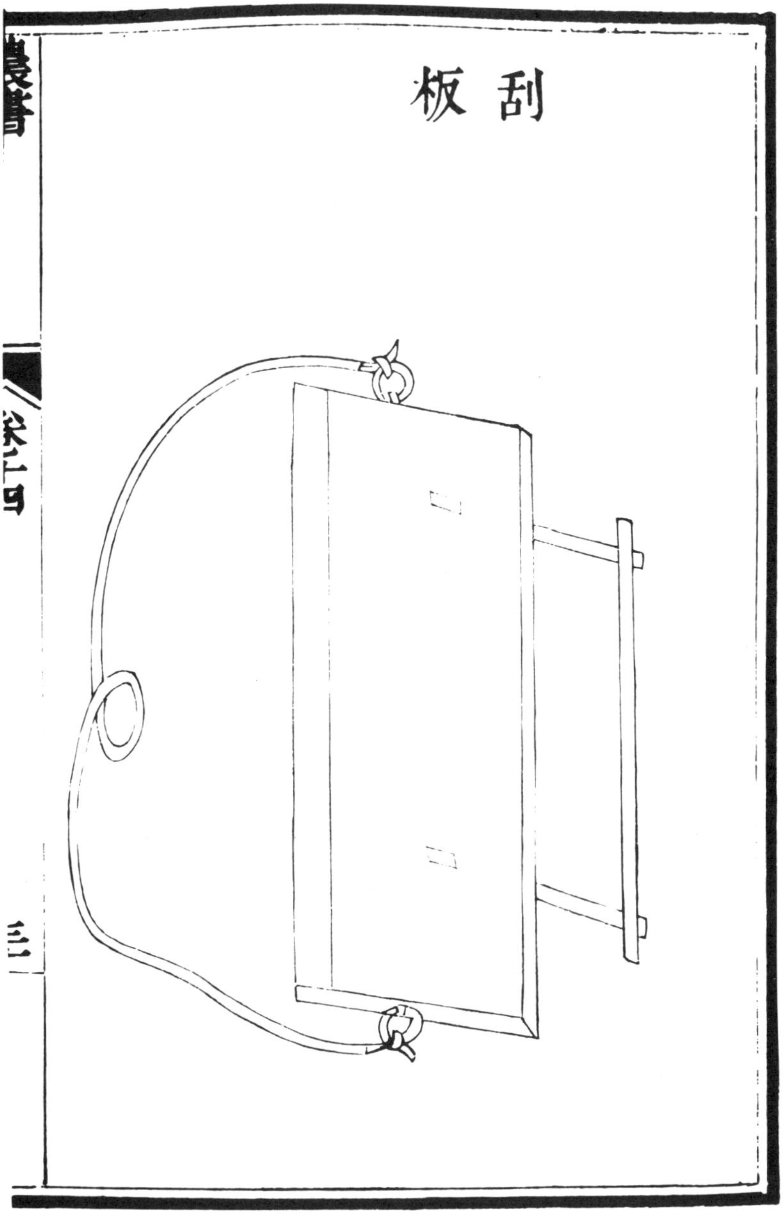


Fig. 90. 'Scraping board' (*kua pan*) for levelling seed-beds; *WCNS* 14/30*a*.

In the winter when it stops snowing, immediately press down with an object (*wu*[1]) the snow covering the wheat ... and so it will be drought-resistant and bear a good crop.

It is unfortunate that the object used to press down the snow is missing in both passages, but since *lin*[2][a] usually refers to the ground crushed under chariot wheels it seems reasonable to suppose that some sort of wheel or roller was involved. No field rollers are known from Han iconography, but since rotary rollers were probably used for crushing and grinding as far back as the Chou and were certainly known in the Han,[b] and since at later periods rollers were used interchangeably on field and threshing-floor, Fan Sheng-Chih could well be referring to the use of a roller in these passages. A common term for the roller in later texts was, as we saw earlier (p. 234), *lu thu*[3]. The earliest use of this term is in a commentary in *Chhi Min Yao Shu* on a passage from the late +1st century work *Ssu Min Yüeh Ling*:[c] 'Oats (they are rather difficult to hull, so this should only be done at the height of the summer with a stone roller, *lu thu nien*[4]).' Otherwise *Chhi Min Yao Shu* does not refer to using rollers, and it is not until Wang Chen (p. 234) that we have explicit mention of their use in the fields in North China. Nowadays they are common in Northern districts, used for firming the ground and conserving moisture. Contrary to Wang Chen's generalisation, both stone and wood rollers are found in North China, in a variety of shapes and sizes, and are most commonly known as *nien*[5] or *kun tzu*[6].[d]

The number and variety of implements mentioned in this section gives some idea of the value that Chinese farmers placed on preparing the soil thoroughly. Although fertilisers were often used at this stage, before the seed was sown (see section 3), it is clear that the Chinese had, like Jethro Tull, a firm faith in the nutritive powers of finely worked soil and were prepared to invest great amounts of labour in order to get the crops off to a good start. In Europe the quality of the seedbed was somewhat less important: in a mixed arable-pastoral economy where fallowing was possible, larger quantities of manure were available, and rainfall was frequent, nutrition and moisture could be provided by other means than painstaking communition of the soil. But where fertilisers and manure were scarce or of poor quality and rainfall was uncertain, as was frequently the case in traditional China, the combination of thorough soil comminution and careful crop rotation (see pp. 429 ff.) could be made to compensate.

[a] The homophone of *lin*[7], a type of reed, which clearly has no meaning in this context.

[b] Vol. IV, pt 2, pp. 92, 199. A wall painting of the Western Chin period (+268 to 317) depicting rural scenes that include mills and rollers for grinding grain was found at Turfan in Central Asia in 1964; see Wang Ning-Sheng (*2*), p. 28.

[c] Quoted *Chhi Min Yao Shu* 10.16.1.

[d] Anon. (*502*), pp. 163 ff.

[1] 物 [2] 閵 [3] 磟 or 碌碡 [4] 碌碡碾 [5] 碾
[6] 磙子 [7] 藺

## (2) Sowing

There is a sede, that is called Discretion, and if a husband haue of that sede, and myngle it among his other cornes, they wyll growe moche the better; for that sede wyll tell hym, how many castes of corne euery lande ought to haue. And a yonge husbande, and may fortune some olde husbande, hath not sufficyente of that sede: and he that lackethe, let hym borowe of his neyghbours that haue. And his neyghbours be vnkynde, if they wyll not lende this yonge husbande parte of this sede. For this sede of Discretion hath a wonders property: for the more that it is taken of or lente, the more it is.[a]

### (i) *Planting calendars and the selection of sowing dates*

The tasks of sowing and planting each have their place: if one knows the right time [for each crop] and does not contravene the prescribed order, then they will sprout and ripen in a constant succession to your benefit and profit. There is no day when something is not sown, no month when something is not harvested—as one crop follows another throughout the whole year, how can anyone suffer want?[b]

We have already discussed the importance of agricultural calendars and almanacs in China which prescribed appropriate dates and conditions for the farmer's various tasks. It is worth noting here that, at least from Han times on, dates for sowing generally took pride of place in these works. This is not fortuitous, for while there is no mistaking when the grain is ripe or the fruit needs picking, judging the right time to sow requires experience, complex evaluation of environmental conditions, and a large measure of good luck. It is the symbolic moment when the seed is given life, and any mistake may mean stunted growth and a poor harvest, so naturally much thought went into evaluating the best sowing time for the most important crops.

In ancient times this crucial matter was settled by divination. References to millet planting ceremonies are found in the Shang oracle bones:[c] 'At Fu Ching's millet planting there will not perhaps be libation', reads one inscription. 'The very fact that the subject was divined', says Keightley,[d] 'indicates the degree to which economic and ritual activity were inseparable.' During the Chou calendrical and phenological observations seem to have replaced the state ritual of divination; this can be seen in the *Pin Chhi Yüeh* of the *Shih Ching*, for example, and in the *Hsia Hsiao Cheng*. Neither of these texts in fact contains any explicit references to sowing, though perhaps this may be attributed simply to their very fragmentary state of preservation; but the *Hsia Hsiao Cheng* does mention ceremonies performed to ensure good harvests of millet and soya beans, which presumably were closely followed by the actual sowing of these crops.[e]

During the Han the month, the position of the stars, and the budding or

[a] Fitzherbert (1), 20.
[b] *CFNS* ch. 1, section 5, p. 4.
[c] Keightley (1), 64.
[d] *Ibid.*
[e] *Hsia Hsiao Cheng Su I* 23; 34–5.

Table 7. *Planting calendar according to* Chhi Min Yao Shu (from Li Chhang-Nien (*3*), 93).

| *Planting periods* / *Crop* | *Best* | *Medium* | *Late* |
|---|---|---|---|
| *Setaria italica* | 1st ten-day period *hsün*[1] of 2nd month; hemp flowers and willows come out. | 1st *hsün* of 3rd month; *chhing-ming* solar period; peach starts to bloom | 1st *hsün* of 4th month; mulberry flowers fall |
| *Panicum mil.* | 1st *hsün* of 3rd month | 1st *hsün* of 4th month | 1st *hsün* of 5th month |
| *Spring soya beans* | 2nd *hsün* of 2nd month | 1st *hsün* of 3rd month | 1st *hsün* of 4th month |
| *small beans* | 10 days after summer solstice | just before dog-days | just after dog-days |
| *hemp* | 10 days before summer solstice | summer solstice | 10 days after summer solstice |
| *hemp seed* | 3rd month | 4th month | 5th month |
| *wheat* | early in 8th month | middle of 8th month | end of 8th month |
| *barley* | middle of 8th month | late 8th month | end of 8th, early 9th month |
| *wet rice* | 3rd month | 1st *hsün* of 4th month | second *hsün* of 4th month |
| *dry rice* | middle of 2nd month | 3rd month | 4th month |
| *sesame* | 2nd, 3rd months | 1st *hsün* of 4th month | 1st *hsün* of 5th month |
| *cucurbits* | 1st *hsün* of 2nd month | 1st *hsün* of 3rd month | 1st *hsün* of 4th month |

[1] 旬

blossoming of various plants and trees were all used to decide sowing dates. The *Huai Nan Tzu* for example gives the correct position of the constellations for planting a number of different crops.[a] Furthermore the solar date, a much more precise indication of the season than the lunar date, had also entered into farmers' calculations. Fan Sheng-Chih recommends very precise dates for sowing various cereals:[b]

> Glutinous panicum millet (*shu*[1]) ... should be sown twenty days before the summer solstice (*hsia chih*[2]) when there is rain.
>
> Winter wheat (*su mai*[3]) should be sown seventy days after the summer solstice. If you sow early the ears will be sturdy and the straw strong, but if you sow too late the ears will be small and contain few grains.
>
> Rice (*tao*[4]) should be sown one hundred and ten days after the winter solstice (*tung chih*[5]).

Although these dates may seem far too rigid to be practical, Fan certainly expected that they would be tempered by the farmer's discretion, for he also gives such phenological indications as the fruiting of elm trees at bean-sowing time and the ripening of mulberries when lentils should be sown;[c] furthermore of the most important crop of all he declares:[d] 'There is no fixed date for sowing setaria millet (*ho*[6]), for the season will depend on the type of soil (*yin ti wei shih*[7]).'

[a] Ch. 9/19*a*, quoted *CMYS* 3.16.6.
[b] *FSCS* 4.2, 4.3, 4.5.2, tr. Shih Sheng-Han (2), 15 ff. adj. auct.
[c] *FSCS* 4.6.1, 4.7.1.
[d] *FSCS* 4.1.

[1] 黍 [2] 夏至 [3] 宿麥 [4] 稻 [5] 冬至
[6] 禾 [7] 因地爲時

To the modern eye this appears sweetly reasonable and proto-scientific, yet Fan, like so many of his contemporaries, believed firmly in astrology and interlarded his empirical precepts with recommendations on auspicious and inauspicious days:[a]

Avoid [sowing] lesser beans on *mao*[1] days, rice and hemp on *chhen*[2] days, setaria millet on *ping*[3] days, glutinous panicum millet on *chhou*[4] days, glutinous setaria millet (*shu*[5]) on *yin*[6] days and *wei*[7] days, wheat on *shu*[8] days, barley on *tzu*[9] days, soya beans on *shen*[10] and *mao*[1] days. All the nine cereals have their taboo days, and if you fail to observe these when sowing, the crop will suffer great damage. These are no empty words, but the inevitable consequence of nature.

At almost the same date Virgil encapsulated similar beliefs rather more elegantly:[b]

The Moon herself hath sown good luck and ill
Among her days: avoid the fifth: on that
Pale Orcus and the Eumenides were born . . .
Choose the seventeenth day
For planting vines and taking up young steers
To train, and adding leash to warp; the ninth
Is dear to runaways but bad for thieves.

There were sceptics even among Fan's contemporaries. Ssu-ma Chhien wrote scathingly:[c]

The philosophers of the *yin-yang* school are bigoted and have many taboos. We can only hope to be familiar with the broad outline of such theories and should not attempt to obey them in all their intricacies. The proverb goes: 'The best strategy [when sowing] is to follow season and soil moisture (*i shih chi tse wei shang tshe yeh*[11])'.

This scepticism did not, however, prevent whole books being composed on *fas et nefas* during the Han and even later, amongst which we might mention the *Tsa Yin Yang Shu*, *Ssu Shih Tshuan Yao* and *Chung I Pi Yung*;[d] the calculations and recommendations from these works were regularly quoted in the great agricultural compendia.[e] The late Han writer Tshui Shih was apparently as sceptical as Ssu-ma Chhien, for what survives of the *Ssu Min Yüeh Ling* gives recommended

[a] *FSCS* 2.1 tr. auct; for the Ten Celestial Stems and Twelve Earthly Branches see Vol. II, p. 357. Though nowadays it is easy to dismiss such beliefs as pure superstition, it is important to remember that Han philosophers used the cyclical system to formulate cosmological laws.

[b] Virgil (1), *Georgics* I. 327–40.

[c] Quoted *CMYS* 3.14.4.

[d] Hu Tao-Ching (*4*) p. 63 lists 34 paragraphs out of 242 of the Sung *Chung I Pi Yung* and its Yuan supplement *Chung I Pi Yung Pu I*, i.e. 15 per cent of the total work, which deal with the selection of auspicious days for planting.

[e] E.g. *Nung Sang Chi Yao* 2/1*a* ff.

[1] 卯 [2] 辰 [3] 丙 [4] 丑 [5] 秫
[6] 寅 [7] 未 [8] 戌 [9] 子 [10] 申
[11] 以時及澤爲上策也

sowing dates solely in terms of the lunar calendar, but in the +6th century Chia Ssu-hsieh continued to quote many examples of taboo and auspicious days.

Most personal recommendations for sowing dates or periods Chia gives in terms of the lunar calendar, which he subdivides into the conventional ten-day periods *hsün*[1], and here he makes a curious statement:[a] 'It is generally true of all five cereals that if they are sown during the first ten days of the month you will get a perfect harvest, if they are sown during the middle ten days you will get a middling crop, and if they are planted during the last ten days you will get a poor crop.' This clearly reflects a belief that the growth of plants is affected by the waxing and waning of the moon, sympathetic magic perhaps not entirely unfounded on empirical observation of the degree of insect damage that young crops suffered when sown at different times. Roman farmers believed that grain sown just before the new moon was less likely to be affected by maggots,[b] and there may be some truth in this. Sown at this time the young shoots would emerge when the moon was nearly full, and modern agronomists have observed that insects are less common on bright moonlit nights,[c] so perhaps such timing did protect the crops from insect attack when they were at their most vulnerable.

Chia usually gives sowing periods simply in terms of the lunar calendar (Table 7), but he is well aware that this does not always accurately reflect the passage of the seasons:[d]

> In years where there is an intercalary month the solar terms (*chhi chieh*[2])[e] will fall behind and so you should sow later [in the lunar year]. Generally, however, sowing should be early, for early-sown crops yield much better than late. (Early crops are clean and easy to tend, but late crops are weedy and difficult to look after. The actual amount harvested depends on the year rather than when the crops were sown, but early-sown grain has thin husks and numerous, full seeds while later grain has thicker husks and fewer seeds, many of them empty.)

Chia also correlates the solar and lunar calendars for greater accuracy, as when he refers to planting setaria millet in the first part of the third month up till the *chhing ming*[3] term.[f] This custom has persisted up to the present day. Fei and Chang remarked in their anthropological study of villages in Yunnan that:[g]

> the solar section ... is not [sufficiently] subdivided; so the lunar calendar must be used in conjunction with it as a system of dating. The peasants ... will tell you, for example, 'This year *chhun fen*[4] will fall on the first day of the second moon and *chhing ming* on the sixteenth day of the second moon.'

Although Chia Ssu-Hsieh offers his readers a wide range of criteria, phenological, calendrical and astrological, on which to base their calculation of sowing

[a] *CMYS* 3.14.1.
[b] Pliny (1), XVIII. xlv. 158.
[c] K. D. White (2), 197.
[d] *CMYS* 3.7.2.
[e] See Vol. III.
[f] *CMYS* 3.4.2.
[g] (1), 28.

[1] 旬 [2] 氣節 [3] 清明 [4] 春分

time, the final choice must be dictated by the weather:[a]

Grain is always best sown just after rain. If the rainfall is slight you should sow immediately while the soil is still damp; if it is heavy wait for the weeds to sprout first. (... If the rainfall is slight and you do not sow immediately there will be no moisture to make the grain sprout, but if you do not wait for the soil to turn pale after heavy rain then the dampness will be trapped in the soil and will make the roots sickly.)

This Pliny would certainly have regarded with approval.[b]

There are some who hasten matters on and put forward the dictum that, while sowing in haste often proves deceptive, sowing late always does ... Some people ignore nice points of meteorology and fix limits by the calendar ... Thus these latter writers pay no attention to Nature, while the previous set pay too much, and consequently their elaborate theorising is all in the dark ... And it must be confessed that these matters do chiefly depend on the weather.

Of course, as Fan Sheng-Chih had pointed out earlier, the type of soil also dictated the choice:[c] 'Fertile soil should be sown later than poor soil; whereas fertile soil will give a good crop even if it is sown early, if poor soil is sown late it will invariably yield badly.'

Subsequent agricultural works quoted what the *Fan Sheng-Chih Shu*, *Chhi Min Yao Shu*, and even the apocrypha such as *Tsa Yin Yang Shu* had to say on planting dates, and also usually indicated the periods appropriate to the locality in which they were written, but works such as the *Nung Sang Chi Yao*, Wang Chen's *Nung Shu* and the *Nung Cheng Chhüan Shu* were intended to be read all over China and specific planting dates could hardly be given with any degree of accuracy, so they generally devote more attention to the actual mechanics of sowing than to dates. The attitude of later agricultural writers is typified as follows:[d]

Whatever crop you are growing, the essential is to know in what season it should be sown, whether early or late, in hot or cold weather. If you sow at exactly the right time you will be guaranteed success. You should bring the field to a fine tilth at your leisure while the weather is still cold; then you can just wait for the warm weather and you will not be obliged to work hastily and carelessly. One often sees people today who do not prepare for sowing until the warm weather, then suddenly a heatwave or cold spell strikes and inevitably they lose three or four tenths of their crop.

### (ii) *Preparation of the seed grain*

In spring the seed grain that had been carefully selected and stored the previous year (pp. 390 ff.) was brought out, examined for quality and prepared for sowing:[e]

[a] *CMYS* 3.6.1.
[b] (1), XVIII, lvi, 204–6.
[c] *CMYS* 3.2.1.
[d] Quoted *SSTK* 34/2*b* as from *CFNS*, but probably from a different *nung shu* as this passage does not occur in the surviving text of *CFNS*.
[e] *CMYS* 2.4.1.

Twenty days or so before the seed grain is to be sown ... bring out the grain and wash it in water. (Throw away any light husks that float to the surface and thus any weed seed will be eliminated.) Then dry the grain in the sun and sow it.

The reasons for soaking the seed are plain:[a] 'All seeds should always be soaked and those that float discarded, for seeds that float are empty and fruits that float are rotten.' Although pure water was the usual medium for seed selection in early times, more recently brine came into regular use as its higher specific gravity makes selection more accurate. Our earliest reference to the use of brine comes in a brief sentence of Hsü Kuang-Chhi, who says that testing the seed in brine is a Northern practice.[b] A fuller account of the method is given by Tseng Hsiang-Hsü at the turn of this century:[c]

First construct a deep bucket of 2 gallon capacity. (Separately prepare a fine-meshed bamboo basket slightly smaller than the bucket.) Put in 1 gallon of water and add salt. For ordinary rice use 120 ounces of salt, for glutinous use 100. Stir with a bamboo brush for 5 or 6 minutes until the salt has dissolved ... then insert the basket into the bucket and gradually put in 4 or 5 pints of seed, not too much at once. Skim off the empty kernels that float to the surface with a small bamboo strainer. Repeat several times until no floating seeds remain, then draw out the basket, put it into another vat of water and wash away the salt.

This work must be done quickly for if the seed remains too long in the brine even the empty kernels will sink.

Tseng recommends that the strength of the brine solution be tested after a while by the simple expedient of checking whether the grains rejected earlier now sink (in which case more salt is required) or those accepted now float (when more water must be added). As a measure of economy he recommends that ordinary rice should be tested first, and then the brine can be diluted to the right strength for testing glutinous rice.

Testing the seed in brine, or in some other liquid with a higher specific gravity than water, is a common practice throughout Asia. In Japan the specific gravity of the brine solution is generally 1·13.[d] In Malaya the strength of the brine was tested by floating a duck egg in the water,[e] and in post-Revolutionary China a fresh hen's egg was used to test the strength not of brine but of a thin suspension of clay; modern agronomists recommend specific concentrations of salt, ammonium sulphate or lime as more accurate.[f] This may appear an over-elaborate device since the seed-grain has already been selected at harvest time, but as Ting Ying, the great authority on Chinese rice, points out, there may be significant differences in speed of germination and flowering time even between grains taken from the same ear, and the heavier grains can be relied upon to flower and set more quickly than the light grains.[g] This is of particular importance in a quick-maturing crop like rice of which two or three crops may be grown in a single year.

[a] *NCCS* 25/9*a*.
[b] *NCCS* 6/15*b*.
[c] *Nung Hsüeh Tsuan Yao* 1/18*b*–19*a*.
[d] Grist (1), 145.
[e] Personal observation.
[f] Anon. (*519*), p. 148; Ting Ying (*1*), p. 311.
[g] Ting Ying (*1*), p. 310.

Perhaps less easy to understand is the deep-rooted Chinese belief that the water in which the seed-grain was washed actually benefited it:[a] 'Snow water fecundates (*ching*[1]) the five cereals and will make the grain drought-resistant. Always store snow-water in the winter; when you have a full vessel, then bury it in the ground. If the seeds are treated with this then the crop will always be doubled.' Such beliefs died hard, and in the +17th century they were reiterated by no less an authority than Sung Ying-Hsing:[b]

If [early rice] is put away in the barn while the noon sun is shining, so that the heat is carried into the storage bin and closed in immediately, the rice grains will retain the fiery quality of the weather [resulting in shrivelled seedlings in the spring] ... [So] wait until the cool of the evening before putting the grain in the storage bin, or gather a jar of snow and ice water at the time of the winter solstice (water gathered at the beginning of the spring is not effective) and sprinkle it on the grain at *chhing ming* sowing time to the amount of a few bowlfuls per *tan*[2] ... This water will immediately dissolve the hot properties of the grain so that the young shoots will be unusually handsome ...

Perhaps these beliefs stemmed originally from the discovery of the advantages of pre-germination,[c] commonly associated with wet rice but in China also applied to a number of other crops. Yet although Fan Sheng-Chih extolled the virtues of snow-water in the −1st century, the first descriptions of pre-germination are given by Chia Ssu-Hsieh in the +6th century. Chia recommends pre-germinating both wet and dry rice:[d] '[After washing the seeds] soak them for 5 nights, then drain them and put them in a grass basket, keeping them warm and damp. After another 3 nights they will begin to shoot. Broadcast when the seeds have grown to 2 inches ...' For wet rice the advantages of pre-germination are considerable, for although both wild and cultivated species will germinate under water, if the water is too deep or too muddy germination is greatly impaired. Additional advantages of pre-germination are that the growth of the rice is accelerated, helping it to compete better with weeds, and the young shoots are more regular and dense than those grown from dry seed, so that less seed is required. Modern experts, however, recommend that the rice should be sown when germination has just begun and long before the radicle has reached 2 inches;[e] indeed the Chinese agricultural books written after rice had become the dominant crop advise sowing as soon as the tiny shoots have turned white[f] (Fig. 91).

The advantages of pre-germination for other crops, in particular small-seeded

[a] *FSCS*, quoted *CMYS* 3.18.5.

[b] *TKKW* 1/7–8; tr. Sung Ying-Hsing (1), pp. 8–11; it is also mentioned in the contemporary *Chhün Fang Phu*, *Ku Phu* 17*b*, quoted *SSTK* 34/7*a*, as a treatment for setaria.

[c] Shih Sheng-Han (2), p. 61 suggests that in North China where river and well water were likely to be full of mineral salts, snow water offered the advantage of purity; but would this explain the enthusiasm of Sung, a native of Kiangsi?

[d] *CMYS* 11.3.1.

[e] Grist (1), pp. 141, 218.

[f] *WCNS* 2/11*b*; *Chhün Fang Phu*, *Ku Phu* 25*b*, quoted *SSTK* 34/5*a*.

[1] 精 [2] 石

Fig. 91. Soaking the rice seed in baskets; an empty chain-pump stand can be seen on the bank between the irrigation canal and the fields; *Keng Chih Thu*, Franke (11), pl. XII.

crops such as vegetables, have recently been discovered in Europe, and commercial market gardeners now frequently sow seed that has been germinated and suspended in a gel. It is interesting therefore to see that in the +6th century Chia advises germinating hemp if the fields are very wet at sowing time (hemp was sown in midsummer during the rainy season, and the seeds were liable to rot if not pre-treated):[a]

> If you soak them in rain-water then the hemp seeds will germinate very rapidly, but if you use well-water the process will be slower. Method: put the seed in water for twice the time it takes to steam a bushel of rice, then drain it and take it out. Spread on a mat to a depth of 3 or 4 inches, and stir frequently so that all the seed is equally exposed to the earth's *pneuma*. After one night the sprouts will appear. But if you leave the seed too long in the water you may keep it for ten days and it still won't sprout.

Pre-germination of field crops other than rice does not seem to have persisted much after Chia's time, though his instructions for sprouting hemp are quoted in later works.[b] The Sung writer Shen Kua mentions that in Hai-tung[1], where

[a] *CMYS* 8.2.6.

[b] E.g. the *Nung Tshan Ching*, quoted Li Chhang-Nien (*2*), p. 176.

[1] 海東

extra-fine hemp was produced, the seed was put in a bag and dipped in boiling water until the water cooled, hung overnight in a well-shaft just above the surface of the water, and then dried off in the sun.[a] Pre-germination of vegetable seeds continued in a tradition recorded in such works as Wang Chen's *Nung Shu*, the *San Nung Chi* and the *Nung Hsüeh Tsuan Yao*, and has survived right up to the present day.[b]

Sunlight was held to be beneficial to the seed. We have seen that most authors recommended drying the seed in the sun after washing or soaking it. Chia Ssu-Hsieh advised sun-drying vegetable seeds such as mallow seeds before sowing[c] and the practice was continued in later times, but few took their enthusiasm as far as the +18th century Szechwanese writer Chang Tsung-Fa:[d]

> To facilitate the germination of vegetables: Dry the seed in the sun during the hottest month of the summer ... In the time that seed sun-dried one year grows one inch, seed dried two years running will grow two inches and seed dried for several years will grow several inches tall.

To us perhaps the most fascinating of all the processes applied to seeds in China is their treatment with various substances intended to nourish the young plant and protect it from insects and disease. In its most complex form the seed was actually pelleted, though often it was merely mixed with fertilisers and insecticides. We must remember that in Europe prior to the development of scientific farming the only treatment accorded to the seed before sowing was to thresh it and select the largest grains. Treatment of seed with pesticides is a recent idea in the West, and seed-pelleting was only developed in the second half of this century, yet in early Han times elaborate methods of seed-pelleting had been evolved. The *Chou Kuan* is the earliest text to mention seed treatment:[e]

> The plants officer (*tshao jen*[1]) is in charge of methods of soil improvement (*thu hua*[2]). He examines the needs of [each particular] soil and fertilises the seeds (*fen chung*[3]) accordingly. Seed fertilisation: for hard red soils (*hsing kang*[4])[f] use ox [bones]; for reddish orange soils (*chhih thi*[5]) use sheep ...

Here follows an enumeration of several soil types and their appropriate treatment; some of the more elaborate decoctions proposed by the *Chou Kuan* and its commentators require the bones of badgers, foxes, or even shrews.[g] The whole notion might seem rather fanciful did not Fan Sheng-Chih, writing only a short while after the *Chou Kuan* was compiled, convincingly fill in the practical details:[h]

a *Meng Chhi Pi Than*, quoted Li Chhang-Nien (*2*), p. 176.
b *WCNS* 2/13*a*; *San Nung Chi* 9/1*b*; *Nung Hsüeh Tsuan Yao* 2/23*a*; Sinkiang Agr. Inst. (*1*), pp. 390–1.
c *CMYS* 17.2.1.
d *San Nung Chi* 9/2*a*.
e *Chou Kuan* 4/32*b*; tr. auct.
f See Vol. VI, Section 38 on soil classification.
g See *CMYS* 2.6.1
h Quoted *CMYS* 3.18.4; the text given in *FSCS* 3.2 is more complex but probably contains later interpolations.

1 草人　2 土化　3 糞種　4 騂剛　5 赤緹

> Take horses' bones, break them up, and boil one *shih*[1] with three times the weight of water. When they have come to the boil three times strain them and throw away the sediment. Then soak five aconite heads (*fu tzu*[2]) in the decoction (*chih*[3]). After three or four days throw away the aconites, add to the decoction equal quantities of silkworm droppings and sheep's dung and stir until the mixture is like thin gruel. Twenty days before sowing sprinkle this onto the seed grain: the mixture should be of the consistency of boiled wheat. Whenever the weather is hot and dry sprinkle the seeds and put them out to dry ... Sprinkle them six or seven times and then stop. Dry them at once in the sun and put them away carefully: do not let them get wet again. When the right time for sowing comes, sprinkle the seeds with what remains of the decoction before sowing. The grain will not suffer from locusts or other insects.

As an alternative to the horse-bone soup, Fan recommends a mixture of silkworm droppings and snow-water or soured rice-water,[a] both of which he claims protect the seed from insects and strengthen its resistance to drought. Shih Sheng-Han[b] points out that the silk-worm dung is hygroscopic and contains freely available potassium, nitrogen and phosphates, as well as microbes that might be stimulated into activity by the nutrient collagen to raise the ambient temperature around the seed and hasten germination. Li Chhang-Nien[c] reports that experiments carried out at the Botanical Institute in Nanking in 1958 showed that Fan Sheng-Chih's pelleting methods produced slight increases in yields from most crops but noticeably promoted germination and growth in wheat; the pelleting process proved very time-consuming and made sowing more difficult which, Li suggests, is why the practice was later discontinued.

Although few Chinese farmers can actually have taken the trouble to pellet their seed in collagen, it was quite usual to treat the seed with some simple fertilising or insect-repellent substance. Hsü Kuang-Chhi, quoting Fan's method, comments:[d] 'Nowadays in the district of Chang-ming[4] near Chengtu [in Szechwan] the peasants grow large quantities of aconite for no other purpose than to treat seed grain.' All members of the aconite family contain the highly toxic alkaloid aconitine, 'one of the most formidable poisons ... yet discovered ... One fiftieth of a grain of aconitine will kill a sparrow in a few seconds.'[e] Aconite is said to be the poison that Medea prepared for Theseus (though Gerard[f] recommends it for its anthelmintic properties—if taken in sufficiently small doses).[g]

Other seed treatments included the addition of salt to melon seed, said to prevent disease,[h] and of the ground root of *Sophora evanescens* (*khu tshan*[5]) mixed with lime to vegetable seed,[i] a decoction of eels' heads to radish and cabbage

[a] *FSCS* 3.1.1; *CMYS* 10.11.2.
[b] (2), p. 60.
[c] (*3*), p. 99.
[d] *NCCS* 6/15*a*.
[e] M. Grieve (1), p. 9.
[f] John Gerard (1), p. 976.
[g] Grieve (1), p. 10, remarks that in the 1920s 55,000 lb. a year of *A. Wilsoni* were produced for drugs in Szechwan.
[h] *CMYS* 14.4.1.
[i] *WCNS* 2/13*a*.

[1] 石 [2] 附子 [3] 汁 [4] 彰明 [5] 苦參

seed,[a] and arsenic or ashes to wheat seed as insecticides;[b] Hsü Kuang-Chhi recommends the addition of cotton-seed oil to barley seed to deter insects and increase drought resistance, but says that most farmers simply mix the seed with ashes.[c] The *Nung Hsüeh Tsuan Yao*[d] advises soaking seed in kerosene because, it says, this speeds germination; in fact its more probable effect was to deter mice and birds from eating the seed.

It is possible that this long tradition of seed treatment is linked to continuous cropping. Where fields are fallowed every one or two years, as was usual in Europe before the +18th century, crop pests and diseases have less chance of building up in the soil than if the fields are under crop year after year. Although most Chinese agriculturalists from Chia Ssu-Hsieh on advocated crop rotations that would help reduce the build-up of disease, many smallholders must have been obliged to plant their fields for years at a time with the staple crop: though they would probably try to obtain new seed or to plant different varieties the crop would still be more disease-prone than if grown on fallowed land. Particularly when we remember that Fan Sheng-Chih speaks of fallowing as a last resort when the land is really exhausted[e] it is interesting to speculate how far the elaboration of methods of seed treatment in the early Han is related to the intensification of land use.

### (iii) *Sowing methods*

In the West up to the +18th century seed was always sown by hand, field crops being broadcast and vegetables and other small-scale cultures being sprinkled in drills. Chinese sowing methods were more elaborate, and different methods would be employed according to the crop, size of field, soil or weather conditions. The general rules were as follows:[f]

> When sowing one must pay attention to the type of seed as well as the soil and climatic conditions. In dry or limey soils [the seed] should be thickly covered, in moist or clay soils it should be thinly covered, and in sandy soils [the seed] should lie even a little shallower. In winter you should sow deep, in summer shallow, in spring, slightly more deeply than in autumn because the cold remaining [from the winter] still lies in [the soil].

The Chinese themselves distinguished three methods of sowing, namely broadcasting, sowing in rows, and sowing individual seeds (what Gervase Markham referred to as 'setting').

#### *Broadcasting*

Broadcasting was called *sa chung*[1], *sa po*[2], *man chih*[3] or *man chung*[4]:[g]

[a] *CFNS* 1/5.
[b] *TKKW* 1/14.
[c] *NCCS* 26/11a.
[d] 1/21*b*.
[e] *FSCS* 1.7.
[f] *Nung Hsüeh Tsuan Yao* 1/21*a*.
[g] *WCNS* 2/7*b*–8*a*.

1 撒種 2 撒播 3 漫擲 4 漫種

For broadcasting (*man chung*) one holds a basin containing the seed grain under the left arm while with the right hand one takes the seed from the basin and scatters it (*sa*[1]). Regularly every three paces or so one takes another handful, taking care that the seed is spread evenly so that the young plants are neither too crowded nor too sparse.

Wang Chen's brief description is not very instructive, but few have described the intricacies of broadcasting successfully. As Pliny rather lamely concludes,[a] 'there is a certain art in scattering the seed evenly', and this art only comes with practice, the more so as fine judgement is required to estimate the rate of sowing which differs for various crops and soils: the only guide the sower has is the rate at which his seed-basket empties. Usually he projects the handful of seed in a wide arc as he walks (Fig. 92):[b] 'And in your castynge, ye must open as well your fyngers as your hande, and the hyer and farther that ye caste your corne, the better shall it sprede, excepte it be a greatte wynde.'

Broadcasting of field crops seems to have been regarded as rather an inferior practice in North China, where sowing in rows by drill was much preferred as it saved seed and economised on soil moisture. If a crop was sown broadcast it was usually because the seed was too small to be sown properly by drill, like hemp, sesame or coriander, all of which must lie very shallow, in which case the land was usually furrowed first and the seed scattered and covered with a light bush-harrow.[c] Here almost the same effect as sowing in rows was produced. In some cases the seed was simply scattered and allowed to grow as it fell, but this was usually only applied to fodder crops, green manures, or crops grown on newly cleared land,[d] though in some Western areas in early times it seems that most grain was sown this way (Fig. 82) and was covered not with a harrow but with a beetle (Fig. 93).

In Southern China most dryland crops were broadcast, including millet, beans, hemp and wheat,[e] and the sower often simply stamped the seed into the earth with his feet[f] (Fig. 94). Wet rice was sown broadcast in both North and South China, and this presented a variety of problems. Often the seed was simply left lying on the surface of the mud,[g] in which case the birds had to be scared away until the rice had rooted.[h] The alternative was to sow the rice in shallow water, which hastened growth and deterred the birds, but if a wind blew up it might sweep all the seed into a pile in one corner of the fields, so if the weather was windy the nursery field had to be drained before sowing.[i] Sometimes the seed-bed was treated with sharp sand or ashes to help the rice seed settle.[j]

[a] (1), XVIII, p. 197.
[b] Fitzherbert (1), p. 19.
[c] *CMYS* 8.2.7, 13.2.3, 24.3.3.
[d] Li Chhang-Nien (*3*), 100.
[e] *WCNS* 2/12*a*.
[f] *TKKW* 1/14; Sung Ying-Hsing (1), p. 14.
[g] *Nung Sang Chi Yao* as quoted *SSTK* 34/4*a*.
[h] *CMYS* 11.3.2.
[i] *CFNS* as quoted *SSTK* 34/4*a*, but not included in surviving text of *CFNS*. *TKKW* 1/8 also mentions this problem.
[j] *Nung Hsüeh Tsuan Yao* 2/3*b*.

1 撒

Fig. 92. Broadcasting seed in 14th-century Europe, after the Luttrell Psalter, and in Southern China, *Keng Chih Thu*; Franke (11), pl. XXII.

Fig. 93. Broadcasting seed and covering it with a beetle; Wei/Chin mural from Holingol, Mongolia; Hayashi (*4*), 6–34.

*Sowing in rows*

Sowing in straight rows (*hang chung*[1] or *thiao chung*[2]) was an ancient practice in China that, as we have seen (p. 105), dates back at least to late Chou times. The −3rd century *Lü Shih Chhun Chhiu* says:[a] 'If the crops are grown in rows they will mature rapidly because they will not interfere with each other's growth. The horizontal rows must be well drawn, the vertical rows made with skill, for if the lines are straight the wind will pass gently through.' It also recommends that the plants should be spaced half a foot apart, and that they should be sown in broad ridges separated by deep, narrow furrows.[b] This system has the advantages of leaving plenty of space for the root systems to develop, cutting down competition for water and nutrients, facilitating weeding and hoeing and providing drainage in case of heavy rain; in fact it was ideally suited to the spring-sown crops of North China.

At first the seed must have been sown along the ridges by hand, either sprinkled regularly a few at a time, if the seeds were small, or dibbled in (*yen*[3]) one at a time, at regular intervals, if the seeds were large. The first method continued to be used for vegetable seeds,[c] while the second remained common for cereals that tillered well: Chia Ssu-Hsieh recommends it for wheat, barley, and dry rice, saying in each case that this method encourages the growth of tillers.[d] But from Han times on the most usual way of sowing rows was to use a seed-drill:[e]

> The rich folk sow a bushel,
> For the poor a few pints do,
> But both must sow together
> In an even, steady flow.

[a] *LSCC* 67.
[b] *Ibid.* 65, 72.
[c] *CFNS* 4.
[d] *CMYS* 10.2.2, 12.2.3.
[e] Poem by the Sung statesman Wang An-Shih[4], known as Wang Ching-Kung[5]; quoted *WCNS* 12/17*b*.

[1] 行種 [2] 條種 [3] 掩 or 稚 [4] 王安石 [5] 王荊公

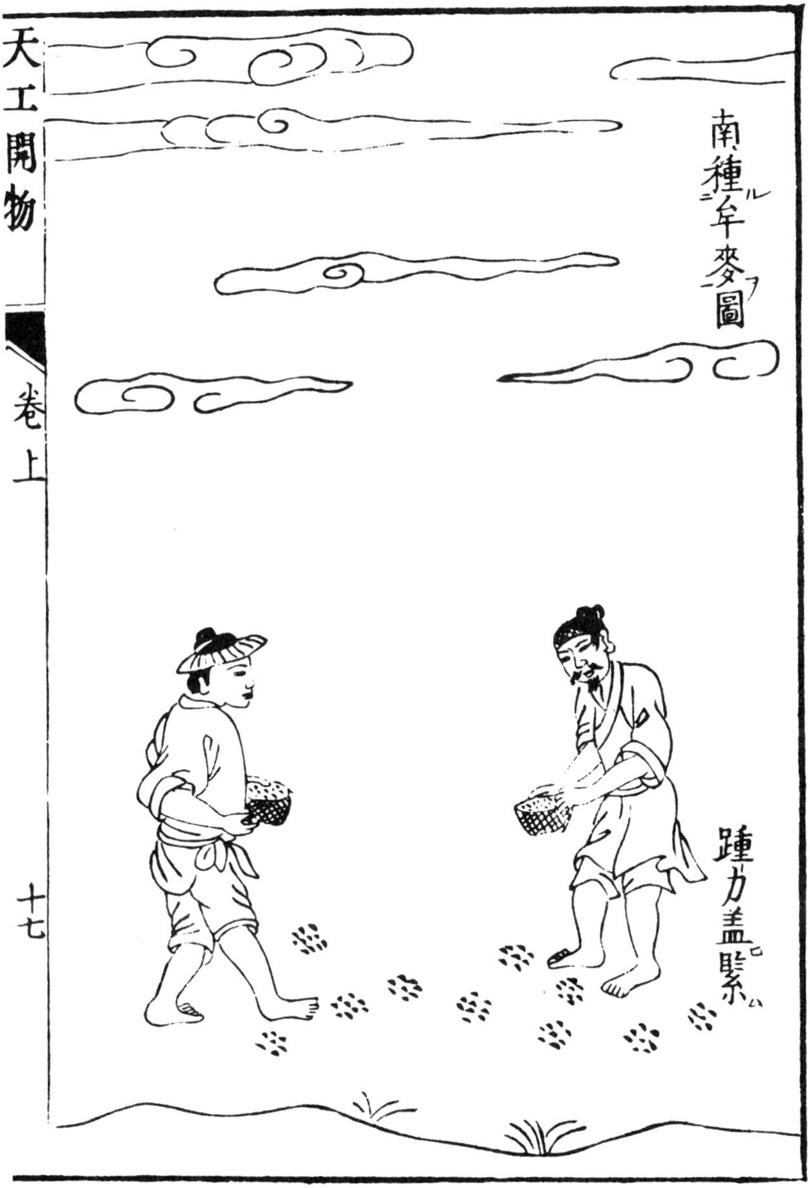


Fig. 94. Broadcasting wheat in South China and covering the seed with the feet. Ming illustration to the *Thien Kung Khai Wu* 1/17*a*.

On thousands of ridges how straight the rows stand,
A few seed-drills covering acres of land.
Though pulled by mere oxen, there's no need for shame:
The seed-drill can justly be proud of its fame!

The seed-drill was variously known as *lou*[1], *lou chhe*[2], *lou li*[3], *chiang tzu*[4] and *chung shih*[5]; it was, and still is, in common use all over North China:[a]

> Drills vary in their construction, some having a single foot, some two and some three. In modern Yen[6], Chao[7], Chhi[8] and Lu[9] [North and Northeast China] the two-footed drill is common; West of the passes [Kansu and Central Asia] there are four-footed drills: they only use one extra ox and work even more speedily. The drill is used in all the central lands but in other regions they have never even seen it.
>
> Unfortunately it is difficult to construct [Fig. 95]. The two handles are curved at the top and should be 3 feet tall. The two feet are hollow in the centre and a furrow's width apart. 4 round poles are set horizontally [between the handles]. In the centre is placed the seed-bin (*lou tou*[10]) which holds the seed-grains, which are sown individually through the hollow feet. To either side there are 2 shafts for the ox. One man is needed to lead the ox and one to hold the drill, shaking it as he walks so the seed falls by itself.

This is the basic principle of the Chinese seed-drill.

The history of the seed-drill is long, complex and rather mystifying. The earliest known representations of the seed-drill occur on Sumerian seals of the −3rd millennium and later (Fig. 96), which show a double-handled stilt with a seed-tube, running behind the share, into which a second man drops the seed. The drill has a long beam to which several oxen are yoked. This Mesopotamian instrument, known in pre-Sargonic times as *apin*, was used in November and December for sowing and covering the seed. It was quite distinct functionally and linguistically from the true plough, called *numun*, which was used in September and October for breaking the ground.[b] We find a similar situation in Vedic India, where the plough used for breaking ground was called *lā́ṅgalam* and the sowing plough *sī́ram*:[c] 'Attach the *sīrā* [plural], spread apart the yokes, sow the seed into the prepared womb.' There are references in the Pali and Sanskrit literature to *sīrā* which could sow several furrows at a time.[d] Drills similar to the Sumerian single-tube type are still in use in Iraq, Iran, the Indus Valley and Northwestern India,[e] while the multi-shared variety were to be found in Malabar, Mysore and the other southern princedoms of India[f] (Fig. 97*a*). The multi-shared type usu-

[a] *WCNS* 12/18*a*–*b*.
[b] Salonen (1); Puhvel (1).
[c] *Rig-Veda* 10.101.3–4, tr. Puhvel (1), p. 187; the first scholar to point out this distinction was Jules Bloch (2).
[d] Puhvel (1), p. 88.
[e] H. E. Wulff (1), p. 265; A. Memon (1), figs. 7–9, 14–15; D. G. Graham (1), p. 52, fig. 3.
[f] Francis Buchanan (1), fig. 73; Thomas Halcott (1); Alexander Walker (1).

1 耬 2 耬車 3 耬犁 4 耩子 5 種蒔
6 燕 7 趙 8 齊 9 魯 10 耬斗

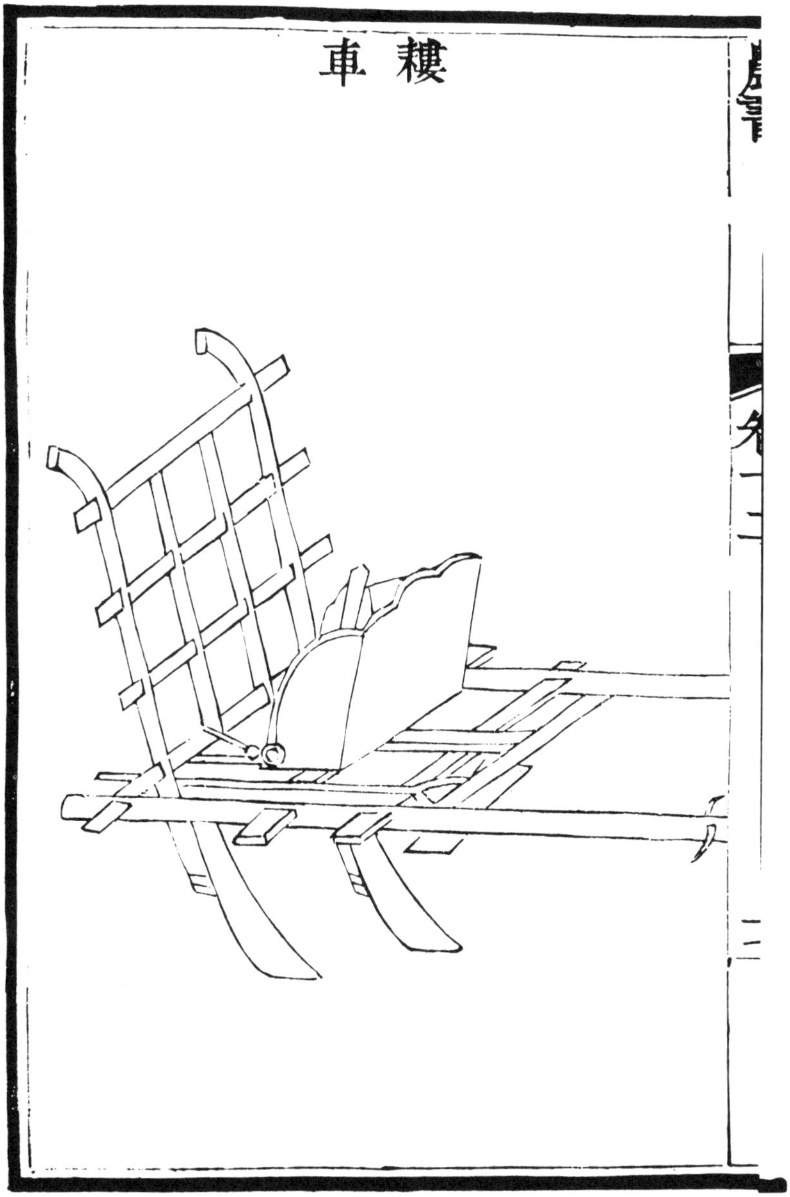


Fig. 95. Chinese seed-drill (*lou chhe*); *WCNS* 12/17*b*.

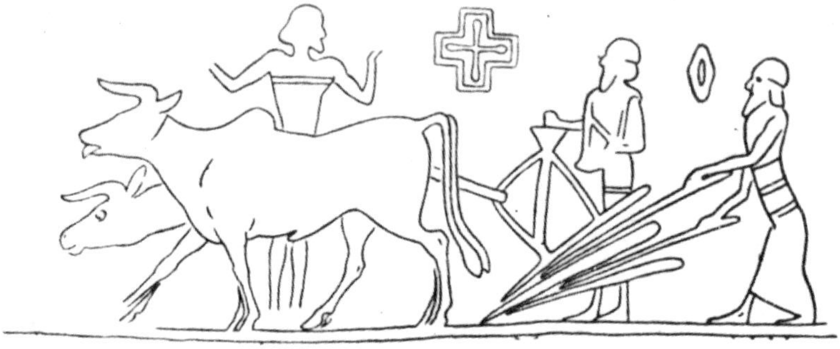

Fig. 96. Babylonian single-tube drill; Anderson (2).

ally had a single seed container from which bamboo tubes led down to each share, but sometimes where intercropping was practised a different seed was sown through each tube.[a] The seed was gradually fed into the container by a woman who walked beside the drill, and the seed was covered after the passage of the drill by an ox-drawn device, like a broad Dutch hoe the width of the seed-drill (Fig. 97*b*), which 'operates by agitating the earth so as to make the sides of the drills fall in, and cover the seed grain, which it does so effectually as scarcely to leave any traces of a drill'.[b] In late 18th-century Karnataka, this drill was used successfully for wet rice as well as for dry crops, but only by the richest *ryots* who could afford at least three yoke of stout oxen to plough and drill the miry soil.[c] Visitors from England were amazed to find that these 'primitive' Indian seed-drills sowed the seed much more evenly than their newly invented mechanical drills.

In Europe the seed-drill was unknown until the 16th century. The first patent for a sowing-machine was granted to Camillo Torello by the Venetian Senate in 1566, but the earliest drill of which a description survives is the one constructed

[a] Alexander Walker (1), p. 187.
[b] Thomas Halcott (1), p. 210.
[c] *Ibid.* 214.

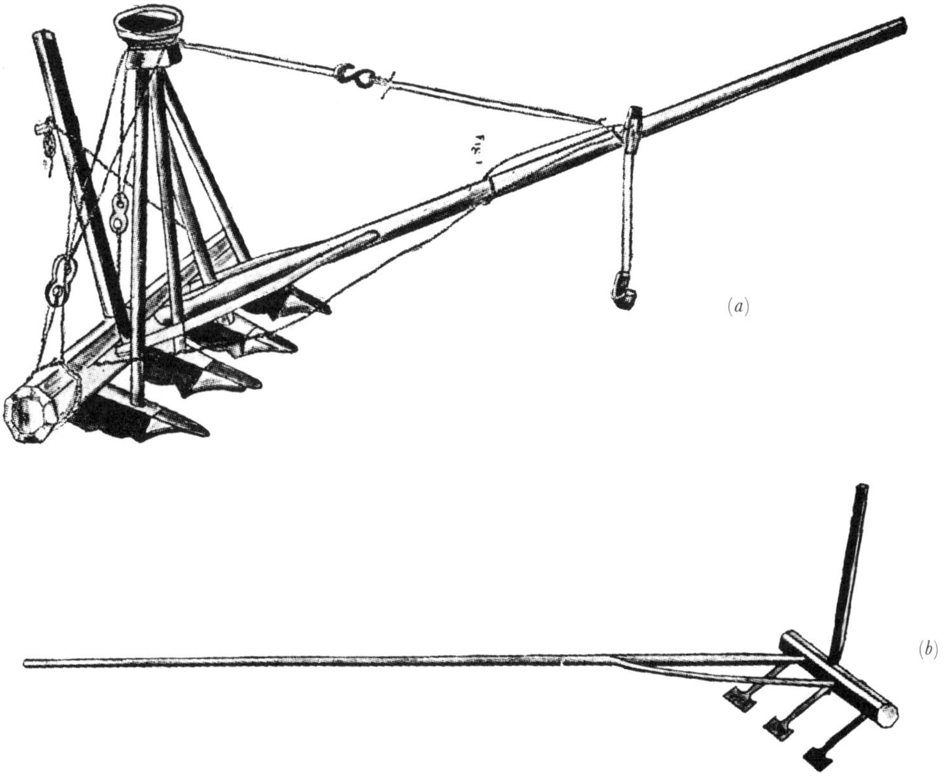


Fig. 97. Multiple-tube drill and covering implement from South India; Halcott (1).

by Tadeo Cavalina of Bologna, described by Canon Battista Segni in 1602:[a]

> Its construction resembles that of a flour sieve carried on a small, simple carriage, with two wheels and a pole. Part of the body holds the grain to be sown and part is constructed under the sieve and is perforated, and to every hole there is fitted an iron tube directed toward the ground and terminating in an anterior knife-blade of sufficient length to make a furrow into which the sifted grain at once passes through the tube and where it is so completely buried that none of it is damaged. It is then immediately covered, by means of another iron implement, with the earth that has been excavated in the making of the furrow.

As Fussell points out, all the elements of the modern Western seed-drill were incorporated in this device, yet it had a long way to go before it was accepted as practical and efficient by agronomists or farmers. Many men of ingenuity set to constructing improved seed-drills,[b] one of the most famous of whom was certainly Jethro Tull, the first Englishman to design a seed-drill that actually worked

[a] Quoted Fussell (2), p. 94. In Section f, pp. 566 ff. we discuss in greater detail whether there were any links between the development of the seed-drill in Europe and Asia.

[b] For a detailed account of the many inventors, see Fussell (2), ch. 3.

Fig. 98. A reconstruction of Jethro Tull's seed-drill; Anderson (2).

(Fig. 98). Tull's *Horse Hoeing Husbandry*, containing pictures and descriptions of drills, was first published in 1733, thirty-odd years after he produced his famous drill, the chief excellence of which was the dropper unit. This

> consisted of the case at the bottom of the seed-box, and the notched axle which passed through it [Fig. 98*a*]. The axle, with its notches and cavities in the periphery, turned with the wheels, received the grain from the boxes above, and dropped it into the furrows below. The passage of the grain past the notched dropper was controlled by a brass cover and an adjustable spring.[a]

This was the ultimate in the control of the sowing rate until the 1880s when John Bailey of Chillingham '[did] much to render drills more perfect by adding a piece of flat iron which could be adjusted to make the slot in the hopper entrance larger or smaller'.[b] But however sophisticated the feeding device, often the holes in the feeding mechanism would clog, and European seed-drills continued to behave erratically and inefficiently until well into the 19th century.[c]

As we mentioned before, English visitors to India were extremely impressed by the efficiency of the apparently primitive local seed-drills. In England until the 1860s or so, few farmers actually used seed-drills. Instead they sought to obtain the advantages of row-sowing by alternative means such as 'sowing under furrow' (broadcasting the seed and then ploughing furrows to cover it),[d] or 'setting', whereby evenly spaced holes were made in the ground with the aid of a special

[a] *Ibid.* 102.
[b] *Ibid.* 109.
[c] See Arthur Young's criticisms of the unwieldiness and fragility of Tull's drill in A. Young (1).
[d] Gervase Markham (2), p. 47.

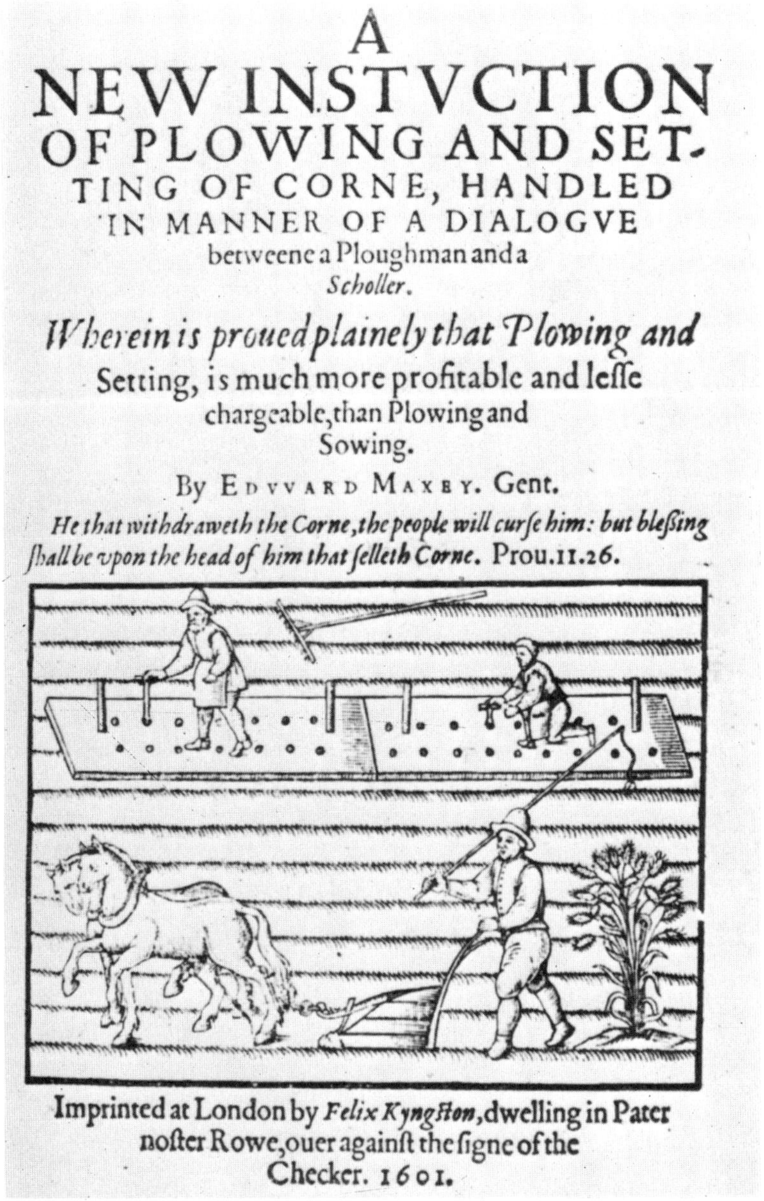

A
NEVV INSTVCTION
OF PLOWING AND SET-
TING OF CORNE, HANDLED
IN MANNER OF A DIALOGVE
betweene a Ploughman and a
*Scholler.*

*Wherein is proued plainely that Plowing and*
Setting, is much more profitable and leſſe
chargeable, than Plowing and
Sowing.

By EDVVARD MAXEY. Gent.

*He that withdraweth the Corne, the people will curſe him: but bleſſing ſhall be vpon the head of him that ſelleth Corne.* Prou. 11. 26.

Imprinted at London by *Felix Kyngſton*, dwelling in Pater
noſter Rowe, ouer againſt the ſigne of the
Checker. 1601.

Fig. 99. Setting board, as shown on the title-page of Maxey (1).

board (Fig. 99) or spiked roller, and the seed was dropped in by hand.[a] Sometimes the holes were simply made by a man holding a double-pronged 'dibbling-iron' in either hand, who walked backwards along the furrow making rows of holes that were filled by small children.[b] Only on more progressive farms had the

[a] On setting corn see Hugh Plat (1); Edward Maxey (1); Gervase Markham (2), pp. 100 ff.

[b] Such 'debblers' were still used by some Suffolk cottagers in the 1950s; George Ewart Evans (1), p. 119.

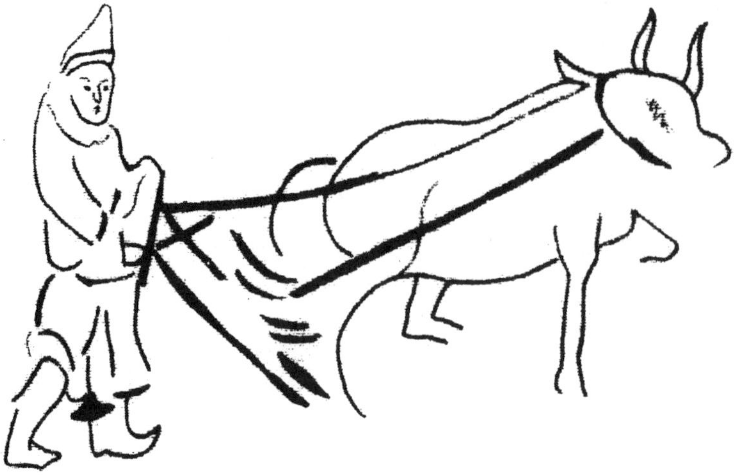

Fig. 100. Han seed-drill from Phing-lu, Shansi. Apparently a two-tube drill. Note the long shafts and the sturdy construction of the stilts. Anon. (*512*), pl. 6.

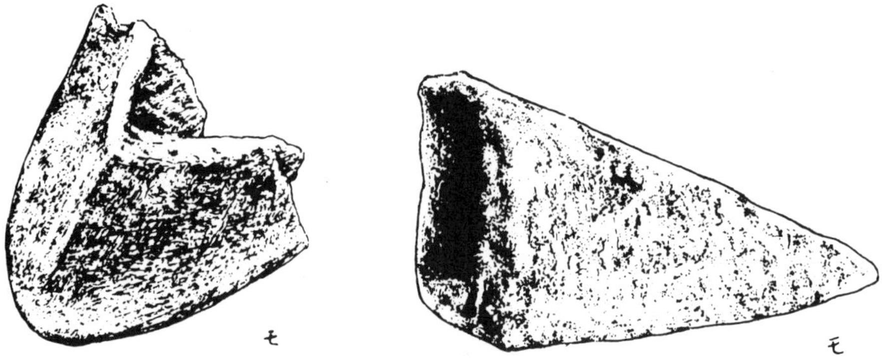

Fig. 101. Han seed-drills. The drill from Peking, on the left, is flanged; the Shensi drill on the right is similar to the *huo* described by Wang Chen and shown in Fig. 103. Hayashi (*4*), 6–22, 23.

mechanical seed-drill become a standard piece of equipment by the end of the 19th century.

Let us return now to the Chinese seed-drill. The earliest picture of such an implement is found in a mural from an early +1st century tomb at Phing-lu[1] in Shansi (Fig. 100), but small iron shares, apparently belonging to seed-drills, have been found in Western Han sites in Hopei and Shensi (Fig. 101); the flanged share from Peking is similar not only to one excavated from a Sung site in Honan[a] but also to the shares still used on many types of drill in China today[b] (Fig. 102).

[a] Liu Hsien-Chou (*8*), p. 32, fig. 59.
[b] Anon. (*502*), pp. 250, 255, etc.

[1] 平陸

Fig. 102. Modern seed-drill from Shantung, double-tubed with markedly flanged shares. The curved bamboo rod at the back pulls the earth over the seed; Hommel (1), fig. 66.

The Western Han official Chao Kuo[1], who was put in charge of government grain supplies by Wu-ti in −85, is credited with introducing the seed-drill to the metropolitan area:[a]

Three plough shares (*li*[2]) were all drawn by one ox, with one man leading it, dropping the seed and holding the drill (*lou*[3]), all done simultaneously. Thus 1 *chhing*[4] [100 *mu*] could be sown in a single day ... (In my opinion[b] three shares all drawn by a single ox is the same thing as the modern three-tube drill (*san chüeh lou*[5]). Nowadays from Chi-chou[6] [modern Shantung] westward they still use ... a two-tubed drill ... But with a two-tubed drill the rows are too close together and it is not as successful as a single-tubed drill.)

Soon after Chao Kuo took up his post Fan Sheng-Chih wrote his agricultural treatise, and although the *Fan Sheng-Chih Shu* does not refer specifically to the seed-drill at any point (or indeed to any other sowing method for field crops), a comparison of the sowing rates in *Fan Sheng-Chih Shu* and *Chhi Min Yao Shu* does suggest that Fan expected the seed-drill to be used.[c]

[a] *Cheng Lun* 11*a*; quoted *CMYS* 1.19.1.

[b] This appears on stylistic grounds to be Chia Ssu-Hsieh's +6th-century comment.

[c] On sowing rates, see Li Chhang-Nien (*3*), p. 106.

1 趙過 2 犂 3 耬 4 頃 5 三脚耬
6 濟州

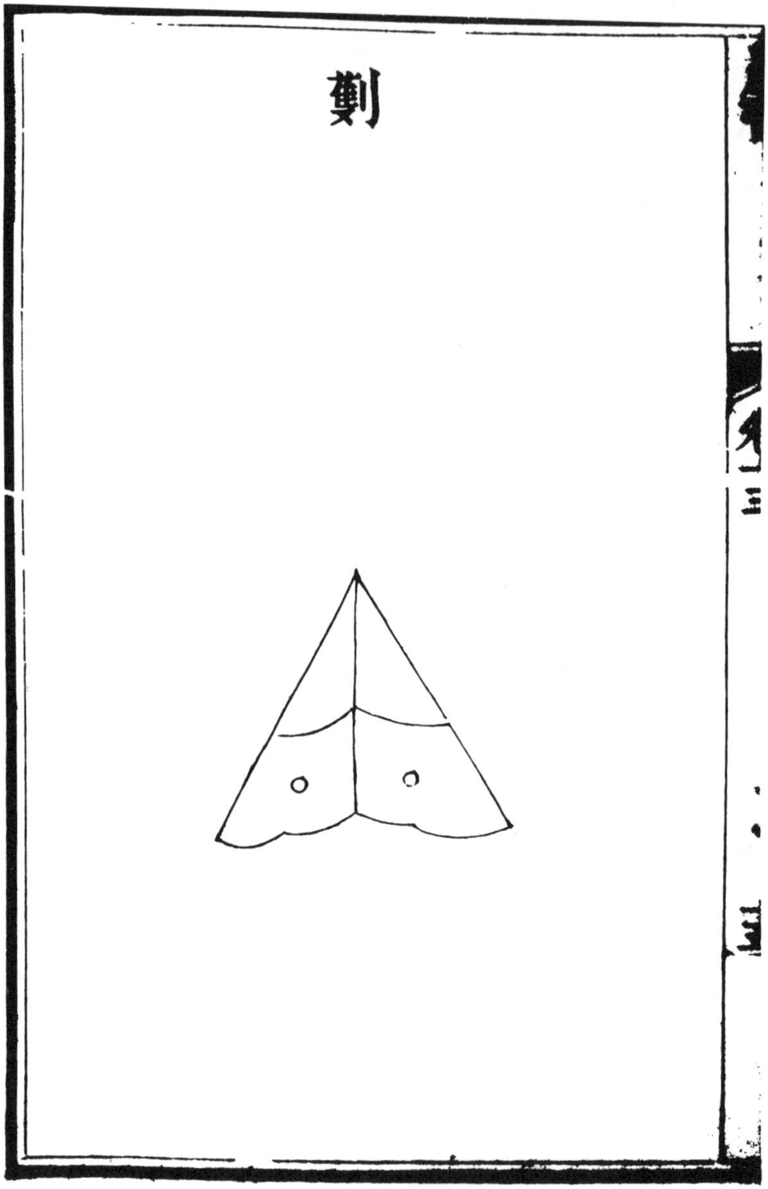


Fig. 103. Drill share (*huo*), Ming, *WCNS* 13/15*b*. Although Wang describes the *huo* as similar to the *chhan* ploughshare (Fig. 56), i.e. triangular, this drawing appears to show a flanged share. The two holes are presumably for binding or pinning the share to the hollow leg of the drill.

Wang Chen describes how the special drill-shares functioned:[a]

> The *huo*[1] ... or sowing shares (*chung chin*[2]) are the shares ... that trace the drill, like a triangular ploughshare (*chhan*[3]) but smaller, with a high ridge down the centre, 4 inches long and 3 inches wide [Fig. 103]. They are inserted into the two holes at the back of the seed-drill's feet and bound tightly to the cross-piece. The share bites 3 inches or so deep into the soil, and the seed dribbles down through the foot of the drill, so it is sown very deep in the soil and the yield is improved. Soil tilled with a seed-drill looks as if it had been gone over with a small plough.

The early Chinese material gives us few indications as to the nature of the seeding mechanism, which unlike the European devices must have been fairly simple, yet reliable and efficient, since the use of the drill was almost universal in North China. Wang Chen unfortunately does not elaborate beyond stating that the man who holds the drill tips it from side to side 'so the seed falls by itself'. But if there was no device within the seed-bin for regulating the flow of grain into the hollow legs, then the flow of seed would be far too rapid. The illustration in the 1783 edition (Fig. 95), which may be based on a +15th-century original, shows the tip of a handle or stick protruding from the bin which may be part of a regulating device, as well as a shuffle which must have been used to push out the last few grains.

Some further illumination comes from the +17th-century *Thien Kung Khai Wu*:[b]

> The wheat-seed is contained in a small bin, the bottom of which is pierced with holes arranged in a plum-blossom pattern [i.e. a five-pointed star]. As the ox moves he shakes the seed which falls through the holes to the ground. If the farmer wishes to sow thickly then he whips up the ox and more seeds will fall, but if he wishes to sow more thinly he reins the ox back and fewer seeds are sown.

Though somewhat elliptical, Sung's description clearly implies that the flow of grain was regulated by sieves between the seed-bin itself and the hollow legs; if the drill was regularly tilted from side to side, as Wang says, then small clusters of grain would fall to the ground first down one tube and then down the other, automatically spacing the clumps of wheat diagonally (Fig. 104).

Yet it is hard to believe that this rather simple device represented the acme of Ming sowing technology, and the *Nung Shu* illustration (Fig. 95) hints at the existence of some more ingenious mechanism, perhaps of the type described by Hommel in the 1930s (Fig. 105*a*, *b*):[c]

> The seed-bin has an ante-chamber, which, instead of a bottom, has two square chutes leading downwards to the shares ... The seed-bin has a passage-way which communicates

[a] *WCNS* 13/16*a*.
[b] *TKKW* 1/13; tr. auct; see also Sung Ying-Hsing (1), p. 15.
[c] (1), pp. 45–7.

[1] 劐 [2] 種金 [3] 鑱

Fig. 104. The caption says this drill is used for sowing wheat, barley, or setaria millet. *TKKW* 1/15*b*.

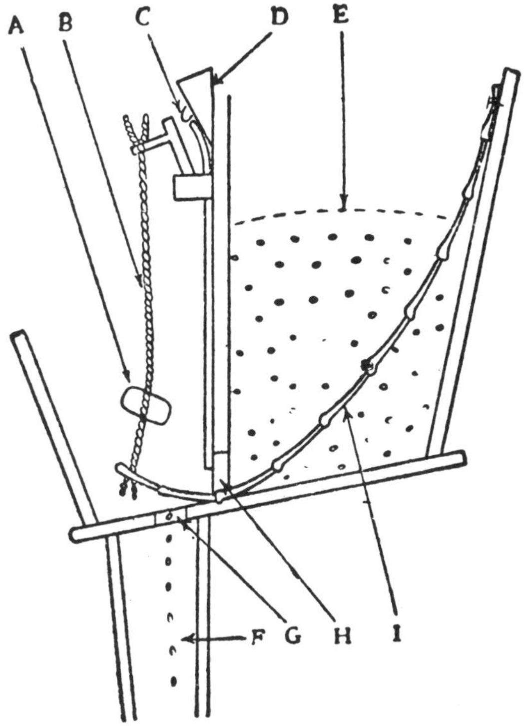


Fig. 105 (*a*)

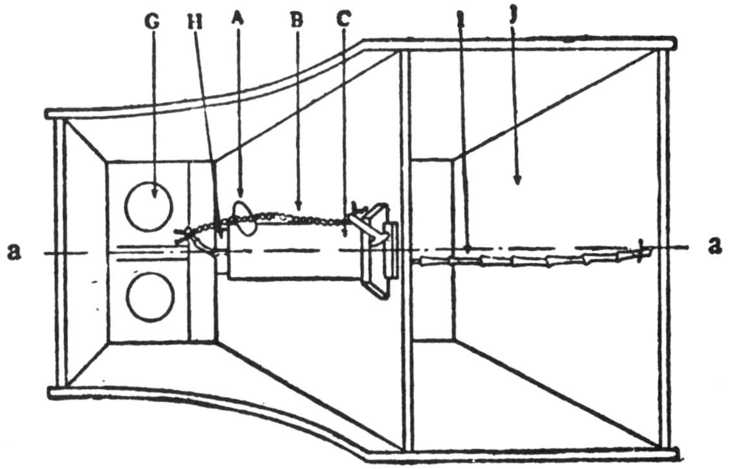


Fig. 105 (*b*)

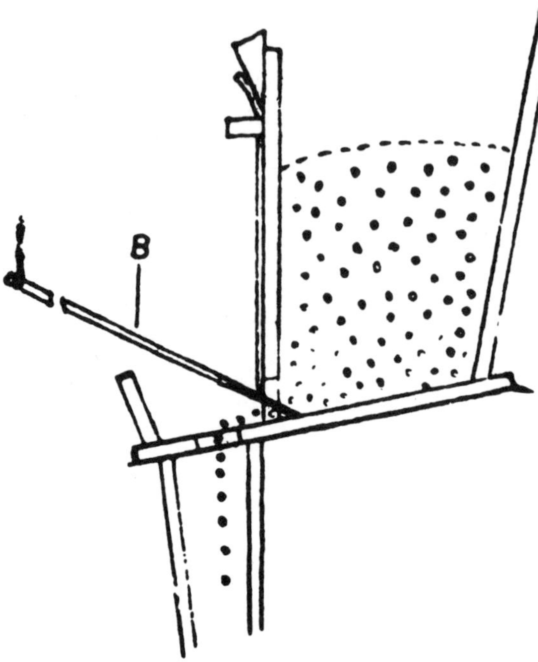


Fig. 105 (*c*)

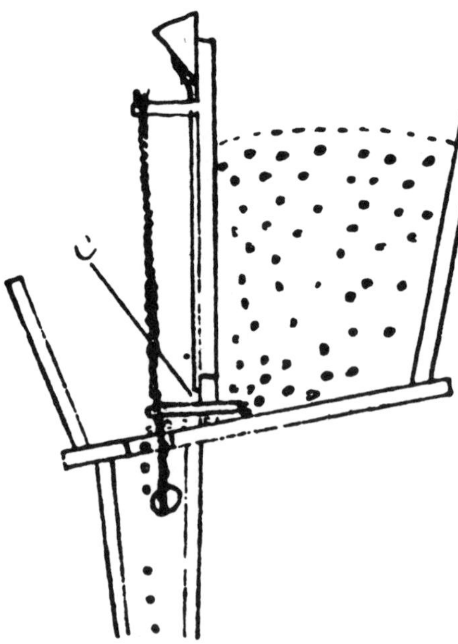


Fig. 105 (*d*)

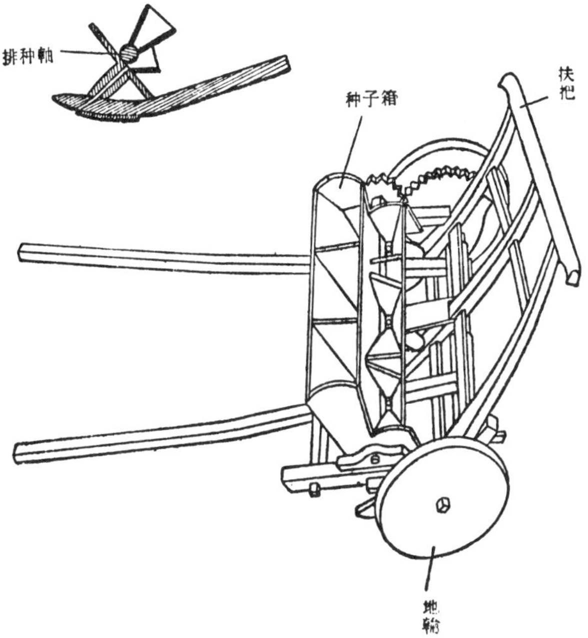


Fig. 105 (*e*)

Fig. 105. Chinese seed-dropping mechanisms; (*a*, *b*) Liu (*8*), p. 35; (*c*, *d*) Amano (*4*), p. 792; (*e*) Anon. (*502*), p. 257.

with the ante-chamber where a stone dangles. The stone is fastened to a flexible bamboo rod, and when the peasant serving the plough swings the plough by means of its handles gently from side to side, the suspended stone swings and imparts a back and forth motion to the bamboo rod which half-blocks the passage for seed from the seed-bin into the side-by-side chute holes. The function of the rod is to retard the flow of the seed, and at the same time throw alternately a few seeds towards the left into the left chute, and then again a few to the right into the right chute, according to the rhythm of the dangling stone.

The passage-way for seed from the bin to the ante-chamber can be varied in size with the slide board or seed-gate held in place by a wedge ... This seed-gate has at one end a semicircular notch and a larger one at the other end. For small seeds the gate is placed with the small notch against the communicating hole to the seed-bin, and for larger seeds, as of wheat and barley, the board is reversed.

Simply yet effective, and as so often in China, bamboo is an essential component of the device. A close paraphrase of Hommel's description is given by Liu Hsien-Chou,[a] who presents it as the standard regulating device, but Amano[b] illustrates two more regulating devices common on traditional Chinese drills (Fig. 105*c*, *d*). One quite simple way of regulating the flow to a variable number of tubes is to attach to the back of the drill a small wheel with cog-wheels on the axle, that

[a] (*8*), pp. 34–5.
[b] (*4*), p. 792.

advance and retract a shutter between the seed-bin and the chute (Fig. 105*e*). This is quite a common device nowadays,[a] but we can find no evidence of its existence in pre-Revolutionary China.

Most Chinese seed-drills sow not only seed but also manure. The idea was almost as old as the seed-drill itself in Europe, and it was common in India too. In China the practice was probably an ancient one, but our earliest evidence comes again from Wang Chen:[b]

> Close in construction [to the simple seed-drill] is the fertiliser-drill (*fen lou*[1]). Behind the seed-bin is placed a separate sieve containing fine compost, sometimes mixed with silk-worm manure, which covers the seed as the drill goes along the row. It is even more ingenious and efficient than the usual drill.

Another cunning improvement is to fix to the drill a container of water which sprinkles the seeds as they are sown; again this seems to be a modern development, perhaps dependent on the availability of light and adaptable containers such as the kerosene can.[c] Often a device for covering the seed is incorporated into the drill. Hommel[d] (Fig. 102) shows a bent bamboo rod curved behind the legs of the drill close to the ground, which dragged soil into the freshly sown furrows; modern drills are sometimes equipped with a pair of heavy wheels similar to the *tun chhe*[2] (see p. 272).

The question now arises of the origin of the Chinese seed-drill. Until recently when concrete archaeological evidence of Han seed-drills was first discovered, many scholars believed that the reference in the *Cheng Lun* to Chao Kuo's introduction of seed-drills in the metropolitan area had been misinterpreted, and that the seed-drill was first introduced to China from Central Asia by Huang-Fu Lung[3] when he was Prefect of Tun-huang[4] *c.* +250. The text on which this theory is based runs as follows:[e]

> In Tun-huang they were not familiar with the seed-drill (*lou li*[5]), so that they wasted both seed and the labour of men and oxen in return for poor harvests. Huang-Fu Lung taught them to make seed-drills, and this reduced their labour by half while increasing their yields by five-tenths.

However, this passage actually implies not that Huang-Fu Lung discovered a new Central Asiatic device in Tun-huang that subsequently spread to the rest of China, but on the contrary that he introduced a traditional Chinese implement to a frontier region where techniques were comparatively backward. The archaeological as well as the literary evidence shows clearly that the seed-drill was known in Western Han China.

[a] Anon. (*502*) shows a very simple single-tube type on p. 223 and a more complex three-tube variety on p. 257.

[b] *WCNS* 12/18*b*.

[c] Anon. (*502*), pp. 251 ff., 270 ff.

[d] (1), p. 45.

[e] *CMYS* 0.7, paraphrasing from *Wei Lüeh*.

1 糞耬 2 砘車 3 皇甫隆 4 燉煌 5 耬犁

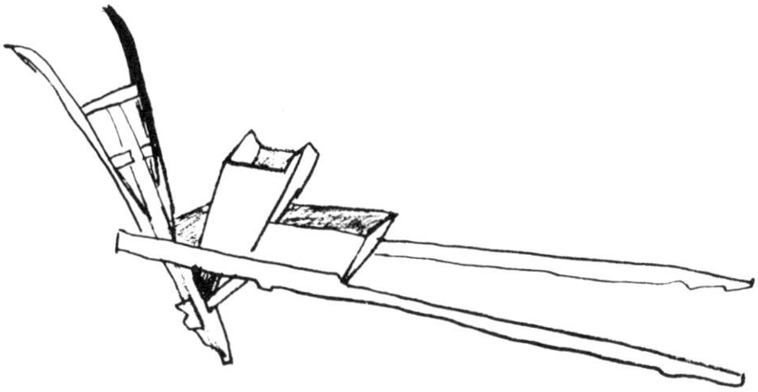

Fig. 106. Chinese single-tube seed-drill from modern Shantung; Anon. (*502*), p. 224.

The earliest seed-drills known anywhere are the −3rd-millennium drills of Mesopotamia. It is interesting to recall that the Han dictionary *Shuo Wen Chieh Tzu* mentions altogether three terms which can be taken to refer to the seed-drills, including the term *lou*[1] with which we are already familiar:[a]

> *Hui*[2]: a six-pronged plough (*liu chha li*[3]).
> *Hsi*[4]: a sowing drill (*chung lou*[5]).[b]

This diversity of terms could well indicate that even in Han times the Chinese seed-drill had a certain antiquity. Although the drill was used to sow many different crops in China it is particularly associated with wheat and barley,[c] both crops of West Asian origin; furthermore the Chinese single-tube drill even today (Fig. 106) bears a close resemblance to the ancient Mesopotamian drill (Fig. 96) in the assemblage of the handles and share. In the absence of contrary evidence we would venture to suggest that the single-tube seed-drill is closely associated with the cultivation of wheat and barley, that it originated in Mesopotamia and the Fertile Crescent where these crops were first domesticated and spread with them South to Iran and the Indian subcontinent, and East to Central Asia and eventually to China, where it arrived in pre-Han times.[d] The diversification of names associated with the drill in Han China would testify to its comparative antiquity. In China as in India, more complex seed-drills with several tubes were subsequently developed, and the two- or three-tube drills already familiar in Han times have continued in use in China up to the present day.

[a] Hayashi (*4*), p. 272; see Amano (*4*), p. 744 on the interpretation of the first definition.
[b] Note the use here of a different radical for *lou*.
[c] See *TKKW*, ch. 1.
[d] If this explanation were correct we should have to account for the lack of seed-drills in +3rd-century Tunhuang either by assuming that the place had fallen on hard times and lost its agricultural expertise, perhaps as a result of continuous warfare and invasions by nomads, or else perhaps that wheat and the associated technology had followed a more northerly route into China through Dzungaria and Mongolia; the first explanation seems the more likely.

1 耬 2 樟 3 六叉犂 4 椺 5 種樓

As for the European seed-drill, its appearance in the +16th century soon after the great voyages of discovery suggests strongly that the idea was borrowed from the East, though whether from the Middle East, India or China is impossible to say. Venice, where the first model was patented, received merchants and travellers from all over the Old World. The complex and clumsy nature of the early European machines shows that their inventors had no very clear idea of how the oriental prototypes functioned. Basically the Europeans tried to build their machines on too large a scale, with the result that they were unwieldy, unreliable and easily damaged. This bias towards engineering rather than craftsmanship was of course natural, given the prevailing trend towards enclosures and large-scale agriculture. One of the most efficient early seed-drills, and probably the closest to its oriental counterparts, was the late 18th-century James Sharp's drill-plough for single-row sowing,[a] which worked perfectly but was designed on too small a scale to attract much attention.

Once the seed-drill had sown the seed it had to be covered and the soil firmed down:[b]

> Spring sowing should always be deep, so draw a weighted bush (*tha*[1]) over the seed. Summer sowing should be shallow, so just sow the seed directly and leave it to sprout on its own. (In spring the soil is cold and germination slow. If you do not use the weighted *tha* the roots will spread into empty cracks [in the soil] and even though the plant germinates it will soon die. In summer the air is hot and germination rapid. If you use the weighted *tha* and then it rains the soil will become compacted ...)

Wang Chen describes the *tha* as a bunch of branches 3 or 4 feet long and a couple of feet wide, weighted down with stones[c] (Fig. 107), similar to the European device of a thorn-bush weighted with a log,[d] which was of some antiquity even though Gervase Markham hailed it as a new invention in the 17th century.[e] Some time before the Yuan a slightly more sophisticated implement was invented for compacting the soil, the *tun chhe*[2], *tun tzu*[3] or *kun tzu*[4], consisting in a number of trimmed stone wheels corresponding to the number of tubes on the farmer's seed-drill, and fixed at exactly the distance between furrows[f] (Fig. 108). These stone rollers are still very common in North China,[g] though nowadays they are not infrequently incorporated into the drill itself.[h] The use of the *tha* seems to have been dying out in Yuan times. Though contemporary agriculturalists recommended that the man driving the seed-drill tie a *tha* around his waist and drag it behind him to make sure that the seed was well covered before the *tun chhe* was used,[i] one can imagine that such uncomfortable advice usually fell on deaf

[a] Fussell (2), p. 104.
[b] *CMYS* 3.5.1–2.
[c] *WCNS* 12/12*b*.
[d] Fussell (2), p. 68.
[e] Gervase Markham (1), p. 70.
[f] *WCNS* 12/19*b*.
[g] Wagner (1), p. 204; J. L. Buck (1), pl. 12; Anon. (*502*), pp. 165–8.
[h] Anon. (*502*), p. 241.
[i] *WCNS* 12/19*b*.

[1] 撻 [2] 砘車 [3] 砘子 [4] 磙子

Fig. 107. Weighted bundle of branches (*tha*) used for covering seed; *WCNS* 12/12*a*.

Fig. 108. Roller wheels for covering seed; *TKKW* 1/16*a*.

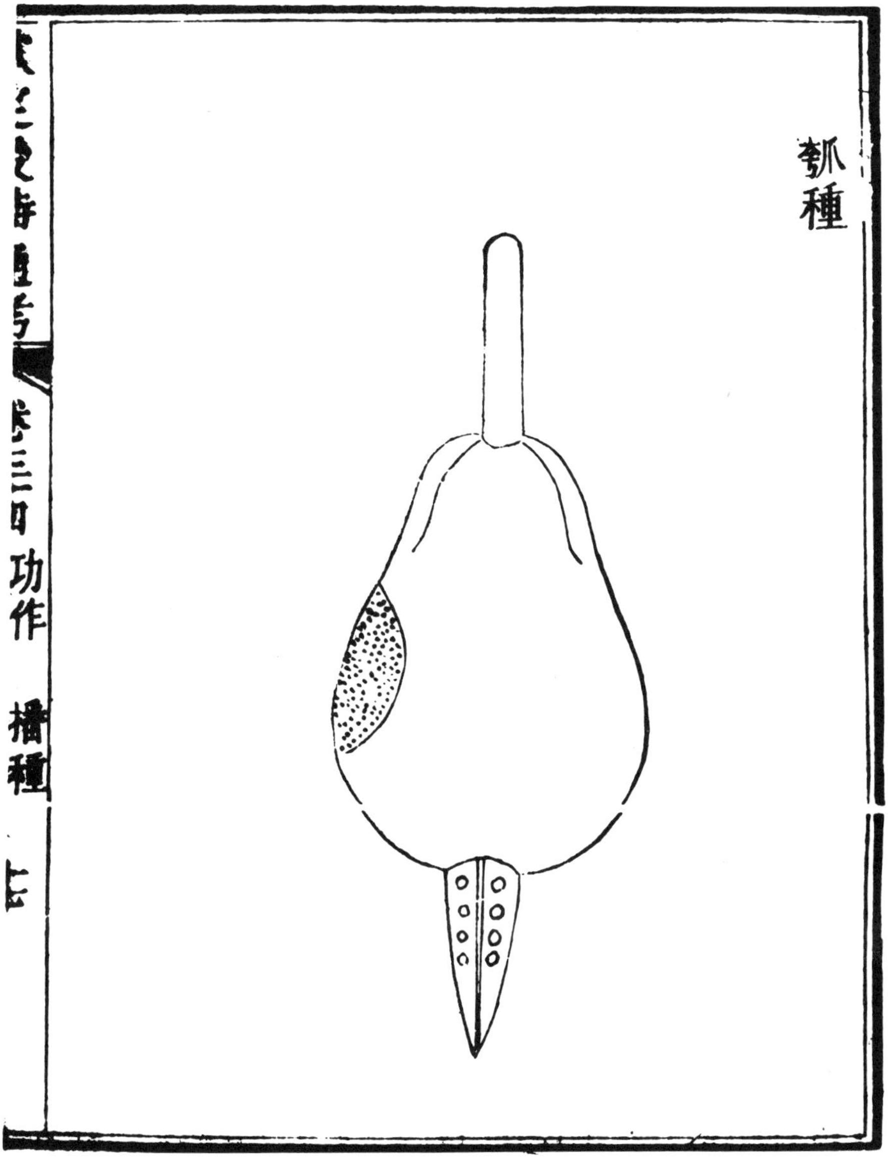


Fig. 109. Sowing calabash (*hu chung*). Chhing drawing; *SSTK* 34/17*a*. Note the lid at the left for inserting the seed, and the small holes in the beak-like tube below for regulating the flow of seed.

ears. After all, as Wang Chen said, the virtue of the *tun chhe* was its practicality and extreme simplicity.

Wang also recommended an alternative to the seed-drill for families who were short of labour or draught animals, the sowing-calabash (*hu chung*[1])[a] (Fig. 109). This was a large hollow calabash with a capacity of a gallon or so, pierced with a long wooden tube one end of which served as a handle, while the other was filled with hollow grass stems through which the seed was scattered. The sowing-calabash is also mentioned by Chia Ssu-Hsieh[b] in a passage which is far from clear but which apparently advises the ploughman to drag the calabash along the ground behind him by a string tied to his waist, a procedure which Wang stigmatises as clumsy.

*Sowing individual seeds; transplanting*

The individual planting or dibbling of seeds, *yen chung*[2], *tan yen chung*[3], or *tien chung*[4], is often considered to be a primitive planting method associated with shifting cultivation,[c] but in China it was often rather the extension of planting in rows. Crops that develop many tillers, like wheat or upland rice, for example,[d] or that are unsuited for drill sowing because of their shape or size, like ginger, were often planted individually on ploughed-up ridges in which shallow furrows had been made with the seed-drill. Chia Ssu-Hsieh refers to this process as *lou chiang yen chung*[5].[e] It was still used on hillsides in Shensi in the 18th century,[f] and dibblers for wheat are still found in China today.[g] The *ou thien*[6] or pit-cultivation system (see p. 127 above) was also a method of dibbling, whereby a checkerboard arrangement of small pits of carefully manured soil were planted with a few seeds of one or several species. Fan Sheng-Chih claimed that wheat grown in this fashion produced a hundred times the normal yield,[h] and Markham said of the similar method of setting wheat in 17th-century England:[i]

> Now for the profit which issueth from this practise of setting of Corne, I must needs confesse, if I shall speak simply of the thing, that is, how many foulds it doubleth and increaseth, surely it is both great and wonderfull: and whereas in general it is reputed that an Aker of the Corne yeeldeth as much profit as nine Akers of sowne Corne, for mine owne part I have seen a much greater increase.

It is difficult to say whether the ancient practice of pit cultivation in China is a development of the system of planting in rows, or whether it is indeed a relic of

[a] *WCNS* 12/20*b*.

[b] *CMYS* 21.2.3.

[c] For example D. Freeman (1) describes this method of planting hill rice among the Iban of Sarawak.

[d] Dryland rice is still dibbled in *permanent* fields in parts of Malaya to encourage free tillering; the process is known as *menugal*; A. H. Hill (1); Bray & Robertson (1).

[e] *CMYS* 10.2.2, 12.2.3.

[f] *Chih Pen Thi Kang*, p. 9.

[g] Anon. (*502*), p. 209.

[h] *FSCS* 7.3.1.

[i] Gervase Markham (2), p. 101.

1 瓠種 2 掩 or 稀種 3 單稀種 4 點種 5 耬耩稀種 6 區田

horticultural swidden techniques like the *citamene* of East Africa or the yam gardens of highland New Guinea.[a]

Individual planting is a technique suited not only to seeds but particularly to the vegetative reproduction of roots and tubers, and it is perhaps not too far-fetched to regard the transplanting of rice seedlings and other young crops as an elaboration of this method. Since transplanting is generally associated with rice it might have seemed more appropriate to include descriptions of the technique in the section on rice rather than under planting techniques but for two reasons. First, transplanting is not confined solely to rice: the dye plant bastard indigo (*lan*[1]) (*Polygonum tinctoreum*) is also grown as an irrigated crop in China, and references to transplanting indigo are just as ancient as those for rice.[b] Secondly, transplanting is the method *par excellence* of economising on seed: a seed-bed of 1 *mu* sown with a few pints of rice should provide enough seedlings for 25 *mu* of irrigated fields.[c] Thus transplanting can be seen as a logical extension of the Chinese techniques of economical sowing.

Here a brief description of the principles of transplanting would doubtless be welcomed by those unfamiliar with the rice farms of Asia. The rice-seed, usually pre-germinated, is sown in seed-beds of various types (see the section on rice, pp. 502 ff., for further details) and then transplanted into the main field after anything from 2 to 8 weeks, depending on the variety of rice.[d] In China transplanting is called *tsai*[2], *chha*[3] or *shih*[4]. The seedlings should normally be transplanted at their period of maximum rate of growth, by which time they will have reached a height of 15–18 cm.[e] The sturdiest seedlings are pulled up from the seed-bed by hand[f] and tied into small bundles which are transported immediately to the main field as they must be transplanted the same day[g] (Fig. 110). The roots of the seedlings are usually washed and the top few inches of the leaves trimmed to reduce evaporation and damage in handling. They are then transplanted into the main field, a few seedlings to each hill, the hills being anything from 10 to 30 cm. apart. The plants should be set shallow and upright in the mud:[h]

Shallow transplanting is recommended, for if the seedlings are planted deeply the roots fail to develop normally ... Roots near the surface are subject to higher temperatures

[a] W. Allan (1), p. 77; Jim Allen (1).
[b] *SMYL* 5.5; *CMYS* 53.2.4; see also the reference in the *Hsia Hsiao Cheng* on irrigating indigo, quoted *CMYS* 53.2.4.
[c] *TKKW* 1/3.
[d] Grist (1), p. 149; Ting Ying (*1*), pp. 344 ff.
[e] Chhen Liang-Tso (*1*), p. 540.
[f] In Hong Kong and parts of Southern China a specially designed hoe is used for the purpose; Grist (1), p. 155; Herklots (4).
[g] In Malabar the seedlings are laid in the sun with their roots exposed for 3 or 4 days, the object being to destroy insect eggs harboured in the roots; Grist (1), p. 155.
[h] Grist (1), p. 156.

[1] 藍 [2] 栽 [3] 插 [4] 蒔

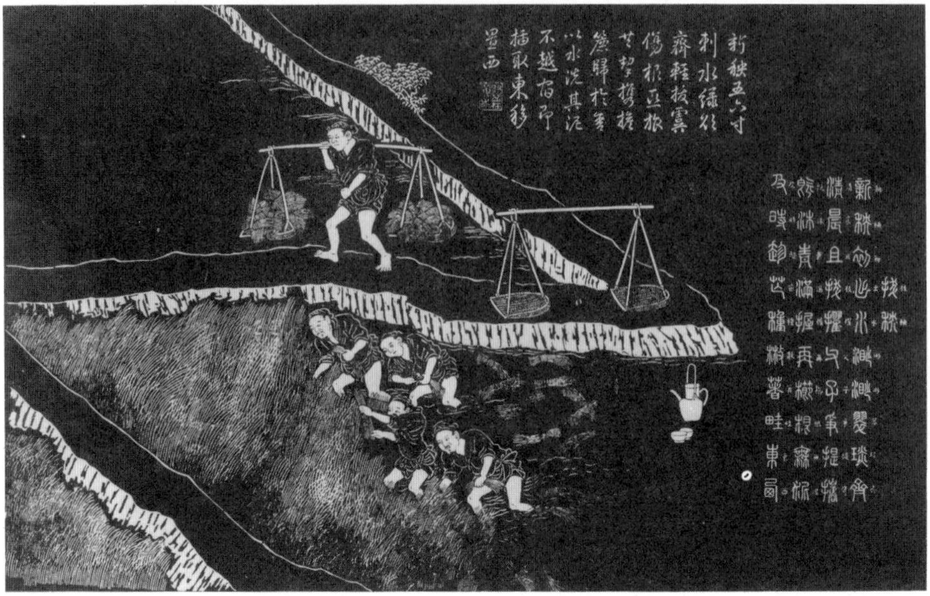

Fig. 110. Pulling up rice seedlings and transporting them in bundles to the field; *Keng Chih Thu*, Pelliot (24), pl. XVIII.

during the day and lower temperatures at night and this encourages tillering ... Investigations in India show that if seedlings are planted in a slanting position there is a setback in tillering, for the seedlings so planted develop at 25 to 35 mm below the surface a short rhizome sideways which delays the establishment of the plant and results in a reduction in the number of tillers.

The physiological factors involved in transplanting are still not fully understood:[a]

The conclusion at present, based on a number of experiments, is that transplantation acts in a similar way to root pruning, the injury to the root system stimulating growth of the sub-aerial portion and resulting in increased tillering. The root system of transplanted rice is developed from the lower nodes of the stem, the first or seedling root system dying completely in most cases.

Transplanting not only encourages tillering but also facilitates weed and pest control and crop management; transplanted rice generally gives substantially higher yields than direct broadcasting.[b] But it is a comparatively complex procedure requiring considerably greater investment of time and labour than the simple broadcasting method, and it is interesting to speculate on the conditions that first prompted the development of such an intensive system. Unfortunately very little is known of the history of transplanting anywhere but in China, so we have no comparative material to draw on in our attempt to unravel the Chinese evidence.

[a] H. W. Jack (1).

[b] Grist (1), pp. 149 ff.

The earliest textual evidence for transplanting in China is a sentence in the +2nd-century *Ssu Min Yüeh Ling*:[a] 'In the fifth month you may transplant (*pieh*[1]) rice and bastard indigo.' Though other interpretations of *pieh*, such as 'thinning out', have been suggested, most scholars are agreed that Tshui was actually writing of transplanting,[b] more especially as the task took place six weeks or so after the rice was first sown, that is the age at which many traditional long-ripening varieties of rice are transplanted in Asia today. Some scholars, relying on references in *Huai Nan Tzu* and *Lü Shih Chhun Chhiu* to weeding growing rice and to planting crops in straight rows, believe that rice was transplanted even earlier than the Han.[c] It is hard to imagine that the *Lü Shih Chhun Chhiu* would give much useful information on rice cultivation, and the implications of the *Huai Nan Tzu* passage are far from clear, but there is more positive evidence for transplanting in the Han. An Eastern Han model of a rice field found in Szechwan bears the inscription *chha yang khung hen*[2], literally 'holes for transplanting rice seedlings': *chha yang* is the term usually used for the process of sticking the seedlings into the mud.[d] Recently excavated tombs probably of early Eastern Han date, one on the Shensi-Szechwan border south of the Chinling Mountains and one in South-western Kweichow on the borders of Yünnan and Kwangsi, have yielded two well-preserved and detailed models of irrigated rice fields, both of which show well-spaced rice seedlings arranged in straight rows[e] (Fig. 111), strong indications that rice was transplanted in Southern China at that time.

The +6th century *Chhi Min Yao Shu* describes two methods of cultivating rice, the first, *sui i*[3],[f] consisted in fallowing the field every other year,[g] while the second required that the seedlings be pulled up and transplanted when they were 7 or 8 inches tall:[h] *chi sheng chhi pa tshun pa erh tsai chih*[4]. It also recommends that upland rice be dibbled in rows to increase tillering, but says that if the clumps of young shoots grow thickly some seedlings should be pulled up after heavy rain and transplanted elsewhere.[i] The commentary then goes on to say:[j]

When transplanting the seedlings should be set shallow so that the hair-roots can branch fully in all directions and the plants will grow thick and luxuriant. If they are

[a] *SMYL* 5.5.
[b] Amano (*4*), pp. 183 ff.
[c] *Huai Nan Tzu* 20/18*a*; *LSCC*, p. 76; Chhen Liang-Tso (*1*), p. 542.
[d] Amano points out, however, that the character *chha*[5] is unclear and could in fact be a mistake for *tao*[6], rice; (*4*), p. 185; see also Okazaki Satoshi (*1*).
[e] Chhin Chung-Hsing (*1*); Anon. (*520*), pp. 31 ff.
[f] *CMYS* 11.2.1.
[g] Shih Sheng-Han (*3*), p. 116; Amano (*4*), p. 194.
[h] *CMYS* 11.6.2.
[i] *CMYS* 12.2.3, 12.3.3. Perhaps these seedlings were transferred to wet rice fields? This was certainly common practice in parts of Kelantan, Malaysia, where upland and lowland rice were treated as complementary crops (Raymond Firth, pers. comm.).
[j] *CMYS* 12.3.3.

[1] 別 [2] 插秧孔痕 [3] 歲易 [4] 既生七八寸拔而栽之
[5] 插 [6] 稻

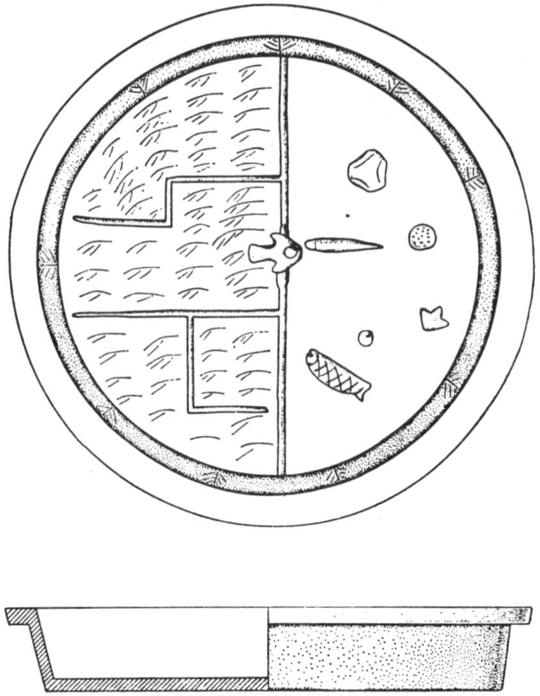

Fig. 111. Han model of a rice field from Kweichow. The seedlings are set in rows. The right-hand section represents the irrigation tank and contains fish, water caltrops and other aquatic plants; Anon. (*520*), fig. 17.

planted too deeply so that the roots grow straight down the tillers will not develop. If the shoots are too tall pinch off the upper few inches, taking care not to damage the heart of the plant.

So to *Chhi Min Yao Shu* falls the honour of first advocating shallow transplanting and trimming the seedlings. Upright planting is not specifically mentioned until the *Shu Hsü Pu* of +1504.[a] One assumes that the transplanters took care from very earliest times to set their rows reasonably straight as this would facilitate subsequent weeding; the Han models certainly show ordered rows of seedlings. But the first explicit reference comes from Lou Shou, the author of the *Keng Chih Thu* poems first published in +1145[b] (Fig. 112): 'To left and right the straight rows never waver (*tso yu wu luan hang*[1]).' The early +14th-century *Nung Sang I Shih Tsho Yao*[c] also exhorts farmers to make sure their rows are straight.

It has been said that the transplanting aid known as the 'seedling horse' (*yang ma*[2]) naturally moved in straight lines and so helped regular transplanting. With-

[a] 11/1*b*; see Chhen Liang-Tso (*1*), p. 543.
[b] *KCT* 1/10*b*.
[c] 1/16.

[1] 左右無亂行 [2] 秧馬

Fig. 112. Transplanting rice-seedlings; *Keng Chih Thu*, Pelliot (24), pl. XIX.

out personal experience it is difficult to judge the merits of this case; nevertheless the *yang ma*, first mentioned by the Szechwanese poet Su Tung-Pho[1] in the +11th century, must offer distinct advantages for it is still in common use in China today (Fig. 113). The modern versions, called 'transplanting boats' (*chha yang chhuan*[2]), are often enhanced by the addition of a parasol, but otherwise have not changed greatly in the nine hundred years since Su described them (Fig. 114):[a]

When I was travelling in Wu-chhang[3] [modern Hupei] I noticed that the farmers all rode 'seedling horses': their bellies were made of elm or jujube wood for smoothness, their backs of *chhou*[4] [*Mallotus japonica*] or *wu*[5] [*Firmiana platanifolia* R.Br.] for lightness. The belly was like a small ship raised at the head and tail, while the back was [convex] like a ridge-tile. This horse allows the rider to hop through the mud as sprightly as a sparrow, the bunches of seedlings being fastened to the head with a straw rope. By day thousands are to be seen going up and down the rice fields, their riders hunched over at their task ... I have composed a poem on the 'seedling horse':

From the low spring clouds fall curtains of rain.
The spring seedlings have reached their prime, the bright green blades of a height.
Father calls to son, gliding through mud and water,
At dawn separated by a single bund, at dusk by a thousand.
Belly like a lute and head like a chicken's beak,
His sinews strain, his bones ache, he neighs in anguish,

[a] *Yang-ma ke*[6], *Su Tung-Pho Chhüan Chi* II, ch. 4, p. 499; quoted *WCNS* 12/24*a*–*b*.

[1] 蘇東坡 [2] 插秧船 [3] 武昌 [4] 楸 [5] 梧
[6] 秧馬哥

Fig. 113. 'Seedling horse' (*yang ma*); *WCNS* 12/23*b*.

Fig. 114. Sketch of modern seedling horse made in Kwangtung in 1958 and published in *Jen Min Jih Pao*; Amano (*4*), p. 239.

My elm-wood horse that I can lift with one hand.
Head and rump are proudly raised, ribs and belly lie deep,
His back is barreled like a ridge-pole tile.[a]
My own two feet are his four hoofs
On which he hops and slides like a drake ...
The young lord with his brocade saddle-cloth, riding from the palace gate,
Mocks me who spend my life plodding behind the plough.
Little does he know that I too own a high-mettled steed—of wood.

The 'seedling horse' eases the backbreaking task of transplanting as well as speeding the work: on a *yang ma* a man can transplant over 1/15 hectare a day, twenty per cent more than he would normally manage.[b] But one imagines that it would actually be more difficult to plant straight riding a 'seedling horse' than standing in the field 'not moving your feet too often, for your hands can easily reach to transplant six hills at a go; then you gradually retreat, taking care to make straight rows'.[c] Perhaps this is why the 'seedling horses' were not recommended in most agricultural treatises, for as the *Nung Hsüeh Tsuan Yao* points out,[d] 'it is better to transplant slowly but skilfully than quickly but clumsily'.

Indeed, more recent developments in transplanting technology all seem to have been directed towards precision first of all. The agricultural manuals advise using ropes stretched between the bunds to ensure straight planting,[e] and indicate precise numbers of seedlings and distances between hills (Table 8).[f]

[a] All signs of a fine steed; see *CMYS* 56.12 ff.
[b] Anon. (*502*), p. 389.
[c] *Chhün Fang Phu, Ku Phu* 26*a*, quoted *SSTK* 34/5*a*.
[d] 2/4*b*.
[e] *WCNS* 14/21*b*; *Nung Hsüeh Tsuan Yao* 2/4*a*.
[f] E.g. *Nung Sang I Shih Tsho Yao* 1/16; *Chhün Fang Phu, Ku Phu* 26*a*, quoted *SSTK* 34/5*a*.

Table 8. *Transplanting distances for rice given in* Nung Hsüeh Tsuan Yao (2/4b)

| | Good soil | | | Medium soil | | | Poor soil | | |
|---|---|---|---|---|---|---|---|---|---|
| | Early rice | mid-season | late | early | mid-season | late | early | mid-season | late |
| Number of hills per pace | 45 | 40 | 35 | 55 | 50 | 45 | 65 | 60 | 55 |
| Number of seedlings per hill | 5 | 4 | 3 | 6 | 5 | 4 | 7 | 6 | 5 |

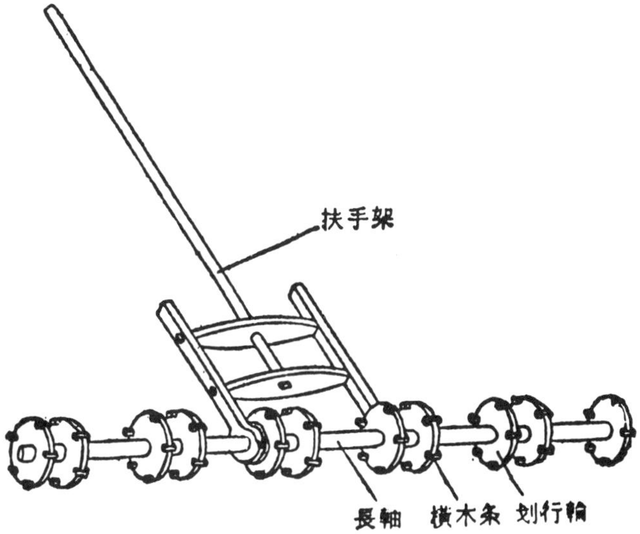


Fig. 115. Adjustable marker for transplanting rice seedlings, from Kiangsi; Anon. (*502*), p. 386.

Modern developments in China and Japan even include transplanting machines, but these are often expensive and require the seed to be sown in special trays if they are to function efficiently,[a] and often elaborate spacers or mechanical line-drawers, for example wide rollers with regularly spaced spikes, are more popular (Fig. 115).[b]

We have described, albeit briefly, the chief technical stages of development of transplanting in China, but we are still no nearer understanding where the technique was first developed or under what stimulus it spread. Various factors have been advanced to account for its adoption: the wish to reduce the number of weeds, the need to use water economically at sowing time before the summer rains arrive, increased pressure on land that made it impossible to fallow rice-land in alternate years, and eventually the absolute necessity imposed by the introduction of double-cropping, when there was simply no time to let the seedlings grow in the main field.[c] To these we would add the need for higher yields and the wish to economise on seed.

The generally accepted conclusion is that although transplanting was known in China as early as the Han it was not widely adopted until much later, when pressure on land made fallowing impractical. According to this theory transplanting was first developed in North China where there was little land suitable for wet rice available,[d] and it gradually spread to the Huai valley and further south as

[a] Grist (1), pp. 221 ff. gives diagrams and details.

[b] Anon. (*502*), pp. 384–8.

[c] All these arguments are discussed in Amano (*4*), pp. 198–201.

[d] The *Chhi Min Yao Shu* speaks of transplanting as a technique necessary in the North where natural rice-lands are rare; *CMYS* 11.6.1–2.

these areas became more densely populated, gaining in popularity with the introduction of winter wheat as a second crop.[a]

Although this model is attractive, we do not find it totally convincing. There is an increasing body of evidence to show that transplanting was practised in parts of South China in Han times. While it is true that there was little land naturally suited to wet rice cultivation in North China, it is equally true that rice remained a luxury rather than a staple in the Northern provinces until comparatively recently, whereas in the South, where rice was the staple food, pressure on irrigable land would be more acute. There are references to rice being cropped twice a year in Cochinchina, Chiao-chih[1], in Han times,[b] and considerable evidence to show that irrigation techniques and wet rice technology were well developed in Ling-nan and other parts of mainland Southeast Asia (see p. 228 above). We would therefore suggest tentatively that transplanting was developed in the rice-dependent cultures of Southeast Asia and first adopted by Chinese in the southernmost territories of Ling-nan, Yunnan and Szechwan. The technique was quickly adopted by those farmers in North China who grew rice, accustomed as they were to intensive cultivation techniques, but was probably largely ignored in the thinly populated areas of Central China around the Huai valley until the period of intensive migration southward from the Yellow River valley in the late Thang and Sung.

### (iv) *Sowing rates*

A general principle for Chinese farmers was that the later you sowed the more seed you should use, and that on poor soil you should use less seed than on fertile land.[c] The second rule reflects the care that went into tending the crops: where weeds are not systematically controlled it is usual to sow poor land more thickly than good to ensure that at least a few plants survive,[d] but where all crops are carefully tended one sows on the assumption that fertile soil can support more plants than poor. On the other hand, seed sown late in the season will generally have a lower rate of germination, and so one has to use more seed than when sowing early.

Precise figures for sowing rates, as for yields, are rare in Chinese literature, and unfortunately it is even rarer to find both given for any one crop, so that it is difficult to calculate the rate of return. The figures that do exist, however, even if they can only be taken as approximate, show that Chinese farmers generally achieved rather high rates of return for traditional agriculture. This is not

[a] Nishiyama Buichi (*1*); Amano (*4*), p. 201.
[b] *I Wu Chih* and *Yü I-Chhi Chien*, quoted *CMYS* 92.2.1, 92.2.2.
[c] Li Chhang-Nien (*3*), p. 102; *CMYS* 21.2.3; 8.2.4.
[d] K. D. White (2), p. 180.

[1] 交趾

altogether surprising when we reflect that millets and rice both produce many more grains on each plant than, say, wheat or barley. The story of the Eastern Han farmer who planted three or four double seeds of black panicled millet (*phei*[1]) and harvested 39 gallons (*tou*[2]) of grain is surely apocryphal.[a] But it is certainly true that Chinese crop characteristics, together with the Chinese farmer's attention to economical sowing and careful tending of each plant, generally produced far higher rates of return than were normal in Europe before the Agricultural Revolution.

Let us make one or two rough calculations. The Chhin statesman Li Khuei[3] reckoned an average yield of the staple crop setaria millet on a peasant's farm to be $1\frac{1}{2}$ bushels (*shih*[4]) per *mu*,[b] which in modern terms is roughly equivalent to 700 kg/ha. *Chhi Min Yao Shu* advises sowing setaria at 5 pints (*sheng*[5]) per *mu*, the equivalent of 2 kg/ha.[c] This would give a return of over 1 to 300, though here the sowing rate may well be too low as modern returns are usually closer to 1 to 100.[d]

Hsü Kuang-Chhi complained in the mid-17th century that farmers sowed their rice too thickly, at over 1 gallon (*tou*[6]) per *mu*.[e] At roughly the same period Sung Ying-Hsing said that 1 *mu* of rice-seedlings was sufficient to transplant 25 *mu* of rice land.[f] Rice yields varied enormously from region to region, but in the early 16th century medium land near Shanghai would yield $1\frac{1}{2}$ *shih* of husked grain, equivalent to at least 2 *shih* of unprocessed padi.[g] Assuming a sowing rate of approximately 1 *tou* per *mu*, this gives a return of roughly 1 to 500, which compares quite reasonably with modern returns.[h]

For wheat and barley, as one would expect from the physiology of these crops, returns were much lower. The late 12th-century *Chiu Huang Huo Min Shu*, trying to encourage the cultivation of wheat and barley as a winter crop in the Yangtze area, quoted the *Ssu Shih Tsuan Yao* as saying that these crops returned tenfold (*shih phei chhüan shou*[7]),[i] a normal figure for traditional cultivation techniques in Asia today.[j]

Although the rates of return given for millet, rice and wheat in China are very far from being precise figures, they do present an interesting comparison with the more accurate figures available for Europe in medieval times and later. On the

[a] *Thu Ching Pen Tshao*, quoted Hu Hsi-Wen (*3*), pp. 411–12.
[b] *CHS* 24*A*/6*b*; Swann (1), p. 140.
[c] Li Chhang-Nien (*3*), p. 106.
[d] Purseglove (2), p. 258.
[e] *NCCS* 25/10*b*.
[f] *TKKW* 1/3.
[g] On rice yields in the Ming see Amano (*4*), p. 332; E. Rawski (1), p. 39. The rate of conversion from padi to milled rice varies considerably; here we have taken it to be 70 per cent, though 50 per cent would not be surprising; Grist (1), p. 432.
[h] Purseglove (2), pp. 181 ff.
[i] Quoted Hu Hsi-Wen (*3*), p. 82.
[j] Purseglove (2), p. 294.

1 秠 2 斗 3 李悝 4 石 5 升
6 斗 7 十倍全收

French royal domain at Annapes in a poor year during the 8th century, for example, 54 per cent of the spelt harvest had to be used for seed, 60 per cent of the wheat, 62 per cent of the rye and *all* the barley.[a] In 13th century England that most efficient manager, Walter of Henley, still talked of a harvest of three times the amount of seed-grain as common, if not desirable,[b] and yield/seed ratios for wheat and barley were usually well under 1 : 10 all over Europe until rotations were improved in the Netherlands and England in the 18th century.[c]

### (v) *Conclusions*

It is clear from our survey of Chinese sowing techniques that they were far more elaborate than those in the West. Chinese farmers did their best to economise on seed, to practise the 'Discretion' that Fitzherbert preached. They selected the seed with care, often treated it with fertiliser or insecticide before sowing, and sowing techniques were designed so that each plant could be carefully tended throughout its growth. Since most Chinese cereals bear a comparatively large number of seeds, the net result was that only a very small proportion of the crop had to be set aside as seed grain.

There is a clear distinction between Northern and Southern Chinese sowing techniques. In the drier North, sowing techniques, like tillage methods, were designed to ensure that each plant received the maximum amount of soil moisture. The seed was sown in straight rows and carefully spaced, a procedure made possible by the use of the seed-drill which has been common in Asia for centuries though it was not perfected in the West until very recently.

In South China the need to conserve moisture was not acute, and dryland cereals were often simply sown broadcast. But for rice, the staple crop, a much more sophisticated method was developed. The technique of transplanting permits efficient use both of land and of water, and since it promotes tillering it gives higher yields than broadcasting; if transplanting is done carefully then all the benefits of straight-row planting, particularly easier weeding, are also conferred.

Both Northern and Southern sowing techniques were labour-intensive compared with the simple broadcasting methods of contemporary Europe, but they were self-sustaining in that they contributed significantly to the comparatively high yields of pre-modern China. This careful approach is no recent phenomenon. 'Husband your seed so that the plants are well-spaced', and 'The horizontal rows must be well-drawn, the vertical made with skill, so that the plants do not interfere with each other's growth' are both adages taken from the *Lü Shih Chhun Chhiu*, written more than two thousand years ago.

[a] G. Duby (1), pp. 38 ff.
[b] *Walter*, c. 59; Oschinsky (1), pp. 324–5.
[c] For a much more detailed account and excellent tables, see Slicher van Bath (1).

## (3) Fertilisation

Growing crops exhausts the mineral content of the soil and damages its structure. From very early times farmers have been aware that in order to repair the damage, and to enable them to grow more and better crops upon the land, nutrients must be returned in some form to the soil. The collection, preparation and application of fertilisers and manures is time-consuming and often expensive, and where space permits farmers often prefer to evade the issue altogether by practising shifting cultivation. Most settled farmers are, however, obliged to take fertilisers very seriously:[a]

> [European] farmers of the [15th to 17th centuries], both high and low, had one main worry, manure ... They dared not neglect any source of supply, however minute, for the success of every crop they grew depended very largely on the amount they could accumulate for use. They were willing to undertake the labours of Hercules to build a sufficient dunghill.

Chinese farmers were equally concerned to maintain their soil's 'strength' (*thu li*[1]), and were equally industrious in seeking out and preparing fertilising substances, as F. H. King's[b] account of the 'tremendous labour of body and amount of forethought required' makes plain, but the range of fertilisers used by Chinese farmers differed considerably from that familiar to Europeans. The European farming economy relied heavily upon stock-raising; as well as providing meat, wool and dairy products, animals provided the manure which was the European farmer's main source of fertiliser, supplemented to some extent by vegetable waste, marl, and (in coastal areas) seaweed and sand.[c] Animals played a much more minor role in the Chinese agricultural economy so that, while animal manures were never wasted, they provided only a very small proportion of the total amount of fertilisers used. The Chinese availed themselves of a much wider range of materials than was traditionally used in Europe, including one at least which repelled foreign visitors:[d]

> And because the Subject I now treat of dispences me to speak of all, I will relate, that which we further observed there, and whereat we were much abashed, judging thereby how far men suffer themselves to be carried by their Interests, and extream avarice; you must know then that in this Country there are many of such as make a trade of buying and selling mens Excrements, which is not so mean a Commerce among them, but that there are many of them grow rich by it, and are held in good account; now these Excrements serve to manure grounds that are newly grubb'd, which is found to be far better for that purpose than the ordinary dung: They which make a trade of buying it go up and down the streets with certain Clappers, like our Spittle men, whereby they give to understand what they desire without publishing of it otherwise to people, in regard the

[a] Fussell (3) p. 61.
[b] (1) p. 161.
[c] Fussell (3), passim.
[d] Pinto (1), ch. 31, §1.

1 土力

thing is filthy of it self; whereunto I will adde thus much, that this commodity is so much esteemed amongst them, and so great a trade driven of it, that into one sea port, sometimes there comes in one tyde two or three hundred Sayls laden with it: Oftentimes also there is such striving for it, as the Governours of the place are fain to enterpose their authority for the distribution of this goodly commodity, and all for to manure their grounds, which soyled with it, bears three crops in one year.

The use of 'nightsoil' or human manure immediately springs to Western minds when Chinese agriculture is mentioned, and although it may occasion considerable repugnance to us, to the Chinese it was an obvious and simple way to return fertility to the soil, the validity of which emerges clearly in modern chemical analyses:[a] 'From the analyses of mixed human excreta made by Wolff in Europe and by Kellner in Japan, it appears that, as an average, these carry in every 2000 pounds 12·7 pounds of nitrogen, 4 pounds of potassium, and 1·7 pounds of phosphorus ...' Although much Chinese farmland suffers from deficiencies in these elements,[b] it is clear that the use of human manure must over the ages have done much to mitigate the problem.[c] King describes most human waste in China as being stored in stoneware receptacles, hard-burned, glazed terra-cotta urns ranging in capacity from 500 to 1000 pounds.[d] The Sung agricultural writer Chhen Fu[1] says:[e]

Beside every farmer's dwelling there should be a manure house (*fen wu*[2]), with low eaves and posts to keep out the wind and rain; besides, if exposed to the sky the manure will lose its fertility. Inside the house a deep pit is dug, lined with bricks to prevent seepage. Into it go sweepings, ashes, husks and chaff from winnowing, chopped straw and fallen leaves, all of which are accumulated in the pit burned and enriched with liquid manure (*fen chih*[3]),[f] and kept there for as long as possible [before use].

The initial collection of the manure from which the *fen chih* was extracted presumably took place elsewhere, in the privy (*tshe*[4]), which was often combined with the family pigsty for the sake of convenience and efficiency (Fig. 116). There it was allowed to 'ripen' or 'cook' (*shu*[5]) before being used, for the direct application of fresh manure (whether human or animal) was recognised as harmful, as it burned the plants. During the process of composting, the heat generated kills

[a] King (1), p. 171.

[b] See p. 26.

[c] The practice is not in fact totally unknown in Europe, for Varro recommended the addition of servants' privies to compost heaps ((1) I. 13.4; see K. D. White (2) p. 133), while more recently human excreta were used as manure by market gardeners in France. Although generally most sewage in the West is simply swept out to sea, growing water shortages have recently encouraged the development of sewage farms where the water is purified and the sewage extracted and treated with micro-organisms to produce a sludge that can be used as agricultural fertiliser; E. J. Russell (1), pp. 208 ff.

[d] (1), p. 175.

[e] *CFNS*, *Fen Thien Chih I*, p. 6.

[f] Also known as 'clear liquid' (*chhing chih*[6]) or 'gold liquid' (*chin chih*[7]). This is the liquid that runs off, or is drained off, a manure heap; the French call it *purin*; the more fastidious English have no special word for it.

[1] 陳旉 [2] 糞屋 [3] 糞汁 [4] 廁 [5] 熟
[6] 清汁 [7] 金汁

Fig. 116. Han model of combined pigsty and privy; Laufer (3), fig. 12. Laufer misidentifies the privy as a grain store.

many of the harmful micro-organisms present in the manure, so that the application of human excreta to the fields is in fact less of a health hazard than has often been supposed.[a] The resulting liquid manure contains a high proportion of the total nutrients, especially of the potassium, and is therefore a very powerful fertiliser,[b] while the solid fraction, being mainly organic, is useful for improving soil structure and retaining moisture, just like the chaff, ashes and fallen leaves with which the liquid manure was to be mixed in Chhen Fu's 'manure house'. In South China wetter manures were preferred, and nightsoil was often used alone, diluted with water before being applied to the rice-fields; in the North nightsoil was always mixed with other organic matter, as Chhen Fu describes, and the composted manure kept as dry as possible.[c]

It is difficult to tell how far back the use of human manure goes in Chinese history, particularly as the Chinese frequently neglect to distinguish between

[a] See J. C. Scott (1). The results obtained by Scott and his colleagues in improving the elimination of pathogens are being actively used in China today; Anon. (*535*).

[b] Anon. (*535*).

[c] T. H. Shen (1), p. 36.

animal and human manure, simply calling it all *fen*[1]. (The combining of privies with pig-pens might explain this (Fig. 116).) It seems likely to be an ancient practice; there is even some slight evidence for its existence in the Shang-Yin period:[a] 'There is a character [in the Shang oracle inscriptions] showing the act of defecation (later modified into *shih*[2]), which may in its context be interpreted as 'manuring the fields'.'

Animal manures were also used as and when available, including ox, sheep and pig manure and chicken droppings. Silkworm droppings (*tshan shih*[3] or *tshan sha*[4]) were a particularly potent fertiliser, mentioned quite frequently as early as the –1st-century *Fan Sheng-Chih Shu*.[b] Where sheep or cattle were kept in stables their litter was carefully collected:[c]

> To prepare bedding straw (*tha fen*[5]): once the fields have been harvested in the autumn, all the stubble and trash left in the fields should be collected and stored together. Every day some should be spread 3 inches deep in the cattle byre, and early each morning this should be collected and composted. Then more should be spread as before, and after a night has passed it should be added to the compost heap. I have calculated that during the winter one ox will use 30 loads of bedding straw. At the time of the 12th or 1st month it should be carted to the fields and used as fertiliser. If it is reckoned that 1 small *mu* requires 5 loads, then 1 ox will provide fertiliser for 6 *mu*.

The composting and preparation of manure could be a complex process. The late Ming scientist Hsü Kuang-Chhi[6] was extremely interested in fertilisers and manures, and his manuscript notebooks in the Shanghai Municipal Library contain several pages on the subject. One passage relates to the ancient practice of pit cultivation (*ou chung*[7]):

> The old methods for pit cultivation say that you should use one pint of 'cooked' [or 'treated'] (*shu*[8]) manure per pit, but they do not give the method of preparation. However, I found the method elsewhere: one must boil (*chu*[9]) the manure on a fire to prepare it, and then it will [help the crop to] resist drought. The *Chou Li* commentary contains indications on the subject: boiled manure, when treated and used in the fields, is one hundred times as effective [as ordinary manure]. Each type of manure receives an addition of bones, ox or horse bones for ox or horse manure, human hair for human manure ... The soil of the pit must first be thoroughly dried out if it is to be drought-resistant, then the ashes from three herbs, 'goose bowels' (*o chhang*[10]) [*Stellarium media*], yellow artemisia (*huang hao*[11]), and *Xanthium strumarium* (*tshang erh*[12])[d] are put in, the treated manure added and watered, and once the pit has dried out the seed is sown and

[a] Lu Gwei-Djen (3), commenting on Hu Hou-Hsüan (5).
[b] E.g. *CMYS* 10.11.2.
[c] *CMYS* 00.12–13.
[d] All these plants were common weeds all over China, and artemisia and *Xanthium* were often used as insect repellents when sowing or storing grain, so perhaps their ashes were supposed to protect the seeds from insect attack.

1 糞 2 屎 3 蠶矢 or 屎 4 蠶沙 5 踏糞
6 徐光啓 7 區種 8 熟 9 煮 10 鵝腸
11 黃蒿 12 蒼耳

> covered with soil with which a small amount of the manure has been mixed. I myself have experimented with this method and [have found that] if I sowed the seed in this manner then 1 *mu* might yield 30 bushels (*shih*); if I only used the treated manure without the herb ashes then I got 20 bushels or so; if I used untreated manure and no ashes the yields were normal and could not be increased. So we see that ancient methods should not be neglected.
>
> In my opinion 'boiled' manure is about equal in strength to liquid manure (*chin chih*[1]), only liquid manure must be applied for several years to take effect, while the boiled manure will work straight away.

Another very ancient form of fertilisation in China was the use of green manures. These are crops which are grown solely for the purpose of improving the soil for a following crop: they not only return to the soil the nutrients which they have absorbed during growth but also add to it minerals and other elements that they have taken in, and improve the soil structure by the addition of organic matter. In China green manures were at first called, appropriately enough, 'embellishers' (*mei*[2]), though later the term 'green fertiliser' (*lü fei*[3]) was adopted. It is supposed that the first step in their use was the discovery that cutting down the weeds in a field, allowing them to rot or ploughing them in, improved the crop. Such natural green manures may have been known in late Shang times if a reference in the earliest part of the *Shih Ching* to smartweed (*thu liao*[4]) rotting in the fields is correctly interpreted; the late Chou work *Li Chi Yüeh Ling* also contains references to the fertilising value of rotting weeds.[a] But the use of cultivated green manures seems to have begun later in China than in the West, for the value of green manure crops was already well known to the Greeks, and they are mentioned by all the Roman authorities, leguminous crops such as the lupin and vetch being preferred.[b]

The earliest Chinese references to the cultivation of green manures occur in the +6th-century *Chhi Min Yao Shu*, which advocates the ploughing-in of adzuki beans (*lü tou*[5]) before cultivating melons, mallows and other vegetables.[c] However, it was the opinion of the Chhing scholar Ma Kuo-Han[6] that these sections of the *Chhi Min Yao Shu* were taken from an early Han work, the *Yin Tu Wei Shu*; the modern scholar Chhen Liang-Tso (*2*) concurs in this opinion and gives a complete list of the very wide range of green manures mentioned in early Chinese texts. These included wheat, barley, peas, lucernes, brassicas and radishes, but the majority of green manures were leguminous plants with nitrogen-fixing properties. Although the legumes' fertilising powers are greatest if they are ploughed in when still young, they add greatly to the soil's fertility even if they are allowed to mature and their beans are harvested before the straw is ploughed in.

a Chhen Liang-Tso (*2*).
b K. D. White (2), p. 135.
c *CMYS* 17.5.10, etc.

1 金汁 2 美 3 綠肥 4 荼蓼 5 綠豆
6 馬國翰

Fig. 117. Wheels of bean-waste manure; King (1), fig. 120.

The role of legumes, and of crop rotations generally, in maintaining soil fertility is discussed in the section on *Crop Systems* (pp. 429 ff. below).

The fertilising methods mentioned so far, like the process of seed-pelleting already described in the section on *Sowing*, can all claim considerable antiquity in China. But as the population grew and cropping practices became increasingly intensive, a whole new range of fertilisers was added to the Chinese repertoire in the Sung and Ming. Some of them the farmer could quite easily make or procure for himself:[a]

> The best fertiliser [for rice seedbeds] is hemp waste (*ma khu*[1]), but hemp waste is difficult to use. It must be pounded fine and buried in a pit with burned manure. As when making yeast, wait for it to give off heat and sprout hairs, then spread it out and put the hot fertiliser from the centre to the sides and the cold from the sides to the centre, then heap it back in the pit. Repeat three or four times till it no longer gives off heat. It will then be ready for use. If it is not treated in this way it will burn and kill the plants. Neither should you use nightsoil, which rots the shoots and harms human hands and feet, causing sores that are difficult to heal. Best of all is a mixture of burned compost with singed pigs' bristles and coarse bran rotted in a pit.

Other fertilisers, such as oil-cake (*yu ping*[2] or *tshai ping*[3]), or the cake left over from making bean-curd (*tou ping*[4]), were industrial by-products and had to be purchased (Fig. 117). The expense was not usually begrudged, for a little oil-cake went a long way (pounded up fine and sown with the seed, a single cake would

[a] *CFNS*, *Shan Chhi Ken Miao*, p. 5.

[1] 麻苦 [2] 油餅 [3] 菜餅 [4] 豆餅

fertilise a whole *mu* of young rice[a]), and the effect was longlasting.[b] A thriving trade in fertilisers developed, not only in the more industrial fertilisers like oil-cake but also in lime and in mollusc shells (used for their lime content),[c] river mud, silkworm waste and human manure, all of which were sometimes transported over considerable distances.[d]

By Ming times the Chinese made use of a very wide range of fertilising materials. Hsü Kuang-Chhi[1] in his notebooks mentions eighty, mainly conventional manures and waste products like oil-cakes, hemp waste, bean-curd liquor, animals' bones and hoofs, chicken feathers and the like. Local variations in their use might be considerable, as in 17th-century Kwangtung:[e]

> Above Canton they use cattle bones as fertiliser, and below Canton they use the husks of tea-seeds or hemp. In the other, more mountainous areas they use lime, because the heat of the fire element in lime eliminates the cold influence of the water.[f] They may also use frogs killed in brine as a fertiliser, but then the grain is always sparse. These manures are not used in the tidal fields (*chhao thien*[2]), which are the more fertile the lower they lie; the most fertile of all are those newly reclaimed from the sand, where the sea covers them with mud and silt that accumulates at every tide.

The Chinese did not use fertilisers indiscriminately, for they were well aware that soils varied in their requirements and that different fertilisers had different properties, so that the thoughtless application of the wrong type or quantity of fertiliser might do more harm than good.[g] If possible all fields were manured at ploughing time, as this both increased the soil's fertility and improved its structure,[h] and seeds were usually sown mixed with compost or manure to get them off to a good start. Wheat benefited from the application of fertilisers such as river mud, banked up around the roots during its growth.[i] Rice, being an irrigated crop, was rather more difficult to cope with and most fertilisers were applied at the seedling stage. Fig. 118 shows tubs of fertiliser, probably well-composted and diluted human manure, being applied to the young plants with a long-handled ladle such as Chinese farmers still use today. The Neo-Confucian writer Chu Hsi[3] (1130–1200), when he was serving as magistrate in Kiangsi in the 1170s, advised the local peasants to get their rice off to a good start by sowing

[a] *Shen Shih Nung Shu*, p. 236.

[b] *Wu Hsing Chhüan Ku Chi*, cited Chhen Tsu-Kuei (*2*), p. 106.

[c] King (1), p. 155; Wagner (1) p. 233; Amano (*4*) p. 309; this form of lime began to be prominent as a fertiliser from the Sung dynasty on; Yang Min (*1*).

[d] See Pinto (1), as cited above, and King (1), passim. The development of commercial fertilisers was an important factor in the increases in agricultural productivity during this period (Perkins (1); Golas (1); Elvin (2), (3)), as it was also in Tokugawa Japan (T. C. Smith (1); Pauer (1)).

[e] *Kuang Tung Hsin Yü*, p. 375.

[f] A common belief, see *KCT* 1/8*a*; Amano (*4*), p. 228. The addition of lime to rice-fields tends to coagulate the clay particles, preventing them from forming colloidal solutions in the water; Buck (2), p. 156.

[g] *CFNS, Fen Thien Chih I*, p. 6.

[h] E.g. *CMYS* 00.13; *CFNS, Keng Nou Chih I*, p. 3; etc.

[i] *NCCS* 26/13*a*; *Pu Nung Shu* 2/31.

[1] 徐光啓 [2] 潮田 [3] 朱熹

Fig. 118. Fertilising the rice seedlings; *Keng Chih Thu*, Imperial ed., 1/8*b*.

it in fertiliser pellets:[a] 'Make the pellets in the slack seasons of autumn and winter. First, gather grass and roots (from waste patches of ground), dry them in the sun and combine them with nightsoil. Push the seeds into the middle of the pellets and sow the pellets.' By a judicious combination of different types of fertiliser it was possible to keep the rice plants well supplied with nutrients throughout their growth:[b]

> You should not put down manure too early or its strength will not last ... Only at sowing time must river mud be applied as a base, and although its strength is lasting and dissipates slowly, by midsummer you should apply a little potash or oil-cake, which also dissipates slowly and is long-lasting. Only at the end of the summer or the beginning of autumn should you apply nightsoil, by which time it will have double the effect, so that the rice panicles will grow very long.

The Chinese had a fine empirical grasp of the practical principles of fertilisation, and understood that the addition of organic manures improved the soil structure and increased water retention as well as nourishing the crop. Deeper than this their knowledge could not go, for only with the development of botanical science and analytical chemistry in the last two centuries have we come to understand the processes of plant nutrition and the chemical composition of soils.

In Europe in the 17th and 18th centuries, many natural scientists believed that plants nourished themselves by absorbing minute crumbs of soil, while such eminent scientific authorities as Bacon and Boyle still adhered to Thales' view that water alone was the essential plant food.[c] Agricultural writers of the period held somewhat different views, believing that plants derived their nourishment from the 'salt of the earth':[d] 'The Juices of the Earth enter into the Roots of the Plants. The Rain as other Waterings dissolve the Salts of the Earth; this puts the Juices in Motion, and then the Subterranean Heat drives them upward, after this comes the Heat of the Sun which dilates the Pores of the Plants, and opens a Passage for the Juices to mount up into the Stem and into the Branches.' This notion of salts in fact corresponds closely to our modern understanding of the processes of plant nutrition, but at the time the state of chemical knowledge was too rudimentary to discern the roles of minerals in plant nutrition. The late 18th century was a turning point for plant sciences in Europe: Cavendish isolated hydrogen, Priestley discovered oxygen, Lavoisier demonstrated the composition of water and air, de Saussure (1) showed that green plants cannot live without carbon dioxide and that plant life is dependent upon the presence of nitrogen in the soil. Such discoveries paved the way for Liebig, the founder of agricultural chemistry, whose *Organic Chemistry in its Applications to Agriculture and Physiology* (1) was first published in Germany and Britain in 1840. It was this analytical,

[a] *Chu Wen-Kung Wen Chi* 99, *Kung I*, cited Amano (*4*) p. 228; tr. Philippa Hawking.
[b] *Wu Hsing Chhüan Ku Chi*, quoted Chhen Tsu-Kuei (*2*) p. 106.
[c] Fussell (3), p. 83.
[d] Anon. (161).

scientific approach to crop fertilisation which made possible the large-scale industrial synthesis of fertilisers, of the ammonia compounds and super-phosphates which are now produced and consumed in vast quantities all over the world.[a]

In China chemical fertilisers were unknown until they were imported from the West, but although they did not make their appearance until the early 20th century, they quickly became popular.[b] Chinese peasants had probably applied more fertilisers to their fields than farmers almost anywhere else, but nevertheless centuries of intensive cultivation had resulted in general deficiencies in nitrogen, potassium and phosphorus, as well as other minerals.[c] These deficiencies the traditional fertilisers were unable to make good. But traditional methods of fertilisation had proved their worth and were not abandoned on the advent of the new chemical products (as has often been the case elsewhere): most of them are still in common use in China today, being regarded as an essential complement to the chemical fertilisers which are low in organic content. Indeed some traditional Chinese fertilisers are acquiring still greater importance as synthetic products become increasingly costly, and scientific methods are being sought to improve their strength and efficacy.[d] It is unlikely that the 'honey-cart' will ever disappear from the streets of Chinese cities.

### (4) Weeding and Cultivation

Until well after the First World War, bands of women and children engaged in the tasks of hand-weeding or picking up stones were a common enough sight in the fields of England, but today many Western farmers do their weeding by aeroplane, spraying powerful chemical herbicides and pesticides over vast areas. Now that we are accustomed to such capital-intensive methods, only those of us with large gardens will appreciate how much protection growing plants need and what a valiant struggle weeds put up against extinction:[e]

> The may wede doth burne, and the thistle doth treate:
> The tine pulleth downe, both the rie and the wheate.
> The dock and the brake, noieth corne very much:
> But bodle for barley, no weede there is such.

It is easy to observe that a plant which is weeded and tended throughout its growth gives much better results than one which is abandoned to survive as best it may; indeed, it has often been suggested that the weeding and tending of certain

[a] The development of agricultural chemistry is treated in detail by Fussell (3) and Rossiter (1).

[b] T. H. Shen (1), pp. 32–9.

[c] *Ibid.*

[d] See Anon. (*535*). An interesting by-product of this research is the development of biogas production, especially from human manure (*ibid.* pp. 72 ff.). Szechwan is the world pioneer in biogas production; a UNESCO conference on biogas was held in Chengtu in 1979, and it is claimed that most villagers in the Chengtu plain now cook and read by biogas.

[e] Thomas Tusser (1), §80.

useful wild plants was the first step towards domestication. Weeds inhibit growth, depriving the crop of moisture, light and nutrients. It has been estimated, for example, that the amount of water saved by eliminating weeds in a maize field is equivalent to providing an entire course of irrigation at the time of maximum need,[a] while the hand-weeding of irrigated rice has been shown to increase yields by up to 45 per cent.[b] Weeds can also harbour pests and diseases[c] and sometimes bear poisonous grains which contaminate the crop.[d]

It is clear why farmers should regard weeds as an unpleasant challenge and devote large amounts of time and energy to their extirpation, but we should point out that there are one or two things to be said in favour of some weeds. Just as 'dirt is matter out of place',[e] so weeds are plants in the wrong bed, and it is important to remember that many modern crops were originally weeds of other crop-plants. Cultivated oats were developed from the common weed of corn fields, and it has been suggested that domesticated rice was evolved from a weed of taro gardens.[f] The very weeds that farmers find most difficult to eradicate, namely the wild relatives of crop plants which thrive in the same environment, can cross-breed with the crops, and although this is frequently regarded as a source of degeneration it may equally well serve as a means of genetic enrichment, enabling the crop to develop, adapt and compete more successfully than 'true strains' bred and maintained in laboratory conditions.[g] In some parts of the world traditional farmers seem to be aware of the advantages of such interbreeding: in Mexico teosinte plants (*Euchlaena mexicana* or *Zea mexicana*) are referred to as 'the maize father', while in West Africa wild sorghum is similarly revered and is not weeded from the crop; as for China, that land of inveterate weeders, it is still possible today, even in experimental agricultural stations, to see the small heads of wild foxtail millet, *Setaria viridis* L. (known variously as *kou wei tshao*[1] or *ku yu tzu*[2]), nodding among the much larger panicles of its cultivated descendant *Setaria italica*, an accepted interloper.[h]

This said, it is undeniable that most weeds are better eliminated, whether by hand, by hoe, or by some more elaborate device, and the resultant stirring of the soil has further beneficial effects on the crop. Hoeing or stirring of dry fields

[a] P. C. Mangelsdorf (4).

[b] International Rice Research Institute (1).

[c] A notorious example is the common barberry, *Berberis vulgaris*, which acts as host to stem-rust of wheat; the Arabs popularised the barberry, whose berries were used to make medicines and sweetmeats, throughout the Mediterranean area in early medieval times, and it was almost certainly the spread of this bush which was responsible for the disastrous wheat-harvests and numerous famines characteristic of the period; Carefoot & Sprott (1), p. 34.

[d] Corncockle, *Agrostemma githago* L., is one familiar English example; another is darnel, *Lolium temulentum* L., a very common weed of wheat before the introduction of modern herbicides; it has been shown that flour prepared from grain contaminated with a mere 0.5 per cent of darnel seeds is toxic; L. J. King (1).

[e] Mary Douglas (1), p. 40.

[f] Haudricourt (13), p. 95; (14), p. 41.

[g] A. Haudricourt, pers. comm., 1980.

[h] G. Métailié, pers. comm., 1980.

[1] 狗尾草 [2] 谷莠子

improves moisture retention and allows the roots of the plants to develop:[a] 'A plant is almost as imperfectly nourished by tillage without hoeing as an animal body would be without gall and pancreatic juice: for roots pass along the soil, as the soil or mass passes along the guts...' In areas such as North China, where the climate is dry so that centuries of manuring have produced a thin, impermeable crust of salts on the surface of the soil, it is only by hoeing immediately after every rainfall that plants can really benefit from the moisture.[b] If they are not properly hoed they will be stunted and cereals will run to straw rather than grain:[c] 'Hoeing ... keeps the soil at a good tilth and will give full ears with thin husks that do not shatter. If you hoe your field ten times you will get "eighty per cent grain" [i.e. grain that gives eight measures of hulled grain for every ten measures in the husk].' Hoeing the growing crops not only keeps them free from weeds and improves the tilth of the soil, it also permits the farmer to thin the crop, to weed out sickly plants and replant bare patches, to earth up cereals so that their roots spread and tillers develop, and to dig in manure or other fertilisers. In very few circumstances can farmers afford to neglect weeding or hoeing, but they have been brought to a particularly fine art in the labour-intensive cultivation systems of China and Japan. The Chinese have a number of terms for the process. The earliest recorded, *yün tzu*[1] (literally 'weeding and earthing up'), is found in the *Shih Ching*;[d] other common terms are *hao yün*[2], *chhu chih*[3][e] and *chung keng*[4], a term still in general use today. The tools and methods adopted in dryland farming systems and irrigated areas naturally present many differences, and we shall accordingly treat them under separate headings.

### (i) *Dryland agriculture*

Weeding and hoeing are particularly important in dryland areas such as North China where the conservation of soil moisture is one of the farmer's primary concerns: 'There are three inches of moisture on the end of a hoe', says the old Chinese proverb.[f] Once the young plants had been thinned out with a small, pointed hoe and bare patches replanted,[g] the farmers set to the long-term task of hoeing between the crowns of cereals, removing weeds, earthing up (*tzu*[5] or *phei*[6]) such plants as wheat or beans, and maintaining a dust mulch that prevented evaporation:[h] 'You should not mind how many times you hoe; once you have been right round the field, start again, and do not stop [even] for a short time just

[a] Jethro Tull (1), p. 118.
[b] W. Wagner (1), p. 265.
[c] *CMYS* 3.10.2.
[d] See Legge (8), vol. II, p. 583.
[e] These are used as section headings respectively in Chhen Fu's and Wang Chen's treatises; *CFNS* 1/p. 6; *WCNS* 3/1*a*.
[f] *CMYS* 00.24.
[g] *LSCC*, p. 86; *CMYS* 3.8.1–2; 3.9.2.
[h] *CMYS* 3.10.2

1 耘耔 2 薅耘 3 鋤治 4 中耕 5 耔
6 培

because there are no weeds.' The earliest weeding-hoes were probably made of a shell or bone blade hafted onto a wooden handle. By Warring States times the blades were often tipped or made entirely of iron. One type with a simple oblong blade was known variously as *chhu*[1], *nou*[2], *chhu chu*[3] or *ting*[4], and ranged in size from the tiny hand-hoes used in vegetable gardens (Fig. 119) to large, heavy field-hoe (Figs. 120, 121). These simple hoes have formed an essential part of the Chinese farmer's equipment for millennia.

An early variant was the drag-hoe. The earliest examples are hexagonal and of cast iron, set on the handle at an acute angle; these first appear in the Warring States and continued to be used through the Han and at least up till the Sui[a] (Fig. 71). But even in Han times the archaeological evidence indicates that they were being replaced by the swan-necked hoe, *po*[5] or *yu chhu*[6] (Fig. 123). This can hoe right round the individual plants without damaging them, and has the additional advantage that a whole range of interchangeable blades can be fitted to the long tubular shank (Fig. 124).[b] Wang Chen was a great enthusiast of the swan-necked hoe, to whose use he attributed the high yields achieved on poor soils in Hsing-chou[7] (modern Hunan) and Yang-chou[8] (the Yangtze Delta); the swan-necked hoe was, he says, used throughout North China too, but in the dry fields of the Huai area as well as in the deep South it was unknown, and Wang felt this was responsible for their poor standard of husbandry.[c]

Other variants of the drag-hoe were the stirrup-hoe (*teng chhu*[9]) (Fig. 125), which Wang Chen said could be used to advantage in any soil, and its simplified form, the weed-scraper (Fig. 126), widely used today in all parts of China on both wet and dry land.[d]

A type of iron spade, known variously as *chhan*[10], *yao*[11] or *chien*[12],[e] was used to chop the weeds off just below the surface. It appears to have been highly esteemed for its efficiency up to the Thang dynasty,[f] but by Wang Chen's time its popularity had apparently waned for he does not give any account of its contemporary distribution or dialectal variants. A more versatile and widespread implement was the grubbing fork or 'universal tool', *thieh tha*[13] or *feng*[14]. We have already discussed its role as a tillage implement (p. 209 above), and it was also much used for tending the growing plants, both in wet and dry fields.[g] It seems to have been particularly popular in Japan.[h]

[a] Anon. (*508*), p. 49.
[b] *WCNS* 13/23*a*; F. H. King (1), p. 214; W. Wagner (1), p. 263.
[c] *WCNS* 13/20*a*; 23*a*–*b*.
[d] Anon. (*502*), vol. 2, pp. 16 ff., 98 ff.
[e] *WCNS* 13/19*a*.
[f] It is praised in the *Tsuan Wen* (quoted *WCNS* 13/19*a*) and its use recommended for pit cultivation by Chia Ssu-Hsieh (*CMYS* 3.19.13).
[g] E.g. *CMYS* 3.10.4; *WCNS* 13/7*b*–8*a*.
[h] Iinuma and Horio (*1*), pp. 124 ff.

1 鋤，耡 or 鉏　2 耨　3 斫钃　4 定　5 鎛
6 櫌鉏　7 荆州　8 揚州　9 鐙鋤　10 鏟
11 銚　12 錢　13 鐵搭　14 鋒

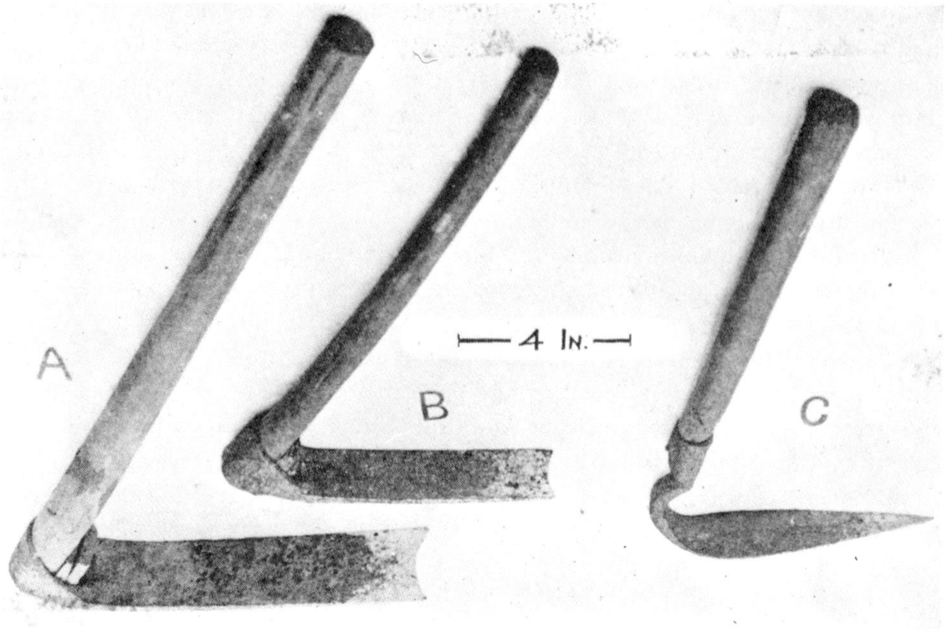


Fig. 119. Modern hand hoes; Hommel (1), fig. 91.

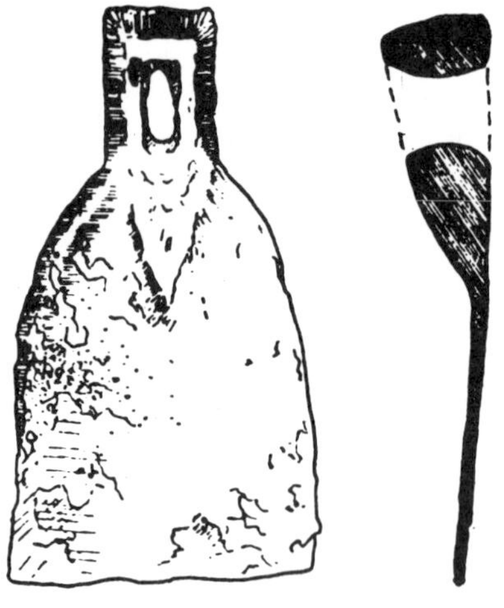

Fig. 120. Heavy iron hoe of the Han period; Hayashi (*4*), 6–29.

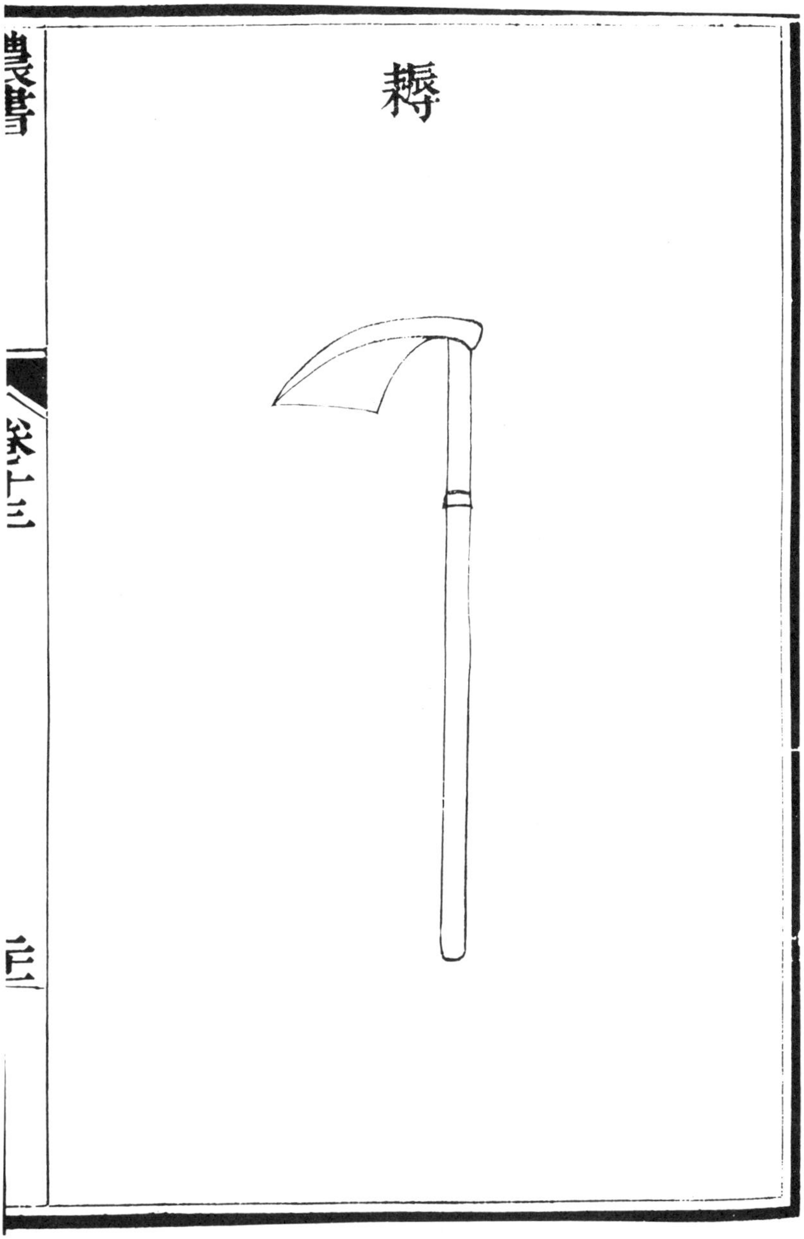


Fig. 121. Ming hoe; *WCNS* 13/21*a*.

Fig. 122. Swan-necked hoes shown in an engraved stone sarcophagus of the Sui dynasty; Nelson Gallery of Art, Kansas City, U.S.A.

Fig. 123. Han swan-necked hoe; Hayashi (*4*), 6–24.

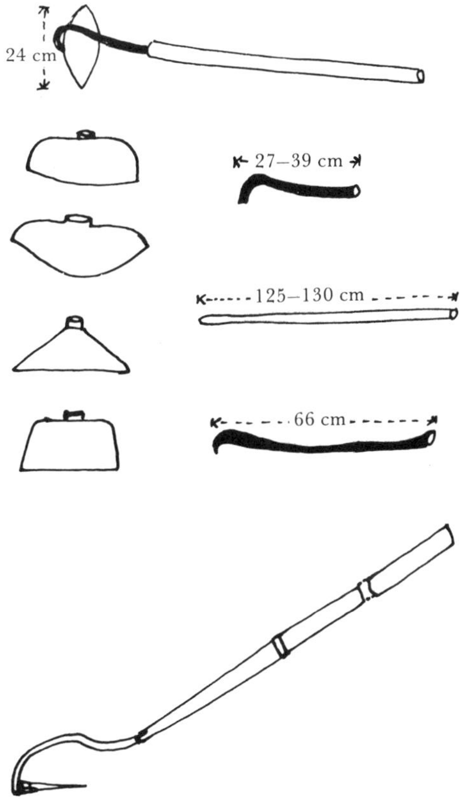


Fig. 124. Swan-necked hoe with interchangeable blades; Amano (7), fig. 50.

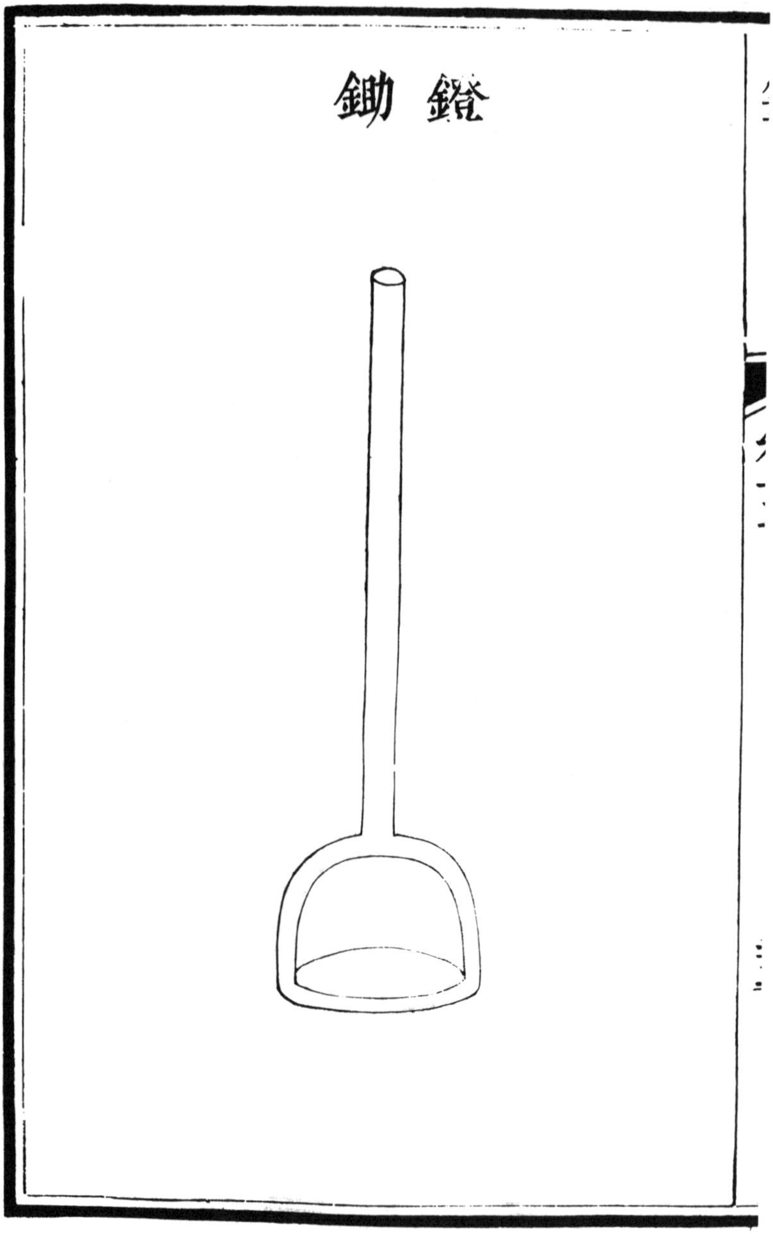


Fig. 125. Stirrup hoe; *WCNS* 13/25*b*.

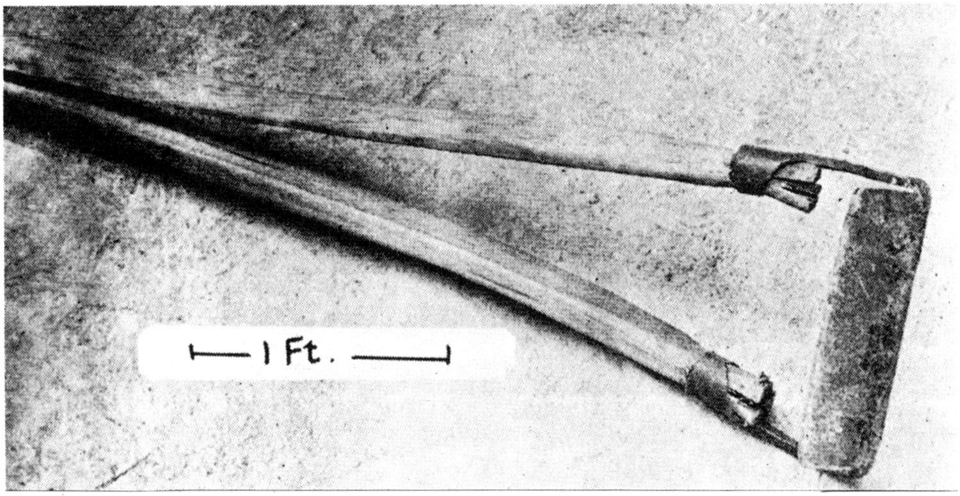


Fig. 126. Weed scraper; Hommel (1), fig. 98.

The task of hoeing was burdensome and dull, and peasants often hoed in groups so as to while away the hours a little more pleasantly:[a]

In the villages of the North they frequently form hoeing societies, generally of ten families. First they hoe the fields of one family which provides all the rest with food and drink, then the other families follow in turn over the ten-day period ... This is a quick and pleasant way of performing the task of hoeing, and if one family should fall ill or meet with an accident the others will help them out. The fields are free from weeds and so the harvests are always bountiful. After the autumn harvest [the members of the society] contribute bowls of wine and pigs' trotters for a celebratory feast.

### (ii) *Horse-hoeing husbandry*

In the early 18th century, the English agriculturalist Jethro Tull developed his famous system of 'Horse-hoeing Husbandry', on the principle that only if crops were thoroughly cultivated could they extract maximum nourishment from the soil:[b]

The earth is so unjust to plants, her own offspring, as to shut up her stores in proportion to their wants; that is, to give them less nourishment when they have need of more; therefore man, for whose use they are chiefly designed, ought to bring in his reasonable aid for their relief and force open her magazines with the hoe, which will thence procure them at all times provisions in abundance, and also free them from intruders; I mean their spurious kindred, the weeds, that robbed them of their too scanty allowance.

[a] *WCNS* 3/3*b*.

[b] Tull (1), p. 118.

Hoeing and weeding were not neglected in European husbandry. They were among the labour services regularly demanded of tenants, and farmers provided their labourers with such tools as weed-hooks or weeding-tongs for the task,[a] yet a veritable army of 'sarclers' or weeders (sixty at a time on one 14th-century Suffolk manor) was required to keep the weeds at bay.[b] Only if the fields were not broadcast but sown in rows, as was rarely the case in Europe before the 19th century, could each plant receive the individual care and attention it required to thrive. Tull realised that the only way in which crops could be more effectively tended was through the elaboration of a system in which they would first be sown in evenly spaced rows so that they could subsequently be thoroughly weeded and hoed. In 'Horse-hoeing Husbandry' Tull proposed that the seed be sown with a seed-drill and then it could be hoed repeatedly, quickly and efficiently, with a horse-hoe, namely a horse-drawn frame to which a number of hoe-blades were attached, spaced according to the width of the furrows. Considerable opposition was raised to Tull's proposals, which were not at first widely taken up, but the objection was not to the theory as such but to the inadequacy of the early European seed-drills (see p. 575 below). The system itself was adjudged sound and as soon as improved seed-drills were developed in the early 19th century so too were horse-hoes (known variously in English as horse-hoes, cultivators, tormentors, etc.).[c] They became standard equipment on the mechanised farms of Northwestern Europe and the New World, especially in the case of root crops, and remained so until rendered largely superfluous by the introduction of chemical herbicides in the 1950's and 60's.[d] It took over a century for Jethro Tull's system to gain due recognition in Britain; one wonders whether he would have been more gratified or disgruntled to know that his revolutionary system had been brought to a fine art in the Far East as early as the +6th century.[e]

We have already seen (p. 254 above) that sowing in rows had been practised well before Han times, that seed-drills were in use in Han China and that by the Northern Wei Chia Ssu-Hsieh recommended that all cereals be sown by drill. With crops sown in rows hoeing could be carried out quickly and efficiently, and there was scope for the development of animal-drawn ridgers and hoes. (The close link between seed-drills and horse-hoes is found not only in China but also, for example, in Southern India (Fig. 127).) Perhaps the first step in such a development was a ridger, used to earth up the crops after they had been hoed. In the Western Han, Fan Sheng-Chih says that after hoeing barley a 'brushwood

[a] Markham (2), p. 92.

[b] Ernle (1), p. 11.

[c] Illustrations of commercial brands may be seen in works such as Malden (1), Watson & More (1).

[d] As an illustration of radical technological change in this connection it is instructive to compare the treatment of weed-control in, say, the 1942 edition of Watson and More (1) with that in the 1977 edition of Fream (1), where cultivation plays only a very small ancillary role and weeds are generally eliminated by the application of a combination of selective herbicides.

[e] See, for example, Amano (7); Wan Kuo-Ting (*6*); Li Chhang-Nien (*3*), pp. 68 ff.

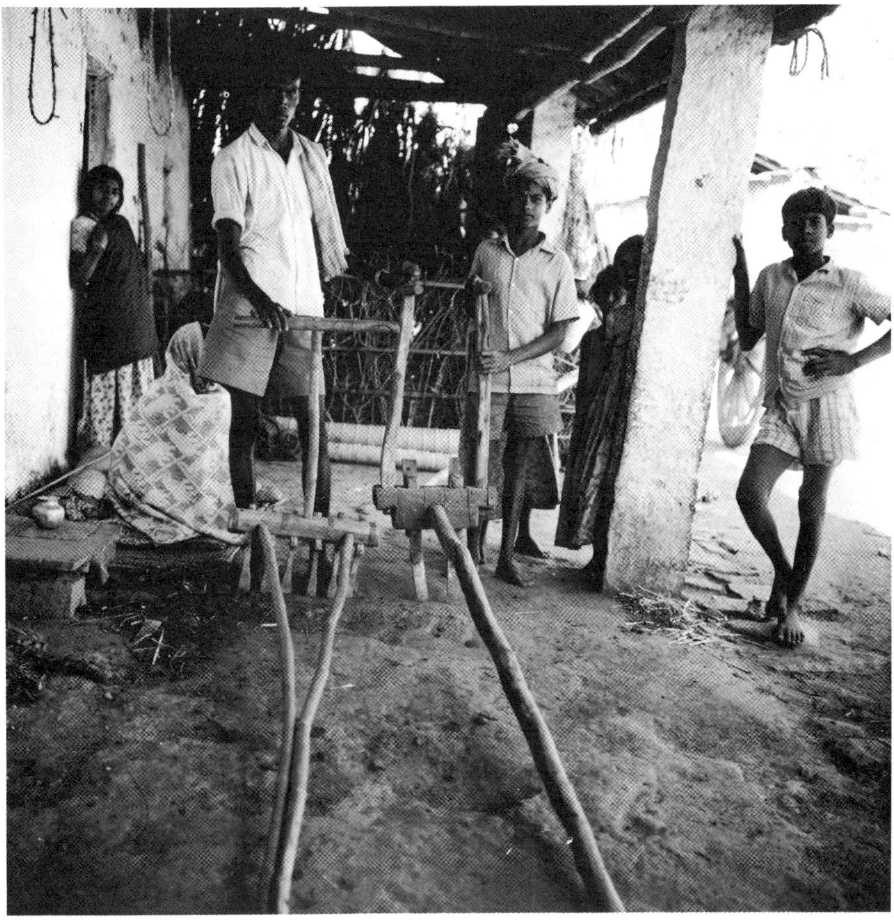

Fig. 127. South Indian horse-hoe; orig. photo Axel Steensberg.

seed-drill' (*chi chhai lou*[1]) should be used to bank up (*yung*[2]) the roots.[a] Now the term *lou*[3], which used alone or in combinations such as *lou chhe*[4] or *lou li*[5] refers to the seed-drill, can also be used in other combinations such as *lou chhu*[6] (see below) to signify an ox-drawn frame, similar to that of a seed-drill, to which hoe-blades or ridging-boards can be attached. It seems possible, then, that the *chi chhai lou* was some sort of ox-drawn ridger, consisting of a frame to which bundles of brushwood were attached in order to sweep the loose, newly hoed soil back up onto the barley ridges.

[a] Quoted *CMYS* 10.11.4.

[1] 棘柴耬 [2] 壅 [3] 耬 [4] 耬車 [5] 耬犁
[6] 耬鋤

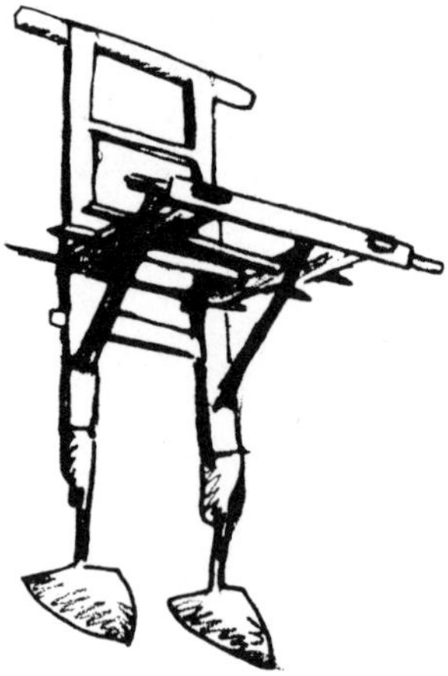

Fig. 128. Modern Chinese horse-hoe; Anon. (*502*), p. 52.

The +2nd-century *Shih Ming* refers to an implement called a *chiang*[1] which was used to cut away the weeds on ridges, at the same time banking up the soil around the roots of the plants and creating an irrigation trench between the ridges.[a] Since the character is formed with the metal radical, Hayashi assumes that this *chiang* is some kind of spade; however, it could equally well apply, according to the functions specified by the *Shih Ming*, to a metal-bladed ridger or horse-hoe. Some four centuries later Chia Ssu-Hsieh refers to the use of an implement also called *chiang*[2], but this time incorporating the plough radical, which is used to earth up such crops as leeks which require blanching.[b] Wang Chen later glosses the identical term *chiang*[2] as 'a plough without a mould-board', and the implement referred to by Chia must have been fairly heavy, for he mentions that its use has a tendency to compact the soil, making subsequent ploughing difficult. It seems likely, therefore, that the term *chiang* as used in both Han and later texts refers not to a spade but to a ridger similar to the *hu tzu*[3] described below. By Wang Chen's time it seems that such implements were interconvertible with horse-hoes and seed-drills, sharing a common frame; this may well have been the case in earlier times too. Wang Chen states that the contemporary *chiang* frequently had two heads, and goes on to say that the *chiang tzu*[4] is the same thing as a seed-drill (*lou chhe*[5]). It is easy to see how a Chinese seed-drill can be converted to a horse-hoe by replacing the narrow, pointed drills with wider blades (Fig. 128),[c] and Wang gives another name for the horse-hoe as *lou chhu*[6]:[d]

[a] Quoted Hayashi (*4*), pp. 275–6.
[b] *CMYS* 20.3.3.
[c] See also Liu Hsien-Chou (*8*), fig. 87 which shows both single- and double-bladed horse-hoes in use in modern China today.
[d] *WCNS* 13/24*b*, quoting the *Chung Shih Chen Shuo*.

1 鏤 2 耩 3 劐子 4 耩子 5 耬車
6 耬鋤

It is very like a seed-drill but without a hopper, and the metal handles [of the hoes] pass through the horizontal struts of the frame; the hoe-blades, shaped like apricot leaves, point downwards and scrape up the weeds. Latterly the horse-hoe has been drawn by a muzzled donkey, but at first it was pulled by a man, though this is no longer common. The driver leans down lightly [on the frame] to push the blades 2 or 3 inches into the soil, that is three times as deep as an ordinary hoe, and in a day it will cover not less than 20 *mu*.

Wang's illustration (Fig. 129) shows a horse-hoe with only a single blade, but as we have seen, the two-bladed type was also common. He tells us that in Yen and Chao (the Hopei area) the horse-hoe was known as *hu tzu*[1], and was convertible from horse-hoe to ridger. The first time it was used simply as a hoe, for at that time the roots of the plants were not fully developed, but the second time a wooden V-shaped attachment called a 'goose-wing' (*yen chhih*[2]) was attached above the hoe-blade, serving to deepen the trench made by the blade and pile the soil up around the roots of the plants.[a] A modern example of the 'goose-wing' is shown in Fig. 130; nowadays the ridging device is usually fixed to a plough-frame rather than a horse-hoe, and the ridger shown in Fig. 131 may well be similar to the 'plough without a mould-board' referred to by Wang Chen. Modern terms for ridger are *hu tzu*[1] or *ho tzu*[3], while the two-bladed horse-hoe, still common in North China, is known as *lou chhu*[4] or *thang thou*[5].[b]

The advantage of horse-hoes and ridgers over hand-hoes is that they save man-power. A horse-hoe drawn by one animal and driven by a single man can, Wang Chen states (and this agrees with the estimates given for modern equivalents), cover over 10 *mu* in one day, several times as much as a man with a hand-hoe.[c] Though it may be necessary for someone to hoe the last few remaining weeds manually, a single round of horse-hoeing can be substituted for several rounds of hand-hoeing.[d] Such implements are clearly advantageous if the animal power is available to use them, and they seem to have been common in most of Northern China right through medieval times, but in Central and Southern China livestock were in short supply and dry fields there seem to have been worked by hand. More recently the generally low ratio of livestock to farm-holdings led to an increasing reliance on hand implements all over China.[e]

### (iii) *Irrigated agriculture*

Although the number of species of weeds that infest irrigated fields is less than those of dry fields, they make up in persistence for what they lack in variety. Perhaps the most invasive weed of Asian rice-fields is 'barnyard millet',

[a] *WCNS* 13/24*b*.
[b] Liu Hsien-Chou (*8*), p. 43.
[c] *WCNS* 3/2*a*–*b*; Anon. (*502*), vol. 2, pp. 52 ff.
[d] *WCNS* 3/2*b*.
[e] F. H. King (1), fig. 115; W. Wagner (1), pp. 263 ff.

1 劐子 2 雁翅 3 耠子 4 耬鋤 5 蹚 or 蹚頭

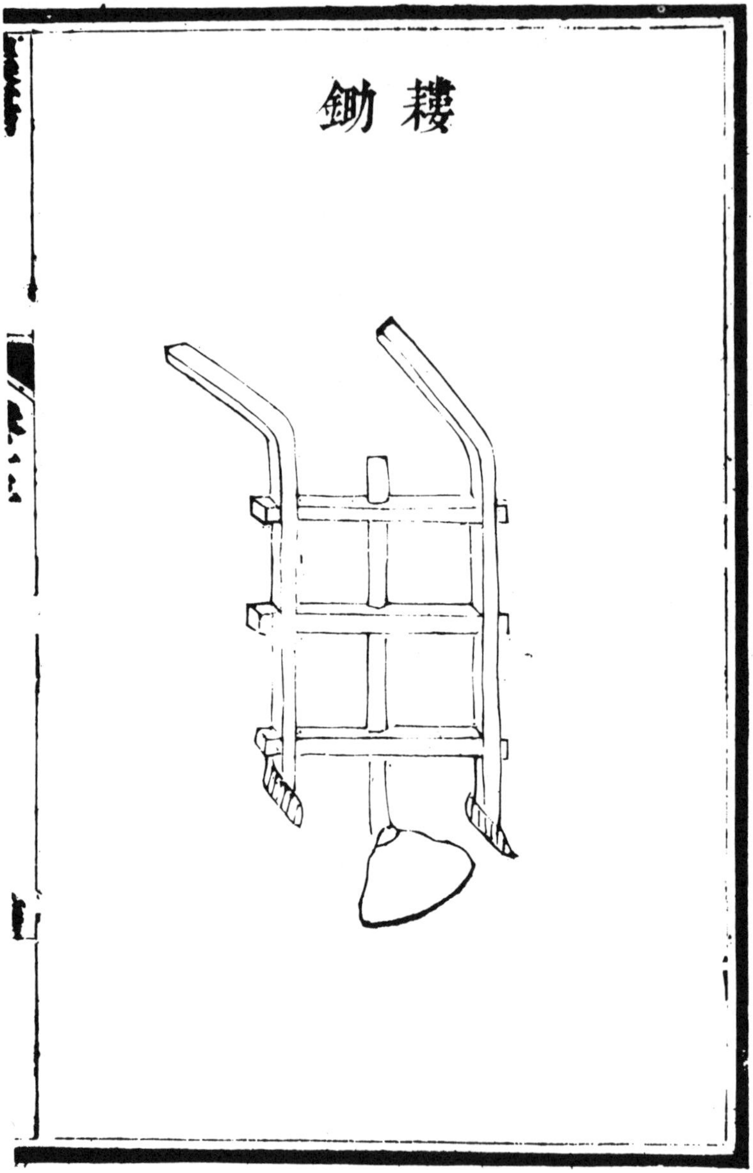


Fig. 129. Ming horse-hoe; *WCNS* 13/24*a*.

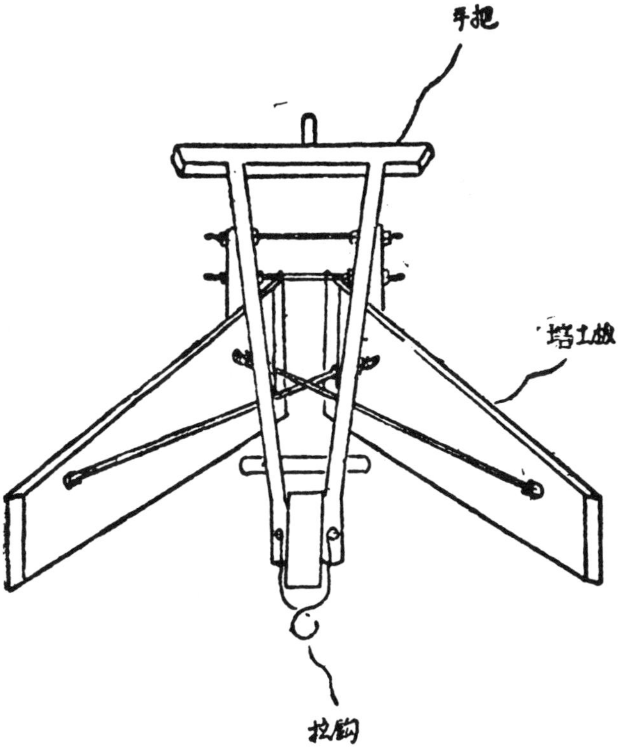


Fig. 130. 'Goose wing' for ridging soil; Anon. (*502*), p. 80.

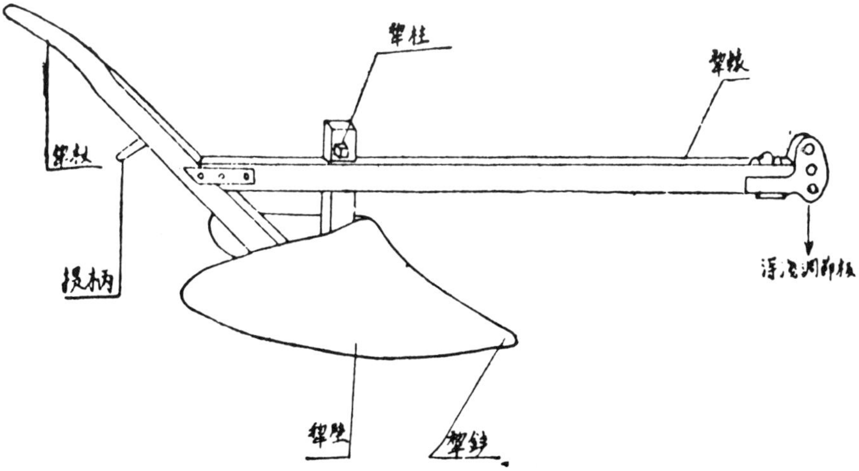


Fig. 131. Ridger; Anon. (*502*), p. 82.

Fig. 132. Hand weeding of rice; *Keng Chih Thu*, Pelliot (24), pl. XXI.

*Echinocloa crusgalli* Beauv., but many types of rush and marsh plant are also widespread pests.[a] Once such weeds become established in wet fields they are almost impossible to eliminate, which is one reason why it is best to transplant rice:[b] 'If you do not transplant the rice seedlings from the seedbed then weeds and barnyard millet will grow up simultaneously so that even cutting cannot kill them. It is therefore necessary to transplant the rice so that the weeds can be taken out by hand.'[c] Hand-weeding of irrigated fields was and remains general practice in China. The *Keng Chih Thu* recommends three weedings in between transplanting and harvest (Fig. 132),[d] and it was usual to weed before applying fertiliser, for obvious reasons. Often the weeders would protect their fingers with bamboo stalls known as 'weeding claws' (*yün chao*[1]) or 'crow weeders' (*wu yün*[2]), and grub the weeds out at the roots (Fig. 133).[e] These claws were still used in parts of China and Japan earlier this century.[f]

Hand-weeding was backbreaking work, and some Chinese farmers perfected the technique of uprooting the weeds with their toes instead of their fingers,

[a] For a comprehensive list see Grist (1), pp. 278 ff.

[b] *CMYS* 11.6.2.

[c] Hand-weeding is called *hao*[3]. Since the young weeds will be several inches shorter than the rice seedlings, it is easy to identify them and remove them. It is, however, almost impossible to remove any weeds that grow up actually within the clump of rice shoots.

[d] *KCT* 1/11*b*–14*a*; O. Franke (11); P. Pelliot (24).

[e] *WCNS* 13/30*a*.

[f] F. H. King (1), p. 258.

[1] 耘爪 [2] 烏耘 [3] 薅

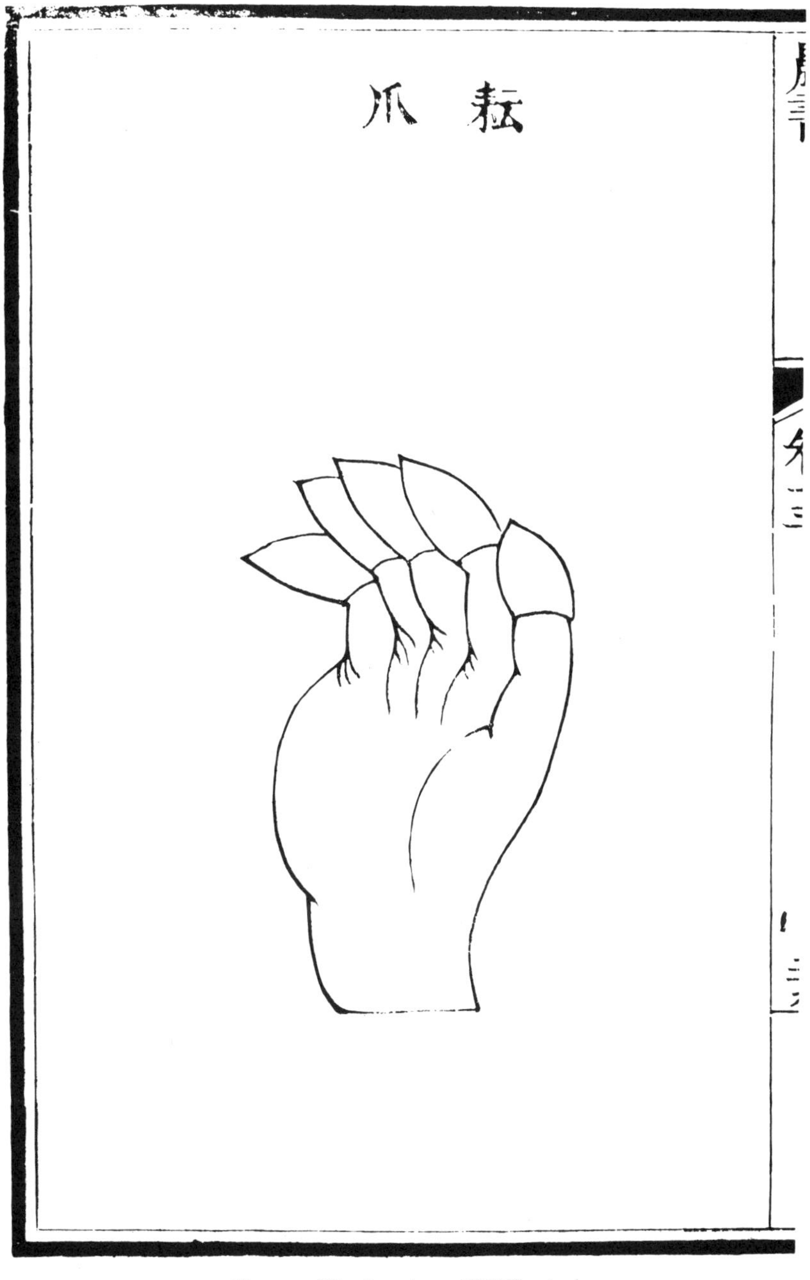


Fig. 133. Weeding claws; *WCNS* 13/29*b*.

Fig. 134. Foot weeding in Szechwan; Anon. (*524*), fig. 2.

steadying themselves on a staff as they went. The earliest depiction of this is on a stone stele of Later Han date from Szechwan (Fig. 134), and it is also mentioned by Wang Chen.[a]

Chinese farmers often turned their weeds to good use by burying them in the mud under the rice roots as they went along, so that they would rot down and act as a form of manure. The Sung agriculturalist Chhen Fu complained that in his day the practice had become neglected, but it was certainly still common in the 20th century.[b] An interesting method of incorporating the weeds into the mud was reported by Wang Chen:[c]

> The *kun chu*[1] [Fig. 135] is used to roll down both weeds and young rice shoots ... Between the Huai and the Yangtze all rice fields are broadcast and so rice and weeds sprout together; this roller is then used to roll both weeds and shoots into the mud. After two nights the rice shoots re-emerge, but the weeds do not grow again.

[a] *WCNS* 3/3*a*.

[b] *CFNS* 1/p. 6; F. H. King (1), p. 258.

[c] *WCNS* 14/20*b*–21*a*.

[1] 輥軸

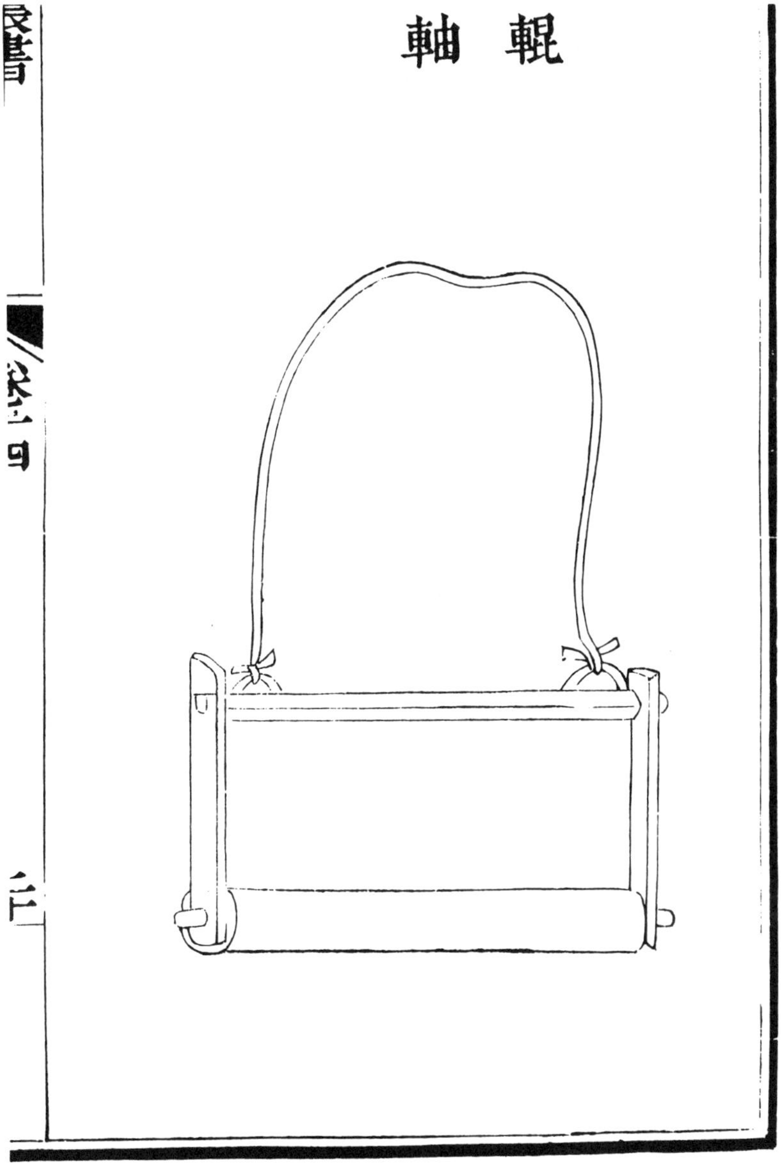


Fig. 135. Weed-roller (*kun chu*); *WCNS* 14/20*a*.

One doubts whether the *kun chu* really eliminated weeds quite as effectively as Wang Chen maintains, but the rolling would have the additional benefit of encouraging the young plants to tiller.

It was also recognised that draining the fields once or twice after transplanting not only allowed more efficient weed control but also aerated the soil and encouraged the development of healthy root systems. It is common in most countries where rice is grown to drain the fields two or three times during growth;[a] Chia Ssu-Hsieh recommends it at an early stage to strengthen the roots,[b] while Chhen Fu proposes that the fields should be so thoroughly dried out during weeding that the mud parches and cracks open, since this is just as beneficial as an application of manure.[c]

Labour-saving devices for weeding transplanted fields seem first to have been developed in the Yangtze Delta in the Sung period (clearly the introduction of double-cropping of rice greatly increased the amount of necessary weeding and stimulated the development of labour-saving methods). Wang Chen refers to an implement called a *yün thang*[1], newly developed in the Chekiang area. Although his illustration[d] apparently depicts a rake, the description makes clear that he is in fact speaking of what Hommel calls a 'hand-harrow' (Fig. 136):[e]

> The *yün thang* ... is shaped like a wooden patten, a foot or so long and roughly three inches broad, spiked underneath with a dozen or so nails in rows ... The farmer stirs together the mud and weeds between the rows of crops so that the weeds are buried in the mud; the field is kept clean and in good tilth ... In Chiangtung and other areas I have seen farmers weeding their fields by hand, crawling between the crops on their hands and knees with the sun roasting their backs and the mud soaking their limbs—a truly pitiable fate [Fig. 132] ... and so I have described the *yün thang* here in the hope that philanthropists may disseminate its use.

These hand-harrows are still very common in the rice-growing areas of China; on the Lower Yangtze they are called *tao thang*[2],[f] and further to the Southwest they are simply known as 'seedling harrows' (*yang pa*[3]).[g] Nowadays many of them have been adapted so that the spikes are mounted, not on a flat frame, but on one or more small rollers.[h] Although the prototype hand-harrow was apparently not adopted in Japan until the +18th century, the wheeled form seems to have been developed rather rapidly there (it appears under the name of *jozōki* (Fig. 137) in a +1900 farming manual),[i] and it is probable that this wheeled form was first introduced into China from Japan earlier this century.

a Grist (1), p. 43.
b *CMYS* 11.4.1.
c *CFNS* 1/p. 7.
d *WCNS* 13/28*a*, see Fig. 18.
e *WCNS* 13/28*b*–29*a*.
f The second element is presumably an alternative form of the *thang*[4] used by Wang Chen.
g Anon. (*502*), vol. 2, pp. 90 ff.
h *Ibid.* pp. 104 ff.
i Pauer (1), p. 143.

1 耘盪 2 稻耥 3 秧耙 4 盪

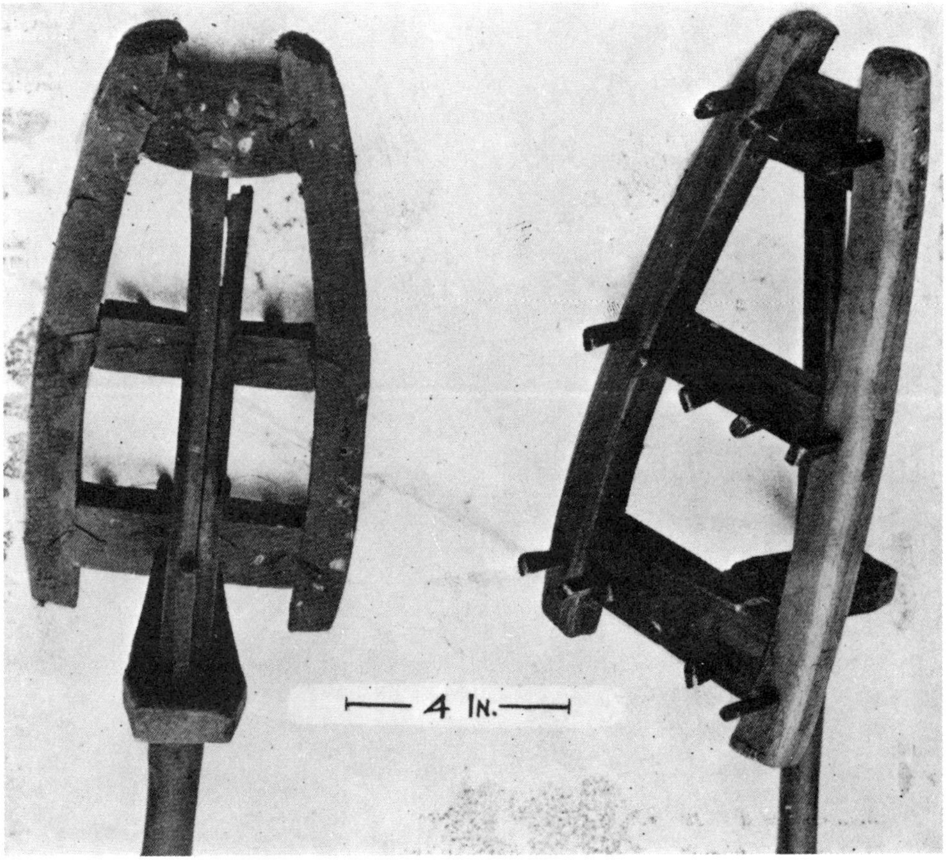


Fig. 136. Modern Chinese hand-harrows; Hommel (1), fig. 97.

## (5) Harvesting, Threshing and Winnowing

### (i) *Harvesting*

At harvest time the farmers,
Bent low over their sickles, vie to fill the grange.
Thick frost cracks their hands into mosaic;
After the endless day they are broken by toil.
The rows of children binding up the sheaves,
Chilled by the wind piercing their scanty garb,
Shout for joy as they reach the warmth of home
By the light of the full moon, high over roof and hill.[a] (Fig. 138.)

Harvesting is, as Gervase Markham said,[b] 'the end, hope, and perfection of the labour, and both the merit and incouragement which maketh the toyle both light

[a] Lou Shou's poem on reaping for the *Keng Chih Thu*, composed *c.* +1145.
[b] (2), p. 112.

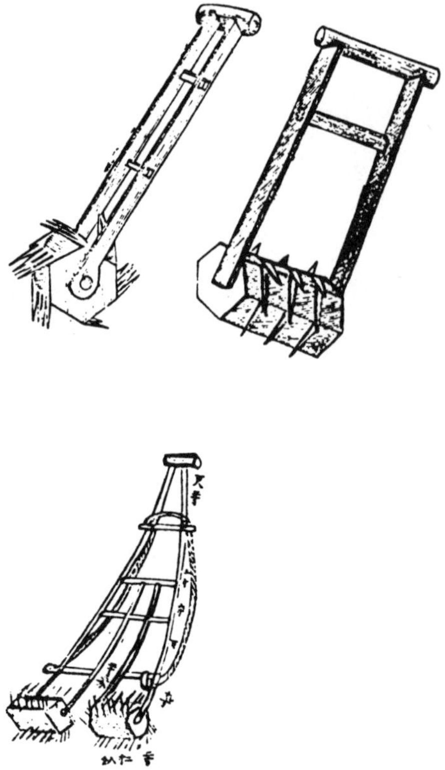

Fig. 137. Japanese wheeled hand-harrow; Pauer (1), fig. 71.

and portable'. It is also a race against time to bring the ripe grain safely in before it is eaten by birds or flattened by a summer storm. If the grain is reaped too soon it will be damp and will not keep, if it is left too long in the field the over-ripe ears will shatter and spill the seed. In the West farmers have tried from earliest times to overcome these difficulties by using faster and more efficient reaping tools: first the sickle, then the scythe, and finally the mechanical harvester. In East Asia, as we shall see, the course of development has been somewhat different.

Man's earliest reaping tool, used for harvesting cereals long before their domestication, was of course the human hand. Indeed it is still used in many parts of the world today, not only by hunter-gatherers but also by settled agriculturalists.[a] The eminent American archaeologist and botanist Jack Harlan, experimenting with stands of wild wheat in Anatolia, found that he could pluck the heads by hand almost as fast as he could reap them with a stone or steel sickle.[b] Another method of hand-harvesting is simply to tear the plants up by the roots, sometimes done in dry areas where plants grow thinly[c] and still found in areas of China

[a] E.g. among some Central Mountains tribes of Taiwan today, where setaria millet is cultivated; Fogg (1).
[b] B. Bender (1), p. 127.
[c] Kraybill (1), p. 127.

Fig. 138. Harvesting rice with sickles: *Keng Chih Thu*, Pelliot (24), pl. XXIV.

Fig. 139. Natufian sickle; Singer, Holmyard & Hall (1), vol. 1, p. 503.

where inter-cropping is practised.[a] But reaping with a knife or sickle is generally less painful than hand-harvesting.

The earliest known sickles were used to harvest wild grasses in the Nile valley *c.* −12,000 to −10,000, but are not found elsewhere earlier than the neolithic Natufian sites of Jordan (*c.* −8000 to −7000). These early sickles consisted of a bone or wooden haft inset with small stone or flint blades (Fig. 139); their form has led to the suggestion that they were modelled on an animal's jawbone.[b] Such sickles are characteristic of Natufian sites and have been found in several, though not all, early Near Eastern sites; they also occur in European sites as late as −2000,[c] but were quite soon replaced in Western Asia and the Mediterranean area by sickles of baked clay, bronze, and eventually iron.[d]

[a] See F. H. King (1), p. 300 on the harvesting of wheat in the Peking-Tientsin area.
[b] Curwen & Hatt (1), p. 107.
[c] Singer, Holmyard & Hall (1), vol. I, p. 503.
[d] K. V. Flannery (1), p. 90.

One tends to think of the sickle as the universal reaping tool, yet it did not appear in many parts of the world until late in the neolithic period or even the metal age. In the pre-Columbian American cultures sickles were unknown, and it is only in the past decade that they have superseded the harvesting-knife in many parts of Malaysia and Indonesia. The reason for this lies, we believe, not so much in different levels of technological competence as in the nature of the crops cultivated. The early West Asian domesticates, wheat and barley, are characterised by small seed-heads containing few grains. Although the grains of wheat and barley are large compared to most millets or even to rice, the total weight of grain contained in a single ear is much less. This is one factor that may have encouraged the early development of the sickle, which could be used to cut several heads simultaneously. Another is that, like their wild ancestors, the early domesticated varieties of wheat and barley had an extremely fragile rachis which was prone to shatter and shed the grain as soon as the ear was ripe. Harvesting was thus a matter of extreme urgency: 'Harvest as if robbers were after you', goes the old Chinese saying.[a]

The first domesticated cereals of China and Southeast Asia, rice, setaria millet and Job's tears, had large seed-heads or panicles and were less prone to shattering. Like wheat and barley, rice and setaria are naturally free-tillering, but in the more primitive varieties the tillers ripen very unevenly.[b] In such conditions it is not desirable to reap all the heads from a single plant simultaneously: it is better to gather the heads individually as they ripen, for which purpose a small knife is preferable to a sickle. So it is not really surprising to find that the sickle was a rather late introduction to this part of the world. On the basis of archaeological evidence for Egypt and the Near East Reed[c] suggests that 'perhaps the transfer of the cultural association between reaping and the sickle accompanied a natural movement of the large-seeded grasses (barley and wheats?) northward into [West Asia from Egypt] in the late Pleistocene'. One might go even further and suggest that it was only with the adoption of wheat and barley cultivation that the sickle was introduced from West Asia into more distant areas such as India and China, for as yet no early sickles have been found in India except in association with wheat-growing, while in China the earliest known sickles are to be found in Lungshanoid sites in Shantung and Northern Kiangsu,[d] in an area and period where wheat and barley were probably already under cultivation.[e]

[a] *C/HS* 24A/3*b*.
[b] Fogg (1).
[c] (3), p. 548.
[d] Liu (*8*), p. 61.
[e] The few neolithic finds of wheat and barley so far are of doubtful reliability, and indeed one find of wheat grain in a supposedly Lung-shan context in Anwhei (Chin Shan-Pao (*1*)) has recently been $C_{14}$-dated to the Eastern Chou (Anon. (*522*)), thus confirming the doubts thrown upon the stratification by ceramics experts (Yang Chien-Fang (*1*)). Nevertheless the frequency with which wheat is mentioned in the Shang oracle bones suggests that it had been cultivated in China for some time previously; Ho Ping-Ti (5), p. 74 suggests that it was introduced at the very end of the neolithic.

Fig. 140. Stone harvesting knife; Higuchi (*1*), p. 107.

*Harvesting knives*

The reaping tool of earliest neolithic China was the harvesting knife, a simple yet practical tool known throughout East and Southeast Asia. A small flat knife, often with a curved blade, it is held in the palm of the hand, usually between the middle and ring fingers. The heads of grain are cut just below the ear by drawing the stem across the blade with the index finger (Fig. 140). The harvesting knife has survived up to the present day not only in Java (where it is called *ani-ani*), Malaya (*tuai*, or *ketaman* in Kelantan), Sarawak (*ketap*) and the Philippines (*yatab*), but even in parts of North China.

Ethnographers have frequently linked the use of the harvesting knife with the cultivation of rice, for its use is often accorded ritual significance by rice farmers,[a] but it is also used by farmers growing millets, both setaria[b] and panicum.[c] The cultivation of setaria probably long predated that of rice in much of the Far East,[d] and the harvesting knife is particularly suitable for reaping primitive varieties of setaria since the heads on the side tillers of the plant generally ripen considerably later than the head on the main stem.[e] We might deduce that the use of the harvesting knife was more ancient than rice cultivation in many parts of Asia. On the other hand the harvesting knife (*hōchō*[1]) apparently reached Japan only with the introduction of rice-farming in Yayoi[2] times, *c.* −300,[f] although it now seems increasingly certain that the mid to late Jōmon[3] economy included swidden cultivation of such cereals as buckwheat and setaria millet,[g] which in other cultures were reaped with harvesting knives.

Harvesting knives of stone, shell and pottery have been found in neolithic sites all over China, and come in a variety of shapes. An Chih-Min (*3*) distinguishes between three principal forms, 'notched', 'crescentic' and 'rectangular' (though perhaps 'trapezoid' would be more appropriate), while the Japanese scholars Iinuma and Horio[h] adopt a more functional approach and subdivide the knives into five categories, 'notched', 'rectangular', 'crescentic with straight cutting edge', 'crescentic with curved cutting edge', and 'shuttle-shaped' (Fig. 141). Attempts have been made to chart the geographical distribution of the different types in order to deduce patterns of cultural contact, but such generalisations usually produce mutually contradictory results.[i] It is possible that a careful reassessment of the material, using recent $C_{14}$-dates to provide a chronological as well as a geographical perspective, would yield interesting results; however that is beyond the scope of this study.

Another important distinction that can be made between harvesting knives as well as their form is the way in which they are held. Either the fingers are passed through a loop of cord, leather or metal (Fig. 142), or else the knife is held by a longish haft mounted perpendicularly to the blade (Fig. 143). The Chinese neolithic notched stone knives, and those pierced with two holes, must belong to the first type, those pierced with a single hole to the second.

Knives with double holes were found all over China in neolithic times and continued in use through the Shang and Chou periods; the first iron forms appear in Warring States times (Fig. 144), though stone and shell models were still in

a A. H. Hill (1), p. 70.
b Fogg (1).
c *WCNS* 14/6*b*.
d Kano Tadao (1), vol. I, p. 291; Barrau (1), p. 6.
e Fogg (1).
f Higuchi Seishi (*1*), pp. 13, 107.
g Furushima Toshio (*2*), p. 6; Sasaki Kōmei (*1*), pp. 34 ff.
h (*1*), pp. 33–4.
i W. Watson (6), figs. 13–15, based on An Chih-Min (*3*); Sasaki Kōmei (*1*), fig. 21, p. 296.

1 庖丁 2 弥生 3 繩文

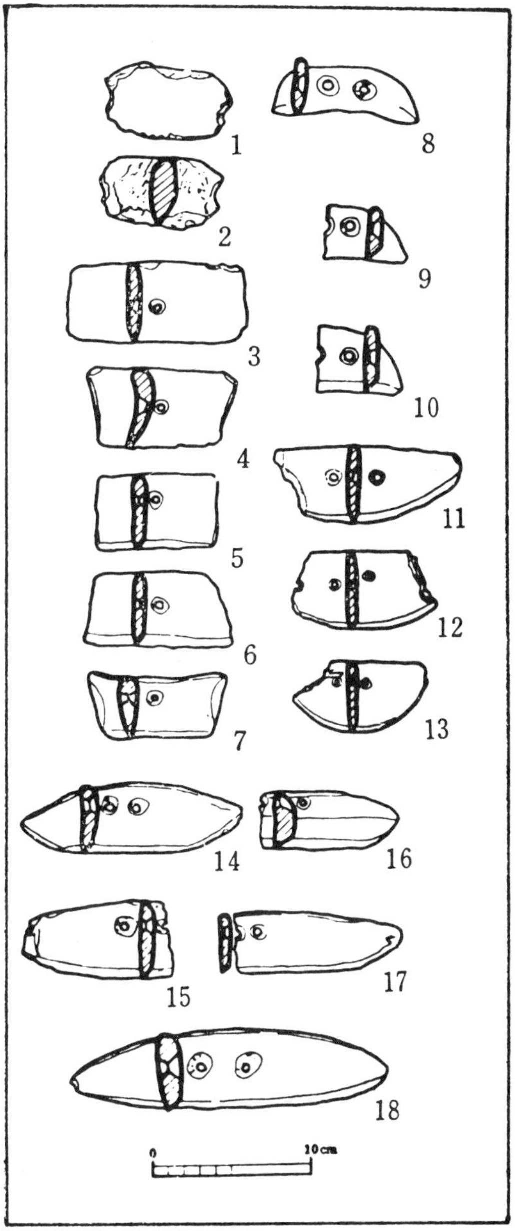


Fig. 141. Typology of Chinese neolithic reaping knives; Iinuma & Horio (*1*), p. 34.

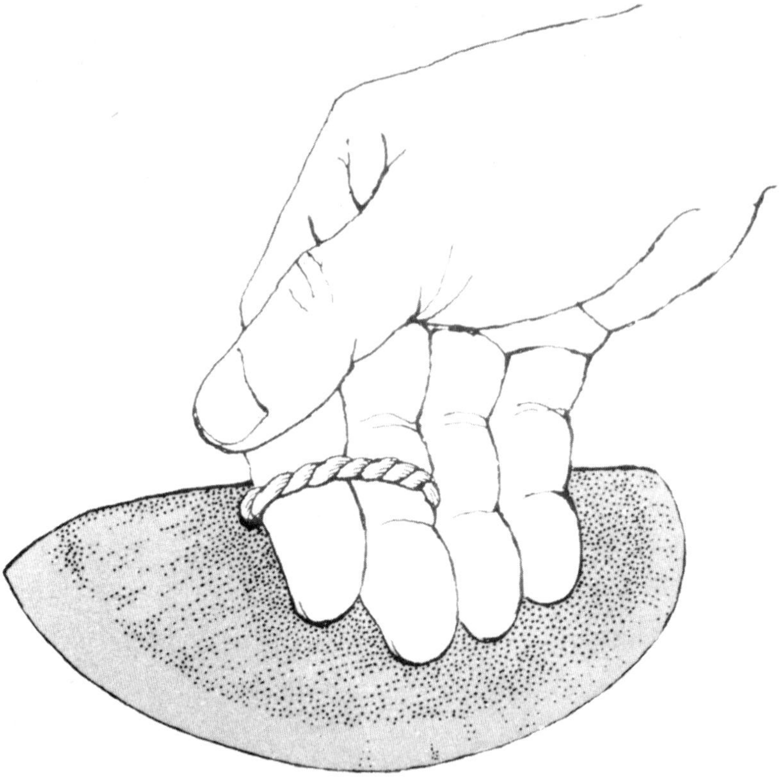

Fig. 142. Looped reaping knife; Higuchi (*1*), p. 107.

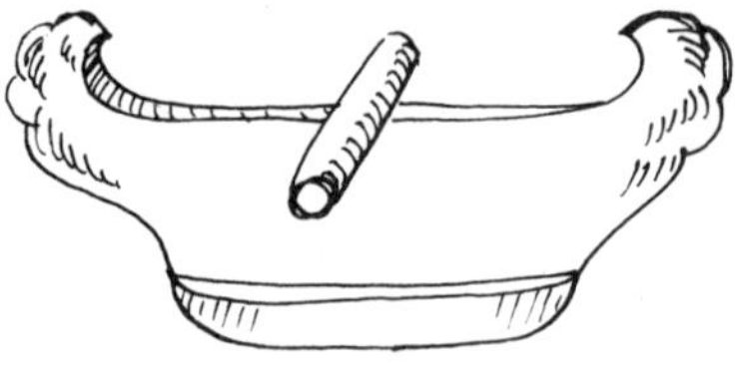

Fig. 143. Hafted knife, as used in Malaya; after A. H. Hill (1), fig. 5.

common use during the Han.[a] The loop on such knives would have been made of some sort of cord or leather, as it still is in Hopei and Lianoning today,[b] but by Yuan times a new type of harvesting knife had been developed where the index finger was passed through a metal loop on the top of the knife;[c] this form too is still common in North China now (Fig. 145).

[a] Liu Hsien-Chou (*8*), p. 58.
[b] *Ibid.*
[c] *WCNS* 14/6*b*.

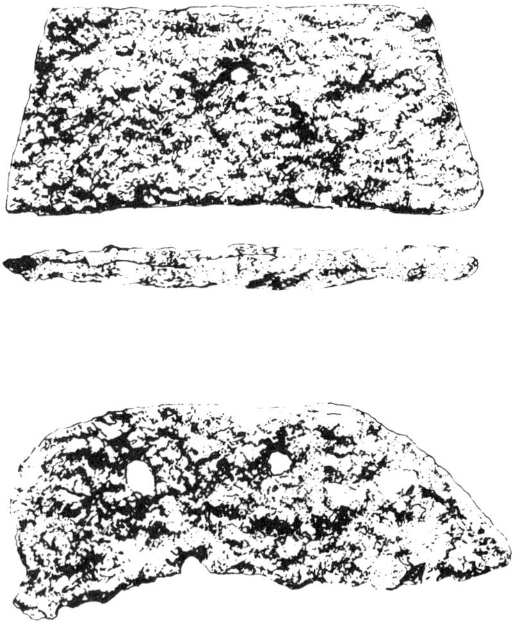

Fig. 144. Han iron reaping knives; Hayashi (*4*), 6–42*b*.

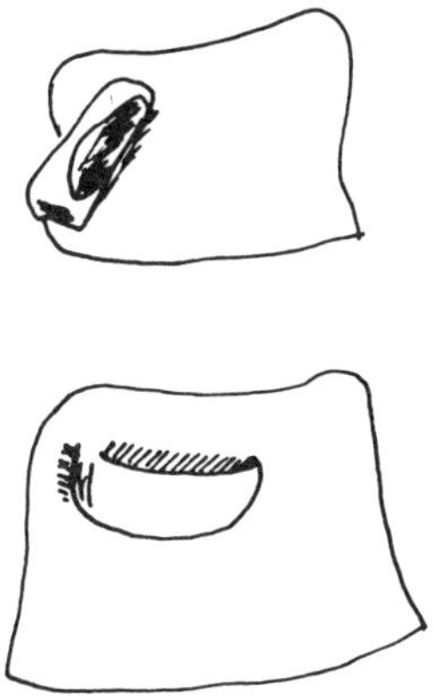

Fig. 145. Modern Chinese iron reaping knife; after Liu (*8*), p. 116.

The hafted knives were apparently known only in the Yang-shao area in neolithic times, though Han iron examples have been found in Liaoning as well as Honan.[a] We know of no later Chinese examples, but hafted knives are still used to good purpose throughout Southeast Asia: the modern form is light and economical, made of wood and bamboo with a small iron blade easily removed for

[a] Hayashi (*4*), figs. 6–42, 6–43; Fig. 144.

sharpening or replacement (Fig. 143). There seems no intrinsic reason why the use of the hafted knife should have died out in China while the looped harvesting knife survived. The *Wang Chen Nung Shu* refers to two types of harvesting knife which differ in shape,[a] the looped knife and another type, the *chih*[1], which 'has been an essential tool ever since ancient times and remains so today',[b] but the artist was clearly unacquainted with the implement and the accompanying illustration[c] is most unhelpful.

Whether the term *chih*[1] applied to a hafted harvesting knife or to harvesting knives in general, it was certainly a term of considerable antiquity, for it occurs in both the *Yü Kung* and the *Odes*.[d] It appears once again in the great Han dictionaries[e] and was apparently still current more than a thousand years later when Wang Chen was alive, though by this time the term 'setaria knife' (*su chien*[2]) was applied to the looped harvesting knife.[f] Nowadays the looped knife is called a 'finger-nail sickle' (*chao lien*[3]; *nien tao*[4] in Liaoning).[g]

It may be asked why the harvesting knife has survived for so long, especially in countries where the sickle is also known. Its late survival in Malaya has been attributed to its ritual properties: unlike the sickle the harvesting knife is almost completely concealed in the palm of the hand, and this is said to avoid frightening away the 'rice soul', *semangat padi*, without which the seed grain will not grow.[h] This may be one important factor in its survival, though some doubt has been cast on the Asian peasant's blind belief in ritual.[i] But though the Chinese apparently invested the reaping knife with no such ritual properties,[j] it has nevertheless survived in China for millennia, alongside the sickle, and presumably as a result of its strictly practical advantages.

The harvesting knife, as we mentioned above, is ideal for reaping crops that ripen unevenly. Since the stem is cut just below the ear one does not need to stoop too low. Most of the straw is left in the field; though it can later be cut for fodder it is usually turned in or burned as fertiliser, and in many cases it is the only fertiliser applied to the field at all.[k] Grain cut in this manner is usually not threshed before storing but made into bundles (Fig. 146) and stored in the ear, where it keeps much longer than threshed grain. But perhaps the most important advantage of

[a] *WCNS* 14/2*a*; 6*b*.
[b] *WCNS* 14/2*a*.
[c] *WCNS* 14/1*b*.
[d] As quoted in *WCNS* 14/2*a*; see also Legge (8), vol. II, p. 583.
[e] Hayashi (*4*), p. 280.
[f] *WCNS* 14/6*b*.
[g] Liu Hsien-Chou (*8*), p. 59.
[h] E. H. G. Dobby (1), p. vii; A. H. Hill (1), p. 70; D. Freeman (1), p. 154.
[i] Raymond Firth (1).
[j] Nothing of the sort is mentioned even in the *Shih Ching*, but of course this may simply be due to Confucian editing-out of popular superstitions.
[k] E.g. Bray (2).

[1] 銍 [2] 粟鑒 [3] 爪鐮 [4] 捻刀

Fig. 146. Rice-harvesting with reaping knives and scythes; stone relief in Szechwan Provincial Museum.

reaping with a harvesting knife is that it permits individual examination of each head of grain and so facilitates the development of new strains. Fogg (1) describes how Central Mountains aborigines in Taiwan carefully pick and set aside the seed from any unusual setaria millet plant and sow it in a separate plot the next season. If it does well or has particularly desirable characteristics they will keep it for cultivation, and in this way they have acquired a wide range of setaria cultivars.

The process of individual inspection not only allows the deliberate breeding of new varieties but also facilitates the maintenance of true stock. In his chapter on the selection of seed-grain the +6th century agriculturalist Chia Ssu-Hsieh writes:[a]

For foxtail and panicled millets, glutinous and non-glutinous, the seed-grain should be harvested separately every year. Select fine heads of even colour, cut them with a harvest-

[a] *CMYS* 2.3.1.

ing knife [?] (*chhiao i*[1])[a] and hang them up in a high place. In the spring prepare the grains and sow them separately to provide the seed-grain for the next year.

Such meticulous selection of seed-grain is not possible if the grain is cut with a sickle. In the Mediterranean, where sickles were used, the Roman agricultural writers complained frequently of the degeneration of wheat and barley stock despite the care taken to select the best grain for seed after threshing,[b] and less than a dozen distinct varieties of wheat, the staple cereal, are named by the classical authors.[c] Compare this with Chia Ssu-Hsieh's list of ninety-eight named varieties of setaria millet, a list which he describes as selective. (Even for rice, a crop little grown in North China at the period, Chia produces a list of thirty-seven varieties.)[d] Chia breaks the list of setarias down into different types: drought-resistant, quick- or slow-ripening, well-flavoured, or disease-resistant. Asian farmers usually sow several different varieties of their staple cereal in any one season, either in separate plots or in different parts of the same field. This not only guarantees some harvest whatever the weather, but also means that both the planting and the harvest are spread over a longer period than if a single variety was grown. There is therefore plenty of time to reap with a harvesting knife. But this leisurely though stable system is disrupted as soon as the harvest becomes a matter of urgency, as when a single variety is grown on a large scale, or multi-cropping is introduced. More rapid reaping methods must then be adopted, and the harvesting knife will be abandoned. In parts of Southeast Asia this change is still in progress under the impact of the 'Green Revolution'.[e]

We have seen that the harvesting knife is still used today in some parts of China, chiefly where setaria millet is grown. In other areas it may have survived as a gleaning tool, for Wang Chen describes a 'gathering knife', *chün tao*[2],[f]

with a blade perhaps 5 inches long and nearly 2 inches wide, pierced by a cord above and below and so attached to the wrist. As the hand moves the ears are cut most

[a] *Chhiao*[3] is a rare term generally defined simply as 'to cut', and nowhere is it glossed as 'harvesting knife'. On the other hand when he speaks of reaping with a sickle Chia simply uses the word *i*[4], 'to reap', so *chhiao i* clearly denotes a special method of cutting. The only other use of *chhiao* in the *CMYS* is in the chapter on barley (10.4.5) where Chia describes the preparation of *tsamba* or parched barley, which he calls 'cut barley' (*chhiao mai*[5]). (The more usual Chinese name for *tsamba* is 'roast flour' (*chhao mien*[6]) (Trippner (1)).) Now *tsamba* is prepared by lightly toasting or roasting the grain while still in the ear; it is therefore not wholly illogical to assume that by *chhiao* Chia Ssu-Hsieh meant the cutting of individual heads with a knife. In Asturia, as in parts of Georgia and Nepal, awned cereals such as spelt are sometimes harvested by pulling the individual heads from the stems with a pair of sticks (see p. 353 below). Before threshing with a flail the heads of grain are briefly burned to remove the awns; Sigaut (1).

[b] K. D. White (2), p. 188.

[c] *Ibid.* p. 189; Moritz (3), (4).

[d] *CMYS* 3.1.3, 3.1.5; 11.1.2, 11.1.6.

[e] For a more detailed account see Bray (2). The organisation of labour is of course also affected; Bray & Robertson (1), I. Palmer (1).

[f] *WCNS* 19/24*b*.

[1] 劁刈 [2] 捃刀 [3] 劁 [4] 刈 [5] 劁麥
[6] 炒麵

Fig. 147. Balanced Italian *falx messoria*; after K. D. White (1), p. 54.

conveniently. Sometimes if wheat or setaria are not harvested as soon as they ripen, the stalks and ears become all entangled and it is not possible to reap the whole crop with a sickle. Then poor people may use this knife to cut whatever is left.[a]

But despite these isolated examples the sickle probably replaced the harvesting knife as the chief reaping tool in many areas of North China as early as the period of agricultural intensification that began with the Warring States, and the development of winter-summer crop-rotations must have assured its ascendancy in many more southerly parts by the Sung dynasty.

*Sickles and scythes*

As we have seen, the sickle came comparatively late to China, and its introduction may be linked to that of wheat and barley. Early sickles were made of stone or shell; they were only slightly curved, and were set at right angles to the handle; they were probably secured with tightly bound thongs.[b] Later bronze and iron were substituted for stone and shell, but the shape of the Chinese sickle remains essentially unchanged to this day.

Most European and many Asian sickles, especially the larger varieties, are balanced, that is to say, the sickle blade is heavily curved and set so that it extends behind the handle as well as in front, balancing the weight around the handle so that it is less tiring to use (Figs. 147, 148). The balanced sickle is not found in either China or Japan (Figs. 149, 150); there seems no rational explanation for this.

When iron tools first came into general use in the Warring States and Han, large numbers of sickles were cast in iron (Fig. 151). They were gently curved with a thick ridge along the back for strength and a smooth blade; a thin tab of iron at the back of the blade served as a tang; the blade was usually 20–30 cm. long. Though cheap and easily produced, these cast-iron sickles cannot have been very durable or sharp, and wrought iron sickles were already common in Han times.[c]

[a] In the 1920s Wagner (1, p. 270) reported that the poor had gleaning rights all over China. Not only were they permitted to glean the remaining ears and straw from grain fields, they might also dig up the stubble, gather up any fallen bolls of cotton, and carefully collect every damaged vegetable or leaf of bean or sweet potato that remained in the field. F. H. King (1), p. 300 describes a curious sight in Chihli, where he found women cleaning the broken plants from a crop of lodged wheat *before* the reapers had passed.

[b] Liu Hsien-Chou (*8*), figs. 121, 122.

[c] Hayashi (*4*), figs. 6–35 to 6–38.

Fig. 148. Balanced Iranian sickle; Lerche (3), fig. 12.

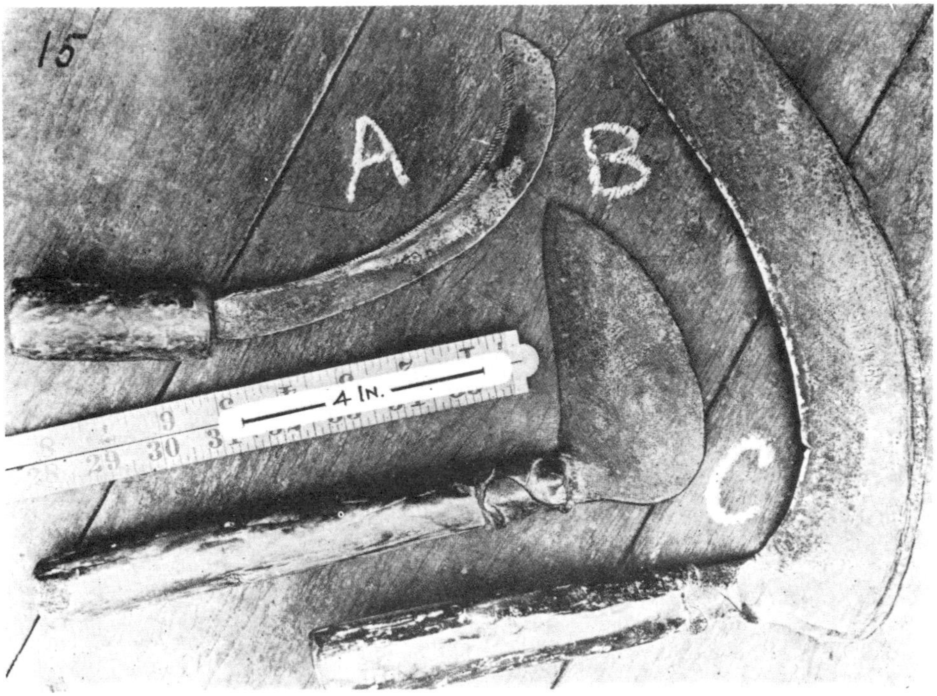


Fig. 149. Unbalanced Chinese sickles; Hommel (1), fig. 103.

The name *lien*[1], still given to sickles today, was current in Han times but several dialectal variants also existed,[a] including the Chinese equivalent of 'reaping hook' (*i kou*[2]). Wang Chen refers to several different sickles,[b] socketed, hooked and hafted, as well as a 'belt sickle' (*phei lien*[3]) (presumably the common type where a thin tab of metal projecting from the back of the blade was wrapped around the handle), and an intriguing 'double-bladed sickle' (*liang jen lien*[4]), which was perhaps a short-handled scythe (see below). Further details Wang does not give. Yet unlike cast-iron sickles, their wrought-iron or steel counterparts were capable of considerable refinement, and in Japan at least much care went into producing them (though of course it never reached the mystic heights of the sword-maker's art). Thao Hung-Ching wrote at the turn of the +5th century that like sabres, sickles were made of steel (*kang thieh*[5]); the process of combining layers of wrought iron and steel to produce damascened blades is attested, at least for swords, in both China and Japan in early times.[c] The *Hyakushō Denki*[6] (Peasants' Chronicle) of +1684 explicitly refers to the damascening process being used for both sickles and hoes, while other Japanese sources say that these implements were reforged every year.[d]

[a] *Ibid.* p. 279.
[b] *WCNS* 14/4*a*.
[c] Needham (32), 42–3.
[d] Furushima Toshio (*1*), p. 304.

1 鎌 or 鐮　2 刈鈎　3 佩鎌　4 兩刃鎌　5 鋼鐵
6 百姓傳記

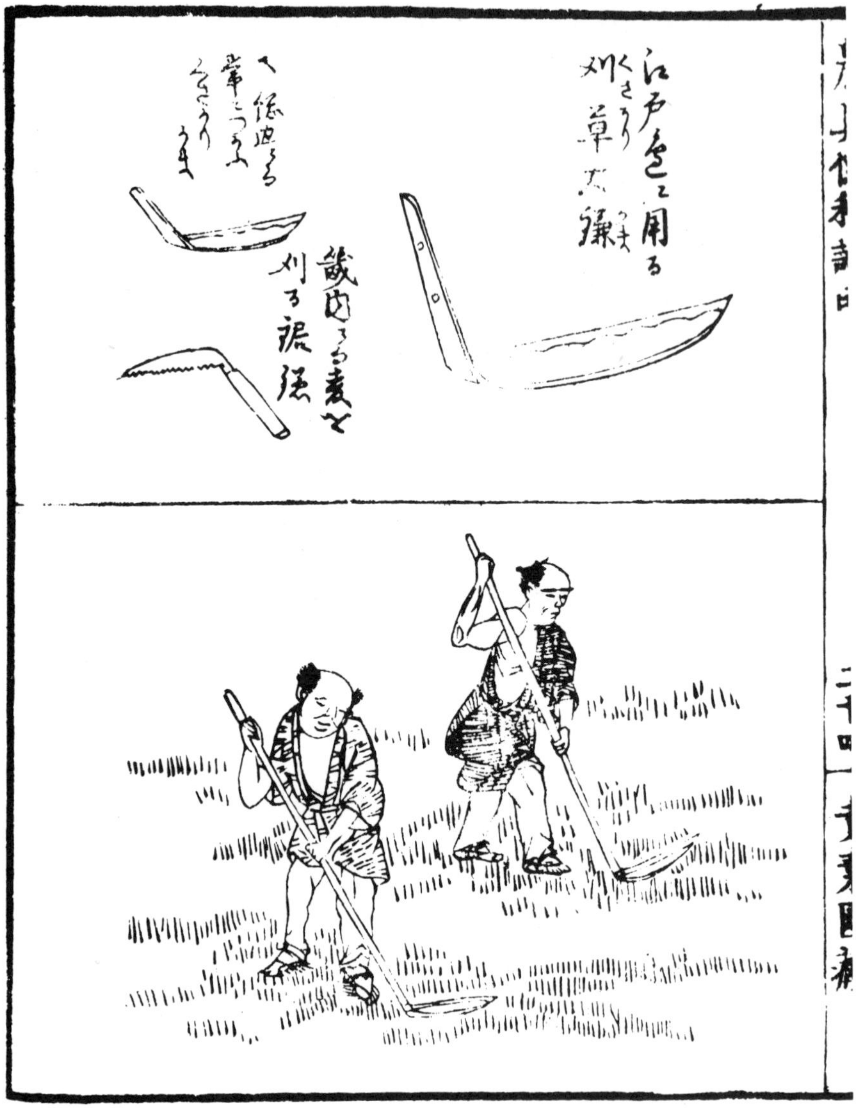

Fig. 150. Unbalanced Japanese sickles: *Nōgu Benri Ron* 2/24*a*.

A steel or wrought-iron sickle can be tempered to razor sharpness, or cut with a cold-chisel to produce a serrated edge. The Chinese literature does not distinguish between smooth and serrated sickles,[a] but a serrated sickle of Han date has been excavated in Anhwei[b] so presumably the two forms have long co-existed, though little further information is available. According to Hommel, serrated sickles are

[a] In English a terminological distinction can be made between the serrated sickle ('sickle') and the smooth ('reaping hook').

[b] Hayashi (*4*), fig. 6–38.

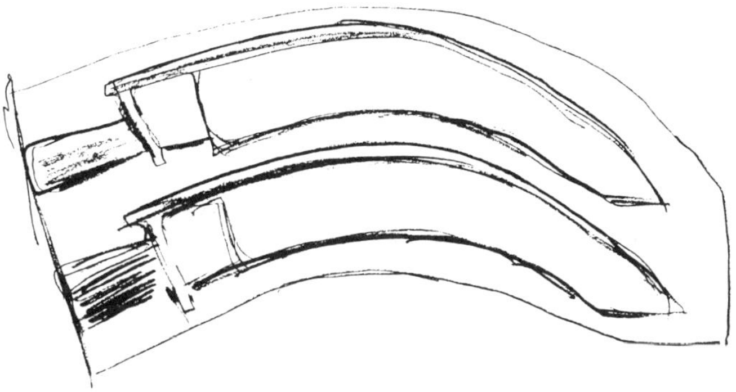

Fig. 151. Mould for casting iron sickles; Han, from Hopei; after Liu (*8*), fig. 125.

commonly used for harvesting rice while smooth sickles are used for cutting grass and fodder.[a] However, the expert on Roman agriculture, K. D. White, tells us:[b] 'It is well known that the smooth edge ... works most effectively when the dew is on the standing crop, since the edge will then bite into the straw and not slide along the surface. Studies of the present distribution of the two types show the serrated sickle predominating in the drier areas of the Mediterranean and in the sub-Saharan regions ...' Since the rice-growing areas of China are humid even at harvest time one would expect, if White is correct, to find smooth sickles in use there, but we are insufficiently informed to say whether Hommel was describing an exception rather than the rule.

When using the sickle the reaper takes a bunch of grain in the left hand and cuts it with a sawing or slashing motion of the sickle (depending on whether it is toothed or smooth). In order to protect the hands and grasp at more stalks at a time, reapers sometimes wear leather or bamboo finger-stalls[c] (Fig. 152), but though these were known in China they were apparently used only for weeding and not for harvesting (see p. 314 above). On the other hand the Chinese did make use of a fagging-stick (*ho kou*[1]) to hold the grain steady in small bundles as it was cut.[d] In Central and Southern China the bundles of grain were usually dried before threshing, either on small bamboo tripods (*chhiao kan*[2]), or on large frames (*hang*[3]) 'constructed like a roof from bamboo and wooden beams'[e] (Fig. 153). The sickle was not suitable for all conditions. Crops that had been sown broadcast, particularly wheat and fodder crops, were often cut with a scythe (*pho*[4]).

The European scythe was first developed in Roman times, probably as a response to an increase in stock-raising and the subsequent demand for fodder.[f]

a (1), p. 67.
b (1), p. 80.
c Lerche (3); Rasmussen (1).
d *WCNS* 14/26*a*.
e *WCNS* 14/24*a*.
f K. D. White (1), p. 102.

1 禾鉤 2 喬扦 3 笐 4 鏺

Fig. 152. Stitched leather finger-stalls used by Iranian reapers; Lerche (3), fig. 13.

The simple straight-handled mowing-scythe was improved in the +12th century by the addition of bars to the handle, permitting a smoother, fuller swing. Scythes came to be used for reaping as well as mowing, probably in medieval times,[a] and to catch the cut grain a simple cradle of bent wood (Fig. 154) or of sacking stretched on a frame was attached to the back of the blade; in the late 18th century a more sophisticated cradle of curved metal bars was invented[b] (Fig. 155). By the 18th century the scythe had superseded the sickle as the reaping tool of Flanders, and in the 19th century it replaced the sickle over most of Northwest Europe.

In China the scythe also began life as a tool for cutting fodder (or weeds):[c] 'The

[a] Steensberg (5); Lynn White (7), p. 155. Both Fitzherbert ((1), pp. 35–6) and Markham ((2), pp. 112 ff.) say that wheat should be sheared with a sickle, but barley and oats mown with a scythe.

[b] L. J. Jones (1), p. 105.

[c] *Shuo Wen Chieh Tzu*, quoted Hayashi (4), p. 279:

Fig. 153. Hanging-frames (*hang*) for drying grain before storing; *Keng Chih Thu*, Pelliot (24), pl. XXV.

Fig. 154. Harvesting with scythes; Brueghel the Younger.

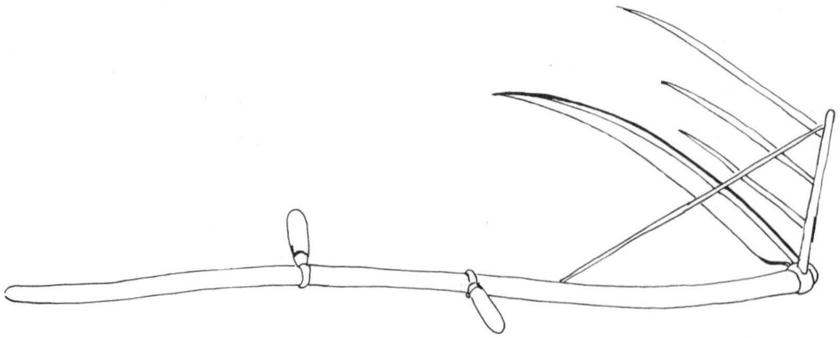

Fig. 155. English reaping or cradle scythe; after Partridge (1), p. 135.

scythe is a tool with a double blade and a wooden handle, used for cutting grasses (*pho, liang jen mu ping, kho i i tshao*[1]).' The Chinese scythe differs considerably from its European counterpart: it is used with a slashing rather than a swinging action (Figs. 146, 156), and consists of a shortish wrought-iron blade, rather similar to a double-edged sword, fixed to a long handle at an obtuse angle. It can vary considerably in size: one Han example is only 9 cm long, while Wang Chen refers to the blade as 'over 2 feet in length'.[a] The Chinese scythe never ousted the sickle as the principal reaping tool for rice or millets, but in wheat-growing country it acquired greater importance and sophistication. The scythe,[b] with a whole range of other implements, was used to harvest wheat in the Central Provinces, where:[c] 'the lands are broad and the grain abundant, so [the wheat farmers] must construct special tools to facilitate harvesting'. Wang Chen devoted a special section to these tools in the hope that their use might spread to other regions, but he wrote in vain, for in the 20th century their use was still confined mainly to Honan.

The shape and mounting of the blade means that most Chinese scythes are used with a slashing movement rather than the swing of a Western scythe. But Wang Chen describes a long-bladed scythe with a cradle (*liao tshao chang*[2]) which, if his illustrator is to be relied on, rather resembled the normal European scythe (Fig. 157). This was probably the scythe used for reaping wheat. Another more complex scythe was the *mai hsien*[3], known in modern Honan as a 'long-handled sickle' (*chhang ping lien*[4])[d] or 'cutting knife' (*shan tao*[5])[e]. Wang Chen and Wagner both describe this implement in virtually identical terms[f] (Fig. 158): a pole is bent so as to form a long-handled bow, and across the mouth of the bow is set a razor-sharp blade mounted on a wooden bar. The bar and the bow form the semi-circular

[a] Hayashi (*4*), p. 279; *WCNS* 14/8*a*.
[b] *WCNS* 19/25*b*.
[c] *WCNS* 19/21*a*.
[d] Wagner (1), p. 268.
[e] Anon. (*502*), p. 163. A photograph of the implement in use in Shensi in the 1930s can be seen in Amano (*11*), p. 188.
[f] *WCNS* 19/26*b*; Wagner (1), p. 268.

[1] 鏺，兩刃木柄可以乂艸 [2] 掠草杖 [3] 麥釤 [4] 長柄鎌 [5] 刪刀

Fig. 156. Chinese scythe, wielded in a slashing motion; Hommel (1), p. 106.

frame for a shallow wicker basket (*mai chho*[1]), which serves as a cradle to hold the cut grain. The reaper holds a grip at the end of the handle in his right hand, and in his left a small wooden reel from which lead two cords, one attached to the handle, the other to the back of the blade. The wheat is cut too short to be bound into sheaves, so it is emptied from the cradle into a light, wheeled wicker basket (*mai lung*[2]), which the reaper pulls along behind him[a] (Fig. 159).

[a] *WCNS* 19/22*b*; Wagner (1), fig. 87.7.

[1] 麥綽 [2] 麥籠

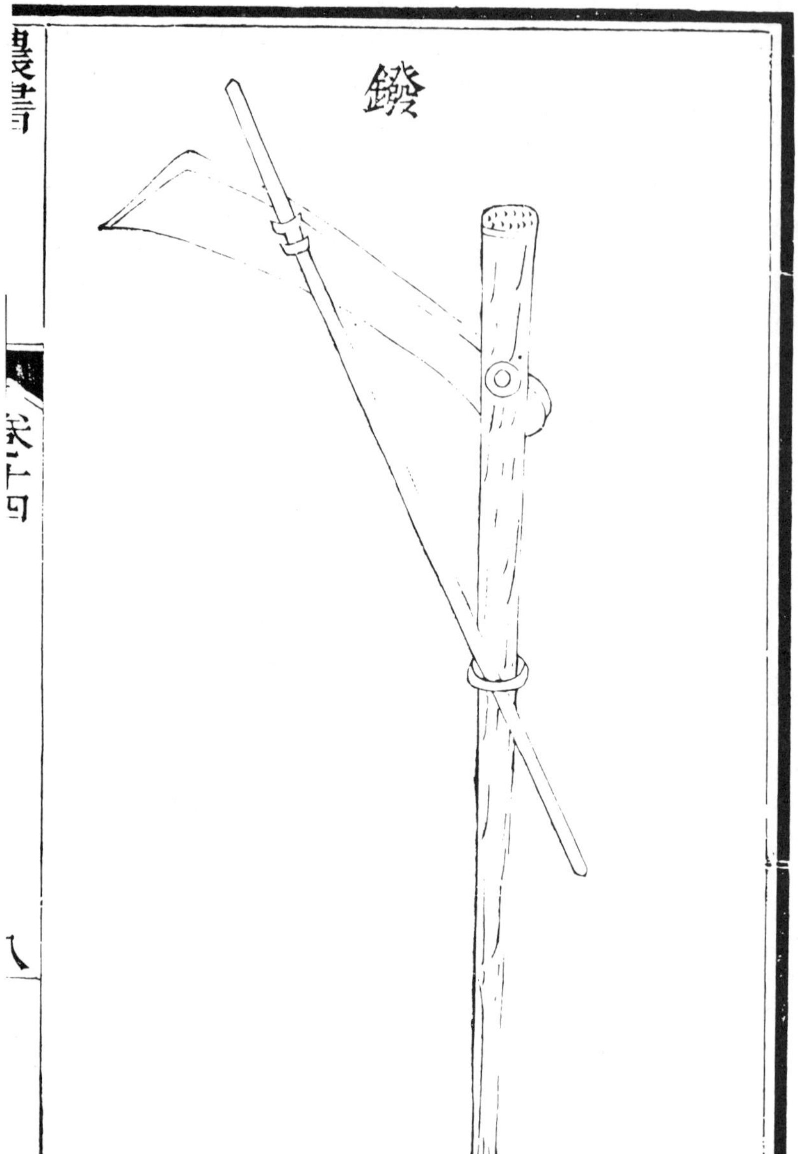


Fig. 157. Chinese cradled scythe (*pho*); *WCNS* 14/8*a*.

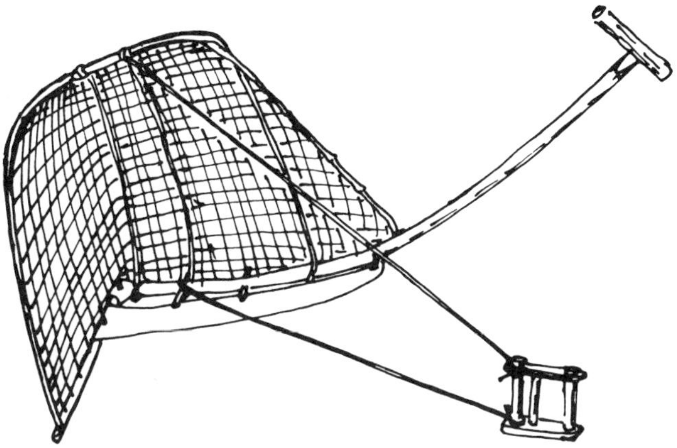

Fig. 158. Modern cradled scythe, corresponding to Wang Chen's *mai hsien*; after Hopfen (la), fig. 89.

Fig. 159. Ming depiction of *mai lung*; the construction of the *mai hsien* has clearly been misunderstood by the artist; *WCNS* 19/21*b*–22*a*.

Fig. 160. Roman stone relief showing a *vallus*, or reaping machine; from Montauban-Buzenol, Belgium; White (1), pl. 15.

*Mechanical harvesting*

The mechanisation of reaping only becomes necessary where a number of factors are combined. A large market for grain and a shortage of labour were the stimuli responsible for the invention of the *vallum*, a reaping machine used in the early centuries of the Christian era in Roman Gaul.[a] According to Pliny:[b] 'On the vast estates in the provinces of Gaul very large frames fitted with teeth at the edge and carried on two wheels are driven through the corn by a team of oxen pushing from behind; the ears torn off fall into the frame' (Fig. 160).[c] After the fall of Gaul to Frankish invaders the *vallum* was forgotten for centuries, and only in the late +18th century, with the consolidation of large estates and the consequent 'rationalisation' of agriculture, did an interest in mechanised reaping revive in Europe. As the demand for wheat grew in the industrial towns of England and other European countries, the wheat-belts of America and Australia were opened up and their huge and often under-manned farms provided the ideal experi-

[a] J. Mertens (1); K. D. White (1), p. 183; L. J. Jones (1), p. 111; J. Kolendo (1).

[b] (1), vol. v, p. 375.

[c] A more detailed description is given by Palladius (1) 7.2.2–4, while a modern reconstruction is shown in L. J. Jones (1), fig. 6.

mental environment for the development of reaping-machines; the repeal of the protectionist Corn Laws in Britain in 1846 must have been an additional stimulus to farmers to invest in the new machines, which by the later part of the 19th century had replaced manual harvesting all over Australia and America.[a]

One might have imagined that the highly developed grain markets of Sung and post-Sung China, in conjunction with Chinese mechanical ingenuity, would have resulted in the invention of Chinese reaping machines. A device such as the *vallum*, for example, was hardly beyond the competence of a Chinese village carpenter, nor would it represent a considerable financial investment, while any farmer who had an ox for ploughing could have spared it to push a machine at harvest time. But only one example of a mechanical harvester is known from the literature: a device described by Wang Chen as a 'push-sickle' (*thui lien*[1])[b] (Fig. 161), like a primitive lawn-mower but with a single, fixed blade, which was apparently used in some parts of North China for the buckwheat harvest. It is only mentioned by Wang Chen, and is the sole example of a mechanical harvester in the traditional agricultural literature.[c]

But there are excellent reasons why the Chinese failed to develop mechanical reaping. First, labour was generally abundant and cheap. Secondly, the most important crop was irrigated rice. Even where the terrain is sufficiently regular to construct fairly large irrigated fields, mechanised reaping is rarely economic, for a significant proportion of the crop will be inaccessible to a mechanical reaper.[d] In irregular terrain most rice fields are of necessity too small for any form of mechanisation. Chinese wheat fields were usually larger and more regular than the padi fields and apparently more appropriate for mechanisation, especially since there was always a ready market for wheat and wheat flour. Yet no more sophisticated harvester than the cradled scythe was developed prior to the 20th century, and it is perhaps typical that the late 19th-century agricultural writer Tseng Hsiang-Hsü, although an enthusiastic advocate of such Western ideas as iron ploughs and winch-ploughing,[e] made no mention of mechanical harvesting in his book although it was already widespread in the West. A crucial point is that the Chinese farmers (and Asian farmers generally) practised not extensive monoculture but intensive polyculture. One farmer would sow several different varieties of rice, wheat or millet each season, and this diversification not only served as an insur-

[a] For an excellent account of the development of modern reaping machines suited to various conditions (i.e. cutters, headers and strippers), see L. J. Jones (1).

[b] *WCNS* 14/5*a*–*b*.

[c] Anon. (*502*), vol. 2, p. 162 shows a small single-wheeled implement with two blades fixed V-shape to either side of the central shaft, used in modern Honan for harvesting two rows of wheat simultaneously. This could well be a survival of the *thui lien*, for Wang Chen's rather vague description would fit, and as we know the *Nung Shu* illustrator frequently mis-drew implements with which he was not personally familiar.

[d] In corners or too close to the bunds. Malay farmers on a modern irrigation scheme complained to the author in 1977 that mechanical reapers were wasteful and caused considerable damage not only to the bunds but to the soil of the padi fields.

[e] (*1*), 1/28*a*–31*b*.

[1] 推鎌

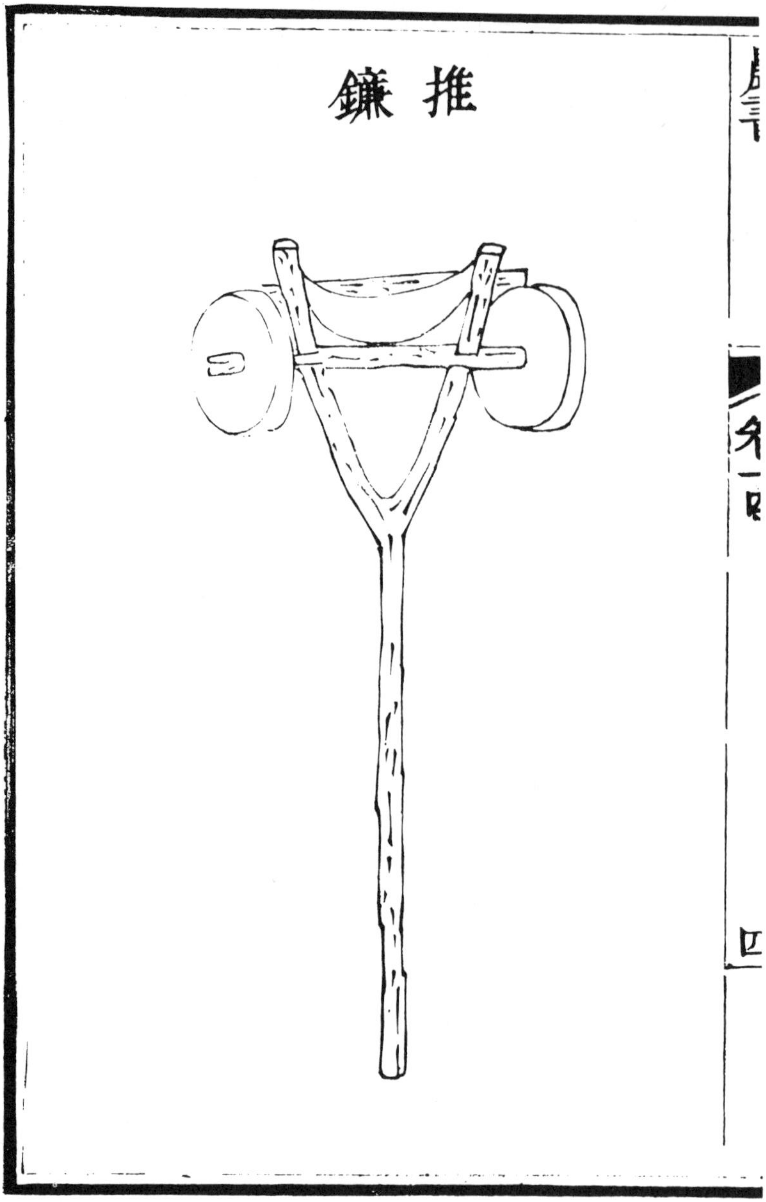


Fig. 161. 'Push scythe' (*thui lien*); *WCNS* 14/4*b*.

ance against total crop failure but also spread the harvesting period over several weeks so that (admittedly by dint of long and hard labour) the crops could all be cut by hand without spoiling. It is significant that even today the only effectually mechanised area of modern China is Manchuria, where extensive wheat monoculture *à l'américaine* is practised; in other areas the combine-harvester is still acknowledged as uneconomic, and where reaping machines of a kind have been developed to meet local requirements they usually resemble a Roman *vallum* rather than a reaper-binder.[a]

## (ii) *Threshing*

*Threshing floors, tubs, mats*

Threshing removes the grain from the straw, and for purposes of convenience grain is usually threshed before storing. But not always, for cereals keep better in the ear and so seed-grain, for example, is frequently stored unthreshed. Grain reaped with a harvesting knife (which is cut close to the ear with only a short straw) is also stored in sheaves and threshed as required.

There are many methods of threshing, but the prerequisite for most of them is a smooth, hard, clean-swept threshing-floor. The Chinese term for threshing-floor is *chhang*[1]. The 'Seventh Month' poem from the Pin section of the *Shih Ching* refers to making *chhang* in the vegetable garden the month before the harvest:[b] 'In the ninth month we pound flat and hard the threshing area in the vegetable garden; in the tenth month we bring in the harvest (*chiu yüeh chu chhang phu, shih yüeh na ho chia*[2]).' That the *chhang* was also used for growing vegetables can be seen from another poem in the *Shih Ching*,[c] which speaks of a fine white colt nibbling the tender spring shoots in the *chhang*. Temporary threshing-floors are still found: in modern Bangalore the threshing-floors are prepared yearly in the *ragi* fields, and once the monsoon rains have fallen the floor is ploughed up again and sown with grain.[d] Hommel[e] reports a similar practice in modern Anhwei. But it cannot have been very convenient to pound one's fields or vegetable patch to the consistency of concrete every winter only to dig them up again each spring. By Han times permanent threshing-floors were found, often in the courtyard of the farm-house, with a hulling-mill and a trip-hammer for polishing the grain conveniently placed nearby (Fig. 162).

Great care goes into the construction of a good threshing-floor. In Bangalore the surface is sealed with liquid cow-dung,[f] while in Calabria the mud surface is polished to burnished smoothness.[g] The +16th-century *Maison rustique* recom-

[a] Anon. (*502*), pp. 164 ff.; recently developed *vallum*-like machines are shown from Szechwan, Honan, Anhwei and Shansi; the combine-harvesters used in Heilungkiang are based on Russian prototypes.

[b] Karlgren (14), p. 99.

[c] *Ibid.* p. 128.

[d] Steensberg (6), p. 248.

[e] (1), p. 73.

[f] Steensberg (6), p. 248.

[g] Rasmussen (1), p. 99.

[1] 場 [2] 九月築場圃，十月納禾稼

Fig. 162. Han model of threshing-floor with quern and frame for trip-hammer adjacent; Laufer (3), fig. 7.

Fig. 163. Threshing on to a mat, using double-headed flails; *Keng Chih Thu*, Franke (11), pl. XLIII.

mended sprinkling the floor with a mixture of ox-blood and olive-oil, before rolling it well to kill any pismires.[a] Chinese methods seem to have been less picturesque:[b]

> The threshing-floors about Canton are made of a mixture of sand and lime, well pounded upon an inclined surface enclosed by a curb; a little cement added in the last coat makes it impervious to the rain; with proper care, it lasts many years, and is used by all the villagers for thrashing rice, peas, mustard, turnips, and other seeds, either with unshod oxen or flails. When frost and snow come, the ground requires to be repaired every season; and each farmer usually has his own.

If animals or heavy sleds or rollers are used for threshing, then clearly a proper threshing-floor is necessary, but with other methods a finely-woven mat can be substituted. This was, and is, a common procedure in China (Fig. 163). As Wang Chen points out:[c] '[The mat] not only avoids the grain being contaminated with dust or sand but also prevents losses. The mat may also be used for sun-drying cereals or rolled up to make a cylindrical basket (*thun*[1]),[d] so that it is really most useful.' Of course such mats also came in useful for winnowing (Fig. 164).

In many parts of Asia today rice is threshed into tubs in the padi-fields. Although this is a simple and very practical method, the earliest mention of a threshing-tub that we can find is in the *Thien Kung Khai Wu*;[e] the +1637 illustration (Fig. 165) shows a man beating a sheaf of rice against the edge of a heavy, round, wooden tub (*thung*[2]), which is unprotected from the wind. Nowadays the tub, which can be round or square, even taking the shape of a light sled on wooden runners in the Mekong Delta,[f] is protected from the wind with a screen of matting;[g] this has recently been replaced by plastic fertiliser bags (Fig. 166). A wooden ladder is placed inside the tub (Fig. 167), and the sheaves are beaten against the rungs.

*Treading the grain; threshing sticks*

In Kelantan, Malaysia, the threshing-tub has only come into use since the introduction of new, hard-grained varieties of rice. Malay farmers said that such violent threshing had a tendency to damage the grains and so they still prefer to thresh the traditional, more fragile varieties of rice by the old method of treading by foot (*mengirek*).[h] This is a very ancient and widespread method of threshing in the Far East. The earliest known Chinese term for it, *jou*[3], occurs in the poem *Sheng Min*[4] from the *Ta Ya*[5] section of the *Shih Ching*, which has been dated on

[a] Stephens & Liebault (1), p. 546.
[b] S. Wells Williams (1), vol. II, p. 9.
[c] *WCNS* 15/33*b*.
[d] *Thun* were frequently used for storing grain; see p. 388 below, and D. Kuhn (1).
[e] 1/53*b*; Sung Ying-Hsing (1), p. 81.
[f] Hickey (1), p. 146.
[g] Grist (1), p. 165; Hommel (1), p. 70.
[h] Bray (2).

1 篅 2 桶 3 蹂 4 生民 5 大雅

Fig. 164. Winnowing on to mats, using winnowing baskets and shovels; *Keng Chih Thu*, Franke (11), pl. XLIX.

linguistic evidence to the −9th or −8th century:[a] 'Some pound the grain, some bale it out, some sift it, some tread it (*huo chhung*, *huo yü*, *huo po*, *huo jou*[1]).' The Han commentary says that *jou* refers to the treading out panicum millet (*shu*[2]),[b] while the Thang commentary explains that the grain was trodden out (*chien jou*[3]), before being pounded in a mortar. The *Chhi Min Yao Shu* in its chapter on panicum millets recommends that they should be threshed by 'treading out while still damp' (*chi shih chien*[4]);[c] it is certainly preferable to tread grain while it is damp for then it is less injurious to the feet. Treading out of rice and millet was common until quite recently in poorer households which did not own oxen, and it was usually women's work (Fig. 168). Copeland (1) describes an ingenious method used in Leyte in the Philippines which combined both threshing and winnowing:

[a] Karlgren (14), p. 201; the dates for the *Ta Ya* are based on linguistic evidence as adduced by W. A. C. H. Dobson (1).

[b] Although the interpretation of the term *shu*[2] in a Han gloss of a Western Chou poem does present some difficulties (see section e, p. 437).

[c] *CMYS* 4.5.1; Shih Sheng-Han (*3*), vol. I, p. 71 for some reason explains *chien*[5] as threshing with a roller, but this is certainly mistaken.

[1] 或舂或揄或簸或蹂 [2] 黍 [3] 踐蹂 [4] 卽溼踐 [5] 踐

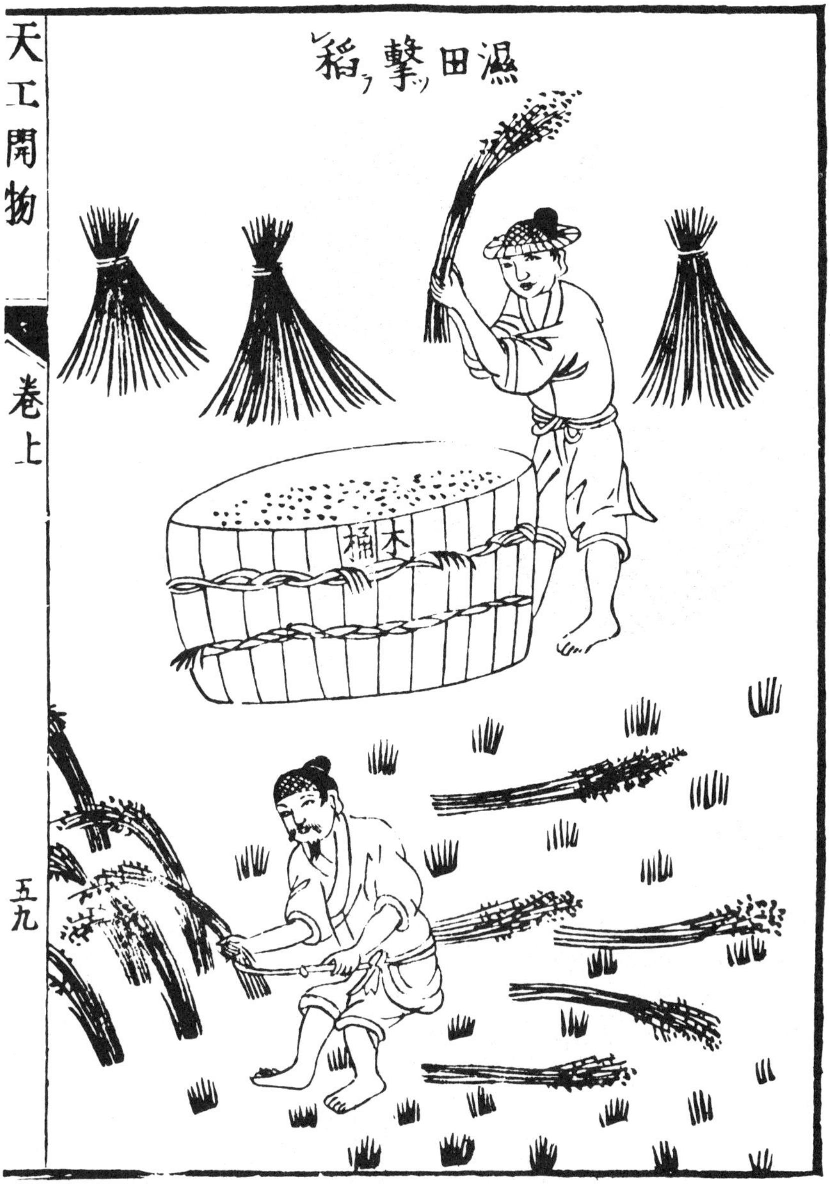


Fig. 165. Threshing tub; *TKKW* 1/59*a*.

Fig. 166. Threshing tub screened by plastic fertiliser bag; FB orig. photo.

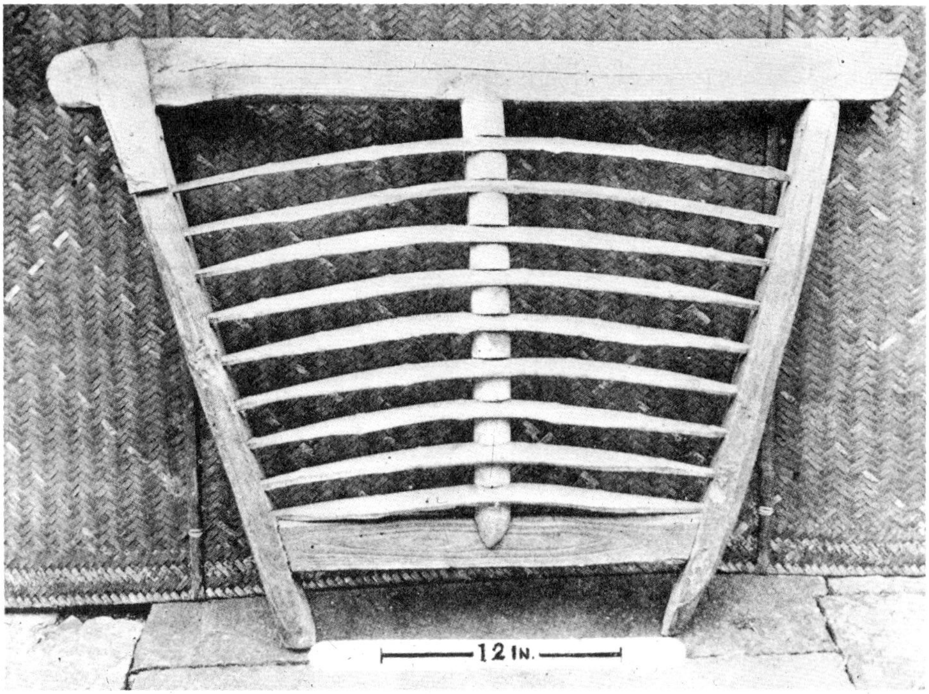


Fig. 167. Wooden ladder for threshing grain into a tub; Hommel (1), fig. 108.

Fig. 168: Vietnamese rice treading; after Huard & Durand (1), p. 129.

Fig. 169. Stripping rice with the fingers: *Nōgu Benri Ron* endpiece.

A platform, eight or ten feet high, with a bamboo floor with cracks between the slats, is erected. A rope is stretched above the platform; to this the workers hold with their hands while they work the rice with their feet. The grain falls through the cracks, is winnowed by the wind, and collects, clean, on a mat on the ground. An efficient worker separates as much as fifty *cavanes* [about 2250 kg] a day in this way.

Still simpler methods of threshing than treading out the grain apparently survived till quite recently in both China and Japan. An illustration in the *Nōgu Benri Ron*, published in Tokyo in 1822, clearly shows women stripping rice grains from the stem with their fingers (Fig. 169), and the + 1676 Japanese reproduction of the +1462 Ming illustrations to *Keng Chih Thu* also shows women who appear to be stripping the grain with their fingers;[a] we can find no written allusions to this, however. A slight refinement on this method was stripping the grain with a pair of thin sticks joined at the top by a short cord. In Japan these were known as *kokihashi* and were commonly used to strip both rice and dryland cereals up until the +18th century (Fig. 170). Although they are not specifically mentioned in the Chinese agricultural literature, women are shown using such sticks in the Sémaillé scroll illustration of the *Keng Chih Thu* (Fig. 171).[b]

*Beating the grain; flails*

A more common Chinese threshing method was simply to beat the sheaves of grain, either into a tub as described above, or onto a mat against a large stone[c], or a stone slab or wooden board (*chhuang*[1]) supported at an angle on a low wooden frame (Fig. 172).[d] In Vietnam the sheaves were twisted in a cord stretched between two sticks (similar to the *kokihashi*) so as to add to the momentum of the swing.[e] Beating out the grain is very exhausting work and is usually a man's task.

One can also crush the sheaves beneath an ox-drawn stone roller, sometimes cylindrical, sometimes ribbed, and sometimes tapered so that it may easily be dragged in a circle[f] (Fig. 173). Although rolling requires only one third the human labour of beating out the grain, it is more damaging; rolling often breaks off the germ, so seed-grain must never be rolled.[g] On the other hand rolling improves the quality of the straw for fodder;[h] the same cannot be said of the threshing-boards (heavy wooden boards with flints or iron teeth embedded in the underside) or threshing-wains (wooden frames fitted with three or four axles on which are fixed iron discs or beaters) of West Asia and the Mediterranean, which chop up the straw almost into dust.[i] Neither of these implements seems to have been known in China or the Far East.

[a] O. Franke (11), pl. XLVIII.

[b] We do not know of anywhere else where such sticks are used for threshing; maybe their use is restricted to the area where chopsticks are used. A similar device is used, however, for harvesting spelt and other husked grains in Asturia, Georgia and Nepal; Sigaut (1).

[c] *WCNS* 15/33*a*.

[d] *TKKW* 1/53*a*; the *Sung Chiang Fu Hsü Chih*[2] (quoted Liu Hsien-Chou (*8*), p. 67) says that the wooden board was ribbed with horizontal strips of bamboo.

[e] Huard & Durand (1), p. 128.

[f] Hommel (1), p. 74.

[g] *TKKW* 1/53*b*.

[h] Hickey (1), p. 145.

[i] Weulersse (3), p. 148. The wain was known in Roman times as a North African machine, *plostellum poenicum* (K. D. White (1), p. 155), but the Carthaginians probably brought it with them from the Middle East, for it is mentioned by the prophet Isaiah (Isaiah 41.15: '... for I have made thee a new threshing-board with teeth like a saw'). Both wain and threshing-board were common throughout the Mediterranean and West Asia at least as early as Roman times (K. D. White (1), pp. 150–156), and are still in use in the Middle East today. Weulersse (3), p. 148 gives a vivid picture of Syrian villages swimming in a haze of golden chaff thrown up by the wains; H. E. Wulff (1), pp. 274 ff. documents both wains and threshing-boards throughout Iran.

[1] 床 [2] 松江府續志

Fig. 170. *Kokihashi*: stripping rice with chopsticks; *Nōgyō Zensho* 1/5*a*.

Fig. 171. The same technique depicted in the *Keng Chih Thu*, at the bottom right hand of a winnowing scene; Pelliot (24), pl. XXVII.

Fig. 172. Threshing board; *TKKW* 1/59*b*.

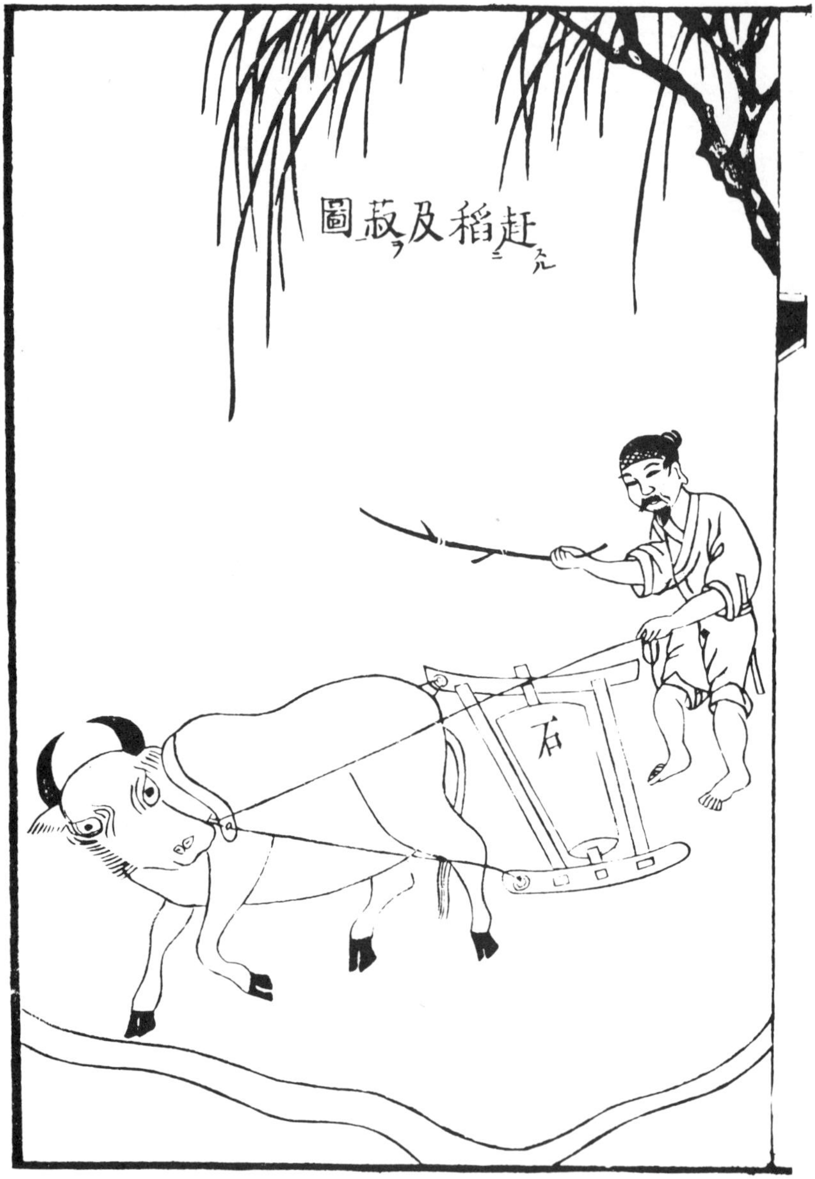


Fig. 173. Stone threshing roller, tapered to move in circles; *TKKW* 1/63*b*.

Fig. 174. Single-headed, hinged flail, on a Han mural from Kansu; Hayashi (*4*), 6–44.

As in Europe, the hinged flail (*lien chiao*[1]), was often used for threshing in China:[a]

> A clear frosty day
> When a strong wind tears the leaves from the trees
> Is best for threshing.
> The irregular beat of the flails rings out,
> The brown hens peck at the fallen grain
> While the crows chatter gleefully ...

The flail was used all over China in Han times, as can be seen from the wide range of dialectal terms given in the *Fang Yen*.[b] Wang Chen says that in Yuan times its use was confined to South China,[c] but modern examples are found in the northern provinces too.[d] The heads or swingles of European flails consist of a single piece of wood, but Chinese swingles are usually made of several strips (Fig. 374a, Vol. IV, pt 2, p. 70). Han terms existed for double-headed flails (*ya*[2]) and triple-headed flails (*lo chia*[3]),[e] but the number of strips may vary from one (Fig. 174) to as many as seven or eight (Fig. 175); where several strips are used they are bound together with rope or rawhide, though the double-headed flails shown in the *Keng Chih Thu* illustration (Fig. 163) were left unjoined. The pole of the Chinese flail is roughly 90 cm. long, the swingle about 30 cm. in length,[f] joined to the pole by a wooden axle projecting at right-angles from the top of the swingle and passing through a loop at the top of the pole, in other words a hinge fastening. In most European flails the swingle was attached to the pole by a link fastening of leather

[a] Lou Shou's poem on threshing for the *Keng Chih Thu*, tr. auct.
[b] Cf. *WCNS* 14/29*a*.
[c] *Ibid.*
[d] Flails from Kansu, Shensi and Liaoning are illustrated in Anon. (*502*), p. 188.
[e] Definitions are given in the *Shih Ming*; see Hayashi (*4*), p. 281.
[f] *WCNS* 14/29*a*; Hommel (1), pp. 73–4.

[1] 連枷 [2] 丫 [3] 羅枷

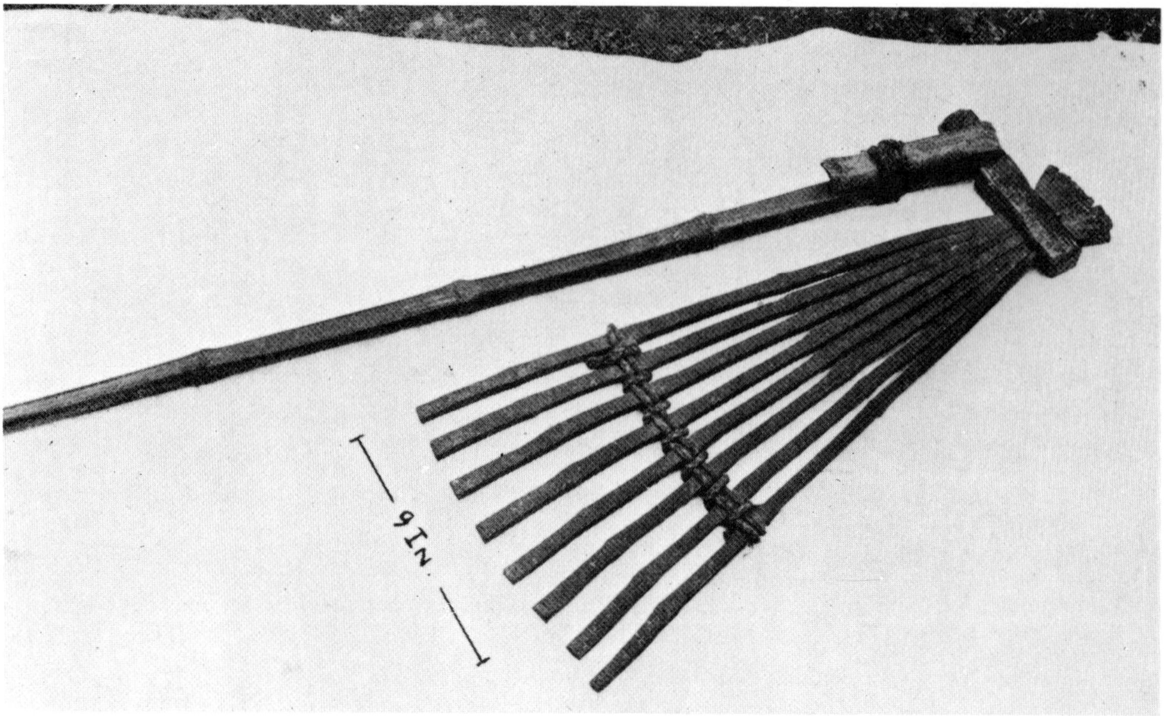


Fig. 175. Eight-stripped modern flail; Hommel (1), fig. 113.

thongs or metal loops;[a] the link flail existed in China in the form of a war flail,[b] but we know of only one Chin wall-painting (Fig. 174) which shows a link fastening in a farmer's flail. The hinge fastening somewhat reduces the swing of the Chinese flail and thus the force of the blow, and Wagner[c] dismisses it as a clumsy, impractical tool when compared to the German flail. Hommel[d] further complains that 'When several natives are threshing together, we miss the rhythm that we are wont to hear when European peasants thresh with the flail. No matter how many or few Chinese swing the flail, one person will always bring down his flail after the rest have come down with theirs together.' This clearly offended Hommel's European sense of order, but perhaps the Chinese found a certain charm in what Lou Shou called 'the flail's irregular beat (*luan sheng*[1])' (see above, p. 357).

[a] Partridge (1), p. 160.
[b] Vol. IV, pt 2, p. 70; Needham, Wang & Price (1), p. 56.
[c] (1), p. 274.
[d] (1), p. 73.

[1] 亂聲

*Threshing combs and mechanical threshing*

One further type of threshing device, Japanese in origin, is of some interest as it embodies the principle on which modern Japanese and Chinese mechanised threshers were developed. This is the threshing-comb, *ina koki*[1], first mentioned in the *Hyakushō Denki* of +1684. It consisted of a row of close-set teeth inserted in a wooden board set on a trestle; the teeth pointed upwards at an angle of about 45°, and the sheaves were pulled through the teeth to release the grain (Fig. 176). The threshing-comb must have been invented prior to +1680, and by the 18th century its use was practically universal in Japan. Combs for threshing rice had close-set iron or steel teeth, while those used to thresh wheat and other dryland cereals had bamboo teeth that were somewhat more widely spaced.[a] These threshing-combs were very effective and were nicknamed 'widow-killers' (*goke taoshi*[2]), for previously much of the threshing, particularly of the wheat crop which was harvested just before the main ploughing season, was done by elderly women or widows using the *kokihashi* or threshing-sticks to earn themselves a meagre living. But old women were not strong enough to use the threshing-comb and lost their employment; hence the sinister name. There is no indication that the Chinese were familiar with the threshing-comb. It is not mentioned in any of the classic agricultural works, nor do any Western visitors mention it. Only when it had been developed into a mechanical thresher did it come into use in China.

In Europe the idea of mechanical threshing was first taken up in the 18th century.[b] Some time before 1735, a Scottish mechanic named Michael Menzies invented a machine consisting of a number of flails attached to a beam rotated by a water-wheel. The flails soon broke when used at high velocity, however, and subsequent inventions were based on the principle of the flax-mill: the grain was rubbed out between rollers. By the mid-19th century, when threshing-machines had become sufficiently reliable and cheap to replace the flail in most parts of Scotland and England, mechanical threshers fell into two basic types. The first was the beater, of which a typical example is the horse-driven machine shown at the Great Exhibition of 1851:[c]

> The drum has five straight iron blades as beaters, and the 'concave' is of ribbed iron plates separated by spaces covered with wire screens. The corn is threshed out of the straw by the rapidly revolving beaters, which knock out the grain as the straw is being carried round in the small space between the drum and the concave, the empty straw finally being ejected at the lower end of the drum. The grain threshed out drops through the concave into the space at the bottom of the machine, and is removed at intervals through the side doors provided.

[a] This brief account of the threshing-comb is based on a much more detailed history given in Pauer (1), pp. 148 ff.

[b] Details of the history and evolution of the mechanical thresher in Europe are given by Fussell (2), ch. 5, on which work this account is based.

[c] Spencer & Passmore (1), p. 69.

[1] 稲扱 [2] 後家倒し

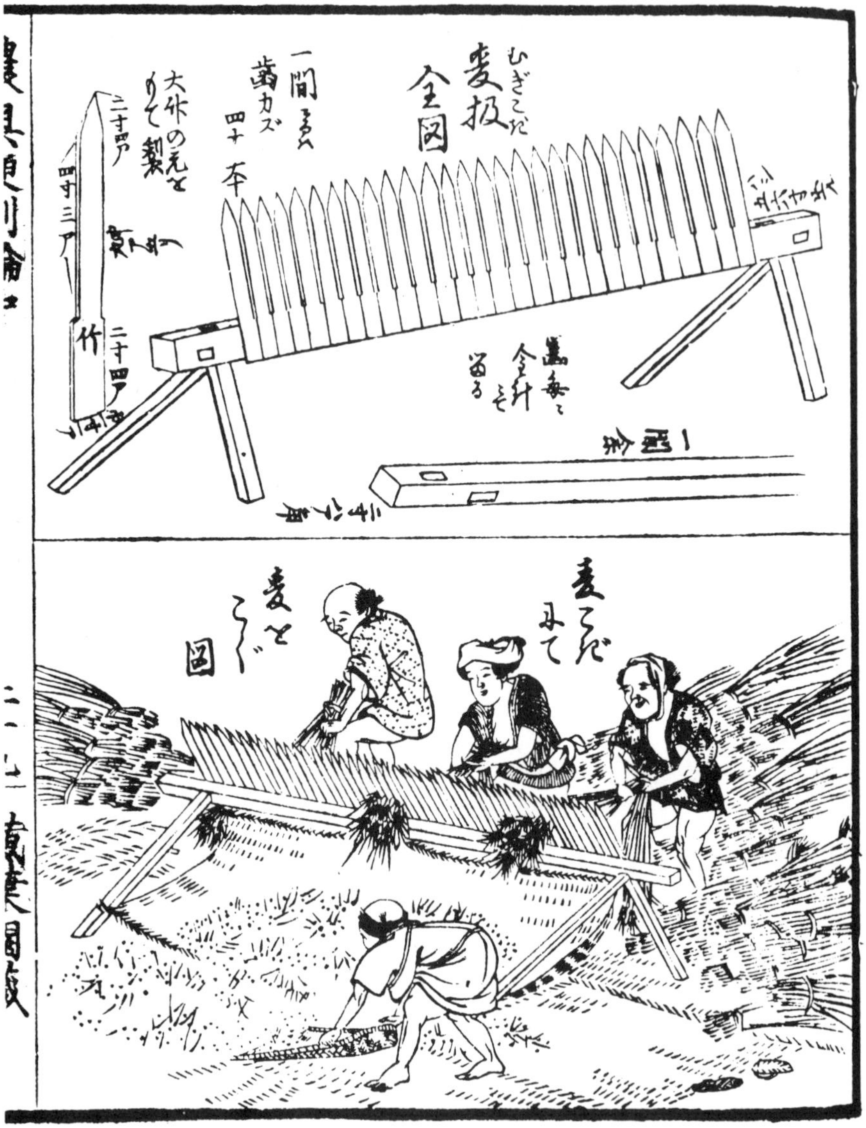


Fig. 176. Threshing-comb or 'widow-killer'; *Nōgu Benri Ron* 2/29*a*.

The clearance between the beaters and the concave could be adjusted on most beater-drum threshers, but the adjustment was delicate and there was a tendency for such machines to damage the grain and bruise the straw.

The second type was the peg-drum thresher, which differed from the beater in the construction of the cylinder: the sheaves were fed into the machine through a sort of mangle which held them firm, while the grain was beaten off by pegs attached to a rapidly revolving cylinder. Though the peg-drum thresher did not separate the grain as efficiently as the beater-drum, it required less attention and

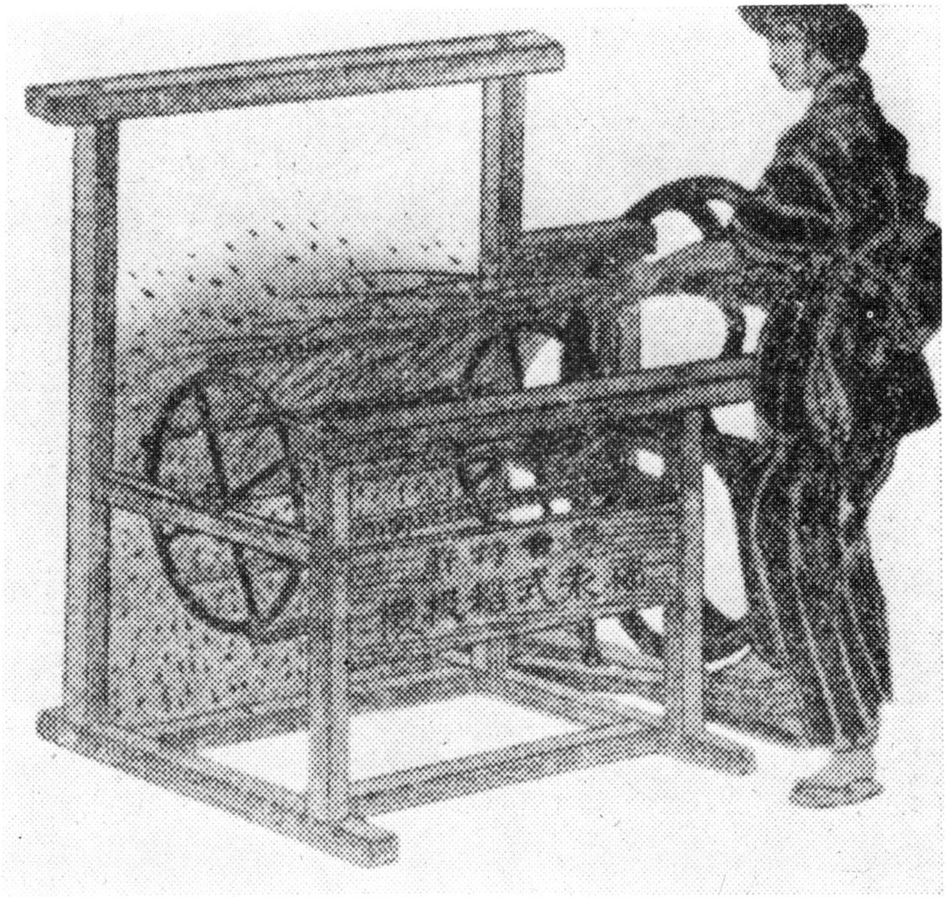

Fig. 177. Early Japanese mechanical thresher; Iinuma & Horio (*1*), p. 197.

was safer to use, so that many farmers preferred it. But it was the beater-drum that was gradually developed and incorporated into the combined threshers and winnowers of the late 19th century, from which came the elevators with which we are familiar today. We shall return to these combination machines in the section on winnowing.

Neither of the aforementioned principles is used in the Far Eastern threshing machine, although since this was first developed in Japan in the late Meiji era, it is likely that it was in some way inspired by Western examples. Almost all threshing machines in use in the Far East today are directly derived from the 18th-century Japanese 'widow-killer' or threshing-comb. The simple threshing-comb was used throughout Japan in the 19th century, and at the end of the Meiji period, in 1911, the first treadle-operated revolving toothed thresher was invented in Yamaguchi Province in Southern Honshu[a] (Fig. 177). These threshers

[a] Iinuma & Horio (*1*), p. 196.

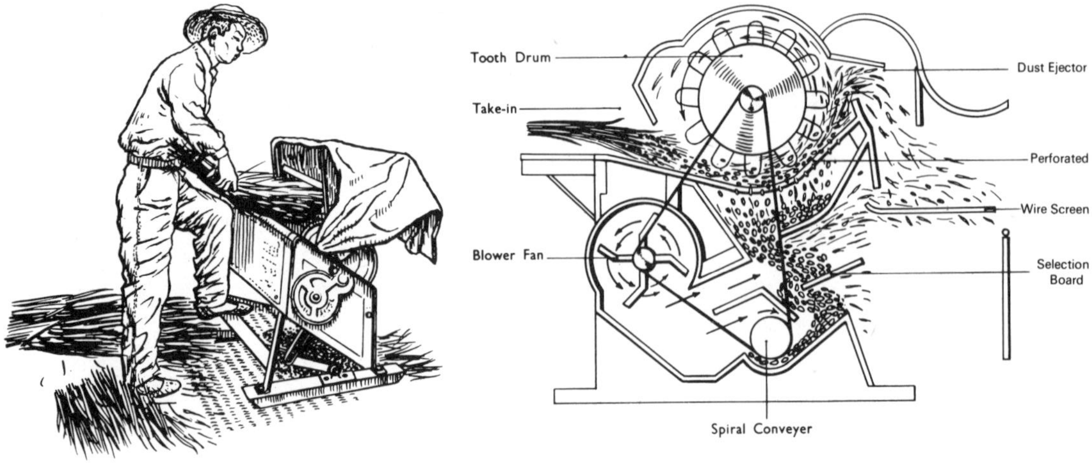


Fig. 178. Modern Minoru thresher; Grist (1), p. 166.

quickly spread throughout Japan, and presumably were introduced somewhat later into China through the Japanese possessions in Taiwan and Manchuria. The earliest (but not conclusive) evidence for their existence is a drawing in Wagner's book of 1926,[a] which unfortunately provides no information as to where and when he saw this machine. In its simplest form, namely a treadle-operated revolving drum with parallel rows of steel teeth or loops, the machine is common all over China today;[b] in some places it has been adapted to animal- or water-power.[c] A Japanese machine widely used in Southeast Asia is the Minoru thresher (Fig. 178), very similar to the Chinese models but incorporating a small fan to winnow the grain.

These Oriental threshing-machines may appear primitive in comparison with their Western counterparts. They are nevertheless efficient, capable of threshing anything from 1 to 6 tonnes of grain in a day,[d] and have the advantage of cheapness and simplicity. Western farmers generally hire threshing or combined harvesting and threshing machinery, or employ agricultural contractors to carry out these tasks for them; this has been the case ever since such machines first became popular.[e] The much smaller threshers of China and Japan are cheap enough for individual farmers or work-groups to purchase, they are simply constructed, easy to maintain and cheap to run, and perfectly appropriate to small-scale intensive farming, just as the giant combine-harvesters are an inevitable concomitant of extensive wheat-farming on the American prairies.

[a] W. Wagner (1), p. 272, fig. 91/13.
[b] Anon. (*502*), pp. 190–210.
[c] *Ibid.* pp. 210–17.
[d] *Ibid.* pp. 190 ff.
[e] Fussell (2), p. 173.

### (iii) *Winnowing*

*Baskets, trays and sieves*

The use of baskets and sieves for winnowing grain probably predates the actual domestication of cereals,[a] and winnowing baskets were certainly used in China in neolithic times.[b] The character *chi*[1], meaning a winnowing-basket, is found in very early texts;[c] the *Shuo Wen* calls the winnowing-basket *po*[2].[d] But despite numerous literary allusions in Chou and Han texts, it is not until much later that we find any illustrations (Figs. 179, 164). These show, as one might expect, that there was considerable variation in the shape and size of Chinese winnowing-baskets.

Winnowing-baskets and trays require dextrous manipulation: a rhythmic flick of the wrists separates the heavy grain from the chaff which is gradually tipped away over the edge of the tray. 'This method provides a very clean sample ... and is not arduous. It is, however, extremely slow, the average rate being 45 kg. per hour.'[e]

The process can be speeded up by substituting for the winnowing-basket a large sieve (*shai*[3]) (Fig. 180). A further improvement is to suspend the sieve from a tripod or the branch of a convenient tree (*shai ku kuai*[4])[f] (Fig. 181); the tripod arrangement is still common both in China and Southeast Asia.[g] A much more recent development is the sieve on rockers from Liaoning, exhibited in Peking in 1958.[h]

*Forks and shovels*

The advantage of both baskets and sieves is that they can be used even in still weather, but if the farmer can rely on a good stiff breeze after the harvest he will probably prefer to winnow the grain by tossing it, either with a fork or with a shovel (*chu yang hsien*[5] or *yang lan*[6]) (Figs. 164, 171, 182):[i]

> Lightly we toss the grain into the wind;
> Out of the clear blue sky comes a sudden rain-like patter.
> Already the wind has swept away chaff and straw;
> What need of a mechanical fan to complicate our lives?

[a] See, for example, Kraybill (1) on winnowing amongst hunter-gatherers of Australia and elsewhere.
[b] D. Kuhn (1).
[c] Karlgren (1), 1952, a–f; Keightley (2), table 26, no. 5; Legge (8), vol. II, pp. 356, 471.
[d] Quoted *WCNS* 15/25*a*.
[e] Hopfen (1), p. 126.
[f] *WCNS* 15/29*b*.
[g] W. Wagner (1), p. 276; Hommel (1), p. 76; Loofs (1), p. 20.
[h] Anon. (*18*), fig. 9; D. Kuhn (1) discusses the various forms of basket and sieve in some detail.
[i] *WCNS* 13/10*a*.

[1] 箕 or more anciently 其 [2] 簸 [3] 篩 or 籭 [4] 篩穀箉
[5] 竹揚杴 [6] 颺籃

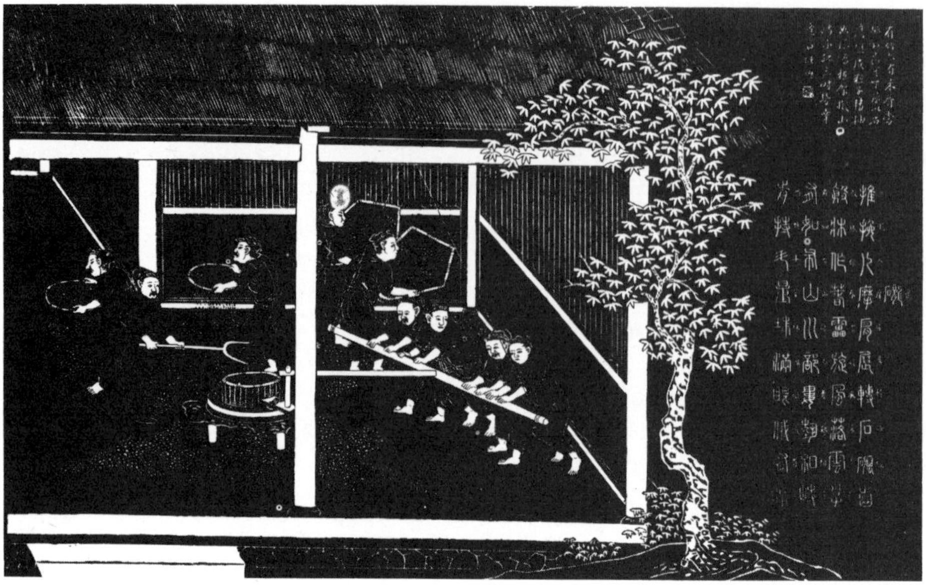

Fig. 179. Winnowing with baskets, preparatory to husking the grain in a bamboo mill; *Keng Chih Thu*, Pelliot (24), pl. XXVIII.

Fig. 180. Winnowing sieves; *Keng Chih Thu*, Franke (11), pl. XLVII.

Fig. 181. Winnowing sieve suspended from the branch of a tree; *WCNS* 15/29*a*.

Fig. 182. Winnowing fork, from a Han mural at Chia-yü-kuan, Kansu; Anon. (*512*), pl. 144.

*Winnowing fans*

The disparaging remark on mechanical fans which Wang Chen imputes to his farmers probably reflects their economic status rather than their true feelings, for in fact even the simplest of fans made winnowing much easier, enabling grain and chaff to be separated in the absence of the slightest breeze. The earliest Chinese winnowing-fans were probably the double fans described in the section on mechanical engineering:[a] pairs of rectangular woven bamboo leaves mounted on vertical poles, which one man worked back and forth while another poured the grain in a steady stream in front of him (Fig. 183).[b] Although such fans are not specifically mentioned in the Chinese agricultural literature they continued in use, at least in South China, up to the 20th century and were also much used in Japan and Vietnam; more recent forms were made not of plaited bamboo but of pleated paper (Fig. 184).[c] Sometimes the ends of a mat were used instead of a proper fan (Fig. 185).

We have already discussed mechanical winnowing in some detail in a previous volume (Vol. IV, pt 2, pp. 151 ff.), but here we should like to take up the theme again, not only to consider in somewhat greater detail the evolution of the winnowing-machine in China but also to investigate its adoption in other parts of the world.

In the section on mechanical engineering (Vol. IV, pt 2, p. 154) we mentioned the possibility that rotary-fan winnowing-machines were in use in China in the Former Han, citing in support both philological evidence and the existence of a Han grave-model which looked remarkably like a crank-operated enclosed winnowing-machine (Fig. 415). But we had to admit that the philological evidence could be considered ambiguous, while the model was not sufficiently de-

[a] Vol. IV, pt 2, p. 154.

[b] On the basis of their hairstyle, Liu Chih-Yuan (*3*), p. 52, suggests that the workers manning the fans in Figs. 436 (Vol. IV pt 2) and 183 were probably enslaved tribal peoples. In Vol. IV we debated whether the Han fans might have been worked in a rotatory rather than an oscillatory fashion, but we have since rejected the possibility of rotation: the requirements of the task and the nature of the apparatus both indicate oscillatory motion, and ethnographic evidence supports this view.

[c] Their use in South China and Japan is referred to by W. Wagner (1), p. 277; F. H. King (1), p. 268; Pauer (1), p. 156; in Vietnam by Truong Van Binh (1).

Fig. 183. Han double fans for winnowing; Hayashi (*4*), 6–54; orig. photo from Szechwan Provincial Museum.

Fig. 184. Modern Japanese double fans; King (1), fig. 159.

tailed to constitute incontrovertible proof. Since then, however, several Han working models of winnowing-machines have been found. In 1969 two models were found in a late Western Han tomb at Chi-yüan in Honan,[a] and in 1971 a clay model with all the wooden working parts preserved except the fan was found in a Later Han tomb in Lo-yang (Fig. 186).[b] So we now have conclusive proof that the enclosed rotary-fan winnowing-machine (known as *shan thui*[1] or *shan chhe*[2 c]) was in use by the −1st century in China.

The Han grave-models show the winnowing-machine as a fixture, built into the wall of a courtyard, with tilt-hammers and querns placed conveniently nearby. A comparatively small, crank-operated fan, with a large air-inlet immediately behind it, was set at the broad end of a sloping tunnel, just in front of

[a] Anon. (*523*).
[b] Hsü Fu-Wei & Ho Kuan-Pao (*1*), pp. 57–9.
[c] See Vol. IV, pt 2, p. 154.

[1] 扇隤 [2] 扇車

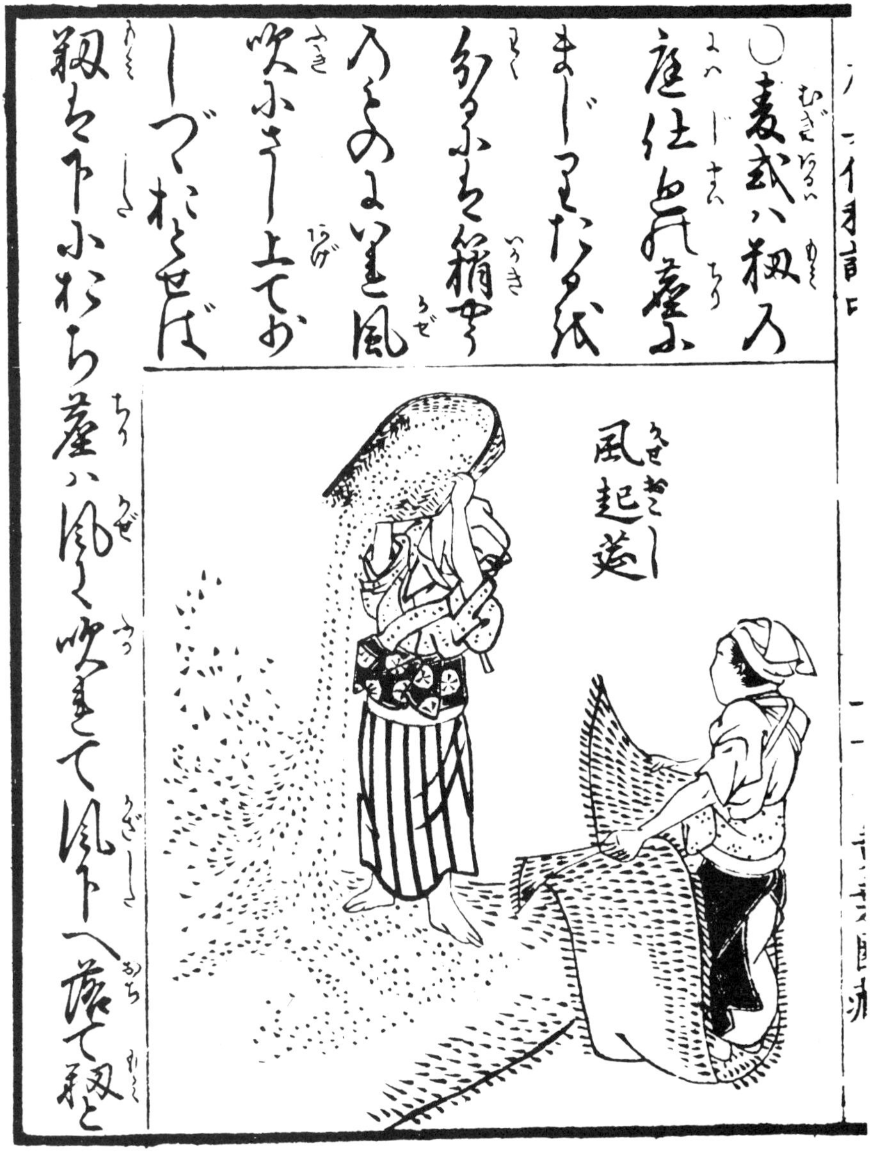

Fig. 185. Fanning with a mat; *Nōgu Benri Ron*, 2/11*b*.

the hopper through which the grain was poured. The heavy grain fell out through an aperture immediately below the hopper, but the lighter chaff (*khang*[1]) and still lighter bran (*sai*[2]) were blown away down the tunnel. If the tunnel was sufficiently long, bran and chaff could also be separated,[a] and one of the Han models (Fig. 415) shows the two side-apertures typical of many later machines.

[a] *Ibid.*

[1] 糠 [2] 粞

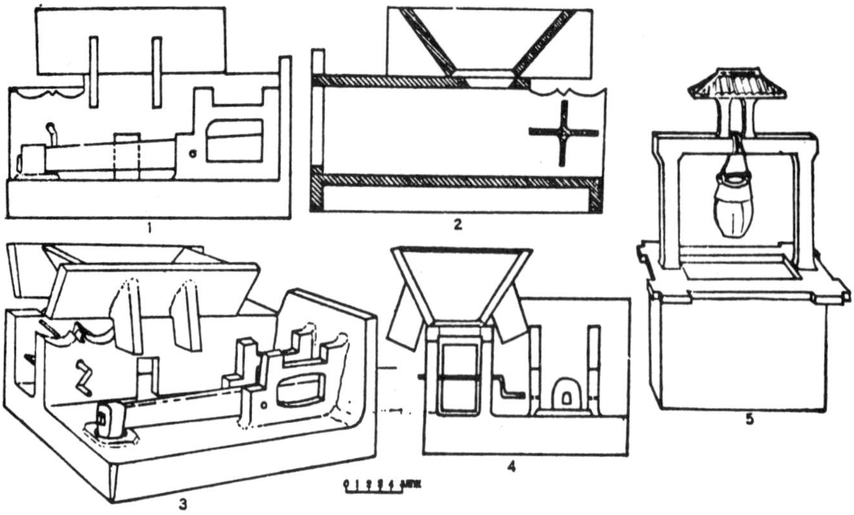


Fig. 186. Han pottery model of rotary winnowing fan, the fan being worked by a crank; Hsü Fu-Wei & Ho Kuan-Pao (*1*), fig. 4.

The winnowing-machine underwent several important modifications between the Han and Yuan, but it is impossible to say when these modifications first occurred. Apart from Yen Shih-Ku's brief gloss on the *Chi Chiu Phien* entry[a] and poems describing a winnowing-machine at work by the Sung writers Wang An-Shih[1 b] and Mei Sheng-Yü[2],[c] the machine is neither mentioned nor depicted again until the publication of Wang Chen's *Nung Shu* in +1313, which provides us with both description and illustration (Fig. 188). The *Nung Cheng Chhüan Shu*[d] and the *Shou Shih Thung Khao*[e] both quote Wang Chen's description, and other texts such as the *Thien Kung Khai Wu*[f] and the *Nung Chü Chi*[g] also mention winnowing-machines. The Ming illustrations of both the *Nung Shu* (Fig. 188) and the *Thien Kung Khai Wu*[h] show a machine with a large fan enclosed in a circular hood; such machines are still in use all over China and the Far East, and the *Shou Shih Thung Khao*[i] and various Japanese texts[j] (Fig. 189) show similar machines in use in Chhing times.

But here a discrepancy creeps in. The *Nung Cheng Chhüan Shu*, although it faithfully quotes Wang Chen's description, apparently illustrates a machine with an unenclosed fan (Fig. 187), and this drawing is copied in the *Shou Shih Thung*

[a] See Vol. IV, pt 2, p. 154.
[b] *Wang Lin-Chhuan Chi*, vol. 2, p. 73.
[c] Mei Yao-Chhen[3], died +1060. His poem is quoted by Wang Chen, and the whole *WCNS* entry is translated in Vol. IV, pt 2, p. 153.
[d] 23/11*b*.
[e] 40/14*b*.
[f] Ch. 1, sect. 4, p. 53*b*.
[g] Quoted in *Nung Ya*, 4.89.
[h] Ch. 1, sect. 4, p. 61*a*.
[i] 40/15*a*.
[j] See Pauer (1), figs. 53, 55 where illustrations from the *Nihon Eitai Gura* and the *Shika Nōgyō Dan* are reproduced.

[1] 王安石 [2] 梅聖俞 [3] 梅堯臣

Fig. 187. Apparently unenclosed winnowing fan; *NCCS* 23/11*a*. The Ming drawing in fact shows a cross-section of the drum enclosing the fan. The fan is worked by a treadle, bottom right.

*Khao*[a] as well as in the additional Chhing illustrations to the *Thien Kung Khai Wu* (Fig. 435, Vol. IV, pt 2, p. 152). This illustration has been the cause of confusion over the years, leading to the conclusion (accepted by ourselves as well as others[b]) that in medieval China there were two different types of winnowing-machine, the open and the boxed-in fan, only the latter of which had survived to the present day.

[a] 40/14*a*.
[b] Vol. IV, pt 2, p. 151; W. Wagner (1), p. 277; O. Franke (11), pp. 156–7; T. Thilo (1) p. 148 and (2), is the only scholar we know to cast doubts on this interpretation. Liu Hsien-Chou (7), p. 39, does not discuss the problem.

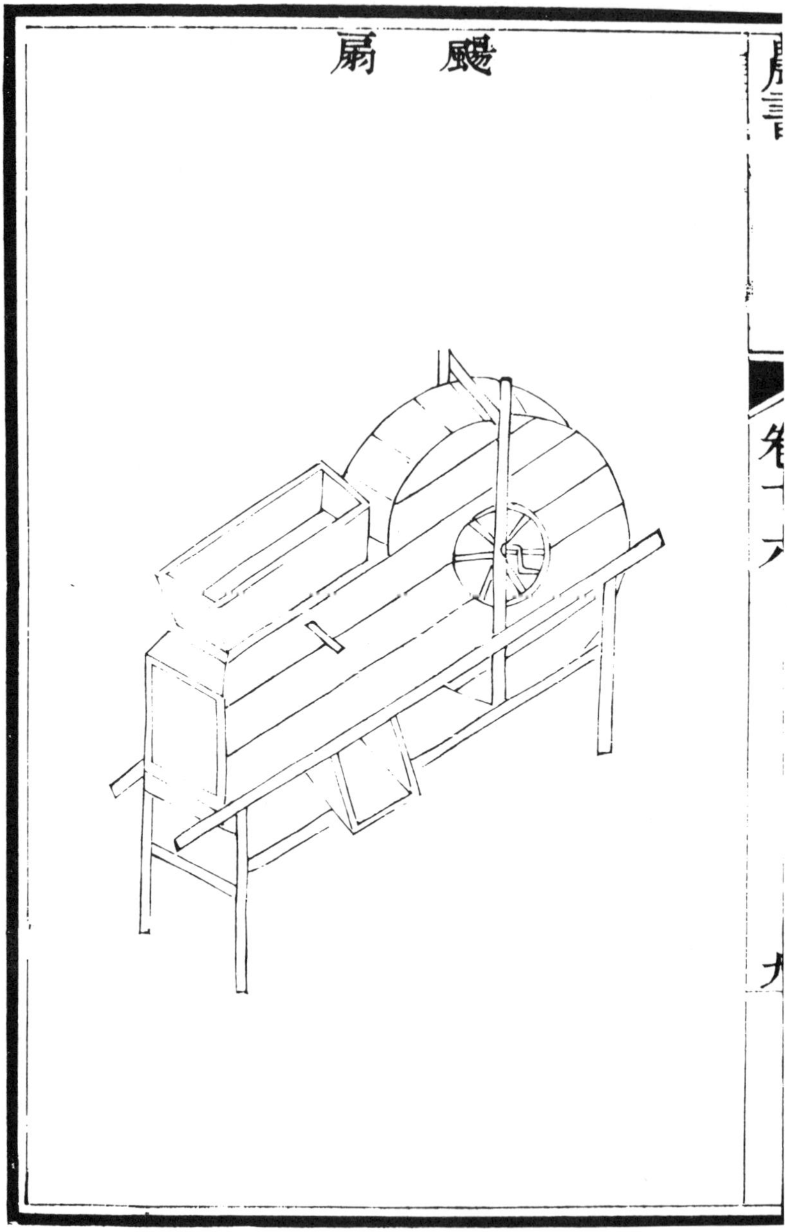


Fig. 188. Drawing of enclosed winnowing fan; *WCNS* 16/9*b*. Note the carrying poles extending from either end of the frame.

Fig. 189. Japanese enclosed winnowing fan (top left), with two apertures; from the *Nihon Eitai Gura* of 1688; repr. Pauer (1), fig. 53.

(No example of an 'open' fan has ever been seen except in the above-mentioned illustrations, all copied from the *Nung Cheng Chhüan Shu* original.) This idea was, however, based on a misinterpretation of the illustrations of the so-called 'open' machine. If we look closely at Fig. 187, we see that the artist is in fact attempting to draw a cross-section of an ordinary winnowing-machine. The two arcs at the top represent the sides of the circular hood encasing the fan; in the foreground the operators are adjusting the flow of grain through the hopper by manipulating a separator (*pien feng*[1]),[a] and the clean grain falls down the chute at the left, out of the machine and into a basket, while the chaff is blown out at the end of the machine (at the bottom of the picture). The *Shou Shih Thung Khao*, which appends an explanation of the illustrations of winnowing-machines,[b] leaves no room for doubt that this is the correct explanation of the 'open' fan.[c]

We must therefore modify somewhat our earlier translation of the passage from the *Nung Shu*. Where we wrote:[d] 'Some people raise the fan high up (without

[a] *WCNS* 16/9*b*; the separator or regulating device for the grain flow can also be seen quite clearly in Hommel (1), fig. 118 and Grist (1), fig. 8.8.

[b] 40/15*b*.

[c] If the drawing did truly represent an open fan, it is difficult to see how it could work, given the peculiar arrangement of fan and grain-inflow.

[d] Vol. IV, pt 2, p. 153.

[1] 匾縫

enclosing it) and so winnow; this is called the *shan chhe*[1]', we would now substitute the translation: 'There are also winnowing-fans which can be carried out on poles (*yü*[2]) to be used on the threshing-floor (*yu yu yü chih chhang phu chien yung chih che*[3]), and these are known as *shan chhe*[1].' Otherwise our earlier translation stands unchanged.

So let us now look again at the characteristics of post-Sung winnowing-fans (*shan chhe*[1] or *yang shan*[4]). First, they were more efficient than the Han forms: the fan was much larger with a central air-intake, and was enclosed in a circular hood which concentrated and directed the draught towards the flow of grain. Secondly, the whole structure was made of wood, and although the illustrations usually show winnowing-machines being used inside a shed it is clear from what Wang Chen says that some machines, the *shan chhe*[1], were sufficiently mobile to be carried out onto the threshing-floor. While some winnowers still had hand cranks (Fig. 188), others were treadle-operated (Fig. 187). Though the Chinese illustrations show machines with only one chute for the cleaned grain, early Japanese depictions, drawn soon after the machine was introduced from China, show machines with double chutes (Fig. 189),[a] and a Swedish visitor to Canton in +1773 reports that Chinese winnowing-machines separated the grain into two grades;[b] presumably the grain was passed through a sieve as it fell. Machines identical to those described in the *Nung Shu* and *Shou Shih Thung Khao* are still to be found in China in the 20th century.[c]

It is interesting that Sung Ying-Hsing wrote in +1637 that the winnowing-machine was used extensively only in South China where rice was the principal crop; in the North, he said, farmers generally used shovels to toss (*yang*[5]) what little rice they grew as well as their wheat and millet; but tossing, Sung felt, was not suitable for rice.[d] Since several of the European regions into which the Chinese winnowing-fan was later introduced were barley- or rye-producing areas (for example parts of Sweden and Austria), it has been suggested by Dr François Sigaut that the winnowing-fan is best suited to awned or husked cereals.[e] Yet the winnowing-machine was also adopted in wheat-growing areas of Europe such as Flanders,[f] and in China Wang Chen specifically refers to its use for wheat and millet;[g] furthermore, as we have seen, three Han models of winnowing-machines have been found in Honan where millets and wheat were the principal crops. So the explanation of 'awned' versus 'naked' cereals does not seem to tally, and it could well be that the answer is one of simple economics. The winnowing-

[a] See also Pauer (1), fig. 55, after the *Shika Nōgyō Dan* of +1788.
[b] Barchaeus (1), pp. 19 ff. cited G. Berg (3), pp. 27–8.
[c] Hommel (1), pp. 74 ff.
[d] *TKKW*, ch. 1, sect. 4, p. 53*b*.
[e] Pers. comm., 1978.
[f] G. Berg (3), p. 32.
[g] *WCNS* 16/19*b*.

[1] 扇車 [2] 舁 [3] 又有舁之場圃間用之者 [4] 颺扇
[5] 颺

machines were many times more efficient than shovels or baskets,[a] but they must have been costly; Hommel[b] reports that in the areas of South China that he visited only the wealthier farmers owned winnowing-machines, while their poorer neighbours used baskets or shovels. It seems probable then that winnowing-fans were used on the large agricultural estates that flourished in North China for some time after the Han, but that as the region became impoverished and relatively backward in the centuries after the Thang,[c] farmers reverted to cheaper if more laborious methods of winnowing, until by the +17th century the winnowing-fan was almost unknown in the North.

The advantages of the winnowing-machine were, however, manifest to those who could afford them, and the machine gradually spread to other Asian countries. In Japan the earliest mention of the 'Thang winnowing-basket' (*tōmi*[1]) occurs in Ihara Saikaku's *Nihon Eitai Gura* of +1688, where a machine with two grain-chutes is shown (Fig. 189); thereafter the winnowing-machine was illustrated in several agricultural books.[d] It spread through most of Japan in the early 18th century and even now is still common in the countryside, though now it is used for beans more often than for grain.[e] The winnowing-fan is also common in Southeast Asia, where it was probably introduced in the early 17th century by the Chinese rice-farmers who settled in Java and elsewhere.[f]

Let us now briefly consider the history of winnowing in the West. The only winnowing implements referred to in the classical texts are the shovel (*pala lignea* or *ventilabrum*)[g] and the winnowing-basket (*vannus*),[h] made of wicker and of a rather different shape from the Chinese *chi*[2] (Fig. 190). These remained the chief winnowing implements of Europe up till the 19th century, and are still in use in some regions today.[i] The arrangement of a large sieve suspended from a tripod, widespread in China, is reported from Southern Europe in the early 19th century by Lasteyrie, but there it was used only for the final sieving of the grain, not for the actual winnowing process.[j]

The use of fans of any kind was apparently rare before the late 18th century, though both Fitzherbert's *Husbandry* of 1523[k] and the English version of *La Maison Rustique*[l] refer to the fanning of corn in some districts of England. Fitzherbert gives

[a] *WCNS* 16/10*a*. An 18th-century Swedish report on a machine imported to Gothenburg from China says that it could easily clean up to 17 barrels of grain a day; Hårleman (1), p. 63, quoted G. Berg (3), p. 27.

[b] (1), p. 76.

[c] Bray (3).

[d] In Japanese translations of the *Keng Chih Thu* and *San Tshai Thu Hui*, as well as original Japanese works such as the *Shika Nōgyō Dan* of +1788.

[e] Pauer (1), pp. 156 ff.

[f] V. Purcell (1). In some areas they are a comparatively recent introduction, but in others they have been in use for a very long time; we have seen venerable hard-wood winnowing-fans in the Malacca Museum which may well date back to the 18th century.

[g] K. D. White (1), pp. 31 ff.

[h] K. D. White (3), pp. 75 ff.

[i] E.g. Rasmussen (1) on Calabria.

[j] Lasteyrie (1), cited Leser (2), p. 437.

[k] (1), p. 41.

[l] Stephens & Liebault (1), p. 548.

1 唐箕 2 箕

Fig. 190. Roman using a winnowing basket (*vannus*); stone relief from Moguntiacum (Mainz); Mainz Zentral-Museum.

Fig. 191. Welsh winnowing fan; Spencer & Passmore (1), pl. IX.

no details, but the *Maison Rustique* describes two types of fan, one 'made of Wickers, of a great compasse, being the one halfe plaine without an edge, the other halfe hauing an edge almost a foot deepe, which being turned to the bodie of the man, and casting the corne to and fro in the same, it disperseth and driueth the chaffe from the corne', the other 'a fan with loose clothes like sailes, which being turned swiftly about, gathereth a wind that will disperse the corne from the chaffe'. This second type was quite widespread in England, where it continued in use in remote country districts until the end of the 19th century (Fig. 191).[a] Although according to *La Maison Rustique* these fans were not as efficient as the time-honoured method of sifting grain in a windy passage, it may be that their presence accounts for the relatively slow adoption of the Chinese-style fan in Britain after its introduction elsewhere in Europe.

The first models of Chinese winnowing-fans were apparently brought to France in the early 18th century by the Jesuits,[b] and by 1720 working machines are reported from Flanders and Silesia; several examples were also brought back to Sweden by scientific visitors to Canton, and the machines were adapted to winnow European cereals by such enterprising technicians as Jonas Norberg, who in +1722 invented an improved machine, for which he says 'I got the initial idea ... from three separate models brought here from China ... In addition to the fact that they do not suit our kinds of corn, all three lack the most essential feature, namely a device for ensuring that the corn is evenly released.'[c]

As we have seen, there is documentary evidence that the early Swedish winnowing machines were imported directly from South China, but there is some debate as to the origins of the machines that were introduced to Flanders some time earlier, *c.* 1700–20. Berg[d] believes that they came from Japan, for the Dutch had by that time well-established trading links with the Japanese through Nagasaki, and he feels that the local tradition reported by Watson,[e] that the winnowing-fan was introduced to Holland from the Dutch East Indies, may well have been no more than vague speculation. We ourselves, however, would give preference to the second view, for as we have seen, the Chinese winnowing-fan was only introduced to Japan at the very end of the 17th century, and its use did not become common there until the mid-18th century. But rice-farmers of Southern Chinese origin were already established in Java when the Dutch first arrived, and immediately after the foundation of Batavia in +1619 they began to clear land for their rice-farms around the city walls.[f]

In China and East Asia the mechanical winnowing-fan was used for cleaning the grain both after threshing and after husking; it was even used sometimes to

[a] G. Berg (3), pp. 37 ff.; Spencer & Passmore (1), p. 68; G. Jekyll (1), p. 231.
[b] Hårleman (1), p. 63.
[c] Quoted Berg (3), p. 40; Berg gives an exhaustive account of the adoption of the machine in most of Europe.
[d] *Ibid.* p. 37.
[e] J. A. S. Watson (1), p. 47.
[f] V. Purcell (1), p. 395.

dry the grain.[a] But it never occurred to the Chinese to incorporate a threshing device into the machine, hardly surprisingly as the only Asian method of threshing that would have been at all suitable was the Japanese threshing comb. But the arrival of the mechanical winnowing-fan in Europe coincided with a growing interest in the development of agricultural machinery. Inventors in Flanders, France and Scotland were working on mechanical threshers, and they soon realised the advantage of incorporating a fan to clean the grain as it was threshed. In Britain by the 1870s multiple-blast, combined threshing-and-dressing machines were in general use, and elevators were being incorporated into these machines to stack the straw as the grain was threshed and winnowed.[b] The standard of winnowing achieved by the original Chinese machine was by then regarded as quite inadequate to meet the requirements of the miller. Yet it is interesting to note that although the Chinese-style winnowing-fan has completely lost favour in the West, its use is still spreading today in many Third World countries where its economic and practical advantages make it an effective competitor with technologically more advanced machinery.[c]

## (6) Grain Storage

### (i) *The importance of storage methods; their place in the literature*

Agricultural writers in both Occident and Orient have always been preoccupied with methods for improving yields, yet sometimes neglect to point out that efficient storage of the crops after the harvest is every bit as important. Even today, losses of rice in the barn may be as high as fifty per cent,[d] and Pliny's numerous and exotic suggestions for preserving wheat in storage, one of which was to hang a toad up by its hind leg at the threshold of the barn before carrying in the corn, show that in Roman times efficient storage was equally problematic.[e] In seventeenth-century England Gervase Markham deprecates the practice of stacking corn in the open, directly on the ground, 'which commonly doth rot and spoyle at least a yard thicknesse of the bottome of the stack'.[f] When the ratio of crop yield to seed sown was frequently well under ten to one, as was the case in most of Europe till the Agricultural Revolution,[g] poor storage meant that many farmers had either to go hungry or else to eat their precious seed grain.[h] It is not surprising, therefore, that both the Roman and the European agricultural treatises contain detailed advice on the construction and maintenance of granaries.[i]

Early Chinese statesmen and writers all acknowledged the importance of keep-

[a] *Nung Chü Chi*, quoted *Nung Ya* 4. 89.
[b] Fussell (2), pp. 173 ff.
[c] Grist (1), p. 168.
[d] Grist (1), p. 401.
[e] K. D. White (2), p. 189; Pliny (1), XVIII. 301.
[f] Markham (1), p. 83.
[g] Slicher van Bath (1), tables 328–33.
[h] Duby (1); W. Abel (1), p. 95, quoted E. R. Wolf (1), p. 5.
[i] Varro (1), I, lvii; Columella (1), I, vi; Stephens & Liebault (1), pp. 14–18, 547–8; G. Markham (1), pp. 83–6, 101–17, etc. On Roman granaries see also G. Rickman (1).

ing large quantities of grain in store, not only for running the state efficiently and maintaining large armies, but also for relieving the populace in times of famine. 'On average the state should be able to save a year's surplus from every three years' crops ... so that despite natural disasters the people will not suffer misery or destruction. A state which does not have [at least] nine years' stores is said to have insufficient food, one which does not have six years' supplies is said to be in peril, and one which does not have three years' food is said to be in desperate straits.'[a] Yet although grain was obviously stored on a considerable scale at least as early as the Spring and Autumn period,[b] even the early Chinese agricultural treatises such as *Fan Sheng-Chih Shu* and *Chhi Min Yao Shu* give very few technical details on grain storage, and it is not until the Sung dynasty that we find the first fuller accounts of the construction, maintenance and management of granaries.

Most of the later material deals with large-scale grain storage in public granaries, and similar methods were doubtless employed in the Buddhist and Taoist monasteries, clan estates and great families.[c] The storage methods used by small peasant families can be deduced partly from the archaeological evidence,[d] partly from the definitions of terms describing storage given in dictionaries and encyclopaedias and their use in historical and geographical documents, and partly from the illustrations and explanations contained in the agricultural treatises and encyclopaedias.

Perhaps one reason why the early Chinese sources devote less attention than their Western counterparts to efficient grain storage is that the principal grain crops of early China (millets in the North and rice in the South) bear a very large number of seeds on each plant, so that the ratio of yield to sowing rate (the amount of grain harvested divided by the amount of grain sown) is very much higher than for wheat and barley, the principal cereals of the West. Even in the late twentieth century, the ratio of yield to sowing rate for wheat varies between twenty and six to one,[e] and in medieval Europe yields sometimes barely surpassed the sowing rates.[f] Twentieth-century ratios for *Setaria italica*, the main crop of early North China, are of the order of one hundred to one, and although the ratios found in the +6th-century work *Chhi Min Yao Shu* were certainly exaggerated,[g] the real ratios for that time were certainly at least ten times higher

[a] *Huai Nan Tzu* 9/18*a*.

[b] See, for example, the grain storage policies of the Legalist statesman Kuan Chung[1], supposed to have lived in the −7th century, described in *Kuan Tzu*, ch. 24; Than Po-Fu *et al.* (1), esp. p. 187.

[c] Again precise information is scanty. There are some references to granary management in clan documents (e.g. Twitchett (9)), and the plan of great Chinese mansions usually shows considerable space allotted to storage and granaries (Liu Tun-Chen (*4*)).

[d] The Han pottery grave models and stone bas-reliefs are particularly informative; e.g. Finsterbusch (1), vol. II, passim.

[e] Purseglove (1), vol. I, p. 294.

[f] Slicher van Bath (1), table II, pp. 328–9.

[g] *CMYS* 3.19.12, etc.

1 管仲

than for contemporary Europe. The ratio for rice nowadays averages fifty to one, so again it is considerably higher than for wheat or barley.[a] Another factor which might help explain the lack of technical references to grain storage in the early Chinese texts is the dryness of the North Chinese climate and soils, which favours the preservation of cereals. Moreover, millets and rice are comparatively resistant to mildews and other insect pests.

Grain was stored in enormous quantities in the Chinese capital and in large provincial centres from very early times, but it is probable that the granaries were constructed, maintained and managed by trained specialists who would have passed their knowledge on by word of mouth rather than in written form to their subordinates or apprentices. The posts of *tshang jen*[1] ('granary man'), *lin jen*[2] ('grain barn man'), and *she jen*[3] ('house [stores] man') mentioned in the *Chou Kuan* indicate the possible existence of such specialised offices before the Han.[b] The *Han Shu* contains the titles of such official posts as Granary Chief (*tshang chang*[4]), Assistant (*tshang chheng*[5]) and Prefect (*tshang ling*[6]) in the court of the Empress, as well as provincial Granary Inspectors (*tshang nung chien*[7]) and Clerks of the Department of Granaries (*tshang tshao shih*[8]),[c] but it does not enter into the duties of the officers involved, which were probably supervisory rather than executive.

By the Sung dynasty not only were the capital and provincial granaries growing both in size and in number, but also it had become common practice to maintain smaller public granaries at village level as a precaution against grain shortages or famine. If these smaller granaries were to be efficiently run, the technology of government granaries had to be publicised, and methods of managing small-scale granaries without waste and corruption had to be evolved. The Southern Sung statesman Fu Pi[9] (1004–83) wrote a series of memorials and recommendations on the running of public granaries in sub-prefectures and villages,[d] and the famous neo-Confucian statesman and philosopher Chu Hsi[10] composed an essay on granary management at village level in 1182.[e] But although these two works, extensively quoted by later writers, contain a wealth of managerial detail, just like the numerous references to granaries and grain storage in the dynastic histories and other textual sources, they contain no technical information on construction, maintenance or storage methods.

The earliest surviving text specifically devoted to granary construction is the relevant section of the *Ying Tsao Fa Shih*, an architects' manual first published in

[a] Purseglove (1), vol. 1; Grist (1), p. 458.
[b] *Chou Li*, 4/37*b*, 26*b* and 27*a*, tr. Biot (1), vol. 1, p. 390, 384 and 388 respectively.
[c] See de Crespigny (1), p. 27.
[d] Quoted *NCCS* 43/12*a* ff.
[e] *Chu Tzu She Tshang Fa*[11], quoted *SSTK* 56/4*a*–6*b*.

[1] 倉人 [2] 廩人 [3] 舍人 [4] 倉長 [5] 倉丞
[6] 倉令 [7] 倉農監 [8] 倉曹史 [9] 富弼 [10] 朱熹
[11] 朱子社倉法

1097. In fact it is really more of a quantity surveyor's handbook than an architectural treatise, and although it enumerates the type and number of pieces of stone and timber required it gives no details of how they should be assembled.[a] Wang Chen's *Nung Shu* (Agricultural Treatise) of +1313 is the first agricultural work to devote a section specifically to grain storage; it not only quotes definitions from the classical dictionaries but also describes contemporary construction methods, uses and distribution, concentrating on small-scale rather than public storage methods.[b] One of the technically most detailed texts comes down to us from the Ming official Chang Chhao-Jui[1] (*fl. c.* 1570), who wrote a very full account of how a district magistrate should choose the site for a public granary, finance the project, order materials, finance grain purchases and levy contributions, and manage the whole enterprise.[c] Sung, Ming and Chhing administrators furnish numerous details of granary management in their essays and memorials, and both Hsü Kuang-Chhi[2] of the late Ming and the compilers of the early Chhing work *Shou Shih Thung Khao* devote considerable space to the historical development, construction techniques and management of various types of public granary, in which they quote at length from earlier works and, in Hsü's case, add many pertinent comments based on personal experience. Hsü, who was keenly aware of the importance of famine prevention and relief, includes the chapters on public granaries at the beginning of the section of the *Nung Cheng Chhüan Shu* devoted to famine relief (*huang cheng*[3]), the later part being devoted to a classified enumeration of edible wild plants that could be gathered if the supplies of grain ran out.[d]

### (ii) *Storage technology*

The principal requirement of grain storage is that the grain should be kept dry and cool. If the grain is not properly dried before storing, or if it is stored in a damp place, the moisture content will cause 'spontaneous heating', resulting in 'loss of viability, loss of material and chemical changes in the proteins, carbohydrates and oil'.[e] Micro-organisms and insect pests also increase the temperature as well as attacking the grain, and these pests are less likely to infest grain kept dry and at low temperatures.[f] 'The granaries ... should receive ventilation through small openings on the north side; for that exposure is the coolest and the least humid, and both these considerations contribute to the preservation of

[a] *Ying Tsao Fa Shih* 19/4*b*–7*a*.
[b] *WCNS*, ch. 16.
[c] *Tshang Ao I*, quoted *SSTK* 57/2*a*–5*b*.
[d] *NCCS*, ch. 45–60.
[e] Grist (1), p. 386.
[f] Tests in Japan have shown, for example, that below 15°C there is no insect damage to rice stores; Grist (1), p. 400.

1 張朝瑞 2 徐光啓 3 荒政

stored grain', says Columella[a], while Kuan Tzu remarks more tersely:[b] 'It is essential to store grain in a granary that does not leak.'

Most grains keep better if stored in the husk, or even in the ear. The Thang granary statutes allow unhusked grain (*ku*[1]) to be kept for nine years, hulled grain (*mi*[2]) and 'miscellaneous grains' (*tsa ku*[3]) for only three.[c] The Southern Sung writer Shu Lin[4] allowed very similar periods for keeping rice, only four to five years for hulled but eight or nine years for unhusked.[d] It is standard practice in peasant holdings all over the world to store the grain in the husk and to hull and mill it a little at a time as required; often it is kept in the ear for even greater protection. In Europe this practice seems to have been associated particularly with millets and spelt (*Triticum spelta* L.), in Southeast Asia, more especially in areas where the reaping knife is still in use, with rice, and in China with millets and wheat.[e] But although grain in the ear or husk is far more resistant to attacks by insects and micro-organisms than milled grain, it is of course much bulkier to store. Grain stored in large quantities is therefore usually husked before storage, and sometimes milled too or even ground to flour. In China most grain was milled and polished before storing, as can be seen from the arrangement of the chapters in the agricultural encyclopaedias, where milling is invariably treated before storage methods.[f] Rice keeps badly if husked but not polished, as the fats in the bran quickly turn rancid; in India, Africa and the West Indies rice is usually parboiled beforehand, as it facilitates husking, reduces the proportion of broken grains, and improves the keeping qualities of unpolished rice.[g] Rice in China was usually well polished before being stored in bulk,[h] so that parboiling was unnecessary, indeed the distinctive flavour imparted by the process of parboiling is disliked throughout East Asia. But the *Chhi Min Yao Shu* refers to panicled millet being parboiled before storage in order to prevent mildew and facilitate hulling,[i] and also advises that wheat, which is particularly prone to insect damage,[j] should be burned while still in the husk to improve its resistance.[k]

Although most preparatory processes such as parboiling were eschewed in

[a] (1), I. vi. 10.

[b] *Sheng Ma* section, 2/13.

[c] *Thang Liu Tien* 19; Twitchett (4), p. 191 interprets *ku* as millet and *mi* as rice, but in this context the interpretation as 'unhusked' and 'hulled' grain seems more satisfactory. *Tsa ku* probably refers to such non-cereal crops as sesame or beans.

[d] Shiba Yoshinobu (1), p. 55.

[e] In Northern Iberia stilted granaries used to store panicles of millet were known as *espighieros*, a term derived from the Latin *spicarium*, 'granary' (from which the German *Speicher* is also derived); Gomez-Tabanera (1). *Spicarium* is not a classical form; it first appears in the Salic Laws of the +6th century, and it has been suggested on etymological grounds that it referred specifically to the storage of grain in the ear, practised particularly in Germanic areas where spelt was more commonly grown than ordinary wheat; F. Sigaut (2). In Southeast Asia rice is invariably stored in the panicle in areas where the reaping-knife is still in use; where the sickle has been introduced, the rice is usually threshed before storing.

[f] *WCNS*, ch. 16; *NCCS*, ch. 23; etc.

[g] Grist (1), p. 425.

[h] *TKKW* 1/4; Sung Ying-Hsing (1), p. 81.

[i] *CMYS* 4.5.2.

[j] Pliny says this is because wheat's density makes it get hot; modern authorities refute this explanation but offer no alternative; K. D. White (2), 196.

[k] *CMYS* 10.4.5; spelt is similarly treated in Asturia; F. Sigaut (1).

1 穀 2 米 3 雜穀 4 舒璘

China, the quality of milled grain (usually rice) in government stores was usually maintained by rapid turnover. The old rice in government granaries was commonly sold off to the lower classes and replaced with fresh stores each autumn.[a] Nevertheless, grain frequently deteriorated in store. The Thang granary statutes regarded a diminution in volume of 2 % in three years or 4 % in five years as permissible,[b] but these figures seem overly optimistic. The grain was likely to be attacked by birds, rats, weevils and mildew, if not by robbers or dishonest runners. The +18th-century handbook of administrative terms *Liu Pu Chheng Yü*[1] lists a variety of forms of deterioration commonly encountered in official granaries: 'rat damage' (*shu hao*[2]), 'red rot' (*hung hsiu*[3]), 'wet rot' (*i lan*[4]) and mildew (*mei shih*[5]), not to mention adulteration of the grain with lime or chemicals by granary runners to make it expand.[c] Earlier texts refer with distaste to 'insects' (*chhung*[6]), of species usually unspecified, rodents, small birds, rotting and mildew. Rodents could sometimes be kept at bay by building granaries on stilts, while birds were quite easy to keep out if all the lights were finely latticed;[d] Chia Ssu-Hsieh recommends steaming for non-glutinous panicum and careful sun-drying for glutinous panicum and wheat as a specific against mildew.[e]

Generally, however, prevention was considered better than cure in China, and every care was taken to prevent the grain becoming overheated and damp, and thus prey to insects and micro-organisms, in the first place. The detailed attention paid to waterproofing and ventilating granary buildings is described in detail later. There are two main differences between the Chinese and the Western approach to these problems. First, whereas in Europe ventilation of granaries was usually achieved by the direct inflow and circulation of cool air from outside, the Chinese relied on the principle of convecting away the warm vapours exhaled by the grain through lanterns (*chhi lou*[7], literally 'ventilation towers') built in the roof. The convection process was often aided by an ingenious device known as a 'grain cup' (*ku chung*[8]) or 'ventilation basket' (*chhi lung*[9]). The origin of this device is not known; it is first mentioned in the *Chi Yün* of 1037 when it seems already to have been in common use. It was defined as a cylinder of plaited bamboo, perhaps 30 cm. in circumference and 6 m. long, slightly wider at the base to prevent it from falling; sometimes three or four such tubes were joined together (Fig. 192). Several *ku chung* were placed in each section of the granary to siphon off the warm vapours exhaled by the grain, thus preventing rot or mildew.[f]

[a] Shiba Yoshinobu (1), p. 57; E-Tu Zen Sun (1), items 787, 788.

[b] Twitchett (4), p. 191.

[c] E-Tu Zen Sun (1), items 757–60, 790; governments of course regarded the adulteration of their grain with disfavour, but farmers often took the opposite point of view: a French author recommends that the corn be sprinkled with 'Sal-nitrum and its scum, well powdered and mixed with very fine earth' so as to increase its bulk; Stephens & Liebault (1), p. 547.

[d] *SSTK* 57/6*a*–*b*.

[e] *CMYS* 4.5.2; 10.4.4.

[f] *WCNS* 16/21*b*.

1 六部成語 2 鼠耗 3 紅朽 4 浥爛 5 霉濕
6 蟲 7 氣樓 8 穀盅 9 氣籠

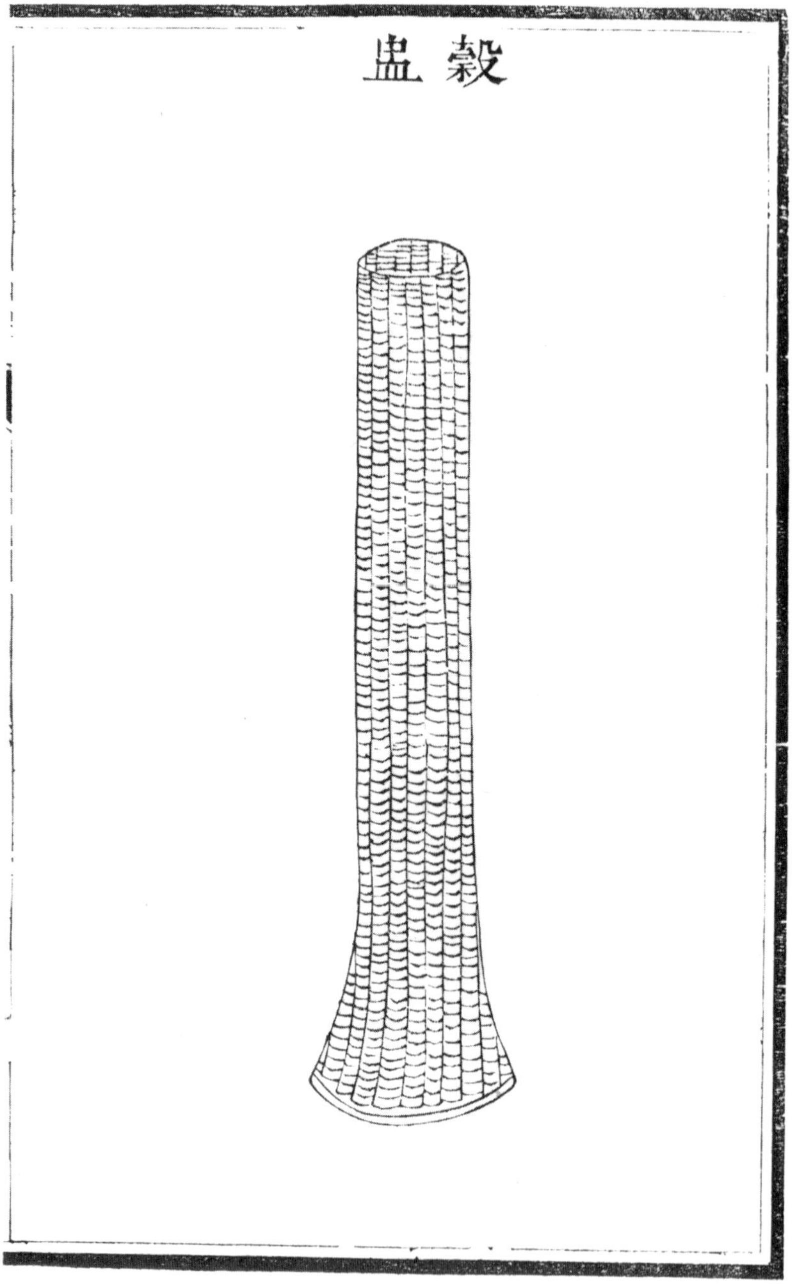


Fig. 192. Ventilation tube of wickerwork (*ku chung*); *WCNS* 16/21*a*.

Secondly, the Chinese made great use of the absorbent properties of reed or grass matting in their granaries and storage pits; the grain was never allowed to touch the bare brick or earth of the walls, but was always insulated with at least one layer of matting[a] which absorbed both the moisture exhaled by the grain and any seepage through the walls. The efficacy of matting in preserving grain carried in ships was recognised in the West,[b] but it was little used on land. Markham recommended that garners and lofts should be covered 'with a smooth layer of at least two inches of plaster inside and out', for plaster kept the grain dry and cool whereas 'boords are too hot, and clay is too apt to breed vermine', but this was perfectionism, for most European farmers simply smeared the walls with lime which, though good against weevils, was 'sharpe and so consequently very vnwholesome for all manner of Graine'.[c] It would be most interesting to compare the efficacy of grain storage in China and England, say in +1600, but unfortunately all our information is purely qualitative.

The problems involved in storing grains such as rice and millet, usually consumed whole, were less acute than those encountered in Europe, where most cereals were ground as flour for consumption, and so had to be kept very dry; this is perhaps the reason why there is no mention in the Chinese sources of the grain-drying kilns commonly used in the West;[d] sun-drying was usually sufficient. Sun-drying was also a remedy as well as a preventative: if grain in the Chinese government granaries was attacked by mildew it was washed, steamed and then dried again in the sun.[e]

Every culture has developed its own complex of insect repellents. The Serer of Senegal line baskets with a special aromatic leaf to protect their millet;[f] modern Nigerians use chili peppers;[g] the Romans laid their faith in coating the walls and floors of their barns with olive lees (*amurca*), with a special earth found at Olynthus or Corinthus in Euboea, or with common wormwood;[h] the French liked to protect their corn by threshing it on a floor 'watered with vineger, the walls ... dressed ouer with mortar tempered in water, wherein hath been steeped the roots and leaues of wild Cucumber: or with Lime tempered with Sheepes Vrine, which shall be of much vse against all kind of shrewd beasts that vse to eat the corne.'[i] Once the corn was threshed and laid in the garner, they protected it from weevils and other vermin with 'wild Organie or the dried leaves of Pomegranat-trees, or Wormewood, or drie Southernwood', or best of all by adding one part of clean millet to ten parts of wheat, for the millet was believed to possess a coolness that deterred insects, and could easily be separated from the wheat by sieving.[j]

[a] *NCCS* 27/8*a*–*b*; *SSTK* 57/5*b* ff.

[b] E.g. G. Markham (1), p. 134.

[c] *Ibid.* pp. 107 ff., p. 113.

[d] Markham (1), p. 103; Stephens & Liebault (1), p. 546.

[e] E-Tu Zen Sun (1), item 461; Markham recommends similar treatment for wheat that has been attacked by mildew, smut or blast; (1), p. 104.

[f] De Garine (1).

[g] K. D. White (2), p. 197.

[h] Varro (1), vol. I, lvii, 2; Columella (1), vol. I, vi, 14.

[i] Stephens & Liebault (1), p. 547.

[j] *Ibid.*

Markham tells us that the lees of sweet olive oil were still used in his time in France and Spain: not only did they kill weevils, but 'if corne by casualty be tainted or hurt, it doth recouer it againe, and brings it to the first sweetnesse'; he also mentions the use of powdered chalk to absorb excess moisture, and of wormwood.[a]

The common property of most of these specifics is their strong scent. Olive lees have 'an exceedingly strong and nauseous odour',[b] peppers are notoriously pungent, and oregano is strongly aromatic. The Chinese generally relied on the use of artemisia species (*ai*[1] or *hao*[2]), all strongly scented, as insect repellents in granaries. The vermifugal qualities of artemisia when taken internally are clearly reflected in the common English name for *Artemisia vulgaris*, 'wormwood'; its bitter taste is famous as the flavouring of absinthe, and in China, apart from its medicinal uses, it is commonly employed as an ingredient in mosquito coils.[c] Apart from their aromatic qualities the artemisias have the advantage of growing widely throughout China, and of having flexible stems well suited to basketry. The earliest mention of their use as an insect repellent in grain stores is in the −1st century, when Fan Sheng-Chih recommends adding a handful of dried *ai* to each bushel of wheat in store;[d] Chia Ssu-Hsieh recommends that wheat should be stored in baskets woven from artemisia,[e] and bins and baskets for grain storage continued to be made from artemisia for centuries afterwards.[f]

Apart from the depredations of mildews and insects, another hazard to be guarded against in granaries was fire. The danger of combustion was very low if grain was stored in airtight pits, but much higher if it was kept in wooden buildings. Peasants in China often plastered their granaries with clay and lime to protect them from fire;[g] on a grander scale the merchants of Sung Hangchou built their warehouses in the form of square stone towers called *tha fang*[3], surrounded on every side by water.[h] As a precaution the use of lamps with open flames was strictly forbidden in all large granaries; instead small brushwood stoves were used.[i]

## (iii) *Storage facilities*

### *Baskets, earthenware jars, and wooden containers*

These types of grain container are all of very ancient use. They have the advantage of being cheap, easy to construct and readily portable; they are suitable for grain in the ear, in the husk, milled or even ground to flour, but they are

[a] Markham (1), p. 110.
[b] K. D. White (2), p. 197.
[c] Burkill (1), vol. I, p. 244.
[d] Quoted *CMYS* 2.8.2
[e] *CMYS* 10.4.4.
[f] *WCNS* 15/18*b*.
[g] *WCNS* 16/16*a*.
[h] Moule (11), also (5) and (15); see also Vol. IV, pt 3, p. 90.
[i] *SSTK* 57/6*b*.

[1] 艾 [2] 蒿 [3] 塌坊

not suitable for storing large quantities, nor will the grain keep for very long periods.

The most ancient of all containers for grain is the basket. Impressions of basketwork are found in potsherds from the very earliest neolithic sites, and although no traces of palaeolithic or mesolithic basketry have survived, the craft certainly predates the development of pottery and agriculture, for the variety and sophistication of basketry techniques that can be seen in neolithic impressions points to a craft already of considerable antiquity.[a] Certainly baskets are an essential item in all modern hunter-gatherer societies. The many references to baskets in the Chinese agricultural encyclopaedias show that they were probably the most widely used form of grain storage in peasant households throughout China, as in many other parts of the world.[b]

Several different types of basket are mentioned in the *Book of Odes*, in particular the *khuang*[1] and *chü*[2],[c] and definitions of most of the basket types illustrated in the later agricultural works are given in Han dictionaries such as the *Fang Yen* and *Shuo Wen Chieh Tzu*, together with local variations in orthography and pronunciation which point to a considerable antiquity for most forms. One of the earliest pictorial representations of baskets in China is a charming relief on a bronze drum from Shih-chai-shan[3] in Yunnan, depicting files of villagers filling their granaries with the contents of large hemispherical baskets carried on the head, and bearing away smaller quantities in spherical baskets similar to those later used for seed grain (Fig. 196).[d] The surviving pre-Sung agricultural works are vague on the subject of baskets, or indeed any other type of grain storage, but the *Chhi Min Yao Shu* states that:[e] 'Rice must be stored in baskets, because it is a water cereal (*shui ku*[4]), and if it is buried in pits it will be affected by the exhalations (*chhi*[5]) of the soil and rot.'[f] The lack of other references to baskets in *Chhi Min Yao Shu* might be thought to indicate that most North Chinese peasants of that period preferred to keep their grain in pits; more likely, however, it reflects the conditions on large estates where the amounts involved were much greater than those kept by a normal peasant family. The number of dialectal terms for storage baskets which are said by the early dictionaries to come from Chhi[6][g] and other parts of North China seem to prove that baskets were as commonly used then as they were in the Sung dynasty and later.

The first agricultural work to devote a section to grain storage is Wang Chen's

[a] See Section 31, and D. Kuhn (1).

[b] E.g. in Hungary, Füzes (1), and Senegal, de Garine (1).

[c] *Shih Ching* 2/10*b*; Legge (8), p. 25.

[d] See p. 390 below.

[e] *CMYS* 11.7.1.

[f] The *chhi* of the soil was humid, and the reaction between likes, the humid element earth and the water cereal rice, was sure to lead to disaster.

[g] Shantung.

[1] 筐 [2] 筥 [3] 石寨山 [4] 水穀 [5] 氣
[6] 齊

*Nung Shu* of +1313,[a] which gives a comprehensive account of the various types of basket in use in medieval China. Most of the terms still in use are already found in the Han dictionaries. The descriptions given in the *Nung Shu* are mostly based on the definitions given in the Sung dictionary *Chi Yün*, to which Wang Chen appends a catchy jingle for each type:[b]

> Our present basket shapes are ages old,
> Square *khuang*[1] and round *chü*[2] still remain the same.
> They can be used for cresses from the stream,
> Or else to store the golden millet grain.

The later encyclopaedias, *Nung Cheng Chhüan Shu* and *Shou Shih Thung Khao* for example, include Wang's prose descriptions but omit the verse.

The most popular storage baskets in China were the large type known as *thun*[3], or sometimes as *shuan*[4] or *chu*[5].[c] Wang Chen says that *thun* are often placed out of doors and used to store food grain, while *shuan* and *chu* are used indoors to store seed grain; but it is unlikely that the distinction was so rigidly defined, as in several works the drawings of *thun*, *shuan* and *chu* are used interchangeably. The *Chi Yün* does, however, mention a more important structural difference: in the North *thun* are round and made of woven withies (*ching liu*[6]) or artemisia,[d] while in the South they are made of bamboo wicker-work, or else of coarse bamboo matting rolled into a cylindrical container.[e] The illustrations (see Fig. 193) often show the *thun* as having a square base and a round top, a particularly strong and stable construction still much used in the Far East today (Fig. 194). Although this shape is not specifically ascribed to the *thun* in the text, it is applied to another type of basket called *lo*[7], which strangely enough always appears with a circular base in the illustrations.

Wang Chen describes the *thun* as indispensable in the peasant household for storing grain in the husk. Other types of basket used to store grain in the husk are the *lo* just mentioned, the *thiao*[8], and the *khuei*[9], straw or wicker baskets of round and cylindrical form respectively, and smaller in size than the *thun*. None of these baskets is ascribed any particular capacity; however, the baskets used to hold milled grain were usually described as containing so many pints or bushels. The *khuang*[1] referred to in the jingle was a square lidless basket used to hold a variety of goods; those used for milled grain came in two sizes, the larger holding 5 *hu*[10]

[a] *WCNS*, ch. 15, passim.

[b] *WCNS* 15/16*a*; both *khuang* and *chü* are mentioned in the *Shih Ching* (see p. 387 n. c above) as being used to hold a variety of objects including water-cresses and millet.

[c] In Han times these terms were also applied to earthenware bins, see p. 392 below.

[d] As well as being a very common plant in Northern China, artemisia has, as we have seen, the additional virtue of repelling insects.

[e] A very similar type of wickerwork container, the *cista*, was used in Rome, but principally for storing fruit; K. D. White (3), pp. 63 ff.

1 筐 2 筥 3 笹 4 篅 5 䉷
6 荆柳 7 籮 8 篠 9 蕢 10 斛

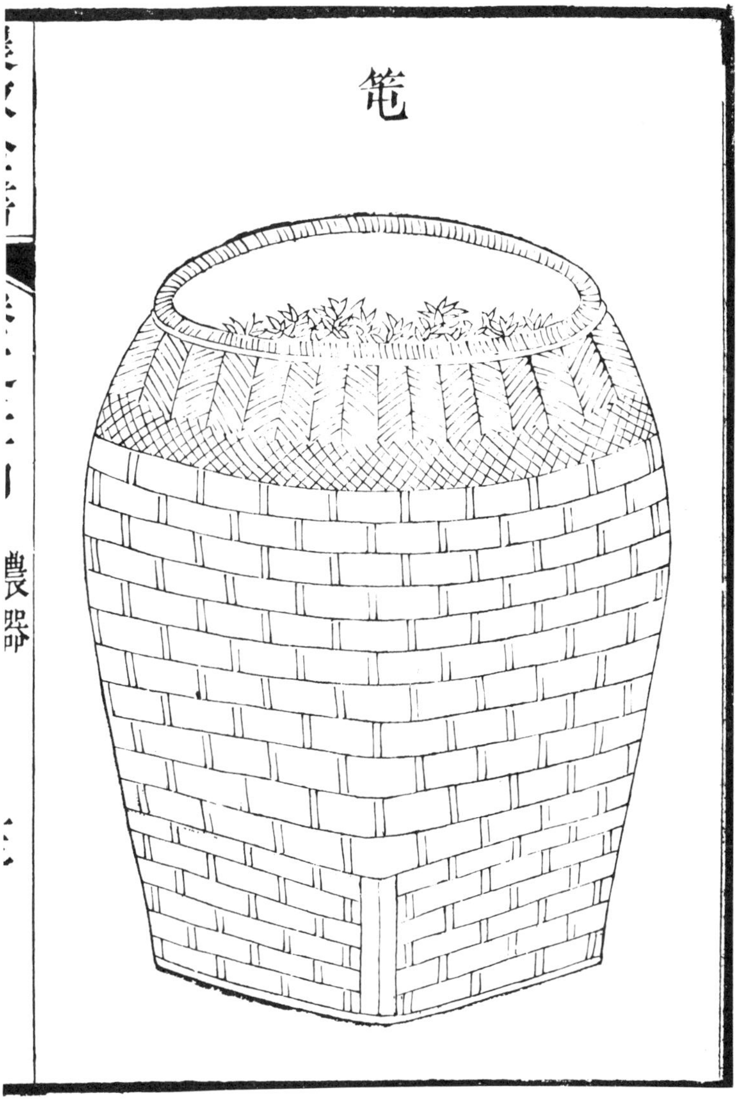


Fig. 193. Storage basket (*thun*), with square base and round top; *NCCS* 24/13*a*.

(very approximately 50 gallons) and the smaller 5 *sheng*[1] (about 10 gallons in Yuan times). According to Wang Chen, *khuang* were very common in peasant households, and were used not only for storing grain but also for conveying wedding gifts of food; this ceremonial purpose perhaps accounts for the rather elaborate shape of such baskets in later illustrations.[a] The *chü*,[2] also mentioned in the *Book of Odes*, by Sung times was defined as an oval bamboo basket containing

[a] *WCNS* 15/14*b*–15*a*; *NCCS* 24/11*a*.

[1] 升 [2] 筥 or 簴

Fig. 194. Similar construction in a modern basket; *Eastern Horizon* (1978), XVII, *3*, p. 27.

5 *sheng*; the *lo* could be used to hold milled grain as well as grain in the husk, and had a capacity of 1 *hu*.

A final category of basket is the type used to hold seed grain. Fan Sheng-Chih recommends that wheat seed should be kept in the ear, either in earthenware or bamboo containers, well mixed with artemisia,[a] but does not describe the baskets used. Wang Chen mentions a type of bamboo basket called *chung tan*[1] as being specifically used for storing seed grain; the basket is 'shaped like a round jar with a closely woven lid'.[b] The accompanying illustration shows the basket first with the stems of the grain protruding, and then with the lid closed, a clear proof of the fact that grain was stored in the ear (Fig. 195). The *chung tan* is obviously the same type of basket as was used in the Tien culture in Yunnan before −200, for the bronze drum from Shih-chai-shan[2] shows exactly the same type of basket, again with the stems of grain protruding (Fig. 196).[c] Wang Chen says that the *chung tan* holds several *tou*[3] (a *tou* in Yuan times was about 2 gallons), and since it is airy the

[a] *CMYS* 2.8.2. [b] *WCNS* 15/35*b*.
[c] Anon. (*28*).

[1] 種簞 [2] 石寨山 [3] 斗

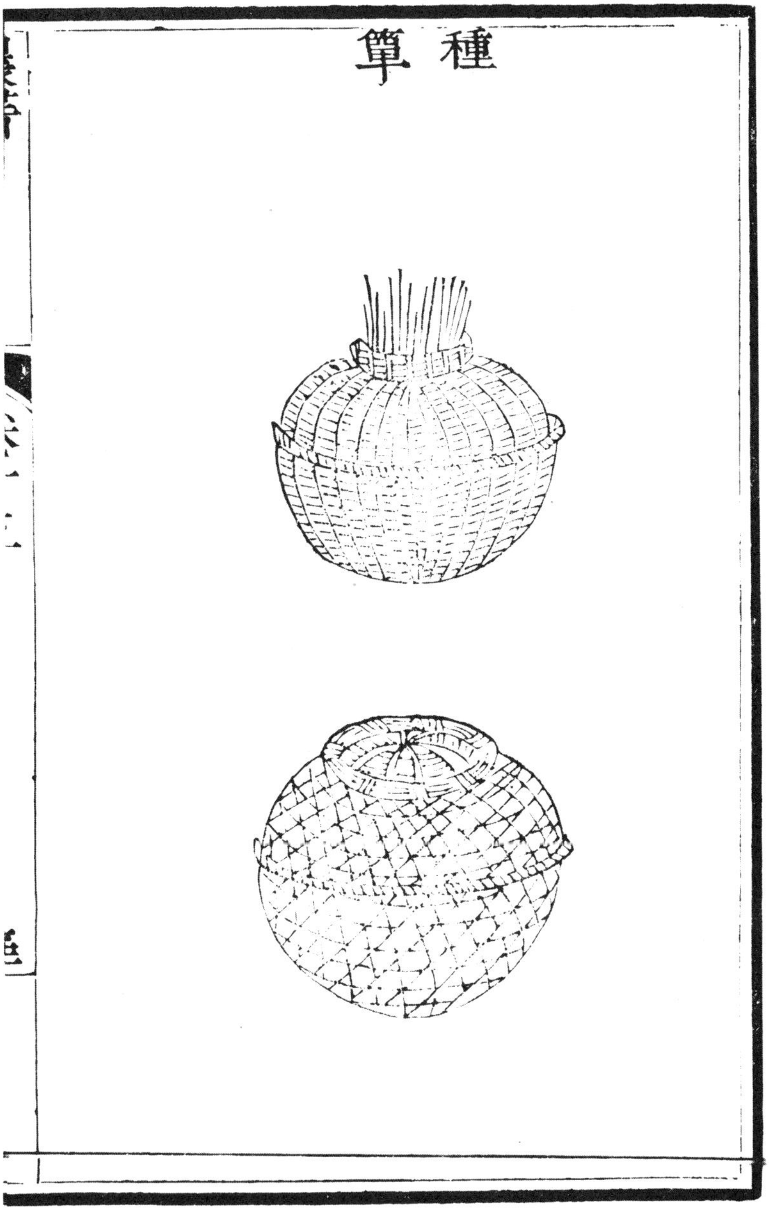


Fig. 195. Seed-grain container (*chung tan*); the upper drawing shows the stems of the seed-grain, stored in the ear, protruding; *WCNS* 15/31*a*.

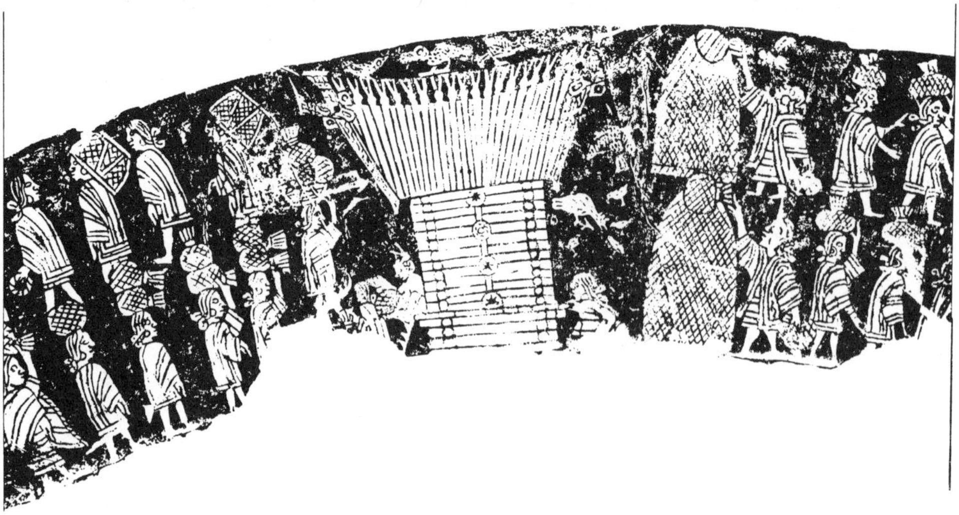

Fig. 196. Stilted granaries of the Tien culture, being filled with bundles of grain in the ear, carried in small, round baskets; bronze drum decoration from Shih-chai-shan, Yunnan; Anon. (*28*), pl. 21.

seed grain will not become damp or rot; for this reason it is preferable to pit storage.[a]

Earthenware jars are another extremely ancient form of storage, more versatile than baskets in that they can be used for liquids as well as solids. Large storage pots of a variety of shapes are found in most neolithic sites in China, and some of them were probably used for storing grain, as remains or impressions are often found in them.[b] Modern archaeologists usually classify these large jars as *kuan*[1], but in Han times they were known by a variety of names. In the central area the most common term was *ying*[2], but further south in Honan and Anhwei they were known as *yu*[3], while in Shantung the common term for the smaller size, holding only 2 *hu*[4] (approximately 9 gallons in the Later Han), was *tan*[5]; in the Yangtze area they were known as *yü*[6] or *shu*[7], while the North Koreans called them *chhang*[8].[c] The Shantung word *tan* later came to be used as the standard term, and Wang Chen says that in Yuan times *tan* were very commonly used in the Yangtze and Huai areas, especially by poorer folk, for storing hulled grain.[d] Although the illustrations in the agricultural encyclopaedias all show a standard storage jar barely flared in shape (Fig. 197), the Han description of jars with wide bellies and

a *WCNS* 15/35*b*.
b K. C. Chang (1), p. 95; Anon. (*501*); Wu Shan-Chhing (*1*); Anon. (*503*).
c See the quotations from *Fang Yen*, *Erh Ya*, etc. in *WCNS* 15/23*a*.
d *WCNS* 15/22*b*–23*a*.

1 罐 2 罌 or 甖 3 缶 4 斛 5 儋 or 甔
6 瓵 7 瓶 8 瓺

Fig. 197. Earthenware storage jar (*tan*); *WCNS* 15/22*b*.

narrow mouths sounds closer to the neolithic Yang-shao model (Fig. 198), but there was undoubtedly always a wide range of shapes and sizes.[a]

In Han times cylindrical earthenware bins for storing grain were also common, and several fine examples have survived (Fig. 199).[b] They were called *thun*[1] or *shuan*[2] like the basketwork bins, and were used to store hulled grain. The *Huai Nan Tzu* describes them as having a large opening at the top for pouring the grain in, and a small hole at the base for taking it out;[c] the illustration shows the lid on the top and the bunged small hole, similar to the spigot-hole in a beer barrel, very clearly. These earthenware *thun* were often inscribed with such phrases as 'ten thousand bushels of millet' (*su wan shih*[3]) or 'ten thousand bushels of wheat' (*hsiao mai wan shih*[4]): not, of course, an indication of their capacity but an invocation of good harvests; the Chhing scholar Tuan Yü-Tshai[5] (+1735 to 1815) mentions that bins of this type with a capacity of several bushels had been reported in Kiangsu province.[d] These earthenware bins were apparently little used after the Han, and there are no references to them in the agricultural encyclopaedias.

Another form of grain storage much used in Europe until very recently was the wooden chest or vat. It was easily obtained, durable, and (perhaps its greatest advantage) mobile. Markham mentions that in Ireland and 'in other Countries where warre rageth' grain was often stored in the ear in wooden chests, byngs or hutches,[e] while for convenience in transport brine barrels well caulked with plaster were used.[f] The Chinese made far less use of wooden containers than the Europeans, principally because wood was a scarce and valuable commodity in most regions, but one ingenious exception called 'grain drawers' (*ku hsia*[6]) is illustrated in the *Nung Shu* (Fig. 200). It consisted of a square framework of four posts, mounted on a raised platform and covered with a roof; a series of wooden drawers was inserted in the frame, and the grain was stored inside. It could be used indoors or out (in the latter case the roof was tiled), and had the advantage of being more mobile than a granary, more capacious than wicker baskets, and resistant to both mildew and rats.[g] It sounds rather similar in conception to the little granaries on rollers used by Hungarian shepherds well into the +20th century, which could be brought inside in winter or quickly moved if fire broke out.[h]

*Subterranean pits*

In the dry loess lands of Northern China, pits were dug for storage in earliest neolithic times, and have remained one of the most important methods of storing

[a] But Laufer (3), pl. XII, fig. 1 shows a Han jar remarkably similar to the +14th-century *tan*.
[b] Hayashi (*4*), pl. 3 and figs. 4–15.
[c] *Huai Nan Tzu* 7/9*b*.
[d] Hayashi (*4*), p. 164.
[e] (1), p. 106.
[f] *Ibid.* p. 135.
[g] *WCNS* 15/19*b*–20*a*.
[h] Füzes (1), p. 591.

[1] 笹 [2] 篅 [3] 粟萬石 [4] 小麥萬石 [5] 段玉裁
[6] 穀匣

Fig. 198. Yangshao storage jar; Medley (3), fig. 8; Stockholm, Museum of Far Eastern Antiquities.

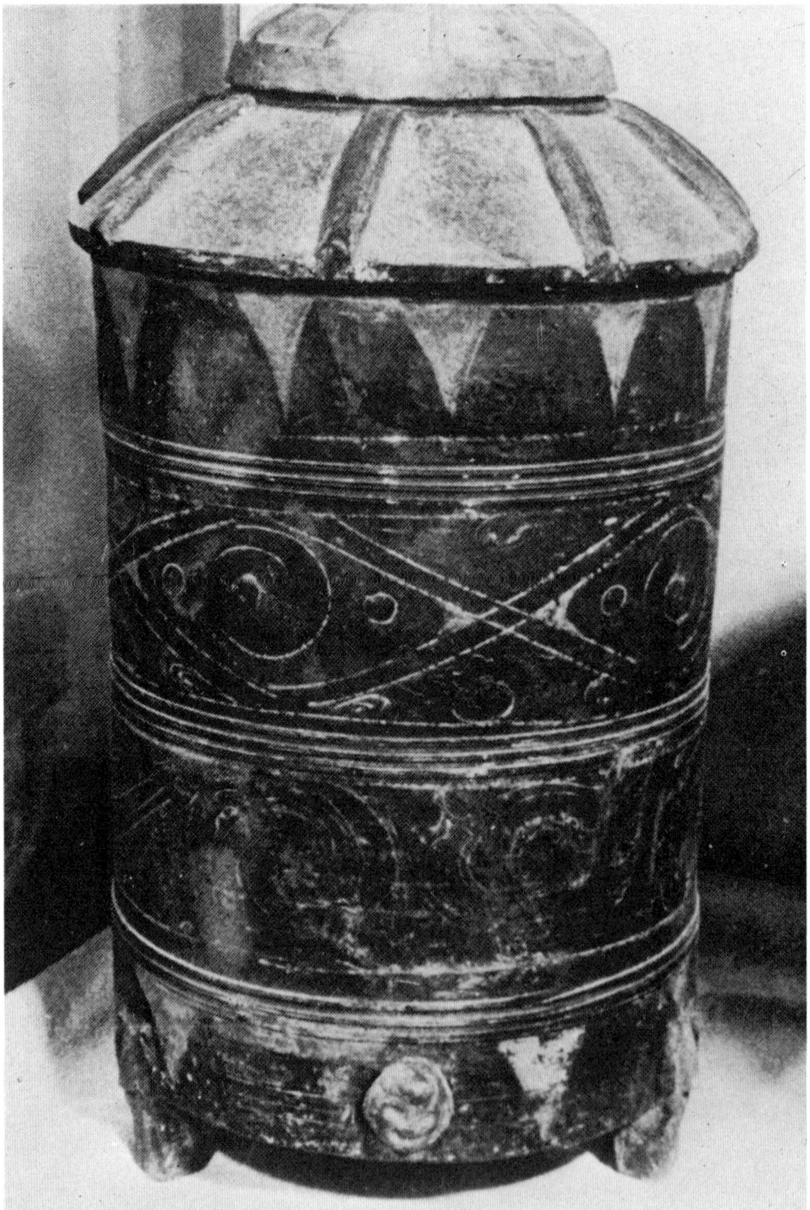

Fig. 199. Han earthenware storage bin; Hayashi (*4*), pl. 3; Shensi Provincial Museum.

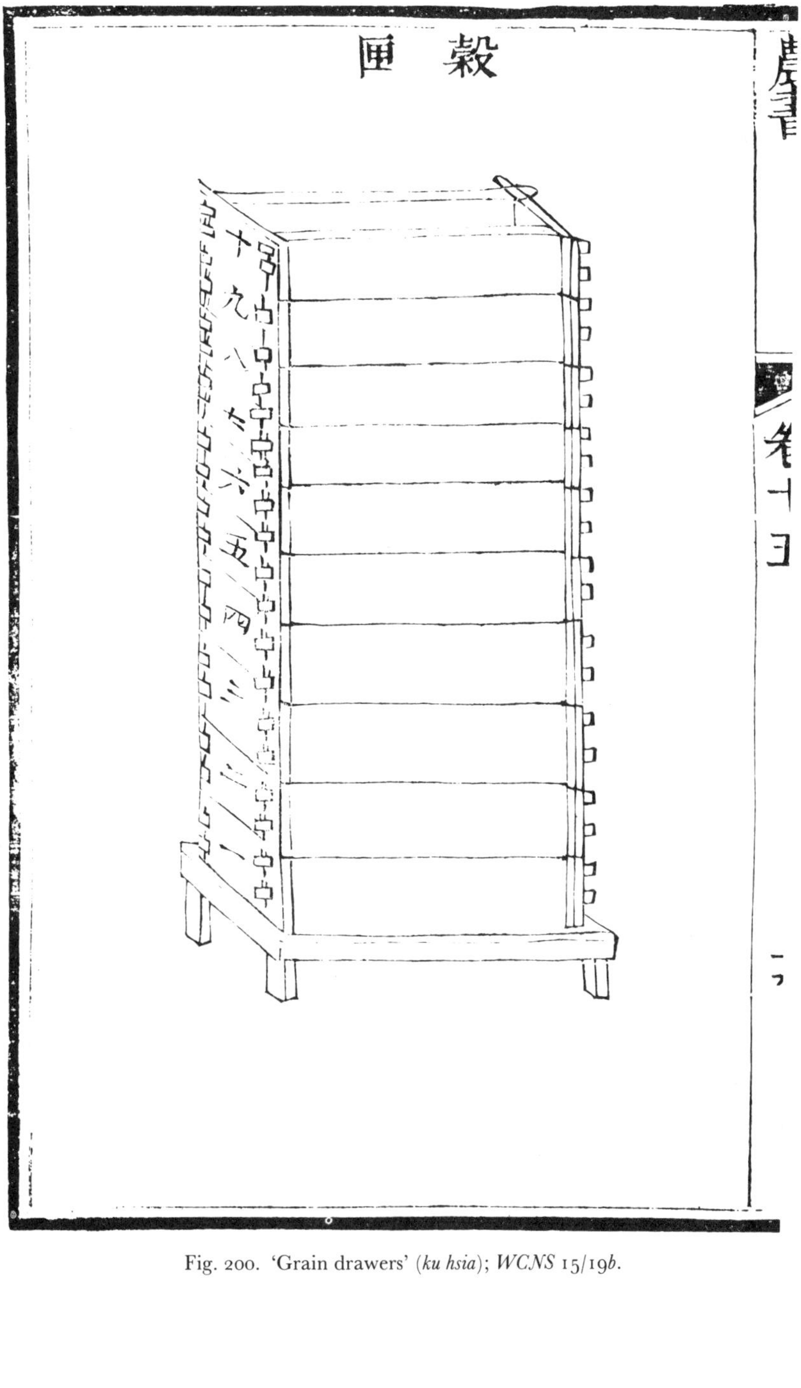


Fig. 200. 'Grain drawers' (*ku hsia*); *WCNS* 15/19*b*.

grain for millennia; indeed they are still found in some provinces today. Pits are a very effective way to store grain, providing the soil is sufficiently porous or the pit is lined to make it watertight, and the system has recently been revived in modern wheat-producing countries under the name of 'silo'.

All but the most primitive forms of pit storage work on the principle of the airtight container. The pit is filled with grain and tightly sealed, the carbon dioxide emitted by the grain kills any insects and larvae, and the grain cannot become overheated.[a] Modern silos are usually lined with concrete or even welded metal, but the simple fired earth pit was sufficiently effective to survive for millennia in a number of countries.

Pits for grain storage are found all over the world in dry areas. Varro mentions the use of underground granaries, *siri*, in Cappadocia, Thrace and North Africa,[b] and Pliny claims that in a *sirus* wheat will keep for fifty years and millet for one hundred.[c] Pits have always been used for grain throughout the loessial areas of Central Europe. Markham, who had come across them in the Azores, hesitated to recommend their use in England except in hot, high, sandy or gravelly areas such as Norfolk, Middlesex or Kent, but claimed that the method would be effective even in clayey areas if the pit was well lined with a six-inch layer of tiles and three inches of plaster.[d] In fact unlined pits were used in Britain in the Iron Age, on the chalky uplands of the South Downs.[e]

In North China linings of tiles and plaster were quite unnecessary. The earliest pits found at Pan-pho[1] village in Shensi date back to nearly −5000; one section of the earliest level at the site contained 22 houses with 43 storage pits, cup-shaped and rather shallow, which presumably were covered with straw or a lid.[f] By the Lungshanoid period the pits had been elaborated, the characteristic form being 'pocket-shaped ... with a bottleneck opening and an enlarged chamber about 4 meters in diameter.'[g] This is exactly similar to the grain pits found in Czechoslovakia up till the Second World War (Fig. 201). In Shang times the shape of storage pits changed again; an excavation at An-yang[2] revealed pits dug in the floors of semi-subterranean houses, some round, some rectangular, and mostly several meters deep.[h] Similar forms are also found at Chou sites.[i] Thus when the Han scholar Cheng Hsüan[3] (+127 to 200) glossed the terms used in the *Li Chi Yüeh Ling*, *tou*[4] and *chiao*[5], respectively as 'round' and 'square pits', he was not merely indulging in the Chinese scholar's passion for contrasting categories.[j] Both round and square pits actually existed, but whether the names really fitted quite so neatly is difficult to say; Wang Chen later understood the difference

[a] Grist (1), p. 403.
[b] Varro (1), vol. I, lvii, 2.
[c] (1), XVIII, 306.
[d] Markham (1), p. 112.
[e] P. Reynolds (1).
[f] Vol. IV, pt 3, p. 121; K. C. Chang (1), p. 99.
[g] K. C. Chang (1), p. 174.
[h] K. C. Chang (1), p. 247.
[i] *Ibid.* p. 300.
[j] On the dichotomy between round and square in Chinese architecture see Vol. IV, pt 3, p. 122.

[1] 半坡 [2] 安陽 [3] 鄭玄 [4] 竇 [5] 窖

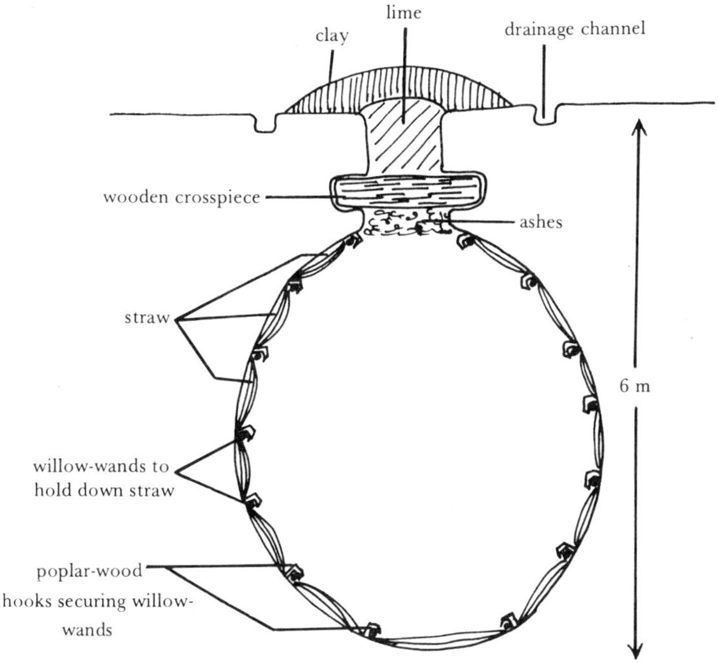


Fig. 201. Cross-section of a grain-pit as constructed in the loessial regions of Central Europe; based on Kunz (1), Füzes (2).

between *chiao* and *tou* to be that while the former were dug vertically into the ground and had quite large openings, the latter were bottle-shaped like the Lungshanoid pits, and were often dug sideways into banks (Fig. 202). Wang described grain pits as being suitable chiefly to the North of China, but said that they were also found in certain elevated spots in the Huai and Yangtze areas, especially where the topsoil was deep.[a]

Although by Yuan times the use of grain pits seems to have been confined mainly to village level, in earlier times many rich grain merchants stored enormous quantities of grain in pits. The *Shih Chi* refers to one Master Jen[1], who made a fortune by hoarding grain in pits during the civil war that brought about the fall of the Chhin dynasty, selling it for hundreds of thousands of cash the bushel when all other supplies were exhausted.[b] Pits were also widely used in official granaries. Although they are not specifically mentioned in the Han histories, the *Yü Hai* tells us that the granaries built by the Sui emperor Yang-Ti[2] (+604–617) all consisted of pits,[c] and in the chief Thang granaries the grain was also kept in storage pits, sometimes of huge dimensions; by the Thien-pao[3] period (+742–756) the pits of

[a] *WCNS* 16/22*a*–24*a*.
[b] *Shih Chi*, ch. 129.
[c] Quoted *SSTK* 54/8*b*.

[1] 任氏 [2] 煬帝 [3] 天寶

Fig. 202. Storage pit dug in a bank of earth; *WCNS* 16/22*a*.

the Han-chia[1] granary at Loyang had a total capacity of six million *shih*[2].[a] But subsequently, as the Chinese state shifted its economic dependence from the northern-produced millets to rice from the Yangtze valley, pit storage declined in importance, for rice unlike millet did not keep well underground.

Excavations carried out at the Thang imperial granary of Han-chia not surprisingly show a higher level of technical sophistication than Wang Chen mentions.[b] 259 pits have been found so far in the Han-chia granary, varying in size from 8 m. across by 6 m. deep, to 18 m. across by 12 m. deep. They are set in rows, about 5 m. apart, over an area of just under 1 km. square. As each pit was filled with grain, an inscribed brick was also placed inside, identifying the granary, the location of the pit, the year, district and type of tax, the amount of grain, the date of deposit, and the name of the officials concerned, a witness to the intricacy of the accounting system. The pits are cup-shaped, smoothly dug, with a floor of tamped earth 7 to 30 cm. thick; the surface was fired to make it waterproof, and some pits had an additional impermeable layer of a mixture of fired clay sherds, charcoal and cinders. The sides and bottom were then lined with wooden boards, many of which still survive, and covered with grass or rush matting; after the pit had been filled it was covered with a lid which seems to have consisted of a spoked wooden frame covered with matting and bundles of grass; the whole was finally sealed with plaster. This elaborate lining was necessary to keep the grain cool and dry: the wood and matting provide a layer between the grain and the non-porous wall that absorbs the moisture and heat exhaled by the grain and thus prevents rotting. A similar use of absorbent matting is found in the grain pits of Fayum, dating from between −4800 and −4000, which are lined with coiled, woven corn-straw,[c] and in the twentieth-century grain pits of Central Europe.[d]

Wang Chen, in his account of village pit storage, merely mentions lining the pits with chaff before filling them with grain.[e] This simple lining may have been very effective, however:[f]

> ... your great Corn-masters and hoarders of corn, when they want roome to lay their Corne in, will thresh up their oldest store, and then keepe it in the chaffe till they have occasion to vse it ... whilest it lyes therein, it will euer keep sweet; and it is a most certaine rule: for nothing is a greater preseruer of Corne than the owne chaffe, except it be the eare it selfe.

Certainly Wang claims that millet stored in this fashion would keep for many years, but he says that pits are not suitable for the storage of any other cereals than millets,[g] and this is borne out by the earlier *Chhi Min Yao Shu*, which insists

[a] Anon. (*525*); Loyang Museum (*1*).
[b] Anon. (*525*); the excavations were begun in 1971.
[c] Caton-Thompson & Gardner (1).
[d] Kunz (1). Füzes (1), (2).
[e] *WCNS* 16/22*b*.
[f] Stephens & Liebault (1), p. 546.
[g] *WCNS* 16/22*b*.

1 含嘉 2 石

that rice must be kept in baskets, while wheat and barley are better kept in artemisia bins.[a] However, Chia does recommend that all seed-grain except rice should be kept in pits rather than jars,[b] a practice borne out by modern experiments at Butser Iron Age Farm in Hampshire, where germination rates of over 90 % were achieved by storing the seed in unlined pits.[c] Chia also recommends the use of pits for storing grapes, pears and other fresh fruit over the winter months.[d]

In the +6th century Chia Ssu-Hsieh regarded pits as the most efficient way to store cereals, but by the +14th century Wang Chen recommends them primarily as a means of concealing one's stores from the eyes of robbers or marauding troops: if one seals the mouth of the pit with a mott of turf, then a stranger will never notice it; better still, plant a small tree over the pit and this will not only conceal it but also indicate the state of the grain underneath, for should the grain start to rot the trees leaves will turn yellow and wither.[e] The decline in importance of pit storage was the result of the shift from millet to other cereals; even in the Thang dynasty many people preferred to eat rice rather than millet, and by Sung times wheat had replaced millet in many areas of North China. These crops did not keep well under ground, and it was only in the arid Northwest, where the poorer peasants still depended on millet for their livelihood, that pit storage survived.

*Storage buildings*

'Granaries (*tshang*[1]) are stores for grain ... the state grain stores have lanterns (*chhi lou*[2]) above for ventilation ... and a verandah at the front; the buildings in which farmers store their grain may be on a smaller scale, but the terms and the principles of construction are the same.'[f] Despite the immense volume of grain kept in government granaries in China, the greater proportion was kept in private stores, ranging from the huge go-downs of the Yangtze merchants to the simple straw granaries of poor farming families. Although Wang Chen seems to imply that the small peasant barns had taken the great government granaries as their model, in fact the reverse is true: the largest constructions were just an elaboration of the simple forms found in the countryside since time immemorial. Wang Chen goes on to say:[g]

There is a difference between round and square granaries. In North China the land is high-lying and dry, so [it is possible to] plant wooden stakes directly in the ground and to plait wicker-work round them to make grain bins. This technique results in circular structures known as *chün*[3]. But in South China the land is low-lying and damp, so [it is

a *CMYS* 10.4.4, 11.7.1.
b *CMYS* 2.3.3.
c P. Reynolds (1).
d *CMYS* 34.27.1; 37.8.1.
e *WCNS* 16/22*b*.
f *WCNS* 16/16*a*.
g *WCNS* 16/20*b*.

1 倉 2 氣樓 3 囷

Fig. 203. Round granary of coiled straw, in a North Chinese village; JN orig. photo.

necessary to] join planks raised from the ground to make an enclosed structure. This technique results in rectangular structures known as *ching*[1].

The geographical boundaries of round and square granaries are not nearly as rigidly defined as Wang Chen suggests, but the distinction in terms of form and material is perfectly valid, and worthy of elaboration.

*Round granaries*

It is certainly true that round granaries are more common in Northern than in Central or Southern China, that they are usually made of straw or wicker, and that they are often built directly on the ground. Although the earliest definitions and models of such structures are no older than the Han dynasty, the form itself is undoubtedly very ancient, and it has survived with few modifications up to the present day (Fig. 203).

One might perhaps think of the most ancient form of the granary as a simple stack on the ground. Although this is a very inefficient form of storage (the dangers of losses to birds, rats and other vermin are obvious), it is extremely undemanding to make, and this charm has ensured its long survival. Markham, deploring its use in +17th-century England, exhorted his readers at least to raise the stack on a platform out of the reach of rats, and to turn the ears of the corn

[1] 京

inwards so that they would suffer less damage from wind and rain.[a] In ancient China stacks of millet were known as *yü*[1], a term first used in the *Book of Odes*,[b] speaking of abundant harvests where 'myriads of stacks stand in line' or 'stand up like eyots and hills'. These *yü* may simply have been stooks of grain drying out before being stored in more permanent granaries, for the *Odes* also refer to granary buildings, *tshang*[2] and *lin*[3]; certainly stacks were a common form of short-term storage before threshing and milling in Yuan times and later.[c] Although Wang Chen simply quotes the *Odes* and the Han glosses on *yü* and appends a little poem, his illustrator shows conical lids of coiled basketwork or straw thatch that were presumably used to protect such stacks.[d]

The next step up from a stack was the round, roofed wickerwork bin called *chün*[4]. As Wang Chen says, *chün* were usually constructed on a pole framework directly on the ground, though the modern version shown in Fig. 203 is raised on a stone and cement platform. The framework is covered with matting or coiled straw, which is then plastered with mud and covered with a roof of thatch or more coiled straw.[e] The door, made of matting or boards, is high off the ground (Fig. 204). These granaries have always been extremely common throughout North China. In Han times they seem to have been built in a more solid form, roofed with tiles and with walls presumably of bricks or clay, for pottery tomb models of such buildings have been found,[f] but there is no mention of tiled round granaries in the later agricultural encyclopaedias. Several Han models of round granaries from South China, raised on stilts, survive; these also seem to have been made mostly from straw or matting.[g]

Round granaries are cheap and easy to build, though they do not provide very efficient protection against rodents or insects. But their structure, being circular, is such that little elaboration or enlargement is possible, and they remain essentially a peasant form of storage. Only a square frame construction is sufficiently flexible to be adapted to large-scale storage.

*Square granaries*

The generic term for granary, *tshang*[2], is also used in geometry to describe that basic solid, a parallelopiped with no square faces,[h] an indication of the fundamental importance of rectangular granaries. With a square timber frame as the basis for construction, granaries of any size or shape can be built, and even as early as the Han dynasty there is a marked contrast between a peasant's small storehouse and the huge and elaborate granaries of the rich.

The term usually given to the rectangular equivalent of the round straw gra-

[a] (1), p. 83.
[b] *Shih Ching* 8/20*a*, 20/37*a*; Legge (8), pp. 157, 368.
[c] *KCT* 1/16*b*–17*a*.
[d] *WCNS* 16/18*a*–*b*.
[e] *WCNS* 16/19*a*–*b*; *SSTK* 57/9*b*–10*a*.
[f] Laufer (3), fig. 12.
[g] Hayashi (*4*), figs. 4–14; Anon. (160), pl. 104; Anon. (*42*), pls. 57–60.
[h] Vol. III, p. 98.

[1] 庾 [2] 倉 [3] 廩 [4] 囷

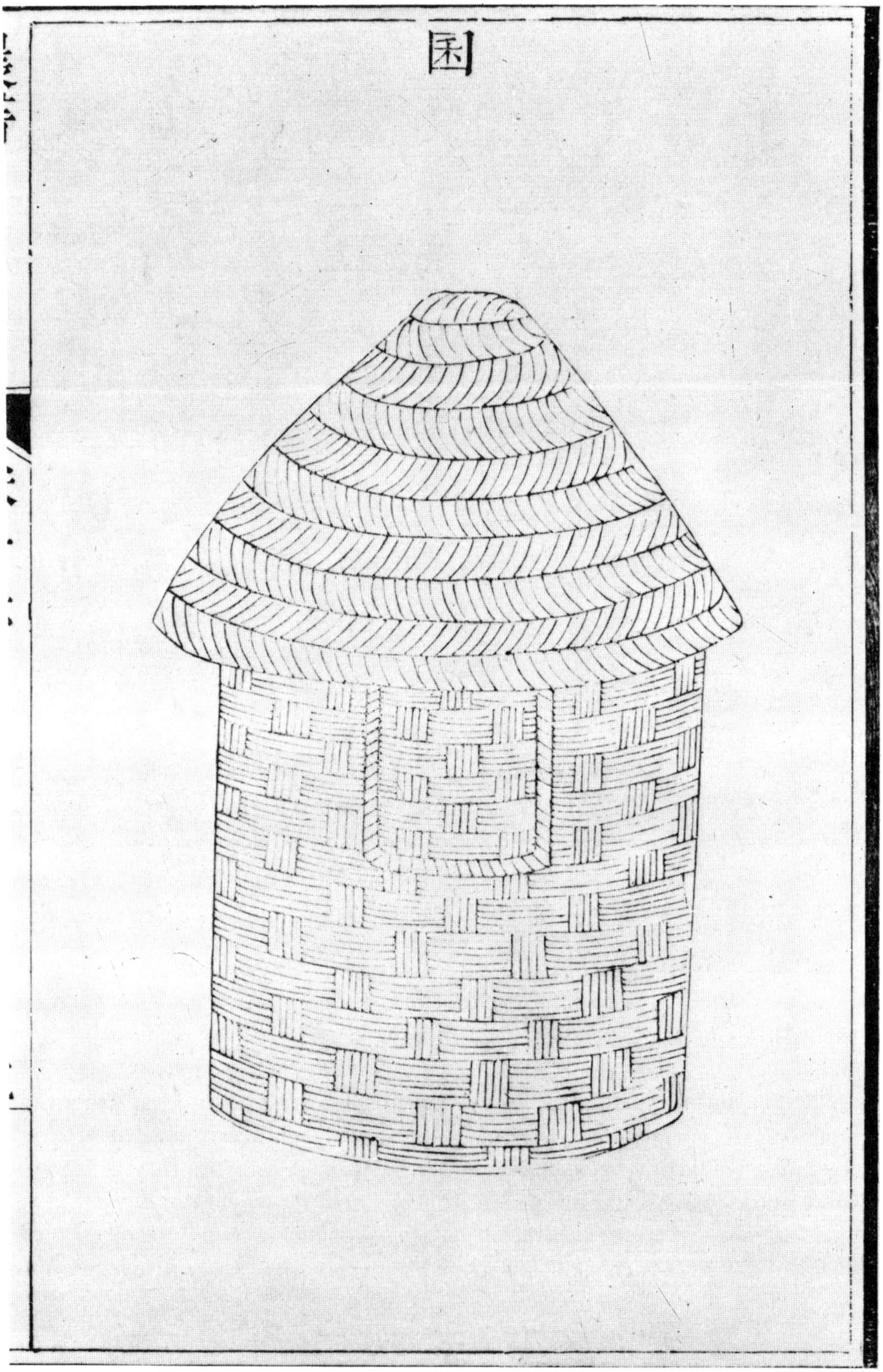


Fig. 204. Round granary of matting and wickerwork; *WCNS* 16/19*a*.

Fig. 205. Han model of a Kwangtung granary on stilts; Anon. (*42*), pl. 48.

nary was *ching*[1].[a] Han models in wood, pottery and even bronze survive;[b] most of them are from the Yangtze area, Kwangtung and Kwangsi, and show granaries on stilts. The stilts are thick and tall, four or more in number, and often have overhanging stones placed at the top to deter rodents (Fig. 205). The granaries usually have a simple gable roof, but sometimes it is underhung and extended on one side to form a verandah; the roof is invariably depicted as tiled. The walls may be in planks, lattice-work, matting, or even brick, pierced by one large door and often one or several lights for ventilation. The door or verandah is reached by a log ladder like a Southeast Asian *tangga*, indeed the whole structure is reminiscent of the rice barns still found in Malay villages. In Southwest China in

[a] *WCNS* 16/20*b*.
[b] Anon. (*42*), pp. 59 ff. shows clay and wooden models; a particularly striking bronze model from Kwangsi is shown in Anon. (*527*), pl. 3.

[1] 京

pre-Han times, the tribes of Tien[1] stored their grain in stilted granaries built of horizontal logs, with wide curved roofs like the old American waggons in a Western (Fig. 196);[a] exactly similar granaries are found in the Celebes today.[b] Over twelve centuries later the agricultural encyclopaedias show small, square granaries on short stilts, taller than they are long, built of elaborately joined planks and with tiled roofs. Wang Chen says that these *ching* were widely used in Southern China, and were of similar construction to the larger granaries, which were called *tshang*.

The *tshang* had already attained a high degree of technical sophistication throughout China by the Han dynasty, as is evident from the numerous stone-engravings and paintings depicting country mansions and their outbuildings. The granaries are large, solid buildings raised on platforms of earth or brick a meter or so thick, broken by one or even a pair of flights of steps (fig. 206). The frame of the building is of heavy timber, the roof of closely laid tiles, and the solid door—often there are two—has double leaves and is fastened with a heavy timber beam. It looks impregnable enough to withstand a siege by the Hsiung-nu. Ventilation is assured by the conspicuous lanterns, like little towers on the roof; sometimes the lantern is so elaborate as to resemble a second storey. The courtyard in front of the granary is often shown busy with kitchen staff pounding grain in mortars, while poultry fight over the chaff.

There is a considerable gap in the evidence, both pictorial and literary, between the Han and the Sung, but the descriptions and drawings of granaries, their siting and construction given in the *Ying Tsao Fa Shih* of +1097, in Wang Chen's *Nung Shu* of +1313, and in later works tally exactly with the Han illustrations, and one may safely assume that most of the technical details given in the *Nung Cheng Chhüan Shu* and elsewhere were already familiar in the Han.[c] Although the charming Han drawings and models of granaries give no indication of interior design or of construction details, it is probable that the techniques quoted below from the *Nung Cheng Chhüan Shu* and the *Shou Shih Thung Khao*[d] were known well before the Han.

Most of the directions quoted below were given with large-scale public granaries in mind (e.g. an average county granary holding 15,000 *shih*[2] or approximately 45,000 bushels).[e] The storehouses of clan estates, monasteries and manors, though somewhat smaller, would have been constructed and run along very similar lines, and even the small barns or storage rooms of peasant farms, though

[a] Von Dewall (2), fig. 7.

[b] Ronald Lewcock, pers. comm.; a similar style of roof is found in the Yayoi drawings and +5th century Haniwa houses of Japan, W. Alex (1), pls. 3, 5.

[c] The most important principles of Chinese architectural design and technology were widespread from at least the −13th century onwards; Vol. IV, pt 3, pp. 126 ff.

[d] *NCCS*, ch. 45, *SSTK*, ch. 57.

[e] *NCCS* 45/8*a–b*.

[1] 滇 [2] 石

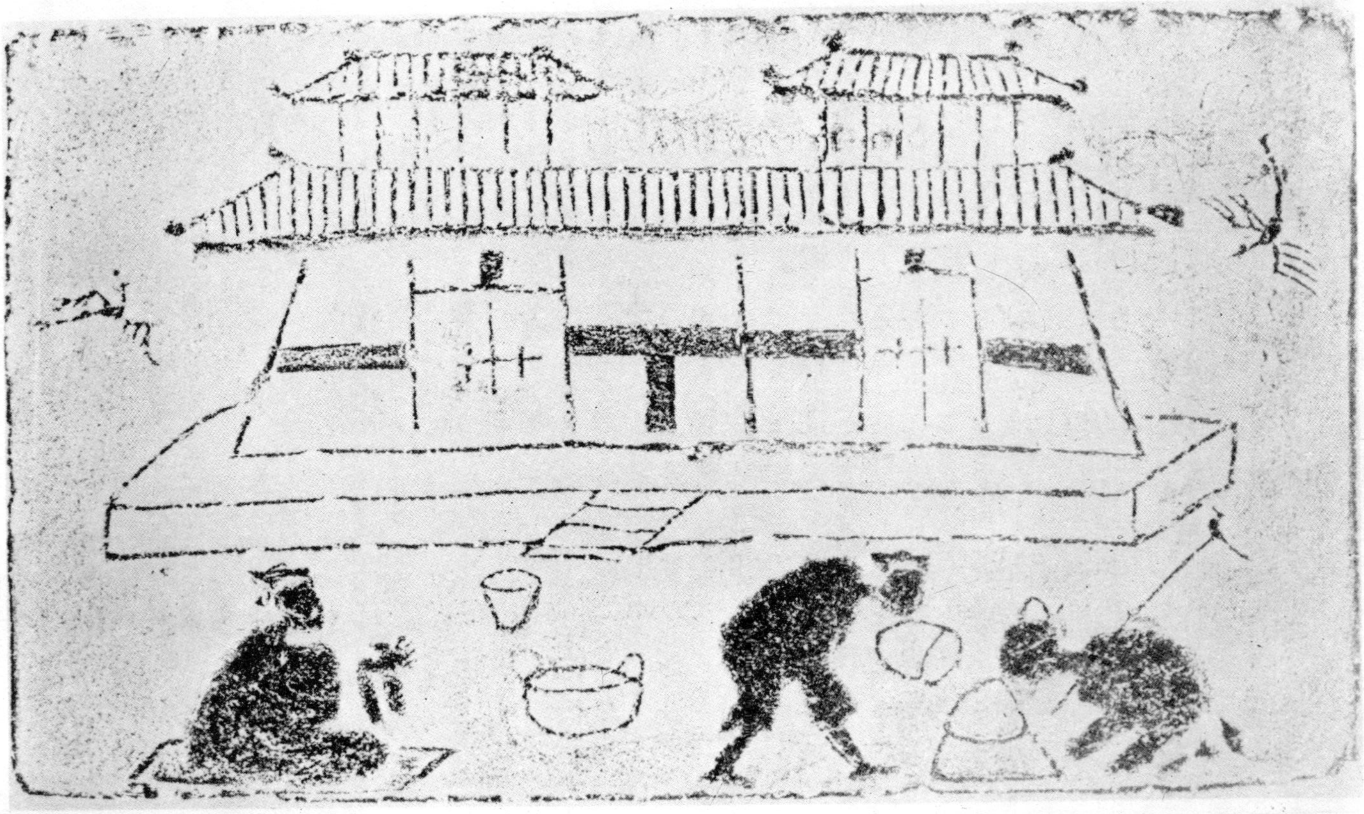

Fig. 206. Granary of a wealthy Han household in Szechwan; note the ventilation towers, the heavy barred doors and the solid platform upon which the building is constructed. Finsterbusch (1), vol. II, fig. 188.

considerably smaller and less extravagant in their use of materials, would certainly have received similar attention to water-proofing and ventilation. Although a first-class granary was extremely costly to build, requiring large quantities of good timber, dressed stone and fired bricks, the initial heavy investment yielded long-term dividends in both money and labour, and it was accepted that one should be prepared to spend as much money on one's granary as on one's house.[a]

The main things to avoid in a granary are damp ground, leaking walls, birds and rats.[b] Where possible the granary should be built on high ground, the site first cleared, levelled and well drained. In China granaries, like all buildings of importance, were designed to face South. In Europe farmers were generally advised to build their granaries so that the lights faced North or Northeast, 'for that exposure is the coolest and least humid'[c] and the least exposed to frost.[d] These considerations were important when ventilation was provided by a direct inflow of air, but in China the system of ventilation by convection allowed the building to be oriented according to ritual rather than meteorological demands.[e]

Once the site had been levelled and prepared, a platform of up to three feet thick was built as a foundation; it could be made of stone slabs, of fired bricks with edges specially ground for a tight fit, or of tamped earth, paved with stone slabs in the verandah area, and with bricks inside the actual granary. The pillars and beams were of excellent wood, thick and solid, cladded inside with thick planks of pine (*sung*[1]) or cryptomeria (*shan*[2]) and then hung with bamboo mats. The square ridge-pole and the purlins were covered with planks to which a bamboo frame was attached; this frame was then filled with earth for insulation, and the roof was finally closely tiled with tiles soaked in a solution of alum (*pai fan shui*[3]) to make them watertight. (In village granaries the roof was often just thatched (Fig. 207), though the lantern and the roof ridge might be tiled for extra protection.) All air vents (*feng chhuang*[4]) had to be protected with bamboo slats to keep out the sparrows.

If the granary was built in a dry area, then all four exterior walls could be built of brick with an inner cladding of acacia (*hsiang ssu*[5]) planks. If the area was damp, the exterior wall at the front was to be of logs protected by a verandah six feet wide. Once the brick walls had dried out, the granary floor received its final dressing of 6 inches of coal ash and the same of wheat chaff, covered with a layer of thick bricks pointed with a mixture of glutinous rice and thick lime, and finally

[a] Lü Khun[6], *Chi Chu Thiao Chien*, quoted *SSTK* 57/6*a*.
[b] The remarks that follow are taken from the *Tshang Ao I*, quoted *SSTK* 57/2*a*–5*b*.
[c] Columella (1), vol. I. vi. 9.
[d] Stephens & Liebault (1), p. 547.
[e] Similarly in Moslem Southeast Asia the rice barns, like the houses, face Mecca.

[1] 松 [2] 杉 [3] 白礬水 [4] 風窗 [5] 相思
[6] 呂坤

Fig. 207. Thatched village granary; note the verandah to keep off the rain, and the shallow, brick-edged platform on which the granary is constructed. *Keng Chih Thu*, Imperial ed., 1/22*b*.

Fig. 208. Horizontal planks used to close the granary door; this is the Semallé Scroll version of the previous figure, and the granary depicted here is altogether on a grander scale, reminiscent of the Han granary shown in Fig. 206. *Keng Chih Thu*, Pelliot (24), pl. XXXI.

with a layer of wooden planks, or of thick matting if wood was in short supply. The doors were the one feature that had changed significantly since Han times. Instead of two solid panels of heavy timber which could only be closed when the granary was full, the Yuan and later granaries were closed with planks laid horizontally across the entrance as the level of the grain rose inside. If the planks were counted, the amount of grain in the granary could be calculated at a glance (Fig. 208). In large public granaries the site usually contained several separate granary buildings, as well as accounting sheds and other ancillary buildings all surrounded by a thick wall twelve feet high. The space between the buildings was paved with stone so that the grain could conveniently be laid out to dry in the sun.

Although the *Nung Cheng Chhüan Shu* and the *Shou Shih Thung Khao* describe granary construction in considerable detail, even giving the precise dimensions and colours of the bricks and tiles to be used and the characters with which they were to be stamped,[a] neither unfortunately describes the construction of the lantern (*chhi lou*[1]). This term occurs both in Wang Chen's *Nung Shu*[b] and in the Chhing commentaries on the *Keng Chih Thu*,[c] but is not found elsewhere, although the *Shou Shih Thung Khao* refers to 'wind vents' (*feng chhuang*[2]). The lanterns were

[a] *NCCS* 45/10*b*–12*b*.

[b] *WCNS* 16/16*a*.

[c] O. Franke (11), p. 125.

[1] 氣樓 [2] 風窗

just as universal a feature of granaries in the Yuan, Ming and Chhing dynasties as they were in the Han, and they were an indispensable adjunct to the interior construction of the granary, expressly designed to maximise ventilation of the grain. The building was divided internally into bays separated by lintel-high partitions; a bay holding 4000 *shih*[1] (about 12,000 bushels) was approximately 4·25 m. high by 3·5 m. wide by 5 m. deep, and each bay had its own door. A public granary might have as many as seven bays, and the central bays were sometimes subdivided into three sections. At least one end bay was always left empty, and every six or seven months the grain was moved up one bay, and cooled and aired in the process.[a] To ensure that no grain was spoiled in between moves, several *ku chung*[2][b] were placed in each section of the granary. If these tubes were marked off in sections they performed the further function of a gauge indicating the depth of the grain.[c]

Not all granaries were as elaborate and costly as this; many grain stores were not even separate buildings. Chinese domestic buildings are usually single-storey, and the roof is not concealed as in the West by a flat ceiling; Chinese farmers do not therefore store their grain in lofts as was commonly the case in Europe.[d] But rooms and outhouses akin to European garners were very common in China.[e] Sometimes grain was stored in an extension of the farmhouse itself (Fig. 209); in larger farms or townhouses built around one or more courtyards, the granary buildings often formed part or the whole of a separate wing (Fig. 210). The materials were of course less lavish than those recommended for public granaries, and Wang Chen tells us that peasant granaries were often built of wood plastered with a mixture of clay and lime to prevent fires and deter woodworm.[f] In the North, where millet, wheat and barley were the main crops, the peasants often stored their grain on a thatch-roofed platform called *lin*[3]. This term is very ancient and occurs in the *Book of Odes*, where it is glossed by a Han commentator as 'a store for millet in the ear';[g] the *Thang Yün* enlarges on this, explaining that the *lin* is a granary with a roof but no walls. This type of building seems to have been common in North China at least since Chou times; the Ming illustrations show a sturdy construction of thick wooden pillars on a stone platform, with a thickly thatched gable roof, containing sheaves and baskets of grain (Fig. 211).[h]

[a] *Chi Chu Thiao Chien*, quoted *SSTK* 57/6*b*.

[b] See p. 383 above.

[c] *WCNS* 16/21*a*–*b*.

[d] Markham recommended spreading corn in thin layers in a well-plastered loft, and turning it every few days with a wooden shovel; this, he said, would keep the corn in perfect condition; (1), p. 113.

[e] Anon. (*42*), p. 39; Liu Tun-Chen (*4*), passim.

[f] *WCNS* 16/16*a*.

[g] *Shih Ching* 27/13*a*; Legge (8), p. 586.

[h] Yen Shih-Ku, commenting on *C/HS* 7/8*a*, says that *lin* were government depots for relief grain (quoted Swann (1), p. 56), but this seems to be a less common usage.

[1] 石 [2] 穀盅 [3] 廩

Fig. 209. Granary as an extension of the farmhouse itself; Han model from Kwangtung; Anon. (*42*), fig. 31*b*.

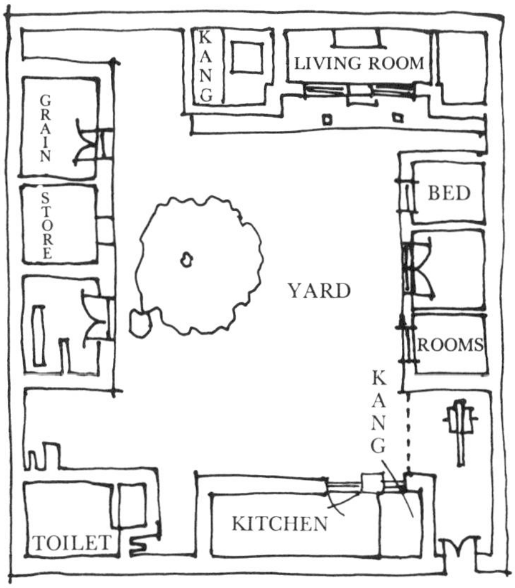


Fig. 210. Ground plan of a Hopei farmhouse, showing the granary wing at the right; Liu Tun-Chen (*4*), fig. 94.

Fig. 211. Open grain store (*lin*); *WCNS* 16/17*a*.

### (iv) *Public grain storage*

There were three main categories of public grain storage in China. The first was concerned with the storage of tax grain, the second with the stockpiles used to control price fluctuations and shortages (the famous Ever-Normal Granaries), and the third with national and local famine relief.

#### *Tax grain*

The Yin oracle bones contain lists of the tribute grains received by the emperor Wu-Ting[1] from his vassal kings in the −14th century,[a] and ever since that remote period the Chinese have paid their feudal dues and state taxes principally in grains and textiles.[b] Under the empire the taxes paid by the peasants at village level were collected by sub-prefectural officials, who retained a proportion for local bureaucratic expenses and forwarded the rest to the provincial centre. There a further proportion was taken for the expenses of the provincial government, and the balance was sent on to the capital.[c] Not only was grain needed to provision the imperial household and the armies stationed at the capital or on the frontiers, but also it was an important form of currency: government salaries, from the Prime Minister's to the lowest *yamen* runner's, were usually paid in grain.[d] The quantities involved were enormous; as early as the Thien-pao[2] period of the Thang dynasty (742–56) the combined total of land and poll tax for the four years is estimated at about 25 million *shih*[3] or 45 million bushels of unhusked grain,[e] while by the middle of the +11th century the Sung government, who calculated in cash equivalents, had an annual budget of between 125 and 150 million strings of copper cash, equivalent to about 370 to 450 million bushels of grain;[f] by Ming times the state budget had declined, and in 1578 the government assessed its total provincial tax quotas at 26, 638, 642 *shih*[3] of grain, or about 80 million bushels.[g] The government built huge granaries to store these enormous amounts of grain, and the number and dimensions of the granaries increased from dynasty to dynasty. The great granary at Hsiao-ho[4], built outside the capital Chhang-an[5] in the Former Han dynasty, was said to have 120 pillars along each wall.[h] In +607 the Sui emperor Yang-Ti[6] built two granaries near the capital, the Lo-khou[7] granary, whose walls were 12 *li*[8] in circumference and which con-

[a] Hu Hou-Hsüan (*3*), 83*a*–89*b*.

[b] The grain was the farmer's contribution, the cloth his wife's. Taxes were sometimes commuted into cash (Ray Huang (3), p. 46, Wang Chih-Jui (*1*), p. 135), but the system of grain contributions has survived up to the present day: a large proportion of the taxes paid by the communes in the People's Republic are still paid in kind (Aubert, Maurel & Pairault (1), p. 10).

[c] See R. Huang (3), fig. 3, p. 83.

[d] Salaries were reckoned in terms of the tax grain paid by a fixed number of households; thus one might be promoted from a five-hundred household position to a thousand-household position.

[e] Anon. (*525*).

[f] Wang Chih-Jui (*1*), p. 135.

[g] Ray Huang (3), p. 46.

[h] *San Fu Huang Thu*, quoted *SSTK* 54/6*a*.

1 武丁 2 天寶 3 石 4 蕭何 5 長安
6 煬帝 7 洛口 8 里

tained 3000 storage pits, and the Lo-hui[1] granary, with walls 10 *li* in circumference and containing 300 pits, each of which held 800 *shih*. The Lo-hui granary thus held 240,000 *shih*, the equivalent of 350,000 bushels of grain.[a] The great Han-chia[2] granary of the Thang dynasty contained nearly six million *shih*, or over nine million bushels;[b] it was used as a temporary store for the tax grain from East China, transferred subsequently to the central Shih-ching[3] granary.[c] The *Sung Hui Yao* records a total of twenty-three granaries in the Northern Sung capital of Khai-feng[4];[d] in +1189 the Minister of Agriculture and other members of the government met to discuss the problem of the Feng-chhu[5] granary, which had for the first time reached its maximum capacity of 1,500,000 *shih* (about 2,750,000 bushels).[e] The Ming dynastic history records that before +1370 there were fewer than twenty granaries in the capital, Peking, but by +1450 there were 85, while extra granaries had also been built in all the provinces and border regions.[f]

Most of the tax grain was transported by boat along rivers and canals, and the construction and maintenance of an efficient network of waterways was a prime concern of the state from the Chhin dynasty on. In pre-Thang times, when the capital and the chief grain-producing areas were both situated in North China, the problem was of manageable dimensions, but as the Chinese state came to rely more and more heavily on the rice-growing areas of Central and Southern China the logistics of grain transport became increasingly complex, large quantities of grain were lost, damaged or stolen *en route*, and by the Ming the total quantities of 'wastage rice' (*hao mi*[6]), 'porterage rice' (*chiao mi*[7]), and other extra levies designed to compensate for losses on the journey to the capital were often even greater than the amount of actual tax grain.[g]

### *Ever-Normal Granaries* (chhang phing tshang[8])

In China as in Europe grain prices suffered considerable fluctuations, not only from year to year but from season to season and region to region.[h] In China, however, the government early felt the need to maintain some kind of control over grain prices, particularly in times of shortage, when inflation was rapid and large numbers risked starvation as much from lack of funds as from lack of actual supplies. Their solution came to be known as the Ever-Normal Granary (*chhang phing tshang*), a term famous throughout Chinese history but unknown in the West until Henry A. Wallace introduced the institution (with several modifi-

[a] *Yü Hai*, quoted *SSTK* 54/8*b*.
[b] Anon. (*525*).
[c] *Thang Liu Tien*, quoted *SSTK* 54/8*b*; also Anon. (*525*).
[d] *Sung Hui Yao* 53/1*a*–5*a*, quoted *SSTK* 56/10*b*.
[e] *Hsü Wen Hsien Thung Khao*, quoted *SSTK* 54/10*b*–11*a*.
[f] *Ming Shih, Shih Huo Chih* 79/11b, quoted *SSTK* 54/13*a*–*b*.
[g] Hoshi Ayao (1), pp. 54–61.
[h] On grain prices see Twitchett (4), Shiba Yoshinobu (1), Ray Huang (3), etc.

[1] 洛回 [2] 含嘉 [3] 實京 [4] 開封 [5] 豐儲
[6] 耗米 [7] 脚米 [8] 常平倉

cations) to the United States in 1938.[a] But the concept, if not the name, was not unknown in Western societies: Wallace cited as a precedent not only China but Joseph's policies in Egypt, and similar systems were practised, for example, by Friedrich Wilhelm I of Prussia (1713–40) and in colonial Mexico.[b] Only in China, however, has this system of government price control been practised on such a scale and over such a long period.

The earliest proposal that the government should control grain prices was made in the *Kuan Tzu*. The statesman Kuan Chung[1], who is described as making this proposal to Duke Huan of Chhi[2], is supposed to have lived from −710 to −645, but the work now known as the *Kuan Tzu* is usually dated to around −300. Kuan Chung advised that the government reduce the grain taxes and purchase large quantities of grain when it was abundant and cheap, to sell it back to the people in times of shortage and high prices.

> By this process of collecting grain when it is cheap and distributing it when it is dear, the sovereign can secure ten times his outlay as profits, and the economy can be stabilised ... [He] should take the storing of grain under his control, to overcome and reduce the rise and fall of the value of grain in terms of other commodities. Then (because he will make a revenue in this way) he may exempt the people from taxes.[c]

There is no indication in the historical records, however, that such a policy was ever practised in Chhi.

The next statesman to whom a policy for price control is attributed is Li Khuei[3], a minister at the court of Marquis Wen of Wei[4] (−403 to −387), who pointed out that ordinarily in a bad year grain was so dear that the people starved, while in a good year it was so cheap that the farmers were ruined; it was therefore necessary for the state to purchase a proportion of the grain harvest in every good year, the amount varying with the abundance of the harvest, and to put the grain up for sale to the public in years when the harvest was poor. The policy was designed both to control prices and to prevent famine, but this time was not intended as a substitute for direct taxation. The *Han Shu* implies that this policy was actually practised with some success in the state of Wei.[d]

The Legalist economists of Han Wu-Ti's[5] reign (−140 to −86) favoured government regulation of the prices of a number of important commodities. Sang Hung-Yang[6] (−152 to −80) introduced the systems of 'equable transport' (*chün shu*[7]) and 'equalisation and standardisation' (*phing chün*[8]) in −115 and −110 respectively, whereby the key economic products of each region were purchased by the government and redistributed at fixed prices in other regions. Unfortunately, we

[a] D. Bodde (24).
[b] *Ibid.* p. 426; Chester L. Guthrie (1).
[c] *Kuan Tzu*, ch. 22; tr. Than Po-Fu *et al.* (1), p. 118–19.
[d] *C/HS* 24A/6*a*–7*b*; see also Swann (1), pp. 136–44.

1 管仲 2 齊桓公 3 李悝 4 魏文侯 5 漢武帝
6 桑弘羊 7 均輸 8 平均

are not told whether grain was one of the commodities handled by the government.[a] But it was not long after that the first Ever-Normal Granaries were established by Keng Shou-Chhang[1], an officer of the second rank in the Ministry of Agriculture, in −54. These granaries were set up in the frontier provinces; grain was bought and stored when prices were low and sold at reduced rates when prices were high. Although the local population seems to have found these Ever-Normal Granaries very useful, a series of natural disasters prompted the Confucians at court to abolish a whole series of Legalist-inspired government monopolies and price-controls, and so the first Ever-Normal Granaries were abolished only ten years after they had been set up.[b]

Although Ever-Normal Granaries were set up again on several occasions in the Later Han, Chin, Chhi and Sui, in the capital cities as well as on the frontiers,[c] it was not until the Thang that the institution met with real success. The founder of the dynasty, Kao-Tsu[2] (*r.* 618–27), appointed official inspectors for the Ever-Normal Granaries in the very first year of his reign; during the Chen-kuan[3] period (627–50), Ever-Normal Granaries were established in eight provinces, mainly in North China,[d] and in 655 they were set up in both the Eastern and Western capitals.[e] During the rebellion of An Lu-Shan[4] in the late 750s the whole system fell into disarray with terrible results for the population, who starved or resorted to cannibalism, and in the reign of Te-Tsung[5] (780–805) it was proposed that the Ever-Normal Granaries should be re-established on a new basis, using cash capital as well as stores of grain to regulate the price; the cash capital for various Ever-Normal Granaries in the capitals, the Yangtze regions and Szechwan was to vary between one hundred thousand and one million strings of cash, and this capital was to be raised by taxing the local merchants at the rate of 2% on all monetary transactions and 10% on sales of bamboo, timber, tea and lacquer.[f] The proposal was accepted but soon had to be abandoned as impracticable, as military requirements were at that time draining all the state's resources. Nevertheless, the idea of maintaining reserves of cash capital as well as grain caught on, and was common practice during the Sung and later periods.[g]

If the Ever-Normal Granary was to function efficiently it was essential to purchase grain immediately after the harvest when prices were low. The Sung statesman and historian Ssu-Ma Kuang[6] complained that

officials and their subordinates cannot be bothered to buy rice at harvest time [when it is cheap], and prefer to rely upon merchants and [rich] families who stockpile grain. The

[a] Swann (1), pp. 314–16; *C/HS* 24B/17*b*–18*b*; Bodde (24), p. 414.
[b] *C/HS* 24A/17*b*–18*a*; Swann (1), pp. 195–7.
[c] See the quotations from *Hou Han Shu*, *Chin Shu* and *Wen Hsien Thung Khao* in *SSTK* 55/1*a*–3*a*.
[d] *C/Thang Shu, Shih Huo Chih* 49/11*b*, quoted *SSTK* 55/11*a*.
[e] *Thang Hui Yao* ch. 88 p. 1612, quoted *SSTK* 55/10*b*.
[f] *C/Thang Shu, Shih Huo Chih* 49/12*a*, quoted *SSTK* 55/2*a*.
[g] *Sung Shih, Shih Huo Chih* 176/14*a*, quoted *SSTK* 55/3*a*.

1 耿壽昌 2 高祖 3 貞觀 4 安祿山 5 德宗
6 司馬光

result is that the peasants, who have to sell quickly, must sell at low prices, while the officials, who buy later, must buy at high prices, and it is the middlemen who pocket the profits.[a]

One of our most detailed sources of information on the Ever-Normal Granaries is the Ming official Chang Chhao-Jui[1] (*fl.* 1570), who in a long essay entitled *I Chhang Phing Tshang Ao Shen Wen*[2] [b] not only describes in some detail how best to site and construct a granary, but also recommends methods of funding and administration.

The materials should be purchased out of the fines awarded in the local circuit, to which should be added all the non-quota income (*wu ai kuan yin*[3]) brought in by judicial investigations ... Every year half the proceeds from the fines and redemption money paid to the circuit and prefectural courts should go towards buying grain, and the authorities may also use funds from dispossessed monasteries and non-quota income ... A large sub-prefecture should buy 5000 *shih* [about 15,000 bushels] of grain, a medium one 4000 *shih* and a small one 3000 *shih*, but one should not compel or order the common people [to sell grain]. Each district should select two wealthy and business-like persons from the vicinity to administer the granary ... and for each 100 *shih* of grain that is bought and later sold, each administrator should receive the average price of 3 *shih* as a fee for his work. Every year the administrators should be replaced. On no account must they be re-appointed.[c]

The local population was divided into the usual units of ten families (*pao chia*[4]) for the purpose of purchasing grain at government prices, each family in turn being permitted to purchase a strictly regulated quantity until supplies ran out. Chang's system was based on the use of official cash capital to control grain prices within a single agricultural year. This permitted a rapid turnover of grain, and had the advantage of requiring fewer storage facilities and involving less waste than if long-term stock-piles of grain were used instead of cash. The use of official funds as capital for the Ever-Normal Granaries seems to have been common throughout the Ming and Chhing.[d] However, considerable stocks of grain continued to be stored, because the government granaries had to serve not only as price-regulating institutions but also as storehouses for famine relief.

*Famine relief: charitable granaries* (i tshang[5]) *and community granaries* (she tshang[6])

Providing relief for the hungry in famine years was one of the fundamental tasks of the Chinese government. The organisation of famine relief in early times was apparently quite haphazard: when local supplies ran out, grain was imported from other areas if possible, but if not the government quite often resorted to

[a] *Chhang Phing Fa*[7], in *Wen Hsien Thung Khao*, quoted *SSTK* 55/4*b*.
[b] Report of Discussions on Ever-Normal Granaries, quoted *SSTK* 55/6*b*–10*a*.
[c] *SSTK* 55/7*a*.
[d] The *SSTK* (55/10*b*, 11*a*, etc.) quotes examples from the 5th year of Ming Hsuan-Te[8] (+1430) and from the 43rd year of Khang-Hsi[9] (+1704), among others.

[1] 張朝瑞 [2] 議常平倉廒申文 [3] 無礙官銀 [4] 保家
[5] 義倉 [6] 社倉 [7] 常平法 [8] 宣德 [9] 康熙

transporting those among the stricken peasants who had not already fled of their own accord.[a] By Han times relief had become more systematic. In times of famine the government granaries distributed grain either as gifts, as loans at reduced rates of interest, or for sale to those who could afford it, special commissioners were delegated to distribute seed-grain in stricken areas, and the rich people in the locality were called upon to contribute relief grain from their own private stores. Regional stores and supplies, however, were often insufficient to cope with real disaster; when possible grain was imported from other regions, yet resettlement from the crowded areas of North China to the less populous Huai and Yangtze regions was still sometimes necessary, as in −115 when the whole of the Yellow River plain was flooded.[b] The migration of thousands of starving peasants in search of food was a phenomenon which recurred with horrifying regularity until the Communist Revolution, but as the imperial bureaucracy perfected its relief machinery the tragic exodi were usually confined to periods of foreign invasion or civil unrest; in peace time the government usually managed to avoid the onerous task of large-scale official resettlement, except in the frontier colonies.[c]

The Ever-Normal Granaries were used for famine relief as well as price regulation until quite recent times: the *Chhing Hui Tien* for example mentions that half the grain accumulated in the Ever-Normal Granaries each year was loaned to poor people in distress.[d] But as early as the Sui dynasty it had become obvious that the Ever-Normal Granaries were too centralised to be able to deal efficiently with local crises. What was required was famine relief granaries organised at the local level.

In the 5th month of 585 Chang-Sun Phing[1], Secretary of the Board of Works, petitioned that the common people and the armed forces throughout the empire should be encouraged to form societies to establish charitable granaries. At harvest time they should be encouraged to contribute quantities of millet (*su*[2]) and wheat (*mai*[3]) proportional to their yields; they should build granaries or pits in which to store the grain, and appoint a society manager (*she ssu*[4]) to administer the stores. Each year the harvest should be stored and none should be wasted or spoiled, and should the harvest fail, anyone in the society who was in need of food would be provided for out of the communal grain.[e]

In 596 the proposed granaries were set up in more than thirteen provinces, and also at the sub-prefectural level in the capital area. Contributions were fixed at three levels: 1 *shih* (approximately 1/2 bushel) from rich families, 7 *tou* (1/3 bushel) from middle-level families, and 4 *tou* from poor households.[f] The Sui

[a] E.g. *Meng Tzu* Ia/3.
[b] *CHS* 24B/15*b*–16*a*; Swann (1), pp. 60–1; 301–2.
[c] Han Kuo-Pan (*1*), p. 73.
[d] Ch. 191, quoted *SSTK* 55/10*b*.
[e] *Sui Shu, Shih Huo Chih* 24/15*a*, quoted *SSTK* 56/1ff.
[f] *Ibid.*

[1] 長孫平 [2] 粟 [3] 麥 [4] 社司

granaries, whether at provincial or sub-prefectural level, were all called 'charitable granaries' (*i tshang*[1]), but later on this term was usually reserved for relief granaries in towns, while those at village level were generally called 'communal granaries' (*she tsang*[2]).[a]

The distinguishing features of the charitable granaries were, first, that they were used exclusively for relief; and secondly, that they were stocked with contributions of grain supplied by the local population over and above the standard taxes. The size of the contribution varied over time. In the early Thang (628) an official from the Secretariat, Tai Chou[3], proposed that the three grades of contribution levied in the Sui should be replaced by *pro rata* contributions of 2 *sheng*[4] per *mu*[5] of all cultivated land (roughly 0.22 bushels/acre), which probably represented not more than 2 % of the total yield.[b] This basic figure was modified locally according to soil fertility, and furthermore 'if four-tenths of the crop was lost only half the amount was levied, and if seven-tenths was lost the farmer avoided the levy altogether'.[c] Merchants also paid contributions graded according to nine ranks, and only the very poor and barbarians were excused altogether. The grain was distributed in times of hardship either as gifts, or as loans of seed-grain to be repaid in the autumn.[d] Despite their special status these contributions were soon misapplied and used by the government for current expenditure; they became known as 'land tax' (*ti shui*[6]), and by the middle of the +8th century constituted the government's most important source of income.[e] Undeterred by this inauspicious example, later officials revived the charitable granaries on numerous occasions in subsequent dynasties, and although like the Ever-Normal Granaries their initial success usually crumbled into mismanagement or corruption, the obvious need for such an institution guaranteed their reappearance.

In the early Sung (+963) an imperial decree ordered the establishment of charitable granaries in every sub-prefecture, and twice-yearly contributions of 10 % of the harvest were levied, but the dynastic history tells us that 'this system was subsequently discarded as being too troublesome to administer'.[f] But as early as +1035 the system had been revived, although based as in the Sui on graded contributions with five different categories of household, and it reappeared and disappeared with some regularity over the next nine hundred years.

In +1182 the great neo-Confucian philosopher Chi Hsi[7] wrote a long essay entitled *She Tshang Fa*[8],[g] which described in some detail how to run such an

[a] E-Tu Zen Sun (1), §740 and passim.
[b] The grain tax in early Thang was at a similar level of 2 *shih* per 100 *mu*; Twitchett (4), p. 25.
[c] *Hsin Thang Shu, Shih Huo Chih* 51/4*a*, quoted *SSTK* 56/13*a*.
[d] *Hsin Thang Shu, Shih Huo Chih* 51/4*a*, quoted *SSTK* 56/13*a*.
[e] Twitchett (4), p. 33.
[f] *Sung Shih, Shih Huo Chih* 176/27*a*, quoted *SSTK* 56/13*a*.
[g] The Administration of Charitable Granaries, quoted in *SSTK* 56/4*a*–6*b*.

1 義倉 2 社倉 3 戴胄 4 升 5 畝
6 地稅 7 朱熹 8 社倉法

institution honestly and efficiently, but although reverently quoted in subsequent works his theories seem not to have been put into practice. By Ming times the administration of relief granaries had devolved upon the people themselves. A decree of +1530 declared that:

Communal granaries should be established by every regional inspector (*fu an*[1]), twenty or thirty commoner families constituting a society (*she*[2]) and selecting one person of well-to-do family with an understanding of commerce to be the society chairman, one person known for his honesty in business to be the manager, and one person skilled at writing and figures to be the accountant. A meeting of the society is to be held at the beginning of each month. The member households shall be divided into three ranks, each of which is required to make a different contribution of between 1 and 4 *tou* [roughly 1/4 to 1 bushel] of rice (*mi*[3a]) (plus an extra 5% to make up for wastage), the upper-rank households superintending the contributions.[b]

Official intervention was limited to an annual inspection of the books, and heavy fines were imposed on the chairman if discrepancies were discovered. In times of hardship the grain was distributed free to the lower and middle ranks, and as low-interest loans to the upper-rank families.

The amounts of grain handled by the local relief granaries were large in theory, although not of course comparable to the quantities kept in the tax granaries of the capital. An average subprefectural charitable granary in Sung times probably contained 5–8000 *shih* (approximately 9–15,000 bushels).[c] But in fact the system was so vulnerable to corruption and mismanagement that the granaries were probably rarely full. Very soon after the system of charitable granaries was inaugurated, the *Sui Shu* complains that 'the common people are not provident and take no care to prevent [the grain in relief granaries] being wasted or spoiled, and so supplies run short'[d], and the Ming system of delegating responsibility to the people themselves is described as 'working extremely well—but subsequently [the administration] had no power to implement it'.[e] Only in Northwest China, where long experience of civil disturbances and poor harvests had presumably convinced the peasantry of the virtues of self-reliance, did the system seem to function more efficiently. As early as the Sui we hear that:

The districts on the northern borders are different from other areas ... they even store miscellaneous grains[f] [as well as millet] in their charitable granaries, they distribute grain only in their own district, and when there is a drought and grain is short they first distribute the miscellaneous grains and the millet that has been in store longest.[g]

[a] Although the term *mi*[3] by this period usually refers to hulled grain as opposed to unhusked rice, *su*[4], in this case since it is measured by volume rather than weight it must be assumed to refer to unhusked rice.

[b] *Ming Shih, Shih Huo Chih* 79/13*a*, quoted *SSTK* 56/8*a*.

[c] Figures given for the second year of Ming-Tao[5] (+1033) in the *Sung Shih, Shih Huo Chih* 178/25*a*, quoted *SSTK* 56/13*a*.

[d] *Sui Shu, Shih Huo Chih* 24/15*b*, quoted *SSTK* 56/12*b*.

[e] *Ming Shih, Shih Huo Chih* 79/13*a*, quoted *SSTK* 56/8*b*.

[f] Barley, beans, or hemp seed, for example.

[g] *Sui Shu, Shih Huo Chih* 24/15*b*, quoted *SSTK* 56/12*b*.

[1] 撫安 [2] 社 [3] 米 [4] 粟 [5] 明道

In the Ming we are told again that although most peasants nowadays are improvident and wasteful, 'in Fen[1] and Chin[2][a] it is customary for all dwellings to have their stores of grain so that even in times of distress or famine the people do not emigrate in search of food'.[b] Here it seems that the peasants had given up looking to the government for support and were determined to rely on their own resources.

## (*e*) CROP SYSTEMS

Chinese crop systems emphasise the cultivation of cereals far more heavily than any farming system in the West. As we mentioned in the introductory section, productive livestock play a very small role in the Chinese farming economy, and the proportion of arable land occupied by pastures is minute.[c] Furthermore, almost all the remaining arable land is devoted to grains: in the early decades of this century 70 % of the total crop area was planted with grains, 10 % with legumes, 3·6 % with oil seeds, the same percentage with fibre crops (including cotton), 3·3 % with tubers and roots, 1 % with fruit and 1 % with vegetables.[d] Though the proportions may have changed somewhat over the centuries, the balance has not altered, as a glance at the chapter headings of the Chinese agricultural treatises will show. Grain was the essential ingredient in the traditional Chinese diet, millets and wheat in the north and rice in the south providing almost all the carbohydrates and part of the protein. Legumes, especially soybeans and their products, provided the supplementary proteins, and the cultivation of these crops had the additional advantage of restoring the soil's nitrogen content; they could also be eaten as famine foods if the grain crop failed, as could tubers and roots. Together with cooking oils (mostly vegetable) and a good supply of green vegetables,[e] this made up a well-balanced and often enjoyable diet, in fact it has been said that the Chinese population, especially during periods of economic prosperity like the Sung, was among the best-fed in the world.[f]

Our treatment of Chinese crops will reflect the importance of their role in the traditional Chinese economy, concentrating above all on cereals and devoting much briefer sections to legumes, oil crops, and so on. We have chosen to follow

[a] The Shansi/Hopei area.

[b] *WCNS* 4/8*b*.

[c] Under 1 % in the 1920s; Buck (2), p. 173.

[d] Buck (2), p. 209.

[e] Although the area devoted to vegetable cultivation was small, cultivation methods were so intensive that several crops were usually produced in the space of a few months.

[f] Perkins (1); Mote (4). Shinoda (*6*) notes that an ordinary Chinese housewife's shopping list, as portrayed in Yuan drama, included seven indispensables: fuel, rice, oil, salt, soy sauce, vinegar and tea; other items usually found in daily fare included wine, sugar, spices, and so on. But although Perkins maintains that agricultural production *per capita* was maintained until about 1800 (see p. 603), his calculations are based on figures for cereal production, and it is quite likely that levels of cereal consumption were maintained at the expense of dietary diversity, and that in order to produce sufficient staple grains to feed the growing population, the cultivation of vegetables and other less important crops had to be sacrificed. We are grateful to Dr Peter Nolan for raising this point.

1 汾 2 晉

this policy not only because cereal crops were so important in both the Chinese diet and economy,[a] but also because the technology of cereal cultivation, especially of rice, tends to dictate patterns of farm management as a whole.[b]

The general patterns and balance of Chinese crop systems may go back as far as neolithic times. Neolithic sites in north China have yielded evidence of the cultivation of millets, hemp and vegetables, while in the southern neolithic sites remains of rice and aquatic vegetables such as water-chestnuts have been found.[c] The Shang oracle bones and agricultural poems of the *Shih Ching* give ample proof of the importance of cereal crops in China at the time, as do the works of later Chou and Han political economists. Variations in crop systems in different regions of China are mentioned, though in scanty detail, in the *Chou Li* (see p. 22), and in our introductory section we have already described some of the main crops characteristic of China's different agricultural areas. Some of China's most important crops, like rice, millet and hemp, have been cultivated since neolithic times, but over time there were many additions to the repertoire. Some came from within China itself, for example soybeans (probably domesticated in Chou times) and tea (which only became important in the late Thang). Others were introduced from one part of China to another: winter wheat was introduced to the Yangtze valley from north China during the Sung, and many fruits and vegetables were transplanted from their native sod with varying degrees of success:[d]

When plants of northern habitat are moved to the South they generally flourish, but when plants from southern regions are transplanted to the North they readily change—as for example [the famous case of] *chü*[1] orange trees growing south of the Huai River but turning into *chih*[2] limebushes (with sour fruits) when moved to the North. [Contrariwise] rape turnips *ching*[3] flourish in the North, but when they are planted in the South they produce no more [large] roots.

So also the *lung-yen*[4] tree and the lichee (*li chih*[5]) tree are prolific in Fukien and Kwangtung, while the hazel-nut tree (*chen*[6]), the jujube-date tree (*tsao*[7]), and [all sorts of] gourds and melons (*kua lo*[8]) are plenteous in Hopei and Shantung. Plants cannot disobey the appropriate seasons [for their regular development]. How could men force plants [to do the impossible]?

But men often succeeded in coaxing plants into doing the impossible, carefully selecting and breeding until eventually they produced a variety that would flourish in adverse conditions. 'The pomegranate (*shih liu*[9]) is grieved when transplanted to

[a] A great deal of grain was grown for sale rather than for subsistence, and by Ming times it was not uncommon for some highly commercialised regions to import almost all their grain.

[b] We shall elaborate this point in our concluding section.

[c] See, for example, Chang (4); Li Fan (*1*).

[d] *Chhün Fang Phu, Hui Phu* 1/2*b*, tr. Joseph Needham; section 38 contains a section on Chinese phyto-geography and speculations on the correct interpretation of the case of the *chü* and the *chih*.

[1] 橘 [2] 枳 [3] 菁 [4] 龍眼 [5] 荔枝
[6] 榛 [7] 棗 [8] 瓜蓏 [9] 石榴

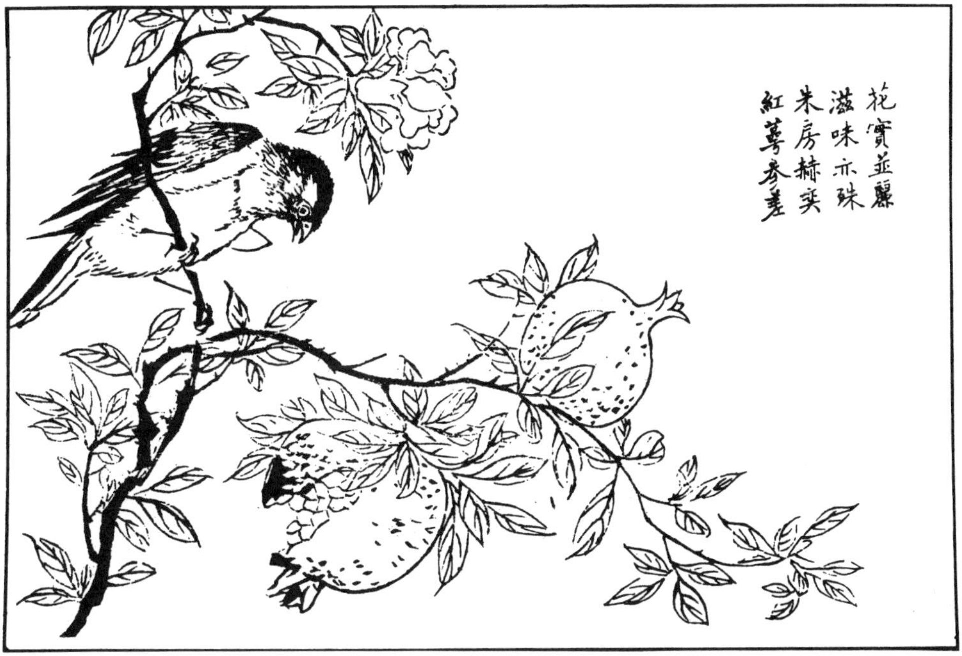


Fig. 212. Pomegranates; *Chieh Tzu Yuan Hua Chuan*, vol. III, p. 175, after the painting by Hsia-Hou Yen-Yu (d. after 965).

the East', the *Huai Nan Tzu* declares,[a] yet a few centuries later the *Chhi Min Yao Shu* devotes a whole chapter to the cultivation of this Persian fruit.[b] Nowadays it is a common sight all over north China (Fig. 212), often planted in groves in wheat-fields, as olives are in Italy and Greece. The tumulus of Chhin Shih-Huang is planted thick with pomegranate trees, which in May transform it into a pyramid of fire rising above the golden fields of wheat.

The Chinese imported a number of useful plants from Central and West Asia over the centuries, many of which were supposed first to have been introduced by the −2nd-century Han general Chang Chhien[1] when he returned to China from long years of captivity among the Western barbarians. This romantic tale has in the main been proved false, though Chang might have brought grapes and alfalfa back to Chhang-an,[c] and perhaps indeed our famous pomegranate.[d] But many valuable plants including peas, broad beans and sesame did come to China from the West in or just after Han times. Perhaps the most important success in naturalising a crop of Western origin was the case of wheat, a winter crop from

[a] The *Chhün Fang Phu, Hui Phu* 1/16 quotes the *Huai Nan Tzu* to this effect, but only part of the relevant passage is to be found in surviving editions of the *Huai Nan Tzu* itself; see Section 38.
[b] *CMYS* 41.
[c] See Laufer (1); Schafer (13).
[d] See Section 38.

[1] 張騫

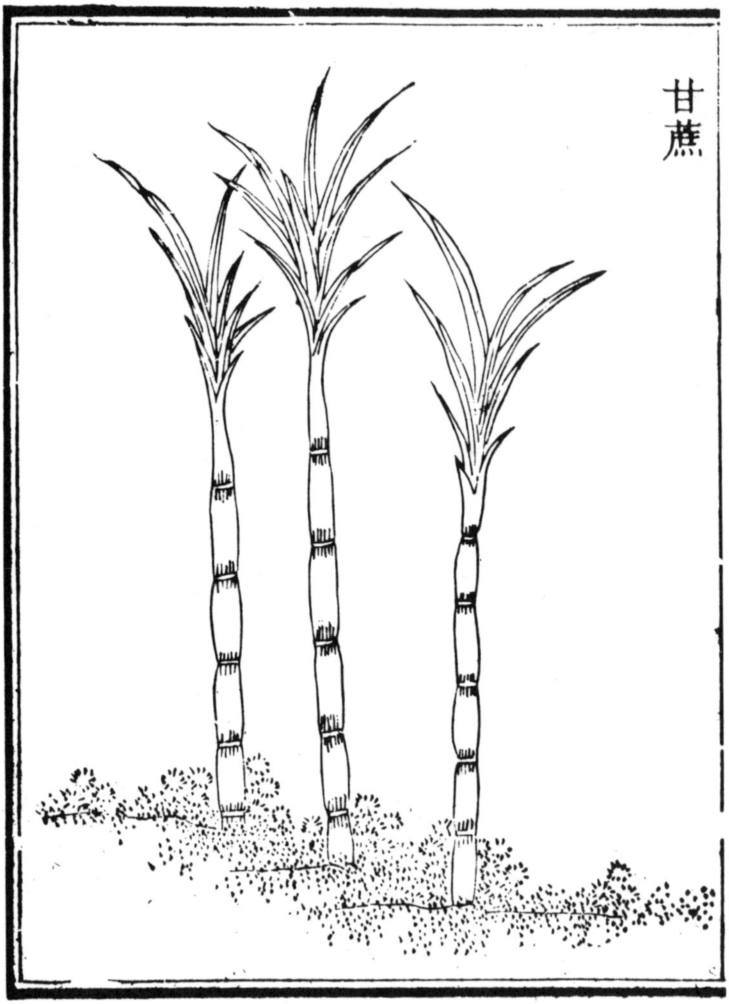


Fig. 213. Sugar cane; *SSTK* 66/8*b*.

West Asia which Chinese farmers persuaded to thrive in a land of summer rainfall maybe as early as neolithic times (see p. 461).

Another important source of Chinese crop plants was the Southeast Asian climatic zone, stretching from the southernmost provinces of China itself right down to Indonesia. It may even be that Chinese agriculture originated in this area: there is certainly a case for situating the domestication of such crucial crops as millets, rice, and yams and taros in this zone.[a] Later important introductions to China from this area included double-cropping rices from Champa, sugar-cane, banana, ginger and some species of citrus fruit (Figs. 213, 214). Other crop plants were introduced (either directly or through mainland Southeast Asia) from

[a] See Origins of Agriculture, above, and pp. 435, 485, 527 below.

Fig. 214. Banana trees were a common shade tree in South China, much appreciated for the sighing of their leaves in the slightest breeze. They were less common as a fruit tree. *Keng Chih Thu*, Imperial ed., 2/11*b*.

India, the most important being cotton, cultivated in Canton by the +8th century, but not widely grown further north until the Mongol dynasty.[a]

A further source of useful plants was the American continent. American crops found their way to China astonishingly quickly after its discovery in 1492. Most were introduced first to the East coast provinces, especially Fukien, probably

[a] H. L. Li (15), Table I, subdivides the vegetation zones (and their cultivated plants) differently, into a Southern Asian zone which includes both India and mainland Southeast Asia, and a southern island zone; however, plants originating in either area usually found their way into China through Annam.

through the intermediary of Chinese settlers in the Philippines and other Pacific islands, and spread rapidly inland from the coast. Another route was overland through Burma and Yunnan. Peanuts were already listed as a local product of Chhang-shu[1] county near Suchou in 1538, while sweet potatoes were in cultivation in Fukien and Yunnan by the mid-16th century.[a] The peanut was much appreciated as a source of oil (and protein) that thrived on otherwise useless sandy soils, while the sweet potato was valued as an auxiliary staple. Both spread rapidly throughout much of China, as did other American plants such as chili peppers, tobacco and tomatoes. Yet the chief American cereal, maize, was very slow to find favour with Chinese farmers, although the hill peoples of the Southwest rapidly incorporated it into their repertoire of swidden crops.[b]

China was not merely a receiver of useful plant species. The native Chinese flora is perhaps the richest in the world:[c] 'In wealth of its endemic species and in the extent of the genus and species potential of its cultivated plants, China is conspicuous among other centres of origin of plant forms. Moreover, the species are usually represented by enormous numbers of botanical varieties and hereditary forms.' Although Vavilov rather overestimated the number of native crops, China has undoubtedly contributed a large number of important crop plants to the rest of the world, some species originating in China itself, others introduced from Southeast Asia or elsewhere and later transmitted from China to other countries. Perhaps the most important example is rice, which may well have been domesticated somewhere in southern China and which was introduced, probably from the lower Yangtze region, to Korea and Japan in the early historical period,[d] and to the Philippines perhaps during the late neolithic. The soybean, first domesticated in north China, found its way to Japan and Southeast Asia, and eventually to the Americas and Europe. China is rich in fruits, including a great variety of apples and pears, of *Prunus* species, and of citrus fruits. It is possible that silk dealers were responsible for the introduction of apricots and peaches to the West, first to Persia (in the −2nd or −1st century) and thence to Armenia, Greece and Rome (in the −1st century).[e] These species were native to north China; more typical of the south were the citrus fruits, and it was Arab traders to the southern Chinese ports who transmitted oranges, pummelos and lemons to the Mediterranean in medieval times.[f] Perhaps the most famous Chinese contribution to the world's crops is tea, which may first have been domesticated in Szechwan as early as Han times, though tea did not become a popular beverage in the rest of China until the late Thang. The fine teas of the

[a] Ping-Ti Ho (1).
[b] See pp. 455–9 below; also Ping-Ti Ho (1), (4).
[c] Vavilov (2), p. 26.
[d] Sasaki (*1*); Furushima (*2*).
[e] Laufer (1), p. 539.
[f] See Section 38; also Lefebvre (1); Cameron & Soost (1).

[1] 常熟

Southeastern coast were later to become China's principal export in its trade with the West, and it was only in the late 19th century that tea plantations were established in Assam, South India and Ceylon which began to rival the Chinese product on western markets.[a]

## (1) Crop Rotations

One of the striking features of traditional Chinese agriculture is the intensive use made of the land. Population densities in some parts of China have been high from early times,[b] and the size of holdings has traditionally been limited not by shortages of labour or livestock but by the exiguous supply of arable land. In medieval Europe farm productivity was dependent upon the number of livestock kept, for these were the principal source of manure. Since supplies of manure were limited, yields were low and the land had to be fallowed one year in every two or three, with the result that 2 or 3 hectares of land were required to feed each person.[c] Fallow rotations like those of medieval Europe were supposedly practised in China in the early Chou period, though even then not, apparently, on the most fertile land. The *Han Shu* gives the following description of the system of land distribution traditionally attributed to the Sage Kings:[d]

> The distribution of land to the people was as follows: best land, 100 *mu* for each family head; medium land, 200 *mu* for each family head; and the poorest land, 300 *mu* for each family head. Land that was ploughed and sown every year was called non-changing land and was the best land; land that was fallowed for one year was called single-change land and was the medium land; land that was fallowed for two years [in a row] was called double-change land and was the poorest land ... Once the peasant household had received its [basic] allotment of land, additional grants were made for the extra adult males in the family; thus the allotment was proportional to the number in the household.

By the Warring States period, writers like Mencius and Hsün Tzu assumed that a household of eight persons (the couple, their aged parents and four children) could survive on a land allotment of 100 *mu* or something under 2 ha.[e] In the Metropolitan Provinces land must have been in really short supply by the early Han, for the *Fan Sheng-Chih Shu* describes fallowing as a last resort to raise diminishing yields rather than a common practice, and we know that at the same period large numbers of peasants were encouraged, if not obliged, to leave the North China plains to open up new land in the Northwest or down in the Yangtze Delta.[f] By western standards Chinese land-holdings were minute even as early as Han times. A measure of the contrast with Europe is that in Sung China *landlords*

[a] The origins and development of the tea industry will be treated in detail in Section 42.
[b] Ping-Ti Ho (4).
[c] Slicher van Bath (1), pp. 18ff.
[d] *C/HS* 24A/2*b*–3*a*, tr. auct.; see also Swann (1), p. 118.
[e] Cho-Yün Hsü (1) p. 9; the *mu* was rather smaller during the Eastern Han than in Sung times, for example.
[f] Cho-Yün Hsü (1), (2); F. Bray (3).

in the Lake Thai area owning more than 100 *mu* (then about 15 acres or 6 ha.) had part of their estates confiscated; actual holding sizes in the more productive rice-growing areas were likely to be as little as 1 acre, though in the 1930s the *average* holding size was $5\frac{1}{2}$ acres in North China and 3 acres in the south.[a] In the most fertile and therefore most densely populated parts of medieval Europe, for example in Flanders or Eastern England *c.* 1300, a farmholding of under 3 ha. or $7\frac{1}{2}$ acres was too small to support a family.[b]

The small size of Chinese farms was made possible by their high productivity—and conversely. In the section on Tools and Techniques we have seen how intensively the soil was tilled and the plants cared for during their growth. The careful spacing of crops, the transplanting of rice, and repeated hoeing and weeding of the growing crops did much to guarantee high yields, as did rigorous seed selection. Only a small proportion of the crop was required for seed grain, and none for feeding draught animals which ate mainly straw or forage; thus almost all the harvest was available for human consumption. The fertility of the fields was maintained, not by allowing livestock to graze on (and thus manure) the fallow land, but by the application of such fertilisers as human excrement, river mud or oil-cake, and by the use of judicious crop rotations including legumes and other soil-enriching crops.

Continuous cropping seems to have been well-established in many parts of China by Han times.[c] The earliest systematic account of crop rotations comes down to us from the early +6th-century *Chhi Min Yao Shu*, which makes a point of suggesting the most suitable preceding crop (*ti*[1], 'base') for a number of cultivars (Table 9).[d]

Green manures also played a very important role in Chinese crop rotations, apparently as early as Chou times;[e] they are repeatedly referred to in the *Chhi Min Yao Shu* and subsequent agricultural treatises. The careful maintenance of soil fertility, together with the breeding of quick-ripening or hardy varieties, permitted the extension of the cropping seasons and the intensification of cropping patterns from continuous cultivation to multi-cropping. In northern provinces where winters were very cold it was often impossible for climatic reasons to grow more than one crop a year, but in milder regions like Szechwan or Kwangtung it was quite common to grow two or three main crops in a year, plus a few catch-crops of vegetables or other quick-ripening plants:[f] '[In 17th-century

[a] Buck (2), p. 272; holdings in Hunan or Yunnan were, of course, much larger than in Fukien or Chekiang.

[b] Slicher van Bath (1), p. 134; in the late 13th century a high proportion (up to half) of the holdings actually fell into this category. Poor farmers with small holdings who were not bound to feudal estates often eked out their living by working as labourers, but still went in permanent fear of starvation. This was, of course, one of the periods when many peasants left the land for the towns.

[c] Yu Yü (*3*); Cho-Yün Hsü (1), p. 13.

[d] See Amano (7), p. 482.

[e] Chhen Liang-Tso (*2*).

[f] *Kuang Tung Hsin Yü* 14, p. 372.

1 底

Table 9. *Crop rotations in the* Chhi Min Yao Shu

| *Crop to be sown* | *Recommended preceding crop* (ti[1]) (*in order of excellence*) |
|---|---|
| Foxtail millet *Setaria italica* (*ku*[2]) | Green gram *Phaseolus aureus* (*lü tou*[3])<br>Adzuki bean *P. angularis* (*hsiao tou*[4])[a]<br>Cucurbits (*kua*[5] or *lo*[6])<br>Hemp *Cannabis sativa* (*ma*[7])<br>Broomcorn millet *Panicum miliaceum* (*shu*[8])<br>Sesame *Sesamum indicum* (*hu ma*[9])<br>Rape turnip *Brassica rapa* (*wu ching*[10])<br>Soybean *Glycine max* (*ta tou*[11]) |
| Broomcorn millet *Panicum miliaceum* (*shu*[8]) | Soybean; foxtail millet; newly opened land |
| Adzuki beans *Phaseolus angularis* (*hsiao tou*[4]) | wheat or barley (*mai*[12]); foxtail millet |
| Hemp *Cannabis sativa* (*ma*[7]) | adzuki beans |
| Cucurbits (*kua*[5]) | adzuki beans; late foxtail millet; broomcorn millet |
| Wheat or barley (*mai*[12]) | broomcorn millet |
| Rape turnip *Brassica rapa* (*wu ching*[10]) | wheat or barley |
| Coriander *Coriandrum sativum* (*hu sui*[13]) | wheat or barley |
| *Lithospermum officinale* (*tzu tshao*[14]) (a dye plant) | wheat or barley; non-glutinous broomcorn millet (*chi*[15]); newly opened land |

[a] These two legumes have recently been reclassified from *Phaseolus areus* Roxb. and *Phaseolus angularis* (Willd.) W. F. to *Vigna radiata* (L.) R. Wilczek and *Vigna angularis* (Willd.) Ohwi & Ohashi respectively; see Maréchal, Mascherra & Stainier (1).

1 底　2 穀　3 綠豆　4 小豆　5 瓜
6 蓏　7 麻　8 黍　9 胡麻　10 蕪菁
11 大豆　12 麥　13 胡荽　14 紫草　15 穄

Kwangtung] they grow two crops of rice in the early fields and then plant brassica (*tshai*[1]) to make oil or indigo for dyes, or grow turmeric or barley, rape or sweet potatoes ... On flat hills and ridges, reeds [for basketry], sugar cane, cotton, hemp, beans, herbs, fruits and melons are also grown in quantities.' This passage describes cropping patterns in the vicinity of Canton, where there was a ready market for such products. Access to markets naturally influenced the farmer's choice of crops—there was no point in sowing large quantities of perishable goods like vegetables unless one farmed near a town; on the other hand if one did own land in a suburb it could often be more profitable to abandon cereal cultivation altogether in favour of market gardening. The 6th-century *Chhi Min Yao Shu* recommends suburban landholders to plant large acreages exclusively

1 菜

with vegetables for market, giving calculations for as much as 100 *mu* of rape turnips, for example, and indicating the very high profits on such ventures.[a] A limiting factor to such diversification, though, was the general level of agricultural productivity in the region: if the climate was harsh and cereal yields were low, farmers would have to concentrate on growing grain for subsistence and could not afford to spare land for commercial crops. Buck reports that in the 1930s in the Spring Wheat region of North China, where conditions were very harsh indeed, over 90% of all the crops grown were seed products (grains, legumes and oil-seeds), that is to say staple foods.[b]

Chinese crop rotations are so varied as to defy any practical classification even within a single region.[c] Buck and his collaborators identified 574 systems of crop rotation in China as the most common.[d] The situation is further complicated by the common Chinese practice of interplanting; thus rows of late rice are grown between early rice,[e] broad beans in trenches between rows of wheat,[f] squash between barley,[g] sweet potatoes on mounds between rice rows or peanuts on sandy soils with sugar-cane.[h] Most traditional Chinese rotations, however, included the basic staples of cereal, legume and oil-crop, though the variations on this theme were extremely complex and many rotations ran over as much as five years.[i] Table 10 gives a few examples of Chinese rotations, culled at random.

Now let us consider those crop species which have played a fundamental role in Chinese agriculture, beginning with the cereals. The Chinese term for grain, *ku*[1], was applied not only to the main cereal crops but also to such field crops as hemp and beans, also cultivated for their grains. Thus *wu ku*[2], the 'five grains', an expression commonly found in the classical texts, was understood to comprise *chi*[3] (setaria millet), *shu*[4] (panicum millet), *tao*[5] (rice), *mai*[6] (wheat and barley) and *shu*[7] (legumes), though some commentators substituted *ma*[8] (hemp) for rice.[j] Other classifications referred to the 'six grains' (*liu ku*[9]) or the 'nine grains' (*chiu ku*[10]), making up the numbers by using more specific names for wheat, for barley, or for large or small beans.[k] Here we shall give precedence to the true cereals.

[a] *CMYS* 18.5.1.

[b] Buck (2), p. 208.

[c] 'An attempt was made to discover whether there was any generally adopted plan of crop rotation, but the best we could do was to secure a long list of three-year programmes', says Gamble (1), p. 219 of the district he studied in Hopei.

[d] (2), Statistical Volume, Table 18.

[e] The earliest reference to this practice appears to be in the 14th-century *Nung Thien Yü Hua*; it was still widespread in China in the 1930s; see p. 508 below.

[f] *NCCS* 26/13*a*.

[g] Gamble (1), p. 219.

[h] Food and Fertiliser Technology Centre (1), pp. 27, 47.

[i] This gives some idea of the security of tenure enjoyed by Chinese tenant farmers.

[j] Bretschneider (1), vol. II, p. 137.

[k] *Ibid.* p. 139; see also the *Chiu Ku Khao* and Liu Pao-Nan (*1*).

[1] 穀 [2] 五穀 [3] 稷 [4] 黍 [5] 稻
[6] 麥 [7] 菽 [8] 麻 [9] 六穀 [10] 九穀

Table 10. *Some common Chinese crop rotations*

| Locality | Period | Year | Winter/early spring crop | Summer crops |
|---|---|---|---|---|
| Yangtze Delta[a] | Sung | 1 | Wheat, beans, or rape | Rice |
| Yangtze Delta[b] | C17th | 1 | Wheat or rape | Rice |
| Canton region[c] | C17th | 1 | Rape, sweet potatoes, vegetables, spices, etc. | Early rice/Late rice |
| Taiwan[d] | C18/19th | 1 | Various vegetables | Early rice/Late rice |
| Chihli (low land)[e] | C20th | 1 | Wheat | Indigo |
| | | 2 | Wheat | Cabbage |
| Shansi (terraced hillside)[f] | C20th | 1 | Irish potatoes | — |
| | | 2 | Spring wheat | — |
| | | 3 | Setaria millet | — |
| | | 4 | Kaoliang | — |
| | | 5 | Millet | — |
| Winter wheat-millet area[g] | C20th | 1 | Winter wheat | Soybeans |
| | | 2 | — | Kaoliang |
| | | 3 | Winter wheat | Soybeans or black beans |
| | | 4 | — | Millet |
| | | 5 | — | Kaoliang |
| Anhwei (medium land)[h] | C20th | 1 | Wheat | Tobacco |
| | | 2 | Barley | Cowpeas |
| | | 3 | Wheat | Buckwheat |
| | | 4 | Vegetables | Vegetables |
| Kiangsu (high land)[i] | | 1 | Wheat or barley | Soybeans & sesame |
| | | 2 | Broad beans or peas | Sweet potatoes |
| | | 3 | Cotton | — |
| Yunnan[j] | C20th | 1 | Broad beans | Rice |
| South China[k] | C20th | 1 | Wheat interplanted with peas | Cotton interplanted with turnips |

[a] *CFNS* 1/3 and other Sung sources; see Amano (*4*), p. 232.
[b] *Shen Shih Nung Shu*, p. 225.
[c] *Kuang Tung Hsin Yü* 14/p. 372.
[d] Food & Fertiliser Techn. Centre (1).
[e] Buck (1), p. 170.
[f] *Ibid.* p. 173.
[g] T. H. Shen (1), p. 145.
[h] Buck (1), p. 169.
[i] *Ibid.* p. 175.
[j] Fei & Chang (1), p. 21.
[k] Wagner (1), p. 261.

## (2) Millets, Sorghum and Maize

'Millet' is a general term for a wide range of small-seeded cereals, including foxtail millet (*Setaria italica*), broomcorn millet (*Panicum* spp.), finger millet (*Eleusine corocana*), bulrush millets (*Pennisetum* spp.), barnyard millets (*Echinocloa* spp.) and a number of other cereals. As well as their large ears containing many small seeds, millets have in common their heat-tolerance and drought-resistance. They are widely cultivated in the tropics, but they can also be successfully grown in more temperate latitudes, and they seem to have been among the earliest domesticated cereals in several quite separate parts of the world. Millets are generally well balanced nutritionally, with a relatively high protein-content,[a] and are often highly valued by rural populations as food for small children, nursing mothers and old people, even when their cultivation has largely been abandoned in favour of improved varieties of such large-grained cereals as wheat, maize or rice. Foxtail and broomcorn millet, *Setaria italica* (L.) Beauv. and *Panicum miliaceum* (L.) Beauv.[b] have been cultivated in China since neolithic times when they apparently formed the staple food of the population in much of the north, and setaria in particular is still economically important in North China today. Since medieval times Chinese millets have faced increasing competition from sorghum and maize, both of which have very similar environmental and cultivation requirements to millets, and indeed have generally been classified by Chinese peasants and botanists alike as belonging to the class of millets. For this reason it seemed appropriate to discuss Chinese millets, sorghum and maize in the same section, even though botanically they are no more closely related than setaria to wheat or maize to sugar-cane.

### (i) *Foxtail millet* (Setaria italica) *and broomcorn millet* (Panicum miliaceum)

Setaria was found in large quantities at the early Yang-shao site of Pan-pho[1] in Shensi (*c.* −5000), as well as in several other northern neolithic sites, and it is generally accepted that it formed the staple diet of early farming communities in the north.[c] Panicum has so far been reported at only one neolithic site, Ching-tshun[2] in Southern Shansi, and must, since it is a unique occurrence, remain suspect.[d] Both panicum and setaria were extremely important in Shang and

[a] Usually 10–12% or more (Purseglove (2)), superior to maize, rice or tubers and roughly equivalent to wheat (Whyte (1), p. 38, table 13).

[b] These are the botanical names generally accepted today; the nomenclature of *Gramineae* generally, and of cereals in particular, has varied enormously at different periods and under different systems; some of the complexity is conveyed by a brief perusal of Ames (1), who lists some of the different Latin names given to important crop plants.

[c] K. C. Chang (1), (4); Ping-Ti Ho (5); Li Fan *et al.* (*1*); etc.

[d] K. C. Chang (1), p. 95, commenting on C. W. Bishop (4).

[1] 半坡 [2] 荆村

Chou China, as can be seen from the numerous references to both crops in the oracle-bone inscriptions, the *Shih Ching*, and other early texts:[a]

> The Descendant's stacks
> Are high as cliffs, high as hills.
> We shall need thousands of carts,
> Shall need thousands of coffers
> For broomcorn and foxtail millet, rice and spiked millet.

Throughout the historical period setaria and panicum have been considered as northern rather than southern crops: 'Northerners rely on setaria for their daily food', says the Ming author Wang Hsiang-Chin[1].[b] In Ling-nan[2] there is very little millet, either setaria or panicum, says the 17th-century *Kuang Tung Hsin Yü*.[c] The importance of millets in the north of China, their scarcity in the south, and their suitability for cultivation in semi-arid environments have led modern scholars to postulate that both setaria and panicum were domesticated in the Yellow River area. De Candolle (1) saw *Panicum miliaceum* as originating in Egypt but thought that one of the original centres for *Setaria italica* was in China, while Vavilov (2) believed that both types of millet had their centre of domestication in the north of China; Ping-Ti Ho (5), (6) stressed the drought-resistance of Chinese millets in support of this hypothesis.

An interesting counter-argument has been developed by Fogg (1) who points out that, while it is true that setaria and panicum tolerate drought better than many other cereals, they in fact thrive best in semi-tropical conditions such as those of southern China and Southeast Asia, or meso-America where millet may have been domesticated earlier than maize.[d] Fogg contrasts the ease of setaria farming among the aborigines of modern Taiwan with the labour-intensive techniques of moisture conservation necessary to produce a crop in the loess lands of China's Nuclear Area and concludes:[e] 'The farming of foxtail millet is difficult in the loess region and logic is against it as the place where highly evolved varieties of *Setaria italica* were domesticated.' In further support of his hypothesis he points out:[f]

> It is known that *S. italica* is almost 100 % self-fertilising with very little out-crossing. Normal out-crossing rates are 0·59 % to 1·11 % (Li, Li & Pao (1)). However one study by Li, Meng & Liu (1) reported a relatively high outcrossing figure of 7·63 %. The researchers reasoned that in the semi-arid region of their test site located near Kaifeng, the

[a] *Fu Thien*, tr. Waley (1), p. 170 adj. auct.
[b] *Chhün Fang Phu, Ku Phu*[3], 15*b*.
[c] *Ibid.* p. 98.
[d] E. O. Callen (1); but see also Charles A. Reed (4), p. 927.
[e] (1), p. 20.
[f] (1), p. 27.

[1] 王象晉 [2] 嶺南 [3] 穀譜

pollen was blown from one plant to the stigma of another plant more easily than in more humid areas. They concluded that out-crossing in *S. italica* is higher in habitats with dry atmospheric conditions and lower in those with relatively more humid conditions. Therefore it appears that initial domesticators of foxtail millet could have had greater success in selecting and maintaining new stable genotypes in regions with relatively more humid atmospheric conditions than that [sic] of North China. Some hybridisation does occur, and genetic introgression between the wild foxtail and the cultivar will occur unless certain farming techniques are performed to prevent it.

Later, as we shall see, the northern Chinese farmers took full advantage of the hybridisation potential of millets in arid areas to develop an enormous range of varieties, but this does not invalidate Fogg's argument as to the most favourable conditions for initial domestication. According to the study by Kano Tadao (1) (Fig. 215), setaria millet appears to be the oldest of the annual cereals domesticated in East Asia, followed by barnyard millet (*Echinocloa crus-galli*) and then panicum millet. Fogg suggests that setaria, and probably panicum too, were first domesticated in Southeast Asia by Hoabinhian horticulturalists whose main staple was taro: Hoabinhian farmers gradually moved up into north China, into drier environments unsuitable for taro cultivation, and as they did so, says Fogg, they devoted greater efforts to developing millet varieties that would supplement their diet in these less favourable conditions.

Fogg's hypothesis is attractive, though whether population movements rather than cultural contacts should be considered responsible for the diffusion of crop plants is debatable.[a] However, many botanists and archaeologists would argue that it is precisely under conditions of stress, where traditional food resources are in short supply, that domestication takes place, and this would point to the north rather than the south of China as the area where millets were first domesticated. On the other hand, it is contestable that the Nuclear Area did in fact have a semi-arid climate at that period, as Fogg and Ping-Ti Ho (5), (6) believe. It has been suggested that the climate of North China in the period with which we are concerned, namely from about −8000 to −500, was considerably warmer and more humid than it is today, probably approximating to the climatic conditions under which Fogg's recent fieldwork in Taiwan was carried out.[b] The annual wild ancestor from which foxtail millet evolved, *Setaria viridis*, is common all over East Asia, north and south,[c] so that *a priori* domestication could have taken place anywhere within this region. For such crops as rice or wheat it is possible (up to a point) to identify amongst archaeological remains forms intermediate between wild ancestors and domesticates, but such identification usually depends on changes in size and shape of grain or modifications in the glume pattern. In setaria the chief difference between wild and domesticated forms lies in the size of

[a] An interesting attempt has been made by Ammerman & Cavalli-Sforza (1) to construct mathematical models for stimulus and demic diffusion of agricultural advance in early Europe, but as yet no model has been developed that evaluates the relative importance of each mode of diffusion.

[b] K. C. Chang (5); Anon. (*530*); Chu Kho-Chen (*11*).

[c] Fogg (1), pp. 4–5.

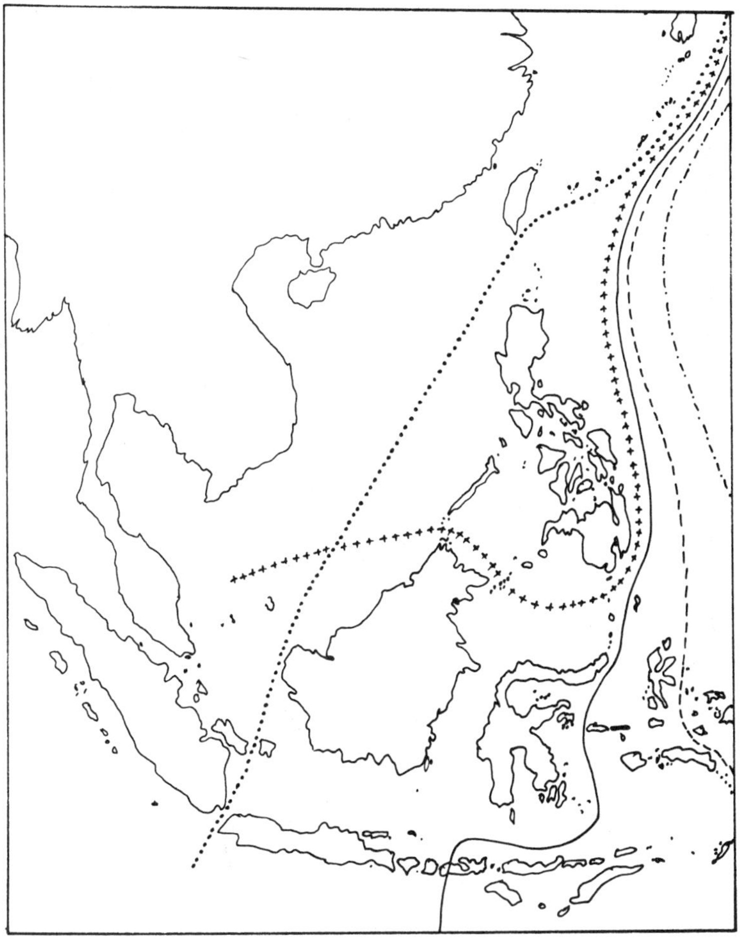

· · · · : *Eleusine corocana*, finger millet
+ + + : *Panicum miliaceum*, broomcorn millet
——— : *Echinocloa cruss-galli*, barnyard millet
– – – – – : *Setaria italica*, foxtail millet
–·–·–· : *Coix lacryma-jobi*, Job's tears

Fig. 215. Map showing the distribution of Coix, Setaria and Panicum in East and Southeast Asia; after Kano Tadao (1).

the panicles rather than the size or shape of seed;[a] this only makes the precise process of domestication the more difficult to establish.

We now come to another thorny problem, namely that of terminology. The confusion begins with the Shang oracle-bone inscriptions, persists throughout the historical period, and has survived triumphant to the present day. Although the

[a] Wild setaria panicles in Taiwan are only 3–6 cm. long, whereas the panicles of domesticated varieties may be from 40 to 60 cm. in length; Fogg (1), p. 39.

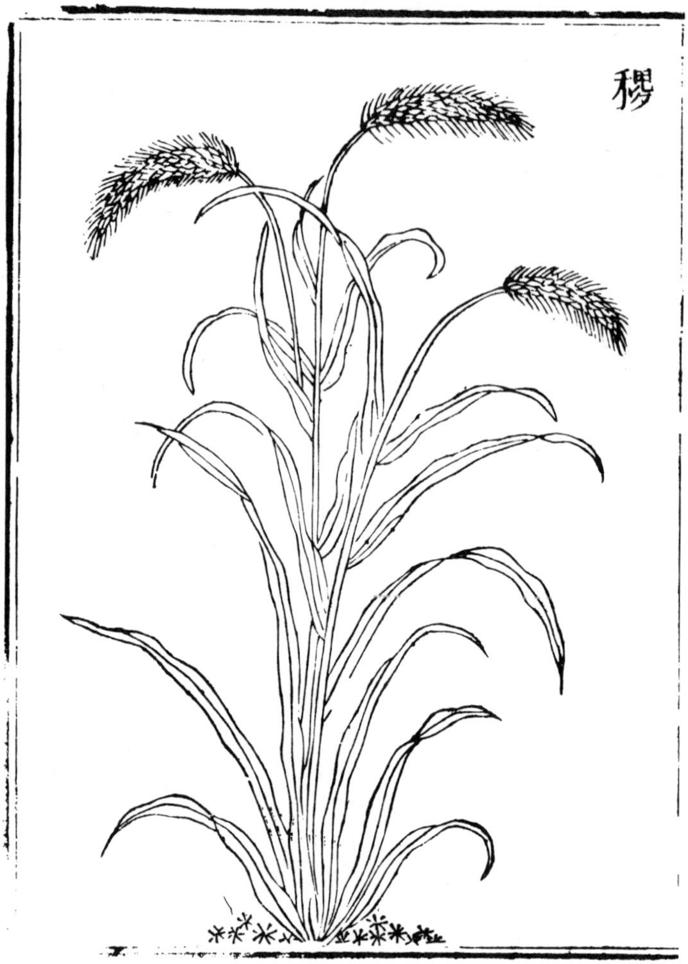


Fig. 216. Setaria millet; *SSTK* 23/5*b*.

Chinese have never been in any doubt that setaria and panicum are separate species, terminological confusion and inconsistency between the two, as well as between general terms and names for glutinous and non-glutinous varieties, has been rampant ever since the Han if not earlier, exacerbated by regional as well as chronological variations in usage (Figs. 216, 217). *Chi*[1], for example, as used in the expression Hou Chi[2], 'Lord Millet', the founding ancestor of the house of Chou, is generally accepted as referring to setaria; it was used in this sense in the *Shih Ching* as well as in the *Chhi Min Yao Shu* and the *Nung Cheng Chhüan Shu*, and according to the *Chhi Min Yao Shu* it was a very common term for setaria throughout north China in those times;[a] but the *Pen Tshao Kang Mu* quotes the +5th century Thao

[a] *CMYS* 3.1.1.

[1] 稷 [2] 后稷

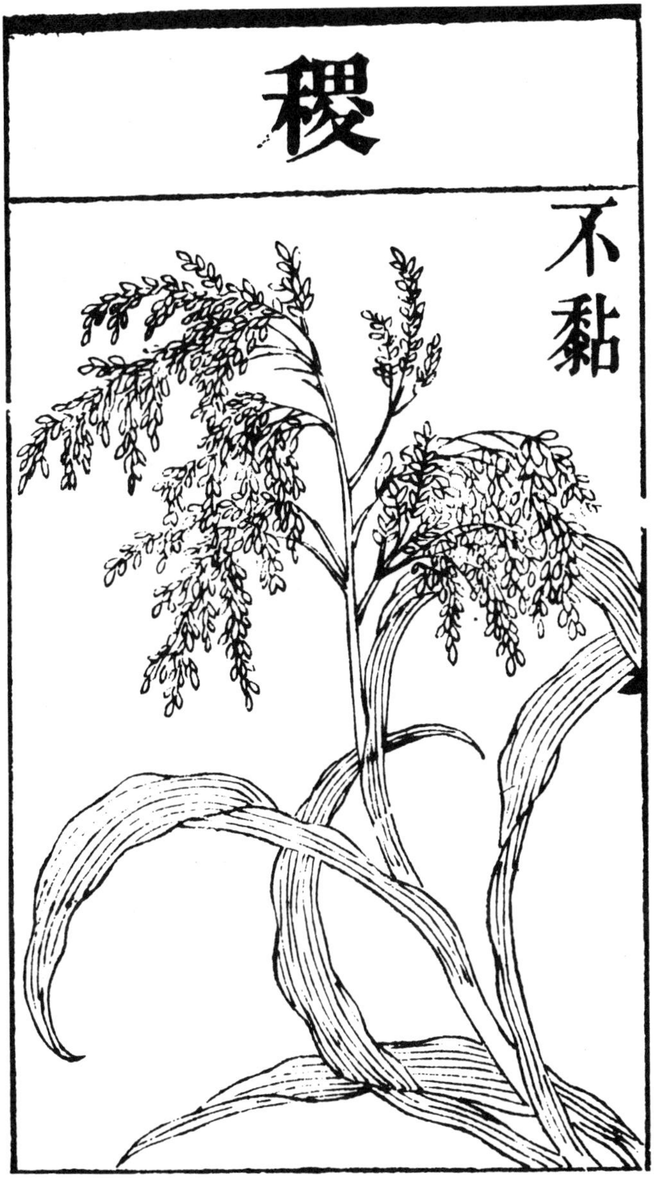


Fig. 217. Panicum millet; *PTKM* (1885 ed.) 2/24*a*. Note that this and the previous figure are both labelled with the same Chinese character *chi* 穄.

Table 11. *Terminology of Chinese millets*[a]

| | |
|---|---|
| General terms: | 禾 *ho:* possibly used in the general sense of 'millet' in the oracle-bone inscriptions; common literary usage in later texts e.g. *Shuo Wen*, *Shou Shih Thung Khao*, etc. |
| | 穀 *ku:* used e.g. in *Nung Cheng Chhüan Shu*; also a general term for cereals or unhulled grain. |
| | 粟 *su:* a general term for unhulled millet-grain as given in the *Shuo Wen*, etc. |
| *Setaria italica (L.) Beauv.:* | 穀/谷 *ku:* used e.g. in *Chhi Min Yao Shu*, *Nung Cheng Chhüan Shu*; modern term for setaria in North China is *ku tzu* 谷子. |
| | 粟 *su:* used in *Fan Sheng-Chih Shu*, *Kuang Chih*, *Wang Chen Nung Shu*. |
| | 稷 *chi:* used in *Shih Ching*, several classical texts, *Nung Cheng Chhüan Shu* and the name Hou Chi 后稷, 'Lord Millet', to denote setaria; otherwise used for panicum (see below). |
| | 粱 *liang:* earliest mention in *Shih Ching*; appears to denote a large-grained, fine-flavoured sub-species of setaria (cf. *CMYS*, ch. 5), probably what is known as 'German millet'. |
| | 秫 *shu:* glutinous setaria according to early lexicographical works such as *Erh Ya* or *Shuo Wen*, but sometimes used as a general term for glutinous cereals, and subsequently incorporated into terms for sorghum and maize; the modern term for glutinous setaria is *nien ku* 黏谷. |
| Less common terms: | 粢 *tzu:* defined by *Erh Ya* as setaria. |
| | 虋 *men:* defined by *Erh Ya* as red-stemmed setaria. |
| | 芑 *ssu:* defined by *Erh Ya* as white-stemmed setaria. |
| Hulled grain: | 小米 *hsiao mi* |
| *Panicum miliaceum (L.) Beauv.:* | 黍 *shu:* usually glutinous panicum, but may also be used as general term for panicum; earliest usage occurs in oracle-bone inscriptions; still current today. |
| | 穄 *chi:* non-glutinous panicum, as used for example in *Chhi Min Yao Shu*, ch. 4, *Wang Chen Nung Shu*, ch. 7; this term is homophonous with the following one, and the two have persistently been confused. |
| | 稷 *chi:* non-glutinous panicum, used in this sense at least since +5th century and incorporated into the modern botanical classification of panicum; it has often been used, however, to refer to setaria. |
| Less common terms: | 秬 *chü:* defined by *Erh Ya* as black, glutinous panicum; used for wine, *chhang* 鬯, as referred to in early bronze inscriptions. |
| | 秠 *phei:* defined in *Erh Ya* commentary as black, glutinous panicum with two grains in each husk. |
| | 糜 *mei:* the Liang work *Yü Phien* quotes the Chhin work *Tshang Hsieh Phien*, defining *mei* as non-glutinous panicum, but the +1st-century *Ssu Min Yüeh Ling* defines it as the most glutinous, sticky kind of panicum. Today the term *mei tzu* 糜子 (usually pronounced *mi tzu*) is used in parts of North China such as Shensi as a general term for panicum. |
| Hulled grain: | 黃米 *huang mi* |

[a] Based principally on the discussions in Amano (*4*), pp. 3–52, Hu Hsi-Wen (*3*), pp. 3–9, and Pernès, Belliard & Métailié (1), as well as the relevant sections of *WCNS*, ch. 7, *Thung Chih*, ch. 75, *NCCS*, ch. 25, etc.

Hung-Ching[1] as saying that many literary works define *chi*[2] as resembling panicum, *shu*[3],[a] and this some decades before the *Chhi Min Yao Shu* was written; in modern botanical works *chi shu*[4] is the term used for the *Panicum* species. Table 11 represents our attempt to unravel some at least of the terminological tangles surrounding Chinese millets.

Having attempted to define which millets are spiked and which are panicled, which are sticky and which are not, we are now in a slightly better position to investigate the variations within these categories. Over centuries and millennia, the Chinese have evolved and selected an enormous number of millet varieties, characterised by differences in yield, flavour, drought or flood resistance, growth period, etc., so that there is a suitable millet for almost every conceivable circumstance. There is an oft-cited proverb to the effect that even an eighty-year old farmer cannot count all the varieties of millet, or as Wang Chen puts it in a pithy understatement: 'There is more than one name for millet (*su chih wei ming pu i*[5]).'[b] One naturally chose the variety that one would plant to suit the conditions in which it was to grow:[c]

> There are all sorts of millet: they ripen at different times, they vary in height and yield, in the strength of their straw, in flavour, and in the ease with which they shed their grain. (The varieties that ripen early have short stems and give a good yield; those which ripen late have longer stems and yield less grain. The strong-strawed varieties belong to the class of short yellow millets, and those with weaker straws belong to the class of tall green, white and black millets. Those which yield a light crop are delicious but shed their grain easily; the large yielders are unpalatable but productive.) Soils vary in fertility. (Fertile land should be sown late and poor land early. Fertile land need not necessarily be sown late: if it is sown early no harm will be done; but poor land must be sown early or the crop will certainly be a failure.) Hilly and marshy land suit different crops. (Hilly land should be sown with hardy varieties that can withstand wind and frost, and marshy land should be sown with weaker varieties to obtain larger heads.)
>
> If you follow the natural seasons and accurately gauge the land's potential, then you will reap large rewards for little labour, but if you are wilful and oppose the natural way then however hard you work you will get no harvest. (If you dive into a spring for timber or climb into the mountains for fish you will always return empty-handed. If you try to sprinkle water against the wind or to roll a ball uphill, then the circumstances are against you.)

The Chinese developed careful selection techniques enabling them to isolate and maintain millet varieties with desirable characteristics at a relatively early stage. We know, for example, that wines fermented from glutinous panicum were an important feature of Shang ceremonial, and that the glutinous starch endosperm is a feature only of recessive mutants of the non-glutinous varieties, caused by a single recessive gene.[d] Fogg (1) concludes that farmers who are able to

[a] *Pen Tshao Kang Mu* 23/2*a*.
[b] *WCNS* 7/1*a*.
[c] *CMYS* 3.2.1–2.
[d] Watabe (1), p. 5.

[1] 陶弘景 [2] 稷 [3] 黍 [4] 稷屬 [5] 粟之爲名不一

maintain the glutinous type are at an advanced stage of cereal cultivation and are able to maintain any variety that is the result of some stable genetic change; we may therefore assume that Shang farmers had a wide range of millet varieties at their disposal. The 'Ode to Hou Chi' in the *Shih Ching* says:[a]

> ... He planted the yellow crop ...
> It nodded, it hung ...
> The black millet, the double-kernelled,
> Millet pink-sprouted and white.

The +6th century *Chhi Min Yao Shu* lists nearly a hundred varieties of non-glutinous setaria alone, including red millet (*chu ku*[1]), highland yellow (*kao chü huang*[2]), hundred day grain (*pai jih liang*[3]), people's prosperity (*min thai*[4]), mountain salt (*shan tsho*[5]), bamboo root yellow (*chu ken huang*[6]), otter-tail dark green (*tha wei chhing*[7]), and merciful yellow (*kho lien huang*[8]). It names fourteen varieties that ripen early and are drought-resistant and insect-free, twenty-four varieties that are awned, wind-resistant and are not attacked by birds, and ten varieties that are flood-resistant, as well as individual varieties with particularly fine or poor flavours or that are easy to hull.[b] The author emphasises the necessity for careful seed selection: 'If you need a particular grain to suit particular conditions, you cannot afford to be careless.'[c] The late 19th-century treatise on famine prevention, the *Chiu Huang Chien I Shu* by Kuo Yün-Sheng,[d] provides a truly wondrous calendar listing types of setaria that can be sown at almost any season throughout the year, some of which will ripen in sixty, fifty, or even forty days; it also includes a list of salt-, flood-, insect- and frost-resistant varieties. Panicum, as the *Thien Kung Khai Wu* says, also comes in many varieties but less numerous than setaria (*liang su chung lei ming hao chih to shih shu chi tu shen*[9]).[e] In the 1950s and 1960s two thousand types of setaria, subsequently reclassified as 600 distinct varieties, were collected in Shantung province alone, even though millet is now a comparatively unimportant crop in this province.[f] The astonishing range of traditional setaria types may be due in part to the relatively high natural incidence of sterile male forms, which increases the chances of hybridisation in this homozygous plant.

The Chinese farmers had at their disposal varieties of panicum and setaria millet suited to almost any conditions of soil and climate. In an emergency, if the spring crop was a total failure, they might be able to avert disaster by planting a fifty-day variety after the summer rains; other types would survive in saline soils, in marshy areas, or on mountainsides. Known difficulties could thus be overcome

[a] Waley (1), p. 242.
[b] *CMYS* 3.1.5.
[c] *CMYS* 2.2.2.
[d] (*1*), quoted Hu Hsi-Wen (*3*), pp. 279ff.
[e] *TKKW* 1/18*b*; one might also compare the list of panicum varieties in Kuo Yün-Sheng (*1*), cited Hu Hsi Wen (*3*), pp. 463–5 with that given for setaria, *ibid.* pp. 279–85.
[f] Pernès *et al.* (1), p. 29.

1 朱穀 2 高居黃 3 百日粱 4 民泰 5 山鹺
6 竹根黃 7 獺尾青 8 可憐黃 9 粱粟種類名號之多視黍稷獨甚

by a judicious selection of the appropriate variety, but however carefully one chose, one still had to reckon with the capriciousness of the Chinese climate, particularly in the North. To a certain extent one could protect oneself against climatic vagaries by growing more than one variety and staggering the sowing times;[a] in some cases, as for example in the cold and arid mountain zone of Northern Shensi, farmers traditionally planted several varieties of setaria simultaneously in the same field, though only one single variety of panicum was sown. There was even a proverb to this effect: 'Red panicum, mottled setaria (*hung mi tzu, hua ku tzu*[1]).'[b] Presumably not all varieties of setaria were likely to succeed simultaneously under the extreme conditions, so that the problem of harvesting varieties that ripened at different times in the same field did not arise. Panicum, on the other hand, was hardier, and so such precautions were not necessary:[c]

> In remote districts in the North only glutinous paricum (*shu*[2]) will grow: one might say that it is sown in the heat and harvested in the heat.[d] The stem is short and the ear small (the local people call it *lin tzu*[3]), and it can be used either for brewing wine or for making gruel; it is gelatinous and sweet ...

Not only were there fewer varieties of panicum than of setaria, but despite its comparative hardiness it was grown less often and over a far smaller area. Wagner (1), reporting on the situation in the 1930s, says that while setaria was cultivated almost all over China, expecially in hilly areas, panicum cultivation was largely restricted to Shantung, where it was chiefly used in the production of spirits. Neither type, he points out, was important in the chief rice-growing areas where millets were practically unknown.[e] Generally speaking, panicum was regarded as a delicacy or as raw material for making wine, while setaria, the non-glutinous variety especially, was everyday food, consumed mainly in the form of gruel made from the hulled grain (Fig. 219).

Millets are gross feeders requiring plenty of soil nutrients. They are often sown after a leguminous crop,[f] though according to the *Chhi Min Yao Shu* glutinous setaria (*shu*[4]) and the fine-flavoured but comparatively low-yielding German millet (*liang*[5]) (Fig. 218) preferred poor soils.[g] The seed was thinly sown, usually by drill, any time between spring and early summer, and the soil compacted around the small seeds with a light roller (*thun chhe*[6]) (see p. 272 above), or simply

[a] E.g. *CMYS* 3.7.1.
[b] G. Métailié, pers. comm.
[c] *WCNS* 7/10*a*.
[d] That is to say, the growing season is very short, presumably the interval between the solar terms *hsiao shu*[7], small heat, and *chhu shu*[8], term of heat, an interval of about six weeks.
[e] Buck's surveys ((2), p. 211) confirmed this: in North China 19·4 % of the cultivated area was sown with setaria, 7·7 % with panicum, while in the south only 1·6 % was sown with setaria and 0·3 % with panicum.
[f] *CMYS* 3.3.1; W. Wagner (1), p. 305.
[g] *CMYS* 5.3.1.

[1] 紅糜子花谷子 [2] 黍 [3] 林子 [4] 秫
[5] 粱 [6] 砘車 [7] 小暑 [8] 處暑

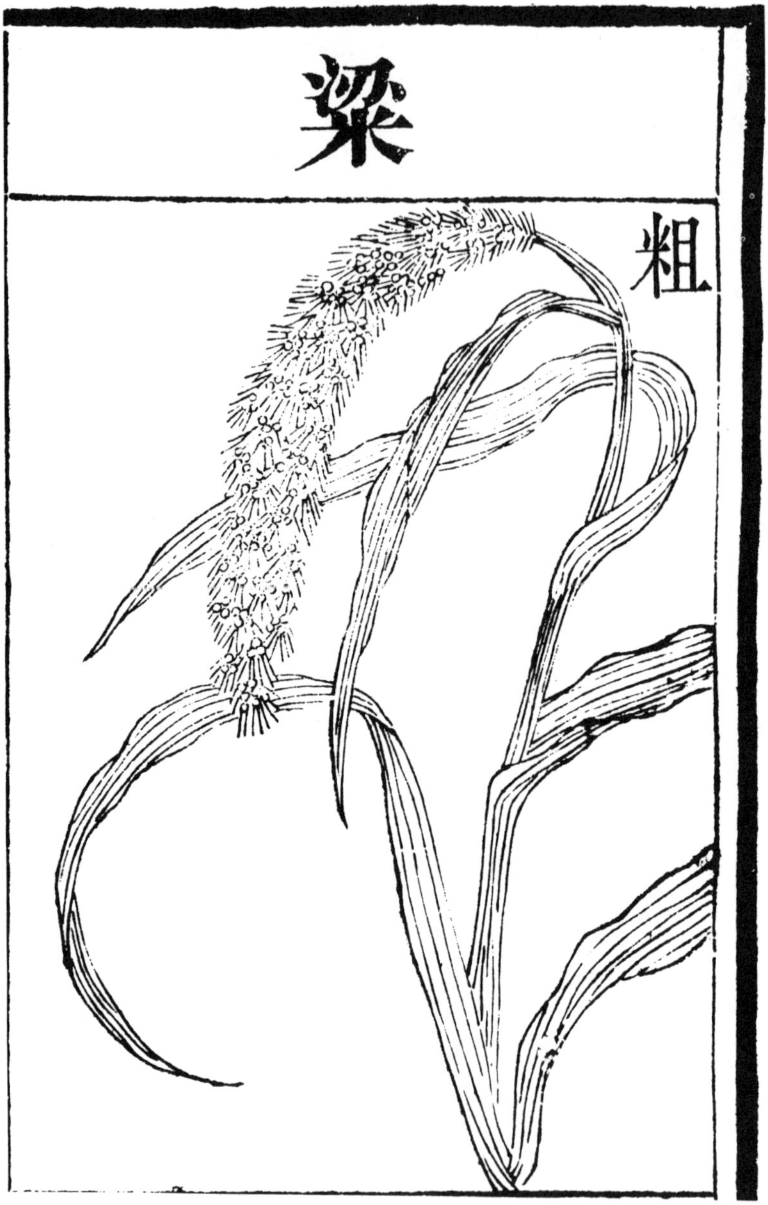


Fig. 218. Large-grained setaria millet (*liang*); *PTKM* 2/25*a*.

Fig. 219. Hulling millet with a hand-roller; *TKKW* 1/65*a*.

by stamping the seed into the furrow. An interesting belief connected with the sowing of panicum is recorded in the *Chhi Min Yao Shu*:[a]

> If you always remember which were the 'frozen tree days' (*tung shu jih*[1]) when sowing panicum, your crop will never fail. ('Frozen tree days' occur when a hard frost completely covers the branches of the trees. If the third day of such a [winter] month is a 'frozen tree day', then you should sow panicum on the third day of the corresponding [spring] month ...)

This rather complicated formula for calculating sowing dates has a remarkable parallel in the classical world, noted by Pliny:[b]

> As a result of careful observation it is said that [root vegetables] give a wonderfully fine crop if they are sown on a day that is as many days after the beginning of the period specified [late July to late August] as the moon was old when the first snow fell in the preceding winter.

The parallel is the more inexplicable since it is highly improbable that there was any factual basis whatsoever to either of these formulae.

Once the millet seedings were established they were carefully thinned. The *Chhi Min Yao Shu* recommends leaving 1 foot between setaria plants in good soil.[c] The *Fan Sheng-Chih Shu* says that panicum plants should be slightly more widely spaced than setaria, but the *Chhi Min Yao Shu* questions the wisdom of this:[d]

> In my opinion, although the plants are bigger when the panicum is widely spaced, the grain is yellow and there are many half-full or empty husks. It is true that the present method of close-spacing makes for smaller plants, but as the grain is white and ripens evenly, and there are no half-empty husks, it is superior to wide spacing ...

If the millet plants were properly spaced, they not only gave higher yields but also depleted the soil less. The *Chhün Fang Phu* quotes a proverb: 'Wide-spaced setaria has large ears and next year the land is good for wheat (*hsi ku ta sui, lai nien hao mai*[2]).'[e] The crop was carefully hoed throughout its growth as this not only increased yields but improved milling rates:[f]

> You should not mind how many times you hoe; once you have been right round the field start again, and do not stop even for a short time simply because there are no weeds. (Hoeing does not just get rid of weeds, it keeps the soil at a good tilth and will give full ears, with thin husks, that do not shatter. If you hoe your field ten times you will get 'eighty per cent grain'.)

It is unlikely that many farmers were willing or able to hoe their millet fields ten times, even if this meant that husks would only account for twenty per cent of their grain yield, yet the +18th-century author Phu Sung-Ling[3] said that when

[a] 4.3.3.
[b] (1) XVIII. XXXV. 132.
[c] *CMYS* 3.9.2.
[d] *CMYS* 4.8.4.
[e] *Ku Phu* 16*b*.
[f] *CMYS* 3.10.2.

[1] 凍樹日 [2] 稀穀大穗來年好麥 [3] 蒲松齡

possible one should try to hoe as many as six or seven times, and pointed out that tenant farmers were often required to hoe their landlord's millet fields as part of their rental contract.[a] Although 'eighty per cent grain' was probably rare, it seems that 'sixty per cent grain' was considered normal in the Han.[b]

As is natural in a region to which a crop plant is native, Chinese millets were subject to a wide variety of pests and diseases, chief among which were army worms (*Cirphis unipuncta*) (*tzu fang*[1]), smut (*Ustilago crameri* Körn.) (*su nu*[2]), and 'green ear', a downy mildew (*Sclerospora* spp.) (*lao ku sui*[3] or *su pai fa ping*[4]).[c] Army worms could be got rid of by the simple procedure of smashing them to death:[d]

> The grubs fear light and stay underground during the day, but at night they all come out. You should hire numbers of people to smash the grubs at night. Not only is it cooler then, so that nobody suffers from the heat, but also you will save the three meals [that you would have had to give] the hired men [by day], and furthermore the grubs will easily be wiped out.

Against millet diseases, however, there were no remedies in traditional China, and even today they continue to ruin millet crops in many regions.[e] To a certain extent resistance to disease can be increased if the cultivated millet varieties are allowed to interbreed with their more resistant wild relatives; since the wild ancestor of setaria millet, *S. viridis* (*kou wei tshao*[5]), is a common weed throughout China, this may have contributed to the vigour of the traditional Chinese varieties.[f]

Harvesting took place in the late summer or autumn. Setaria and non-glutinous panicum had to be harvested quickly to avoid danger from wind or rain and the shedding of grain. 'Once the awn has grown long and the leaves have turned yellow you should harvest setaria at once. Setaria should be harvested as soon as half of it is ripe.'[g] If it rained near harvest time the grain should be cut immediately, even if it was not fully ripe: 'If there is a sudden rainstorm when the setaria is a third or half ripe, then you should cut it rapidly as soon as the rain stops. It should all be cut in the space of one or two days, for if you leave it any longer the setaria will lodge or turn brown and not a single grain will you have

[a] *Nung Tshan Ching*, cited Hu Hsi-Wen (*3*), p. 219.

[b] Lien-Sheng Yang (9), p. 154; Ying-Shih Yü (1), p. 73; on the traditional milling techniques of China see Vol. IV, pt 2, pp. 183 ff. Modern milling ratios are hard to come by, but the Han figures compare quite favourably with the ratio of 7 to 10 given in Purseglove (2), p. 200.

[c] Hu Hsi-Wen (*3*), p. 15.

[d] *Nung Tshan Ching*, quoted Hu Hsi-Wen (*3*), p. 250.

[e] Pernès *et al.* (1), p. 6 report a serious outbreak of *Sclerospora* in Shantung in 1977, for example.

[f] Modern introduced hybrids are noticeably less resistant to disease than the traditional varieties. For a discussion of the role of *S. viridis*, and the possibly intermediate form *ku yu tzu*[6], in the evolution of traditional varieties of setaria in China, see Pernès *et al.* (1).

[g] *Fan Sheng-Chih Shu*, quoted *CMYS* 3.21.1–2.

[1] 虸蚄 [2] 粟奴 [3] 老穀穗 [4] 粟白髮病 [5] 狗尾草
[6] 谷莠子

left.'[a] Glutinous panicum, on the other hand, had to be left until it was really ripe before harvesting: 'Non-glutinous should have a green neck, glutinous a drooping head (*chi chhing hou, shu che thou*[1]),' said the proverb.[b]

Average millet yields are difficult to estimate, as figures are so rare. The Warring States politician Li Khuei[2] referred to 1·5 *shih/mu* as an average yield for setaria, which in modern terms is roughly equivalent to 6–700 kg/ha, a rather low figure. According to Buck, millet yields in early 20th-century China varied between 400 and 1200 kg/ha, while Wagner gives figures of 800 to 1000 kg/ha of grain, plus 1300 to 1600 kg/ha of straw.[c] The straw was very important as it was reckoned highly nutritious fodder. Modern improved varieties are said to yield as much as 5000 kg/ha in experimental conditions, and anything between 2200 and 4500 kg/ha in the field.[d] Compared with rice, or even with wheat and barley, traditional millet yields were rather low, but very little of the crop had to be reserved as seed-grain. Each millet-spike contains several thousand seeds: one Chinese proverb said that a setaria spike contained three thousand grains[e] but the number might well be far greater. Speaking of the miraculous harvest of 1834, the author of the *Ma Shou Nung Yen* tells us that one spike of grey setaria (*hui ku*[3]) was found to have 76 spikelets and altogether 8989 grains, while one spike of small yellow setaria (*hsiao huang ku*[4]) contained altogether 9835 grains; such a number in a single spike, he said, had never been surpassed.[f] Some husks would be empty or infertile, however, so that the ratio between the amount of seed sown and grain harvested usually varies between 1 : 50 and 1 : 100.[g]

Although regarded as poor men's food, rich Northerners preferring to eat wheaten breads or even rice, millet is still valued for its pleasant flavour and nutritious qualities even today, and despite competition from kaoliang and maize, its cultivation remained widespread in northern China until very recently. It is unlikely that maize became a serious competitor until this century, but kaoliang has occupied an increasingly important place in North China ever since Yuan times, and has gradually replaced millet in many regions. The grain yields of the two crops are roughly equivalent, but the chief attraction of kaoliang as compared to millet is its greater resistance to flooding and its very high yields of straw, used for fuel, which eventually overcame the Chinese farmer's reluctance to abandon the more appealing traditional food-crop.

[a] *Nung Tshan Ching*, quoted Hu Hsi-Wen (*3*), p. 250.
[b] *CMYS* 4.5.1.
[c] Buck (2), pp. 222 ff., W. Wagner (1), p. 306.
[d] Shansi Agricultural Institute (*1*); Shansi (Chin-Tung) Technical Institute (*1*); Kansu Agricultural Institute (*1*).
[e] *Chhün Fang Phu, Ku Phu* 16*a*.
[f] Hu Hsi-Wen (*3*), p. 265.
[g] Purseglove (2); Wagner (1), p. 306.

[1] 穄青喉黍折頭 [2] 李悝 [3] 灰穀 [4] 小黃穀

### (ii) *Kaoliang or sorghum*

Cultivated sorghum (*S. bicolor* or *Andropogon sorghum*) is an African domesticate.[a] Archaeological evidence suggests that it reached India soon after −2000,[b] but the date of its introduction to China remains uncertain. Remains of sorghum have been reported from several neolithic, Chou and Han sites in North China, but their identification as sorghum is not universally accepted,[c] and possible references in early texts are far from unambiguous.

Chinese sorghum is nowadays generally referred to as *kao liang*,[1] literally 'tall millet'; this name appears for the first time in the *Wang Chen Nung Shu* of +1313,[d] but the plant was more commonly known in pre-modern times as *shu shu*[2], 'Szechwan millet' (Fig. 220), a name still in common use today in the South and West of China.[e] The term *shu shu* first appears in the +3rd-century *Po Wu Chih*, which reads: 'If a field is planted with *shu shu* for three years, then for seven years after there will be many snakes (*ti san nien chung shu shu, chhi hou chhi nien to she*[3]).'[f] But even if the text is authentic (the *Po Wu Chih* has long since been lost except in quotation, and is full of later interpolations), the identification of *shu shu* with sorghum is not certain.[g] The *Chhi Min Yao Shu*, in its section on exotic plants, quotes the late +3rd-century *Kuang Chih* which refers to a cereal called *ta ho*[4], 'great millet', introduced to China from Su-the-kuo[5] (probably Sogdiana), over ten feet tall and with seeds like mung beans, as well as to a cereal called 'willow millet' (*yang ho*[6]), as tall as rushes, which it says was the same as the 'Szechwan millet' (*pa ho*[7]) or 'tree millet' (*mu chi*[8]) of the Central States.[h] These plants do sound similar to kaoliang, which is characterised by its tall stems ten feet high, huge panicles and comparatively large seeds. In Wang Chen's description of *shu shu*[9], written in 1313, which definitely refers to sorghum, he says: 'the stalks are over ten feet high and have panicles as large as a broom. The grain is black as lacquer and like frogs' eyes.'[i] Sorghum was known in both India and the Arab world well before its introduction to China, and coming from either place Szechwan would be a likely point of entry; the name *shu shu*[9] means 'Szechwan millet', as does the term *pa ho*[7] given in the *Kuang Chih*, which adds some weight to Wang Yü-Hu's proposal that *yang ho*[6] and *pa ho*[7] be identified with sorghum.[j]

[a] Wild sorghum is widespread throughout Africa, but domestication probably took place in the savannah zones of Eastern Africa (Harlan (1), p. 374; de Wet, Harlan & Price (1)), though it may have occupied a much wider zone or noncentre, a long belt right across Africa (Doggett (1), p. 114).

[b] Doggett (1), p. 115.

[c] Amano (*4*), pp. 23, 927; Ping-Ti Ho (5), p. 380.

[d] Ch. 7/13*b*.

[e] Wagner (1), p. 298.

[f] Cited Hu Hsi-Wen (*3*), p. 531.

[g] Hagerty (17), among others, contests this interpretation and is convinced that the cultivation of sorghum was only developed in China in the Yuan.

[h] *CMYS* 92.3.3–4.

[i] *WCNS* 7/13*b*.

[j] (*6*), p. 12.

[1] 高粱 [2] 蜀黍 or 蜀秫 [3] 地三年種蜀黍其後七年多蛇
[4] 大禾 [5] 粟特國 [6] 楊禾 [7] 巴禾 [8] 木稷
[9] 蜀秫

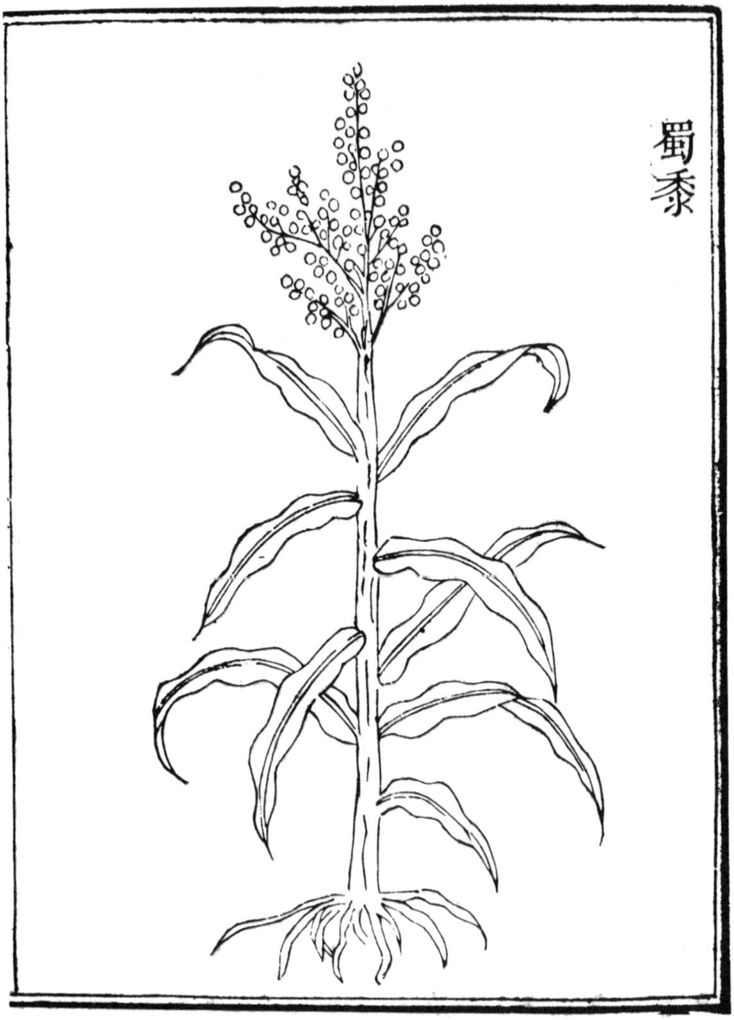


Fig. 220. Sorghum; *SSTK* 24/11*a*.

Even Hagerty, who doubts any introduction of sorghum into China proper before the Southern Sung, concedes that it may have been known in parts of Szechwan at a much earlier period.[a]

Hagerty says that the earliest unambiguous references to sorghum are to be found in the Yuan texts, though he believes on linguistic grounds that the grain may have been introduced into Northern China during the Southern Sung, while Ping-Ti Ho says that the first unmistakable botanical description is given in the +1175 edition of the *Hsi-An Chih*, the history of Hui-Chou[1] prefecture in

[a] (17), p. 259.

[1] 徽州

Southern Anhui written by the famous natural historian Lo Yuan[1].[a] On the other hand, both Amano and Wang Yü-Hu refer to a Northern Sung text, the *Pei Meng So Yen*, which describes the general Chu Wen[2] on campaign in North China in *c.* +910, coming with his troops to a deep channel. They thought that their way forward was barred until they saw that *shu shu*[3] stems had been piled up in the channel to make a passage (*hu chien kou nei shu shu kan chi i wei tao*[4]), which Chu and his troops were able to cross on horseback. The *shu shu* in this passage almost certainly refers to kaoliang, the stems of which were often used for such purposes, and Wang deduces that kaoliang was already cultivated in North China in the early +10th century.[b] A still earlier introduction into China or its border regions is suggested by the fact that the kaoliangs are morphologically related to *Sorghum bicolor*, the most primitive and least specialised of the early major races of sorghum, and appear to represent a Chinese variant of some early *bicolor* race.[c]

Be that as it may, whenever sorghum was introduced into China proper, we can say for certain that it was reasonably familiar to Yuan and Ming writers. Wang Chen describes its uses as follows:[d]

> The grain can be hulled and eaten, and anything left over fed to livestock. It is a famine food. The tips of the stems can be made into brooms and the straw woven into trays, plaited to make fences or used to provide fuel. No part of the plant need be thrown away.

However, we have already seen that Wang Chen described the grain as being 'black as lacquer', and in modern China the dark varieties of sorghum, red, brown or black, are considered too bitter for human consumption and are used only for fodder. White and yellow sorghum, on the other hand, are considered reasonably palatable and sweet.[e] Although sorghum is mentioned in the *Wang Chen Nung Shu*, *Nung Cheng Chhüan Shu* and other post-Yuan treatises, it is not until 1760 that the *San Nung Chi*, a Szechwanese work, gives a detailed account of its cultivation techniques, which are generally very similar to those of millets,[f] and one suspects that sorghum occupied only a very small fraction of the cultivated area until population pressure mounted in the 18th and 19th centuries. Sorghum gives good yields on poor soils and was therefore usually grown on land not suitable for wheat and millet.[g] Hsü Kuang-Chhi remarked on its flood-resistant qualities and its suitability for low-lying land, saying that once autumn had begun, even if the water rose in the fields to a height of ten feet the crop remained undamaged.[h] Sorghum is therefore an ideal crop for marginal land, providing a coarse but edible staple as well as abundant supplies of straw for fodder, fuel and handicrafts. The grain can also be used to distil wine and a fierce spirit, and Wagner claims that up to 90 % of the kaoliang crop was used for this purpose in

[a] Hagerty (*17*), p. 259; Ping-Ti Ho (*5*), p. 382.
[b] Amano (*4*), p. 22; Wang Yü-Hu (*6*), pp. 12–13.
[c] Harlan & Stemler (*1*).
[d] *WCNS* 7/13*b*.
[e] Wagner (*1*), p. 302.
[f] *San Nung Chi* 7/18*b*.
[g] *NCCS* 25/14*b*.
[h] *Ibid.*

[1] 羅願 [2] 朱溫 [3] 蜀黍 [4] 忽見溝內蜀黍秆積以爲道

Yunnan and Szechwan.[a] Grain yields are comparable to those of Chinese millets, in the region of 800 to 1000 kg/ha,[b] though the early 18th-century *Nung Tshan Ching* claimed yields of 2 *shih/mu*, equivalent to approximately 1900 kg/ha,[c] but the yields of straw are double those of millet, varying from 1500 to 3000 kg/ha,[d] a very important consideration in impoverished areas.

It seems probable, then, that sorghum was a relatively uncommon crop in China in the Yüan and Ming, but that as population pressure mounted during the Chhing it came to occupy an increasingly large proportion of the cultivated area. Not only could it be used to reclaim areas of marginal land too poor for millets or wheat, but eventually it even replaced millet in many areas, for although it was less esteemed as food, its straw was more abundant and extremely valuable; Wu Chhi-Chün was already deploring the replacement of millet by kaoliang in the Northwestern provinces in the mid-19th century.[e] Kaoliang was cultivated principally in the Northeastern provinces, where marshy areas were common and the climate was less severe than in the Northwest; there millets could best withstand the aridity and low temperatures.[f] Even so, when Buck carried out his agricultural surveys in the 1920s and 1930s, kaoliang occupied only 4·7 % of the cultivated area of China, whereas 9·4 % was planted with millet.[g] In recent years, particularly in the period 1960–76, kaoliang has replaced traditional Chinese millets in large areas of North China, presumably as a result of policies emphasising the expansion of rice cultivation for human consumption and that of kaoliangs and maize for animal fodder, but already there is a visible tendency to revert to the cultivation of millet, which the Northern Chinese much prefer as food to kaoliang.[h]

### (iii) *Maize*

'Maize is of American origin, and has only been introduced into the old world since the discovery of the new. I consider these assertions as positive, despite the contrary opinion of some authors.' De Candolle wrote these words in 1855, and as one of the outstanding modern experts on maize points out, his conclusions are still valid today.[i] This is not to deny that numerous alternative hypotheses have been proposed in the meantime. Though few people would deny the Middle Eastern origins of wheat or the African origins of sorghum, for some reason the origins and dispersal of maize, which one would expect, given the very recent discovery of America, to be well documented and unquestionable, have been the

[a] (1), p. 303.
[b] Wagner (1), p. 302; Buck (2), pp. 222 ff.
[c] Cited Hu Hsi-Wen (3), p. 535.
[d] Wagner (1), p. 302.
[e] (2), p. 134.
[f] Buck (2), p. 27 classifies the Northeastern provinces as the 'Winter Wheat-Kaoliang Area'.
[g] (2), p. 213.
[h] Pernès *et al.* (1), p. 20.
[i] De Candolle (1), p. 388; Mangelsdorf (5), p. 201.

subject of dispute ever since the 16th century, and the disputes continue to this day.

The first bone of contention was the geographical origin of maize. Domesticated maize (*Zea mays*) is a highly evolved form which bears little immediate resemblance to any other American *Gramineae*; on the other hand it does bear a certain superficial likeness to Job's tears (*Coix lacryma-jobi*), a very ancient Old World domesticate; from this fact some botanists have deduced that maize too must have originated in the Old World. Since then two close relatives of maize, *teosinte* and *Tripsacum*, have been identified in America; some botanists think that *teosinte* is the ancestor of cultivated maize, others (Mangelsdorf in particular) believe that the ancestor of domesticated maize was wild maize, a form that has since become extinct. In any case, the American origins of domesticated maize are beyond dispute.[a]

The second question that arises is the date at which maize was introduced to the Old World. The possibility of a pre-Columbian introduction has appealed to many people. Stonor and Anderson (1) in 1949 described a 'primitive maize' cultivated by hill tribes in Assam which they assumed must be pre-Columbian in origin, while Thapa (1), on the basis of an illustration in an early edition of the *Pen Tshao Kang Mu* (Fig. 221) and some ritual practices common in the Eastern Himalayas of offering corn-cobs to a deity before the harvest, considered it clear that maize had been known in Asia in pre-Columbian times. There is also some dubious archaeological evidence for the early existence of maize in Asia.

One of the earliest causes of confusion as to the origins of maize, or the date of its introduction to the Old World, was the fact that it was often known by names like 'Turkish wheat', but as de Candolle argued, such names were based on popular error rather than fact:

> The names *blé de Turquie*, Turkish wheat (Indian corn) given to maize in almost all modern European languages [Fig. 222], no more prove an Eastern origin than the charter of Incisa.[b] These names are as erroneous as those of *coq d'Inde*, in English *turkey*, given to an American bird. Maize is called in Lorraine and in the Vosges Roman corn; in Tuscany, Sicilian corn; in the Pyrenees, Spanish corn; in Provence, Barbary or Guinea corn. The Turks call it Egyptian corn, and the Egyptians, Syrian *dourra*. This last case proves at least that it is neither Egyptian nor Syrian. The widespread name of Turkish wheat dates from the sixteenth century. It sprang from an error as to the origin of the plant, which was fostered perhaps by the tufts which terminate the ears of maize, which were compared to the beard of the Turks, or by the vigour of the plant, which may have given rise to an

[a] The exact process of domestication, however, is still under debate. Mangelsdorf (5) gives a historical account of maize and the various controversies surrounding it, a detailed description of the present state of genetic and archaeological knowledge, and a comprehensive bibliography. Mangelsdorf's hypothesis that domesticated maize is descended from a (long extinct) form of wild maize is forcefully criticised by Beadle (1), who advances impressive genetic arguments for accepting *teosinte* as the ancestor of modern maize. Cf. also Harlan (5).

[b] A +13th-century document which mentioned a golden grain, similar to maize, brought back from Anatolia by the Crusaders. This charter was subsequently shown to be a modern forgery; de Candolle (1), p. 388.

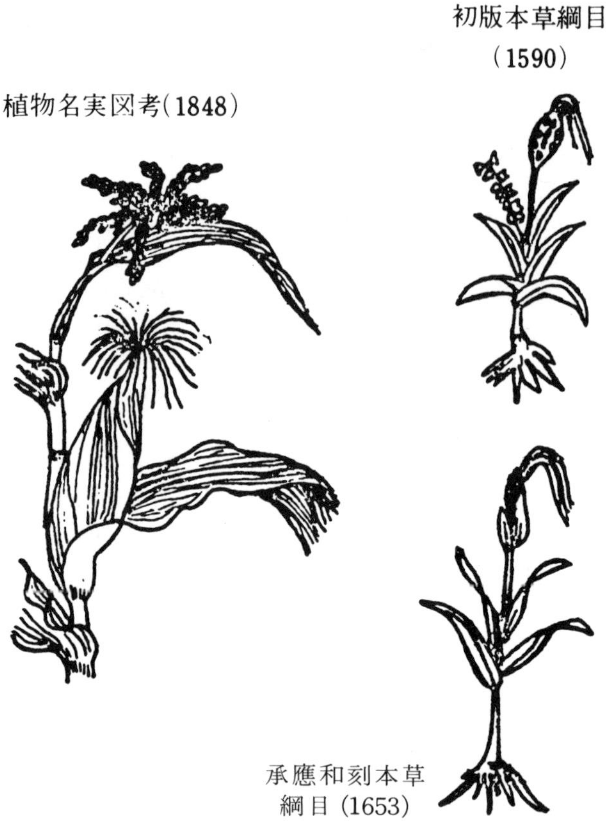


Fig. 221. Illustrations of maize from three editions of the *PTKM* (1590, 1653 and 1848). The early date of the first illustration was taken by some scholars to indicate a pre-Columbian origin for Asian maize.

expression similar to the French *fort comme un turc*. The first botanist who uses the name, Turkish wheat, is Ruellius, in 1536. Bock or Tragus, after giving a drawing of the species which he calls *Frumentum turcicum*, *Welschkorn*, in Germany, having learned by merchants that it came from India, conceived the unfortunate idea that it was a certain typha of Bactriana, to which ancient authors alluded in vague terms. Doedens in 1582, Camerarius in 1588, and Matthiole rectified these errors, and positively asserted the American origin. They adopted the name *mays*, which they knew to be American.[a]

But the un-American names given to maize in so many parts of the world have continued to intrigue philologists, and although they now accept the American origins of domesticated maize, scholars like Jeffreys (1) have evolved intricate linguistic arguments to prove that maize was known in Africa and Europe before 1492. Mangelsdorf (5) disposes of these arguments, in our opinion quite convincingly.[b]

[a] De Candolle (1), p. 389.

[b] We should also briefly allude to the works of diffusionists such as Carter (2) and Heyerdahl (7), who are convinced that there were substantial trans-Pacific contacts in pre-Columbian times. H. L. Li (1) has even gone so far as to postulate a developed Arab trade with the Americas some time before the Conquest.

Fig. 222. 'Turkish wheat' (maize); Fuchs (1).

The question which concerns us most closely here is, of course, the introduction of maize into China: where it was first introduced, how rapidly it spread, and its importance in the Chinese economy. The Chinese have many names for maize, including *yü mai*[1], 'tribute wheat', popularly known as *yü mai*[2], 'jade wheat', *yü mi*[3], 'jade grain', *pao ku*[4], 'wrapped grain', and *yü shu shu*[5], 'jade sorghum', all of which refer exclusively to maize.[a] 16th-century works such as the *Pen Tshao Kang Mu*, or the slightly earlier *Liu Chhing Jih Cha*, written by the Hangchou scholar Thien I-Heng[6] in 1572, point out that maize was first known in China proper as

[a] Except for *yü mai*[2], which in one single instance, an early 18th-century Yunnan gazetteer, was found to refer to buckwheat; Ping-Ti Ho (1), p. 199.

[1] 御麥 [2] 玉麥 [3] 玉米 [4] 包穀 [5] 玉蜀黍
[6] 田義蘅

*fan mai*[1], 'Western barbarian wheat', because it came to China from the territories to the West. This was one of the bases for Laufer's (36) hypothesis that maize was introduced to China by an overland route from India and Burma, though as Ping-Ti Ho (1) points out Laufer was not familiar with the several 16th-century Yunnan gazetteers that mention maize: the *Ta-Li Fu Chih*[2] and *Yün-Nan Thung Chih*[3][a] record maize as having been grown in six prefectures and two counties in northern and western Yunnan, on the upper reaches of the Yangtze, Mekong and Salween rivers.

On the other hand, still earlier references to maize are to be found in gazetteers of eastern provinces, Anhwei, Honan, Kiangsu, Chekiang and Fukien, for example.[b] The earliest of all is dated 1511 and appears in the *Ying-Chou Chih*[4]; Ying-chou is in Northern Anhwei. Wan Kuo-Ting[c] believed maize must have come to Anhwei from the coast, and Amano[d] agrees with him, but Wang Yü-Hu[e] rejects this evidence on the grounds that another Anhwei gazetteer, the *Chhien-Lung Huo-Shan Hsien Chih*[5] of 1776 which describes an area only two or three hundred miles distant from Ying-chou, says that maize was first introduced to the region some 40 years previously, being accidentally sown in a vegetable garden. Wang says it is impossible that maize should have taken two centuries to spread from Ying-chou to Huo-shan, but for reasons stated below we disagree.

Laufer (36) believed that maize spread very quickly through China after its introduction, becoming an important economic crop. He cites evidence from Spanish sources, indicating that enormous quantities of maize were paid as tax grain in the late 16th century.[f] On the other hand the important agricultural works of this period pay little attention to maize. The *Nung Cheng Chhüan Shu*[g] dismisses it in a footnote, while the *Thien Kung Khai Wu* fails to mention it altogether; the first work to give a brief account of its cultivation techniques is the *San Nung Chi* of 1760, a Szechwanese work:[h]

It should be grown on mountain slopes. Dibble in the seed in the 3rd month, allowing 3 feet or so between each hill and 2 or 3 seeds per hill. Once the shoots are six or seven inches high, hoe away the weeds and remove any weakly plants, leaving one healthy one per hill. Maize sown in the 3rd month can be harvested in the 8th or 9th month. Bring the

[a] *Ta-Li Fu Chih*, 1563 ed., 2/24*a*; *Yün-Nan Thung Chih*, 1574 ed., ch. 2, 3, 4 passim; see Ping-Ti Ho (1).

[b] For a complete list of all the references to maize found so far in Chinese gazetteers before 1720, see Amano (*4*), p. 930.

[c] (*9*), pp. 29–34.

[d] (*4*), p. 929.

[e] (*6*), p. 18.

[f] Laufer's source was the Augustinian monk Martino de Herrada, who claimed that in 1577 some 30 million bushels of maize were paid in tax; as Ho (1), p. 196, points out, the official Ming tax returns make nonsense of Herrada's figures. Other European writers, e.g. Gonzales de Mendoza (1), I, p. 15, refer to extensive cultivation of maize on Chinese hillsides under pine trees, but as Lach (5), I, pt 2, pp. 742–94, in his discussion of Gonzales' work points out, the accuracy of his information has yet to be established.

[g] *NCCS* 25/14*a*.

[h] *San Nung Chi* 7/12*b*.

[1] 番麥 [2] 大理府志 [3] 雲南通志 [4] 潁州志 [5] 乾隆霍山縣志

cobs back to the farm and put them on wooden racks to dry, preferably indoors with the doors and windows closed, until they ooze no liquid when struck. Then dry them once more in the sun before storing.

The author Chang Tsung-Fa[1] recommends that maize should be grown on hill-sides, and indeed maize has a deep rooting system which is very suitable for hillside or swidden cultivation, since the plants are not likely to be washed away by the rain; for the same reason maize prospers better when the soil is deeply cultivated with hand-tools such as swidden farmers are likely to possess, rather than the ploughs of lowland farmers.[a] Furthermore, cultivation on slopes minimises the danger of frost-damage, to which maize is very susceptible.[b] The principal inhabitants of the mountainous areas of South China have been non-Han peoples, Miao, Yao, I and many others, most of whom live by shifting cultivation. To them the introduction of maize, with its extremely high yields,[c] was a god-send. The fact that maize contains relatively little useful protein[d] was far less important to these mountain-dwellers, who supplemented their diet by hunting, than it was to the plain-dwelling Chinese, who relied on cereals for a large proportion of their protein intake. These tribal groups developed quite a range of maize varieties (the *San Nung Chi* mentions black, white, red and dark green varieties, as well as glutinous and non-glutinous), which they prepared in a number of different ways.[e] Glutinous varieties were probably favoured as food as well as for their superior brewing qualities: the mountain peoples of Southeast Asia generally prefer glutinous to non-glutinous cereals, unlike the valley dwellers.[f] Stonor and Anderson (1) seized upon the glutinous quality ('waxiness') of upland Assamese maize as proof of its independent origin, but as Mangelsdorf says:[g]

> The fact that waxy maize occurs so commonly in a part of the world that also possesses waxy varieties of rice, sorghum, and millet can be attributed to artificial selection. The people of Asia being familiar with waxy varieties of these cereals and accustomed to using them for special purposes recognised the waxy character of maize after it was introduced into Asia following the discovery of America and purposely isolated varieties pure for

[a] The adoption of maize cultivation in Southwest France brought about a striking increase in the manufacture of spades (F. Sigaut, pers. comm.). In post-Conquest Mexico, where the Indian population had been reduced to a fraction of its former numbers by war and disease, lack of manpower forced native farmers to adopt the plough for maize cultivation, but although this permitted the extension of the area under cultivation, yields per acre were greatly reduced (A. Warman, pers. comm.).

[b] This is one of the reasons why maize was exclusively cultivated on mountain slopes in South America, the other being that terracing permitted a more efficient use of water; Donkin (1).

[c] Maize is well known for being the cereal that produces by far the highest yield of carbohydrate calories per acre.

[d] The total protein content of maize varies between 6 and 15 %, but it is deficient in tryptophane and lysine; Purseglove (2), p. 316.

[e] Amano (*4*), p. 937, fig. 2.

[f] Moerman (1).

[g] (5), p. 143.

[1] 張宗法

waxy endosperm. But the fact that waxy endosperm came to their attention in the first place is probably due to genetic drift. The gene for waxy endosperm, which has a low frequency in American maize, apparently attained high frequency in certain samples of Asiatic maize. Indeed the practice reported by Stonor and Anderson of growing maize as single plants among other cereals would result in some degree of self-pollination and in any stock in which the waxy gene was present would inevitably lead in a very short time to the establishment of pure waxy varieties with special properties that people accustomed to the waxy character in other cereals could hardly fail to recognise.

Han Chinese not only consider other cereals (or even tubers) superior to maize, they positively dislike it. Wagner points out that in the 1930s, while maize was a staple in the mountains of West Szechwan, Yunnan and Kwangsi, all areas where minorities abound and where the maize was consumed as flour products, elsewhere maize flour dishes were unknown and maize was consumed only as a vegetable, the half-ripe cobs being roasted and eaten.[a] The *Chhün Fang Phu* recommends the addition of a *small quantity* of maize flour to wheat flour to give it whiteness and bulk, but otherwise only quotes the *Pen Tshao Kang Mu* on the use of maize leaves and roots, boiled in soups, as a remedy for diseases of the urinary tract.[b] Today maize flour is one of the ingredients in the buns known as *wo thou*[1], but nobody eats *wo thou* if they can avoid it, and a Mexican agricultural delegation visiting China recently was astonished to be asked if they knew of any method for removing the smell from maize pancakes (the aroma of tortillas which Mexicans find so appetising), as it made the average Chinese feel sick.[c]

It is clear that for the Han Chinese maize was very much a last resort, and we would suggest that this is one reason why it is mentioned so rarely in agricultural works in the three centuries following its first introduction. Its cultivation was confined chiefly to the ethnic minorities,[d] Li Shih-Chen refers to its cultivation as 'rare',[e] and many of the references in early gazetteers probably refer to its cultivation by non-Han whose contacts with the Han Chinese were limited. This could be the explanation of the discrepancy between the earliest dates for maize in Anhwei, remarked upon by Wang Yü-Hu.

Maize cultivation, then, seems to have spread very slowly among the Chinese population, contrary to Laufer's hypothesis. The first example of extensive maize cultivation by Chinese farmers comes from the 18th century, when migrants from the over-crowded Yangtze valley took refuge in the hills and mountains of Szechwan, Yunnan and the upper Han River (Southern Shensi, Western Hupei and Southwest Honan). Maize, together with sweet potato, were the crops that could be most successfully grown in these conditions, and according to Ping-Ti

[a] Wagner (1), p. 310.
[b] *Chhün Fang Phu, Ku Phu* 11*a*; *Pen Tshao Kang Mu* 23/8*a*.
[c] A. Warman, pers. comm.
[d] It is still very widely grown among them today; we ourselves have seen extensive maize cultivation in swidden fields in Southwest Yunnan in 1980.
[e] *Pen Tshao Kang Mu* 23/7*b*.

[1] 窩頭

Ho, 'in the whole Han River drainage, which was one of the leading maize-producing areas in China, although sweet potatoes were also grown, maize ruled supreme'.[a] Given modern Chinese attitudes to maize-consumption one suspects, *pace* Professor Ho, that sweet potatoes were the preferred food in these remote areas. Maize is, however, a useful and productive fodder crop,[b] and the area under cultivation slowly increased throughout the late 19th and 20th centuries, especially in regions like Manchuria where extensive production techniques could be used.[c] Nevertheless, Buck's survey shows that in 1930 only 3·7 % of the total cultivated area was planted with maize, less than any other cereal crop.[d] Yields varied between 400 and 4000 kg/ha.[e] Since 1949 the area under maize has increased enormously and a great deal of attention has been given to producing high-yielding hybrids;[f] this has resulted in quasi-miraculous increases in production figures, as at Tachai, but as the average Chinese refuses to eat maize and more is produced than is required for stock-raising purposes, the Chinese government is now faced with the problem of disposing of the surplus.

## (3) Wheat and Barley

Wheat and barley are known collectively in China as *mai*[1], wheat being 'lesser' or *hsiao mai*[2] and barley 'greater' or *ta mai*[3]. Although neither crop is native to China, together they have played an increasingly important economic role and have eventually come to supersede more typically Chinese crops such as setaria and panicum millets in many areas. In pre-Han times, wheat was a delicacy even among the aristocracy; in the mid-17th century, Sung Ying-Hsing[4] estimated that 50 % of the staple food of the northern Chinese consisted of wheat, though only 5 % of southern farmers grew it; today wheat is the second most important cereal in China after rice, accounting for two thirds of the total cereal production in the North China Plain and one third in Central China.[g]

Both wheat and barley are Near Eastern in origin. Their wild ancestors are found throughout the Middle East and are particularly concentrated in the areas of the Levant and the Tauros and Zagros Mountains (Figs. 223, 224). The process of domestication had probably begun in this area by about −8000, and the earliest examples of domesticated wheat and barley known so far are two-rowed barley (*Hordeum vulgare*) and einkorn (*Triticum monococcum*) from Pre-

[a] (4), p. 188.

[b] In North America and Northern Europe maize is now produced in great quantities, almost exclusively for use as animal feed.

[c] Ping-Ti Ho (4), p. 189.

[d] Buck (2), p. 213.

[e] *Ibid.* pp. 222 ff.

[f] Harlan (4), p. 309.

[g] *TKKW* 1/13*b*; R. Myers (2); for present-day acreages and yields, see Lai-Yang Agricultural College (*1*), R. O. Whyte (1).

1 麥　2 小麥　3 大麥　4 宋應星

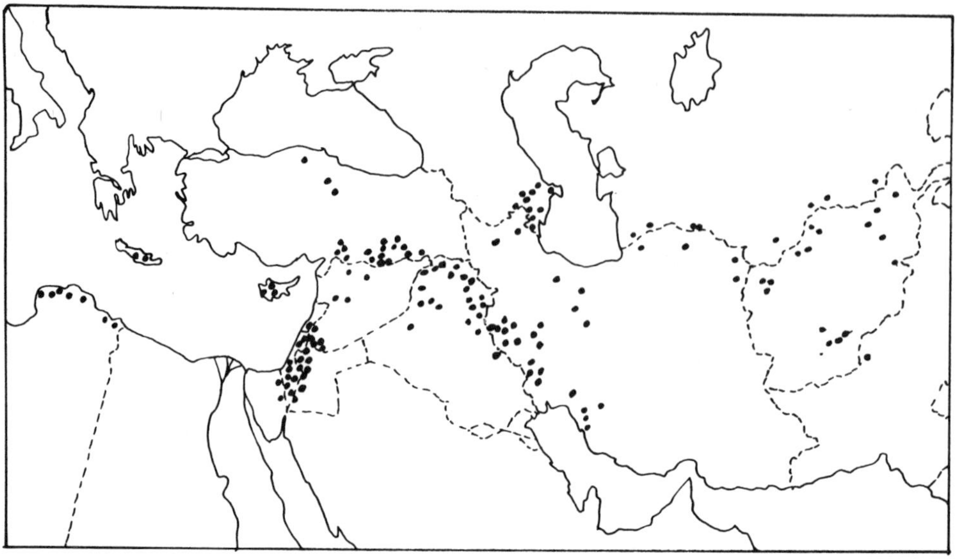

Fig. 223. Map showing the distribution of wild and weed barleys; based on Harlan & Zohary (1).

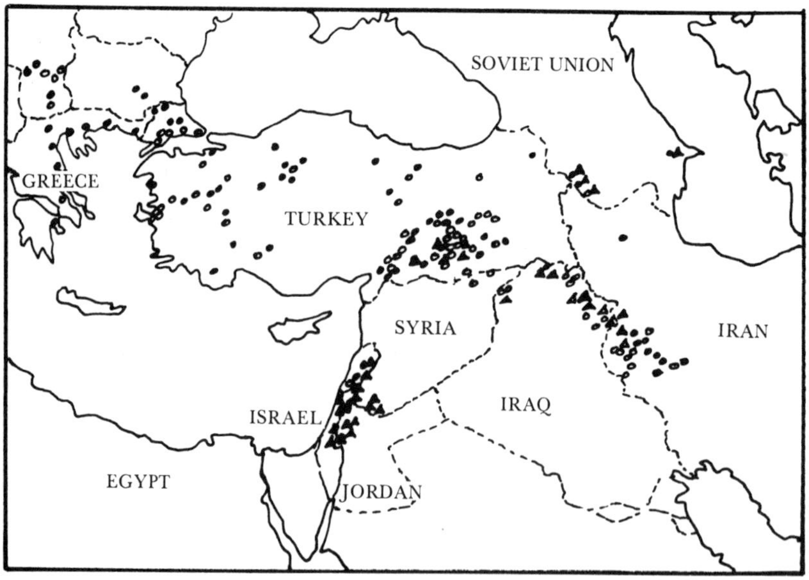


o : einkorn

Δ : emmer

Fig. 224. Map showing the distribution of wild and weed einkorn and weed emmer; based on Harlan (1).

Pottery Neolithic sites in Jericho (*c.* 9000 BP), einkorn, emmer (*T. dicoccum*) and barley from Ali Kosh in Southern Iran (*c.* 9500–8750 BP), and einkorn and emmer from Çayönü in Southern Turkey (*c.* 9500–8500 BP). The cultivation of both crops spread rapidly throughout the Middle East to Egypt and North Africa, Crete and the Balkans, and south through Afghanistan, reaching Pakistan probably by about 5000 BP; the early Indian civilisations of Mohenjo-Daro and Harappa both depended on wheat and barley.[a]

There are various theories as to the origins of Chinese wheat and barley. There have been as yet no authenticated prehistoric finds of either crop, but references are made to them in the late Shang oracle-bone inscriptions and one view is that they were first introduced to China in about −1500.[b] Great interest was aroused by the discovery of 1 kg. of carbonised grains, identified as *Triticum antiquorum*, found in a Lung-shan site in Anhwei, but as early as 1963 Yang Chien-Fang (*1*) had suggested that they were probably a post-neolithic intrusion since the jar in which they were contained was of typically Chou style; this was later to be confirmed by carbon-dating which put the wheat at about −500, that is to say in the Warring States period.[c] Amano's (*4*) hypothesis is that barley arrived in China earlier than wheat, indeed that it may have been domesticated independently there, while wheat only began to be grown in China to any significant extent during the Han. Amano's argument is based on dietary grounds. Barley, he says, is softer than wheat and has traditionally been consumed as porridge, requiring no special apparatus for processing. Wheat, on the other hand, is a hard grain which the Chinese are known to have preferred to consume in the form of flour-based products, bread, noodles and the like, and he sees the cultivation of wheat as being closely associated with the development of hand-mills and the larger animal- or water-powered mills, all of which only became common in Han times (see Section 27).[d]

Against this, many Chinese scholars would argue that the Shang oracle-bone inscriptions refer frequently to an autumn-sown cereal called *lai*[1], which Han lexicographers identified with *hsiao mai*[2], wheat, and that therefore wheat was cultivated in Shang times; however, the identification of certain oracle-bone graphs with *lai*[3], as proposed for example by Yü Hsing-Wu (*1*), has been contested by other experts like Chheng Meng-Chia and Matsumaru Michio[4] [e] (Fig. 225). Be that as it may, the Shang oracle-bone inscriptions do undeniably

[a] The process of domestication of wheat and barley is described at greater length in Feldman (1), Jack R. Harlan (1), (2) and (3), and Zohary (1), (2); Harlan (1) gives details and dates of their spread throughout Eurasia. Further details on the spread of the crops westwards through Europe and South to India are given in R. Tringham (1), F. R. Allchin (1), and Vishnu-Mittre (1) and (2).

[b] Ping-Ti Ho (*1*), (5).

[c] Anon. (*522*).

[d] Amano's opinion is shared by such distinguished Japanese scholars as Shinoda Osamu (*1*), (*7*) and Kitamura Shirō (*1*).

[e] Chhen Meng-Chia (*4*), p. 528; Matsumaru Michio, pers. comm. 1960, cited Amano (*4*), p. 70.

1 來 or 秾 2 小麥 3 來 or 秾 4 松丸道雄

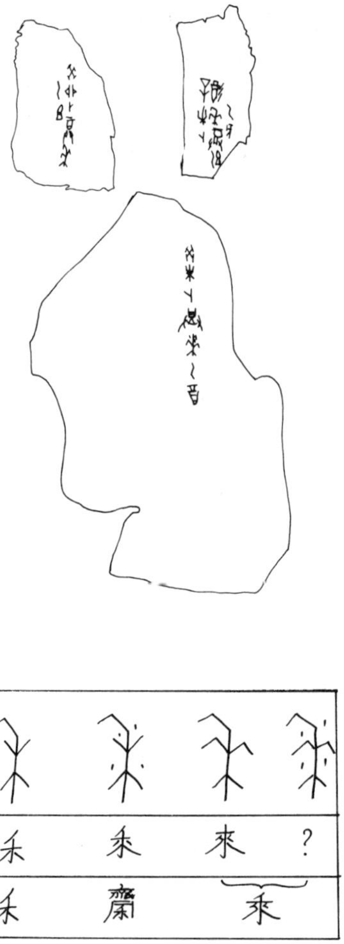

Fig. 225. Oracle graphs possibly representing wheat; the upper line shows the original graphs, the second Chheng Meng-Chia's interpretations, and the bottom line the interpretations of Yü Hsing-Wu.

refer to an autumn-sown, spring-harvested cereal, and that can only have meant wheat or barley. There is no need to assume that the ancient Chinese could only consume wheat after it had been ground to flour: in early Rome wheat was usually consumed as porridge, *puls*,[a] and this could well have been the case in China too before noodles and breads became popular. On the other hand the ancient Chinese might have prepared wheat as a form of *tsamba*, whereby the grains were first roasted and then pounded or milled to flour. *Tsamba* is a Tibetan word; in Tibet this roasted-grain flour, generally made from barley and mixed with water, tea, butter or sometimes even honey, was the staple food of all travellers. It is still widely used in Tibet, Chhinghai and Northwest China, and

[a] J. R. Harlan (1), (3); barley too was usually consumed in this way; L. A. Moritz (2).

the Chinese often borrow the Tibetan term, sinicised to *tsan pa*[1]. But there are also traditional Chinese terms for this food (which must date back to pre-Han times as they are commented on in the *Shuo Wen*), including *chhiu*[2] and *chhao mien*[3]. Trippner (1), writing of Northwest China and the Central Asian provinces in the 1940s and 1950s, says that although *tsamba* was most commonly made of barley in this region, wheat *tsamba*, which was occasionally made, commanded a far higher price, and the *Chhi Min Yao Shu* recommends making wheat into *tsamba* to protect it from insects.[a]

It is just conceivable that barley was domesticated independently in China, for a number of wild barleys have been found in the Yunnan plateau and Tibet. In the 1930s wild six-row barleys were discovered in Hsikang[b] and it was thought that barley might have been domesticated in the region from these varieties: 'Supposing that barley grew wild in Hsikang, we may suppose that it spread from the mountains to the Szechwan Basin and thence North to the Central Plain.'[c] Subsequent genetic research has indicated that such six-rowed wild barleys could not in fact be the ancestors of a domesticated form,[d] but in recent years Chinese and Tibetan botanists have also discovered wild two-rowed barleys (*Hordeum spontaneum*) in Szechwan and Tibet, so that there are some grounds for supposing an independent domestication of barley in the Far East.[e] On the other hand in every other part of Eurasia and Africa wheat and barley appear to have been introduced together as a crop 'complex'[f] and one wonders why China should be an exception. Although there are in Chinese a number of distinct terms for the two crops, they all derive ultimately from the term *mai*[4], which is used collectively for both wheat and barley as well as for other autumn-sown crops of the Middle Eastern complex such as rye (*Secale cereale*) and oats (*Avena sativa*).[g] We should therefore suggest that domesticated wheat and barley came to China together, from the West, probably towards the end of the neolithic period: autumn-sown crops (either wheat, barley or both) are mentioned in the oracle-bone inscriptions as being grown in several different regions of Shang China, which suggests that

[a] 10.4.5; the term used here is *chhiao mai*[5], literally 'cut wheat' (see p. 330 above), but a near homophone of the other Chinese terms. Wheat was also parched, or made into *tsamba*, as far afield as the Yemen; see R. B. Serjeant (1), p. 42.

[b] Also in Japan.

[c] Kitamura Shirō (*1*), tr. Philippa Hawking.

[d] 'Six-row genotypes with fragile ears are known, but do not appear to be truly wild plants and are probably derived from six-row cultivars,' Harlan (2), p. 94.

[e] Shao Chhi-Chhüan & Li-Chhang-Shen Pa-Sang-Tzhu-Jen (*1*).

[f] Harlan's term; (1).

[g] Barley is known not only as *ta mai*[6] but also as *mou*[7] (an archaic term) and *keng mai*[8] (a term variously interpreted by Chinese authors as meaning husked or naked barley; see Amano (*4*), p. 60 for a detailed discussion); oats are called 'sparrow corn' (*chhiao mai*[9]), 'swallow corn' (*yen mai*[10]) or 'green seed corn' (*chhing kho mai*[11]), and rye is called 'black corn' (*hei mai*[12]); see Anon. (*109*); B. E. Read (1); G. A. Stuart (1).

[1] 昝巴 [2] 糗 [3] 炒麵 or 麨麵 [4] 麥 [5] 𠜱麥
[6] 大麥 [7] 麰 or 牟 [8] 穬麥 [9] 雀麥 [10] 燕麥
[11] 青粿麥 [12] 黑麥

they were not a new introduction, and we have already suggested (p. 322 above) that the introduction of these comparatively sparse-growing, small-eared, summer-ripening crops might perhaps be associated with the appearance of the sickle as a harvesting tool, that is to say it might tentatively be dated to the later Lungshanoid period, say the late −3rd millennium. This rather late date, as compared with the introduction of the same Middle Eastern crop complex to Egypt and Europe for example, would tally with the fact that almost all Chinese wheats are hexaploids, the highly evolved bread-wheats (*T. aestivum*).[a]

The most important feature distinguishing wheat and barley from the native Chinese cereals is that they are winter crops, that is to say they are sown in the autumn or winter and harvested in the late spring. Of course spring-sown varieties of wheat and barley also exist, but they were a rather late development in China and have never been grown very extensively. The great attraction of winter wheat and barley to the Chinese farmer was that they were supplementary to, not substitutes for, the more traditional crops, and they were harvested in the lean summer season when stocks of millet and rice were nearing depletion. Since they did not compete for field-space with autumn-ripening crops, wheat and barley permitted the development of highly productive crop rotations. As early as the Northern Wei it seems that the system of growing three crops in two years was well known in North China. The author of the *Chhi Min Yao Shu* says:[b] 'For lesser beans it is usual to use land that has been cropped with wheat or barley (*hsiao tou ta shuai yung mai ti*[1]), though I am afraid that this is leaving it a little late' (beans, he says, should preferably be sown just after the summer solstice). He also recommends sowing crops such as coriander or turnips in the sixth month on land which has previously been used for wheat.[c] This implies that it was then quite common to grow three crops in two years in the north.[d]

One of the earliest texts to mention alternate cropping of wheat and rice is Fan Chho's[2] *Man Shu* of *c.* +860.[e] Writing of the wet-rice farmers of the Yunnan plateau to the west of Kunming, Fan says: 'The wet fields produce one crop [of rice] a year. From the 8th month, when they harvest the rice, to the transition between the 11th and 12th months they sow barley in the rice-fields, which ripens in the 3rd or 4th month. After they have harvested the barley they sow rice again. Wheat they plant on the hills.' Yunnan was known as the land of eternal spring, where crops would grow almost all the year round, but in other parts of the south where conditions were less favourable wheat and barley did not become integrated into cropping rotations until the Sung, when the introduction of the

[a] Hexaploidy seems to be a comparatively late development in wheat; a few tetraploid varieties (*T. turgidum*) are, however, found in parts of Western China; Vavilov (2), T. H. Shen (1), p. 181; Harlan (3), p. 7.

[b] *CMYS* 7.1.1.

[c] *CMYS* 18.4.2, 24.6.3.

[d] Yoneda Kenjirō (*1*) maintains that this was the case, but Nishijima Sadao (*1*) pp. 235–53 presents a number of counter-arguments.

[e] Cited Hu Hsi-Wen (*2*); p. 61.

[1] 小豆大率用麥底 [2] 樊綽

quick-ripening Champa rices allowed more leeway for draining the irrigated fields and preparing them for winter crops. Furthermore, the southward migration of millions of wheat-eating refugees from the north provided a ready market for the winter grains:[a]

> After the fall of the Northern Sung many refugees from the Northwest came to the Yangtze area, the Delta, the region of the Tung-thing Lake and the Southeast coast, and at the beginning of the Shao-Hsing[1] reign (1131–63) the price of a bushel of wheat reached twelve thousand cash. The farmers benefited greatly, for the profits were double those of growing rice. Furthermore, tenants paid rent only on the autumn crop, so that all the profits from growing wheat went to the tenant household. Everyone competed to grow the spring-ripening crop, which could be seen everywhere in no less profusion than to the North of the Huai.

Thereafter wheat and barley were regularly grown in rotation with rice, sometimes even as a third crop after two crops of irrigated rice had been grown.[b] There is some debate as to whether winter crops were in fact exempt from tax and rent, as the *Chi Le Pien* implies. Katō Shigeshi and Amano Motonosuke have suggested that farmers in +16th century Fukien switched from the double-cropping of rice to an alternation of wheat and rice simply because wheat, unlike rice, was exempt from rent (as it appears to have been in South China in the Sung and as it was, according to Katō, in pre-modern Japan). However, there is no evidence in surviving Ming or Chhing rent contracts to support this hypothesis.[c]

In South China wheat and barley were grown in rotation with rice, barley often being preferred to wheat as it matured earlier and was better adapted to humid conditions and poorly drained land.[d] In the North they were often rotated with millet or kaoliang, and even replaced them entirely in some areas.[e] Northern rotations often included soybeans or other legumes and, from the Yuan and Ming on, cotton. Hsü Kuang-Chhi recommends a method whereby the wheat was dibbled in ridges immediately after the winter ploughing and the cotton dibbled in between the wheat plants in the early spring;[f] similar interplanting of wheat and cotton is practised in North China to this day (Fig. 226).

Since wheat and barley were grown in the spring they had the further desirable quality of avoiding the summer floods which so often destroyed standing crops of rice or millet:[g]

> In the north the worst category of land is the submerged. On it natives usually grow sorghum and can secure, on the average, only one crop in several years. For this reason they are poverty-stricken. I have taught them to grow wheat, which may not be affected

[a] *Chi Le Pien*, cited Hu Hsi-Wen (*2*), p. 75.
[b] Rawski (1), p. 201.
[c] Katō Shigeshi (*2*); Amano Motonosuke (*10*); also Fu I-Ling (*1*), pp. 60–2. A discussion of other factors that may have affected the substitution of wheat for rice and vice versa is given in Rawski (1), pp. 32–8.
[d] T. H. Shen (1), p. 208.
[e] R. H. Myers (1), pp. 178–9.
[f] *NCCS* 35/10*b*.
[g] *NCCS* 25/15*b*, tr. Ping-Ti Ho (4), p. 179.

[1] 紹興

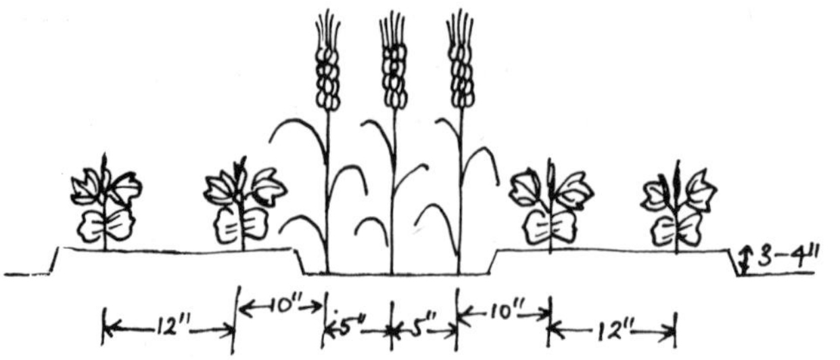


Fig. 226. Interplanting of wheat and cotton; Peking Ag. Coll. (*1*), p. 330.

by [annual floods]. For the flood usually comes in late summer or early fall and can do no harm to wheat. In places where drainage is possible after the recession of the flood the land dries up in autumn and is therefore suitable for autumn wheat. In localities where drainage is not feasible the land dries up in winter and is suited to spring wheat ... This method can practically assure nine crops in every ten years.

The evident advantages of wheat and barley only gradually overcame the Chinese farmers' mistrust of the new crops. Although they may have been grown in some parts of China as early as the neolithic and were certainly numbered among the crops of the Shang and Chou, wheat and barley were long confined to areas such as Shantung and Anhwei, which seem to have been particularly well suited to their cultivation. The early Han text *Fan Tzu Chi Jan*, for example, says of the 'five cereals': '... In the East they grow much *mai*[1] and rice',[a] but it does not refer to *mai* cultivation as a speciality of any other region. The new crops did gain slightly in popularity during the Han. In the −1st century, the statesman Tung Chung-Shu[2] urged the expansion of *mai* cultivation in the metropolitan area of Kuanchung,[b] and the agronomist of the same period, Fan Sheng-Chih[3], described *mai* cultivation in great detail, prefacing his account with the following approving remark:[c] 'If you manage to sow *mai* at the right time it will always give a good crop.' Wheat and barley seem to have been grown quite extensively in the military colonies on the Northwest frontier during the later Han,[d] but perhaps the best proof of its growing popularity was the expansion of flour-production and of the range of flour-based foods at this time.[e]

Despite the increasing popularity of *mai*, there do not seem to have been many varieties available compared with crops such as rice or millet. The *Chhi Min Yao Shu*, for example, which lists literally dozens of varieties even for rice, which was not then a common crop in the north, quotes the late +3rd-century *Kuang Chih* on wheat varieties, but otherwise lists no further names as it invariably does in the case of other cereals:[f]

[a] Cited Hu Hsi-Wen (*2*), p. 29.
[b] *HS* 24A/16*a*.
[c] *FSCC*, quoted *CMYS* 10.11.1.
[d] Cho-Yün Hsü (1), p. 85.
[e] Ying-Shih Yü (1), pp. 73, 81.
[f] *Kuang Chih*, cited *CMYS* 10.1.2.

1 麥 2 董仲舒 3 氾勝之

*Lü hsiao mai*[1], 'slave wheat', has ears the same shape as barley, with tassels. *Wan mai*[2] [unidentified] resembles barley and comes from Liang-Chou[3] [the region of Chungking]. *Hsüan mai*[4], 'immediate wheat' [i.e spring wheat] is sown in the 3rd month and is ripe by the 8th; it comes from the West. 'Crimson wheat' *chhih hsiao mai*[5] is crimson and oily and comes from Cheng-chou[6] [in modern Shensi]; one talks of Hu[7] pork [presumably from Hu-chou[8], in Chekiang] and Cheng wheat. Shan-thi[9] wheat [from the region of Chengtu] is very glutinous and soft and is sent as tribute to the emperor ... there are also 'midsummer wheat' (*pan hsia hsiao mai*[10]), 'bald-awned barley' (*thu mang ta mai*[11]) and black *keng* barley (*hei keng mai*[12]).

As well as citing the *Kuang Chih* on spring-sown wheat, the *Chhi Min Yao Shu* mentions a variety of barley that was sown in the spring (*chhun chung keng mai*[13]),[a] and these are apparently the earliest references in Chinese literature to spring wheat and barley, crops which are now of some importance in the northernmost provinces of China where the climate is too severe for winter wheat to be grown;[b] in all other regions winter wheat was preferred as it fitted better into crop rotations. The *Kuang Chih* distinguishes between 'bald-awned barley' (*thu mang ta mai*[11]) in distinction to *keng mai*[14], presumably implying that the former was naked while the latter was not.[c] Both wheat and barley may be either awned or naked, but the awned varieties were traditionally more common in China, and indeed in early texts *mai* was often defined as an 'awned species' (*mang chung*[15]) (Figs. 227, 228).[d] Similarly in the early texts *keng mai*[14] seems to have referred to awned barley, though in later periods it came to refer to the naked varieties.[e] Today most northern varieties of wheat and barley are awned and most southern varieties are naked: the awns are characteristic of drier areas, improving drought- and wind-resistance and preventing rain-water from falling directly on the flowers, inhibiting pollination.[f]

The Chinese made a distinction between awned and naked, spring and winter varieties of *mai*, but otherwise the range seems to have been limited compared to millets and rice. This may be precisely because *mai* was the summer-ripening crop *par excellence*, that is to say it had to be harvested at great speed, partly to avoid the danger of summer storms, partly to clear the fields in time for the next crop of rice or millet, and partly to avoid the ears shedding their grain as they became over-ripe. 'Harvest *mai* as you would fight a fire (*shou mai ju chiu huo*[16]),' says the proverb.

[a] *CMYS* 10.1.4.
[b] J. L. Buck (2) distinguishes a 'Spring Wheat Area' comprising the dry, rugged, hilly borders of Kansu, Shensi, Hopei and parts of Mongolia and Manchuria.
[c] Thao Hung-Ching[17] in his commentary glosses *thu mang ta mai*[11] as *lo mai*[18], literally 'naked barley'.
[d] As in the commentary of the *Chou Li*, *Ti Kuan*, *Tao Jen* section, *Chou Li* 4/33*b*.
[e] See Amano (*4*), pp. 60–2.
[f] Staff of the Northwestern Agricultural College, Wukung, Shensi, pers. comm.

| | | | | |
|---|---|---|---|---|
| 1 虜小麥 | 2 椀麥 | 3 涼州 | 4 旋麥 | 5 赤小麥 |
| 6 鄭州 | 7 湖 | 8 湖州 | 9 山提 | 10 半夏小麥 |
| 11 秃芒大麥 | 12 黑穬麥 | 13 春種穬麥 | 14 穬麥 | 15 芒種 |
| 16 收麥如救火 | 17 陶弘景 | 18 倮麥 | | |

Fig. 227. Wheat; *PTKM* 2/22*b*.

Fig. 228. Barley; *PTKM* 2/22*b*. Note that both wheat and barley are shown as heavily awned.

*Mai* seems to have been reaped with sickles from very earliest times, and by the Sung northern Chinese farmers had developed a whole array of special scythes and trolleys for harvesting wheat more rapidly; these have continued in use to the present day (pp. 338 ff. above). Despite this relatively advanced technology, the wheat harvest remained one of the busiest times of the year:[a]

> Thresh the wheat, thresh the wheat, the flails whistle and thud,
> Echoes fly afar across the very mountain tops.
> In late spring the sun rises to the northeast:
> As it brushes the seashore peaks the wheat is still bright green
> But at the zenith the crop has turned to gold.
> The pheasant's clarion wakens us at dawn,
> By night we snatch no rest,
> For the partridge clucks to warn of rain, the clouds spread dark as ink.
> The grown women stoop over their sickles, the young follow with baskets.
> In the upper fields we pluck stems that are still green
> While lower down the slope the sheaves already stand in stooks.
> At the cost of such bitter toil we peasants reap our joys—
> Bowed heads seared by a savage sun that burns our faces black.

Under such circumstances the selection of seed-grain was likely to be a hasty affair, and this may be one reason why the range of *mai* was comparatively restricted. Even in the 18th century, the local gazetteers quoted by the *Thu Shu Chi Chheng* list only a few varieties of wheat and barley for most localities, including those where *mai* was the most important cereal crop. The gazetteer for Hsing-thai[1] in Hopei names five varieties of wheat produced in the district, yellow-skinned, red, white, bald, and purple-stemmed white-grained, as well as spring wheat and barley. It goes on to say that *mai* ripens earliest in the northwest of the district, then in the southeast, and lastly in the western mountains, with altogether ten days between the different zones. The gazetteer for Hsien-yang[2] in Shensi lists white wheat, purple wheat, and an early-ripening variety called third-month yellow, while the gazetteer for Thai-ping-fu[3] in Anhwei, today one of China's chief wheat-producing areas, reports seven varieties of wheat and five of barley.[b] We should point out, however, that even in the West, where wheat and barley were of prime importance, very few named varieties are mentioned in pre-modern texts (see also p. 330).

A modern classification of the wheat-regions reads as follows:[c]

> The regions that produce spring wheat, north of the Great Wall, and those in the Yellow River Valley which produce winter wheat may be called the hard-wheat belt. The flour of wheat produced there is best in quality for bread and Chinese noodles. Its quality

[a] Poem by the Northern Sung writer Chang Wu-Min[4], cited Hu Hsi-Wen (*2*), p. 73.
[b] *TSCC* 23/15*a*, 17*a* and 18*a*.
[c] T. H. Shen (1), p. 181.

[1] 邢臺 [2] 咸陽 [3] 太平府 [4] 張舞民

is comparable with that of the high-grade flour of either Canada or the United States. The wheat produced in the Huai River Valley, the northern part of Szechwan, and the southern part of Shensi is medium-hard and not so good for bread and noodles as the hard wheat. The wheat produced in the Yangtze Valley and to the south of that valley is soft wheat. It is low in gluten, absorbs less water, yields less flour, and is greyish white in colour. It is poor for bread and noodles, but suitable for cakes and biscuits.

The +12th-century *Chi Le Pien* remarks of Shensi wheat:[a]

The border regions of Shensi are bitterly cold, and however long it is left, the wheat never ripens properly; it sticks to the teeth and cannot be eaten. For example, the flour of Hsi-chou[1] has to have a handful of lime added to every catty before it can be properly kneaded and cut into noodles. The mutton is also very rank. Only in Yuan-chou[2] are both products good: their noodles are packed in paper and sent everywhere as special gifts.

With exceptions such as these, the climate of North China was far more favourable than that of the South to successful wheat cultivation. Even in the relatively temperate climate of Yunnan the results were poor:[b] 'Noodles made from local wheat are mushy and tasteless, and the barley is usually made into *tsamba* (*chhao*[3]) but is otherwise not used.' But though not appetising, the Yunnan noodles were at least edible. The late 17th-century *Kuang Tung Hsin Yü* comments:[c] 'Ling-nan is a hot region and therefore little *mai* is grown, and more wheat than barley ... The flour made from it is often slightly poisonous.' Wheat flour did have a tendency to go rancid or mouldy in the tropics. The 16th-century author Wang Chi[4] describes how he attempted to persuade the peasants of Heng-chou[5] (in modern Kwangsi) to grow wheat on uncultivated uplands unsuitable for rice. He managed to convince them to give wheat a try, but they stored it before it had been properly dried so it soon went mouldy and tasteless, and those who ate it became sick. Not surprisingly they were reluctant to grow any more wheat: they felt that the soil could not be suitable. Only after an intensive remedial campaign, involving the public reading of a tract on correct methods of wheat cultivation and storage in all the villages around, were a few farmers induced to take up growing wheat again.[d] T. H. Shen points out that even in the 20th century, traditional Chinese stone mills produced flour with a high moisture content and poor keeping qualities.[e]

It is not surprising, then, that *mai* was slow to spread to some of the southern provinces, where the climate was basically unsuitable and a second crop of rice was a viable alternative.[f] In the north the advantages to growing wheat and

[a] Cited Hu Hsi-Wen (*2*), p. 134.
[b] *Man Shu*, cited Hu Hsi-Wen (*2*), p. 61.
[c] *Kuang Tung Hsin Yü*, ch. 14, p. 377.
[d] *Chün Tzu Thang Jih Hsün Shou Ching*, cited Hu Hsi-Wen (*2*), p. 122.
[e] (1), p. 194.
[f] Ping-Ti Ho (4), p. 179 points out that wheat was very slow to spread even to such provinces as Hupei, where wheat production is very important today, remaining an upland crop until well into the 18th century. Increasing population pressure and the development of more resistant varieties eventually tipped the balance.

[1] 熙州 [2] 原州 [3] 麨 [4] 王濟 [5] 橫州

barley were clearer: nevertheless in the +6th-century *Chhi Min Yao Shu* their cultivation is described only in the tenth chapter, after millets, beans and hemp (though before rice and sesame), and it was only in the Thang that wheat and barley really became economically important.[a]

Even in North China wheat and barley offered certain disadvantages. They were comparatively low-yielding crops, requiring careful and labour-intensive cultivation, and they were prone to insect attacks and disease during growth and in storage; they also shed their grains very easily, making harvesting a matter of some urgency. But at least in the North the preparation of the soil for *mai* was straightforward. The field was ploughed into ridges (*lung*[1]) and the *mai* was either sprinkled along the ridges or, more usually, sown by drill (Fig. 229).[b] The crop was hoed and earthed up throughout its period of growth to increase the supply of moisture and improve the quality of the grain: 'Well-hoed wheat yields twice as much: the husks will be thin and it will give large quantities of flour.'[c] 'There is a saying, "If you wish to be rich, bury your barley in gold", that is to say, bank up the roots of the barley while you are doing the autumn hoeing.'[d]

If *mai* was grown in alternation with wet rice, as was generally the case in the South, the preparation of the fields was onerous: barley and especially wheat would not grow well unless the soil had been adequately drained. Wang Chen described the preparation of wet fields for wheat (p. 111 above), and later authors elaborated on the necessity of constructing high ridges (*lun*[2]), gently rounded like a tortoise-shell and separated by deep drains (*kou*[3]) which intercommunicated so that the water never accumulated or stagnated.[e] This is still a key factor in successful wheat cultivation today, except that nowadays to maximise the available soil surface the ridges may be from 4 to as much as 8 feet across.[f] The drains between the ridges had to be kept clean throughout the period of growth:[g]

> In the winter months the drains should be cleaned and repaired so that they are deep and straight and the spring rains will flow through easily and not soak into the roots of the *mai*. When clearing the drains one man should go first with a hoe digging up and loosening the soil in the drain, while a second man follows him with a shovel spreading the soil on to the ridges. Since the mud from the drains is fertile, this is extremely beneficial to the roots of the *mai*.

Usually in the South the soil was simply spread broadcast (Fig. 94),[h] but in the Ming and Chhing the techniques of rice cultivation were sometimes applied to *mai*, that is to say the seed was soaked until it sprouted, sown in a seed-bed and then transplanted into the ridges: 'Each hill should contain fifteen or sixteen plants', says the *Shen Shih Nung Shu*.[i] Transplanting clearly increased yields, and

[a] R. H. Myers (1), p. 178.
[b] *CMYS* 10.2.2, *TKKW* 1/14*b*.
[c] *CMYS* 10.4.3.
[d] *FSCS*, quoted *CMYS* 10.11.4.
[e] *NCCS* 26/13*a*; *Shen Shih Nung Shu* 1/8; *Pu Nung Shu* 2/31.
[f] Hu Hsi-Wen (2), p. 3.
[g] *NCCS* 26/13*a*.
[h] *TKKW* 1/17*a*.
[i] *Shen Shih Nung Shu* 1/9.

[1] 壟 [2] 埨 [3] 溝

Fig. 229. Spring wheat sown by drill in North China; *SSTK* 34/13*b*.

the practice later spread to the northern provinces too; we find it mentioned in the *Chih Pen Thi Kang*, which described agricultural practice in Shensi, where the author claimed that transplanting doubled yields.[a] Transplanting wheat is still standard procedure in parts of China today, though in some areas it has been discontinued as being too labour-consuming.[b]

Both wheat and barley are hungry crops requiring large quantities of soil nutrient: 'If your field is not fertile then you should not sow barley', says the *Chhi Min Yao Shu*.[c] *Mai* was often grown in rotation with soil-enriching crops like legumes, and the fields were well-fertilised before and during growth. When the *mai* was grown in drained rice-fields, the mud cleared from the drainage channels was a useful source of nutrients. In the 17th century, Chang Lü-Hsiang[1] remarked on the development of the root system of *mai* and its implications for fertilisation:[d]

> The roots of *mai* grow straight down but not deep, so the fertiliser[e] need cover the roots only in the early stages of hilling in order to be effective. Similarly for the mud in the drainage channels, which should also be dug and applied early. It is commonly said that digging out the channels in early winter results in golden channels, in mid-winter in silver channels, and in early spring simply in water channels. When I went to Shaohsing I noticed that everyone used oil-cake (*tshai ping*[2]) on their ridges at the rate of 10 catties per *mu*. They put a pinch on every clump just as the wheat was emerging from the soil, and every time it rained the wheat would grow a little more. In my area they use bean-cake mixed with urine during earthing-up, which is even more effective. It is applied at the rate of 2 pints of beancake for every pint of seed, in pinches like the seed itself. However it is preferable if the seed has been sprouted before sowing, for if it is sown dry then the beans soon rot and rot the wheat-seed with them.

On the other hand the *Thien Kung Khai Wu* says that wheat fields must be fertilised before sowing, as afterwards it is impossible.[f] It also says:[g] 'Wheat disasters are only one third as numerous as those pertaining to rice. After the seeds are sown it does not matter whether there be snow, frost, dry weather or excessive rain.' Maybe Sung Ying-Hsing[3] was not entirely familiar with wheat cultivation and the care that went into weeding, watering, and protecting the crops and harvested grain from insects and diseases. Although it is true that specific diseases of wheat and barley are mentioned only rarely in the Chinese sources, this is

[a] P. 22.
[b] The procedure is recommended in several works from the time of the Cultural Revolution, e.g. Lai-Yang Agricultural College (*1*), pp. 168–181, Peking Agricultural College (*1*), pp. 287–91; it is now regarded with less favour, as we discovered when visiting the Northwestern Agricultural College in Wukung, Shensi, in 1980, where none of the wheat in the experimental plots had been transplanted.
[c] *CMYS* 10.3.1.
[d] *Pu Nung Shu* 2/31.
[e] Chang advocates a mixture of potash and compost.
[f] *TKKW* 1/15*a*.
[g] *TKKW* 1/17*b*.

[1] 張履祥 [2] 菜餅 [3] 宋應星

probably because there was very little that the farmer could do, in the days before chemical fungicides and pesticides had been developed, to prevent them. Those diseases that are mentioned, *mai nu*[1] or 'wheat slave', *po ju*[2] or 'thin mould' and *huang tan wen*[3] or 'jaundice epidemic', can probably be identified as smut (*Ustilago carbo*), blight (*Gibberella* or *Fusarium* spp.) and yellow rust (*Puccinia striiformis*) respectively, all of which are common and devastating diseases of wheat and barley throughout the world today.[a] Chinese wheats and barleys were doubly vulnerable to disease in that there were no native wild species present from which the cultivated varieties could acquire resistance.[b] Insect attacks were another matter, since they could be controlled to a certain extent:[c]

> Insects only appear in warm, damp conditions ... How do we know this? We can tell from weevils (*ku chhung*[4]): if grain is [stored] dry then no weevils will appear in it, if wet, it will turn mouldy and there will be no way of preventing weevils from appearing. When storing winter wheat (*su mai*[5]), if it be well-dried on a hot day and stored in dried vessels then no weevils will appear, but if it is not well dried in the sun they will gobble it up: weevils will appear in the wheat in great clouds ... From the gobbling weevils it is clear that insects appear in warm and damp conditions.

Wheat and barley were notoriously the most difficult of all the grains to keep. Some of the remedies against insect attacks in storage included *tshang erh*[6] (*Xanthium strumarium*), powdered hemp leaves, lime and moxa, all of which were sprinkled on the grain in storage.[d] One might imagine that the addition of moxa or other aromatic plants or chemicals would spoil the flavour of the wheat, but clearly there was no alternative if it was to be preserved, and the Chinese additives seem very inoffensive compared with the remedy suggested by the late 14th-century Yemeni author al-Malik al-Afḍal al-'Abbās bin Ali:[e] 'If wheat or other seed be placed in earthenware pots and covered with a hyena-skin in such a way that the smell of the skin can pervade it then it will stay free of all pests'—the bread made from such wheat must have had a rare fragrance. Only in exceptional conditions, as for example in the loess highlands of Shensi, would Chinese wheat keep well:[f]

> The land in Shensi is high and cold, and the soil strata lie horizontally. Preservatives are never used in the official grain stores. Even though wheat is normally extremely difficult to keep for long periods, here it will keep for up to twelve years without a single grain being spoilt by insects. The peasants simply dig holes in the middle of their fields as if they were making wells.

[a] Hu Hsi-Wen (*2*), p. 6; see also Chekiang Agricultural College (*2*); Gair, Jenkins & Lester (1); Leonard & Martin (1).
[b] On this topic see Feldman & Sears (1).
[c] *Lun Heng*, *Shang Chhung Phien*, 16/4*b*.
[d] Hu Hsi-Wen (*2*), p. 7 cites a number of classical texts; see also p. 386 above.
[e] Tr. R. B. Serjeant (1), p. 40; his *Bughyat al-Fallāḥīn* was composed *c.* +1370.
[f] *Chi Le Pien*, cited Hu Hsi-Wen (*2*), p. 74.

1 麥奴　2 薄茹　3 黃疸瘟　4 蠱蟲　5 宿麥　6 蒼耳

The growing crop is also extremely susceptible to pests,[a] in particular to grubs such as wireworms or nematodes which damage the roots. As a counter-measure Chinese farmers often treated the seed-grain with silkworm dung, cotton-seed oil or arsenic.[b]

Was the considerable effort that went into the cultivation of *mai* in China worthwhile? In the West, wheat and barley cultivation was integrated into a mixed farming system where large numbers of livestock were raised and their dung used to fertilise the fields. Even so before the introduction of improved rotation systems yields were generally very low.[c] We have few figures for yields in China before the 20th century, and those that we do possess are far from reliable. Perkins[d] gives half a dozen figures for Ming and Chhing wheat yields in Shensi, deduced from rental figures on the not indisputable assumption that the rent always represented half the crop; the yields he gives for the period from 1600 to 1900 are all of under 500 kg/ha, much lower than average rice yields and somewhat lower than average millet yields for the same region in the same period. But even if Perkins' figures are accurate, Shensi was one of the poorest of all the Chinese provinces and it is reasonable to assume that yields in, say, Shantung or Hopei would have been substantially higher. Buck's figures[e] for wheat yields in the early 20th century ranged between 250 and 3500 kg/ha, with an average of 400 kg/ha; Myers[f] gives a range between 650 and 1100 kg/ha for the period 1917–57, and remarks that China had already achieved fairly high wheat yields in the 1930s according to the world standard of that time, although depending on pre-modern technology. Indeed in 1934–6 China, with an average yield of 1090 kg/ha, ranked fourth in the world for wheat yields.

Given the paucity of data for both China and the West, it is difficult to make any very useful comparisons, apart from pointing out that the yield to seed ratio in pre-modern Europe, that is to say the number of grains reaped for every grain sown, was generally of the order of 3 : 1 or 4 : 1,[g] whereas one Sung writer urges the expansion of wheat production on the grounds that the harvest was fully tenfold.[h] This was clearly a well-rounded figure, but in any case wheat and barley with their rather large grains and small ears naturally gave lower returns than rice, where a ratio of 50 : 1 or 100 : 1 could confidently be expected. Although in a good year millet yields were generally higher, wheat and barley yields were more reliable since they were not subjected, like the spring-sown crops, to drought or flood in the early stages of growth.

[a] Gair, Jenkins & Lester (1), pp. 29–32 give a list of some forty insect pests likely to attack wheat in Great Britain during the various stages of growth.

[b] Hu Hsi-Wen (2), p. 7.

[c] See for example Duby (1), Oschinsky (1), where we find instances of 30 or even 50% of the crop being retained as seed grain.

[d] (1), p. 330, Table G 13.

[e] J. L. Buck (2), pp. 233 ff.

[f] R. H. Myers (2), Table 1.

[g] Slicher van Bath (1), Table II, pp. 328 ff.

[h] *Chiu Huang Huo Min Shu*, cited Hu Hsi-Wen (2), pp. 81–2.

Of late considerable effort has gone into wheat-breeding in China. Often Mexican varieties are crossed with Chinese strains, and spring and winter wheat crosses are common, the aim being to develop a high-yielding intermediate type relatively insensitive to day-length.[a] Yields are now high: we were told at the experimental station of the Northwestern Agricultural College in Wukung, Shensi in 1980 that one of their new varieties, Aifeng[1]-3, produces 4500 kg/ha.[b] Although wheat took many centuries to establish its place in the Chinese economy, there is now no question that it is the most important crop of North China.

## (4) Rice

Rice is the world's most important food crop after wheat; it is the staple in many African and Asian states and throughout most of the Far East, and has played a crucial role in the Chinese economy at least since Thang times.[c] The nutritional value of rice varies considerably according to type, environment and method of preparation, but generally speaking the endosperm is highly digestible and nutritious. Unmilled rice compares favourably with wheat and other cereals in its protein, fat, vitamin and mineral content, but unmilled or 'brown' rice has little sale outside the health-food stores of the West. It takes a long time to cook and is difficult to chew, and most rice-eaters prefer their rice to be not only hulled (removing the husk) but also milled and polished (removing all the coloured pericarp as bran). This leaves the grain white and shining. In polishing rice loses much of its nutritional value: highly polished rice contains only 7 % protein, whereas rice that has simply been husked contains nearly 10 %.[d] Washing and cooking methods often deplete the nutrients further, and deficiency diseases such as beri-beri are not uncommon among consumers of rice too poor to supplement their diet with alternative sources of proteins and vitamins.[e] Just as Europeans traditionally regarded white bread as a luxury more desirable than brown, so

[a] Harlan (4).

[b] In 1974 France had the world's highest average yields, of just over 4500 kg/ha; Myers (2), Table 3. On the future possibilities for the development of more productive and resistant wheat varieties incorporating genetic material from wild varieties see Feldman & Sears (1).

[c] Rice cultivation and consumption have been on the increase in the West in recent years, and also in Africa where intensive efforts have been made to popularise swamp-rice cultivation in countries such as the Gambia, Senegal and Liberia (K. Hart (1), Buddenhagen & Persley (1)). This process often involves not only the substitution of rice for more traditional food-crops such as tubers or dry cereals, but in some areas where African varieties of rice were traditionally grown they have been replaced by Asian varieties grown according to Asian methods (many of the high-cost rice schemes of West Africa are run by Chinese personnel, though whether from the People's Republic or Taiwan of course depends on the political affiliations of the state). Interestingly enough, while rice consumption is increasing in popularity outside Asia, in many Asian countries wheat products such as bread or noodles are now considered to be higher status foods and are beginning to replace rice as the staple amongst the better-off members of the population. The most striking example is Japan which, at great cost to the state, subsidises its rice farmers to produce rice that nobody wants, meanwhile importing large quantities of wheat from abroad.

[d] These figures are taken from Grist (1), p. 451; R. O. Whyte (1), p. 38 gives lower figures.

[e] Grist (1), ch. 19; R. O. Whyte (1), (2).

1 矮豐

most Asians wish their rice to be as highly polished as possible. The problem of nutritional deficiency has been exacerbated by the spread of efficient mechanised mills, for now almost all rice is very highly polished, even in villages. Traditionally the Asian poor used their own hand-mills or bought inexpensive rice that was poorly polished and so they were, despite themselves, protected in some measure against deficiency diseases. They also garnished their rice with soy products, fish sauces and vegetables which combined to make an impressively healthy diet compared with that consumed by the proletarians of rural or urban Europe.[a] So although modern analyses of the nutritional value of rice show it to be poorer in many respects than wheat or other cereals, in fact many traditional Asian rice-based diets were nutritionally more than adequate.

Rice must be the most diverse and adaptable crop known to man. Some varieties flourish on dry mountain slopes while others can survive floods five or six metres deep, growing more than ten centimetres a day as the waters rise.[b] Some varieties take seven months to ripen, others only two. Some are sensitive to day-length, others to temperature, some are so tolerant of salt that they can be used to reclaim marshes along the sea-shore. There are fragrant rices, 'glutinous' rices, red and white rices, and every traditional rice-growing region can count dozens if not hundreds of local varieties. The range of variation in rice is so great that no internationally recognised system of classification has yet been developed, although repeated attempts have been made ever since the Rice Congress at Valencia in 1914 urged the 'formation of a real botanical classification of the varieties of cultivated rice'.[c] There are, however, some very broad distinctions that may be drawn which will facilitate the following discussion of the origins and dissemination of rice cultivation in the Far East.

There are two species of domesticated rice, the West African rices, *Oryza glaberrima*, and the Asian rices, *Oryza sativa*, both of which are generally supposed to descend from a common Gondwanaland progenitor.[d] Here we shall only consider the Asian rices, *O. sativa*; the African rices are not cultivated outside that continent and have had no influence on the development of the Asian species (Fig. 231). Among the Asian domesticated rices, two sub-species are commonly distinguished, *indica* and *japonica*, both of which include glutinous and non-glutinous varieties. 'Glutinous' varieties in fact contain no gluten, and the stickiness of the cooked grain is generally assumed to result from the presence of dextrin and a little maltose as well as starch in the endosperm.[e] It is probable that the 'glutinous' characteristic is governed by a single recessive gene, as is the case with other cereals (see p. 457), and is the result of human selection. Glutinous rice is highly

[a] R. Fortune (4), p. 42.

[b] Grist (1), p. 141.

[c] Grist (1), pp. 101 ff. describes some of the more useful classification systems based on morphology and other criteria.

[d] T. T. Chang (2).

[e] Adair *et al.* (1); Grist (1), p. 100.

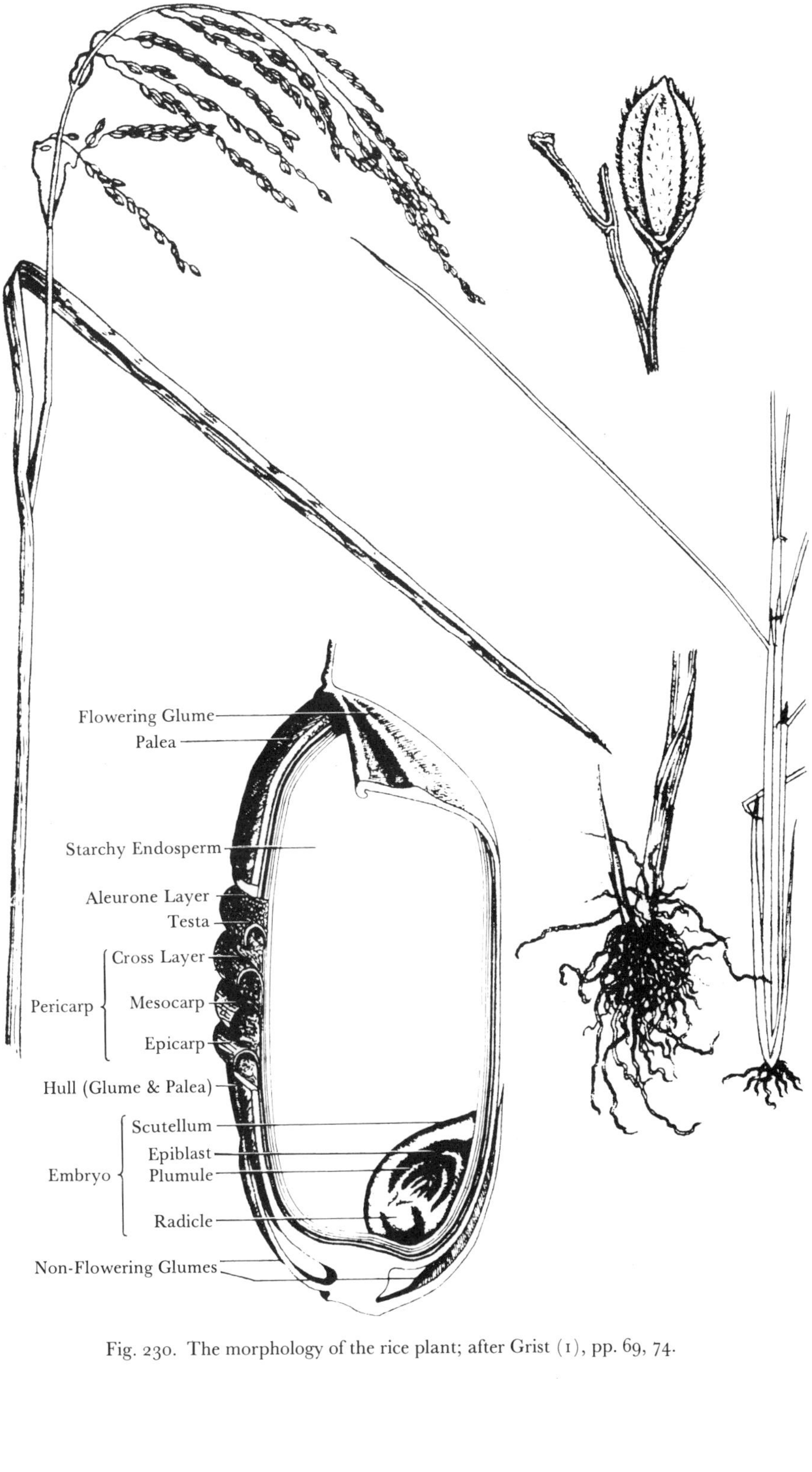


Fig. 230. The morphology of the rice plant; after Grist (1), pp. 69, 74.

valued as a ceremonial food and as the chief ingredient in rice 'wines' or beers; cakes made of glutinous rice are exchanged at weddings and festivals all over Asia.[a] Although most Asians reserve glutinous rice for ceremonial purposes, many mountain dwellers eat it as their staple, regarding it as much more nutritious than ordinary rice (though inclined to make the mind sluggish).[b]

A clear description of the most important differences between the *indica* and *japonica* subspecies is given by Grist:[c]

> The *japonica* forms are typical of the more northerly (and southerly) areas of paddy cultivation and flourish under very long photoperiods. When grown in tropical areas where the day length is short, they respond to the shorter photoperiod by greatly curtailed life-periods and often become so precocious as to be useless. Seedlings of cold water-tolerant *japonica* varieties may grow and develop faster than seedlings of *indica* varieties when the temperature of soil or water is low. Further ground for distinguishing these sub-species is given by the partial sterility of the progeny of *japonica* x *indica* hybrids. Though the chromosome numbers are the same and no morphological differences can be distinguished between the chromosomes of the two sub-species, there is clearly some incompatability between the genes of the two forms. There are other characters which often, though not invariably, separate these sub-species. For instance, the grain of *japonicas* is commonly shorter and broader than is usual for *indicas*; *japonicas* usually have broad leaves, rather hairy glumes and translucent endosperm. These differences are, however, by no means constant and if taken individually would not serve to distinguish one from the other. Another difference of great importance is in the cooking character of the rices; *japonicas* tend to soften rapidly after a certain time of cooking and to become 'mushy' if only slightly overdone; *indicas*, in contrast, tend to resist some overcooking and to give a rice with each grain separating and not sticky. Strong preferences for particular types of rice dictate the forms which will be acceptable in each area among people whose diet is normally rice. These differences are not to be confused with those between the 'glutinous' and 'non-glutinous' types; both of these can be found in *japonicas* and *indicas*, but the above description deals with differences within the non-glutinous types.

Unfortunately the problem is even more complex than Grist allows, for not only do intermediate forms between *japonica* and *indica* occur, as well as long-grained *japonicas* and round-grained *indicas*, but the very terms *japonica* and *indica* are found unacceptable by some experts. In order to understand what is at stake, we shall have to consider the various theories advanced as to the origins of rice domestication in Asia.

[a] One of the best-known Chinese examples must be the *tsung tzu*[1], cakes of glutinous rice steamed in broad bamboo leaves, which are exchanged on the day of the Dragon Boat festival. At Malay weddings the bride is presented with a large cake of glutinous rice, coloured red, yellow or purple, and decorated with eggs and flowers, as a symbol of fertility.

[b] Moerman (1) notes how in Northern Thailand this dietary distinction is felt by the local peasants to be a mark of ethnic if not political identity, and as such is jealously preserved: only effete southerners, soldiers or government officials, eat non-glutinous rice, and though local farmers will grow the higher-yielding non-glutinous varieties for sale, they always take care to grow sufficient of the (lower-yielding) glutinous rice for their own consumption.

[c] (1), p. 94; similar distinctions are drawn by Ting Ying (*1*), p. 185.

[1] 粽子

(i) *The origins of domesticated rice in Asia*

De Candolle, taking as literal the tradition that one of the cereals grown by the legendary emperor Shen Nung[1] was rice, concluded that rice had been domesticated in China by −2800, long prior to its use in India.[a] This was undoubtedly unreliable evidence. The only archaeological support for de Candolle's hypothesis until recently was the identification in the early 1920s, by two Swedish botanists, of imprints on a pottery jar from the neolithic site of Yang-shao[2] in Honan; these imprints, they believed, were made by cultivated rice.[b] But until recently the Yang-shao culture was supposed to have flourished rather late, *c.* −2000, and in any case doubt was cast on the identification of the imprints. Subsequent botanical investigations identified the chief area of distribution of wild *Oryza* species as the Eastern Himalayas and mountainous zones of mainland Southeast Asia (Fig. 231), while archaeological remains of domesticated rice were discovered at Harappan and other Indian sites which were presumed to predate any of the finds of rice in China.[c] Vavilov concluded:[d]

> India is undoubtedly the birthplace of rice ... Even though tropical India may stand second to China in the number of species, its *rice*, which was introduced to China, where it has been the staple food plant for the past thousand years, makes tropical India even more important in world agriculture. That India is the native home of rice is borne out by the presence there of a number of wild rice species, as well as common rice, growing wild, as a weed, and possessing a character common to wild grasses, namely, shedding of the grain at maturity, which ensures self-sowing. Here are also found intermediate forms connecting wild and cultivated rice. The varietal diversity of rice in India is the richest in the world, the coarse-grained primitive varieties being especially typical. India differs from China and other secondary regions of cultivation in Asia by the prevalence of dominant genes in its rice varieties.

Vavilov's opinion that domesticated rice originated in India was shared until recently by most botanists, and when in the late 1920s a group of Japanese workers divided Asian rices into two sub-species, on the basis of geographical distribution, morphology and hybrid sterility, it was natural for them to label the sub-species most commonly found in Japan '*japonica*' and the other, more widespread sub-species '*indica*' after its supposed country of origin.[e] However, some varieties, notably those of Indonesia, did not seem to conform to either category, and in 1958 a third subgroup, *O. sativa* subsp. *javanica* was proposed, to designate the *bulu* and *gundil* rice varieties of Indonesia.[f]

[a] (1), p. 385.
[b] Anderssen (1), p. 366.
[c] Vishnu-Mittre (2), especially p. 572. These have subsequently proved to be misidentifications of impressions of wheat; Charles A. Reed (4), p. 918.
[d] (2), p. 29.
[e] Katō *et al.* (*1*).
[f] Grist (1), p. 93.

1 神農　　2 仰韶

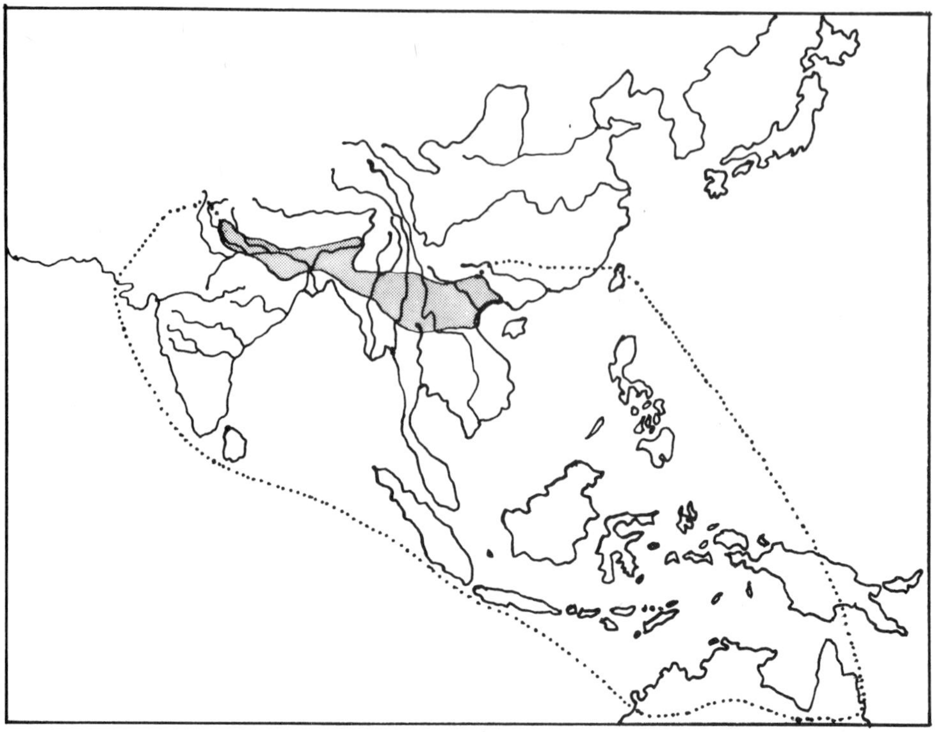


Fig. 231. Map showing the distribution of wild Asian rices; after T. T. Chang (2).

The criteria for division into the two main sub-species, as quoted above, are generally accepted, but Chinese botanists have never been happy with the nomenclature *indica* and *japonica*, which they believe obscures both the origins of rice domestication and the process of sub-speciation.[a] They would point out that Katō's analysis was in fact based chiefly on rice varieties collected in China, where both main sub-species (as well as intermediate forms similar to the '*javanica*') are to be found. The Chinese language has traditionally made a semantic distinction between the two sub-species, the term *keng*[1 b] corresponding to the '*japonica*' rices and *hsien*[2] to the '*indica*' varieties. Furthermore, archaeological discoveries in the 1950s and 1960s indicated that domesticated rice was grown in the Yangtze area much earlier than had previously been imagined,[c] while rice cultivation was known to have been introduced to Japan from China not earlier

[a] Liang Kuang-Shang *et al.* (*1*).
[b] Sometimes pronounced *ching*.
[c] Ho Ping-Ti (5), pp. 61 ff.; (*1*), pp. 140–5.

[1] 稉, 粳 or 秔 [2] 秈 or 籼

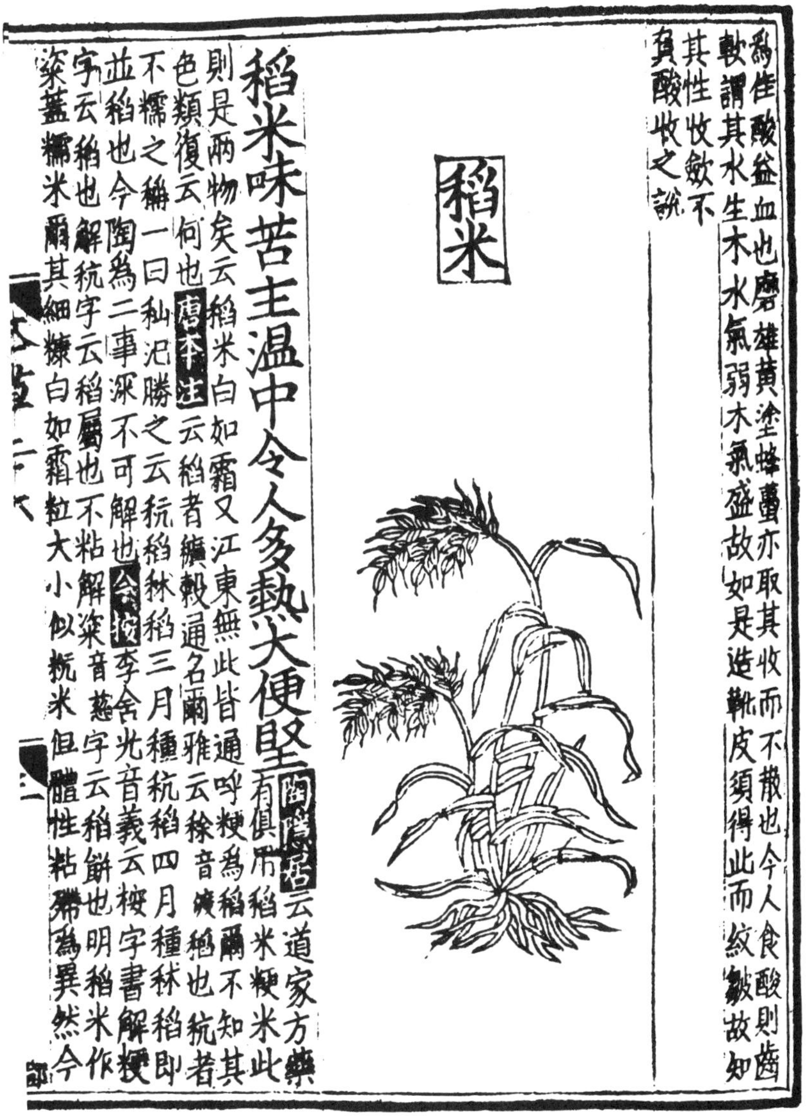


Fig. 232. Rice; *CLPT*, 1249 ed., 26/3*a*.

than the −4th century. A number of wild rices have been identified in southern China, some of which might qualify as possible ancestors of domesticated rice. Three varieties of wild rice, *O. rufipogon*, *O. officinalis* and *O. meyeriana*, have been found in China, in a zone stretching from Hainan to Taiwan and from Northern Kwangsi to Ching-hung[1] on the Upper Mekong in Yunnan.[a] Botanists still disagree as to the exact pedigree of domesticated rice, whether it evolved directly

[a] Yu Hsiu-Ling (*1*).

[1] 景洪

from the perennial form *O. rufipogon*, or through an intermediate wild annual form such as *O. spontanea*,[a] but there is no reason, *a priori*, why rice should not have been domesticated in South or Southwest China from the wild species native there. In 1961 the great Chinese expert on rice, Ting Ying, wrote that, given the extensive geographical distribution of wild rices in China and the comprehensive range of domesticated varieties, it was reasonable to presume that rice had in fact been domesticated in southern China, though he did not exclude the existence of other centres of domestication,[b] and he suggested that the nomenclature of the two sub-species should be changed from *indica* to *hsien* and from *japonica* to *keng*.

As more and more discoveries of early domesticated Chinese rice were made through the 1960s and 1970s, many of which could by then be carbon-dated and were thus shown to be earlier than the known finds in India, Chinese and Japanese scholars, and indeed most sinologists, came to take for granted that Chinese rice had been domesticated in China. The problems they saw as important were the identification of the area or areas where rice was domesticated, the identity of its domesticators, and the process of differentiation between the *keng* and *hsien* sub-species.

Early rice remains have been found at a number of sites in the Yangtze area and southward, some of the principal sites being the Chhing-lien-kang[1] cultural site of Sung-tse[2] near Shanghai (dated to about −4000), where the rice was identified as belonging to the *hsien* sub-species, and the Liang-chu[3] site of Chhien-shan-yang[4], on Thai-hu[5] lake in Northern Chekiang (dated to about −3300), where both *keng* and *hsien* were identified; several other sites on the Lower Yangtze were found to contain rice remains, as well as some Chhü-chia-ling[6] sites in Hupei.[c] Early as these sites were, they were all sufficiently late to be seen as 'Lungshanoid' cultures, and several archaeologists believed that the earliest Chinese farmers were the millet-growers of the Yang-shao culture in the Yellow River area, and that Chinese rice was domesticated by migrants from the North, who introduced their superior culture to the southern regions during a period of rapid expansion under population pressure. Accustomed to eating cereals and unable to grow their millets in sub-tropical latitudes, they domesticated rice either from completely wild species or from varieties growing as weeds in the taro gardens of the more backward southerners;[d] the cultivation of rice was for a long time confined mainly to the Yangtze area, and did not reach the Ling-nan[7] area of

[a] See T. T. Chang (2); H. I. Oka (1); many geneticists, including Chang and several Indian workers, consider *O. spontanea* to be a hybrid form between the perennial and domesticated species. For an up-to-date and comprehensive account of Chinese wild rices, see Huanan Agricultural College, Agricultural Department (*1*).

[b] (*1*), p. 13.

[c] Ho Ping-Ti (5), pp. 61–3; Wu Shan-Chhing (*1*); Mou Yung-Khang *et al*, (*1*).

[d] See for example K. C. Chang (1), (6), and Ho Ping-Ti (5); on rice as a weed of taro gardens see the section on agricultural origins, p. 46 above. The hypothesis of domestication by northern migrants of course ignores the inconvenient fact that Chinese millets grow very well in tropical conditions; see p. 435.

[1] 青蓮崗 [2] 松澤 [3] 良渚 [4] 錢山漾 [5] 太湖
[6] 屈家嶺 [7] 嶺南

Kwangtung, Kwangsi and Annam until it was introduced by Chinese colonists in the Chhin and Han.[a]

A heavy blow was dealt to this particular hypothesis by the discovery in 1976 of extensive remains of domesticated rice at the site of Ho-mu-tu[1] near Ningpo. The Ho-mu-tu site was occupied by neolithic rice farmers for several millennia, the most extensive rice remains being found in the earliest cultural stratum which dates from about −5000, just as early as the first known farming settlements of the Yellow River area.[b] The village of Ho-mu-tu consisted of houses on stilts, built on the edge of a marsh. Even the earliest cultural stratum shows signs of considerable technological sophistication, including well-made, finely decorated pottery and complex carpentry work, and the sheer volume of the rice remains shows that the inhabitants were not proto-farmers but relied heavily on cultivated rice as a food supply. The discovery of an advanced neolithic culture, largely dependent on wet rice cultivation, established on the Southeast coast of China by −5000, naturally leads us to enquire into its origins. The Ho-mu-tu culture has few, if any, affinities with the contemporary cultures of the Yellow River valley, but we may find a plausible explanation for this in the hypothesis, now accepted by many archaeologists in both East and West, that mainland Southeast Asia played a crucial role in early East Asian cultural development.

As Fig. 231 shows, the most likely region for the domestication of rice is a wide area that includes the river valleys of upper Assam and Thailand, Burma and Yunnan, and stretches eastwards to the coast of North Vietnam and South China. This falls well within the boundaries of the Hoabinhian cultural horizon, a pre- or proto-neolithic culture characterised by polished stone tools and cord-marked pottery that many archaeologists would postulate as the common substratum from which the markedly different neolithic cultures of East and Southeast Asia evolved (see p. 42).

Claims for piedmont Southeast Asia, and in particular the zone including Northeast Thailand, as the earliest agricultural centre of East Asia were first advanced in the 1960s following the spectacular discovery of plant remains at the site of Spirit Cave.[c] Difficulties of dating and identification caused many archaeologists initially to reject such claims as unwarranted, but as more and more early sites were discovered in the vicinity it gradually became accepted that agriculture was known in this zone much earlier than had previously been thought, and that it might even qualify as a centre of plant domestication. Two

[a] See Bray (3) for an account of the literature.

[b] The preliminary reports on the site are given in Anon. (*503*), (*504*).

[c] Dated to before −7000, Spirit Cave in North Thailand contained the remains of a variety of plants, including *Piper*, *Areca*, fruits and legumes; on the basis of the Spirit Cave data Solheim (1) postulated a Southeast Asian domestication of plants as early as −10,000. However, the dating, the identification of the plant remains, and their domesticated character have all been disputed, and most Southeast Asian archaeologists, e.g. Gorman (1), would now see the Spirit Cave plant remains as evidence of the plant-tending and incipient cultivation characteristic of pre-neolithic Hoabinhian subsistence patterns.

[1] 河姆渡

sites in Northeast Thailand, Non Nok Tha and Ban Chiang, dating back to about −5000 and −4500 respectively, 'strongly suggest the presence of rice farming in the northeastern Thai plateau prior to −4500';[a] Gorman (1) suggests that the domestication of rice began in naturally marshy areas *c.* −7000 and that, as their skills increased, farmers were able to spread to non-palustrian sites such as Non Nok Tha and Ban Chiang; evidence from later Thai sites suggests that population pressure and the development of iron technology encouraged the spread of rice-farming to the lowland plains some time in the −2nd to −1st millennium.

Gorman's model of an early domestication of rice, say *c.* −7000, in the piedmont zone of Southeast Asia is not only consistent with evidence from Thailand and Vietnam (where developed wet-rice agriculture reached the Red River Delta by −3000 at the latest[b]); the dates also tally with the evidence from Ho-mu-tu and later Lower Yangtze sites which indicate a well-established tradition of wet-rice cultivation by the −5th millennium. Cultivated rice has also been found in South China, for example at Shih-hsia[1] in Kwangtung, which may date back to −2000 or −3000.[c] Interestingly enough, the earliest finds of rice in Yunnan, that is to say the piedmont zone, to date go back only as far as −1300, though local archaeologists are confident that earlier farming sites will be discovered in the region soon.[d]

Further support for a domestication of rice in this extended piedmont zone of Southeast Asia is provided by linguistic evidence. The names of rice in Miao, Japanese, and a number of Southeast Asian languages contain the phonetic *-n-* or *-ni*$^{w}$*-*, as did the ancient Wu[2] dialect names for rice (*nuan*[3] and *nuo*[4]).[e] Benedict, examining the nature of Austro-Thai loan words in Chinese, noted that they included a number of terms for rice, including cooked rice, as well as the words for plough, pestle and mortar, seed, sow and winnow.[f] Eberhard, in his analysis of Chinese regional cultures, connected wet-rice cultivation in China with Thai peoples whose original home was in the Lower Yangtze region;[g] Benedict suggests an origin in Southern China for the Austro-Thai language family, but Chinese ethnographers adduce documentary and archaeological data to show that the Thai did not in fact migrate South and West from the Yangtze area as many Western scholars believe but originated in the border region of Yunnan, North

[a] Gorman (1), p. 344. There are, however, many difficulties associated with these dates; see Charles A. Reed (4), pp. 911–17; Muhly (1), p. 134.

[b] J. Davidson (1).

[c] Yang Shih-Thing (*1*); Su Ping-Chhi (*1*); Kwangtung Provincial Museum (*1*).

[d] Wang Ning-Sheng[5] of the Archaeology Department of the National Minorities Institute, Kunming, pers. comm. 1980.

[e] The contemporary Spring and Autumn period pronunciation of both these terms was *i'nuân*; this is closely related to the Japanese *ine*, the Annamese *n'êp*, Cham *ñióp*, Sedang *ñ'ian* and Buhnar *bānān*; see Andō Kōtarō (*1*); Amano (*4*), p. 93; Sasaki (*1*), p. 288.

[f] (2), p. 316.

[g] (2).

[1] 石峽 [2] 吳 [3] 暖 [4] 稬 [5] 汪寧生

Thailand and Burma.[a] It seems clear that rice cultivation either developed in South China (and probably in other zones of piedmont Southeast Asia too), or was brought there from the Western piedmont zones. In any case, northern intervention seems excluded.

Neither the archaeological, the botanical, nor the linguistic evidence define a centre of domestication more precisely than the piedmont zone of Southeast Asia described above; the uncertainty as to the exact pedigree of *O. sativa* compounds the problem. Few would disagree with the eminent geneticist T. T. Chang's view that both Asian and African domesticated rice are descended from a common Gondwanaland ancestor, but more recent developments are impossible to establish with any degree of certainty: 'Cytogenetic evidence has been well established on the relatively close affinity among the wild perennial, wild annual, and cultivated annual forms of rice in Asia, but it is impossible to critically compare [sic] the affinity between any two members with that of the others because of the deficiencies in describing and designating the parents used in interspecific crosses.'[b] Even archaeological evidence can only be suggestive, not conclusive, when it comes to the history of domestication. Benedict and Ting Ying (*2*) propose a South Chinese origin for Asian domesticated rice, Solheim and the Southeast Asianists a Northern Thai origin, Liu Tze-Ming (*1*) believes it originated in the Yunnan plateau and spread thence to South China, Southeast Asia and India, while Watanabe Tadayo (*1*) thinks domesticated rice had its origin in Assam and spread from there to Yunnan and points East. As a Chinese ethnographer, himself a member of a national minority, said when we were discussing the problem in the National Minorities Insititute in Kunming, the debate really seems to hinge as much on national pride as on any scientific exactitude.[c] So although we shall henceforward conform to the Chinese usage in distinguishing between *keng* and *hsien* rather than *indica* and *japonica*, we do this for purposes of convenience, not because we are convinced that Asian domesticated rice is Chinese in its origins.

The impossibility of precise identification and definition of pure taxa cited by T. T. Chang is also an impediment to our understanding of the subspeciation of *keng* and *hsien* rices. The earliest record of terminological differentiation is in the +100 *Shuo Wen Chieh Tzu*, which distinguishes between *keng*[1], a type of rice (*tao*[2]), and *hsien*[3] or *lien*[3], a non-sticky *tao*[2]; the written form of *hsien*[4] used today appears for the first time in the early +3rd-century *Kuang Ya*. The archaeological record, however, shows that the differentiation is far more ancient. Chinese archaeologists use the ratio of length to breadth of grain to distinguish between *keng* (round-grained) and *hsien* (long-grained); while this is not an infallible criterion it is

[a] Wang Ning-Sheng, pers. comm. 1980.
[b] T. T. Chang (2), p. 429.
[c] Ma Yao[5], pers. comm. 1980.

[1] 稉 or 秔 [2] 稻 [3] 稴 [4] 秈 or 籼 [5] 馬耀

the only one that can be usefully applied to carbonised grains. On this basis, then, the earliest known domesticated rice from the site of Ho-mu-tu near Ningpo (*c.* −5000) has been identified as belonging to the *hsien* subspecies,[a] as have the rice remains from Sung-tse near Shanghai (*c.* −4000)[b] and the Kwangtung rice remains;[c] round-grained *keng*-type rice was however identified at Tshao-hsien-shan[1] in Southern Kiangsu (perhaps as early as −4000), and at some sites dated −3000 to −2000 on the middle and lower Yangtze.[d]

*Keng* rices are sensitive to photoperiod and do not do well in tropical zones. In recent times they have been most extensively grown in Japan and Korea,[e] though they are also grown in China from Shantung northwards and on land from 500 to 2000 m. above sea-level in Kiangsu, Hunan, Fukien, Kwangtung and Kwangsi.[f] *Hsien*-type rices are grown almost exclusively in the plains of South China. A study of cultivated rices in Yunnan showed that *hsien* varieties predominated up to 1750 m. and *keng* varieties over 2000 m., while the zone from 1750 to 2000 m. intermediate varieties were found.[g]

There are two schools of thought as to the process of differentiation between *keng* and *hsien* rices. The first is that the two sub-species evolved independently: *keng* in the region north of the Yangtze where one can find wild rices very similar to *keng*, known locally as 'spontaneous rice' (*lü tao*[2]) or 'tank rice' (*thang tao*[3]); *hsien* in South China where *hsien*-like wild rices occur, known as 'ghost grain' (*kuei ho*[4]) or 'doesn't go home' (*pu kuei chia*[5]). The second is that *keng* evolved from *hsien* as the cultivation of rice spread to higher latitudes and elevations; evidence adduced is the later appearance of *keng* in the archaeological record, the fact that *hsien* varieties shed their grains more easily than *keng*, though not as easily as wild species, and the lower incidence of sterility in *O. spontanea* x *hsien* hybrids than in *spontanea* x *keng* hybrids.[h] The first hypothesis may be objected to on the ground that the wild rices cited are in all probability the result of cross-breeding between wild species and local cultivated varieties, while to the second we may object the almost invariable sterility of *keng* x *hsien* hybrids.[i] Experiments carried out in Jaipur in India indicate that either *keng* or *hsien* types may emerge from the same wild strains during the process of domestication; furthermore, when *O. perennis* was crossed experimentally with *O. sativa hsien*, plants with *keng* characteristics were found among the offspring and vice versa.[j] But these indications of a mono-

[a] Anon. (*503*), p. 20; Yu Hsiu-Ling (*1*).
[b] Wu Shan-Chhing (*1*).
[c] Yang Shih-Thing (*1*).
[d] Yu Hsiu-Ling (*1*); Mou Yung-Khang *et al.* (*1*).
[e] And in Taiwan after it was ceded to the Japanese at the Treaty of Shimonoseki[6] in 1895.
[f] Amano (*4*), p. 108.
[g] Ting Ying (*2*).
[h] Yu Hsiu-Ling (*1*).
[i] Grist (1), p. 115; Oka (1), p. 26.
[j] Oka (1), p. 25.

[1] 草鞋山 [2] 穭稻 [3] 塘稻 [4] 鬼禾 [5] 不歸家
[6] 下關

phyletic origin of the two sub-species do not resolve the problem of how the differentiation occurred, nor do they cover the third sub-species, *O. sativa javanica* which, although it is dismissed as irrelevant by many Chinese workers,[a] is still considered significant by such authorities as T. T. Chang (1), (2).

We have wrestled long enough with the intractable problems of the origins of Chinese domesticated rices. Now let us retreat to firmer ground, and turn to the records to examine the historical development of rice cultivation in China.

### (ii) *Chinese rice varieties and nomenclature*

Rice is largely self-pollinated, but cross-pollination does occur in degrees varying between less than 1 % and as much as 30 %.[b] Crosses between members of the *keng* and *hsien* sub-species are almost invariably sterile, but crosses between cultivated and wild rices show some fertility.[c] Wild rices are common in many of the areas where rice is cultivated, so that in the natural course of events one would expect rice cultivars to evolve and change continuously. Human selection has clearly been crucial in maintaining desirable strains and selecting for new cultivars. We have already discussed (p. 330) the contribution of Asian harvesting techniques, such as the use of the reaping-knife, to crop diversification. Where wet-rice cultivation is concerned, the technique of transplanting provides an additional opportunity for selection and control. As a result of careful tending and selection of seed-grain, most traditional rice-farming communities were able to maintain at least a dozen or twenty rice varieties suited to different soil types, cultivation techniques or climatic conditions, and differing in ripening period, colour or flavour. As most varieties were given local names, it is impossible to calculate the total number of rice varieties known at any one period, for one variety might be given different names in different localities, while the same name might be used for different rices in different places. Furthermore, names changed over time.[d]

There is no doubt that China had an enormous range of rice varieties from earliest times. The *Chhi Min Yao Shu* (written in Northern China at a time when rice cultivation was rare north of the Huai), as well as citing varieties mentioned in earlier texts, names twelve non-glutinous rices (*tao*[1]) and eleven glutinous varieties (*shu*[2]).[e] The *Shou Shih Thung Khao*[f] gives over three thousand names,

[a] See Liang Kuang-Shang *et al.* (*1*), p. 252.

[b] Grist (1), p. 72.

[c] *Ibid.* p. 115. The exchange of genetic material has also contributed to the diversification of weed rices: 'Samples of recently collected weed races indicate that considerable introgressive hybridisation has taken place and that gene flow has been largely from the cultivated races to the wild forms' (T. T. Chang (2) p. 428).

[d] The Sung work *Khai-Chi Chih*[3], ch. 17, lists 56 names of rice varieties, almost all of which were unfamiliar to local farmers by the Chhing, according to the *Shao-Hsing Fu Chih*[4]; see Amano (*4*), p. 213.

[e] *CMYS* 11.1.6–7.

[f] *SSTK*, ch. 21.

1 稻 2 秫 3 會稽志 4 紹興府志

which modern experts reckon represent some thousand or so distinct varieties,[a] but given the immense ecological range of cultivated rice in China it is probable that many more varieties existed than the *Shou Shih Thung Khao* records. Within the *keng* or *hsien* sub-species farmers would distinguish between varieties according to their ripening period, their morphology, water requirements and resistance to disease, as well as to whether or not they were glutinous, fragrant or coloured. Quick-ripening varieties usually yield less than the slow-maturing types, but are less prone to drought-damage; if water is scarce it may be preferable to grow dry rice rather than wet; long-stemmed varieties are less likely to suffer from floods than short-stemmed rices, but are more likely to lodge. These are just a few of the factors the farmer must take into account when deciding which varieties to grow in which fields.[b]

We shall not enter into details of the more exotic Chinese rices, the fragrant rices (*hsiang tao*[1]), one grain of which would perfume a whole pot-full of ordinary rice, or the glutinous rices (*shu tao*[2], *nuo*[3] or *lo*[3]) used for making cakes, wine and spirits.[c] Red rices (*chhih mi*[4]) we shall refer to again when we come to the reclamation of saline land, and the differences between wet and dry rice will be touched upon when we come to cultivation techniques. Readers who wish to know more are referred to the works of Copeland (1), Angladette (1) and Grist (1) for a general treatment; Amano (*4*) devotes a long section to the history of various exotic rice-types in China, based chiefly on textual material which is also to be found in Chhen Tsu-Kuei (*2*), while Ting Ying (*1*) remains the classic work on rice in modern China.

The most important distinctions, as perceived by Chinese farmers, were between *keng* and *hsien* rices, which differed in flavour and were suited to different natural conditions, and between early and late rices, the latter giving heavier yields while the former permitted multicropping. As we have already mentioned, the *Shouo Wen Chieh Tzu* is the first text to distinguish between *keng*[5] and *hsien*[6], while the modern character for *hsien*[7] is first found in the *Kuang Ya*. Since on the whole *hsien* rices ripen earlier than *keng* varieties there has been a tendency to conflate the category of *hsien* rices with that of early rices (*tsao tao*[8]), though in fact both late-ripening *hsien* rices and early-ripening *keng* varieties are fairly common.[d] Most Chinese, though, were quite aware that there was more to the difference

[a] Yu Hsiu-Ling, pers. comm. 1980.

[b] Nowadays many Asian farmers are relieved of this burden of decision-making, since the choice of which variety to plant is dictated to them by the government or local agricultural development authority; the imposition of this uniformity has mixed results; Bray & Robertson (1) gives details for Malaysia.

[c] One of the most unusual rices, floating rice, which can grow 5 or 6m. in only a few weeks, is hardly known in China, though Ting Ying (*1*), pp. 545–55 does describe it; floating rices are typical of the heavily flooded deltaic plains of Bengal and Burma.

[d] See, for example, Amano (*4*), p. 212. or Li Yen-Chang (*1*) on early rices (quoted by Chhen Tsu-Kuei (*2*), pp. 403 ff.).

[1] 香稻 [2] 秫稻 [3] 稬 or 穤 [4] 赤米 [5] 稉 or 秔
[6] 稴 [7] 秈 or 籼 [8] 早稻

between *keng* and *hsien* than just the ripening period. A 12th century scholar from Hsin-an[1] in Anhwei, Shu Lin[2], wrote as follows:[a]

> The greater cereal *ta ho ku*[3] is nowadays known as *keng* rice. Its grains are large and it has awns, and it can only be grown on fertile land. The lesser cereal *hsiao ho ku*[4] is nowadays known as Champa rice (*chan tao*[5] or *hsien*). It has smaller grains and no awns, and can be grown on any land, fertile or not. The grain called *keng* gives low yields and commands a higher price, and apart from being used to pay taxes is eaten only by upper-class people. The small-grained rice [*hsien*] gives high yields and commands a low price, and is eaten by everybody from modest property-holders down.

According to Lo Yuan[6], whose gazetteer of the same district appeared in 1175, Hsin-an was extremely unsuitable for growing *keng*,[b] which would explain the low yields and high prices there. The difference in grain shape and cultivation requirements between *keng* and *hsien* was generally recognised, as were variations in leaf shape and colour and the presence or absence of awns. The confusion between *hsien* and early rices is understandable when we consider the key role played by the early-ripening Champa rices (all of which were *hsien*) in Chinese economic history.[c]

Most rice varieties in early China, whether *keng* or *hsien* (and we know that both had been cultivated since the neolithic period), ripened around the (old) 9th month. The *Chhi Yüeh*[7] ode from the *Shih Ching* speaks of reaping rice in the 10th month;[d] since this rice was to be used for making the spring wine, it must have been a glutinous variety and so would ripen a little later than usual. The *Chhi Min Yao Shu* says rice should be harvested well into the ninth month at 'hoarfrost descends' (*shuang chiang*[8]).[e] The *Kuang Chih* of *c.* +300 mentions early varieties from the south and from I-chou[9] in Szechwan which ripen in the 6th or 7th month, Thao Yuan-Ming[10] wrote a poem in +410 on farming in Kiangsi entitled 'Harvesting the early rice in the Western fields' (*yü hsi thien huo tsao tao*[11]), and the 9th century writer Lu Kuei-Meng[12] also refers to early rice in the Yangtze Delta in a poem entitled 'Harvesting song' (*i huo ke*[13]).[f] Both the *Kuang Chih* and Thao Yuan-Ming are speaking of areas well to the south of the Yangtze, so in these two cases at least the early rices were almost certainly *hsien*, similar to the Champa rices introduced to the Yangtze and Huai areas in the early Sung. Until the arrival of the Champa varieties, though, early rices played a very minor role in China, as did multiple-cropping. The new quick-ripening Champa varieties were made

[a] Quoted Amano (*4*), p. 105.
[b] *Hsin-An Chih*[14], ch. 2, quoted Amano (*4*), p. 105.
[c] Ho Ping-Ti (7) is the key western language text on this subject.
[d] *Shih Ching* 15/3*a*; Karlgren (14), no. 154.
[e] *CMYS* 11.5.1.
[f] Both cited Amano (*4*), pp. 126–7.

[1] 新安 [2] 舒璘 [3] 大禾穀 [4] 小禾穀 [5] 占稻
[6] 羅願 [7] 七月 [8] 霜降 [9] 益州 [10] 陶淵明
[11] 於西田穫早稻 [12] 陸龜蒙 [13] 刈穫歌 [14] 新安志

available at a time when the traditional agriculture of southern China was fast approaching its limits, and their introduction made possible fundamental changes in the southern farming system and important advances in agricultural productivity.

The precarious position of North China during the Five Dynasties and early Sung not only increased the reliance of the state on southern agricultural production but also drove a large number of northerners to migrate to the safety of the Yangtze provinces and other southern regions. This shift in China's economic balance reduced the total area of farmland from which the population as a whole was fed and increased the nation's dependence upon grain production in the south, and the rapid population increase concentrated in the southern provinces exacerbated the danger of famine. In this time of national peril it was clearly in the state's interest to encourage increases in agricultural production by all the means at its disposal, and a whole series of coordinated measures was launched to this end. The key factor in the long-term success of the campaign was the introduction to the Lower Yangtze of quick-ripening rices from Champa.

Champa[1] was an Indo-Chinese state to the south of Annam known for its drought-resistant, early-ripening strains of rice. As early as the Eastern Han the Chinese knew by hearsay of rice varieties so precocious that they could be cropped twice in one year. The *I Wu Chih* mentions their existence in Cochin-China (Chiao-chih[2]), the *Shui Ching Chu* says they were grown in Annam (Chiu-chen[3]). The Annamese apparently possessed such a wide range of wet and dry, early and late varieties that they could grow rice all the year round, but the quick-ripening varieties gave very low yields while requiring just as much labour as the other sorts.[a] These southern rices did, however, have very moderate water requirements: they would grow in poorly watered fields, did well as dry crops when grown in hilly regions, and were generally very resistant to drought; many farmers must have grown them, if not as their main crop, at least as a form of insurance. The Champa rices probably spread gradually northwards during the medieval period, passing from farm to farm, for when the Sung emperor Chen-Tsung[4] decided to popularise their cultivation in Southeast China he sent for 30,000 bushels of Champa seed to be delivered to the Lower Yangtze and Huai provinces from Fukien.[b] This was in +1012. The seed was distributed by officials to the local farmers, particularly those with upland fields, together with precise instructions as to the appropriate cultivation methods.[c] The campaign was a

[a] Amano (*4*), p. 193.

[b] *Sung Shih*, ch. 173, *Nung Thien*[5]; the 14th-century *Wang Chen Nung Shu* (ch. 7/7*a*) and the 17th-century *Kuang Tung Hsin Yü* (p. 374) also say that Champa rice was first introduced to the Yangtze not from Champa but from Fukien; however, the Sung writer Shih Wen-Ying[6] gives a different version of the story in which the emperor sent directly to Champa for 20 piculs of seed (*Hsiang Shan Yeh Lu*, ch. 2).

[c] The instructions published in the *Sung Hui Yao Kao*, *Shih Huo*[7], ch. 1. advised a long and careful pre-germination of the seed before sowing it in an inundated field; since the Champa rices were often grown as dry rices on upland fields, instructions on their cultivation as a dry crop were presumably circulated too.

[1] 占城國, also written 戰城, 金城 or 京城 [2] 交趾 [3] 九眞
[4] 眞宗 [5] 農田 [6] 釋文瑩 [7] 食貨

resounding success. By the mid-12th century it was reported that 70 % of the rice grown in Kiangsi was Champa rice, and by the end of the century we hear that 80 to 90 % of the wet-rice in the Lower Yangtze was the Champa type.[a]

The original Champa imports were early-ripening and drought-resistant, but lower-yielding than the traditional Chinese varieties. In +1192 the official Liu Chiu-Yuan[1] said in a letter that the Lower Yangtze farmers divided their wet-fields into early and late, and while the majority were sown with early-ripening Champa rices, the late fields were sown with *keng* rice which gave higher yields.[b] Chinese farmers worked fast to develop new, more flexible Champa varieties. Even during the Sung, later-ripening more prolific types had been developed, and soon the range was as wide as for any other type of rice. Specially successful local varieties would sometimes find their way all over China; for example, a variety known as 'Honan early' was recorded as far away as Southern Fukien in the early 17th century.[c]

The growth period of most Champa rices varied between 60 and 120 days after transplanting.[d] Though most farmers with a reliable water supply seldom grew the most precocious varieties because of their poor yields,[e] they were invaluable in

> marginal rice areas which called for varieties that would ripen even more rapidly or for special strains that could survive unusually menacing natural conditions. The marshy flats of Kiangsu north of the Yangtze, visited by the midsummer flood and consequently submerged for a greater part of the year, made up one such blighted area. For this reason Kao-yu and Thai-chou, both in the heart of the Kiangsu flats, became virtually the experimental farm for extremely early-ripening varieties. To beat the annual midsummer flood, peasants of Kao-yu developed the fifty-day in the sixteenth century. Some inland districts in southwestern Chekiang and lakeside Kiangsi may also have independently developed this fifty-day variety. It was the fifty-day that saved Kao-yu's peasants from total crop failure during the terrible flood of 1720–1. In the eighteenth century the forty-day was developed, probably independently, in Kao-yu and in Heng-chou in southern Hunan. When Kiangsu was particularly hard hit by the flood of 1834–5, the thirty-day variety, which was said to have been developed in Hupei, was rushed in and distributed to the peasants of the Kiangsu flats.[f]

There was also a need for varieties that could be sown late and harvested quickly; they could be planted after spring crops or early rice as part of a regular rotation, or sown in low-lying areas once the waters retreated after the summer floods. Since they were harvested late in the year they were known as 'cold (*han*[2]) Champa' or 'winter (*tung*[3]) Champa' rices. Such varieties are frequently recorded in gazet-

[a] See Amano (*4*), p. 106; Katō Shigeshi (*3*); Ho Ping-Ti (*7*); Yü Ching-Jang (*1*).
[b] *Hsiang-Shan Wen Chi*, ch. 16, letter to Chang Te-Mao[4], quoted Amano (*4*), p. 106.
[c] Ho Ping-Ti (4), p. 173.
[d] Li Yen-Chang (*1*).
[e] Amano (*4*), p. 214.
[f] Ho Ping-Ti (4), p. 173. However, doubts have been cast by modern experts on the possibility of any rice variety actually ripening in less than two months after transplanting (Yu Hsiu-Ling of the Chekiang Agricultural College, pers. comm. 1980).

[1] 陸九淵 [2] 寒 [3] 冬 [4] 章德茂

Table 12. *Chinese categories of rice*

| | |
|---|---|
| General terms: | *tao* 稻: from Chou texts on; its existence in Shang inscriptions, though often suggested, has yet to be satisfactorily proven.[a] |
| | *tu* 稌: according to the *Erh Ya* commentary this was a term used locally in the Huai area |
| | *nuan* 暖: the ancient term for rice used in southern China; probably derived from an Austro-Thai root. |
| Wet rice: | *shui tao* 水稻 |
| Dry rice: | *han tao* 旱稻 |
| | *lu tao* 陸稻 |
| | *shan tao* 山稻 |
| Early rice: | *tsao tao* 早稻 |
| Late rice: | *wan tao* 晚稻 |
| Glutinous rice: | *nuo/lo* 稬, 穤 or 糯 |
| | *shu tao* 秫稻 |
| *Keng* (*japonica*) rices: | *keng* 稉, 粳 or 秔 |
| | *ta ho* 大禾 |
| *Hsien* (*indica*) rices: | *hsien* 稴, 秈 or 籼 |
| | *hsiao ho* 小禾 |
| Champa rices: | *chan tao* 占, 秥 or 粘稻 |
| | *nien tao* 粘 or 黏稻: the latter especially appears to be a southern usage derived from an original Austro-Thai root |

[a] See Amano (*4*), pp. 128–38.

teers of Chekiang, Fukien, Kiangsi and Hunan.[a] Perhaps one of the most useful types of Champa rice to be developed was the 'red' (*chhih*[1]) or 'peach blossom' (*thao hua*[2]) rice. This was known as early as the Sung: Lo Yuan[3] wrote in 1175 that it was grown in small quantities in Anhwei as it ripened very early and could be used to tide over the lean period before the summer harvest.[b] It was also notable for its salt-resistance, and this was perhaps its most valuable quality:[c]

> Rouge red [rice] (*yen chih chhih*[4]) is soft, fragrant and sweet, and when it is cooked it turns a uniform red colour. It is one of the best of the late rices. One variety tolerates saline conditions and is ideal for brackish fields by lakes or near river-mouths.

It was with similar red rices that the Man clan reclaimed their lands from the sea on Deep Water Bay in Hongkong in the 13th and 14th centuries.[d]

The introduction in the 11th century of the highly adaptable Champa rices thus permitted a quantum leap in the productivity of Chinese rice-farming. Not only did it facilitate the spread of rice cultivation into areas, particularly hilly

[a] Ho Ping-Ti (4), p. 174.
[b] *Hsin-an Chih*, cited Amano (*4*), p. 116.
[c] *Chhün Fang Phu* 8/26*b*.
[d] J. L. Watson (1), pp. 30 ff.; see also p. 123 above.

[1] 赤 [2] 桃花 [3] 羅願 [4] 臙脂赤

areas, where the water supply was insufficient for traditional varieties, and permit the reclamation of marginal land affected by floods or salinity, but the quick-maturing varieties only occupied the fields for three or four months at a time so that multi-cropping became possible. The alternation of rice and winter wheat or the double-cropping of rice quickly became common in the Lower Yangtze provinces, spreading inland to Hupei and Hunan and south down the coast in the Ming and Chhing, while in the southernmost provinces it even became possible to grow three crops of rice in one year.[a]

Different names were used to denote the Champa rices in different regions. In the Yangtze Delta and Huai area they were simply known as *hsien*, while in Honan and Kiangsi they were called *chan*[1] as in Champa[2] or *chan/nien*[3]; in South China and Szechwan they were called *nien*[4]. Now the term *nien*[3,4], when applied to cereals other than rice, means 'glutinous', yet *hsien* rices are well known to be the least glutinous of all the rice types. Yü Ching-Jang (*1*) has suggested that a cereal radical was added to the element *chan*[1] simply as a signifier; yet in that case one would have expected the pronunciation *chan* to be maintained. However, *nien* is by far the commoner pronunciation of both forms of the compound character, and is closely related phonetically to the putative Austro-Thai and to the Southeast Asian and ancient Southern Chinese terms for rice, all of which contain at least one *-n-*. We would therefore suggest that the term *nien*, commonly applied to Champa rices in the southern provinces of China, is, like the rice itself, Indochinese in origin.

### (iii) *Cultivation methods*

The productivity of a particular rice cultivation system is by and large proportional to its labour requirements, and the choice of cultivation techniques is largely determined by population pressure and the available labour supply. Generally speaking, wet-fields are more productive than dry, transplanting is more productive than direct sowing, and the more precisely the water supply is controlled, the higher the yields will be. Increased use of fertilisers, multicropping of rice or the alternation of rice with other crops, the integration of rice-farming with sericulture or pisciculture, all improve overall output but increase the amount of work the farmer must put in. On the other hand a little extra work will usually guarantee an increase—however slight—in yield (which is not the case with all crops), and wet-rice agriculture can maintain higher densities of population than any other farming system.[b] But few people care to work harder than they must,

[a] *Kuang Tung Hsin Yü*, p. 371.

[b] Notorious cases are Bangladesh and Java, where enormous populations eke out a marginal existence on very restricted areas of arable land. The labour-absorbent character of wet-rice agriculture is the basis of Geertz's (1) arguments on 'agricultural involution'.

[1] 占　[2] 占城國　[3] 粘　[4] 黏

which is why farmers familiar with 'advanced' technology will often opt for less demanding methods until driven by economic necessity or, as so often today, official pressure.

Rice can be grown either in wet fields or in dry fields. It is easy to distinguish between the extremes of wet and dry cultivation, between growing transplanted rice in dyked fields constantly supplied with water by irrigation machinery, and growing broadcast rice in an undyked hillside field dependent on natural rainfall alone (Fig. 233). Since every possible intermediate stage exists it is difficult to know exactly where to draw the line between the two, particularly since many dry rice varieties can also be grown under irrigation and vice versa.[a] Perhaps the most useful way for us to distinguish between the two, and that which corresponds most closely to Chinese usage, is to define wet rice as being grown in bunded, irrigated fields. By this definition, dry rice does not seem to have been grown in China in ancient times; all the characters denoting rice in the Chou inscriptions indicate a plant grown in standing water.[b] It was at one time common to postulate that dry rice cultivation preceded wet, on the ground that the techniques involved were less complex,[c] but most botanists reject this on morphological grounds: all the evidence shows that rice, both wild and domesticated, is naturally aquatic. In China at least the historical evidence seems to support the botanists' views. There are extremely brief references to what was probably dry rice in the *Kuan Tzu* and *Chou Li*, but their treatment of wet cultivation is far more elaborate. Some minority tribes, the Miao[1] and Li[2] for example, have traditionally grown hill rice together with other crops in swidden farms in the mountains of South China,[d] but swidden farming was seldom practised by Han Chinese (see p. 100 above). Dry rice was grown by Chinese farmers where the water was insufficient for wet varieties, but even dry rice can only be grown successfully under humid conditions and must have an assured rainfall during its growth period. Although *shan tao*[3] and *lu tao*[4] are used to denote dry rice, by far the most common term is *han tao*[5], which only a single stroke of the pen distinguishes from *tsao tao*[6], early rice, and the confusion often seems to be made in Chinese texts. It is therefore rather difficult to trace the history of dry-rice cultivation in China in any detail.

It seems that the cultivation of dry rice in China was confined chiefly to hillside

[a] For various classifications of dry rice cultivation systems as well as descriptions of cultivation methods see Grist (1); R. D. Hill (2), (3) and (4); J. E. Spencer (4); etc.

[b] Ting Ying (*1*), p. 25.

[c] E.g. R. D. Hill (1), (2).

[d] 'In agriculture, the Miao men and women work together. They have more mountain farms than irrigated fields. Burning the thorny trees and decomposing plants, they plant sesamum, millet, rice, wheat, beans ...' (Yen Ju-Yü (*2*), ch. 8/8*b*, tr. Geddes (1) p. 32). The Miao cultivated their fields several years in a row before clearing new ones, but the Li, according to the *Kuang Tung Hsin Yü* (p. 376), grew only a single crop of rice in the ashes of their swidden fields before moving on.

[1] 苗 [2] 黎 [3] 山稻 [4] 陸稻 [5] 旱稻
[6] 早稻

Fig. 233. Hill rice farm in Sarawak; orig. photo A. F. Robertson.

fields and to areas in the north where irrigation was not possible. The *Chhi Min Yao Shu* recommends growing dry rice in low-lying fields with a tendency to waterlogging, less because they are inherently suited to rice cultivation than because rice is the only crop that will grow there successfully, whereas in upland fields it is preferable to grow barley or millet.[a] The seed is soaked to pre-germinate it, sown by drill or broadcast, in a well-tilled field, and hoed and weeded repeatedly throughout its growth.[b] The whole process as described in *Chhi Min Yao Shu* is so labour-intensive that it is not surprising that northern farmers generally preferred to grow millet or barley instead. According to the *Wang Chen Nung Shu*, the methods prescribed for dry-rice cultivation by the +6th-century *Chhi Min Yao Shu* were still followed in North China in the 14th century; it also mentions that dry rice was grown quite commonly in the hills of Fukien.[c] Dry rice is still grown occasionally in China today,[d] but crops like maize or sweet potatoes will produce far more under the same conditions.

Nearly all the rice cultivated in China is wet rice, grown in bunded fields in which the water stands for the greater length of its growth period (Fig. 234). A plentiful supply of water is the key factor in rice cultivation, more important than the quality of the soil or the amount of fertiliser used. 'Whether the land be good or poor, if the water is clear then the rice will be good', says the *Chhi Min Yao Shu*, adding that ricefields should always be near the upper course of a stream.[e] The point is elaborated by the modern authority Grist:[f]

> Yield of paddy depends in no small measure on the quality of the water used for irrigation. Water may have a considerable fertilising value because of its mineral nutrients, or may cause damage to the crop by poisonous or indirectly harmful substances.
>
> Quality of water is dependent on its origin. River water is generally preferable to that from other sources. In addition to the fertilising elements dissolved in such water, it carries silt and clay. While large masses of silt must be avoided, a reasonable quantity of coarse silt deposited on the land has a favourable effect on the soil. Very fine silt carried to the land in like manner frequently has an unfavourable effect on the plant's growth.

It is significant that rice land is often classified not according to the soil type but according to the nature of the water supply. In the Kelantan plain of Malaysia, when all the rice fields were still simply rain-fed, the rent on a field was determined by whether it was high-, medium- or low-lying land (*tanah darat*, *serderhana*, or *dalam*), low land of course being reckoned the best; since the provision of irrigation facilities in the 1970s, the level of the land has become less important than its proximity to an irrigation channel.[g] In the Mekong plain around Vientiane the wet fields (*na*) are classed according to the water source: *na nam phôn* are fed only by rain water, *na nam houey* by streams in flood, and *na nong* by the

[a] *CMYS* 12.1.1–2.
[b] *CMYS* 12.2–3.
[c] *WCNS* 7/7*a*, quoted verbatim by the 17th-century *Chhün Fang Phu* 8/23*b*.
[d] T. Matsuo (1), W. Wagner (1), p. 284.
[e] *CMYS* 11.2.1.
[f] (1), p. 38.
[g] Bray (2); Bray & Robertson (1).

Fig. 234. Irrigating rice with a chain-pump; *Keng Chih Thu*, Pelliot (24), pl. XXIII.

overflow of a pond.[a] In China, as we have seen (pp. 106 ff.), wet fields were divided into seven or eight different categories according to their construction and water supply; generally speaking hillside fields were supplied by gravity-fed contour canals, streams or tanks, while fields in river valleys or plains relied on pumping equipment to supply water from ditches and canals leading off the main river (also to drain the fields when necessary). The development of Chinese water-raising equipment has been described in detail in Vol. VI, pt 3, as has the history of Chinese irrigation generally, so here we shall only make two points relevant to rice cultivation.

The earliest wet fields of China were in all probability natural swamps like the marshes surrounding Ho-mu-tu; later, bunds would have been constructed around the fields not only in order to retain the water but also to limit the size of field: rice is extremely sensitive to water depth and the water should be of equal depth throughout the field, which is one of the reasons why rice fields are usually very small. Bunded fields were known in southern China by the late Chou (see p. 113), and more complex irrigation systems were developed on that foundation. As irrigation systems become more elaborate the amount of labour necessary to maintain and run them increases, so that labour shortage may prove to be a limiting factor on the number of irrigated crops grown. Potter calculates that as much as one-tenth of the Thai farmer's annual labour is directly devoted to maintaining his irrigation system,[b] while in the low-lying fields of the Yangtze

[a] C. Taillard (1), p. 131.

[b] (1), p. 92.

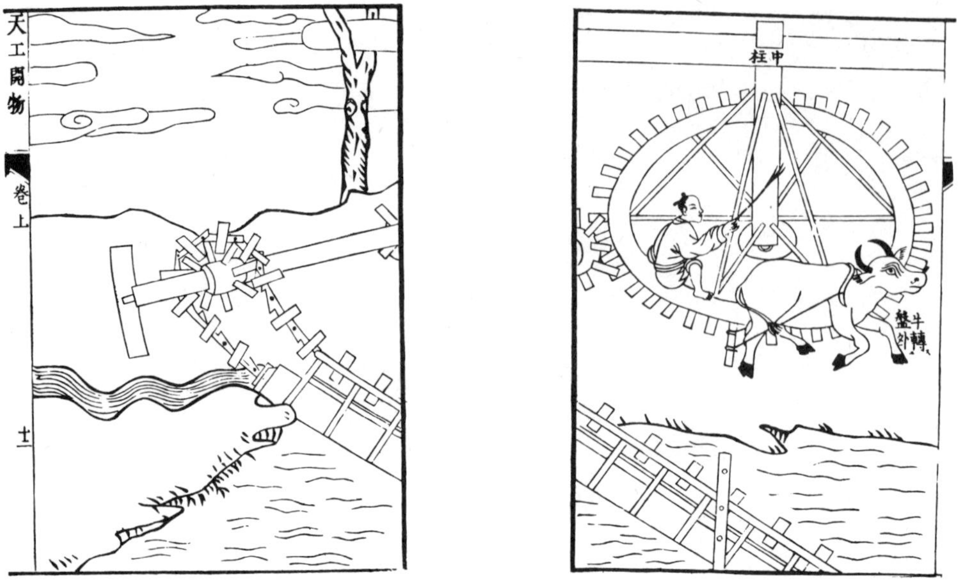


Fig. 235. Ox turning a chain-pump; *TKKW* 1/11*b*–12*a*.

Delta the work of irrigation and drainage combined could absorb as much as 30 % of all agricultural labour.[a] Mechanisation or the substitution of animal for human labour is often a resource open only to the richer farmer (Fig. 235): in South India, where it requires two sturdy bullocks and one strong man to raise irrigation water from a well, poorer land-owners often have to lease their land out rather than cultivate it themselves, and many peasants grow only one crop a year instead of the three that are possible, simply because of family labour shortages.[b]

The ease with which irrigation systems can be established depends principally on natural conditions. In South China streams and rivers are numerous, the soils are leached and sweet water is freely available for irrigation purposes, but in the north irrigation is fraught with difficulties. One basic problem is the salinity of local soils, combined with the very heavy silt content of North China's great rivers: prolonged irrigation under these conditions could produce effects similar to those suffered in ancient Mesopotamia or the modern Punjab. Indeed, in 1957 a Japanese agronomist reported that apart from the newly opened irrigation area around Tientsin, the only areas of North China to grow wet rice were those where clear and sweet water was available (in effect, in mountainous areas near river sources).[c] Further problems were posed by the vast scale of plains and rivers in North China. The high degree of centralised organisation and control necessary to maintain irrigation systems in the north was seldom achieved for more than a few years at a time; even when the central government was sufficiently powerful

[a] Fei Hsiao-Thung (2), p. 161.
[b] Nakamura (1).
[c] Amano (4), p. 195.

and wealthy to initiate such schemes its intervention was often bitterly resented and opposed by local landlords, who saw public irrigation works as a populist threat to the almost feudal organisation of production on their estates.[a]

It was principally the difficulty of providing irrigation which inhibited the development of rice cultivation in North China. Although a little on the dry side, the climate is not inherently unsuitable, and we know that for long periods the climate of North China was distinctly warmer and more humid than it is today (see p. 23). Rice is the principal food crop in Korea and Japan, both of which are at a higher latitude than much of North China, but their terrain is hilly and broken, crossed by many streams, and thus like much of South China suitable for the construction of wet-fields dependent on small-scale irrigation systems.

Some scholars have suggested that in fact North Chinese farmers grew rice in the Shang period or even earlier, either in irrigated fields[b] or in natural swamps.[c] The evidence is hardly conclusive, however, and a scholar as experienced as Amano[d] prefers to reserve judgement on this question. There are certainly references in pre-Chhin texts to northern states growing irrigated rice,[e] but it is extremely unlikely that this formed an important proportion of their agricultural production. Texts such as the *Hsia Kuan*[1] section of the *Chou Li*[f] confirm that the chief rice-growing areas were in the South, along the Yangtze. Repeated attempts have been made, from the Western Han to the present day, to establish irrigation schemes along the Yellow River, in the Metropolitan area, in Hopei and Honan, or as far north as the Great Wall, but until 1949 none of them was particularly successful or long-lasting.[g] When Buck published his survey of Chinese farming in 1937, only 1·3 % of the arable area of North China was occupied by rice, as opposed to 59·8 % of farmland in the south (much of which would produce two or even three crops of rice in a year).[h] It is not surprising, therefore, that almost all the Chinese tools and techniques associated with advanced rice cultivation were developed in the South.

We have already dealt at length with the field types typical of Southern China. We have also described in detail the elements of the tool complex typical of Southern Chinese rice agriculture, consisting of the turn-plough (*li*[2]), two types of tined harrow (*pa*[3] and *chhiao*[4]), a range of weeding implements, the sickle (*lien*[5]) and winnowing-fan (*shan chhe*[6]), as well as numerous mills, sieves and riddles used

[a] See p. 596 below on the collapse of irrigation-based agricultural schemes in North China in the Han; T. Brook (1) examines similar failures in the Ming and Chhing.

[b] Hu Hou-Hsüan (*3*).

[c] Ho Ping-Ti (5), p. 72.

[d] (*4*), pp. 128–38.

[e] Amano (*4*), pp. 177 ff.

[f] 8/25*a*.

[g] See p. 590 on the Han; *Sui Shu*, *Shih Huo Chih* 24/7*a*; *Chiu Thang Shu*, *Hsüan-Tsung Pen Chi* 8/20*a*, etc.; Sung, Yuan, Ming and Chhing examples are cited in T. Brook (1).

[h] (2), p. 211.

[1] 夏官 [2] 犁 [3] 耙 [4] 耖 [5] 鎌 [6] 扇車

Fig. 236. Rice seedbed; *Keng Chih Thu*, Franke (11), pl. XXIII.

for husking, polishing and cleaning the rice, which have been treated in the section on mechanical engineering in Vol. IV, pt 2. We have described the construction and preparation of wet-fields by humans and oxen, but have not yet mentioned the enlistment of fish in this delicate operation:[a]

In such districts as Hsin-chou[1] and Lung-chou[2] [in modern Kwangtung] the fields are all on hillsides; they choose flat pieces of virgin soil and work them to make dyked fields. They observe the rainfall carefully, and once water has accumulated in the hills they first of all buy *huan*[3] fish[b] which they distribute among their fields. After two or three years the little fish have grown large and have quite eaten away the roots of the weeds so that the field is ready to be planted. The farmers can then sell the fish and plant their rice in weed-free fields. This popular method is really excellent.

A great deal of care was devoted to the preparation of the nursery bed (*yang thien*[4]) for the young seedlings (Fig. 236). This had to be carefully chosen on fertile, well-watered soil, at a good distance from any roads or paths so that livestock would not stray in; it should preferably be surrounded by a sturdy fence

[a] *Ling Piao Lu I*, cited Chhen Tsu-Kuei (*2*), p. 54.
[b] A type of *Orthorhyncus*, a round-bodied fresh-water species.

[1] 新州 [2] 瀧州 [3] 鯇 [4] 秧田

to make assurance doubly sure.[a] Chhen Fu devotes a whole section of his *Nung Shu* to this important task of preparing the seed-bed, entitled 'Taking care of the roots and shoots' (*shan chhi ken miao phien*[1]):[b]

When growing any crop the first thing is to take care of the roots and shoots so as to get the plant off to a good start, for it is very seldom that a crop finishes well if it starts badly. If you want healthy roots and sturdy shoots you must sow at the right time, meet the requirements of the land you have chosen, and use your fertilisers sensibly. If these three requirements are all met, and followed up by zealous attention, you will save yourself much worry, and short of some natural disaster the crop will always do well ...

Nowadays the first task when growing rice is to prepare the seed-bed. In autumn or winter it should be deeply ploughed two or three times so that it will be frozen by the snow and frost and the soil will be broken up fine. Cover it with rotted straw, dead leaves, cut weeds and dried-out roots and then burn them off so that the soil will be warm and quick. Early in the spring plough again two or three times, harrowing and turning the soil. Spread manure on the seed-bed [Fig. 118].

The best manure is hemp waste, but hemp waste is difficult to use. It must be pounded fine and buried in a pit with burned manure. As when making yeast, wait for it to give off heat and sprout hairs, then spread it out and put the hot fertiliser from the centre to the sides and the cold from the sides to the centre, then heap it back in the pit. Repeat three or four times till it no longer gives off heat. It will then be ready for use. If it is not treated in this way it will burn and kill the plants. Neither should you use night-soil, which rots the young shoots and damages human hands and feet, producing sores that are difficult to heal. Best of all the fertilisers is a mixture of burned compost, singed pigs' bristles and coarse bran rotted in a pit.

The field should be soaked and brought to a fine tilth, and then sprinkled with chaff and compost. Trample them into the soil, rake the surface quite smooth, and then you can broadcast the seed ... I have often seen farmers use urine poured directly into the irrigation water; the resulting damage is immediately visible.

Generally speaking, seed-beds like fresh, moving water and fear cold and stagnant water; even a thin layer of duckweed will prevent the rice from growing well ...

Despite Chhen's seemingly comprehensive treatment, later authors found still more to add. Both the *Shen Shih Nung Shu* and the *Chhi Min Ssu Shu*, for example, stress the importance of eliminating any seeds of barnyard millet (*pai*[1]) (*Echinocloa* spp.), the most persistent weed in rice-fields and a particular nuisance as it is almost impossible to distinguish from rice:[c]

You must avoid *pai*[2] seed in the nursery bed at all costs. First pare away an inch or so of mud from the surface, sweep it up and wash it clean. Just before you sow the rice, bring the mud in baskets and spread it on the surface of the bed, then broadcast the seed. The old method was to use one sheet of oil-cake, pounded fine and sown with the seed, to each *mu* of seed-bed. Then the seed was covered with ashes so that the roots were loose and the

[a] *So Shan Nung Phu*, pp. 6, 7.
[b] *CFNS*, pp. 5–6.
[c] *Shen Shih Nung Shu*, p. 236; see also Pao Shih-Chhen (*1*), cited Chhen Tsu-Kuei (*2*), p. 459.

[1] 善其根苗篇 [2] 稗

plants easy to pull up [for transplanting]. Nowadays farmers sow rice very closely, saying that they are afraid that [otherwise] weeds will spring up in between, but in fact if they cut away the surface mud all the weed seed will be removed and so it does no harm to sow the rice more thinly if you wish for sturdy plants.

We have already described the selection, pre-germination and sowing of rice seed, and dealt with the question of transplanting as well as the application of fertilisers. The main field was, ideally, fertilised before transplanting took place with river mud, burned compost, hemp, bean-cake or other fertilisers according to soil type;[a] sometimes green manures were used, the earliest mention being in the *Kuang Chih*, which recommends sowing the creeper *Bignonia grandiflora* (*thiao tshao*[1]) in rice fields in the winter.[b] A 16th-century author quotes an old farming adage of Kiangsu:[c]

> You should not put down manure too early or its strength will not last ... Only at sowing time must river mud be applied as a base, and although its strength is lasting and dissipates slowly, by midsummer you should apply a little potash or oil-cake which also dissipates slowly and is long-lasting. Only at the end of summer or the beginning of autumn should you apply nightsoil, by which time it will have double the effect, so that the rice panicles will grow very long.

Of course, one of the simplest methods of fertilising the main field was to burn the stubble after the harvest. This is common practice in many parts of Southeast Asia today, and seems to date back to very ancient times in China. The phrase *huo keng shui nou*[2] ('tilling with fire and weeding with water'), first used to describe the agriculture of Southern China in the *Shih Chi*, has generally been interpreted as referring to swidden cultivation, where trees are burned down to clear the land. We think it is far more likely that it describes wet-rice cultivation, where the rice-plants stood in fields full of water that impeded the growth of weeds, and after the harvest the stubble was burned to enrich the soil (see p. 99). A 17th-century work on Kwangtung makes this explicit:[d] 'Most land, whether high or low, benefits from tilling with fire; 'fire' refers to the ashes of the rice stubble which are thus returned to the soil to fertilise the grain.'

The rice-plants were irrigated during most of their growth (Fig. 234), but it was customary in many places to drain the fields for a short period in mid-growth to toughen the roots. The field was allowed to dry out in the sun until the soil cracked, which took anything between two and ten days.[e] The practice is first mentioned in the *Chhi Min Yao Shu*[f] and is referred to in most agricultural works

[a] *Shen Shih Nung Shu*, p. 235.
[b] Cited Chhen Tsu-Kuei (*2*), p. 35.
[c] *Wu Hsing Chang Ku Chi*, cited Chhen Tsu-Kuei (*2*), p. 106.
[d] *Kuang Tung Hsin Yü*, p. 376.
[e] Hsi Chheng (*1*); *Chih Fu Chhüan Shu*; cited Chhen Tsu-Kuei (*2*), pp. 479 and 199 respectively.
[f] Chhen Liang-Tso (*1*), p. 8.

[1] 苕草 [2] 火耕水耨

thereafter. The *Wang Chen Nung Shu* calls it 'baking the fields' (*khao thien*[1]), the *Nung Sang I Tsho Yao* 'baling out the field' (*hu thien*[2]), the *Shen Shih Nung Shu* 'drying the field' (*kan thien*[3]) and the *Chih Fu Chhüan Shu* 'shelving the rice' (*ko tao*[4]).[a]

Mid-term drainage not only firmed the roots of the rice, it was also recognised as having a fertilising effect,[b] and it provided an opportunity for thorough weeding.[c] The more intensive rice cultivation becomes, the more care and attention is devoted to weeding (Fig. 132). Sung officials remarked with puritanical disapproval that farmers in sparsely populated areas like Kwangtung and Hunan hardly ever bothered to weed or fertilise their fields,[d] whereas in the vicinity of Suchou and Hangchou rice fields were weeded at least three times,[e] by hand, by hoe, and with ingenious toothed wheels which spared the farmer much back-breaking toil (see p. 320 above).

The growing rice was subject to a number of setbacks such as droughts, heavy rains before the seeds had taken root or just before the harvest, or strong winds that lodged the grown plants. The *Thien Kung Khai Wu* enumerated eight separate 'rice disasters', most of which could be remedied only by prayer.[f] The *So Shan Nung Phu* describes the helplessness the peasants felt in the face of such cruel turns of fate:[g]

> There are two disasters that strike rice fields in the mountains, which can be counted neither as flood nor drought although their effect is intensified by either ... The peasants accept these disasters as inevitable; although they weep and curse them there is nothing they can do. Alas, how bitter is such a fate! The first disaster strikes in mid-autumn, when the grain has started well, has flowered, and is just heading. Suddenly a cold spell strikes several nights running—what the locals call 'freezing the cassia flowers'—the rice panicles are blasted by the cold, shrivel, and turn black and mottled ... This is known as the 'dark wind' (*chhing feng*[5]). The other strikes in midsummer, in stifling humid weather when the hot air presses down on the land so that clouds form in the mountains. Frequently this produces rain, and its arrival coincides with a strong southerly wind which carries the rain back and forth with it. Such constantly changing weather, veering from wet to dry, affects the grain in the fields, for leaf-hoppers (*the chhung*[6]) appear and, munching noisily, devour all the leaves completely. The peasants say these two disasters fall from the heavens.

Although little could be done to combat the vagaries of the Chinese climate, there were measures that could be taken to reduce damage by insects. Since rice has been cultivated in China for at least 7000 years all the main pests and diseases are

[a] *Ibid.*
[b] *CFNS*, p. 7; Grist (1), p. 44.
[c] *CFNS*, p. 7.
[d] Shiba (1), p. 53.
[e] As witness the *Keng Chih Thu* illustrations.
[f] 1/4*b*–6*a*; tr. Sung Ying-Hsing (1), pp. 8–12; see also Chang Chieh (*1*).
[g] P. 21.

[1] 熇田 [2] 戽田 [3] 乾田 [4] 閣稻 [5] 青風
[6] 螣蟲

endemic, and they are recognised to affect yields seriously.[a] The *So Shan Nung Phu*[b] recommends running a bamboo comb through the rice plants to kill lurking insects, while the 16th-century *Nung Shuo*[c] describes a method recently developed for killing insect pests, consisting in sprinkling lime or tung oil on the rice leaves. Other writers recommended such preventive methods as ploughing the field up in the winter so that frost would kill off any insects in the soil.[d] Constant weeding and examination of the rice plants also helped to keep down insect damage, but until the introduction of chemical pesticides there was really very little the Chinese farmer could do to protect his crop.[e]

The ripe rice was cut with sickles and either threshed directly in the fields or hung up on frames to dry (see p. 335). In parts of Asia, even today, it is common for awned varieties of rice to be harvested with a reaping-knife, unlike the un-awned varieties which are cut with a sickle and threshed on the spot.[f] It would be interesting to know if there was any association in early China between the use of the reaping-knife and the cultivation of awned *keng* varieties, but the connection would not be easy to trace. The advantage of the reaping-knife, apart from enabling individual selection of the ears harvested, was that the rice could easily be stored still in the ear; that way it keeps much better than if it has been threshed. Provided it is kept reasonably dry, unhulled rice or paddy will keep for a considerable period, but milled rice lasts a much shorter time and has a tendency to turn rancid.[g] The Sung writer Shu Lin[1] said that paddy would keep for eight or nine years, while milled rice rotted after four or five.[h]

Rice yields in traditional China are difficult to estimate. Figures do exist from Sung times on, but one crucial piece of information is often lacking, namely, whether the figure represents the total rice-yield of a piece of land over the whole year (which may mean the yields of two or three crops), or is the figure for a single crop. This is rarely made explicit in the Chinese sources; however, in Chinese communes today one is usually given cumulative figures for the whole year's production, and we may reasonably assume, in the light of the figures quoted for earlier periods, that this was traditional Chinese usage.

[a] 'The main pests in China are rice blast, stem rot, bacterial leaf blight, stem borer, rice borer and plant-hopper. The severe pests in South China are rice blast and rice borer, and those in the central area are blast, bacterial leaf blight, stem rot and borers ...' (Matsuo (1), p. 166); see also Shen (1), p. 200.

[b] P. 21.

[c] P. 9.

[d] Hsi Chheng (*1*), quoted Chhen Tsu-Kuei (*2*), p. 296.

[e] Chemical pesticides are probably the most important innovation in modern Chinese rice cultivation, and have arguably had a greater impact on production figures than the introduction of new heavy-yielding varieties. In the summer months the vast plains of Chengtu or the Yangtze Delta are a shallow green sea of rice, quartered by lonely figures wading thigh-deep through the fields, trailing long white plumes of spray from the insecticide packs on their backs. Sections on the prevention of insect and disease damage figure conspicuously in the modern Chinese literature on rice cultivation: Ting Ying (*1*), pp. 603–24; Wei Ching-Chhao (*1*); Anon. (*532*), (*533*), etc.

[f] H. T. Lewis (1), p. 55; Bray (2).

[g] Grist (1), p. 386.

[h] Quoted Shiba (1), p. 56.

[1] 舒璘

Rice yields throughout the world vary enormously, ranging from as low as 700 kg/ha of paddy in some tropical areas to as much as 7000 kg/ha in Australia.[a] Yields are usually higher in temperate than in tropical zones (where the day length is shorter), wet rice yields more than dry rice, and on the whole *keng* or *japonica* varieties (which respond to photoperiod and fertilisers) give higher yields than *hsien* or *indica* varieties.[b] In his 1937 survey, Buck found that Chinese rice yields ranged from 1000 to over 8000 kg/ha.[c] Modern figures for the improved varieties of rice range from 3500 to 7000 kg/ha for a single crop.[d]

Table 13 gives some rice yields recorded in Chinese works. The conversions from milled rice are based on Amano's calculation[e] that two *shih* of paddy give one *shih* of milled rice; this tallies with Grist's figures for the products of milling and polishing (namely, 50 % whole rice, 17 % points and broken rice, 10 % bran, 3 % meal and 20 % husk).[f] Since the Chinese figures were usually given for the purposes of tax assessment, it is reasonable to assume that they would represent whole rice; on the other hand, traditional Chinese milling methods were rather gentler than modern mechanised techniques, so that the ratio of polished rice to paddy in pre-modern China may have been rather higher than Amano suggests. Our conversion figures to modern metric equivalents are based on Wu Chheng-Lo (*2*); Amano[g] thinks that the Sung *mu* may have been rather larger than is generally assumed, so that the Sung equivalents should be revised downwards.

It is by no means true that double-cropping will always produce double the yields of single-cropping, far less that triple-cropping will triple yields. Attempts to grow three crops of rice a year in many central provinces in the early 1970s were abandoned when it was realised that the farmers were working day and night to achieve rather less than the equivalent of two crop yields. Nor is this a recent phenomenon; 17th-century Canton produced two crops of rice a year, but the yield from the late crop was only two-thirds that from the early, 'because it is a second crop'.[h] Often overall yields are improved if a dry crop is alternated with the wet rice,[i] which may have been one of the reasons for the popularity of the rice-wheat rotation in the Huai and Yangtze regions; sometimes farmers preferred to grow crops of early and late rice simultaneously in the same field:[j]

In the Eastern Man[1] region [the Kwangtung coast] they harvest two crops of rice; as soon as they have harvested one crop they transplant the next. This is a mistake. I have

[a] Grist (1), p. 485.
[b] This of course does not apply to the recently developed heavy-yielding varieties, which are all *indicas*.
[c] (2), pp. 222 ff.
[d] See Anon. (*532*) and other agricultural handbooks.
[e] (*4*), p. 256.
[f] (1), p. 432.
[g] (*4*), p. 256.
[h] *Kuang Tung Hsin Yü*, p. 374.
[i] Most Chinese rotations now alternate two wet crops with one dry.
[j] *Nung Thien Yü Hua*, cited Chhen Tsu-Kuei (*2*), p. 95; this practice was still widespread in the Yangtze Delta and other parts of Southern China in the 1930's; Buck (2), p. 78.

[1] 蠻東

Table 13. *Chinese rice yields in* shih/mu

| Date | Locality | Paddy (*ku*[1]) | Milled rice (*mi*[2]) | Rough paddy equivalent in kg/ha |
|---|---|---|---|---|
| *c.* 1050[a] | Suchou | | 2–3[b] | 2500–4000 |
| S. Sung[a] | Shaohsing | | 2[b] | 2500 |
| 1175[a] | Anhwei | 2 on best land | | 1300 |
| | | 1·5 on medium | | 1000 |
| S. Sung[a] | Yangtze Delta, Fukien | 5–6[b] | | 3200–4000 |
| 16th century[c] | Chang-chou[3], S. Fukien | 4–7[b] | | 4000–7000 |
| | Chien-ning[4], N. W. Fukien | 1·2–3[d] | | |
| 1511[e] | Shanghai | | 3 on best land[b] | 6000 |
| | | | 1·5 on medium[b] | 3000 |
| Late 17th century[f] | Canton | | 4 in good years[b] | 3800 |

[a] Source: Amano (*4*), pp. 255–6.
[b] Probably a cumulative figure for two crops.
[c] Source: Rawski (1), pp. 79–80.
[d] Probably a single-cropping area; see Rawski (1), p. 33.
[e] Source: Amano (*4*), p. 332.
[f] Source: *Kuang Tung Hsin Yü*, p. 374.

1 穀 2 米 3 漳州 4 建寧

followed the advice of a scholar of Yung-chia[1] [in Chekiang] ... who told me that in his district they sowed rice before the Chhing-ming[2] Spring Festival, transplanting it two months later in single rows, widely spaced but with several plants in each hill. First they transplanted the early rice, then ten days later the late, setting it between the rows of early rice. The early rice ripened four months later, and once it had been cut and taken away they would hoe and earth up the late rice so that it grew very vigorously and set much seed, and then later they would harvest the second crop.

Rice is a crop which responds very positively to increased care, and especially to increased inputs of labour. Generally speaking, Chinese rice farming methods have become increasingly labour-intensive as productivity has increased over the centuries. Neolithic farmers probably grew their rice in natural swamps, but late Chou farmers had the additional work of bunding and irrigating their fields; transplanting (a notoriously laborious task) is first mentioned in the Han; the Thang and Sung showed many improvements and elaborations in field construction techniques and irrigation equipment that permitted an expansion of the cultivated area, while the introduction of early rices from Champa in the 11th century led to a gradual spread of double-cropping throughout Southern China;

1 永嘉 2 清明

increases in productivity in the Ming and Chhing were principally due to improvements in the supply of fertilisers and the organisation of water control, again requiring additional work.

The choice of cultivation techniques and of the number of crops grown in a year depended on the supply of labour, the demand for rice and the possibility of alternative economic activities. In districts with easy access to markets, farmers often found it more profitable to direct their energies to growing commercial crops or to the production of manufactured goods, and officials frequently complained that districts with the natural capacity to produce a surplus of rice were in fact reduced to importing supplies from outside.[a] Districts only a few miles distant might differ radically in their choice of crops or occupation:[b]

> Today in the Ling mountains[1] and Annam there is a surplus of land and the population is sparse. Wherever you look there is fertile land, and both elevated and low-lying fields are well-watered. The inhabitants devote themselves to agriculture and stock-raising, and reckon their wealth according to the fruitfulness of their cattle. They produce more grain than they can eat, and so they carry it in great wains to the fairs of Heng-chou[2] [modern Nan-ning in Kwangsi], where it is bought by merchants who transport it down the Wu[3], Man[4] and Than[5] rivers to Canton. The same is true of the western part of the province as for the eastern. The reason for the abundance of grain is ... that the climate in these southern regions is so warm that the land produces three crops in a single year ... They grow two crops of rice in the early fields and then plant brassica to make oil or indigo for dyeing, or grow turmeric or barley, rape or sweet potatoes. Once the main-crop fields have been harvested they soak the straw in sea-water and burn it for the salt. On flat hills and ridges reeds, sugar cane, cotton, hemp, beans, aromatic herbs, fruits and melons are grown in profusion. The people are all extremely industrious and devote themselves so diligently to their farming that truly no patch of land is wasted and no hands are ever idle.

But the author also points out that despite the diligence of the local farmers the chief wealth of the southern provinces lay in the commercial production of luxury goods (perfumes, sugar, wax, rattan, pepper and so on), which were traded all over the Far East. Wealth was restricted largely to the ports, and to remedy their poverty the farmers would often take goods down to Canton, hoping to supplement their income by a little trade.

We shall discuss the relationship between rice cultivation and commercial activities at more length in our concluding section. The amount of land available was also a crucial factor in determining cultivation methods. In newly opened areas where land was in plentiful supply peasants farmed less intensively than in areas of dense population; the Sung writer Chou Chhü-Fei[6] recorded such a lax approach with some disapproval:[c]

[a] E.g. Shiba (1), pp. 50 ff.
[b] *Kuang Tung Hsin Yü*, p. 371.
[c] *Ling Wai Tai Ta*, quoted Chhen Tsu-Kuei (2), p. 69.

1 靈山 2 橫州 3 烏 4 蠻 5 灘
6 周去非

> The farmers of Chhin-chou[1] [southernmost Kwangtung] are very careless. When tilling they merely break up the clods, and the limit of their sowing techniques is to dibble in the seed. Nor do they transplant the rice seedlings. There is nothing more wasteful of seed! Furthermore, after sowing they neither weed nor irrigate, but simply leave Nature to take care of the crop.

In Sung times the careless methods of Hupei and Hunan also presented a striking contrast with the meticulous farming techniques of the Yangtze Delta and Fukien, more striking still in the case of Fukien where the farmers managed to produce good crops despite lack of land and exceedingly poor soils.[a] However, by Chhing times Hupei and Hunan rice farming had developed to the point where they were supplying the national market. Even so, farming techniques in the broad, fertile inland plains never rivalled the intensity of cultivation techniques in the Yangtze Delta: there was no need.[b] The rigour of techniques necessary in crowded districts like the Yangtze Delta was already evident in the 8th-century poem by Tu Fu[2]:[c]

> ... North of the Yangtze River
> Ten thousand acres lie flat as a table.
> By the sixth month the green rice is plentiful,
> In a thousand fields the jade waters mingle.
> The seedlings have already been transplanted as they should
> And the streams led in to increase the irrigation.
> I go over the square dyke
> Where the channel flows in through a gap in the bank.
> Here all your land lies open to view.
> The inflowing waters show no signs of the dry weather.
> The overseer tells one of your servants
> To show me round and explain all the channels and streams:
> Straight as arrows and bright as kingfisher feathers,
> They sparkle and glitter like a reflection of the Milky Way.
> A seagull passes into the mirror
> Which outlines the bordering mountains in snow.
> The autumn grasses have ripened to give black seeds,
> Their inner essence is a snow-white feast.
> The jade-pale grains suffice for our evening meal,
> Rosy and fresh, scattered like the rainbow.

## (5) Legumes

'There are as many kinds of legumes (*shu*[3]) as of rice and millet. Their sowing and harvesting times last through the four seasons, and they have been used daily as

[a] Shiba (1), p. 53.
[b] Rawski (1).
[c] 'Water Fills the Paddy Fields of Circuit Official Chang Wang-Pu (*hsing kuan Chang Wang-Pu tao chhi shui kuei*[4])' quoted Chhen Tsu-Kuei (*2*), p. 47.

[1] 欽州 [2] 杜甫 [3] 菽 or 叔 [4] 行官張望補稻畦水歸

human food since man's need for sustenance was first known.'[a] Legumes belong to the family of *Leguminosae* (one of the largest families of flowering plants) and to the sub-family *Papilionoideae*; they are defined as having a fruit formed by a single carpel and dehiscent by both ventral and dorsal sutures so as to separate into two valves[b]—in this somewhat technical description it is easy to recognise such plants as peas, beans, lentils and peanuts. A remarkable feature of the legumes is their ability to improve soil fertility by adding nitrogen to the soil. Although their value as a fertilising crop had long been recognised, it was not until the 19th century that the process and mechanism of nitrogen fixation was discovered. The roots of legumes attract bacteria of the *Rhizobium* species, which enter the roots through their hairs and cause the cells to divide, producing characteristic nodules. The bacteria inside the nodules are able to fix atmospheric nitrogen, some of which is then available to the host plant and some of which finds its way into the soil as the nodules are sloughed and begin to disintegrate.[c] This process of nitrogen-fixation not only makes legumes valuable as soil-enriching green manures or rotation crops, but also enhances their protein content and makes them highly nutritious.[d]

It is interesting that while legumes were among the earliest plants to be domesticated in West Asia and the Americas,[e] there is no firm evidence that legumes were cultivated in China in prehistoric times, even though several species are native to China, and even though legumes played a very important role in Chinese agriculture from the late Chou onwards. Though remains of leguminous species have been found in one or two Chinese neolithic sites, they have been identified as broad beans (*Vicia faba*) and groundnuts (*Arachis hypogaea*); since the former is believed to have been introduced to China from West Asia in Han times and the latter is a native American species, the authenticity of these identifications is doubtful, to say the least.[f]

There are apparently no references to legumes in the Shang oracle records, and the earliest uncontested evidence for the cultivation of leguminous plants in China is found in Chou bronze inscriptions and in the *Shih Ching*. Both these sources refer to a crop called *shu*[1], and the early form of the character clearly depicts the nodules on the roots of the plants (Fig. 237). The term *shu*, though also

[a] *TKKW* 1/20*a*; tr. Sung (1), p. 24, adj. auct.

[b] Purseglove (1), p. 199.

[c] A number of distinct strains of *Rhizobium* are recognised, each of which is able to infect the roots of any member of a group of related legumes, but not those of other groups; *ibid.* p. 200.

[d] Whyte, Nilsson-Leissner & Trumble (1).

[e] Carbonised peas and lentils, assumed to be domesticated, have been identified at farming sites in the Near East and Europe dating back as far as −7000; D. Roy Davies (1); Zohary (4); Zohary & Hopf (1). Beans of the *Phaseolus* species seem to have been domesticated by the −6th millennium in Peru and Mexico; Alice M. Evans (1); Pickersgill & Heiser (1). Groundnuts are believed to have been domesticated in the Upper Amazon regions, in South Bolivia and North Argentina, well before −3000; Gregory & Gregory (1).

[f] K. C. Chang (1), p. 181; see also the discussion on peanuts below.

[1] 菽

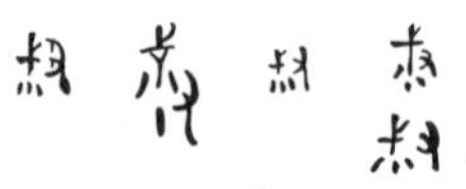

Fig. 237. Oracle graphs denoting legumes; after Ho Ping-Ti (5), p. 80.

used to refer to legumes in general, is primarily associated with the soybean, *Glycine max* (L.) Merrill, and its presence in these writings is usually taken to signify the domestication of this crop.[a] In later lexicographical texts the term *shu*[1] is usually qualified when soybeans are referred to: *jung shu*[2], *jen shu*[3] and *ta shu*[4] are mentioned,[b] and this does suggest that even in pre-Han times *shu* was used as a generic term for legumes rather than a specific term for soybeans. The modern term for soybean is *ta tou*[5], 'greater bean', a term which first appears in the −1st-century *Fan Sheng-Chih Shu*.[c] In classical times the term *tou*[6] was applied only to a type of 'wooden vessel or dish for containing flesh sauces at sacrifices or feasts',[d] but by Han times it seems that it was also used as a general term for pulse crops, as it still is today.

Whether or not the unqualified term *shu*[1] in early texts and inscriptions should be understood to refer to soybeans or to legumes in general, there is no doubt that the soybean was domesticated in China some time around the beginning of the Chou, for texts describing −7th-century events refer to soybeans (*jung shu*[2]) as novel introductions to the Central States, whereas by the time of Mencius they had already become a staple food of the common people.[e] The presumed wild ancestor of the soybean, *Glycine ussuriensis* Regl. & Maack or *G. soja* L., is native to Northeast China and adjacent areas of Manchuria, Korea and Japan,[f] and it may be significant that the cultivated soybean was called *jung shu*[2], for Jung[7] was the name commonly given to the Tungusic tribes of Northeast China in Chou times. The *Kuan Tzu* states that in the −7th century Duke Huan of Chhi[8 g] led an expedition to the territory of the Mountain Jung and brought back 'winter onions and soybeans (*jung shu*[2]) for dissemination throughout the various states'.[h] An alternative gloss of *jung*[7] and of its (then) homophone *jen*[9] is that it simply meant 'large, luxuriant',[i] for the soybean is a large, bushy plant growing up to 6 feet tall; its modern name of *ta tou*[5] is attributable to its habit, and not to the size of its beans which are only about as big as a lentil (Fig. 238).

The soybean's rapid conquest of Chou China is a tribute to its superior qua-

a Hu Tao-Ching (*14*) in fact postulates a much earlier domestication; Ping-Ti Ho (5), p. 70, believes that domestication occurred around −1000.

b In the *Erh Ya* and its commentary for example.

c See *CMYS* 6.6.1 ff.

d Bretschneider (1), vol. II, p. 162.

e Ping-Ti Ho (5), p. 79.

f Hymowitz (1), (2).

g Chhi covered most of modern Shantung.

h *Kuan Tzu* 10/4*a*; tr. Ping-Ti Ho (5), p. 78.

i Bretschneider (1), vol. II, p. 164.

1 菽 2 戎菽 3 荏菽 4 大菽 5 大豆
6 豆 7 戎 8 齊桓公 9 荏

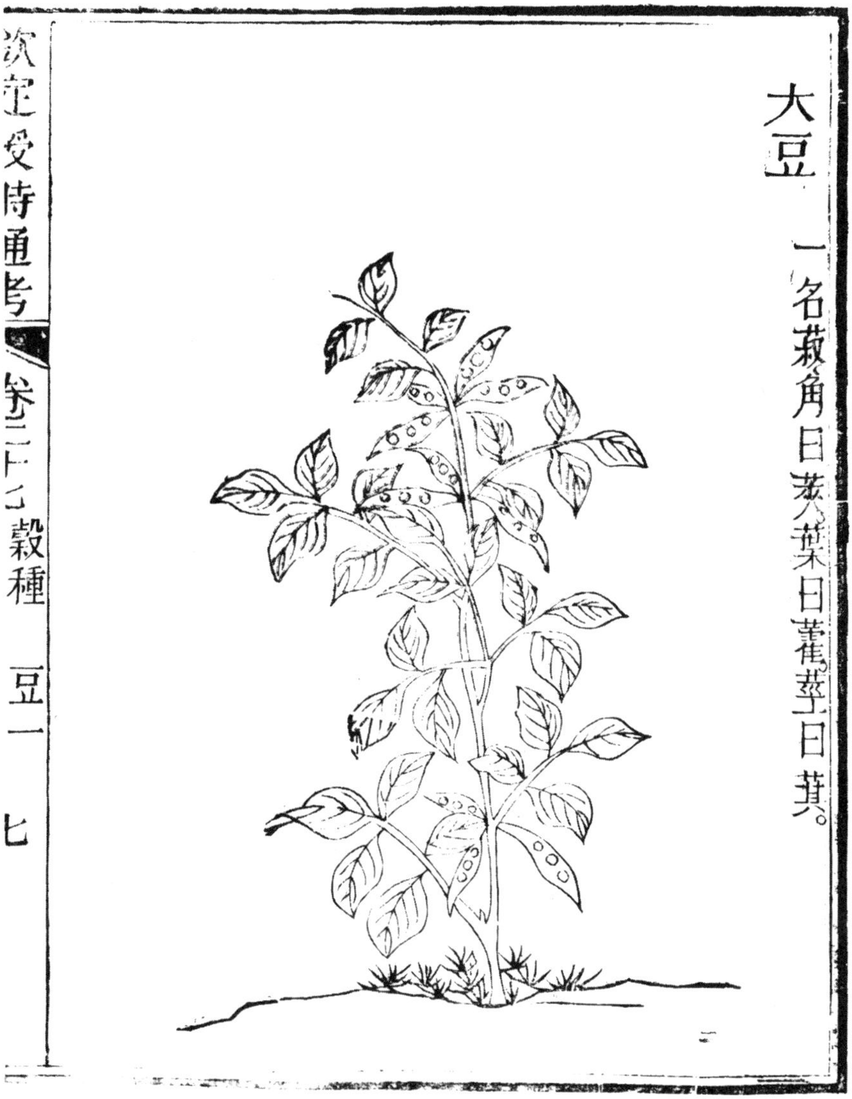


Fig. 238. Soy bean plant; *SSTK* 27/7*a*.

lities, qualities so outstanding that they frequently provoke outbursts of surprising lyricism in modern writers: 'that miracle, that noblest of crops, that wondrous plant, the soybean'.[a] The ancient Chinese were distinctly more moderate in their praise: for them soybeans were simply 'one sort of legume'.[b] However, we may pardon their reserve when we remember that without the benefits of modern nutritional analysis the Chinese were unaware of many of the qualities for which soybeans are most appreciated today; they did not know that a

[a] Anderson & Anderson (1), p. 346.

[b] *TKKW* 1/20*a*.

pound of ordinary dry soybeans contains twice the protein of a pound of beefsteak, that it is high in vitamin A, calcium, thiamine, riboflavin and other B-complex vitamins, but low in calories and cholesterol.[a] As far as the Chinese were concerned, the chief virtues of the soybean were that it produced good crops even on poor land, that it did not deplete the soil, and that it guaranteed good yields even in poor years, so that it made a useful famine crop. Soybeans, it was claimed, could be relied upon to yield between 5 and 10 bushels a *mu*,[b] three or four times the yield from millet, and the *Fan Sheng-Chih Shu* says that in former times[c] it was customary for peasants to plant 5 *mu* a head of soybeans to guard against famine.[d] So soybeans were certainly useful, and they were widely grown in China from Chou times on, but they were not held in high esteem for their gastronomic qualities. When the irrigation works broke down in Ju-nan[1] (modern Anhwei) in the −1st century, the local people composed a song complaining that all they had to eat was soybeans and yams,[e] and in the 14th century Wang Chen writes:[f] 'Black soybeans are a food for times of dearth; they can supplement [cereals] in poor years, and in good years they can be used as fodder for cattle and horses.'

There were, however, several kinds of soybeans,[g] and although some were thought to be fit only for fodder except in times of famine, others were considered to make wholesome porridges and gruels.[h] But the form in which the Chinese most appreciated the soybean was fermented, made into sauces (*chiang*[2]), relishes (*shih*[3]) or beancurd (*tou fụ*[4]). These were generally made from the yellow variety of soybeans.[i] The various fermentation processes were discovered quite early. *Chiang* is mentioned in the *Lun Yü* and was produced on a large scale in Han times,[j] as was *shih*;[k] the earliest beancurd is reported to have been made in the Han.[l] We shall not dwell on soybean products here, as they are treated at length in section 40; suffice it to say that as well as improving the soybean's palatability immeasurably, the various fermentation processes also considerably improve its nutritional qualities.

The soybean was probably the most important legume grown in China, but it

[a] Anderson & Anderson (1), p. 347; Shurtleff & Aoyagi (1), p. 6.

[b] *FSCS*, quoted *CMYS* 7.5.5; since these figures approximate to 2–4000 kg/ha, similar to yields obtained in China and the US today (Heilungkiang Agricultural Institute (*1*); Purseglove (1), p. 270), it is reasonable to assume that they are figurative.

[c] Presumably the late Chou.

[d] *CMYS* 6.5.1.

[e] *C/HS* 84/22*a*; see Ying-Shih Yü (1), p. 76.

[f] *WCNS* 7/11*b*.

[g] *CMYS* 6.1.1–5; *TKKW* 1/20*a*; Li Chhang-Nien (*4*), passim.

[h] *WCNS* 7/11*b*.

[i] Though the *Chhi Min Yao Shu* recommends black beans; 70.1.3.

[j] Shih Sheng-Han (1), p. 84.

[k] *Ibid.* p. 87.

[l] Shinoda Osamu (*7*), p. 110 accepts that the earliest reference is in the *Huai Nan Tzu*, but Ying-Shih Yü (1), p. 81 rejects the textual evidence as too flimsy.

[1] 汝南 [2] 醬 [3] 豉 [4] 豆腐

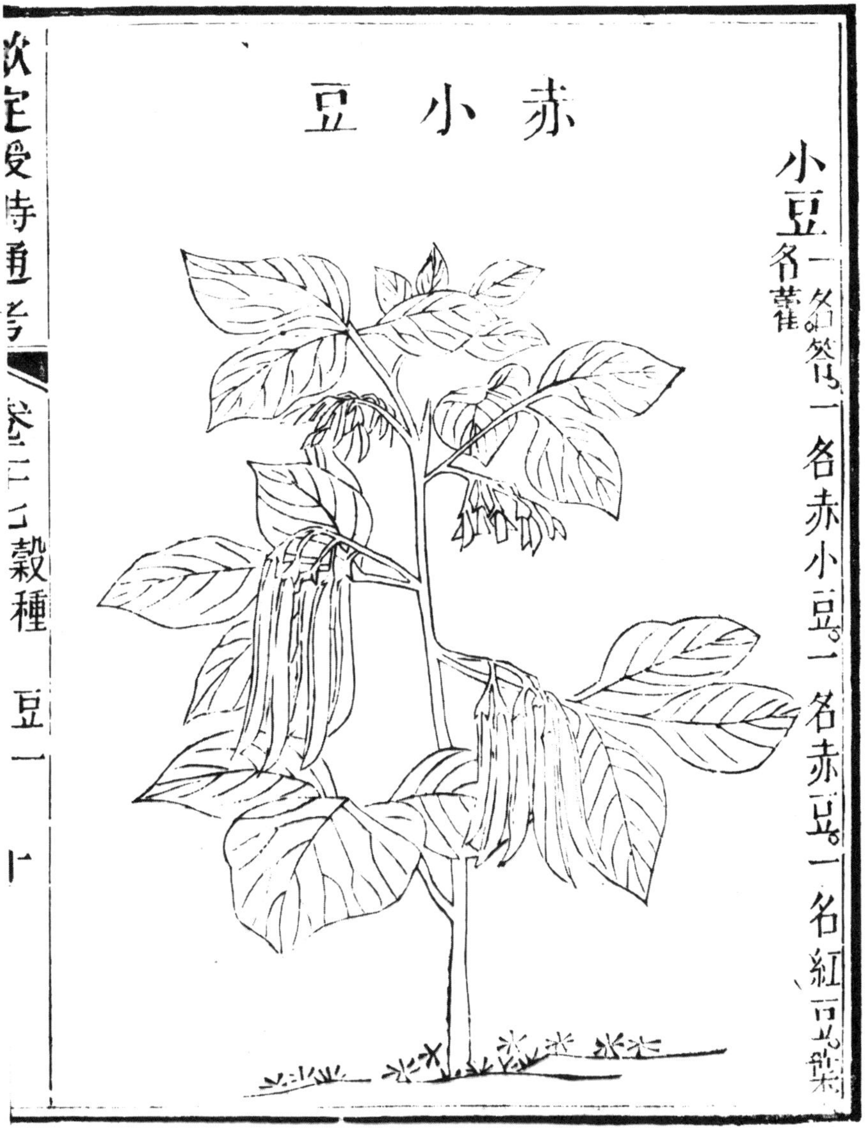


Fig. 239. Red adzuki bean (*chhih hsiao tou*); *SSTK* 27/10*a*.

was by no means the only one. 'Lesser beans' (*hsiao tou*[1] or *ta*[2])[a] came, according to the *Chhi Min Yao Shu*, in three varieties, red, green and white (Fig. 239).[b] The term *hsiao tou* can probably be identified with the adzuki bean, *Phaseolus angularis* (Willd.) Wight,[c] which is native to China and Japan.[d] The 'green bean' (*lü tou*[3]),

[a] *Kuang Ya*, quoted *CMYS* 6.1.2.
[b] *CMYS* 6.1.5.
[c] Purseglove (1), p. 289.
[d] *Ibid.*; Vavilov (2), p. 21.

[1] 小豆 [2] 荅 [3] 綠豆

much valued as a rotation crop for its fertilising properties, has been variously identified as *Phaseolus radiatus* L., a native of India akin to *Phaseolus mungo* L., the black gram,[a] and as *Phaseolus aureus* Roxb., the green gram.[b] This 'green bean' was widely cultivated in North China and was often ground and used as flour to make noodles and cakes[c] (as gram flour is in India also).[d]

Several important legumes came to China from West Asia, including the broad bean (*Vicia faba* L.) and the pea (*Pisum sativum* L.), both of which were said to have been introduced by the Han envoy Chang Chhien[1] on his return to China from Central Asia in −126.[e] Both were sometimes referred to as *hu tou*[2], 'Western bean'; the pea was more commonly known as *wan tou*[3][f] and the broad bean as the 'silkworm bean', *tshan tou*[4], because of its shape (Fig. 240).[g] Both peas and broad beans were used to make noodles, cakes and dumplings as well as being eaten whole,[h] and they were appreciated for their keeping qualities[i] (as they were in Europe where they also formed an important part of the traditional peasant diet).[j] Peas were more common in Northwest China, broad beans in the South and especially the Southwest;[k] today broad beans are still the most important food crop after rice in Yunnan.[l]

Another important pulse grown in China was the cowpea, *Vigna sinensis* (L.) Sair ex Hassk, or asparagus pea, *Vigna sesquipedalis* (L.) Fruw. (*chiang tou*[5]). This crop is presumed to have originated in tropical Africa and was known in India by Sanscritic times,[m] but it probably reached China through Central Asia since it too was sometimes called *hu tou*[2] or 'Western bean' in early texts,[n] and the Northern Sung *Thu Ching Pen Tshao* says that it was much grown in the Northwestern provinces of China.[o]

No account of Chinese legumes would be complete without a brief mention of the peanut. A native of South America, the peanut or groundnut (*Arachis hypogaea* L.) first appears in Chinese literature in the 16th century, in the *Treatise on Tuber Cultivation* (*Chung Yü Fa*[6]) by Huang Hsing-Tseng[7] (1490–1540), a native of

[a] Bretschneider (1), vol. II, p. 166.
[b] Nishiyama & Kumashiro (*1*), p. 5; Herklots (3), p. 245.
[c] *CMYS* 6.5; *WCNS* 7/12*b*.
[d] *Phaseolus angularis*, *P. radiatus*, *P. aureus* and *P. mungo* have recently been reclassified as *Vigna angularis* (Willd.) Ohwi & Ohashi, *V. radiata* (L.) R. Wilczek, and *V. mungo* (L.) Hepper, *Phaseolus aureus* and *radiatus* now being classified together as a single species; Maréchal, Mascherra & Stainier (1).
[e] But see Laufer (1), p. 305.
[f] This term is first used in the early +3rd-century *Kuang Ya*, see *CMYS* 6.1.2; also *WCNS* 7/12*b*; *NCCS* 26/5*b*; etc.
[g] This term does not appear in pre-Sung texts; see Li Chhang-Nien (*4*), p. 351.
[h] *WCNS* 7/13*a*.
[i] *NCCS* 26/6*a*.
[j] Markham (2), (3); K. D. White (2), p. 190; Ames (1), pp. 49, 52.
[k] *WCNS* 7/13*a*; *Pen Tshao Kang Mu* 24/20–1.
[l] Fei & Chang (1).
[m] Purseglove (1), p. 324.
[n] The *Kuang Ya*, quoted *CMYS* 6.1.2, says that the *hu tou*[2] is the same as the *chiang shuang*[8].
[o] Quoted Li Chhang-Nien (*4*), p. 294.

1 張騫 2 胡豆 3 豌豆 4 蠶豆 5 豇 or 跭豆
6 種芋法 7 黃省曾 8 跭䝄

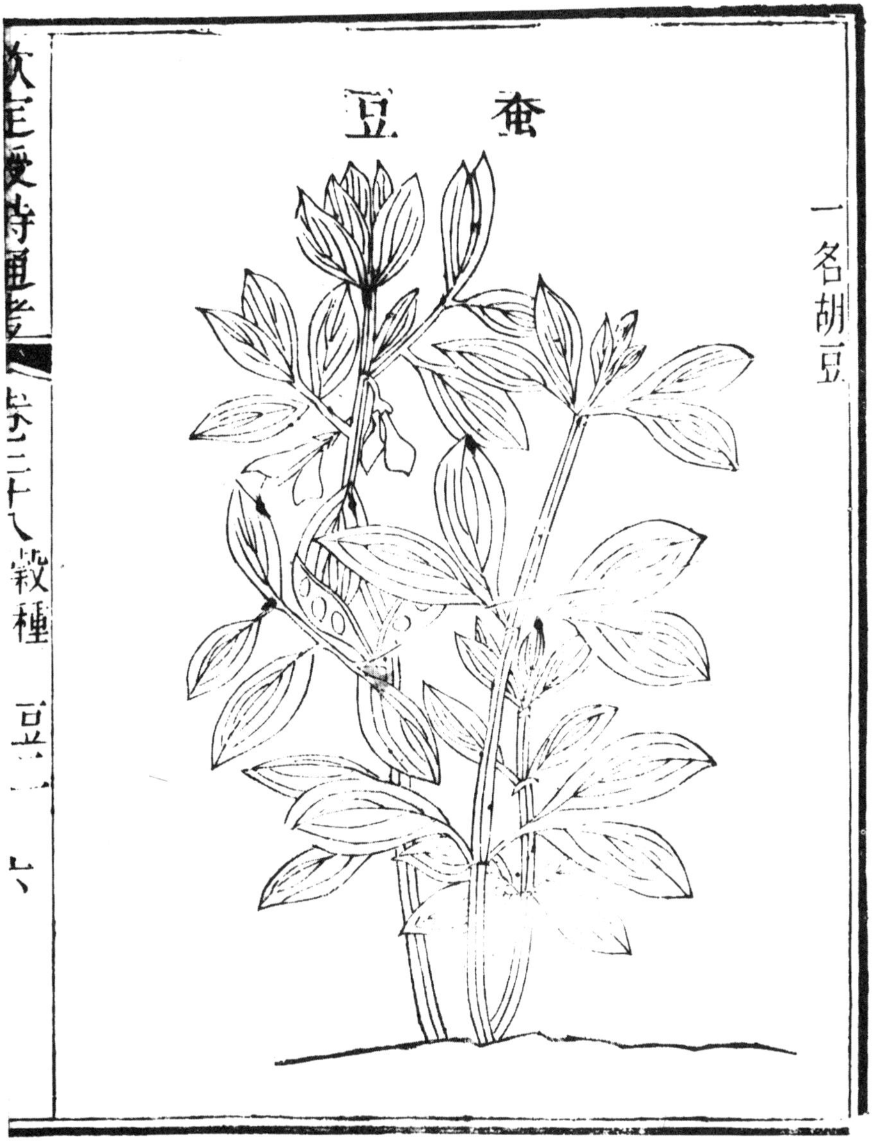


Fig. 240. Broad bean or 'silkworm bean' (*tshan tou*); *SSTK* 28/6*a*.

Suchou:[a] 'There is [a kind of tuber] whose flowers are on the vinelike stem. After the flowers fall, [the pods] begin to develop [underground]. It is called to *lo hua sheng*[1]. [It is] produced in Chia-ting[2] county [near Shanghai].' *Lo hua sheng* literally means that the seeds are born from the flowers which fall to the ground, an accurate description of the plant's peculiar growth. Later the Chinese often abbreviated the name to *hua sheng*[3]. Despite supposedly very early finds of carbon-

[a] Tr. Ping-Ti Ho (1), p. 191.

[1] 落花生 [2] 嘉定 [3] 花生

ised or even fossilised peanuts on the South China coast, finds which have led some scholars to postulate a Chinese origin for the peanut,[a] there can be no doubt that the peanut was introduced to China from America after Columbus, probably imported by the Portuguese to Fukien early in the 16th century.[b] Peanuts were regarded as a delicacy in China at first; by 1700 they were grown for export in a number of southeastern districts, and in the 18th and 19th centuries their cultivation spread to many hitherto underdeveloped areas of the south and southwest: 'Many previously backward areas in Kwangsi and Yunnan [were] transformed by this valuable new crop into prosperous areas of specialised farming.'[c] Since the peanut thrives on poor, sandy soils unsuitable for most other crops, it was a welcome addition to the Chinese repertoire of crops; although its cultivation was slow to spread to the northern provinces, by the 20th century Shantung and Northeast Honan had become the chief peanut-producing areas of the country.[d]

Many other varieties of leguminous plant were grown in China, but often as green vegetables rather than as pulses. The legumes we have just described were grown as field crops,[e] and because of their nitrogen-fixing qualities played a very important role in Chinese crop rotations. It is in large measure due to these leguminous plants that Chinese farmers have been able to practise intensive cropping systems from Chou times to the present day without drastically depleting the fertility of their soils.

## (6) Oil Crops

Master Sung observes that, although Nature has divided time into day and night, man is able to prolong the day artificially in order to carry out his tasks. The reason is not due to his being fond of toil or disliking leisure. If, for example, the weaver must burn wood for light and the student rely on the glow of the snow to read by, how little work would be accomplished in this world!

Stored in the seeds of grasses and trees there is oil which, however, does not flow by itself, but needs the aid of the forces of water and fire and the pressure of wooden and stone [utensils] before it comes pouring out in liquid form. [Obtaining the hidden oil] is an ingenuity of man that is impossible to measure.

For the transportation of goods, and travel to distant places, men must depend on boats and carts. One drop of oil [in the axle] enables a cart to roll and one *shih* of oil used in caulking a ship makes it ready for the voyage. Thus, neither cart nor boat can move without oil. Furthermore, cooking vegetables without oil is like letting a crying infant go without milk. The uses of oil are indeed varied and numerous.[f]

[a] Pheng Shu-Lin & Chou Shih-Pao (*1*); K. C. Chang (1), p. 181 also mentions supposed neolithic finds of peanuts, but without drawing any provocative conclusions.

[b] Ping-Ti Ho (1), (4), p. 184.

[c] Ping-Ti Ho (4), p. 185.

[d] *Ibid.*

[e] Methods of cultivation were very similar to those used for dryland cereals; for further details see Li Chhang-Nien (*4*); Heilungkiang Agricultural Institute (*1*); Wang Chin-Ling (*1*); etc.

[f] *TKKW* 2, section 12, p. 63*a*; tr. Sung (1), p. 215.

In Northern Europe most oils and fats were traditionally of animal origin: tallow and wax for lighting, lard and butter for cooking, and whale oil for lubrication. In the Mediterranean countries olive oil alone served most of these purposes. In China, too, though lard was quite important for some forms of cooking, most oils were of vegetable origin, but they were obtained from a much wider range of plants which included hemp seed, sesame, various brassicas, soybeans, peanuts, cotton seed and tea oil.[a]

Very little is known about the early history of oil extraction in China. The Chinese oil-press (*cha*[1]) is not described in pre-Sung sources, but since it is extremely simple to construct and use, consisting of a hollowed-out tree-trunk in which oil-bearing seeds are pressed between wedges, it may well be of very ancient origin.[b] There are also other, still simpler, methods of extracting oil, such as boiling the oil-seed or simply pounding it in a mortar with a pestle,[c] which were used in ancient Egypt and other early cultures,[d] and which could date back to very early times in China too. It could even be that the brassica seeds found at the early neolithic site of Pan-pho[2] in Shensi[e] were used for oil. Yet even Han archaeological and iconographic material, so informative on most other agricultural and culinary matters, offers no evidence as to how oils were produced, although we know that hemp at least was grown for its oil-bearing seeds at the time.[f] The +6th-century *Chhi Min Yao Shu* advocates planting large acreages with brassicas and other species for oil-seed that was to be sold to 'oil-pressing houses' (*ya yu chia*[3]), but makes no mention of how the oil was extracted.[g]

It seems probable that the earliest oil-crops cultivated in China were brassicas and hemp. Hemp (*ma*[4]), *Cannabis sativa* L., is a tall, bushy, annual, often as much as 6 or 8 metres high (Fig. 241), which the Chinese valued most for its fibres (see p. 532 below). But hemp seeds contain about 30 % oil,[h] and hemp oil seems to have been used quite extensively in traditional China. It was considered to have an offensive smell and was rated lowest of all the cooking oils,[i] but it was a good lamp oil as it produced no smoke and did not harm the eyes.[j] Hemp was grown in China for its fibres in neolithic times, but when it was first used for oil is not known. It is a dioecious plant, and of course the female plants produce the seeds. The traditional Chinese term for the seed-bearing hemp was *tzhu*[5] or *chü*[6].[k] The authors of the *Ssu Min Yüeh Ling* and the *Chhi Min Yao Shu* believed that white seeds

[a] For more comprehensive lists see *TKKW* 2, section 12, tr. Sung (1), pp. 215–21; Chang Wei-Ju (*1*); Anon (*534*). One of the most important oil-crops, rape, *Brassica napus* L. var. *oleifera*, has only recently been introduced to China but is already very widely cultivated; Szechwan Agricultural Institute (*1*).

[b] For details see Volume IV, pt 2, pp. 205 ff.

[c] *TKKW* 2, section 12, p. 64*b*.

[d] Schmauderer (1).

[e] K. C. Chang (1), p. 121.

[f] *FSCS* and *SMYL*, quoted *CMYS* 9.

[g] *CMYS* 18.6.1.

[h] Purseglove (1), p. 42.

[i] *TKKW* 2, ch. 12, p. 63*b*; see also Shiba (1), p. 81, citing the Sung work *Chi Le Pien*.

[j] Shiba (1), p. 82, citing the *Chü Chia Pi Yung*.

[k] *CMYS* 8.1.1; 9.1.1.

1 榨 2 半坡 3 壓油家 4 麻 5 苧
6 苴

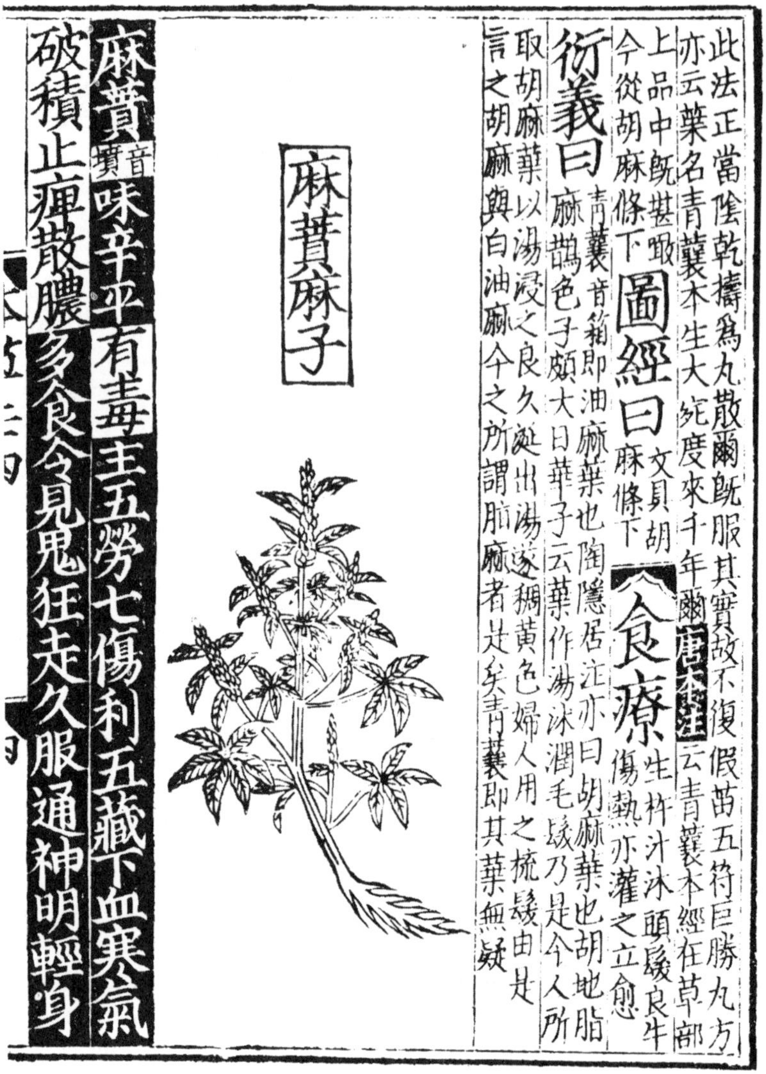


Fig. 241. Hemp plant in seed; *CLPT* 24/4*a*.

tended to produce male plants (*hsiung ma*[1]), and black female.[a] There is no scientific basis for this assertion and (except where modern, genetically monoecious strains are available) seed-raised populations will produce roughly equal numbers of male and female plants.[b] It is possible, of course, that the early Chinese authors mistook a varietal difference for one of gender: fibre varieties of hemp often differ considerably from seed-bearing types, as the former are tall,

[a] *CMYS* 9.2.1–2.

[b] Simmonds (2).

[1] 雄麻

little branched and as little seed-fertile as possible, while oil varieties are short, branched and seedy; in both cases male plants are inferior to female.[a]

Though hemp is still occasionally grown for oil in China today, even traditionally its chief value was as a fibre-crop. Brassicas, which probably have just as long a pedigree, were far more popular oil crops and have long been one of China's chief sources of vegetable oil.[b] The nomenclature of the various brassica species is somewhat confusing: the genus is large, many varieties have a wide range of uses, and the problem of accurate botanical identification is compounded by the existence of dozens of local names for different varieties. Most of the brassicas grown for oil in China belong to the species *B. campestris* L.; the turnip or colza, *B. campestris* var. *rapa*, generally known as *wu ching*[1] (Fig. 242) or *man ching*[2], *lai fu*[3] or sometimes *sung*[4], was grown for its roots and its green leaves as well as for oil. Its close relative *B. rapa* var. *oleifera* DC was also sometimes called *sung*[4], as well as *yün thai*[5] (Fig. 243), 'the greater oil plant' (*ta yu tshai*[6]) or 'white leaf' (*pai tshai*[7]) (*pak choi* in Cantonese): this variety was often grown as a green vegetable, the famous 'Chinese cabbage', but it can also be grown for seed and produces a fine colza-type oil. The Chinese also grew mustards, *B. juncea* Coss., for their oil; they were known as 'Szechwan mustard' (*shu chieh*[8]), 'mustard leaf' (*chieh tshai*[9]), or 'lesser oil plant' (*hsiao yu tshai*[10]). The *Chhi Min Yao Shu* describes the cultivation of all these plants,[c] and it appears that they were often grown on a considerable scale:[d]

One *chhing* [100 *mu*] of colza (*wu ching*[1]) will yield 200 bushels of seed. If you take the seed to an oil-pressing establishment they will bring you three times their weight in hulled grain, that is 600 bushels. This is a better yield than you would get from 10 *chhing* of grain.

The brassica oils were valued for their palatability and relatively high yield. While sesame might yield as much as 40 % of its weight in oil, the *Thien Kung Khai Wai* affirms[e] that colza yielded 27 % and was sweet, palatable and wholesome; Chinese cabbage, though it usually yielded only 30 %, might give up to 40 % if cultivated carefully on fertile soil. The oil obtained from *Perilla ocimoides* L. (*jen*[11] or *su*[12]), a *Labiata*, was also highly esteemed and used for many purposes:[f]

Perilla is by nature extremely easy to cultivate ... it can be grown anywhere. Sow it in a corner of the garden and it will come up by itself year after year ... If you mean to sow large quantities of perilla, the method is the same as for millet (but sparrows have a great taste for it, so it must be grown near human habitation). Harvest the seeds and express the oil, which can be used for cooking cakes. (It is a pretty green colour and has a pleasant,

[a] *Ibid.*

[b] Wagner (1), p. 333 says that they provided three-quarters of all the vegetable oil in central and western China in the early 20th century.

[c] *CMYS* 18 and 23.

[d] *CMYS* 18.6.1.

[e] *TKKW* 2, section 12, p. 64*a*.

[f] *CMYS* 26.2.2–3.2.

[1] 蕪菁 [2] 蔓菁 [3] 萊菔 [4] 菘 [5] 蕓薹 or 芸苔
[6] 大油菜 [7] 白菜 [8] 蜀芥 [9] 芥菜 [10] 小油菜
[11] 荏 [12] 蘇

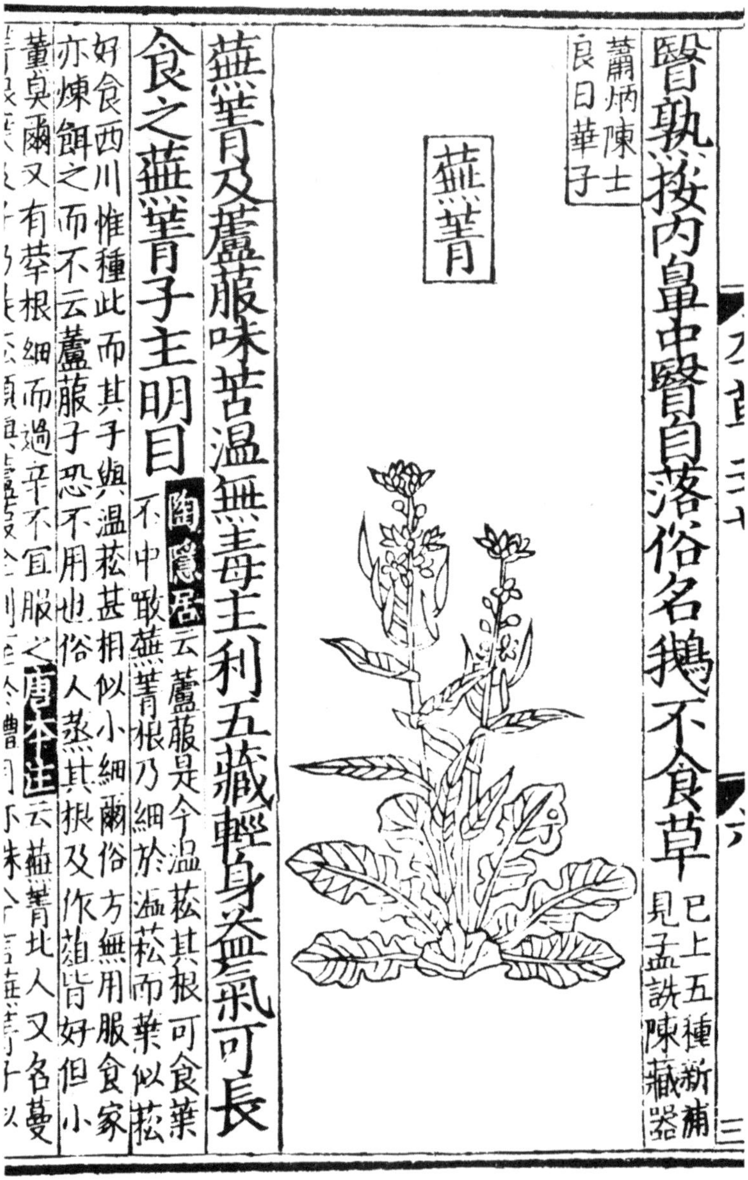


Fig. 242. The oil-seed bearing colza plant (*wu ching*), *Brassica campestris*; *CLPT* 27/6*b*.

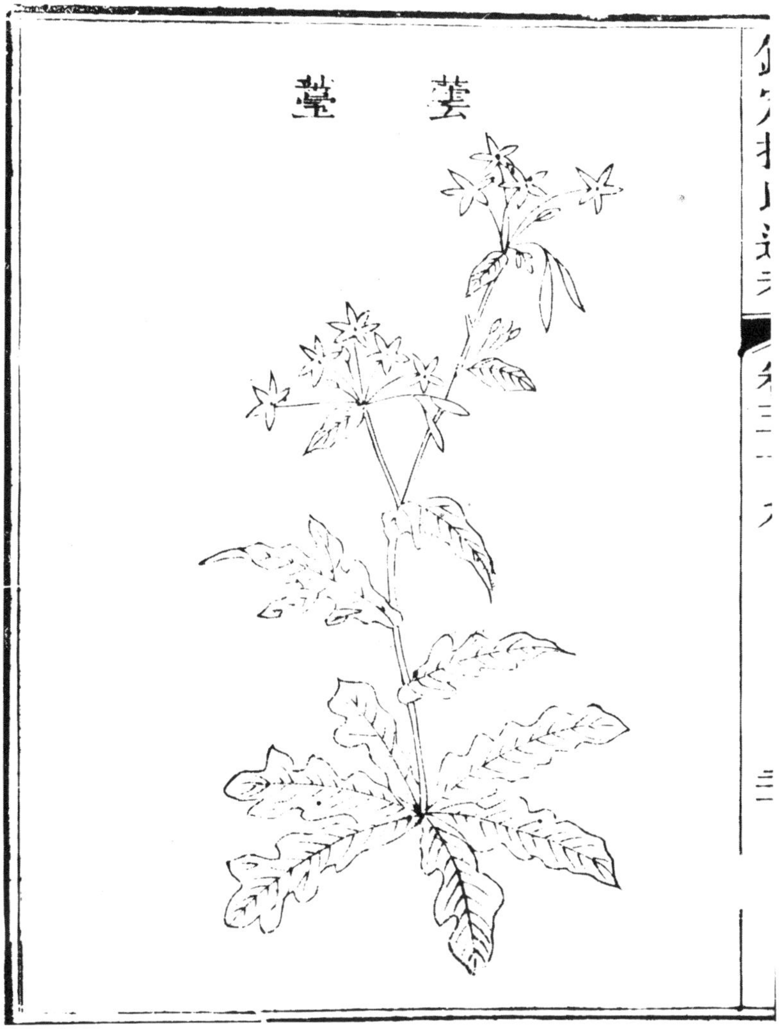


Fig. 243. The oil-seed bearing rape-turnip (*yün thai*), *Brassica rapa*; *SSTK* 59/21*b*.

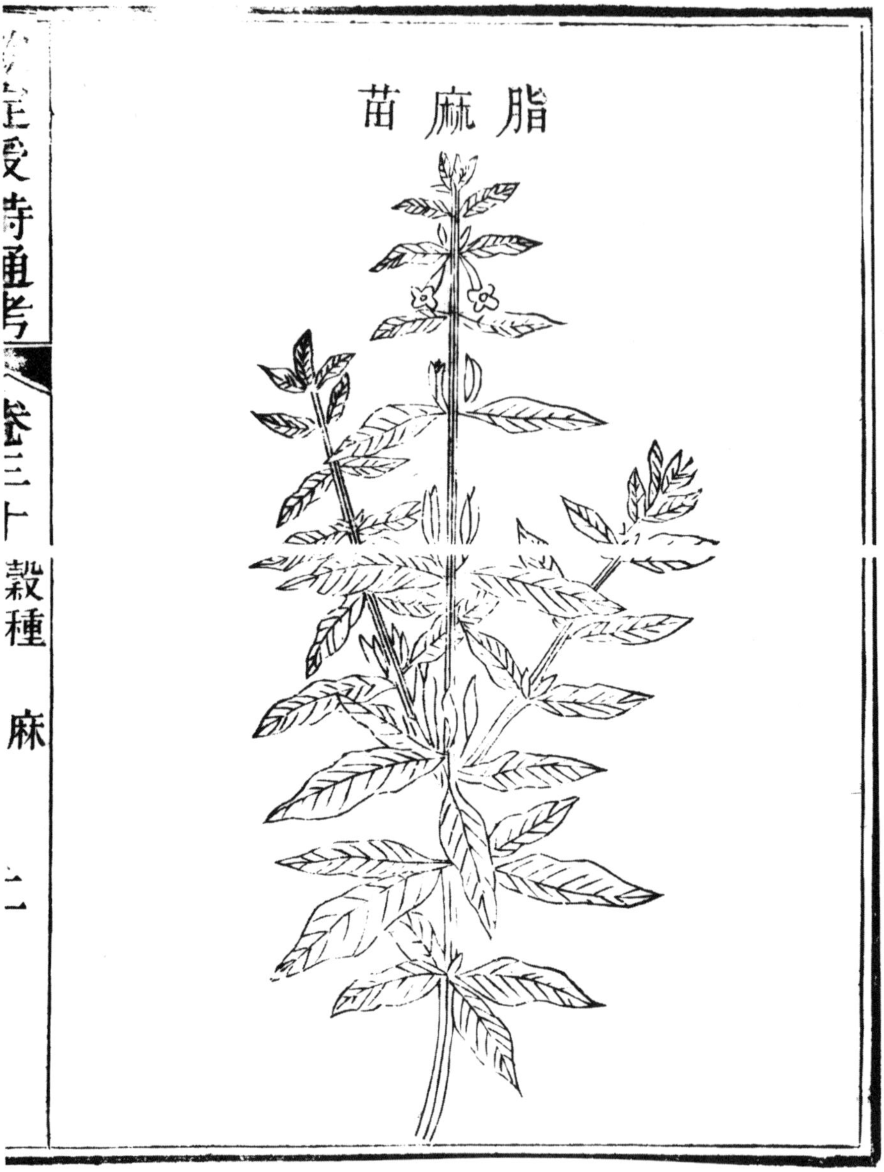


Fig. 244. The sesame plant; *SSTK* 30/2a.

aromatic smell.[a] It is not quite as good for cakes as sesame oil, but it is much better than hempseed grease which smells rancid. However, perilla cannot be used for pomades as it dries out the hair. It is a far better seasoning for broth than hempseed, and can also be used as a lamp oil ...) Perilla oil is excellent for waterproofing cloth.

The king of oilseeds, the Chinese considered, was sesame (Fig. 244). The original home of *Sesamum indicum* L. has not been established beyond doubt. It was known in India, Egypt and West Asia in very early times, and is recorded in Assyrian tablets, while in Sanscrit the word for oil, *taila*, is almost homophonous with that for sesame, *til*.[b] De Candolle suggested on philological grounds that sesame might have been introduced to India from the Sunda Isles in prehistoric times, but this view is now known to be mistaken.[c] Vavilov (2) considered that sesame had more than one centre of origin, in Ethiopia, the Middle East and Central Asia as well as in India; in a more recent study Nayar and Mehra (1) point out that no wild taxa occur in Central Asia or the Middle East and propose that sesame could have originated in either the Ethiopian region or in peninsular India, or even in the two independently.

Wherever sesame is grown, it is highly esteemed for its excellent qualities:[d]

Sesame is both delicious and nutritious; indeed it would be no exaggeration to say that it is the king of all grains ... A few handfuls [of the seed] are sufficient to quell one's hunger for a long time; cakes, breads and sweetmeats, when sprinkled with a few sesame seeds, will have their flavours improved and their values increased. When made into oil sesame can enrich the hair, benefit the intestines, make the strong-smelling [meats] savoury, and dissolve poisonous elements. What enormous profits would accrue to farmers if they devoted more of their land to this crop!

The Chinese call sesame oil 'fragrant oil' (*hsiang yu*[1]). Not only does it have all the qualities mentioned in the *Thien Kung Khai Wu*, the expressed cake is a protein-rich livestock food (sometimes eaten, fermented, by humans in India and Java),[e] and in China it was recognised as an excellent fertiliser.[f] Considering the high opinion in which it was held and the fact that it was grown all over China, it is odd that Chinese botanists had such confused notions about this crop. The problem perhaps lay in its original name of 'Persian hemp' (*hu ma*[2]), a name which was also given to flax (*Linum* spp.) and, in the early *materia medica*, to an indigenous Chinese swamp plant, probably *Mulgedium siberiacum*; early botanical and medical texts, and even the *Pen Tshao Kang Mu*, give extremely confused descriptions of the *hu ma*, which they believed the −2nd-century envoy Chang Chhien[3] to have introduced to China from Ferghana. 'The confusion of *Sesamum*

[a] Very similar to linseed oil according to Burkill (1), p. 1694.
[b] Nayar (1).
[c] De Candolle (1), p. 422; Laufer (1), p. 290; Nayar (1).
[d] *TKKW* 1, section 1, p. 19*a*; tr. Sung (1), p. 24.
[e] Purseglove (1), p. 431.
[f] *NCCS* 26/14*b*.

1 香油 2 胡麻 3 張騫

and *Linum* arose from the common name *hu ma*, but unfortunately proves that the Chinese botanists, or rather pharmacists, were bookworms to a much higher degree than observers; for it is almost beyond comprehension how such radically distinct plants can be confounded by anyone who has even once seen them,' says Laufer,[a] and goes on to point out that while both plants were certainly introduced to China from Central Asia there is not a shred of proof that Chang Chhien had anything to do with it.

The problem of identification is compounded by the fact that flax, like sesame, was grown in China exclusively for its seeds, linen cloth being unknown.[b] Since both sesame and flax had similar cultivation requirements, preferring light, well-drained soils, early agricultural texts are no more enlightening than their botanical counterparts. Both the +2nd-century *Ssu Min Yüeh Ling* and the +6th-century *Chhi Min Yao Shu* describe the cultivation of *hu ma*,[c] but it is impossible to say whether they are referring to sesame or to flax. Only with the introduction of a specific term for sesame, *chih ma*[1], already in common usage by the Sung,[d] is the confusion clearly resolved. But since the Chinese later cultivated sesame far more extensively than flax and clearly preferred its oil to any other, whereas linseed oil seems to have been produced only in Northwest China and was considered smelly and inedible,[e] it does seem reasonable to deduce that most of the references in the early agricultural treatises in fact refer to sesame, not to flax.

The quality of the oil itself was naturally the first consideration when growing oil-crops, but the by-products were often important too. Tea oil from the seeds of *Camellia sinensis* L. (*chha tzu*[2]) was as delicious as lard, but the residual meal was only good for firewood or fish poison, whereas soybean cake could be fed to pigs.[f] The cake left over from making brassica oil (*tshai ping*[3]), or soybean oil (*tou ping*[4]), had become a widely marketed commercial fertiliser by Ming times.[g] As usual the Chinese farmer did his best to ensure that nothing was wasted, and that as much as possible of the goodness extracted was returned to the soil.

## (7) Tuber Crops

It has often been suggested that tubers such as yams and taros may have been the first food plants to be domesticated by man, for they are grown not from seed but by vegetative propagation: simply stick a taro-top or a piece of potato in the ground and a whole new plant will soon appear. It is possible that the yams

a (1), p. 293.
b *Ibid.*
c *CMYS* 13.
d *Chi Le Pien*, cited Shiba (1), p. 80; also *WCNS* 7/14*a*.
e *TKKW* 2, section 12, p. 63*b*; Sung (1), p. 216.
f *Ibid.*
g See the section on Fertilisation.

1 指 or 芝麻 2 茶 or 樣子 3 菜餅 4 豆餅

(*Dioscorea* species) were originally used as poisons rather than as food, for many of the wild forms (and some of the cultivated) contain high concentrations of alkaloids; they are still used by some tribes in Southeast Asia for poisoning fish. Many yams require a long process of scraping and soaking before they are fit for human consumption.[a] Taros (*Colocasia* species), on the other hand, are not poisonous, though in inferior varieties the tissues of both tuber and leaf may contain irritating needle-like crystals.[b]

Taros (Fig. 245) are semi-aquatic plants, yams (Fig. 246) require rather less water, but both are tropical in origin. Taros probably originated in South Asia, while different species of yam were domesticated independently in Asia, Africa and tropical America.[c] It has often been postulated that these tubers were domesticated well before the cereals; it has also been suggested that rice was domesticated from weedy forms growing in taro fields.[d] Although circumstantial evidence (ethnographic, linguistic, and so on) for the early domestication of tuber crops is strong, the hypothesis is unfortunately unlikely to be substantiated, as tubers leave no traces in archaeological sites. The earliest evidence for the cultivation of tubers in China, therefore, is not archaeological but textual.

Interestingly the aquatic taro (*yü*[1]) appears to have been cultivated in North China earlier than the dryland yam (*shu*[2]). The −1st-century *Fan Sheng-Chih Shu* devotes several paragraphs to describing a novel method of taro cultivation:[e]

Taros should be grown in pits 3 feet square and deep. Take dry bean stalks and put them into the pit, treading them down until they are a foot and a half thick. Take the moist soil dug out from the pit, mix it with manure, and put it back in the pit on top of the bean-stalks, making a layer 1 foot 2 inches deep. Water well, then tread it down so that the moisture will be retained.

Take five taro-tops and plant them in the four corners and the centre of the pit. Tread them down. If the weather is dry, they should be watered several times. Once the bean stalks have rotted, the taros will sprout, and the tubers will all grow to be three feet long—each pit will yield three bushels.

The late +3rd-century *Kuang Chih* says that taros were grown as commercial crops in Shu-Han[3], that is to say the mountainous region of Northeast Szechwan and Southern Shensi, and lists fourteen different varieties:[f]

[a] Purseglove (2); Burkill (1). The South American cultivar cassava, *Manihot utilissima* Pohl., also known as manioc or tapioca, requires even more careful treatment as it contains large quantities of hydrocyanic acid in its untreated state.

[b] Burkill (1).

[c] Plucknett (1); Coursey (1), (2); Purseglove (2).

[d] Sauer (1); K. C. Chang (6); Sasaki (*1*); Meacham (2); etc. On the putative connection between rice and taro cultivation, it is worth noting that the *NCCS* 27/14*b* advocates growing taros in rice fields, in the sheltered places beneath trees or near the banks where the rice may not flower.

[e] *CMYS* 16.2.1–2; in *CMYS* 16.3.1–2 Fan describes how taros should be cultivated if a source of water is convenient—this is a much simpler process.

[f] *CMYS* 16.1.3.

1 芋 2 藷，諸 or 薯 3 蜀漢

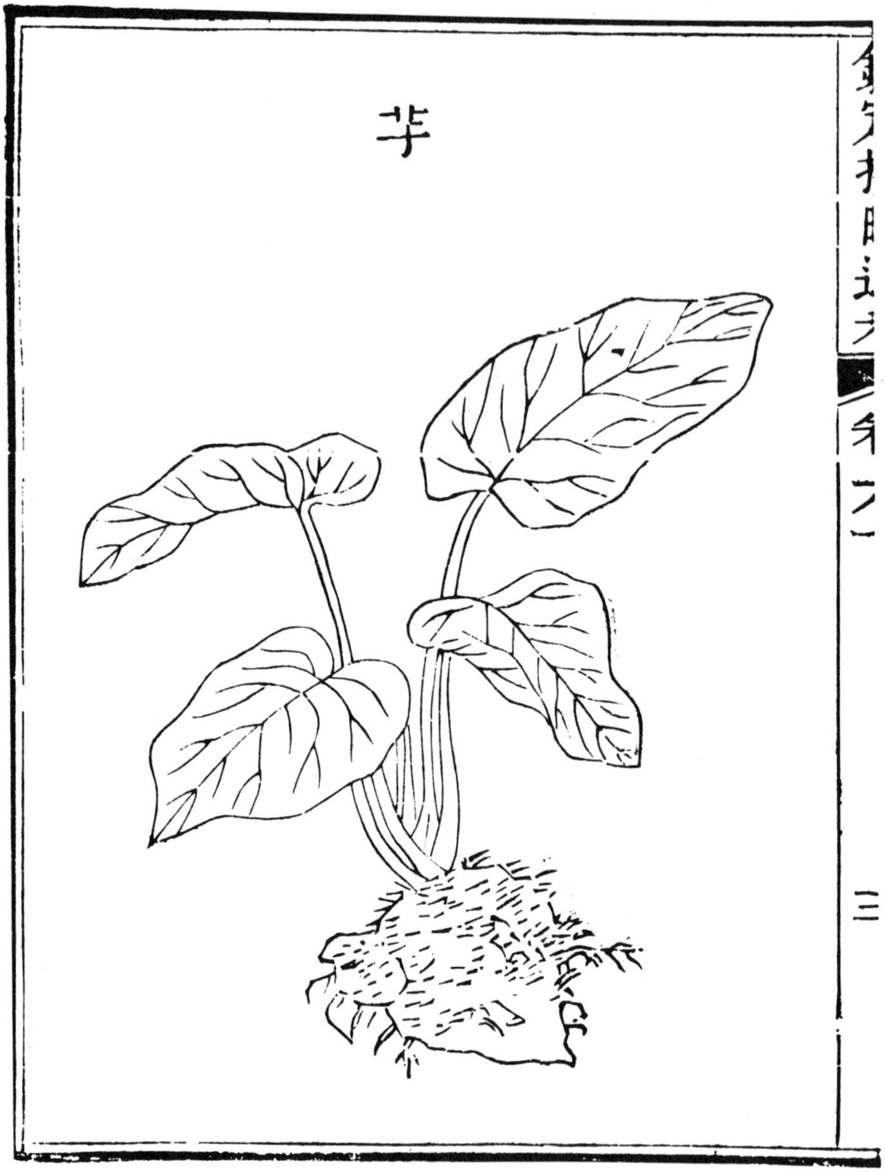


Fig. 245. The taro (*yü*); *SSTK* 60/3*b*.

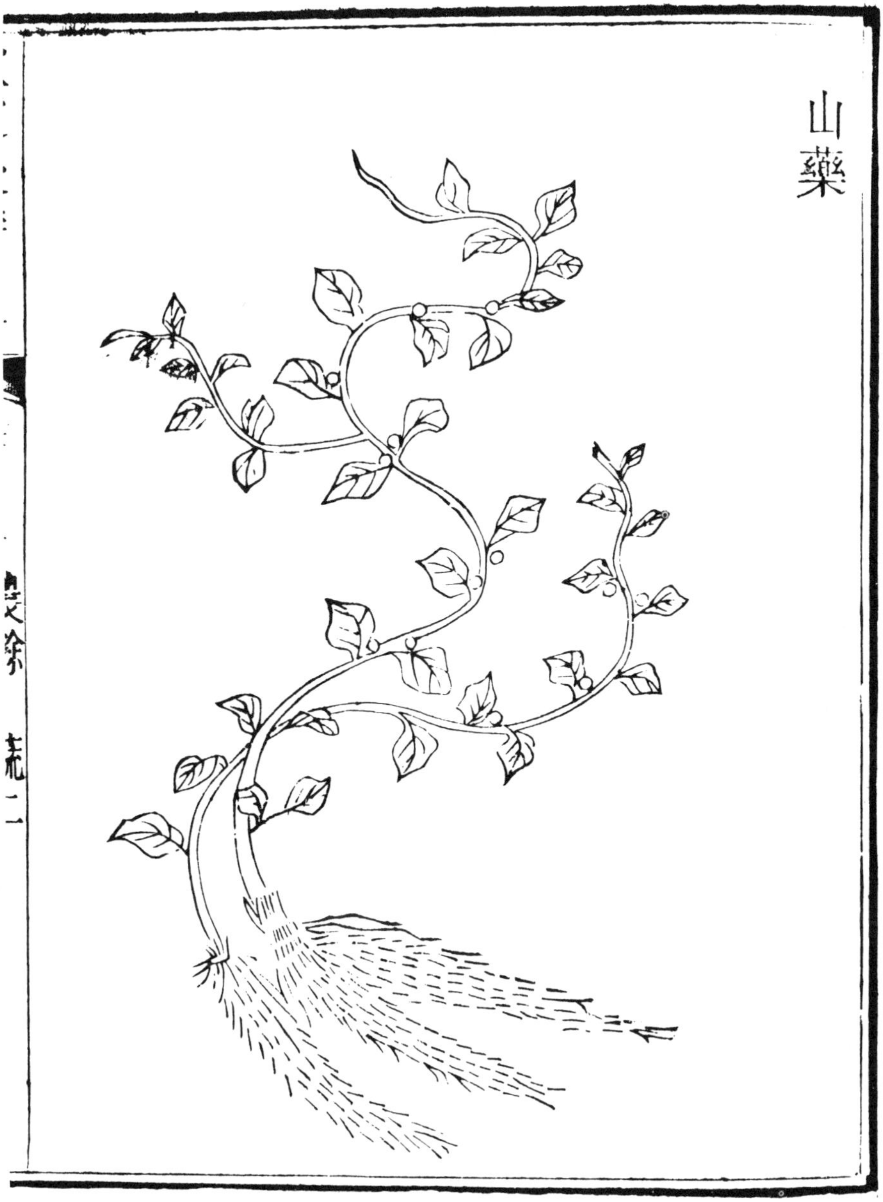


Fig. 246. The yam (*shan yao* or, more commonly, *shu*); *SSTK* 60/2*a*.

... The *than shan*[1] taro has tubers as large as a bottle but few in number, with umbrella-like leaves with a reddish tinge; its stems are purple and grow over 10 feet high. It ripens well and has a good flavour, and is the best of the taros. Its stem can be used to make delicious broth, nutritious and fattening, and a good cure for a hangover ... The 'air-like taro' (*hsiang khung yü*[2]) is large but contains so little nourishment that you could easily starve on it...

While taros were extensively grown in North China by Han times, the yam seems till to have been considered a southern exotic, for the +6th-century *Chhi Min Yao Shu* lists the *shu*[3] or *kan shu*[4] in its final section under non-Chinese plants, saying that it was grown in Annam and Southwest China, and quoting the *Po Wu Chih* to the effect that Southerners ate yams instead of cereals.[a] This may indeed have been true of many hill-dwellers, as it is of the peoples of highland New Guinea and many Pacific islands today.

In themselves tuber crops are not particularly nutritious except in calories; their protein content is very low, less than 2 % in most cases,[b] but yields are enormous[c] and tubers make very good pig food,[d] so carbohydrates and vitamins from the tubers combine with proteins from the pork to provide a more than adequate diet. Grain cultivators usually prefer to use tubers as a side-dish rather than a staple, a tasty ingredient in soups and stews. Tempting the departed soul back with the promise of earthly delights, the *Chao Hun*[5] includes in its list of delicacies 'stewed turtle and roast kid served up with yam sauce'.[e] Some Chinese districts, for example Thai-tshang[6] near Shanghai, specialised in particularly fine-flavoured taros that were exported all over the country.[f]

Tubers were also very valuable famine foods, being less vulnerable than grain crops to floods, locusts and other disasters.[g] By Thang times yams as well as taros were cultivated in most parts of China; the *Pen Tshao Yen I* explains an alternative name for yam, *shan yao*[7], by the fact that the old name *shu yü*[8] was tabooed under Thang Thai-Tsung[9] (r. 763–80).[h] In the 17th century, Hsü Kuang-Chhi mentions the Shantung area as specialising in yam production.[i]

Popular as yams and taros had traditionally been in China, they were rapidly supplanted once the sweet potato was introduced from America. The sweet potato, *Ipomoea batatas* (L.) Lam (Fig. 247), was introduced to China before the mid-

[a] *CMYS* 92.31–32.
[b] Purseglove (2); they are, however, rich in vitamin C and certain minerals.
[c] Yields of over 30 tons/acre have been reported for taro in Melanesia; Purseglove (2), p. 64. Hsü Kuang-Chhi says taro will yield over 2000 tubers per *mu*, each tuber weighing 2 catties; *NCCS* 27/14*b*.
[d] *Kuang Chih* quoted *CMYS* 16.1.3; Jim Allen (1), p. 173; etc.
[e] *Chhu Tzhu*, tr. Hawkes (1), p. 107.
[f] Ping-Ti Ho (4) p. 186.
[g] *CMYS* 16.4.1; *NSCY* 5/3*b*.
[h] Quoted *NCCS* 27/21*a*.
[i] *Ibid.*

1 談善芋 2 象空芋 3 藷 4 甘藷 5 招魂
6 太倉 7 山藥 8 藷藇 9 太宗

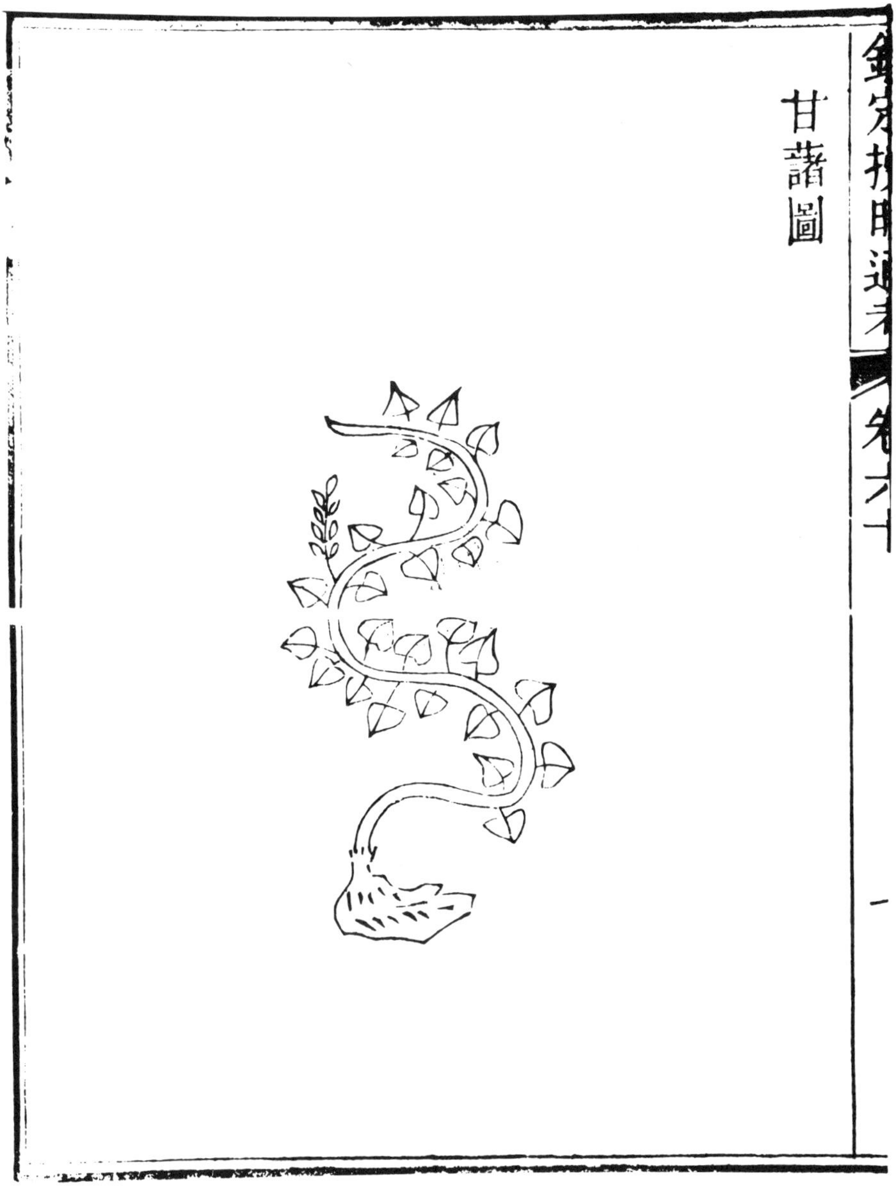


Fig. 247. The sweet potato (*kan shu*); *SSTK* 60/7*b*.

16th century, both overland from India and Burma, and through the seaports of Fukien. It was known as the 'foreign taro' (*fan yü*[1]), the white, red or golden yam (*pai shu*[2], *hung shu*[3] or *chin shu*[4]), or simply as the sweet yam (*kan shu*[5]), like the ordinary yam itself. This last appellation has caused considerable confusion among botanical historians, leading Bretschneider (6), for example, to deduce that the sweet potato was known in South China in pre-Columbian times.[a] The rapidity with which the sweet potato spread throughout China in the 17th and 18th centuries is, however, clear enough proof of its late introduction.[b] The sweet potato had many advantages to offer: it was high-yielding, nutritious, had a pleasant flavour, was more resistant to drought than the native Chinese tubers, and grew well on poor soils.[c] By the 18th century it was grown in all the Yangtze provinces, and Szechwan had become a leading producer; by 1800 it accounted for almost half the year's food supply of the poor of Shantung.[d] The sweet potato did not take long to become the third most important food crop in China after rice and wheat.[e]

## (8) Fibre Crops

Though silk is the fibre most readily associated with China, silk cloth was a luxury product worn only by the rich. The poorer classes dressed in cloth made from plant fibres (wool cloth—unlike in the West—being a rarity). Here we shall briefly discuss the main fibre crops of China and their origins; their special properties and the technology associated with cloth production are dealt with fully in Section 31.

Hemp, *Cannabis sativa* L., which belongs to the *Urticaceae* or nettle family, has been described as 'the most characteristic fibre of warm temperate Asia'.[f] It was widely used all over the Old World in ancient times, and several centres of origin have been proposed for it; in a recent work Schultes proposes a diffuse origin somewhere in the huge area from the Caspian and the Himalayas to China and Siberia.[g] Hemp (*ma*[6]) is perhaps the oldest fibre crop to have been cultivated in China: imprints of what appear to be hempen cloth have been identified on neolithic pots from the site of Pan-pho[7] in Shensi, dating back to perhaps −5000

[a] Ting Ying & Chhi Chhing-Wen (*1*) do much to clarify the terminological confusion; Shih Sheng-Han (*3*), vol. IV, pp. 748–9 is also very useful in this respect, giving names not only for the most common tubers grown in China but also for much rarer species.

[b] Ping-Ti Ho (1), (4).

[c] *NCCS* 27/23*b*; *Min Shu* (1629 ed.), 130/4*b*–6*b*, Liang Chia-Mien & Chhi Ching-Wen (*1*).

[d] Ping-Ti Ho (4), p. 187.

[e] The Irish potato (*ma ling shu*[8]) was not introduced to China until the late 19th century and has not been successful. Nor has cassava, another American tuber, which is now very widely cultivated in Africa and Southeast Asia.

[f] Burkill (1), p. 437.

[g] Schultes (2); see also Simmonds (2); Kirby (1), pp. 46–61.

1 番芋 2 白薯 3 紅薯 4 金薯 5 甘藷
6 麻 7 半坡 8 馬鈴薯

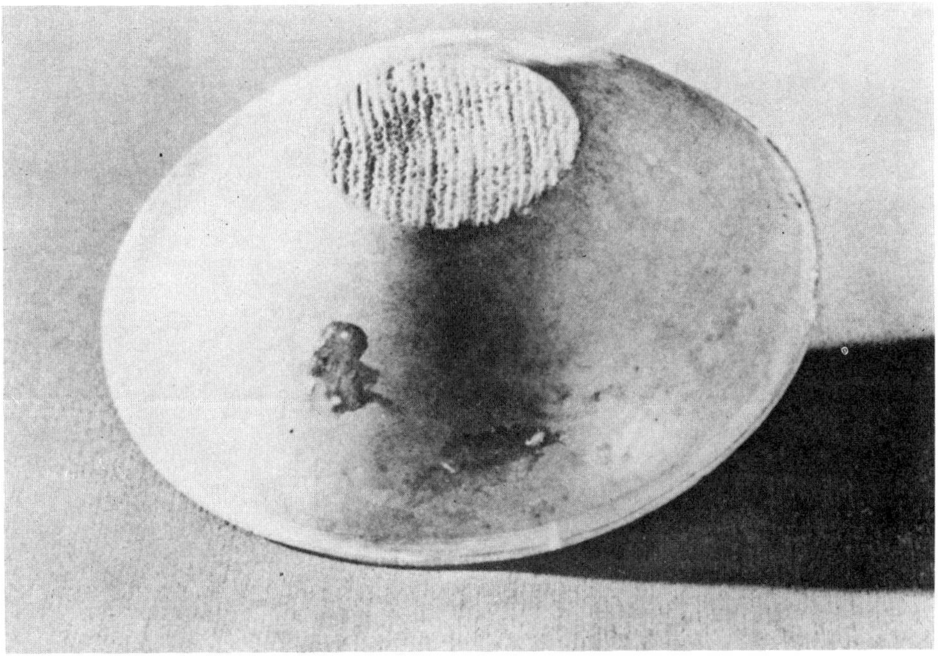

Fig. 248. Imprints of hemp cloth on pottery from Pan-pho; Li (6), fig. 2.

(Fig. 248); the earliest surviving fragment of hemp cloth dates from the Western Chou,[a] and both the plant and the cloth made from it are mentioned many times in the classical texts, including the *Shih Ching*, the *Chou Li* and the *Li Chi*.[b]

Hemp fibres are obtained from the plant by retting the stems (that is to say, soaking them in water until the peel and pith are dissolved by bacterial action). Hemp fibres are very long, often 10 to 15 feet or more, soft and strong. Nowadays hemp fibres are generally reserved for manufacturing yarns and ropes, but in traditional China they were woven into a rather coarse cloth. Hempen cloth was the dress of the common people, worn by the upper classes only for mourning or for special ceremonies. It was called *pu*[1], as opposed to silk cloth *po*[2]. However, the term *pu* later came to be used for all cloths made from plant fibres.

Hemp, as we have already seen, produces an oil-bearing seed as well as fibres, and in China hemp for oil and hemp for fibres were usually grown separately (Fig. 249).[c] Grown for its fibres, hemp is a summer crop requiring well-tilled,

[a] H. L. Li (6), p. 440; Ko Chin (*1*). In 1960 a site belonging to the Chhi-chia[3] culture was excavated in Yung-ching[4] County, Kansu, and among the remains were fragments of coarse hempen cloth adhering to the interior of a jar. The site has been dated to around −2150 to −1780; Anon. (*539*), pl. 6; K. C. Chang (1), p. 488.

[b] Bretschneider (1), II, 205–7 lists the references.

[c] *CMYS* 8 and 9.

[1] 布 [2] 帛 [3] 齊家 [4] 永靖

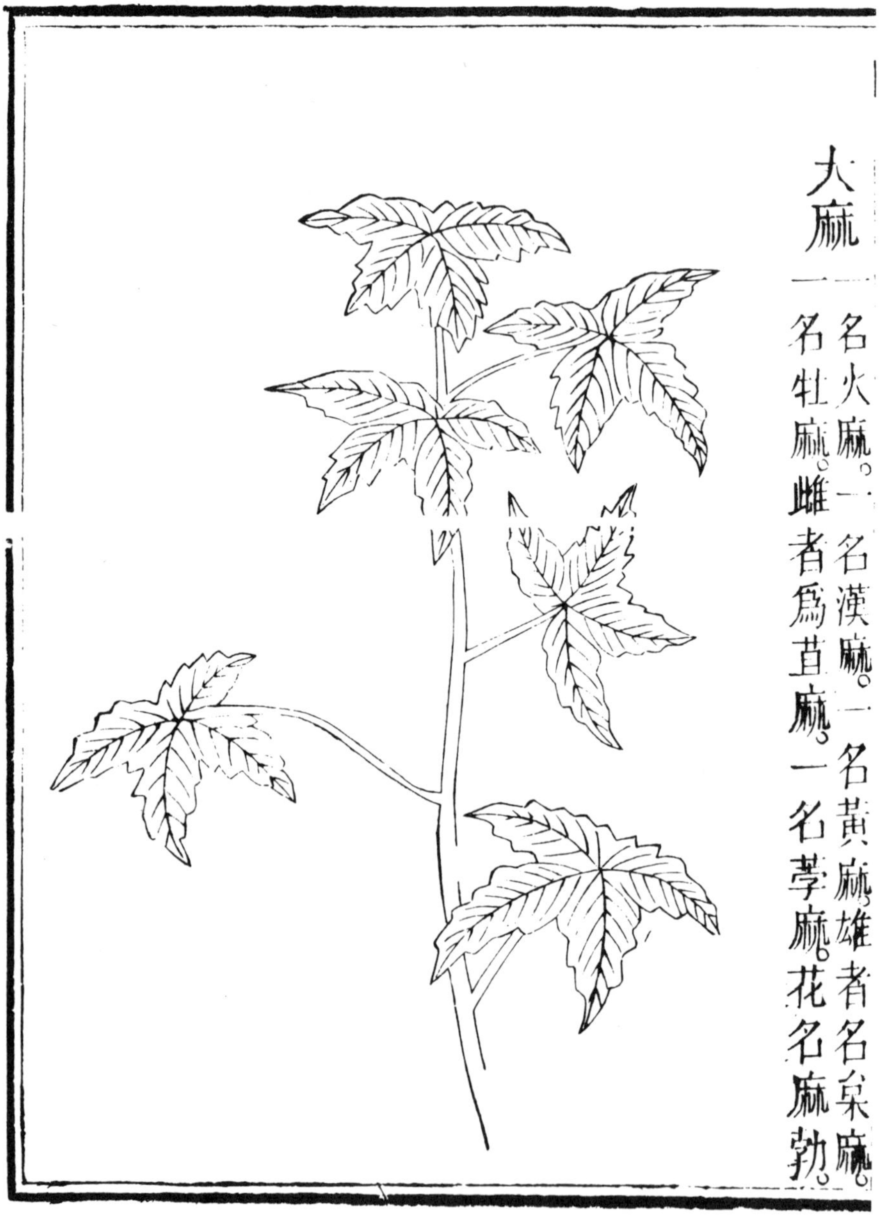


Fig. 249. Hemp grown for fibre; *SSTK* 30/8*b*.

fertile land.[a] It was cultivated all over China in the past, but especially in the North, its cultivation being described in some detail by texts as early as the −1st-century *Fan Sheng-Chih Shu*[b] and the +6th-century *Chhi Min Yao Shu*,[c] while the *Wang Chen Nung Shu* devotes a special section to its preparation and weaving.[d] Once cotton was introduced, however, hemp rapidly fell from favour as cotton made a much finer cloth; the *Nung Cheng Chhüan Shu*, written in the late Ming, has nothing original to say on hemp, and by the 20th century hemp had been almost completely superseded by cotton (for cloth) and jute (for making sacks and ropes).[e]

In South China another indigenous fibre crop was cultivated. Ramie (*chu ma*[1]), *Boehmeria nivea* Gaudich, like hemp, belongs to the *Urticaceae* but is native to the tropical parts of Asia.[f] The qualities of the fibres are remarkable. They are not quite as long as hemp fibres (only 6 to 8 feet), but they are extremely strong: 'It gives the best hempen thread from which cords of every kind are twisted, and nets which last for a very long time'[g]—indeed ramie is perhaps too strong for its own good:[h]

> For the ordinary purposes of life ramie is too good; it outlasts all dyes, and so, for fabrics for wear, it is not fitted unless they are undyed. Also, it is too harsh to take the finish of flax, and the thread, when spun, by reason of the stiffness of its fibres, is hairy with outstanding ends.

It sounds quite unbearable next to the skin, but in fact in China ramie grass-cloth had a reputation as a fine, light cloth especially comfortable in hot weather, and in the 19th century there was a flourishing trade in grass-cloth from Canton, which produced the finest varieties, to Europe.[i] Ramie cloth was worn by most common people in South China, though in the North people were unfamiliar with it and did not know how to prepare it,[j] not surprisingly as it requires quite different treatment from most fibres:[k] 'This fibre differs from flax, hemp, jute and all others in the peculiar pectic substance which holds the strands together and is not easily broken up by bacteria in retting, but must be removed in some other way.' The process of treatment is described by the +3rd-century scholar Lu Chi[2] in his commentary on the *Shih Ching*:[l]

> An iron or bamboo knife is used to strip off the bark. After the thick outer bark has been scraped off, they get to the soft but tough fibres of the inner bark, which are boiled and then twisted and manufactured into cloth. This fabric is used for garments all over southern [most] China (Nan Yüeh[3]).

[a] *CMYS* 8.2.4–5.
[b] Quoted *CMYS* 8.5.1–3.
[c] *CMYS* 8.
[d] *WCNS* 22.
[e] Anon. (*536*); Chhen Hsi-Chhen (*1*).
[f] The name ramie comes from the Malay *rami*.
[g] De Loureiro (1), p. 559.
[h] Burkill (1), p. 343.
[i] Fortune (4), p. 259.
[j] *WCNS* 22/1*a*.
[k] Burkill (1), p. 341.
[l] Bretschneider (1), II, p. 210.

[1] 苧麻 [2] 陸璣 [3] 南越

The special tools required to deal with ramie's tough bark are illustrated in the *Wang Chen Nung Shu.*[a] Though the fibre is difficult to prepare, ramie does have the advantage of being very easy to grow:[b] 'Numerous stems come out from the same root, which is perennial, and in spring young plants shoot forth again without being raised from seed.'

Unlike hemp and cotton, ramie has never been a field crop but has always been grown in very small quantities because of the laborious treatment necessary to prepare the fibres.[c] The Mongol government attempted to introduce its cultivation to North China: the officially compiled *Nung Sang Chi Yao*, published in 1273, contains a detailed description of how to grow ramie from seed,[d] but ramie is not suited to North China's dry climate and the attempt was unsuccessful. In the South, however, the peculiar qualities of ramie have ensured its survival even in the face of competition from cotton, and it is still quite extensively cultivated in southern China today.[e]

Another fibre crop native to China is the kudzu vine (*ko*[1]), *Pueraria thunbergiana* Benth., a leguminous plant which is valued not only for its fibres but also for its starchy roots which may weigh as much as 80 pounds.[f] The *ko* was used in ancient times to produce a grass-cloth finer than hemp, though not as fine as silk.[g] Though kudzu cloth is still much prized in China, the plant is no longer widely cultivated.

The most important addition to the native Chinese fibre crops was, of course, cotton, *Gossypium* L.. The relationship between the various species in the genus *Gossypium* is extremely complex, and for the cytotaxonomic background we would refer the reader to Phillips (1). Briefly, the cottons are basically tropical perennials. Two diploid species, *G. herbaceum* and *G. arboreum*, were domesticated in the Old World; *G. herbaceum* was probably brought under domestication in Arabia and Syria before finding its way to India, where *G. arboreum* developed from it as a cultigen. *G. arboreum* subsequently became the dominant species throughout Africa and Asia, but during the last hundred years tetraploid cottons from the New World have supplanted it everywhere but in India.

Cotton is known both from Mohenjo-Daro and Harappa in India, and so it must have been cultivated there before −2300.[h] From India cotton cultivation and weaving technology spread to Persia (some time before Alexander's invasion in the −4th century), and to Malaya and Indonesia (where all the terms of weaving and for weaving tools are Sanscritic). Cotton is a short-staple fibre (the lint varying from less than 1 inch to $1\frac{1}{2}$ inches in length according to the species), and its spinning requires quite different techniques from other plant fibres, a factor which must have contributed to the rather slow spread of cotton cultivation

a *WCNS*, ch. 22.
b Lu Chi, Bretschneider (1), II, p. 210.
c Burkill (1), p. 343.
d *NSCY* 2/9*b*.
e Chhen Hsi-Chhen (*1*), pp. 14, 18.
f Burkill (1).
g Bretschneider (1), II, p. 208.
h Vishnu-Mittre (1).

1 葛

in the Old World, for without the associated skills the cotton plant itself was of little use. While cotton cloths were known and admired in China in Thang times if not earlier, it was several centuries more before cotton cultivation reached the Celestial Kingdom. Chinese merchants bought Indian calicoes and muslins at the entrepot of Palembang in Indonesia in the Thang,[a] by which time cotton cultivation had reached Indochina. There are frequent references in Thang literature, from the 7th-century *Nan Shih* on, to the cultivation in southern regions of a plant called *ku pei*[1] or *chi pei*[2],[b] which produced a fine, soft white cloth:[c] 'From Lin-i[3] [Cambodia] comes *ku pei*. *Ku pei* is the name of a tree which has flowers like goose-down. They pick the down and spin it to make cloth, which is no different from ramie cloth.'[d] The cloth thus produced was called *mu mien*[4]. *Mien*[5] was originally a term for fine silk cloth, and it is thought that the homophone *mien*[6] (with the tree instead of the silk radical), which later came to be used for the cotton plant itself, was a conflation of *mu mien* probably dating from the late Thang or early Sung.[e]

Although cotton cloth was quite frequently imported into China in earlier dynasties, its cultivation did not reach China until the Sung, when it appears to have reached the border provinces more or less simultaneously by two independent routes.[f] The first was the southern route from Indochina, which brought cotton cultivation to the provinces of Kwangtung and Fukien where it rapidly became a popular commercial crop: 'South of the Min Range[7] there is much cotton [*mu mien*[4]] and the local people compete fiercely in its production', says the *Wen Chhang Tsa Lu* of 1086.[g] The southern cotton was tree-like and was clearly a variety of *G. arboreum*, the Indian cotton,[h] whereas the cotton that reached Kansu and Shensi at about the same time, by the western route across Turkestan, was described as being shrub-like and was a variety of the older diploid cotton *G. herbaceum*.[i] The cultivation of *G. herbaceum* never spread far beyond the Northwestern borders, however, whereas *G. arboreum* came to be cultivated all over China (Fig. 250).[j]

Once established on China's borders, cotton cultivation spread very rapidly indeed. By the late 13th century, or possibly even earlier, cotton had reached the

[a] Burkill (1), p. 1103.
[b] *Chi pei* being a slip of the pen; Chao Ya-Shu (*1*).
[c] *Nan Shih*, quoted Chao Ya-Shu (*1*), p. 226.
[d] Although the description of the tree alone might be taken to describe the kapok tree, *Bombax malabarium* DC, which is native to Southeast Asia, kapok fibres cannot be spun; Burkill (1), p. 345.
[e] Chao Ya-Shu (*1*), p. 222.
[f] Chao Ya-Shu (*1*) and Amano (*4*), pp. 482 ff. review the considerable literary evidence.
[g] Quoted Chao (*1*), p. 231.
[h] Amano (*4*), p. 487.
[i] *Ibid.* p. 494; a clear distinction was drawn between the two varieties by Li Shih-Chen, *Pen Tshao Kang Mu*, 36/71*b*.
[j] Feng Tse-Fang (*1*), p. 21.

1 古貝 2 吉貝 3 林邑 4 木綿 5 綿 or 緜
6 棉 7 閩嶺

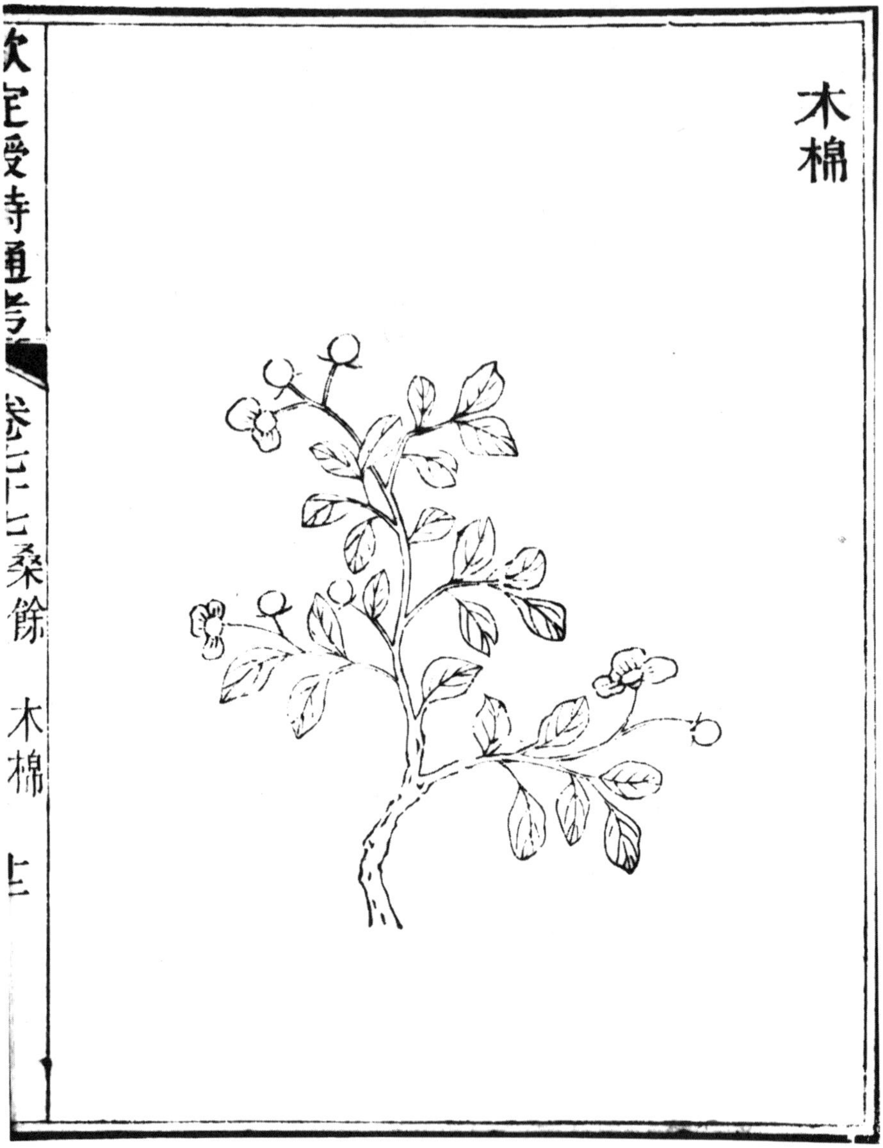


Fig. 250. Chinese cotton plant; *SSTK* 77/12*a*.

Yangtze area,[a] but it was during the Yuan dynasty that the Chinese cotton industry really became established. The Mongol government was so anxious to increase cotton production that it set up Cotton Promotion Bureaux (*mu mien thi chü ssu*[1]) in Fukien, Che-tung, Chiang-tung, Chiang-hsi and Hu-kuang (that is to say the provinces of the Southeast coast and the lower and middle Yangtze) in 1289. These bureaux were responsible for instructing the local peasants in cotton cultivation and weaving techniques.[b] The *Nung Sang Chi Yao*, compiled by a state commission in 1273, gives the first account in Chinese of cotton cultivation methods: it should be sown in raised beds and carefully irrigated during the early stages of growth; when the plant was two feet tall the centre shoots should be pinched out to ensure that the plant became nicely bushy.[c] Although later writers, Hsü Kuang-Chhi for instance, elaborated on the process of pinching out and other details,[d] the laborious cultivation methods have remained largely unchanged since they were first described in the Yuan.[e] Since cotton appreciates a good supply of water, it was often interplanted with rice in South China, while in the North interplanting with wheat was (and is) common.[f]

As a final encouragement to Chiang-nan farmers to cultivate cotton, in 1296 the government declared that they might pay their cloth tax in cotton as well as in the old forms of hempen or silk cloth.[g] By the late Ming, cotton production was firmly established in the Yangtze Delta and the provinces further upstream, as well as in Shantung, and attempts were being made to popularise its cultivation in Hopei;[h] already cotton had become much the most important fibre crop in China.

## (9) Vegetables and Fruits

> ... With her green progeny the parent [Earth] crown,
> Bedeck her hair, in order set her locks;
> Now, let the flowery earth with parsley green
> Be curly, let her joyfully behold
> Herself dishevelled with the leek's long hair
> And let the parsnip shade her tender breast...
> Let the sea-cabbage come; with healthful juice

[a] Chao Ya-Shu (*1*), p. 232.
[b] *Yuan Shih* 15, *Shih Tsu Chi*; Amano (*4*), p. 496; Feng Tse-Fang (*1*), p. 20.
[c] *NSCY* 2/10*b*.
[d] *NCCS* 35/7 ff. Hsü had previously written a monograph on cotton entitled *Chi Pei Shu*[2] or *Chung Mien Hua Fa*[3]; although this work was subsequently lost it is probable that most of it was incorporated into the section on cotton in the *NCCS*; Wang Yü-Hu (*1*).
[e] The *WCNS* has little to add to the *NSCY*'s description of cultivation techniques, but is the first Chinese work to describe and illustrate cotton processing tools and machinery; *WCNS* 22.
[f] Food & Fertiliser Technology Centre (1); Anon. (*537*).
[g] *Yuan Shih* 15, *Shih Tsu Chi*; Amano (*4*), p. 499.
[h] *NCCS*, ch. 35; Amano (*4*), pp. 506 ff. gives a detailed account of the spread of cotton cultivation through the various regions of China during the Ming.

1 木綿提擧司 2 吉貝疏 3 種棉花法

Let lettuce haste to come which can assuage
Sad loathing caused by lingering disease;
One kind grows thick and green, another shines
With tawny foliage, both Caecilian called
After Metellus; a third kind is pale
With dense smooth top and still retains the name
Of Cappadocian from its place of birth;
Then there's my own, which on Tartessus' shore
Gades [Cadiz] brings forth (pale is its curled leaf,
And white its stalk); and that which Cyprus rears
In Paphos' fertile field, with its purple locks
Well-combed, but its stem is white as milk...[a]

Those of us reared on roast beef and Lancashire hotpot may find such outpourings of lyricism over a mere lettuce hard to credit, yet vegetables have a humble yet potent charm, redolent of moderation and virtue, to which even the habitual carnivore will on occasion succumb. The author of 'A Dissertation on Roast Pig' wrote:[b] 'Coleridge holds that a man cannot have a pure mind who refuses apple-dumplings. I am not certain but he is right.'

Where 'cereals stay the hunger and vegetables add the savour', where meat is a luxury reserved for feast days or for the rich, as it was in classical Rome[c] and in China throughout its history, even the poor man's leek may find honourable mention in a poem:[d]

Lucky to own a melon patch by the Green Gate;
Who would envy the marquisate of a hundred square miles?[e]
Watered from the well-pump, leek shoots grow thick and broad.
How Fan Chhih[1] [f] was elated to learn gardening!
Three cups of wine beneath the pear tree,
A mat in the shade of the willows—
What unrestrained freedom!
My meal: a schoolteacher's meagre porridge,
A poor fellow's yellow leek.

Onions, scallions and leeks were the poor man's relish, available all the year round, but cabbages, radishes, egg-plants, mustard greens and many other vegetables were to be found at every country market in season. The Chinese used fruits as well as vegetables to flavour their gruels and porridges; apples and jujubes,

[a] Columella x. 165 ff.; (1) vol. III, p. 21.
[b] Charles Lamb (1), 'Grace Before Meat'.
[c] K. D. White (2), p. 246.
[d] Song by the Yuan poet Ma Chih-Yuan[2], tr. S. S. Sherwin, in Fu, Liu & Lo (1), p. 422.
[e] 'Shao Phing[3] was Prince of Tung-ling[4] under the Chhin. When the Chhin fell he was degraded to the status of commoner and his family became very poor. So he grew melons east of the walls of Chhang-an. The melons were delicious, and the common habit of calling [good melons] Tung-ling melons dates from Shao Phing'; *Shih Chi* 53/4*a*.
[f] A disciple of Confucius.

[1] 樊遲 [2] 馬致遠 [3] 邵 or 召平 [4] 東陵

for example, were used to make a sort of powdered soup that could be added to parched meal, *tsamba*, on long journeys:[a]

To make sour jujube parched grain (*suan tsao chhao*[1]): pick large quantities of the soft red jujube. Sun-dry them thoroughly on a rush mat, and then stew them in a large cauldron, just covering with water. Lift them out with a perforated spoon and then grind them in a bowl. Strain the thick liquid off through a raw silk cloth into a platter or bowl. When the weather is very hot, stand the jujubes in the sun to dry, then gradually crumble them between your fingers to obtain a powder. Use a square ladle to scoop this powder into a bowl of water: when it tastes sufficiently sour you will have a good sauce. On long journeys mix it with parched grain: it is both satisfying and thirst-quenching.

A favourite relish with rice or millet porridge was salted *mei*[2] plums (*Prunus mume* Sieb.), the Japanese *umeboshi*[3], sometimes simply pickled in brine, sometimes smoked too.[b] All sorts of fruits and vegetables were (and are) dried, salted, pickled and preserved to be used as relishes. In Chengtu in early summer before the rains start, every window boasts a washing-line strung with drying cabbage leaves, and some preserves are nationally famous, like the dried vegetables of Tientsin. In many parts of China a prospective bride's beauty counted for less than her skill in pickling.

But vegetables and fruits provided more than savour to enliven a dull meal of cereals:[c]

In Shih Yu's[4] *Chi Chiu Phien* it says: 'Vegetables and fruits from the garden supplement the cereal foods.' The 'Record of Lofty Mountains' (*Sung Kao Shan Chi*[5]) says: 'In the north-east stands Mount Niu[6]. On this mountain grow many apricots; in the 5th month they glisten in golden profusion. Since the Central States have fallen into confusion and the common people have been starving, they have seen these trees as a heaven-sent gift and all have eaten their fill of the fruit.'

In my opinion, if one single species, the apricot, can still help the poor and save the starving, then all the five fruits and the different vegetables together must be still more efficacious in preventing starvation, for they do not merely supplement the cereal foods. There is a saying, 'with a thousand wooden slaves there is never a bad year', which means that fruits of all kinds can be traded for grain.

The *Shih Chi* remarked in the −2nd-century that there were fortunes to be made from orchards and market gardens,[d] and the +6th-century *Chhi Min Yao Shu* harps constantly on the advantages of specialising in horticulture:[e]

Take 1 *chhing* of good land near a town and sow turnips ... 1 *chhing* will yield 30 cartloads of leaves. In the 1st and 3rd month they may be sold for pickling. 3 cartloads will fetch the price of a male slave.

[a] *CMYS* 33.13.1; see also 39.4.1. ff.
[b] *CMYS* 36.4.1–3.
[c] *CMYS* 36.13.1–2.
[d] *SC* 129/31; tr. Swann (1), 432.
[e] *CMYS* 18.5.1–2.

[1] 酸棗麨 [2] 梅 [3] 梅干 [4] 史游 [5] 嵩高山記
[6] 牛山

Hangchou in the Sung was almost surrounded by vegetable gardens, as were most large cities of the time. Some renowned local specialities (certain kinds of bamboo-shoots, melons, mushrooms, ginger, lichees or oranges, for example) were exported to all parts of the empire.[a] There was also a thriving local trade in horticultural produce:[b]

The market at dawn—
Fruits in profusion.
Boxes that brim with loquats, plums
With one red spot on each
And apricots all yellow.
The greengages wait for the summer's warmth,
Quinces evoke the frosts.
The busy peddlers deal
With everything in season.

To keep out thieves, orchards (*yuan*[1]) were surrounded by walls or thick, high hedges, closely interwoven of willows or of thorn-trees. A well-grown, carefully pleached hedge was almost a work of art:[c]

Not only will evil-doers grin with shame and retreat whence they came, but foxes and wolves too will stop of their own accord, stare at the hedge, and turn away. Passers-by when they see it will all sigh with admiration. They will not notice the sun moving westwards, they will forget how far they still have to go, they will walk up and down admiring your hedge, and it will be long indeed before they can tear themselves away.

Vegetable gardens were called *phu*[2]. They were, if possible, situated in fertile land and consisted of rows of raised beds (*chhi*[3]), well fertilised and often irrigated (Fig. 251):[d]

*Phu thien*[4] are fields where vegetables and fruit are grown; the *Chou Li* calls them *chhang phu*[5] or *jen yuan*[6] ... They are encircled by a wall or surrounded by a hedge and moat. Inside the enclosure there should only be 10 *mu*, which is sufficient to support several people. If you are slightly further away from the city you may increase the extent of your garden up to $\frac{1}{2}$ *chhing* [50 *mu*], but no more. Build a small hut at the top and plant the perimeter with mulberry trees for your silkworms. All the area inside should be planted with vegetables. First make one or two hundred beds (*chhi*[3]) of perennial leeks and 20 or 30 varieties of early seasonal vegetables (*shih hsin tshai*[7])[e] taking care to add plenty of manure to give a really fertile base. To guard against drought [such gardens] are best made next to a stream; otherwise you should survey the land and dig a well [Fig. 252].

[a] Shiba (1), p. 85.
[b] Poem by the Sung Writer Hsiang An-Shih[8] (d. +1208), tr. Mark Elvin in Shiba (1), p. 87.
[c] *CMYS* 31.2.1.
[d] *WCNS* 11/13*b*.
[e] What the French, who also know about vegetables, call 'primeurs' (there is of course no word for these in English). Well-to-do Chinese were prepared to pay very high prices for the first vegetables of the season such as egg-plants or asparagus; Freeman (2), p. 155.

1 園 2 圃 3 畦 4 圃田 5 場圃
6 任園 7 時新菜 8 項安世

Fig. 251. Vegetable garden, from a Gardening Album by Shen Chou (1427–1509); Nelson Gallery of Art, Kansas City, U.S.A.

Instead of raised beds small pits were sometimes dug (see p. 127 on 'pit cultivation' (*ou chung*[1])). Horticultural techniques were very intensive, requiring large investments of labour and fertilisers. Even in the Han techniques were very intensive and not an inch of ground was wasted. The *Fan Sheng-Chih Shu* advises growing melons in pits together with shallots or adzuki beans,[a] and the *Chhi Min Yao Shu* adds that five pints of manure should be added to every pit—a quite substantial amount, taking the whole vegetable garden into account.[b] We may get some idea of the hard work that went into market gardening from the 14th-century *Wang Chen Nung Shu*:[c]

> When sowing vegetables you must first dry the seed in the sun. The soil cannot be too fertile, so if it is poor manure it well. You cannot hoe too frequently, and you must water the vegetables whenever the weather is dry. The labour is great, but the profits will certainly be tenfold.
>
> As a rule, leaf vegetables (*tshai*[2]) should be sown in beds (*chhi*[3]) and vine plants (*lo*[4]) in pits (*ou*[5]). The beds should be 10 feet or so long and 3 feet wide. For several days before sowing chop the soil and mix it with straw ash—if you burn this to get rid of any insects it can well be used as fertiliser. Just before sowing add some other type of manure and sow the seed in the prepared beds ...

[a] *CMYS* 14.14.2–3.
[b] *CMYS* 14.11.3.
[c] *WCNS* 2/12*b*–13*a*.

[1] 區種 [2] 菜 [3] 畦 [4] 蓏 [5] 區

Fig. 252. Raised vegetable beds (*chhi*) watered from a well; *TKKW*, illustration from the Chhing ed.

As for the pits, they should be a foot or so deep and in diameter. Before sowing mix equal parts of compost and soil and sow the seeds in this. Once the shoots have emerged, then thin them to the required density.

One can also pregerminate the seeds. Any type of seed may be first washed and placed in a calabash, then covered with a damp cloth. After three days the seed will germinate. When it is about the length of a finger it may be planted. First water the soil in a finely tilled bed, then spread the germinated seed evenly and cover with finely sieved soil and manure to prevent it drying out in the sun. With this method the vegetables all come up at the same time and no weeds grow.

If the vegetables should be attacked by insects, pound *khu shen*[1] [*Sophora flavescens* Ait.] root and mix with lime-water. Sprinkle this on the plants and the insects will die.[a]

Market gardeners were busy all the year round. The +6th-century *Chhi Min Yao Shu* speaks of three crops from each of three plantings of mallow in a single year,[b] and several vegetable crops were grown right through the year even in the open air;[c] greenhouses appear to have been an invention of the Han, though their produce at that time was reserved for the Imperial court.[d] In the southern provinces no special precautions were necessary to grow winter vegetables; indeed vegetables or herbs were the standard winter crop in many southern rice fields.[e]

All the general agricultural treatises contain long sections on horticulture, and Wang Yü-Hu's bibliography of Chinese agricultural works (*1*) lists some 65 monographs on horticulture or on individual fruits and vegetables, beginning with the early Han work *Yin Tu Wei Shu*.[f] Given the large profits to be derived from market gardening, these figures are hardly surprising. Since the investments involved were considerable, most gardens and orchards were small. Around Lake Thai, an area famous for its fine oranges, the terraced orchards belonged to wealthy families who counted their holdings not in terms of individual trees but in *mu*[g]—that is to say a large orchard might be a few acres in size, hardly extravagant by occidental standards. Vegetable gardens were often even smaller. In the 1930s, only 2 % of the total cultivated area was planted with fruit and vegetables,[h] but even this small area was sufficient to keep most Chinese well supplied with green stuffs, at least during the summer months.

Horticulture and horticultural techniques are covered in Section 38, so here we shall simply give a brief list of the more important fruits and vegetables in China. The range has changed constantly over the centuries: some old favourites have reverted to the status of weeds, while new species have been found to replace them. For example the mallow (*khuei*[2]), *Malva verticillata* L., a mucilaginous leaf-vegetable (Fig. 253), was extremely popular in early China,[i] but has since been completely supplanted by greens such as the Chinese cabbage (*pai tshai*[3]), *Brassica chinensis* L. (Fig. 254). Enduring favourites have been various types of melon and gourd (*kua*[4]) (Figs. 255, 256), and alliums such as garlic (*suan*[5]), *Allium sativum* L.,

[a] The root of the *khu shen* is a well-known Chinese drug, extremely bitter but non-poisonous; Bretschneider (1), III, 85.

[b] *CMYS* 17.3.1.

[c] *CMYS* 17.5.1; 24.7.1 refer to the winter cultivation of mallows and coriander respectively; see also M. Freeman (2), p. 155 on winter vegetables in the Sung.

[d] Ying-Shih Yü (1), p. 76; for a detailed description of the elaborate greenhouses of the Chhing court see Amiot (10), pp. 423–37.

[e] *Kuang Tung Hsin Yü* 14/372; Food & Fertiliser Technology Centre (1).

[f] This is not to mention compendia such as the *Chhün Fang Phu*.

[g] Shiba (1), p. 89.

[h] Buck (2), p. 209.

[i] *CMYS* 17; H. L. Li (14).

1 苦參 2 葵 3 白菜 4 瓜 5 蒜

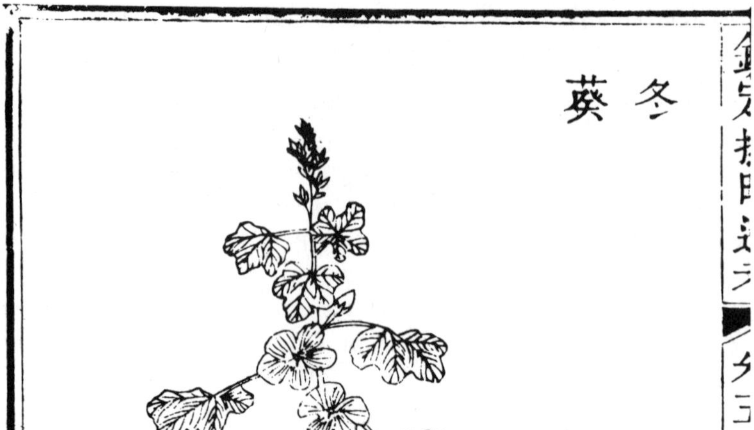

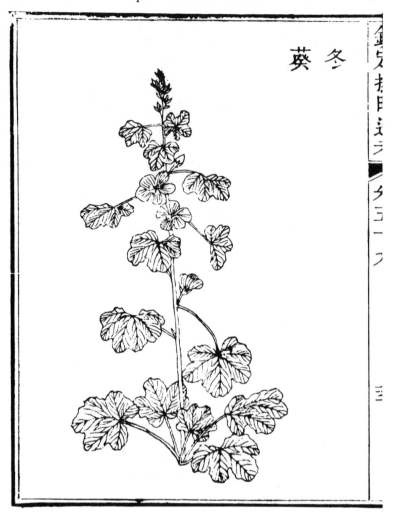

Fig. 253. Mallow plant (*khuei*), once a popular vegetable in China; *SSTK* 59/15*b*.

onions (*tshung*[1]),[a] and the Chinese leek (*chiu*[2]), *A. odorum* L. (all of which—except garlic—are valued as much for their green shoots as for the fleshy bulbs). China also boasted several aquatic vegetables such as the water-chestnut (*chhien*[3]), *Euryale ferox* Salisb., commonly known as 'cock's head' (*chi thou*[4]), the water-caltrop (*ling*[5]), *Trapa bispinosa* Roxb. (Fig. 257), and the beautiful lotus (*lien*[6]), *Nelumbium speciosum* Willd., of which the Chinese not only admired the flowers but also ate the seeds, the stem and the root (*ou*[7]), which can be ground into flour and

[a] The native Chinese onion is what we would call the Welsh onion or shallot, *A. fistulosum* L.; the larger European onion, *A. cepa* L., did not reach China until quite recently.

1 葱 2 韮 or 韭 3 芡 4 雞頭 5 菱
6 蓮 7 藕

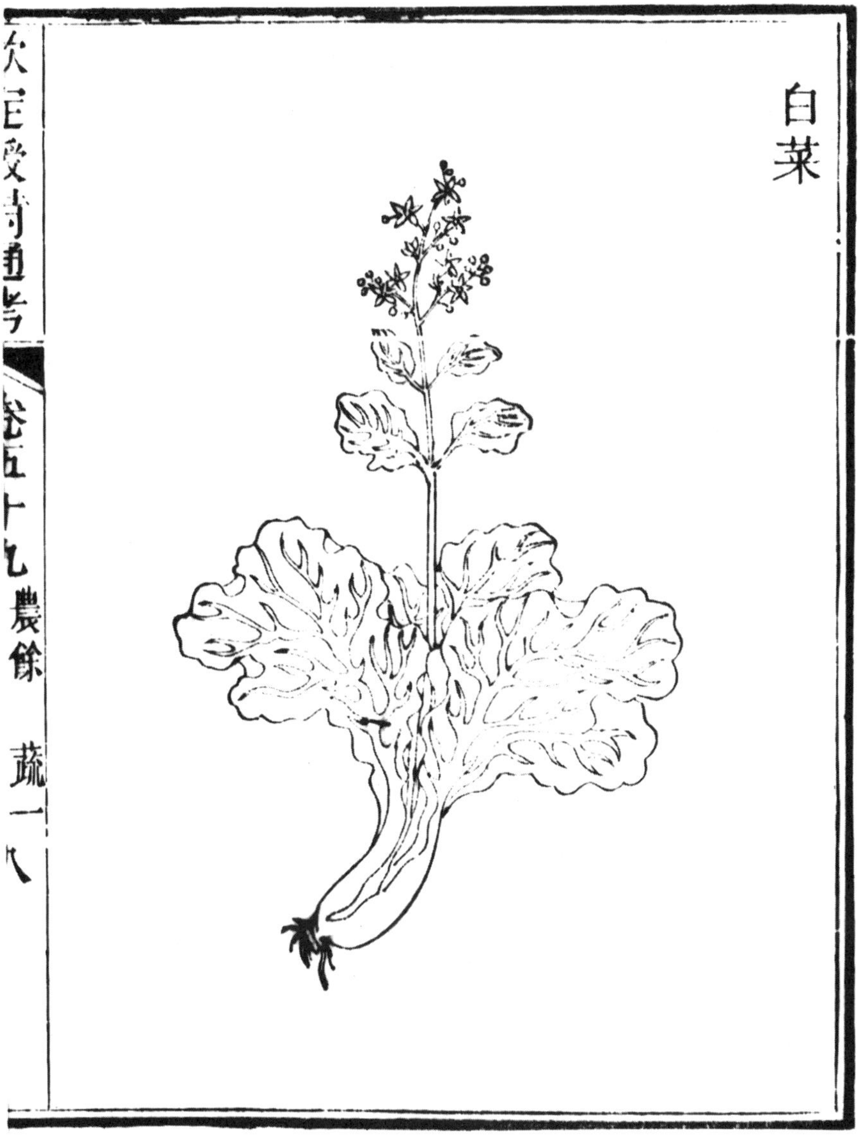


Fig. 254. Chinese cabbage (*pai tshai*); *SSTK* 59/8*a*.

Fig. 255. Water melons, an introduction to China from Central Asia; from the India Office Collection, Archer (1), pl. 12.

used to make a wholesome jelly (Fig. 258).[a] Herbs and spices such as coriander, basil, fagara and ginger (Fig. 259) were also widely cultivated.

The native fruits of North China were peaches (*thao*[1]), *Prunus persica* Batsch.; plums (*mei*[2]), *Prunus mume* Sieb., and (*li*[3]), *Prunus trifolia* Roxb.; apricots (*hsing*[4]), *P. armeniaca* L., the kernels of which were used like almonds; pears (*li*[5]), *Pyrus sinensis* Lind.; crab-apples (*nai*[6] or *thang*[7]), *Malus pumilla* Mill.; jujubes (*tsao*[8]), *Zizyphus vulgaris* Lamb.; and persimmons (*shih*[9]), *Diospyros kaki* L. (Figs. 260, 261). The fruits of the south came in bewildering variety; they included, as well as the plums and peaches of the north, citrus fruits of all sorts, arbutus, loquats, longans, the famous lichee (Fig. 262), and bananas.[b]

[a] Bretschneider (1), II, 217; lotus jelly is still a speciality of Hangchou today.

[b] For further details on Chinese vegetables and fruits we would refer the reader to Section 38 as well as to Bretschneider (1); Buck (2); Herklots (3); H. L. Li (14); Chinese Institute of Agronomy (*1*), (*2*); Anon. (*538*); Hopei Agricultural College (*1*); Li Lai-Jung (*1*); Sun Yun-Yü (*1*).

[1] 桃 [2] 梅 [3] 李 [4] 杏 [5] 梨
[6] 奈 [7] 棠 [8] 棗 [9] 柿

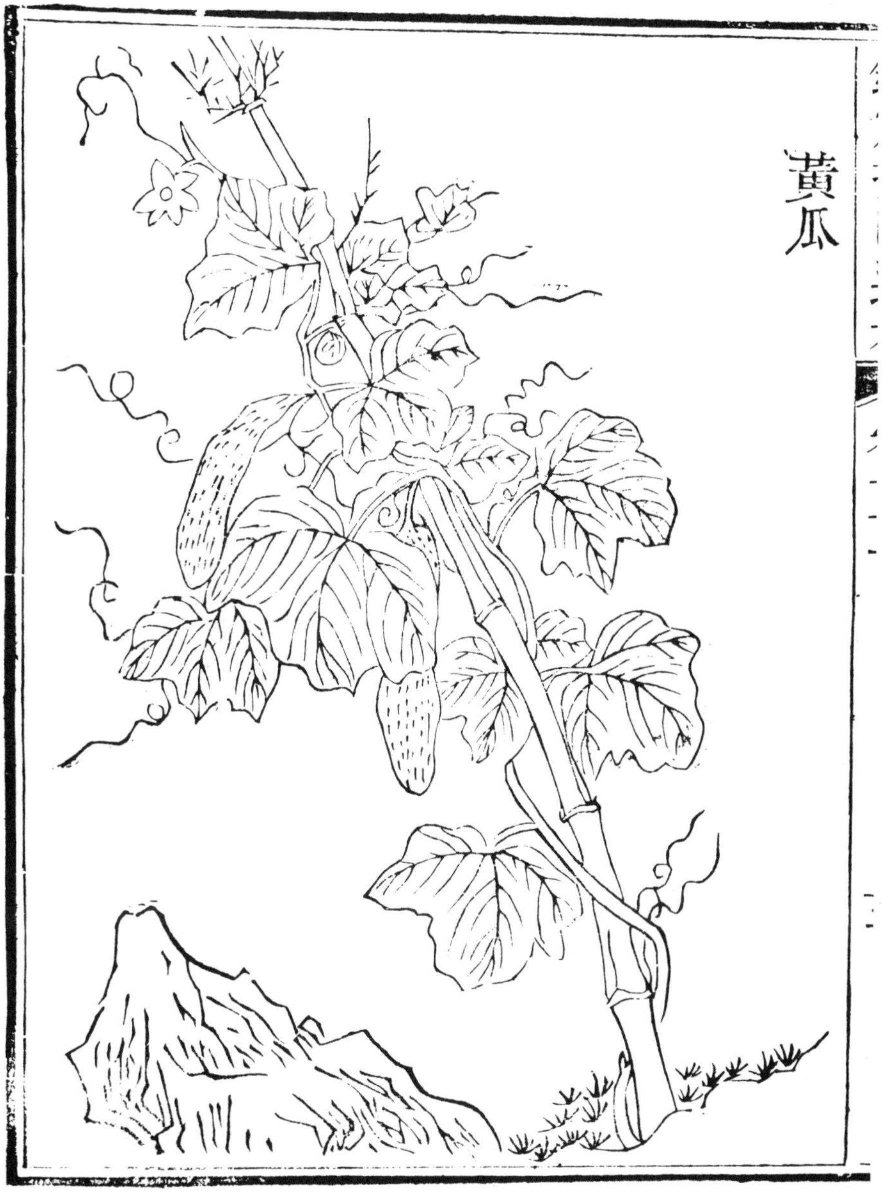


Fig. 256. Cucumbers (*huang kua*) growing up a pole; *SSTK* 61/3*b*.

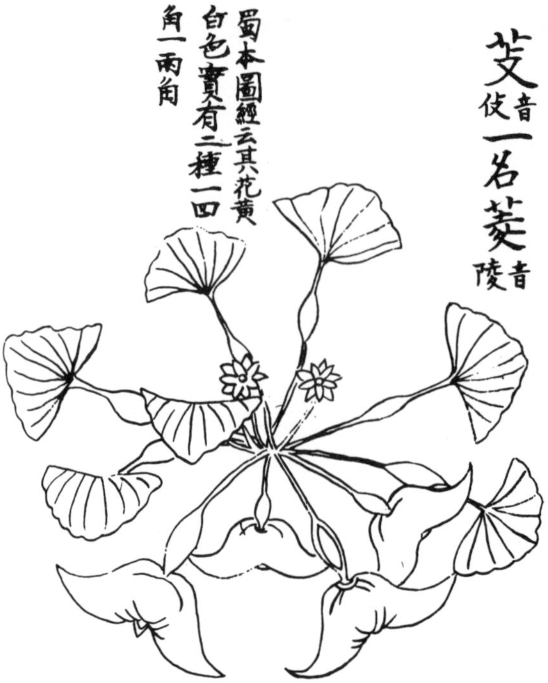


Fig. 257. Water caltrop, a common aquatic vegetable, as illustrated in the +1159 ed. of the *Shao Hsing Pen Tshao* (Karrow (2), p. 55).

Many fruit-trees were valued by the Chinese almost as much for their aesthetic as for their epicurean qualities. The *mei*[1] plum was the first tree to flower in the spring, and many a poet found inspiration in its brave pale blossom:[a]

At the corner of the wall a few plum trees
Open all alone in the hard frost.
From a distance I would take them for snow
But for the dark sweet tang on the air.

Vegetables and fruit were not only an essential part of the Chinese diet, they also symbolised purity and sobriety. They were the food not only of the common people unable to afford meat, but also of pious Buddhists, hermits with magical powers, and scholars living in rustic seclusion; they signalled the renouncement of worldly values:[b]

[a] From a poem by the Sung statesman Wang An-Shih[2], quoted *Chhün Fang Phu*, *Kuo Phu* 1/18*a*.
[b] *CMYS* 36.12.1.

[1] 梅 [2] 王安石

Fig. 258. Lotus plant, showing the edible roots and seed-pods; *CLPT* 23/3*a*.

The 'Tales of Spirits and Fairies' (*Shen Hsien Chuan*[1]) says: 'Tung Feng[2] dwelt on Mount Lu[3] and had no commerce with men. If he cured someone of an illness he would take no money, but if it was a grave illness he would tell the patient to grow five apricot-trees from stones, and if he healed a mild disease he would tell them to plant one cutting. Within the space of a few years there were several hundreds of thousands of apricot trees that had flourished and grown into a forest. When the apricots were ripe he made stores of them throughout the wood. He explained to those who came to buy apricots: "There is no need to come and tell me—just help yourselves. If you bring a jar of grain you can have a jar of apricots." One man brought less grain than he took apricots, and suddenly five tigers appeared and chased after him. The man took to his heels in terror and the apricots he was carrying started to spill to the ground. As soon as he had no more apricots left in his jar than there had been grain before, the tigers turned round and went away. After this occurrence, all those who came for apricots weighed them themselves in the forest lest they should leave with too many. Feng used all the grain he acquired to help the poor and succour the needy.'

1 神仙傳 2 董奉 3 廬山

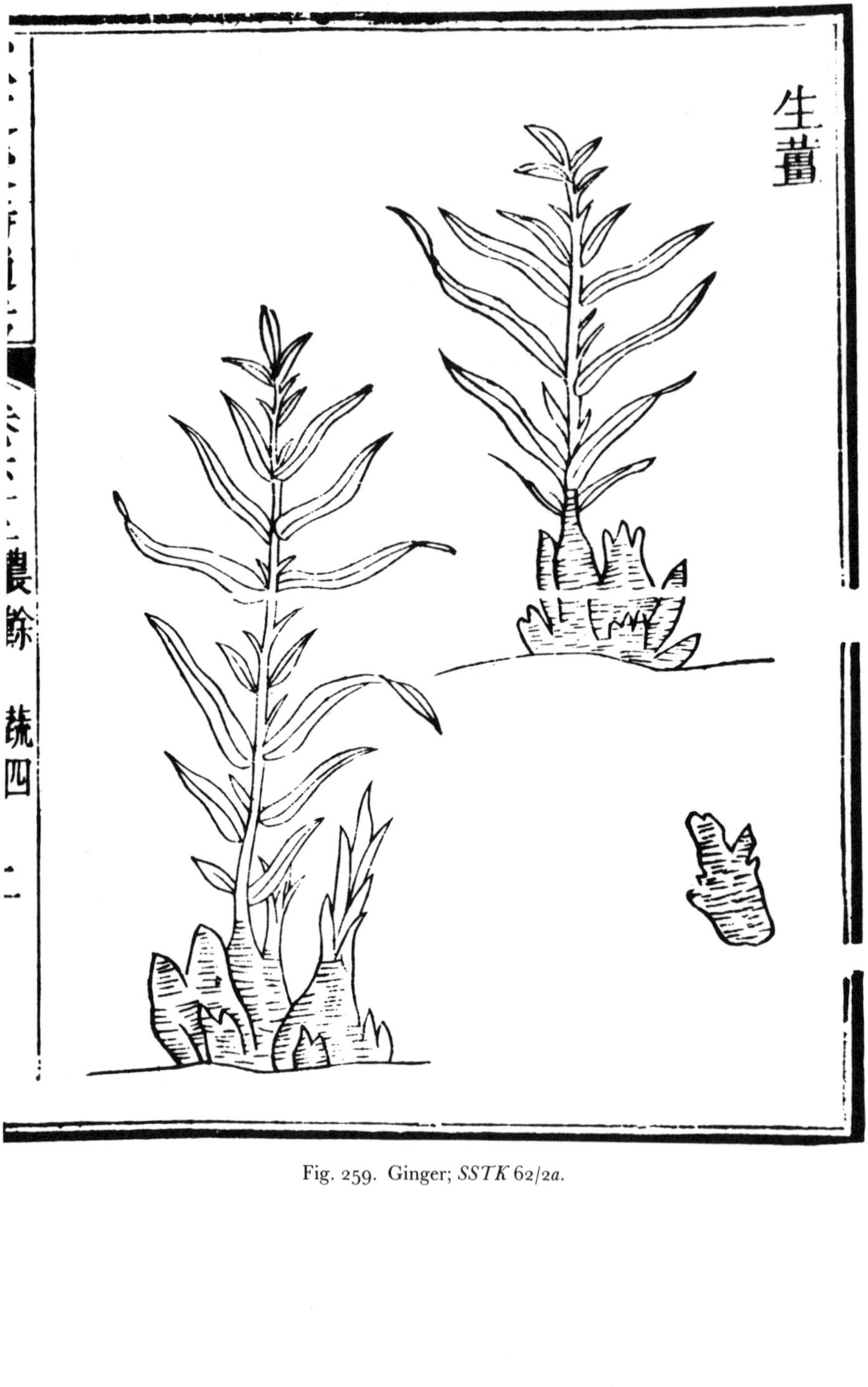


Fig. 259. Ginger; *SSTK* 62/2*a*.

Fig. 260. Jujubes, a typical fruit of North China; *CLPT* 23/7*a*.

## (*f*) CONCLUSIONS: AGRICULTURAL CHANGES AND SOCIETY—STAGNATION OR REVOLUTION?

Paul Leser, one of the greatest historians of agricultural technology, attributed the transformation of 18th-century European agriculture to an infusion of new ideas from the Far East:[a]

> I believe that the curved [plough] mould-board, like many other extremely significant advances in European agricultural practice, originated in East Asia ... These importations stimulated further innovation, bringing the spirit of progress and development into agriculture—all subsequent discoveries and improvements were unthinkable without this foundation. Until the 18th century, European agriculture was caught fast in the grip of tradition; its rationalisation and intensification are ultimately due to East Asian stimu-

[a] Leser (1), p. 456, tr. auct; see also Leser (2).

Fig. 261. Six Persimmons, ink on paper, by the Chhan artist Mu-Chhi (+13th century); Daitokuji, Kyoto.

lus. Without the revolution [in agricultural organisation and technology] that began in the early 18th century, the population increase of the last two centuries and all its concomitant developments would hardly have been possible. I think it is right to say that we owe the very foundation of our present culture to East Asia.

Leser argued that the countries of East Asia, and China in particular, had developed features of advanced agriculture long before they were known in the West, and knowledge of these features had become available to Europeans during the 18th century, through the reports of Jesuits and other travellers, at a time

Fig. 262. Lichee and Gardenia with Bird, handscroll attributed to the Sung Emperor Hui-Tsung (r. 1101–26); British Museum Collection.

when native European agriculture was in stagnation. He concluded that it was the dazzling example of Oriental agricultural expertise that shook European farmers out of a centuries-long torpor and stimulated an agricultural, and thus eventually an economic and industrial revolution.

It is certainly true that Chinese expertise in agriculture, as in many other branches of knowledge, still rivalled and surpassed anything known in the West when the Jesuits arrived at the Ming court. Early visitors marvelled at the high productivity of Chinese farming, its ingenious crop rotations, sophisticated water-raising devices and other equipment, and the rationality and industry of the Chinese peasant farmer. They were also struck by the predominant position accorded to agriculture in the Chinese political economy, where it was known as 'the fundamental' (*pen*[1]). This was a crucial influence on the theories of the French school of political economists, the Physiocrats, who like the Chinese philosophers insisted that agriculture, rather than commerce or industry, was the only source of true and enduring wealth. They felt that the state should be governed according to what they believed to be the Chinese pattern, by a monarch with absolute powers, an enlightened despot who would invest heavily in agricultural development and not in trade or industry (as had been the case under Colbert's ministry in France), for only in agriculture would application of capital and labour give

[1] 本

rise to a net surplus: 'the growth of agricultural output and investment provided the key to investment as a whole'.[a] The influence of the Physiocrats was widespread in late 18th-century Europe. Not only did Quesnay and his colleagues contribute important sections such as *Fermier*, *Grain* and *Culture* [Cultivation] to Diderot and d'Alembert's famous *Encyclopédie*,[b] but they had considerable influence on economic policies in Russia, Sweden and Germany as well as France.[c] Many of their views on the fundamental importance of agriculture to economic development were shared by Adam Smith, one of the founders of modern economics:[d]

> Upon equal, or nearly equal profits, most men will choose to employ their capitals, rather in the improvement and cultivation of land, than either in manufactures or in foreign trade. The man who employs his capital in land, has it more under his view and command, and his fortune is much less liable to accidents than that of the trader, who is obliged frequently to commit it, not only to the winds and waves, but to the more uncertain elements of human folly and injustice ... The ordinary revolutions of war and government easily dry up the sources of that wealth which arises from commerce only. That which arises from the more solid improvements of agriculture is much more durable, and cannot be destroyed but by those more violent convulsions occasioned by the depradations of hostile and barbarous nations continued for a century or two together ...

Smith and the Physiocrats were writing during a period of rapid economic change in Northern Europe. The adoption of new agricultural tools and machinery, of improved rotations and crop varieties, together with a radical reorganisation of farm management, rapidly brought about (particularly in Holland and Britain) an important increase in agricultural production, so marked that it has often been called the 'Agricultural Revolution'.[e] Economic historians, Marx not least among them, generally believed that without the economic stimulus of this Agricultural Revolution, the 19th-century Industrial Revolution (and thus the development of world capitalism) would not have been possible. The exact nature of the relationship between Industrial and Agricultural Revolution is much debated, but if Leser's hypothesis is correct, then Chinese technology made a significant contribution to the transformation of the European economy. We shall therefore investigate Leser's claims with some care in the following section, and show (we hope to general satisfaction) that he was substantially correct.

If, however, it is the case that the introduction of elements of Chinese technology contributed significantly to rapid economic and social transformation in Europe, we are clearly faced with a paradox: why is it that the same elements stimulated no equivalent transformations in China itself? In the preceding sec-

[a] Winch (1), p. 519.
[b] Brandenburg (1), p. 98.
[c] Winch (1), p. 52.
[d] (1), pp. 156, 172. Smith was also impressed by what he knew of contemporary China, where the predominant role of agriculture, he believed, in no way impeded commercial or industrial development; (1), pp. 282–3.
[e] For a more precise definition of this term and its implications, see the introduction in Mingay (1).

tions we have attempted to show how the agricultural systems typical of China, their field patterns, crop plants, and agricultural technology, evolved in response to environmental and practical constraints. We have distinguished two main agricultural traditions in China, the dry-cereal cultivation of the North and the wet-rice agriculture of the South, each characterised by distinctive crops, tools and field-patterns; we have indicated the role of climate and land form in determining the basic principles of these farming systems, and we have shown how they evolved in response to increases in population or the introduction of new tools or crops. We must remember, however, that Chinese agriculture did not develop in the social vacuum to which, for purposes of convenience and length, we have hitherto confined it. To keep our task within manageable proportions we have so far treated agriculture as an isolate, ignoring in so far as was possible the complex and intimate relations between advances in agriculture and other branches of technology, between levels of agricultural productivity, social organisation and political structure, the impact of trade and the growth of markets. Nor have we dealt in any detail with the form and consequences of the persistent official intervention and sweeping agrarian reforms that recur throughout Chinese history. The time has now come to consider these matters, albeit briefly, for only by examining Chinese agriculture in its wider context can we hope to understand why it developed as it did.

The Chinese state was an agrarian state, directly dependent upon agricultural production for the bulk of its revenues. The Chinese population paid poll taxes and gave labour services, as well as paying extra duties on such commodities as salt or tea, but the chief part of the state revenues came from the land tax (*tsu*[1]). This was paid by all land-owners and was proportional to the amount of land owned, though rates varied according to the quality of the land. Why the Chinese state should have been organised in this fashion, whether it should be seen simply as a mechanism of exploitation serving the interests of a dominant class of wealthy landlords, or whether it should be regarded (at certain periods at least) as a more benevolent and disinterested institution, possessed of a certain autonomy and serving what might be seen as the best interests of the nation as a whole, the majority of whom were peasants—these questions are clearly far too complex and important to embark upon as an after-thought to a volume on technology. Nevertheless, given the particular fiscal organisation of the Chinese state, it is clear that, as a general principle, it was in the interests of the state not only to encourage agriculture and to foster improvements in agricultural production, but also to minimise competition for the surplus produce from powerful landlords who, under certain conditions, might be in a position to deprive the state of a considerable proportion of its dues.

Thanks to the richness of the Chinese official documents, it is often possible to reconstruct Chinese agrarian policies and their outcome in some detail. Below we shall examine two of the most important campaigns by the Chinese state to

[1] 租

increase agricultural production, one of which took place in the Han dynasty, the other in the Sung. They offer an instructive contrast, illustrating inherent differences in the development potential of the northern and southern agricultural systems of China and showing how inseparable technological development is from social change. In the first case the nature of the technological change was such that it led to conflict between the interests of large landowners and the state, eventually resulting in stalemate. In the second, South China under the Sung experienced what might well be called an 'agricultural revolution'. Elvin (2) was perhaps the first scholar to characterise it in this way: he linked it with contemporary scientific, technological and economic advances which he also saw as revolutionary, and was then faced with the problem of explaining China's subsequent 'lapse into stagnation', its failure to exhibit transformations of similar scope to those of 18th- and 19th-century Europe. We hope to show in this section that in fact, while the changes in medieval South China were enormous and far-reaching, they were anything but revolutionary: although advances in agriculture stimulated the whole economy, as was the case in 18th-century England, the nature of the Chinese agricultural system was such that these changes reinforced the relations of production rather than upsetting them, so that no radical departures were to be expected.[a] This might be called 'stagnation', except that it was a system in which gross (if not per capita) output continued to increase for several centuries, while the population steadily grew. Only in 1800 or so was the potential for further expansion of production exhausted.

## (1) Did China Contribute to Europe's Agricultural Revolution?

Leser first advanced his hypothesis in 1931, when little had been written on East-West contacts and few Westerners were familiar with the wealth of documentary

[a] It will of course be objected here that China experienced a peasant-based communist revolution. This, however, does not invalidate our argument that in some cases technological development contributes to social instability, whereas in others it tends to maintain the *status quo*—for naturally we do not deny that radical socio-economic changes often result from non-technological causes. The Chinese Revolution is a case in point. Yet the technological base may well have a profound influence on the path of such change.

Peasant rebellions were frequent in China throughout imperial times, but it has been pointed out that agrarian unrest in late imperial China seldom took the form of concerted attacks by peasants against their landlords. Most peasant riots or rebellions previous to the 1930s were aimed against the official agents of the state, and in fact were often led by members of the landlord class (Chesneaux (1); K. C. Hsiao (6); Wakeman & Grant (1)). 'The peasant contribution to the Chinese Revolution resembled much more a mobilised response to a revolutionary elite's initiatives than did the peasant contributions in France and Russia' (Skocpol (1) p. 154). Furthermore, although the first Chinese Soviet was established in South China, in Kiangsi in 1931, the Communist Party's most solid base in the 1930s was in Northwest China, and it was in North China that early land reform was most successful. In the southern provinces social distinctions were less clear-cut: economic exploitation in the rural areas took more diverse forms and it was not possible simply to draw a class divide between landlords and tenants (Philip Huang (2)). Moise (1) says: 'There can be, for instance, little serious doubt that in some societies landlords have been legitimate, in the sense that they have shown at least some minimal consideration for their tenants' needs, so that their tenants did not regard them as enemies and did not leap at the first chance to get rid of them. One has only to look at the difficulties faced by Communist cadres, in Central and South China after 1949 and in North Vietnam after 1954, persuading the peasants of some villages to turn against the landlords.'

evidence (principally in Chinese) available on the history of Far Eastern technology. While Leser's hypothesis was of necessity based chiefly on conjecture, fifty years later it is possible to re-examine the problem with the detailed attention it deserves.[a]

Leser's argument was chiefly based on the course of development of the curved iron mould-board in 18th-century Europe. European ploughs traditionally had wooden mould-boards, usually straight, and it was only after contacts with East and Southeast Asia had been established that curved iron mould-boards were developed in Europe, although in the Far East they had been known for centuries previously (Fig. 263). Another, still more striking parallel, of which Leser was unaware when he advanced his hypothesis, was that between the system of row-cultivation of cereals perfected in North China in the early centuries of our era, and the Northwest European system of 'horse-hoeing husbandry', as it was dubbed in 1731 by Jethro Tull. Tull was the first European to formulate the principles of an integrated system of row-cultivation for dryland cereals, which was the basis for the highly mechanised farming system with which we in the West are so familiar today. Tull's basic principles ran as follows:[b]

> Hoeing is the breaking or dividing of the soil by tillage, whilst the corn or other plants are growing thereon.
>
> It differs from common tillage (which is always performed before the corn or plants are sown or planted) in the times of performing it; it is much more beneficial, and is performed by different instruments ...
>
> The earth is so unjust to plants, her own offspring, as to shut up her stores in proportion to their wants; that is, to give them less nourishment when they have need of more; therefore man, for whose use they are chiefly designed, ought to bring in his reasonable aid for their relief and force open her magazines with the hoe, which will thence procure them at all times provisions in abundance, and also free them from intruders; I mean their spurious kindred, the weeds, that robbed them of their too scanty allowance ...
>
> It is necessary to know how deep we may plant our seed without danger of burying it; for so it is said to be when laid at a depth below what it is able to come up at ...
>
> The proper quantity of seed to be drilled on an acre is much less than must be sown in the common way, not because hoeing will not maintain as many plants as the other; for on the contrary, experience shows that it always will, *ceteris paribus*, maintain more; but the difference is upon many other accounts: as that it is impossible to sow it so even by hand as the drill will do ...
>
> The distances of the rows is one of the most material points wherein we shall find many apparent objections against the truth; which, though full experience be the most infallible proof of it, yet the world is by false notions ... prejudiced against wide spaces between rows.

The principles enunciated by Tull in +1731 were first recorded in Chinese literature in the −3rd century, and later agricultural texts show that the im-

[a] The arguments in section (1) are based on an article written for the special issue of *Chung Hua Wen Shih Lun Tshung*[1] [Collected Articles on Chinese History and Culture] published in 1982 in honour of Joseph Needham's eightieth birthday; Bray (7).

[b] Tull (1), pp. 117 ff.

[1] 中華文史論叢

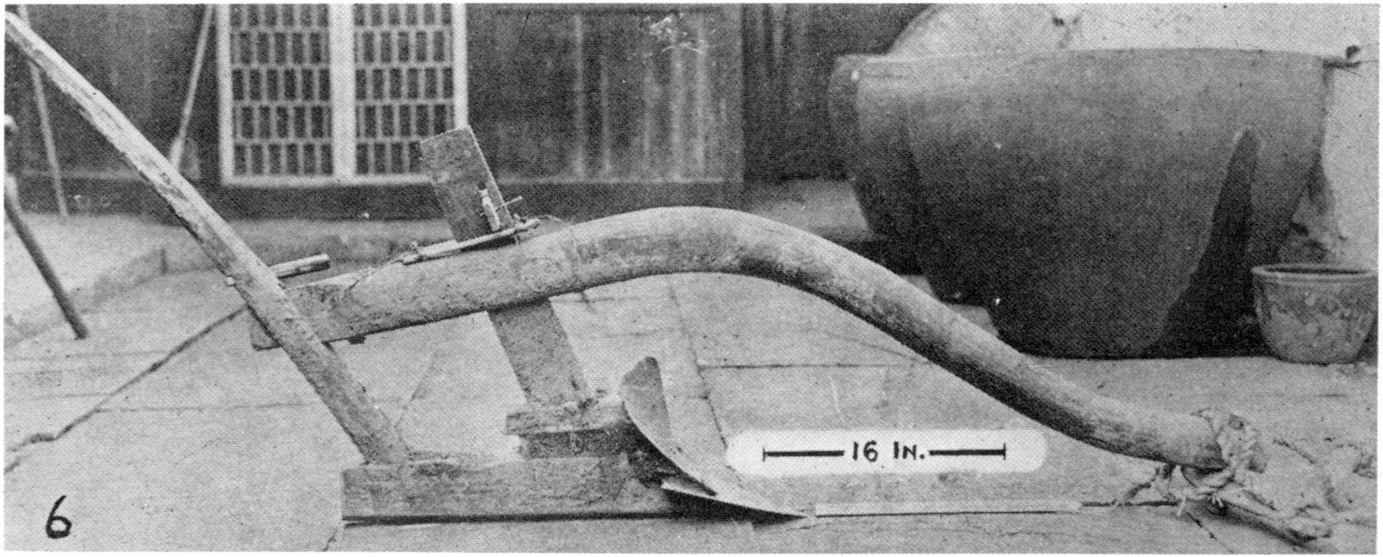


Fig. 263. Modern Chinese plough from Chekiang; Hommel (1), p. 41.

plements traditionally associated with this Northern Chinese system of row-cultivation were precisely those that Tull was trying to develop in England in the +18th century, namely the seed-drill and horse-hoe:

> If you do not sow in rows, then the stems of the grains will not grow sturdy and the young shoots will steal each other's [nourishment].[a]
>
> It is not fitting that farmers should know how to rotate their fields but not how to space their crops.[b]
>
> Take care that the seed is sown neither too close nor too far apart and that it is covered with neither too much nor too little soil.[c]
>
> The plants grow in rows and so they mature quickly; while they are still weak they do not interfere with each other and so they quickly grow tall.[d]
>
> The horizontal rows must be finely judged, the vertical rows made with skill, for if the rows are straight then the wind will blow through without harming them.[e]
>
> When you use a seed-drill to sow wheat, not only is the soil the right depth for easy germination, but also it is more convenient for hoeing.[f]
>
> If you hoe wheat it will yield twice as much; the husks will be thin and it will give large quantities of flour.[g]
>
> The horse-hoe (*lou chhu*[1]) is several times as efficient as a hand-hoe and will cover no less than 20 *mu* [about 3 acres] in a day.[h]

It is clear, then, that Leser's hypothesis is worth investigating. On the other hand it is not true, as he implied, that the period up to 1700 in Europe was one of technological stagnation. On the contrary, the economy of Northern Europe was in a ferment of activity and development. Farmers were anxious to raise productivity, and had identified several key problems that were inhibiting progress; various solutions were being hotly contested both in and out of print, and scientific method as well as practical experience were being called to aid. Clearly the time was ripe for agricultural revolution. So although the distinct similarities between technological developments in East and West lend weight to Leser's hypothesis of transmission, yet it is conceivable that the technological parallels may simply reflect a similarity of circumstance: might one not presume that a similar problem (namely increasing yields of dryland cereals on large farming units in a move from subsistence to commercial farming) would tend to produce similar technological solutions? And even if this is not so, if we rule out independent invention, we must still take care to distinguish between the direct transfer of technology and stimulus diffusion, where ideas rather than concrete objects are transferred between cultures.

a *LSCC*, p. 65.
b *LSCC*, p. 71.
c *LSCC*, p. 73.
d *LSCC*, p. 76.
e *LSCC*, p. 76.
f *CMYS*, 10.7.3.
g *CMYS*, 10.4.3.
h *WCNS*, 3/2*a*.

1 耬鋤

(i) *Pre-modern agricultural technology in Europe*

Quite apart from the socio-economic problems associated with the feudal system of land tenure, there were two strictly technological factors that kept productivity in traditional North European agriculture at a low level. The first was the heavy and poorly constructed ploughs which turned the soil inefficiently and slowly, and which required the labour of large numbers of draught animals. The second was the inefficiency of broadcast sowing, which necessitated the use of large proportions of the crop for seed and frequently gave very poor yields. Only if plough construction and sowing methods were improved could European farmers hope to lower their costs and raise their yields.

The inefficiency of the traditional European turn-plough was largely due to the enormous friction it engendered.[a] This was because of the wide, heavy sole which was thought necessary to turn a wide furrow, the heavy wooden wheels, and the large wooden mould-board, which lay along a different axis from the share so that the soil would not turn smoothly over it (Fig. 40). Many European mould-boards were flat, and could turn the soil only because they were so long that they allowed the furrow-slice to topple over under its own weight; others were slightly curved, but still far from efficient; since the mould-board and share did not fit smoothly together weeds and soil would get caught in the gap between them, and the ploughman would have to stop his team every few minutes to clear the mud from the plough with a stick. The plough was so heavy, and the friction engendered so great, that several pairs of oxen or horses were required to draw it. The exact number depended on the type of plough and the nature of the soil, but the average team was between four and eight; for example, a team of four oxen and two horses could plough one acre a day on a farm in medieval Essex.[b]

> The success of the old traditional ploughs seems to have been bound up with the use of these large teams ... Although very powerful, [the oxen] usually moved very slowly, so slowly that the ploughman could keep his plough steady at its right depth and his furrow-slice turning properly. However awkwardly the plough was constructed, whatever tendency it had to run light or dig deeply or let the furrow-slice flop back into the furrow, the ploughman had complete freedom and time to correct it. This, of course, meant severe labour for him.[c]

Any modification in plough design that reduced friction would allow smaller teams and faster ploughing, though at the same time it would be necessary so to improve the dynamics of the plough that it ran straight and level of its own accord, for the ploughman would have less control over his plough the more the pace of ploughing was increased.

[a] For details of the evolution and different forms of the plough in Europe see Leser (1) and Haudricourt & Delamarre (1).
[b] Fussell (2), p. 36.
[c] F. G. Payne (1), p. 80.

Broadcasting, the only method of sowing cereals known in Europe till the 16th century (Fig. 92), is extremely wasteful of seed:[a]

> Let [the hand] spread the seed never so exactly (which is difficult to do to some seeds, especially in windy weather) yet the unevenness of the ground will alter the situation of the seed, the greatest part rebounding in the holes, and lowest places, or else the harrows in covering, draw it down thither; and though these low places may have ten times too much, the high places may have little or none of it: this inequality lessens in effect the quantity of the seed, because fifty seeds in room of one, will not produce so much as one will do, and where they are too thick, they cannot be well nourished, their roots not spreading to near their natural extent for want of hoeing to open up the earth. Some seed is buried (by which is meant laying them so deep that they are never able to come up ...). Some lies naked above the ground which, with more uncovered by the first rain, feeds the birds and vermin.

Tull gives a fair idea of the wastefulness of broadcast sowing: as much as half or a third of the crop might have to be kept for seed-grain,[b] and even those grains which did germinate properly sprang up so haphazardly that thorough weeding was almost impossible. To some extent this problem could be remedied by covering the seed with a plough rather than a harrow, 'sowing under furrow',[c] yet even this was still wasteful of seed.

Other problems contributed to keeping medieval European agriculture undeveloped and underproductive. Almost the only fertiliser was animal manure, which was usually in short supply as fodder crops were rare and few livestock could survive the lean winter months. It was difficult, therefore, to keep arable land in good heart, and in many places the three-course rotation (whereby a field lay fallow one year in three) gave way to two-course rotations (where a field was only cultivated one year in two). And yields were low, partly because of poor farming practice and partly because few good strains of seed were available. Under the feudal system arable fields were subdivided into long strips sometimes no more than a few feet wide, and individual holdings were scattered over a large area; this also contributed to inefficient farming and low yields.

In a feudal economy, concerned with subsistence rather than profit, it is not surprising that few technological advances were made.[d] Some historians have suggested that the monastic estates of Northern Europe, managed by literate monks and often with considerable sums of capital at their disposal, were more advanced and enterprising than the manors of the feudal nobility,[e] but since the monks often referred their practice to the Latin agricultural treatises, written

[a] Tull (1), p. 120.

[b] In 13th-century England Walter of Henley spoke of a harvest of three times the amount of seed-grain as common; Oschinsky (1), p. 324. Yield: seed ratios were usually well below 10 : 1 all over Europe until the Agricultural Revolution; Slicher van Bath (1), pp. 172–7.

[c] Markham (2), p. 47; Sigaut (4).

[d] Kula (1), p. 34.

[e] Duby (1), p. 22.

several centuries earlier and concerned with the farming practice of the Mediterranean, their literacy may have been more of a hindrance than an advantage. No agricultural writings of innovative consequence were produced in Europe during this period (see p. 87).

But as the feudal system began to break down, as cities grew up and with them manufactures and markets for both foodstuffs and raw materials, so a new trend became apparent in the agriculture of Northern Europe. As villeins lost their servile status, personal ownership of land came to replace feudal relations and a free market in land developed;[a] commodity production superseded subsistence farming in many areas. As early as the 12th century, some feudal lords had perceived the advantages of farming a consolidated holding rather than scattered strips interspersed with common grazing land, and they 'began to withdraw their demesne land from the village farms, to consolidate, enclose, and cultivate them in separate ownership'.[b] By the 14th century this process was well established in England: even peasants were eager to buy out their poorer neighbours as soon as the opportunity arose, and we see the incipient formation of a class of capitalist tenant-farmers, working their holdings with free but landless labourers.[c] Under the stimulus of a growing market for wool, enclosures of land for sheep-raising created a serious problem in Tudor England:[d]

> Sheepe have eate up our medows and our downes,
> Our corne, our wood, whole villages and townes ...

But not all land was enclosed for sheep-runs: experienced Tudor farmers like Fitzherbert and Tusser understood the advantages to be gained by consolidating holdings on which both corn and cattle could be raised for food, and by the late 17th century most enclosures were used for improving mixed-arable farming, where a combination of grain and livestock was produced. On consolidated farms it was possible to improve drainage, fertilisation and crop rotations, and to deploy both animal- and man-power more effectively. Although the enclosures reduced many smallholders to abject poverty, creating a rural proletariat of landless labourers, the apologists of the enclosures argued that the overall benefit to the nation of the higher yields attained on enclosed land far outweighed the adverse social costs.[e]

[a] This process was already apparent by the 13th century in England; Ernle (1), p. 39; Macfarlane (1).

[b] Ernle (1), p. 38.

[c] *Ibid.* p. 48. The same process is documented for other parts of Europe by Duby (2).

[d] Bastard's *Chrestoleros* of 1598, quoted Ernle (1), p. 63.

[e] Opponents of enclosure on social grounds first appeared in the 15th century (Ernle (1), p. 63) and included such distinguished names as Sir Thomas More. Many pamphlets fiercely arguing the pros and cons of enclosure were exchanged in the 17th century (e.g. John Moore (1) and Joseph Lee (1)). By the 18th and 19th century, agriculturalists stated categorically that the preservation of open fields was untenable on agronomic grounds (Arthur Young (2); Ernle (1), pp. 190 ff., (2), p. 68), though recently Dahlman (1) has attempted to show that the open-field system was economically more viable than its detractors would care to admit. Many of the later enclosures were certainly futile and wasteful, as, for example, the attempt to enclose large tracts of Dartmoor on the 1870s (Torr (1), vol. II, p. 30). Apart from the enclosures of arable land, enclosures of woodland or common grazing-land had a deleterious effect on the living standards of rural labourers (Cobbett (1) passim). Some

As feudal relations disappeared and enclosures became common, so a new class of well-to-do farmers emerged, proud of their occupation, anxious to improve their farms, and sufficiently wealthy to invest in new methods or equipment. They were both literate and numerate; many of them committed their experience or their new ideas to paper, and the publishing houses that sprang up with the rapid development of printing from the mid-16th century found a ready market for books on agriculture (see p. 88). Many of these were simply general descriptions of farming methods, but sometimes authors adopted a more specialist approach and published monographs on such topics as crop rotations or fertilisation.[a] The scientific spirit of the age is apparent in such works as Walter Blith's *English Improver Improved* of 1653, which on the title-page declares his methods to be 'all clearly demonstrated from Principles of Reason, Ingenuity, and late but most Real Experiences'. Meanwhile, engineers and mechanics were busy experimenting with various agricultural machines, though fruitlessly for the most part. It was in this atmosphere of development and experiment that new tools and methods were first adopted in Europe which corresponded to a system elaborated in China many centuries before, based, as we have seen, on the principle of growing crops in evenly-spaced rows. This system was typical of Northern Chinese dryland agriculture. It depended for its effectiveness upon the use of a turnplough with a curved iron mould-board, a multi-tube seed-drill with a controlled flow of seed, and a variety of horse-hoes and ridgers to weed quickly and effectively between the young plants. All these elements had probably been perfected by the +6th century, when they formed an essential part of the equipment of a northern agricultural estate (see p. 593 below). They are referred to in the *Chhi Min Yao Shu*, described and illustrated in Wang Chen's *Nung Shu* of 1313, and are still used in North China today.

Before the development of horse-hoeing husbandry in 18th-century Europe, the tool complex of North China (as described above) was unique. Although seed-drills and horse-hoes were to be found in other parts of Asia, the complete range of implements only occurred in North China. Some elements of the system had already appeared by the late Han, but it seems to have reached its full flowering only with the growth of the large agricultural estates typical of the period from late Han to Thang (*c.* +100 to +800). Developments in China during the Han had something in common with the turn of events in Britain during the early modern period (see p. 594 below): an expanding demand for

economic historians maintain that the withdrawal of common rights was so gradual as to have little effect on the incomes of the rural poor (Mingay (1) and Introduction to Hammond & Hammond (1)), but others feel that the substitution of parish relief for economic independence was the final degradation that incited many farm labourers to violence and, when that failed, drove them from the land (Hammond & Hammond (1)).

[a] The earliest of the new agricultural writers was the Englishman Fitzherbert, who published his *Book of Husbandry* (1) in 1523. His example was soon followed by the works of Tusser (1) and (2), published in 1557 and 1573, Gallo's *Dieci Giornata* (1) of 1556, Estienne & Liebault's *Maison rustique* (1) of 1567 and Olivier de Serre's *Théâtre d'agriculture* (1) of 1600, and German works such as Grosser's *Kurze Anleitung* (1) of 1590. These were all general works. More specialised pieces included Markham's *Inrichment of the Weald of Kent* (4) of 1625, Sha's book on fertilisation and manuring of 1657 (1) and Yarranton's essay on rotations including clover (1) of 1663.

agricultural produce stimulated the widespread adoption of technological improvements that raised land productivity and thus the value of land, so that rich farmers gradually dispossessed peasant smallholders, consolidating large estates on which they produced crops for the market. The concentration of landholding permitted the rich farmers to rationalise management and to practise economies of scale such as the substitution of animal-drawn implements for human labour. Wang Chen said, for example, that a one-man one-beast horse-hoe could cover several acres in a day,[a] and the extent to which large northern estates relied upon animal-power is well illustrated in Fig. 264.

Apart from the mould-board plough, the comparatively sophisticated field-implements of North China were unknown in the South. This was not the result of mere ignorance, for writers like Wang Chen made systematic efforts to familiarise southern farmers with northern equipment,[b] but although considerable amounts of dryland crops like wheat and barley came to be grown in the southern provinces from Sung times on, such efforts were fruitless. This was because the cultivation of irrigated rice dominated the southern economy and dictated different patterns of technological development, as we shall see later.

### (ii) *European access to Asian agricultural technology*

The key elements in the 18th-century transformation of European agricultural technology were the introduction or development of a plough with a curved iron mould-board, the seed-drill and the horse-hoe. The only area where all three had previously been used in conjunction was North China, but other elements of the system were to be found separately elsewhere in Asia: single-tube seed-drills in West Asia, multi-tube drills and horse-hoes in South Asia, and ploughs with curved iron mould-boards in South China, Japan, Java and the Philippines where they had been introduced by Chinese migrants. To what extent was information on all or any of these areas available to Europeans in the 16th to 18th centuries?

The earliest European travellers to the Orient, like Marco Polo or William of Rubruck, purveyed little or no reliable information on science or technology in the regions they visited; not until regular contacts were established through merchants and missionaries did more accurate reports begin to filter back to the West. In the 16th century, Portugal established trading bases on the coasts of China and India and in Malacca. The Spanish, crossing the Pacific westwards from their American empire, annexed the Philippines in the 17th century. The Dutch founded Batavia in 1619, ousted the Portuguese from Malacca a few years later, and occupied Taiwan for a brief period at the end of the Ming; they were also allowed to set up a trading base in Nagasaki[1]. The British and French

[a] *WCNS* 3/2*a*.

[b] *WCNS* 12/17; 13/24; 19/passim.

[1] 長崎

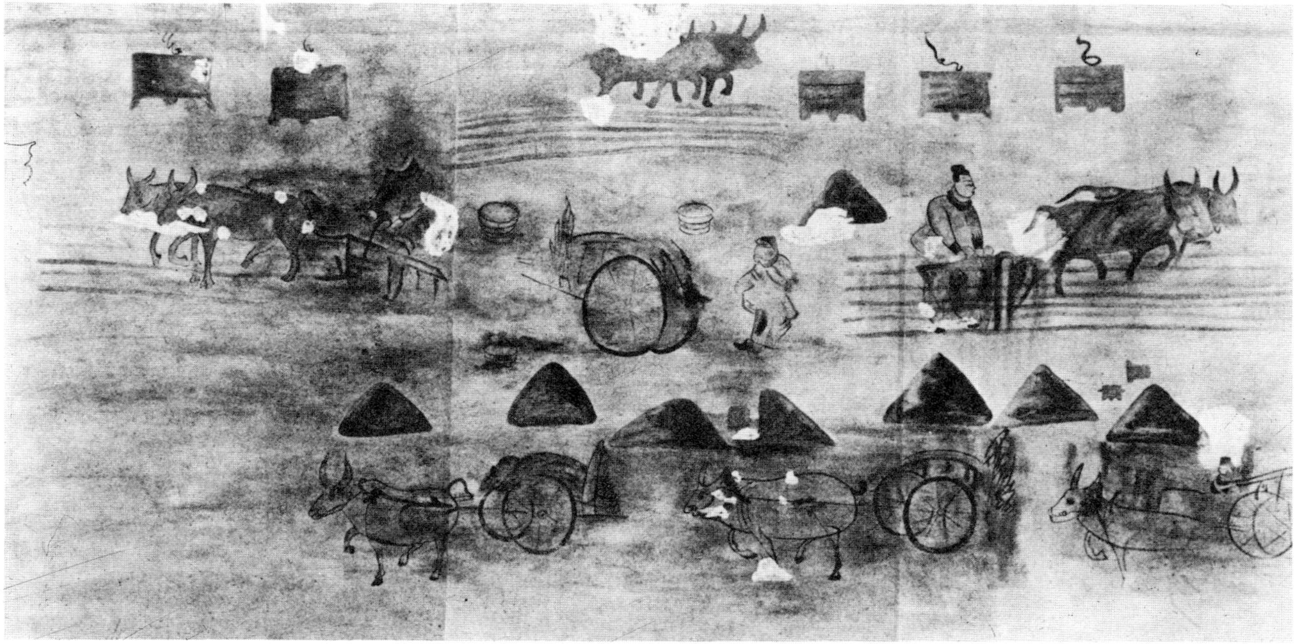

Fig. 264. Estate farm in Mongolia, +2nd century mural from Holingol; Anon. (*512*), pl. 34.

governments, though both were trading with China in the 17th century, confined much of their energy to India, where their rival East India Companies competed for influence throughout the 17th and 18th centuries; by the 18th century the British were also looking for bases in Southeast Asia. The East India Company of Sweden operated along the China coast as well as in Indonesia and along the littoral of mainland Southeast Asia.

European trading contacts with China were chiefly limited to the coasts of Kwangtung and Fukien. Christian missionaries could travel much more widely inside China once the Jesuit mission had been established in Peking in the early 17th century, and it is the voluminous publications and correspondence of the Jesuits which provided 17th- and 18th-century Europeans with the image of a huge, wealthy and highly civilised empire which they might do well to copy. The Jesuits were keen observers, scientifically trained, and prepared to report on the positive as well as the negative aspects of Chinese culture, unlike proselytisers of many other Christian sects. Some were specialists in astronomy, some in engineering or botany, but all were intellectually versatile and accurate observers. Although their spiritual mission naturally took priority, they and their sponsors were aware that there were material advantages too to be gained from contacts with China. Louis XIV of France, for example, sent six Jesuits to China on a specifically scientific mission in 1658;[a] before their departure the French Academy of Science drew up a list of topics that they wished investigated, including questions on history, geography, science, flora, fauna and food production as well as religious and political organisation. Leibniz corresponded extensively with the Jesuits in Peking, as did many other noted philosophers and scientists of the day:[b]

> [In 1689, in correspondence with the Jesuit Grimaldi, Leibniz] stressed the wealth of information available in China to the attentive observer ... [and asked] if certain useful plants and drugs could not be introduced to Europe. These he had learned about by studying Father Michel Boym's *Flora Sinensis*. Leibniz also asked for more detailed information about the manufacture of metals, tea, paper, silk, 'true' porcelain, dyes, Japanese sword blades (*laminarum*), and glass. More geographical details on the insular regions of the Far East he felt might also help in the correction of maps. He urged that closer attention be paid to Chinese agricultural, military and naval machines with an eye to the improvement of European devices. He requested the identification of more Chinese books, particularly relating to history and natural history, that could with value be translated into Latin.

The Jesuits liked to impart to educated Chinese the sciences (such as mathematics and astronomy) which were more advanced in Europe; Leibniz and many of his secular contemporaries were anxious that information should flow the other way, so that European manufactures (especially in the silk industry) could com-

[a] Huard & Wong (5), p. 140.

[b] Lach (1), p. 29; the originals of Leibniz's letters are preserved in a collection in the Niedersächsische Landsbibliothek at Hanover.

pete more effectively with China:[a] 'I besought Grimaldi not to worry so much about getting things European to the Chinese, but rather about getting remarkable Chinese inventions to us; otherwise little profit will be derived from the China mission.' In a letter to the Peking mission in 1707 Leibniz even suggested that the Jesuits should send Chinese animals, machines, models and scholars to Europe as well as their written accounts of Chinese agriculture and industry.[b] The Jesuits did in fact bring several Chinese theological students to Europe and also despatched many Chinese books to France and other countries,[c] including a series of albums illustrating Chinese agriculture, tea production and other rural occupations now in the collection of the Bibliothèque Nationale; significantly one of these is a series of twelve pictures depicting the cultivation of wheat.[d]

Just how much information was available on Chinese agriculture and agricultural policy can be deduced from the works of the Physiocrats, such as Quesnay, though the exact source of their information is not always clear.[e] Chinese agricultural works almost certainly figured among the titles in the Chinese collections brought back to Europe, while most Chinese encyclopaedias included long agricultural sections, and for those who could not decipher Chinese script there was food for thought in the illustrations. In this context it is certainly significant that the Jesuits' most important convert was the Minister of State Paul Hsü (Hsü Kuang-Chhi[1], 1562–1633), author of the *Nung Cheng Chhüan Shu*,[f] which contains copious illustrations of Chinese agricultural implements and machinery. Hsü had very close links with all the leading Jesuits of the time, including Johann Terrenz (Teng Yü-Han[2], 1576–1630), who in collaboration with a Chinese, Wang Cheng[3], had published an illustrated book on Chinese machines.[g]

The Jesuit mission in Peking in the 17th century was constantly pestered for information by European savants. By the 18th century, European publishers had made a real industry out of publishing Jesuits' personal memoirs as well as collections like Halde's *Letters Édifiantes et Curieuses* (2). Furthermore, the Jesuits had begun to send back Chinese models and machines as well as books describing

[a] Lach (1), p. 31.

[b] *Ibid.* p. 65.

[c] Andreas Muller's catalogue of Chinese books in the Berlin Library, prepared in 1679, included 25 Chinese titles or some 300 volumes; there were also collections in Paris, Vienna, and of course Rome; Lach (1), p. 46. Unfortunately many collections have since been dispersed, while others are not open for inspection.

[d] Huard & Wong (5), p. 180.

[e] Quesnay published a long monograph on the Chinese polity entitled *Despotisme de la Chine* (1). He stated that he obtained his information chiefly from the Jesuits as published by du Halde, but said that he also had recourse to original sources as well as consulting reports by Spanish, Italian, Dutch, Russian and British travellers (p. 566). Whether Quesnay did have first-hand knowledge of these sources is in fact debatable (Cummins (1); Maverick (2)).

[f] The final chapter of the section on water management contains a short treatise on European irrigation methods, *Thai Hsi Shui Fa*[4], written by Sabbatino de Ursis (Hsiung San-Pa[5], 1575–1620), which had first been published in six *chüan* in Peking in 1612.

[g] *Chu Chhi Thu Shuo*; his more famous work was on Western machinery, the *Chhi Chhi Thu Shuo*, published in three *chüan* in Peking in 1627.

[1] 徐光啟 [2] 鄧玉函 [3] 王徵 [4] 泰西水法 [5] 熊三拔

them. In the 1720s a Chinese winnowing-machine was sent to France and attracted much attention,[a] and Pierre d'Incarville (Thang Chih-Chung[1], 1706–57) sent home a model of a Chinese seed-drill, which the agriculturalist Duhamel de Monceau was able to examine.[b]

By this time secular channels had been set up to cater for the seemingly insatiable curiosity about China and the Far East. The French Academy of Science had correspondents among the French merchants of Canton,[c] and the Swedish Academy fostered close links with the Royal East India Company of Sweden:[d]

> [Passengers,] mates, ship's captains and others of the Company's servants have in the last few years, as it were, competed with each other to bring with them, on their return from China, some goods suitable for the Royal Academy of Sciences, namely, natural history specimens, discoveries, observations, drawings of machines, models, reports and descriptions of all kinds of subjects, which not only enrich and enlarge the Academy's ... model museum but also provide useful information on ... the customs, trade, economic management and handicrafts of the Chinese people.

Works such as *An Account of the Chinese Husbandry* by Captain Ekeberg, who spent altogether fifteen months in the Canton area during three visits to China in the 1740s, were translated into German and English,[e] and Swedish scientists also brought back copies of Chinese agricultural encyclopaedias.[f] Our information for other European countries is less complete, but we know, for example, that the Dutch brought examples of Chinese-style winnowing-machines back to Holland, probably from Java (see p. 377), in the early 18th century, and many other curiosities must also have been brought back by travellers from Indonesia.

Political and economic conditions in Southeast Asia were very different from China. Westerners found them primitive by comparison and less worthy of interest, and apart from a few naturalists' accounts of the local flora and fauna, very few detailed reports were published by Europeans prior to the early 19th century, while there was little scope for collecting indigenous literature. Stamford Raffles' superb *History of Java* (1), published in 1817, is the first work on Indonesia that can be set against the Jesuits' accounts of China. And although Westerners were intrigued and impressed by Japanese culture, the strict ban on communications with foreigners during the Tokugawa[2] era (1615–1868) prevented them from gaining more than a superficial acquaintance with the country.

Although early travellers to India were dazzled by the splendour of the Moghul court, they were not impressed by Indian agricultural policy. True, many praised the industry of the peasants and the neatness with which the fields were tended, but they did not adopt the same adulatory attitude as they did

[a] Hårleman (1), p. 63.
[b] Gösta Berg (3), p. 37.
[c] Huard & Wong (5), p. 11.
[d] *Proceedings*, 1754, p. 233; tr. Berg (3), p. 26.
[e] Osbeck (1).
[f] Berg (3), p. 27.

[1] 湯執中 [2] 德川

towards Chinese farming. Indian farm equipment was usually dismissed as primitive, not least (as an early 19th-century observer pointed out) 'because the Indian Husbandman can afford to throw away nothing on ornament. The same instrument painted and smoothed by the plane would have given a very different idea of its value.'[a] Again, it was only at the very end of the 18th century that accurate descriptions of Indian farm implements began to reach Europe. Buchanan, Halcott and Walker all give accounts of the Indian seed-drill, which Walker described as 'one of the most beautiful and useful inventions in agriculture';[b] Halcott, who 'until lately [had] imagined the drill-plough to be a modern European invention',[c] was so impressed by the Indian combination of plough, drill and horse-hoe that he sent a complete set to the Board of Agriculture in London, who published sketches of them in the first volume of *Communications of the Board of Agriculture* in 1797 (Fig. 97). But by this time European drill designers had advanced so far along completely different lines that the simple advantages of the Indian drill aroused little interest (see below, p. 576).

There was no dearth of information on Asian agriculture available to Europeans in the 16th to 18th centuries, and we know that European scientists and inventors were eager to learn whatever they could from the East; some specifically requested that details of Chinese agriculture and even Chinese farm equipment be sent to them for inspection. But if we hope to find explicit acknowledgement of such influence in their works we shall be disappointed: Western writers and inventors plagiarised each other's ideas quite shamelessly, and if they dared to flout their near neighbours in such a fashion we may be sure that they had no scruples in passing off as their own ideas that had come from the other side of the world.[d] We must therefore look closely at the technical details of agricultural development in 16th- to 18th-century Europe to determine what, if anything, they owed to the East.

## (iii) *The transformation in European agriculture*

### *The seed-drill*

As we have already seen, the system of horse-hoeing husbandry proposed by Jethro Tull so resembles the farming practice of Northern China that one is tempted to assume that Tull borrowed the system lock, stock and barrel from China. Tull was after all, as Arthur Young pointed out,[e] a very widely-read man;

[a] Alexander Walker (1), p. 186.

[b] *Ibid.* p. 185.

[c] Halcott (1), p. 210.

[d] Most of the work of the prolific 17th century English writer Samuel Hartlib, for example, was pirated from the works of other authors. *Samuel Hartlib, his Legacie* (1) was actually a verbatim reprinting of a manuscript by Weston (1), not even an enlargement as it claimed to be in the title (Ernle (1), p. 426).

The American Jefferson claimed to have discovered the mathematical principles of mould-board construction upon which the Dutch or Rotherham plough was based, even though the plough itself had undergone some years of development in Holland and Britain before being introduced to America (Fussell (2), p. 45).

[e] (1), p. 314.

he was also the first European to invent a seed-drill that actually worked. But Western inventors had been working on seed-drills long before Tull, and we must admit the possibility that they were initially inspired by the notion of 'setting corn', which seems to have been fairly well known in the 16th century.

'Setting' or dibbling corn consisted quite simply in using a stick to make rows of holes into which single seeds were dropped. The method had long been used in kitchen gardens for root vegetables such as carrots and radishes, and Sir Hugh Plat attributes the invention of corn-setting to 'some silly wench' who dropped wheat seed into carrot holes by mistake![a] The practice clearly gained some currency in England in the 16th century, for two pamphlets on improved methods of corn-setting appeared in quick succession in 1600 and 1601.[b] Plat proposed that setting corn could be speeded up and improved by using a board pierced with rows of holes to guide the setter (Fig. 99), but subsequent inventors raised their sights and worked on machines for setting corn.[c] Nothing much is known of these setting-machines except that they rarely worked, yet it could be that the practice of corn-setting was at the root of the later development of the European seed-drill, in which case any Asian influence or inspiration would have to be ruled out.

On the other hand, the first true seed-drill invented in Europe antedates the pamphlets on setting; it is the work of an Italian, and we have no evidence that corn-setting was practised in Italy. In 1566 Camillo Torello was granted a patent by the Venetian Senate for a system of sowing grain that was much commended for saving seed and increasing the yield.[d] But although it must have depended on some sort of drill, no details of Torello's system are known. However, a few years later, in 1580, Tadeo Cavalini applied in Bologna for a patent for a seed-drill, and we possess a description of Cavalini's drill written by Battista Segni in 1602:[e]

> By means of it the corn is planted rather than sown, and there is a great saving of grain in the sowing. Its construction resembles that of a flour sieve carried on a small, simple carriage, with two wheels and a pole. Part of the body holds the grain to be sown and part is constructed under the sieve and is perforated, and to every hole there is fitted an iron tube directed towards the ground and terminating in an anterior knife-blade of sufficient length to make a furrow into which the sifted grain at one passes through the tube and where it is so completely covered that none of it is damaged.

Later scholars have differed in their interpretation of Cavalini's drill: 'It may have been built like a bolting frame with a covering of open material through which the seed dropped as it revolved, as a hexagonal prism with rows of holes from which the seed dropped, or, as seems more likely, as a cylinder in which were cavities to receive the seed as the cylinder rotated under holes leading from the

[a] (1), chapter 1.
[b] Plat (1); Maxey (1).
[c] E.g. Gabriel Plattes, Otwell Worsley and others; for a fuller account see Fussell (2), pp. 95–8.
[d] Russell H. Anderson (2), p. 162.
[e] Tr. Fussell (2), p. 94.

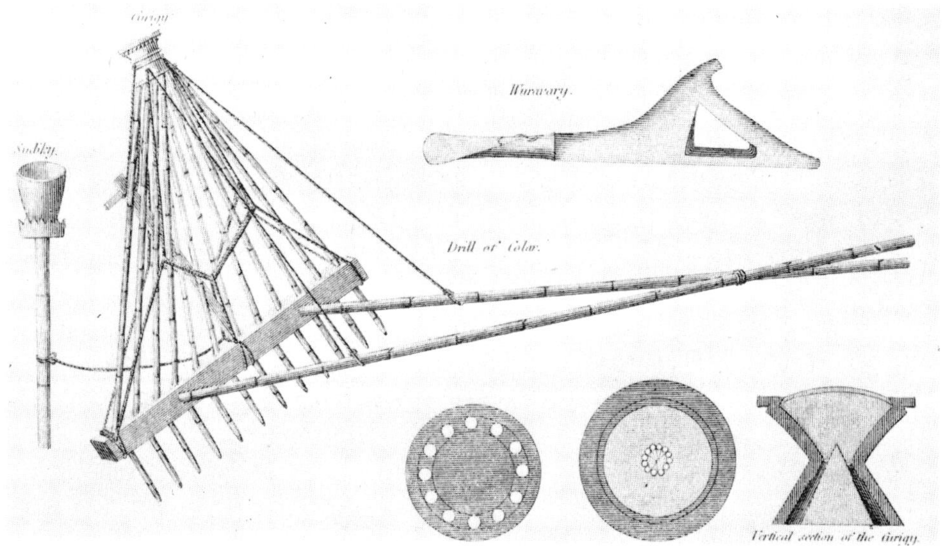


Fig. 265. South Indian drill showing details of feed mechanism; F. Buchanan (1), pl. 11.

seed-box above.'[a] It all hinges on what Segni meant by a flour-sieve. The first of the interpretations cited by Anderson above assumes that he meant a flour-bolter, while the others seem to ignore Segni's description and to interpret Cavalini's feeding mechanism as closely akin to the types found in later European drills; all three assume that the feeding mechanism is operated by the rotation of the wheels. As an alternative explanation we would suggest that the sole function of the wheels was to propel the drill and that by 'flour sieve' Segni meant an immobile sheet of wood or metal, perforated with several holes like a kitchen flour-sifter; to each hole would correspond a sowing tube, and the motion of the drill would shake the seed through the holes automatically. In that case Cavalini's drill would rather resemble the South Indian drill, or the perforated drill described in the *Thien Kung Khai Wu* (see p. 265), but supported by a wheeled frame (Fig. 265 and 104).

We have proposed this interpretation of Cavalini's drill because it seems that the description offers no grounds for assuming that the feeding mechanism rotated.[b] In the South Indian drill the sowing rate was controlled by hand, but if Cavalini's drill did indeed employ a similar sowing mechanism then, since the seed came directly from the seed-box, the farmer would have very little control over the sowing rate—which would explain why Cavalini's drill was not more widely taken up. If on the other hand it did contain a rotating feed-mechanism then, as Anderson points out,[c] his drill would have incorporated all three essential

[a] Anderson (2), p. 163.
[b] Though unfortunately we have not had access to the original Italian text.
[c] (2), p. 163.

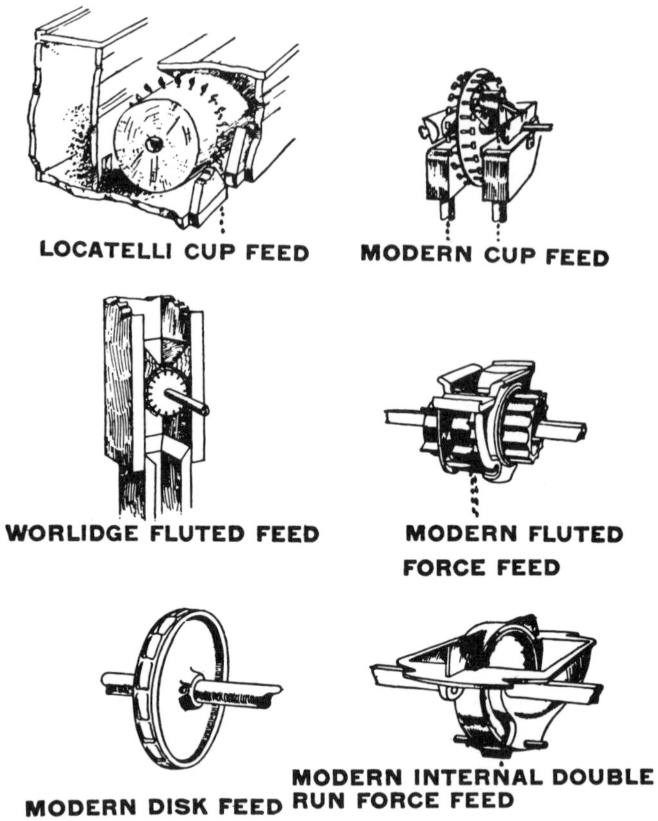


Fig. 266. European seed-drill feed mechanisms; Anderson (2), fig. 10.

elements of a successful drill (shares to open a furrow, sowing-tubes to direct the flow of seed, and a feed-mechanism to control it), and it is difficult to see why it was not more enthusiastically received.

It is just conceivable, then, that one of the earliest European seed-drills was based on an Asian model, especially as it was invented in Italy, which at that time still had vigorous trade links with the Orient. Thereafter, though, European seed-drills were all constructed on totally different principles from anything known in Asia. They usually had sturdy, rectangular wheeled frames supporting a wide, funnel-shaped seed-bin, one or more iron shoes to open a furrow for the seed, and a corresponding number of tubes to drop it into the furrow. The flow of seed into the tubes was controlled by any one of a variety of rotating feed mechanisms (Fig. 266), which might either form part of the axle of the wheels or else be driven indirectly by a chain-drive or gears. It is this consistent use of rotating feeding mechanisms which is the most essential difference between Western and Asian drills. We know that the 18th-century French inventor Duhamel de Monceau was perfectly familiar with the construction of the Chinese drill, for he actually published a drawing of a Chinese drill, almost identical to the *Nung Shu* illustration (Fig. 95), in his long work on drill-husbandry entitled *Expériences et réflexions relatives au traité de la culture des terres* (1). Evidently, though, he felt it was quite irrelevant to his own purposes, for the drill that he developed was built on very much the same principles as Tull's, which was probably the most successful of all the European drills before the 19th century:[a]

[Tull's] drill, which sowed three rows of seed, was drawn by one horse. The three coulters or hoes were narrow and shaped to enter the soil readily. At the rear of the coulters were passages, open behind, which served to guide the seed from the funnels above to the channels in the ground. These coulters, the framework supporting them, and the shafts rested upon the ground and not upon the four wheels of the machine. The two large wheels in front carried the seed-box and dropper unit which fed the centre coulter on a $1\frac{3}{4}$ inch axle. Two smaller wheels at the rear carried the droppers and seed-boxes feeding the two other coulters, placed 14 inches apart and some distance behind the central coulter so there would be no interference between them—an arrangement much used during the last century. The dropper unit consisted of the case at the bottom of the seed-box, and the notched axle which passed through it. This axle, with its notches or cavities in the periphery, turned with the wheels, received the grain from the boxes above, and dropped it into the funnels below. The passage of grain past the notched dropper was controlled by a brass cover and an adjustable spring, patterned after the tongue in the organ mechanism [Fig. 98].

Perhaps the Chinese model that Duhamel examined was too small to contain a feeding mechanism, or perhaps he was unimpressed by its simplicity and apparent clumsiness. It is true that the traditional Chinese seed-drill can hardly compare with the modern mechanical force-fed drill for precision, yet within certain limits it is remarkably effective and reliable, and has the advantage of being cheap, sturdy and foolproof. Duhamel could have done worse than to take a leaf from the Chinese book, for the European rotating feeding mechanisms, despite their apparent sophistication, were to remain the chief stumbling-block in design until well into the 19th century:[b]

A very little attention will discover the causes of the drill husbandry making so slow a progress, even under the supposition of all the merit which the most sanguine of its pursuers assert it to possess. In the first place, the principal reason of all others is the insufficiency, real or imaginary, of all the drill ploughs hitherto invented in performing the complex offices which are requisite in such a machine ...

Suppose ... that the common plough was so complex in its powers as to render simplicity extremely difficult to be preserved in its construction; that the variety of its parts was so great, and had so little firmness or connection in them, as to render the whole machine unavoidably weak; that the same objections which rendered it so complex and so weak made it likewise difficult and expensive to repair; without multiplying these suppositions to a tenth of the extent to which they might be carried, we may venture to determine that husbandry would at once be reduced to infancy if the common plough remained under these three disadvantages.

[a] Anderson (2), pp. 169–70.

[b] Arthur Young (1), pp. 122–3.

Now the drill plough is attended with many other disadvantages, for it is of a high price, very difficult to procure, and notwithstanding the variety invented, not one is of such particular excellency as to be allowed to exceed the rest. In such a situation is it possible that drilling can flourish?

As we have seen, Westerners also brought back Indian seed-drills at the end of the 18th century. Halcott wrote:[a] 'I have reason to think that [the Indian] drill plough, simple as it is, possesses an advantage that the patent drill plough does not; for I remember reading in some publication, that the patent drill plough was defective in not dropping the grain equally; this plough has no defect of that kind.' Nevertheless, although travellers remarked upon the effectiveness of Oriental seed-drills, they had only curiosity value for Western engineers. Even though they were aware of the shortcomings of their more complex drills, they were by now irrevocably committed to mechanical drills incorporating rotating feeding mechanisms.

Well-to-do farmers in both Britain and France were favourably inclined towards the theory of the new husbandry, but few actually practised it, for despite the efforts and ingenuity of numerous engineers, it was not until the mid-19th century that drills became sufficiently reliable, effective and economical to enter into common use.[b] The two centuries or so that elapsed between the invention of the first European seed-drills and their perfection represent not theoretical innovation but progress in the general level of engineering skill. Though modern drills are much larger and mechanically more complex than the early prototypes, they still fall into the two basic categories of the 17th century: the 'Suffolk' drill with a cup-feed and rocking hopper, based on the principle of Locatelli's drill which was certificated by the German emperor in 1663, and the force-feed drill, based on a device first developed by the Englishman Worlidge who published it in 1669 (Fig. 266).[c]

All that we have seen shows that the Western seed-drill was fundamentally different from its Asian counterparts, based on mechanical principles so far removed that, even though Westerners expressed admiration of them in agricultural journals, they had no influence at all on the development of the European drill.

*The curved iron mould-board*

The drawbacks of traditional European ploughs have already been mentioned: they were heavy and engendered enormous friction, so that a large team of animals was required to draw them. By the 16th century, agriculturalists were already casting about for improvements. Naturally the traditional ploughs were

[a] (1), p. 211.

[b] Drills for sowing turnips and fodder-crops were more successful than grain-drills. Britain and America were the first countries to develop successful grain-drills in the 1860s and 1870s; manufacturing techniques lagged behind in other European countries, and few French farmers sowed by drill even as late as the 1890s; Anderson (2), p. 196.

[c] Spencer & Passmore (1), p. 19.

to be found in a wide variety of local versions. In England both Fitzherbert (1) and Markham (2) noted that certain types of plough were suited to particular soils. Markham specified suitable principles of plough construction for a wide range of soil types, for example on white clays the plough should be 'wide and open in the hinder part, that it may turne and lay the furrowes one upon another'.[a] Like many of his contemporaries, Markham believed that a wide sole was necessary to turn heavy soils effectively, yet clearly this added to the friction and to the weight of the plough. Walter Blith highlighted the objections to the broad-soled plough in his formulation of the principles elementary to good plough design, published in 1649:[b]

> First, that whatever moveth upon the Land, or that worketh the Land, and carrieth the least earth or weight upon it, must needs move or work easiest. A wheel the lesser ground it stands upon, the easier it turns, and the lesser the wheel the easier still; so the Plough, the more earth or weight it carries with it, the more strength must be required. The naturall furrow it must carry, but the lesser compass both in height and length it bears upon the Plough, the easier the Plough must go.
>
> Secondly, the more naturally anything moves, the more easily and the more Artificially, the more difficulty.
>
> Thirdly, the sharper and thinner is any tool, the easier it pierceth, and the less strength is required; so contrary, the thicker or duller any tool is the more strength must work it, and;
>
> Fourthly, that which is the plainest and truest to the Rules and admits of least multiplication of work, must needs be the easiest.

By this time, as we can see from Blith's principles, agriculturalists were beginning to conceive of their subject as a science, and were looking to engineers to formulate solutions of many of their problems:[c]

> I wonder that so many excellent *Mechanicks* who have beaten their brains about the perpetual motion and other curiosities, should never so much as honour the *Plough* (which is the most necessary Instrument in the world) by their labour and studies. I suppose all know, that it would be an extraordinary benefit to this *Country*, if that 1 or 2 horses could plough and draw as much as 4 or 6 ... Surely he would deserve well of this *Nation* and be much honoured by all, that would set down exact rules for the making of this most necessary but contemned *Instrument*, and so for every part thereof; for without question there are exact rules to be laid down for this, as for Shipping and for other things.

To increase the efficacy with which the furrow-slice was turned, one alternative to widening the plough was to build the mould-board with a slight twist. The earliest wooden twisted mould-boards may have occurred in Flanders in the 14th century;[d] by the 18th they were fairly common in Northwest Europe.[e] But there was a limit to the twist that could be given to a wooden mould-board, and

[a] (2), p. 57.
[b] Blith (1), quoted Fussell (2), p. 41.
[c] Hartlib (1), pp. 5–7; quoted Fussell (2), p. 39.
[d] F. Sigaut, pers. comm.
[e] Andreas Berch (1) describes them in Sweden, for example, in 1759, and James Small (1), cited Fussell (2), p. 49, mentions them in England in 1784.

furthermore it was difficult to marry it to the share so that there was no gap in which earth and weeds could catch. That is why the iron mould-board was such a revolutionary conception, for it could be constructed with the most pronounced or complex curve, and fitted exactly to the share. In the 17th century Blith, reviewing various British plough types, was signally impressed by the so-called Bastard Dutch plough, made specially for use on boggy land: it had an iron share made in one piece with a plate of iron that covered the mould-board, leaving no gap. The Bastard Dutch plough, as its name implies, originated in Holland and was brought to England by Dutch engineers who at that time were employed in great numbers by the English government to drain the East Anglian Fens.[a] Blith did not believe that the Bastard Dutch plough could be used on ordinary dry land, but the farmers of Holland and East Anglia thought otherwise: in 1707 John Mortimer describes an East Anglian plough 'very peculiar for its Earth-board [mould-board] *being made of Iron, by which means they make it rounding* [italics ours]; which helps to turn the Earth or Turf, much better than any other sort of Plow'.[b] Here we have the first record in Western literature of a curved iron mould-board.

In 1730 the Dutch or Rotherham plough, based on Dutch designs, was patented in England;[c] it incorporated most of the essential elements of advanced plough design (Fig. 267): it was a very light swing-plough (that is to say it had no wheels) with a narrow sole, a coulter and share made of iron and a mould-board covered with an iron plate. It was introduced from England to Scotland and from Holland to America and France, and in the 1770s it was still considered the lightest for draught and the cheapest for general use of any plough then made.[d] The Rotherham plough had a profound influence on all subsequent European plough design, which aimed to produce still lighter and yet stronger frames and to improve the efficiency of the mould-board. In the 1770s the first iron-framed ploughs were introduced, and by the early 19th century ploughs constructed entirely of iron were commonly used by progressive farmers. The substitution of iron for wood permitted still greater flexibility of design. The frame, for example, was often reduced to a single saddle-shaped iron sheet (Fig. 268) on to which interchangeable shares or mould-boards could be bolted.

Now that mould-boards could be accurately cast in the required shape it was possible to conceive of designing them according to mathematical principles. The Scot James Small was the first to attempt their formulation in his *Treatise of Plows and Wheel Carriages* of 1784, where he states that the mould-board should be so

[a] Fussell (2), p. 41 gives a detailed description of the Bastard Dutch plough. Systematic reclamation of the marshlands or Fens of East Anglia began in the 17th century: 'Seventy miles in length, and varying in breadth from ten to thirty miles, the fens comprised an area of nearly 700,000 acres. Now a richly fertile, highly cultivated district, it was, in the 17th century, a wilderness of bog, pools and reed-shoals'; Ernle (1), p. 115. The reclamation of the Fens was undertaken chiefly by Dutch engineers like Vermuyden (1), who were contracted by the local authorities, and many of the settlers on the reclaimed land were French or Dutch Protestant refugees; Ernle (1), p. 119.

[b] Quoted Fussell (2), p. 44.

[c] J. Allen Ransome (1), p. 13.

[d] Matthew Peters (1), cited Fussell (2), p. 46.

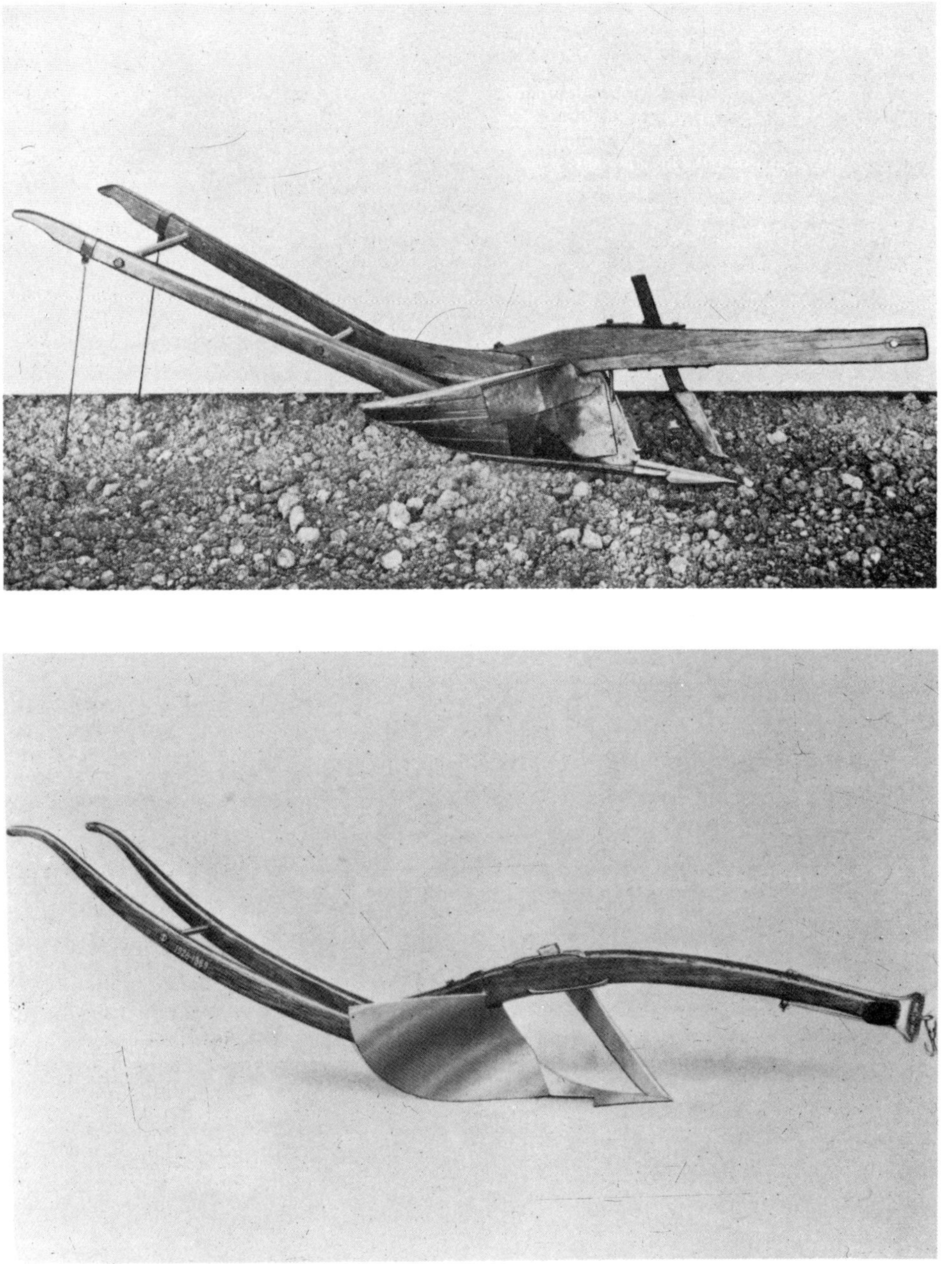

Fig. 267. Rotherham plough and James Small's plough; Spencer & Passmore (1), pl. 3.

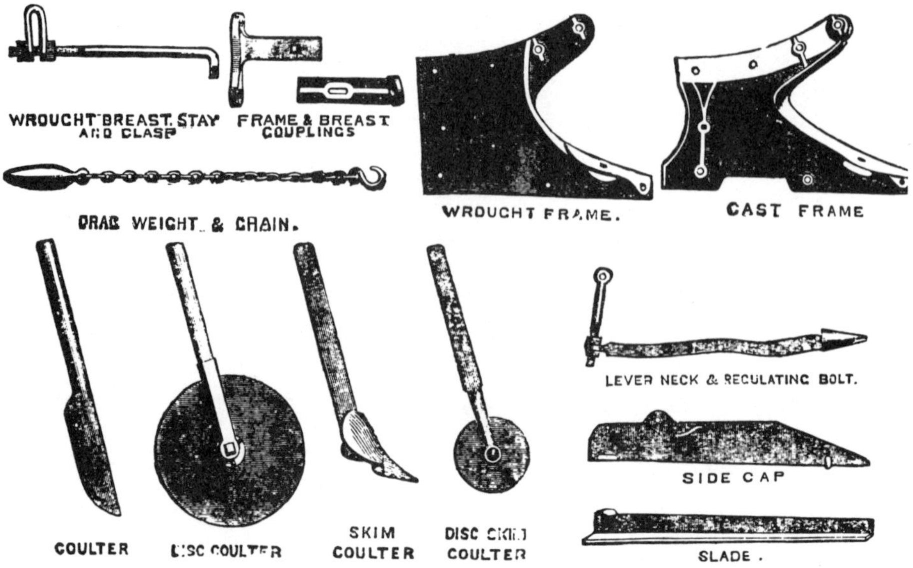


Fig. 268. Plough components, including saddle-shaped frame; Malden (1), p. 119.

constructed that 'the back of the sock [share] and the mould-board should make one continued fair surface, without any interruption or sudden change. The twist, therefor, must begin from nothing at the point of the sock, and the sock and the mould-board must be made by the very same rule.'[a] Over the next fifty years much energy was devoted to discovering the exact mathematical principles of mould-board design, but as J. Allen Ransome, himself the designer of many improved ploughs, said in 1843, no universal rule had as yet emerged, since different soils, for example light sands and heavy clays, might require mould-boards of quite opposite form.[b] But with the steel frame it was now possible to design interchangeable mould-boards for different purposes, such as a long, gently twisted one for laying the furrow-slice on edge, unbroken, at the autumn ploughing, or a short, sharply twisted one for pulverising the clods at the spring ploughing.[c]

All the technical developments which led logically to the modern plough stem originally from the 17th and 18th century improvements in Dutch plough design. Here Leser seems to be completely justified in hypothesising East Asian influence, for though certainly not identical to the East Asian turnplough, the 17th-century Dutch ploughs shared all its key characteristics: lack of wheels, light frame, and curved iron mould-board fitting neatly to the share; all features missing from traditional European ploughs (see Table 6). Almost all previous European ploughs had wheels, as well as heavy frames and broad soles which were con-

[a] (1) Cited Fussell (2), p. 49.
[b] (1) p. 25.
[c] Spencer & Passmore (1), p. 10.

sidered essential for turning the furrow, and the mould-boards were huge rectangular wooden constructions which were not connected directly to the ploughshare. But the typical Chinese turn-plough, which by the 17th century was not only used all over China but had spread to most of the rice-growing areas of East and Southeast Asia, had never had wheels, was light-framed with a narrow sole, and had a curved, cast-iron mould-board that was designed to fit snugly against the share in a single smooth curve. That the most important features of the new Dutch ploughs should also be characteristic of Chinese ploughs but non-existent in the traditional European forms cannot be coincidental: the new Dutch ploughs must have been very heavily influenced by East Asian plough design.

It may be significant that in Holland the new ploughs were first used on wet and boggy land. We would suggest that the Dutch saw Chinese turn-ploughs at work in the wet-rice fields of Asia and immediately recognised their potential for the intractable marshy fields of the Dutch maritime provinces (and the Fenlands of East Anglia). Dutch travellers may first have noticed these ploughs in Java (where they were used by the Chinese rice-farmers who had settled around Batavia), perhaps in Japan, or else in South China, and they probably brought back actual examples of the ploughs to Holland, where their main features were adopted to more traditional designs, resulting in the new Dutch ploughs of the 17th and 18th centuries and providing the basis for far-reaching technological development.

### (iv) *Asian contributions to Europe's Agricultural Revolution*

Until the 17th century, productivity in European agriculture was severely limited by the inefficiency of ploughing, sowing and hoeing methods. The 17th to 19th centuries saw a transformation of North European agricultural technology, based on the development of the turn-plough with curved iron mould-board, the seed-drill and the horse-hoe. Jethro Tull was the first European explicitly to formulate this integrated system of 'horse-hoeing husbandry' in 1731, yet an agricultural system incorporating all the very same elements had existed in North China since Han times, while individual elements of the system were to be found in several other parts of the Far East. The transformation of European agriculture coincided with a growing awareness among Western intellectuals of the civilisations of the Far East and of China in particular; Chinese agricultural policies were taken up by the Physiocrats, and a great deal of specific technical information on Chinese and Southeast Asian agricultural implements, including copies of agricultural treatises, was brought back to Europe by missionaries, scientists and merchants. Can we then conclude that the key technological elements of the European Agricultural Revolution were directly copied from China?

As far as the system of horse-hoeing husbandry is concerned, examination of the development of the Western seed-drill shows that direct copying must be ruled out. Not only were European drills unlike their Chinese counterparts—they were actually based on completely different mechanical principles from any of the Asian drills. It is possible that the two very earliest European drills, patented in Italy in the 16th century, were based on Asian models, but thereafter European types developed along quite different lines, so much so that when concrete examples of Chinese and Indian drills were shown to European inventors they were in no way moved to modify their designs. The horse-hoe is not a complex implement, and its form and use follow logically from the use of the seed-drill; it therefore seems otiose to try and trace any connections between its development in East and West.

There is, however, strong evidence to show that the modern European plough with its light frame and curved iron mould-board was directly influenced by the Chinese plough. But paradoxically, although such ploughs were used as much in the dry fields of North China as in the irrigated rice-fields of the South, it seems to have been the second type that was copied in Europe, even though European farmers did not themselves grow irrigated crops. It appears likely that the Dutch saw Chinese ploughs at work in the rice-fields of South China—or perhaps Java or Japan—and were struck by their potential for use in the marshy lands along the Dutch coast. They adopted the main features of the Chinese plough in a new type which was first used in the boggy lands of West Holland and East Anglia, and soon afterwards came to be adapted for use on drier soils where it proved immensely successful and economical both of human and animal labour.

The Agricultural Revolution thus probably owed one of its most fundamental elements, the efficient mould-board plough, to China—but to South China rather than the North. Is the striking similarity between Northern Chinese farming methods and the horse-hoeing husbandry of Europe then due simply to coincidence, to a similarity of environmental circumstances producing similar solutions to basic technical problems? We feel sure that there is more to it than that. One might argue that the European seed-drill was a logical development from earlier horticultural techniques such as setting, yet it cannot be fortuitous that European inventors suddenly started working on machines to sow several rows of corn simultaneously in straight lines, just like the Chinese machines, precisely at the period when information about Chinese agriculture was becoming freely available. We would suggest that here we have a case of stimulus diffusion. Europeans read about North Chinese farming methods and were impressed by their reported efficiency and economy, but their sources, whether Western or Chinese, were insufficiently detailed for European mechanics to copy the Chinese instruments directly. The descriptions and illustrations in Chinese agricultural encyclopaedias were, as we have seen (p. 256), too imprecise to use as blueprints, and it was difficult at first to send actual models back to the West, as the regions where they were in use were so far from the main seaports of Canton and the Southeast coast. So Western mechanics set to and invented their own versions, which is why Western and Chinese drills have so little in common. The case for the turn-plough was quite different: although it was an essential element of the

North Chinese farming system, it was also widely used in more accessible regions such as the South China coast and Southeast Asia, so there was no difficulty in shipping examples back to the West where they could be copied and modified by local carpenters and blacksmiths. (Other agricultural implements from the same areas, such as the winnowing-fan, were also introduced directly into a number of European countries including Holland, France and Sweden.) In the comparatively accessible areas of Asia, the efficient Chinese turn-plough, with a curved iron mould-board, was used mainly in irrigated rice-fields, and had the Dutch not been faced with the problem of cultivating wet land in their maritime provinces, the advantages of the Chinese plough might never have struck them. As it was, they recognised its potential and imported it to Europe, where local farmers soon realised that it could advantageously be adapted for use in dry fields. So although the new European plough was originally based on a Southern Chinese model, it soon came to be incorporated into a dryland system of drill-husbandry very similar to that of North China.

From the 17th century on, European agricultural inventors began to think in terms of machines rather than implements, and of science rather than skill. It is no coincidence that the motto of the Royal Agricultural Society of England, instituted in 1838 under the patronage of the Prince Consort, was 'Practice with Science'.[a] The scientific approach to agriculture offers a clear contrast with China, where mathematical and scientific concepts were never applied to agriculture until the advent of Western agricultural science in the late 19th century. Although we naturally think of the Western approach as more advanced, the difficulties encountered in working out formulae for the construction of agricultural machinery, plant breeding, or fertiliser response should put us on our guard: even today it is debatable to what extent agriculture should be considered a science rather than a craft.

European agricultural inventors might be accused of having been overambitious at first. There was a long gap between the first introduction or invention of new implements and their mechanical perfection. Indeed, in the case of the seed-drill it was over two centuries before a satisfactory working model was developed, and had the design been simpler a solution might have been reached more quickly. But the scientific approach of the Western inventors gave scope for further elaboration, improvement and enlargement such as was impossible with the simpler Asian equipment.

The purpose of the European inventions was not simply to economise on time and effort but also to replace human labour by machines, a reflection of the relations of production in Europe at the time. Capitalist farmers drew their workforce from a rural class of landless labourers, paid in cash by the hour, so there were very clear economic advantages to replacing human labour by machinery wherever possible. By increasing labour productivity, mechanisation of

[a] Copland (1), p. 118.

agriculture has three possible effects: (i) an expansion of the cultivated area without any increase in the labour force; (ii) a saving of labour on the same cultivated area; and (iii) a reduction in the time required to perform certain tasks (which in turn allows more tasks to be performed). In sparsely populated countries like North America and Australia, the first factor was crucial in stimulating mechanisation; in Europe the two other factors were more important. The chief desire of 19th-century British farmers was to use machinery to reduce cost: Pusey[a] calculated that the mechanisation of an average arable farm with the machinery available at the time (1858) would cut the total running costs by almost one half. The mechanisation of even one simple process could effect considerable savings: on a 19th-century English farm of 730 acres the purchase of a threshing-machine at £500 saved £200 per annum in labour costs.[b] Another much appreciated aspect of such machines as the steam-plough and reaping-machine was that they could be called into action at a moment's notice:[c]

> Machinery has given to farming what it most wanted, not absolute, indeed, but comparative certainty.
>
> One chief advantage [of steam cultivation] is that it renders the farmer comparatively independent of labour; at all times he has an enormous power at his command on a moment's notice, and is thus able to deal with the land when in a fit state for cultivation.

The process of mechanisation in Europe was slow and complex, but given the nature of the relations of production in agriculture, inexorable. It is interesting to note that the initial effect of the Agricultural Revolution in England was actually to increase demands for labour. New methods such as row-sowing permitted more intensive cultivation of the crops during their growth; improved rotation systems intensified land-use; deeper ploughing and more effective crop care increased yields and thus the demand for harvesting labour. In Britain this intensification of production initially guaranteed a high level of rural employment, and the agricultural population continued to grow until it reached a peak in about 1860.[d]

In the early days of machinery profit-margins were small, and only when labour inputs had reached a high level of concentration did mechanisation become worthwhile:[e]

[a] (1), p. 192.

[b] Slicher van Bath (1), p. 306.

[c] Pusey (1), p. 193; John Scott (1), p. 46.

[d] The census of 1851 showed a total rural labour force of 1,788,000 males and 229,000 females; Mingay (1), p. 10. Although this was only 21.5 % of the total labour force, that is to say a decrease of about 5 % from the beginning of the century, these figures in fact represented an absolute increase in numbers. Marx and other economic historians thought that the process of enclosure had depopulated the rural areas of Britain and thus provided a cheap labour supply for industry, while the resulting labour shortage was responsible for stimulating the mechanisation of agriculture. But in fact both urban and rural populations were in expansion until about 1860 (J. D. Chambers (1)), and it was only with the rapid development of agricultural mechanisation after 1850 that the rural population began to fall significantly. In Britain, then, agricultural mechanisation was the cause, rather than the result, of rural depopulation.

[e] Ravenstone (1), writing in 1824; quoted by Karl Marx (1) (vol. 1, p. 556 of the 1976 Penguin edition).

Machinery can seldom be used with success to abridge the labour of an individual; more time would be lost in its construction than could be saved by its application. It is only really useful when it acts on great masses, when a single machine can assist the work of thousands ... It is not called into use by a scarcity of men, but by the facility with which they can be brought to work in masses.

By the early 19th century, these conditions were realised on many British farms, the majority of which covered between 100 and 1000 acres and were thus large enough to benefit from the costly investment of buying or hiring machinery.[a] Opposition from desperate agricultural labourers slowed down the rate at which farmers adopted machinery,[b] but it could not dampen the enthusiasm of the agricultural engineers, who saw very clearly which way the wind was blowing. Their inventiveness was stimulated by competitions run at various agricultural shows and exhibitions. By 1849 at least sixteen different firms in Britain were producing 'combined' threshing, winnowing and dressing machines;[c] steam ploughs, mechanical drills and reaping machines were also produced in a number of versions, and the intensity of competition between engineering firms served both to improve design and to lower prices.[d] In the second half of the 19th century, mechanisation had become a significant trend in British agriculture. Whereas in 1851 more than a fifth of the working force was in agricultural employment, by 1900 less than a tenth of the working population was employed on the land;[e] today the figure is nearer 2 %.

European mechanised cultivation and sowing methods ultimately derive from the Chinese technology and ideas introduced to Europe in the 17th and 18th centuries. But the Asian prototypes, even where they were directly copied, are unrecognisable in the sophisticated and elaborate equipment of modern Western farms, or even of late 19th century enterprises. Who would connect the multi-shared gang-plough (Fig. 269) or the elaborate drilling-plough (Fig. 270) with the wooden implements seen in the fields of North China? Today Western farm machinery is being introduced to China as part of the campaign to modernise

[a] Ernle (2), p. 156.

[b] Active revolt on the part of rural labourers broke out in the notorious Swing Riots of 1830, triggered by the threatened introduction of threshing machines; Hammond & Hammond (1). This revolt was very severely repressed, and thereafter outbreaks of violence were rare. Nevertheless the threat of unemployment continued to deter farmers from adopting machinery in many parts of the country, for the 19th-century British Poor Laws were such that well-to-do rate-payers were obliged to contribute heavily to the support of the unemployed in their parish. Farm labourers regarded parish relief as a degrading pittance, farmers as a drain on their purses which went to support idle layabouts, and so where the Poor Laws were in force many farmers preferred to forego mechanisation; E. J. T. Collins (1), p. 30.

[c] E. J. T. Collins (1), p. 19.

[d] For example the price of the award-winning general-purpose drill at the Agricultural Society of England Show was reduced from £53 in 1849 to £35.12*s*.6*d* only two years later; Vamplew (1), p. 204. S. Nielsen (1) shows how the cost and reliability of reaping-machines affected their adoption in 19th-century Denmark. A big step forward was the formation of contracting firms which hired out expensive machinery, for despite engineering improvements and reduced costs, small machines were generally less economical and efficient in real terms; E. J. T. Collins (1), p. 26.

[e] Mingay (2), p. 29.

Fig. 269. Gang-plough; Scott (1), fig. 36.

Fig. 270. 'Nonpareil' seed-drill; Scott (1), fig. 69.

agriculture, but how many Chinese realise that the new and alien machines are in fact related to their own cultural heritage?

## (2) Agricultural Revolution in China?

As we have already suggested, changes occured in Chinese agriculture almost as fundamental and far-reaching as those of 17th and 18th century Europe, but whether they merit the name of 'revolution' is a moot point. It is impossible to disassociate agriculture from its social context, for the nature of agricultural technology is inextricably linked with the relations of production; where agricultural change does not go hand in hand with social transformation, as in the case of the European Agricultural Revolution, may we really speak of these changes as 'revolutionary'? In the first case we shall consider below, that of North China in the Han, it appears that the technological changes involved did have the potential to transform both agricultural production methods and the organisation of society, but for the inherent contradiction between the fiscal demands of the state and the nature of labour organisation required for efficient production.

### (i) *Agricultural development and agrarian change in North China in the Han*[a]

The Han state economy was based chiefly on the regions of Kuan-chung[1] and Chung-yuan[2], that is the Wei River valley and the lower course of the Yellow River. State revenues derived principally from the land and poll taxes paid by individual peasant households, each provided by the state with a plot of land varying in size according to its fertility; the household was bound to this land and supposedly had only usufructuary rights over it.[b] Aristocrats and officials held much larger grants of land which were farmed by peasant families; the only rent in kind that the grantees were entitled to claim from the peasants was the proportion of the crop due to the government as land-tax (between 1/30 and 1/15 of the crop during the former Han); this the grantee was supposed to pass on to the government, but he was also entitled to claim a small cash sum from the peasant as his own income.[c] But the government knew well that large landowners not only charged their tenants exorbitant rents of half the crop or more,[d] but also withheld large sums due to the government. During the early Han constant efforts were made to reduce the power of the landed classes and to maintain direct control over the peasants.[e]

It was imperative for the Han government to increase agricultural production (and thus its tax receipts), for not only was the population expanding, but also

[a] This section was first published in *Early China* 5, Bray (3), and is reproduced here by permission of the editors.

[b] Ho Chhang-Chhün (*1*), chapter 1.

[c] Duman (1).

[d] *C/HS* 24A/15*a*; 19*b*.

[e] *C/HS* 14/3*b*; 24A/14*b*–15*b*.

[1] 關中 [2] 中原

expensive wars were being fought along the southern borders and in the Northwest and Central Asia, particularly during the reign of Wu-Ti[1]. An intensive campaign was therefore mounted both to improve agricultural methods and to expand the agricultural area. The campaign was specifically designed to benefit smallholders, and to improve peasant agriculture by providing the necessary technical inputs and physical infrastructure. The most famous official involved was Chao Kuo[2], whom Wu-Ti appointed *sou su tu wei*[3] or chief officer for military grain procurement[a] in −87. Chao clearly transformed agriculture in the metropolitan area of Kuan-chung (Shensi), but the basic infrastructure had been laid some years previously. The Chhin government had already built two considerable irrigation projects during the −3rd century, the Cheng Kuo[4] canal in Chhin (Shensi) itself and the Kuan-hsien[5] canal in Szechwan, and some have attributed the meteoric rise of Chhin to the economic success of these irrigation schemes.[b] Wu-Ti was the first emperor of unified China to realise the importance of water control, and he carried out an enormous programme of canal building in Honan and Shansi that irrigated over a million acres of arable land, while lesser projects were realised in Northwest China and the Wei and Huai valleys.[c] By the middle of Wu-Ti's reign the lower Yellow River valley was contributing six million bushels of grain to the exchequer in place of the few hundred thousand it had sent in the first years of the Han,[d] and productivity in the arid areas of the Northwest had also been raised considerably. It is important to remember that while the peasant smallholders undoubtedly benefited enormously from the finished irrigation works, they were also responsible for providing the corvée labour to construct the canals in the first place.

The government also provided peasant farmers with seed-grain, tools and draught animals, sometimes on credit, sometimes on loan, and sometimes as outright gifts.[e] It monopolised the iron industry in −117, putting it under the control of an experienced iron magnate,[f] with the intention of cutting out profiteering manufacturers and middlemen, lowering the price of iron tools and making them available to all.[g] Although opponents of the monopoly protested that such widespread government intervention would result in harmful standardisation of agricultural implements so that they no longer met local needs,[h] the wide variety of tools found in Han iron hoards from different parts of the country belies that argument, while the enormous number suggests that iron tools did in fact become available to almost everyone. The account of Chao Kuo's achievements indicates that considerable effort went into producing tools that were suited to local conditions,[i] though they were not necessarily those with which the farmer was familiar.

[a] Swann (1), p. 184.
[b] Chi Chhao-Ting (1), p. 75.
[c] *C/HS* 24B/9*a*–*b*.
[d] Chi Chhao-Ting (1), p. 77.
[e] *Hsi Han Hui Yao* 48.
[f] *C/HS* 24B/11*a*, 12*a*.
[g] E. M. Gale (1), p. 35.
[h] *Ibid.* p. 33.
[i] *C/HS* 24A/16*a*; Swann (1), p. 186.

[1] 武帝 [2] 趙過 [3] 搜粟都尉 [4] 鄭國 [5] 灌縣

Chao was in fact responsible for introducing a completely new cultivation system to Kuan-chung and the Northwest frontier, namely the *tai thien*[1] system.[a] There has been much debate about the actual nature of the *tai thien*,[b] but the *Han Shu* says that it was an ancient system,[c] and the description it gives, the nature of the implements associated with the *tai thien* system (namely turn-plough and seed-drill),[d] and the ascription of antiquity strongly suggest that it is none other than the *mu chhüan*[2] ridge-and furrow system described in the late Chou work *Hou Chi Shu*,[e] imported into Kuan-chung from lower down the Yellow River (see p. 105). Although primarily suited to heavy soils, the ridge-and-furrow system can also be used in lighter soils to good effect,[f] but in the loess soils of Northwest China erosion would quickly result unless the soil was kept moist and cohesive by irrigation. In −1st-century Kuan-chung the *tai thien* system is said to have increased yields by at least 1 bushel per *mu*, and in some cases almost to have doubled yields.[g] It immediately proved a great success and was practised not only in Kuan-chung but also along the Northwest frontier in the military colonies (*thun thien*[3]), as well as lower down the Yellow River.[h] Judging by the number of Han hoards and other evidence, the *tai thien* system with its adjuncts of mould-board plough and seed-drill remained popular in the Northwest and Kuan-chung for some time,[i] and where it was practised in conjunction with irrigation, erosion was probably minimal. But where irrigation systems did not exist or had fallen into disrepair, the comparatively deep ploughing must have led to a dangerous rate of erosion, and one suspects that the *thai thien* were abandoned as soon as the power of the central government dwindled. Ridging and the use of the turn-plough were not a feature of recent agriculture in the Northwestern provinces,[j] yet the seed-drill continued in use: sowing in rows has obvious advantages in the dry loess lands, for it saves both seed and water. So Chao Kuo and his colleagues did leave the farmers of the Northwest with at least one enduring benefit.

*Pressure on land*

The basic land allotment allowed for poor land to lie fallow two years out of three, and medium land one year out of two; only top quality land was supposed to be cropped continuously. But land was growing short and apparently in Kuan-chung fallowing was considered a last resort even in the early Han.[k] Fertilisers must therefore have been used quite extensively to prevent soil exhaustion, in particular pig and human manure.[l] In the metropolitan province the latter must

[a] *C/HS* 24A/16*a*–*b*; Swann (1), pp. 184–7.
[b] A review of the debate is given in Hara (*1*).
[c] 24A/16*a*.
[d] *CMYS* 1.19.1.
[e] *Lü Shih Chhun Chhiu* 26/8*a*, 9*b*.
[f] After careful research it was introduced into the savannah lands of Northern Gold Coast and Nigeria in the 1930's; Charles Lynn, pers. comm.
[g] *C/HS* 24A/16*b*.
[h] *Ibid.*
[i] Hsü Cho-Yün (2), p. 262.
[j] Anon. (*502*).
[k] *FSCS* 1.7.
[l] Hsü Cho-Yün (2), p. 260.

[1] 代田 [2] 畝甽 [3] 屯田

have been sufficiently plentiful to allow farmers to crop most land continuously. Silkworm, droppings, compost and ploughed-in weeds were also used as fertilisers in the early Han.[a]

The Han government encouraged more intensive cropping; for example, the cultivation of winter wheat in Kuan-chung,[b] which one suspects was a failure for climatic reasons,[c] and experiments were carried out with pit cultivation (*ou chung*[1]).[d] The pits were carefully watered and fertilised, and required no animal traction but only human labour: this must have been a great incentive at times when epidemics frequently deprived the peasants of their draught animals.[e] Pit cultivation could be practised on poor or marginal land, or where steep slopes ruled out conventional cultivation.[f] But although yields were reported to be phenomenally high,[g] pit cultivation seems to have been practised chiefly by gentleman-farmers or generals with plenty of conscript labour to spare rather than by peasant smallholders, presumably because the initial inputs of labour and fertilisers were beyond most small peasants' means.[h]

Despite this intensification the land already under cultivation could not produce enough to feed the growing population, maintain the large bureaucracy and sustain prolonged military campaigns. Land hunger (and the consequent exploitation of poor peasants) was growing in the central states, and a single crop failure could turn the small but steady stream of vagrants into a flood of refugees.[i] Here were ready candidates for opening up new land, and the government sometimes resettled several hundred thousand refugees at a time in sparsely populated areas such as Kiangsu or the banks of the Yangtze.[j] Considerable areas of crown or government land (*kung thien*[2]) in the central states were also made available,[k] but by the middle of the Former Han this supply was largely exhausted and the government generally settled refugees in border areas, particularly in the Northwest. During Wu-Ti's campaigns vast tracts of land in Shensi, Mongolia, Ninghsia and Kansu were recaptured from the Hsiung-nu and extensive areas of Central Asia were brought under Chinese dominion.[l] Colonisation by Han Chinese was desirable both to consolidate Chinese claims to the territory and to provide maintenance for the troops stationed there, and so military colonies (*thun thien*[3]) were set up throughout the Northwest. Originally these colonies were occupied only by soldiers and convicts, but they were soon joined by large numbers of displaced peasants (*liu min*[4]).[m] Irrigation was often provided, for example

[a] Shih Sheng-Han (2), p. 57.
[b] Hsü Cho-Yün (2), p. 260.
[c] J. L. Buck (2), p. 27; however, the Chinese climate during the Han dynasty was less harsh than it is today.
[d] *FSCS*, section 7.
[e] *C/HS* 24A/16*b*.
[f] *FSCS* 7.1.1.
[g] *FSCS* 7.3.1.
[h] Hsü Cho-Yün (2), p. 262 holds that the method was particularly appropriate for small farmers, but there does seem to be evidence to show that only farmers who disposed of a large labour force actually practised this method; *CMYS* 3.19.2.
[i] *Hsi Han Hui Yao* 48.
[j] *C/HS* 24B/15*b*.
[k] Hsü Cho-Yün (2), pp. 257–8.
[l] *SC* 110, 123.
[m] *Hsi Han Hui Yao* 56.

[1] 區種 [2] 公田 [3] 屯田 [4] 流民

in the Shuo-fang[1] area in the northern bend of the Yellow River, and colonists were provided by the government with food, seed-grain, animals and implements either as gifts or on credit, while for several years they were not required to pay taxes. The *thun thien* were found to be both productive and effective, and the institution was regularly revived by subsequent dynasties (see p. 96).

The Han government also encouraged a 'march to the tropics', sending Northern farmers to settle in the Yangtze area and even as far south as Annam. Han writers were usually disdainful of southern agricultural methods and sought to impose their familiar northern ways on their new subjects;[a] yet, as we have seen, there is clear evidence that a quite separate southern tradition, based on the cultivation of wet rice, was already well advanced by Han times.

*Decline of the smallholder economy*

Despite the sophistication of rice agriculture in some areas of Southern China, the northern plains and northern crops remained the basis of the Han economy. Although some new varieties of local crops had been evolved and many new species were introduced in the Han from Central Asia, their impact was small during the early Han. Peasant farmers continued to grow chiefly millet and wheat or barley (the crops in which taxes were paid), various beans (in case of poor grain harvest),[b] and hemp for textile production (for their own use as well as for tax purposes). It is not surprising that the range of crops was so restricted, for the standard allotment of 100 *mu* produced barely enough to feed a farming family once their taxes had been paid.[c] Even though land productivity was raised in many areas, pressure on land due to population growth must rapidly have reduced the size of government allotments (though this is not stated explicitly in the sources), and the peasants' livelihood became increasingly precarious. They could hardly hope to produce a marketable surplus of their normal crops, let alone afford to experiment with new ones. In any case, the early Han government regarded trade as a pernicious drain on the resources of the state and introduced many discriminatory regulations on merchants and petty traders, including heavy taxes on carts and boats, which hampered all but the most determined.[d]

The new farming technology required quite high investment in equipment and livestock which, despite government loans, would certainly deplete the resources of many peasants, particularly of small families who might also need to hire extra labour at ploughing or harvesting time. The government attempted to protect independent peasants by regulating prices[e] and providing relief when crops failed,[f] yet many rapidly fell into debt[g] and were forced to sell their land to work as petty traders, tenant farmers or agricultural labourers.[h] The value of land rose

a Schafer (16), pp. 9–17; *CMYS* 0.7.
b *FSCS* 4.6.
c Chhü Thung-Tsu (1), p. 109.
d Chhü Thung-Tsu (1), p. 118.
e *CJHS* 24B/17*b*–18*a*; Swann (1), p. 314.
f Swann (1), p. 57.
g *Ibid.* p. 392.
h *CJHS* 24B/19*b*; Swann (1), p. 322.

1 朔方

rapidly, not only because of population pressure but also because of its increased productivity under the new technology, and there was no lack of ready purchasers. By the end of the Former Han, despite Wang Mang's[1] efforts at reform, it seems that landed estates had ousted smallholdings as the dominant unit of production in North China.

The policies of monopolists and 'legalists' like Sang Hung-Yang[2], which included measures for prices and currency regulation as well as the famous (or infamous) monopolies on salt and iron, were bitterly opposed by the rural gentry.[a] They claimed that the interventionist policies were inefficient and extortionate, financing useless wars instead of allowing the people to prosper undisturbed, and they accused the ministers responsible of profiteering.[b] It is true that these 'legalist' policies were inspired less by disinterested concern for the people's welfare than for the maximisation of state income, and certainly both profiteering and hardship must often have occurred. Yet it was clearly in the legalists' interest to foster an independent and productive peasantry, and their policies did at least guarantee the peasant farmer a modicum of protection from rapacious merchants and landowners. The Confucian politicians, on the other hand, for all their pious discourses on the honour due to the peasant class and the general peace and prosperity that could be expected to appear spontaneously once interventionist policies were abandoned, belonged to the landed gentry. They claimed to express 'the grievances of the people',[c] but naturally they had their own interests at heart too, and once the 'legalists' fell from favour the way was clear for the gentry to profit from the free market in land and other opportunities. Although there were fortunes to be made in commerce,[d] investment in land was particularly attractive. Productivity was now high and labour was easily found, while with the collapse of government controls the market in agricultural produce was flourishing. Furthermore, in the Confucian ethic it was preferable for a 'gentleman' to make his living by farming, in which case he was devoting himself to the 'fundamental' pursuit and 'contributing to the people's livelihood', rather than by vulgar trade, which was considered 'secondary' if not actually pernicious.

In −44, after a series of floods and crop failures which the Confucians hastened to blame on the government 'striving with the people for profits', the interventionist measures and policies were repealed[e] and the peasants were abandoned to the mercy of the Confucian ethic and the landed gentry. In −7, Shih Tan[3] complained at court that 'the fortunes of overbearing persons of wealth among officers and people have grown to an estimated amount of [as much as] several hundred millions, [while] in contrast the poor and weak have become the

[a] Swann (1), pp. 24 ff.
[b] E. M. Gale (1), p. xxix.
[c] *Ibid.* p. 1.
[d] Swann (1), pp. 425–60.
[e] *C/HS* 24A/18*a*; Swann (1), p. 197.

[1] 王莽 [2] 桑宏羊 [3] 師丹

more straitened'.[a] Various proposals to regulate land ownership or even (under Wang-Mang) to nationalise land and slaves failed: under such threats land prices fell immediately, powerful families protested, and the proposals were shelved *sine die*.[b] Not until the 'equal allotment' (*chün thien*[1]) regulations of the Northern Wei were the peasants, temporarily, to regain their land.[c]

*Growth of large estates*

Large landowners were, potentially at least, better able to profit from the new agricultural technology and methods propagated by the government than the smallholders that the innovations had been designed to help, for the size of their estates and the amount of labour of which they disposed permitted considerable rationalisation. A single peasant would normally dispose only of family labour and would find it difficult to keep even a single ox, let alone the minimum of two required for ploughing in Han times (see p. 180); often he would have to borrow or share his neighbours' animals and hire or borrow labour, and this could cause difficulties during peak agricultural periods. The late Han murals from Mongolia and elsewhere show that estate agriculture suffered from no shortage of draught animals (Fig. 264), so ploughing and sowing would not be held up beyond the most suitable season just for lack of animal power. Similarly labour could be deployed more efficiently. The landlord was not obliged to devote all his acreage to producing crops for subsistence and tax, for as often as not he managed to avoid paying a substantial proportion of his dues. This meant that proper crop rotation was possible: a crop of wheat or millet that exhausted the soil could be followed by sowing a nitrogenising crop such as soya or alfalfa, much to the advantage of overall yields.[d] Marginal land could be planted with timber or bamboo, which looked after themselves,[e] rather than being carefully nursed to produce a few ears of grain. A peasant farmer struggling to make ends meet could not afford to experiment with new crops, but a wealthy landlord with grain in his barns and gold in his coffers could afford not only to try out new crops but to produce them on a large scale if they turned out to be a commercial success. Many Central Asian crops, for example broad beans, sesame and alfalfa, became popular in the later Han,[f] while commercial crops such as indigo were grown in considerable quantities.[g] Rice was still a luxury rather than a staple in North China, and the glutinous varieties (from which wine and cakes were made) seem to have been grown as commonly as ordinary rice; here technological progress

[a] *CJHS* 24A/18*b*; Swann (1), p. 201.
[b] *CJHS* 24A/19*a*; Swann (1), p. 203.
[c] Han Kuo-Pan (*1*), chapter 3. The foreign rulers of the Northern Wei felt severely threatened by the power of the Chinese landed gentry/bureaucracy, and one of the ways in which they sought to reduce their influence was by redistributing land.
[d] *Chhi Min Yao Shu* is in fact the earliest work to specify such rotations systematically.
[e] *SMYL* 1.6.
[f] *SMYL* 1.8, 2.4, 7.3.
[g] *SMYL* 5.5; Hsü Cho-Yün (2), p. 266.

[1] 均田

was made on the estates, for the first reference to transplanting is found in the +2nd-century *Ssu Min Yüeh Ling*.[a] The taste for wheat-flour products such as noodles and dumplings was spreading rapidly, especially in urban areas,[b] and late Han estates were able to take advantage of the trend not only by turning over more of their land to wheat but also by investing in water-mills and other equipment for grinding flour.[c]

Eastern Han estates were clearly geared to market production. Some transactions were probably just local, such as the seasonal speculation in grain advocated by Tshui Shih[1],[d] but some products must have been intended for a much wider market—indeed some scholars have argued for a network over 200 km in radius.[e] Hsü Cho-Yün believes that it was the peasant smallholder enterprise that was integrated into this market network and that the 'manorial economy', which was self-sufficient, only appeared as a defence mechanism in times of unrest.[f] It seems clear to us that the very opposite was true. The independent smallholders of the early Han not only were financially unable to undertake the risks of market production, they were actively discouraged by the government from producing anything but essentials. On the other hand the large estates were in an ideal position to play the market, and did so whenever the political situation allowed, but in times of civil unrest they could afford to draw in their horns and become self-sufficient, diverting some of their labour force from agricultural production to para-military training for defence purposes.[g] The institution was sufficiently flexible to re-adapt quickly to market production as soon as peace returned.

The question still remains of how labour was organised on these estates, and indeed of their layout. If a landlord acquired his property purely by buying out local smallholders as they fell into penury, then a map of his estate might well resemble a patchwork quilt;[h] but if he started off with a substantial grant of several hundred acres of land, as most Han aristocrats and officials did, and by an adept combination of economic pressure and coercion induced his poorer neighbours to sell their land to him, then his estate would cover an increasingly large but continuous area (like the estates of post-Enclosures England). The latter is more desirable if farming is to be organised efficiently in North China, using draught animals and comparatively large pieces of equipment, and Han sources indicate that a large part of most estates was centrally organised, but perhaps the periphery consisted of scattered fields which, being more difficult to reach, were leased out to tenants who farmed them as they wished.

The nature of labour on Han estates has been much debated. It has been

[a] *SMYL* 5.5.
[b] Yü Ying-Shih (1), p. 81.
[c] Vol. IV, pt 2, pp. 189 ff., 391 ff.
[d] Herzer (1), p. 51.
[e] Utsunomiya (*1*), pp. 349–53.
[f] (2), p. 268.
[g] For a brief account of the role of 'guests' (*kho*[2]) and personal troops (*pu chhü*[3]) as seen in the Han sources, cf. Chhü Thung-Tsu (1), p. 132.
[h] Fei & Chang (1), p. 154.

[1] 崔寔 [2] 客 [3] 部曲

proposed that they were worked by slaves,[a] by serfs or bonded peasants,[b] or by free tenant farmers.[c] Positive evidence is scanty. There are references in Han literature to sharecrop tenants[d] and to hired agricultural labourers,[e] as well as frequent mention of slaves whose duties are not usually specified but who are generally assumed to have played no important role in agricultural production.[f] There are also references to tenants on princely estates paying fixed amounts of cash to their lords during the early Han,[g] but there is no mention of cash rents in later texts.

To narrow down the possibilities of how labour might have been organised on the estates, it helps to consider their economic role, which we have argued was geared to market production of a wide range of crops. Farming on at least part of the land must have been organised under strict central supervision (there are references to bailiffs (*jen thien che*[1])[h]), and included many non-subsistence crops. This part of the estate is unlikely to have been farmed by sharecropping tenants: first, sharecroppers are dependent on their own produce for subsistence and could not afford to grow commercial crops; and secondly, the division of the commercial part of the estate into small, separate plots would pose complex problems of management. On the other hand, there is plenty of evidence that the institution of sharecropping was widespread in Han times, so perhaps the poorer or peripheral land on an estate was leased out to sharecroppers whose rent in kind provided the estate with food, while the central or more productive areas were controlled directly by the landlord and worked by serfs and/or hired labourers.

There is far too little evidence to assert that one of these categories of labour was more typical than another. The Han empire lasted for four centuries and covered an area roughly the size of Europe, and the nature and status of the agricultural labour force must have varied greatly according to place and time. In some newly colonised areas, for example, it is certain that tribal peoples were enslaved and made to work in the fields as a matter of course,[i] but although debt slavery was common in Han times in the central areas of China, the Han government regarded the enslavement of Chinese as an unnatural degradation and debt slaves were often freed by royal decree.[j] Obviously the Han was in some respects a period of transition, and in any case the categories of slave, serf and tenant are not necessarily distinct in practice even when terminological distinctions are drawn.[k] However, it seems likely that the rural population in late Han times had considerably less personal and economic freedom than in early Han, and we can be fairly sure that technological and productive levels were generally higher on the central manorial estates than on tenant farms or individual smallholdings.

[a] Yang Lien-Sheng (*1*); Nishijima Sadao (*1*).
[b] Ho Chhang-Chhün (*1*); Duman (1).
[c] Utsunomiya (*1*); Ebrey (1), (2).
[d] *C/HS* 24A/19*b*; 99B.
[e] Chhü Thung-Tsu (1), pp. 347, 356, 368, etc.
[f] Duman (1).
[g] *Hsi Han Hui Yao* 34.
[h] *SMYL* 12.3.
[i] Schafer (16), pp. 33–4.
[j] Chhü Thung-Tsu (1), p. 158.
[k] Weber (4); Yeh Hsien-En (1) discusses the considerable problems of terminological confusion that the Chinese documents present.

1 任田者

*Inherent contradictions*

In his systematic studies of Han agriculture, Hsü Cho-Yün has shown that agricultural advancement was closely related to population growth.[a] Adopting Boserup's inversion of Malthusian theory,[b] he argues that many Han improvements in agricultural technology were known in China previously but were only widely adopted when population pressure forced the government to mount campaigns for agricultural improvement. However, he also assumes that the individual peasant enterprise remained the dominant unit of production throughout the four centuries of Han rule. He maintains that Han peasant farmers were normally integrated into a large-scale market economy, but says that in periods of unrest rural communities were often reduced to self-sufficiency and that at such times 'since such a small community was often under the dominance of local leaders, the self-sufficient economy exhibited the characteristics of a manorial system ... However this isolation and self-sufficiency were only temporary phenomena.'[c] We have argued, on the contrary, that the dominant mode of production changed radically during the course of the Han. During the early Han the government took the independent peasant farm as the basic unit of production on which its tax system was founded, and tried in its own interests to eradicate large estates. It provided an infrastructure which permitted individual peasants to increase production, together with a series of interventionist measures aimed at guaranteeing their independence and security; this system encouraged the production of essential food and fibre crops—indeed peasant marketing was actively discouraged. But with the rise of powerful families during the later Han, the independent peasant farm was superseded as the basic unit of production by the 'manor'. On the technological base already provided by the state, the manorial estates were able to develop a system of consolidated farming suited to commercial production, which brought Northern Chinese agriculture to its technological apogée in the late Han and Northern Dynasties.

During the Former Han, when the government's centralised power was at its height, the productivity of North Chinese peasant farmers must have reached an unprecedented level, stimulated by official investment and aid. But the very success of the system brought about its downfall: the increase in land productivity soon tempted the rich and powerful to build up vast estates at the expense of peasant holdings. The nature of northern agricultural technology encouraged economies of scale, culminating in the sophisticated systems of commercial farming described in the +6th-century *Chhi Min Yao Shu*. Despite repeated attempts by successive governments to re-establish free peasant agriculture, estate farming remained the basis of the North Chinese economy for centuries, even during periods of civil disturbance when the estates abandoned commercial agriculture in favour of self-sufficiency. But for success such estates depended on active and efficient supervision; their decay is little documented and its course is unclear, but

[a] (1), (2).
[b] (1), p. 11.
[c] (2), p. 268.

probably the shift of the economic centre to the Yangtze area in the mid-Thang, with the consequent decline of public and private interest in the less productive North, was just as much to blame as the many wars and invasions that ravaged the area.

Like the estate agriculture, peasant farming in North China also declined in later dynasties. Both reached a level in the Han and Northern dynasties that was not to be surpassed until the creation of the People's Communes: a comparison of the +6th-century *Chhi Min Yao Shu*, or even of the −1st-century *Fan Sheng-Chih Shu*, with northern agricultural treatises of the 18th and 19th centuries[a] shows a marked decline in the later period. The poverty of the Northern regions compared with the more fruitful South was documented statistically earlier this century by J. L. Buck (2), but it struck foreign visitors forcibly as early as the 15th century.[b] With so little left to attract landlords, it is not surprising that the Northern peasants regained much of their independence as the situation deteriorated: in the 1920s, only one-sixth of the land in North China was rented rather than farmed by the owner, compared with nearly half the cultivated area in the South.[c] Yet during the early phase of its expansion, the technological constraints of North Chinese dryland agriculture seemed likely to produce a transformation in social organisation rather reminiscent of the process of enclosure in England, whereby smallholdings were superseded by large estates worked by landless labourers. It is interesting to speculate whether, in the absence of foreign invasions and economic competition from the South, the dominant class of northern landlords would have succeeded in changing the structure of the state sufficiently to resolve the conflict of interests between landowners and government, and thus achieved a permanent transformation of class relations, unbroken by 'equal allotment' land reforms and other official measures aimed at restoring to the peasantry some degree at least of economic independence.

### (ii) *The 'Green Revolution' in South China*

The most striking period of development of Southern Chinese agriculture began in the Sung, when the government initiated a series of development policies so sweeping in scope and in result that they may well be compared to the so-called 'Green Revolution' of Asia today.[d] The similarities are quite striking. 'Green Revolution' is the name given to a technological package (consisting of new high-

[a] E.g. Wang Yü-Hu (2).

[b] Meskill (1), p. 15.

[c] J. L. Buck (2), p. 195.

[d] Though the agencies and governments in question see their policies as aimed primarily at purely technological transformation, the term 'Green Revolution' is the more apt in that the social side-effects have often been profound and very unsettling, indeed in some cases it has even been suggested that the Green Revolution was turning red. F. Frankel (1), and other studies of the social disruption and exacerbation of class tensions resulting from the introduction of the new technology, show that such effects have been much more marked in the wheat-producing areas of Asia; where rice is the chief crop, not only is technological transformation usually slower but also social differentiation as a result of the new policies is less marked, for reasons that we hope to make clear later in this section.

yielding varieties of wheat and rice together with fertilisers, the provision of irrigation facilities, and so on) developed by international scientific organisations and adopted by most non-socialist Asian nations in the 1960s and 1970s, in the hope of achieving rapid increases in agricultural productivity. The governments in question encourage peasant farmers to adopt the new technology by constructing new irrigation schemes and by building agricultural development centres to distribute seed and fertilisers and to provide instruction and credit facilities. Surprising as it may seem, Sung China not only faced similar problems to those of the modern Asian nations, but tackled them in a remarkably similar fashion.

The economic centre of China first began to shift from the dryland plains of the north to the Yangtze regions during the later part of the Thang. Fear of Khitan and other nomadic invaders drove thousands of peasants to abandon their lands in the North, and by the Sung dynasty the greater part of the population lived in the southern provinces.[a] The Chinese government was faced with the double problem of feeding an increased population on a greatly reduced area, and of maintaining large armies to protect its borders.[b] It was quick to perceive the necessity for increasing rice production in the southern provinces, and it approached the problem on several fronts simultaneously.

One of the most famous measures taken was the introduction to the Yangtze Delta by Chen-Tsung[1] in +1012 of new varieties of quick-ripening rices from Champa. We have seen how this transformed production patterns, allowing double-cropping of rice or the alternation of summer rice and winter wheat (p. 492). Seeds of the new varieties were distributed to farmers through the district yamens, and written instructions on their cultivation methods were circulated. These were presumably intended not for the peasants themselves, most of whom would be illiterate, but for the Sung equivalent of a rural extension officer: 'master farmers' (*nung shih*[2]) were local farmers chosen for their skill and experience to fill a minor official post which carried the duty of improving agricultural techniques in their villages. They were to instruct their peers not only in new techniques such as improved sowing and fertilising methods or crop choices, but also in the organisation of mutual aid and so on. They had the right to report peasants who failed in their duties to the district authorities, and they themselves

[a] Sung population figures are quoted in numbers of households rather than individuals, which makes exact interpretation difficult. The census of +1080, that is before the population had been driven south by Jurchen and Khitan invaders, put the population at 14·5 million households, of which 10 million lived in the southern provinces. By +1173, when the Jurchen Chin dynasty ruled North China, the total population of both Sung and Chin was 18·75 million households, of which 12 million lived in the South under Sung rule; Lewin (1), pp. 45 ff.

[b] Golas (1), p. 295 estimates that the total area of cultivated land around +1100 was 7 million *chhing*, while the Thang figure for +755 was over 14 million *chhing*, that is roughly double the amount. The Thang figure was almost certainly too high, as Twitchett (4) points out; nevertheless, it is clear that a great deal of arable land was lost to the northern invaders.

[1] 眞宗 [2] 農師

were absolved from corvée and other services.[a] It was presumably these 'master farmers' who channelled to the ordinary peasants the information contained in the agricultural books newly commissioned or reprinted on government order. *Chhi Min Yao Shu* and *Ssu Min Tsuan Yao* were the first agricultural works printed by imperial order in the early 10th century; other Sung works distributed officially include the *Keng Chih Thu* and Chhen Fu's *Nung Shu*, as well as many other lesser-known titles (see the section *Sources*, pp. 49 ff.). These contained information on better cropping practices, new tools, machines, fertilisers and irrigation methods.

The Sung government offered financial incentives to its farmers to invest in improvements. Most famous among these was Wang An-Shih's[1] Green Shoots Policy (*chhing miao fa*[2]), which for the short time that it was in force provided loans to farmers at far lower rates than they could hope to obtain elsewhere.[b] As Golas (1) points out, 'despite its stormy history, this reform may well have boosted agricultural productivity significantly by providing middle and smallholders with land improvement capital that would otherwise have been unavailable to them'. Not only did the government provide loans but, still more important, it refrained from overtaxing:[c]

> A number of tax policies were designed specifically either to stimulate agricultural production or at least to keep from hindering it. Agricultural products and implements were regularly exempted from commercial taxes. The government sometimes rejected commutation of grain tax payments, on the grounds that this would weaken the people's incentive to practise farming ... Tax breaks were often used to encourage the opening of new or derelict fields. In assessing the fertility of land for tax purposes, efforts were made to distinguish between the natural fertility of the land and improved fertility brought about through the efforts and expenditures of the farmers, so that farmers who tried to improve their fields were not threatened by higher taxes ...
>
> At least in the southeast, official rates of taxation under the biannual tax (*liang shui fa*[3]) were often at or even below what the government saw as the ideal: a modest one-tenth or so of the harvest. Of course, to these rates must be added a whole range of miscellaneous and supplementary assessments (the rates of which were moving inexorably upward), labour service obligations, and all kinds of chicanery on the part of the tax collectors. Still, the remarkable fact that the Sung is the only period in China before modern times when receipts from land taxes dipped below fifty percent of total central government receipts suggests very strongly that, as compared with other dynasties, Sung agricultural taxation on the whole promoted agricultural productivity.

[a] *Sung Shih* 173/4*a*; Yoshioka Yoshinobu (*1*). Lewin (1), p. 73 points out that historical precedents for such an institution are to be found in Han texts, while this particular post was said by Ma Tuan-Lin[4] to have been established in +982.

[b] Higashi Ichio (*1*); James T. C. Liu (2); H. R. Williamson (1) all discuss the fate and implications of Wang's reforms.

[c] Golas (1), p. 311.

1 王安石 2 青苗法 3 兩稅法 4 馬端臨

The Sung government also undertook land reclamation on an extensive scale, aided in large measure by improvements in irrigation techniques and water control.[a] It set up numerous agricultural colonies where new, improved techniques were put into practice. Some were military colonies (*thun thien*[1]) set up in frontier areas, others were civilian colonies (*ying thien*[2]) established on crown lands to settle refugees or landless peasants. On most such schemes irrigation facilities were constructed, and modern equipment and new seed varieties were provided to the settlers free or at very low cost. From +988 on, over 18,000 soldiers were settled on colonies in Hopei, in individual settlements of 200 men; dykes were built along 600 *li*, as were irrigation channels, and high yields of rice were obtained. Similar colonies were set up along the frontier in Shensi, where the fields had to be protected by fortifications. By +1021 it was said that 4200 *chhing* of *thun thien* had been established, but though this figure seems very large it in fact only represents under one-thousandth of the total cultivated area.[b] A similar proportion of land was given over to civilian colonies.[c]

*Peasant response to agricultural development*

The Sung government obviously played a crucial role in stimulating improvements in agriculture, but perhaps its chief success was the way in which the rural population, alive to the benefits of the new technology, were willing to experiment and improve on their own initiative. There was some initial resistance to innovation; for example, some peasants objected to double-cropping because they feared the extra work involved would not be justified by the increases in yields, while landlords feared that it would wear down the fertility of their soils.[d] But as more commercial fertilisers became available and varieties improved, these objections faded away. Peasants bred locally new and improved varieties of rice and other crops, some of which travelled from hand to hand over vast distances (see p. 493 above), landlord and lineage associations reclaimed lakeside marshes, building dyked or poldered fields, while land-hungry peasant families opened up hillside terraces or migrated to the wide, fertile plains of the sparsely-populated middle Yangtze, taking with them improved seeds and advanced technology.

The Sung 'Green Revolution' had its roots in the most populous, agriculturally and economically advanced areas of China, the Lower Yangtze provinces of Kiangsu and Chekiang and the coastal province of Fukien. By the Yuan and Ming the changes had gained their own momentum and the new technology was spreading to less developed areas such as Anhwei, the Yellow River plains and the

[a] An excellent bibliography of the daunting literature on Sung advances in water control is given in Golas (1), p. 298.

[b] *Chhin Ting Hsü Thung Chih* 153/1.

[c] As from +1002 the distinction between civilian and military colonies was dropped and all were known as *thun thien*; Lewin (1), p. 81.

[d] Sudō Yoshiyuki (*4*), pp. 278–9.

[1] 屯田 [2] 營田

deep south, while in the early Chhing agricultural innovation and intensification were proceeding in Hupei, Hunan, and the Southwestern provinces, areas that had hitherto been sparsely populated but now attracted vast numbers of migrants from the agriculturally advanced and overcrowded Southeast and the central plains. It was not until about 1800 that the geographical limits of expansion were reached and population growth overtook agricultural production, leading to the rural impoverishment with which the works of 20th-century authors such as F. H. King (1), J. L. Buck (1), (2), and W. Hinton (1) have made us familiar.

It is clear that, as a result of the innovations described above, there was a rapid upsurge in agricultural productivity in Sung China. Improved yields and multi-cropping of staple grains produced unprecedented surpluses, and as a result it was possible for commercial cropping and rural industry to develop on a scale hitherto unknown. Rice was exchanged for charcoal, tea, oil, wine and other locally produced goods at the village markets that sprang up all over the country, while a vigorous national trade in these and other commodities permitted intensive regional specialisation;[a] indeed, some areas could afford to give up cereal farming altogether, since their needs were amply supplied by the long-distance trade in rice. Chhen Fu describes the importance of sericulture in the neighbourhood of 12th century Suchou:[b]

> Some of the people there rely exclusively upon silkworms for their living. A family of ten persons will raise ten frames of silkworms. From each frame twelve catties of cocoons will be obtained, and from each catty comes 1·3 ounces of silk thread. From every 5 ounces of thread one length of small silk may be woven, and this exchanges for 1·4 piculs of rice. The price of silk usually follows that of rice. Thus supplying one's food and clothing by these means ensures a high degree of stability.

One important commercial crop was sugar, which was especially popular among farmers in Fukien, Szechwan and Kwangtung. The *Thang Shuang Phu*[c] records that in one area of 12th-century Szechwan as many as 40% of the peasants were engaged in growing sugar-cane. Sugar was well established in Fukien by the Sung, and by Ming times it had totally supplanted rice in several districts. Fukien sugar was sold not only in China but also throughout Southeast Asia.[d] Other commercial crops included tea, vegetables, fruit, timber, oil-seeds, dyes and fibre-crops, bamboos and (after 1500) tobacco. There was also a very marked increase in manufacturing, most of which took the form of 'cottage industries'. The farmer's wife had traditionally been responsible for spinning and weaving, not only for her own family's use but to pay that part of the tax dues which was levied in cloth. The silk industry had previously been small and predominantly city based, under official control, but during the Sung it expanded rapidly, especially in the Southeastern provinces and Szechwan; some areas specialised in rearing silkworms or in growing mulberry leaves, others in making a

[a] Shiba (1), (*1*) are the pre-eminent studies of Sung commerce.
[b] *CFNS*, p. 21, tr. Elvin (2), p. 168.
[c] 3*a*.
[d] E. S. Rawski (1), p. 48.

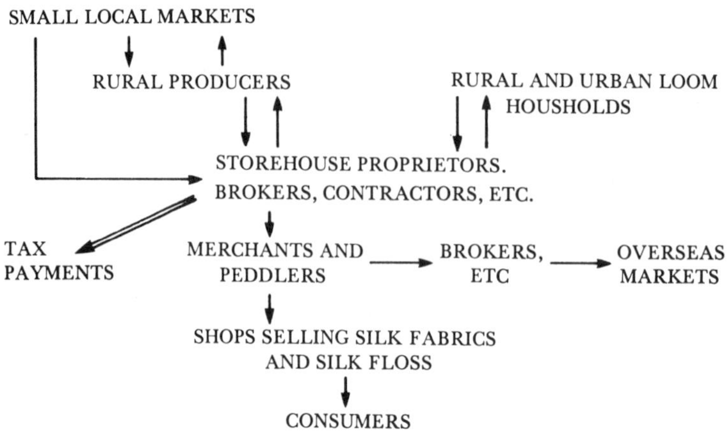


Fig. 271. The structure of the Sung silk industry; Shiba (1), p. 121.

particular type of silk.[a] Much of the silk cloth was woven in peasant households; brokers provided the silk thread, paid the women for their work and marketed the cloth[b] (Fig. 271). Ramie and cotton, especially the latter, first became important in the Yuan and Ming. The cotton industry was run along similar lines to the silk industry: there was a national market for raw cotton, which peasant women bought from traders at local markets, span and wove, and then sold back again, presumably to the same merchants, in the market where they had bought the raw materials:[c]

> It is not only in the country villages that spinning is found, but also in the county capital and market towns. In the morning the village women take the thread which they have spun and go to the market, where they exchange it for raw cotton, with which they return. The following morning they again leave home with their thread, never pausing for an instant.

In 16th-century Fukien merchants ran a flourishing trade in cotton imported from Chekiang (Fukien was ill-suited to cotton cultivation), exchanging it for sugar; some cotton thread used in Fukien was imported from as far away as the Philippines.[d]

[a] D. Kuhn (3) gives details of Sung spinning and weaving technology, and remarks that the rapid expansion of silk production in the southern provinces during the Sung was probably stimulated by the adoption of the more efficient treadle-operated looms, which had been used in Shantung in earlier times but were unknown in the South until the arrival of northern refugees.

[b] Shiba (1), pp. 111–21.

[c] *Shang-Hai Hsien Chih*[1], 1750 ed., 1/21*a*, tr. Elvin (2), p. 273.

[d] E. S. Rawski (1), p. 74.

[1] 上海縣志

Other industries included paper-making (mainly from bamboo), the production of lacquer wares, metal goods, charcoal, and comestibles such as wines, spirits, bean-curd, sauces and pickles. Again, almost all production was on a household scale, and most of the producers were farming families.

*Relations between social and economic change*

Increases in the productivity of wet-rice land, in the absence of such modern inputs as chemical fertilisers, pesticides, or mechanisation, generally require markedly higher inputs of labour, and it has often been said that as land productivity increases, returns to labour diminish. It is certainly true that the transition from direct sowing of rice to transplanting, or from single to double cropping, necessitates increased labour inputs, but this does not necessarily mean that labour productivity decreases. Yields may increase more rapidly than labour demands. Improvements in crop varieties, irrigation technology and fertilising methods may all improve yields without increasing labour requirements; furthermore, since a smaller area of land will produce the same output, the work of tilling and weeding will be cut down. Naturally there does come a point where returns begin to diminish. Geertz (1) describes the extreme case in colonial Java, where so little land was available for food production that farmers were obliged to nurse each plant individually in order to produce sufficient to feed their families from their tiny plots; Geertz gave the name 'agricultural involution' to the process whereby, as the population of Java grew, more and more labour was devoted to refining techniques of rice cultivation in return for proportionately minute increases in yields. Elvin (2) sees a similar process of involution as one of the chief factors in China's failure, despite its economic and technical sophistication in the medieval period, to 'take off'.

Yet while it is true that in the period between 1000 and 1800 China experienced no fundamental historical changes in economic relations or in class structure, no quantum leap from handicraft production to industrialisation, such as we observe in Europe, it would nevertheless be quite wrong to dismiss this period as one of technological and economic stagnation. Not only did agricultural productivity manage to keep pace with rapid increases in population, but local industries and trade expanded, and the prosperity and stability of the Chinese economy continued to impress foreign visitors from Marco Polo to the 18th-century Jesuits. The sustained economic expansion of these eight centuries sprang from the technological innovations of Sung agriculture.

In our view, there are two fundamental questions about the Sung economy that require explanation if we are to make sense of the subsequent course of China's development. The first is how the industry and commerce that sprang up in the wake of the Sung 'agricultural revolution' were organised. The second is why the rapid upsurge in production and economic activity did not result, as it so often has elsewhere, in a severe disruption of existing economic relations, in fact a transformation of the relations of production, a true economic revolution. Why was the form of economic organisation that crystallised in the Sung both so dynamic and so resilient? Notions such as Elvin's 'high-level equilibrium trap'[a] do not, in our opinion, satisfactorily account for the stability of a society in which rapid economic expansion and diversification failed to produce any significant changes in the relations of production, but rather consolidated them. This phenomenon is not confined to medieval China alone. We believe that the nature of China's medieval economy was closely related to the peculiar characteristics of wet-rice cultivation (on which the state economy depended), and that parallels may be drawn between economic organisation and trends in medieval China and those in other countries where the economic staple was (or is) also wet rice.

Wet-rice agriculture has enormous potential for increasing land productivity. As we have seen, most improvements are either scale-neutral and relatively inexpensive (improved seeds, fertilisers such as oil-cakes or river-mud), or involve increasing inputs of labour (transplanting instead of broadcast sowing, multi-cropping, more careful weeding or irrigation). For technical reasons the optimal size for an irrigated rice field is very small, less than one-sixth of an acre (see p. 107), the fields are separated by bunds, and unlike in dryland cereal agriculture there is little scope for introducing animal-powered machines or other economies of scale. As production intensifies and the size of holding necessary for subsistence decreases, a typical farmer's equipment may be reduced to a few hoes and a sickle,[b] but this by no means implies that his techniques are primitive, nor is it necessarily a reflection of poverty; it simply means that animal-drawn implements are not suited to such small-scale, highly-skilled farming techniques. Since efficiency depends less upon the range of equipment than on the quality of labour, a skilled and experienced smallholder or tenant farmer is in just as good a position to raise the productivity of his land as a wealthy landlord. Indeed, as productivity rises the costs of adequately supervising the many minute tasks involved in wet-rice farming become prohibitive: inspecting an irrigated field for weeds is almost as onerous as weeding it oneself. So although prices of rice-land rise as production is intensified and yields increase, and there are often very high rates of tenancy in areas where wet rice is intensively farmed, the difficulty of effective supervision means that landlords find little or no economic advantage (as is usually the case where dry crops are grown) in evicting their tenants to run large, centrally-managed estates. Instead they generally prefer to leave their tenants to manage their small farms independently. There is evidence to show that as the new

[a] Many objections have been raised to Elvin's arguments (2), in particular to his explanatory device of the 'high-level equilibrium trap', according to which any technical improvements to late medieval China's highly developed agricultural and manufacturing technology would run into such sharply diminishing returns as to discourage any further attempts at innovation or improvement; see Sivin (17); J. S. & D. C. Major (1). In our view, one of the major deficiencies of Elvin's explanatory device is that, although it might be applied to economic behaviour in a society that was approaching the limits of expansion, it does not tackle the root causes of the situation and explain why no historical change occurred in the period from 1000 to 1400, when by Elvin's own admission South China was still in a phase of rapid technical expansion and innovation.

[b] Fei & Chang (1).

technology spread through the southern provinces of China the position of tenants *vis-à-vis* their landlords improved: they acquired more managerial and economic independence and tenurial contracts were modified in their favour.

Many large estates were amassed during the economic expansion of the Sung. There is considerable controversy as to how such estates were managed, whether they were consolidated or dispersed, but on balance it seems not only that dispersed holdings were more prevalent during the Sung, but also that they became increasingly so as time went on. It is generally accepted that consolidated holdings were more common in areas of low population density, while widely dispersed estates were typical of areas such as the Yangtze Delta where population density was high and farming techniques advanced.[a] Conditions for tenants seem to have improved steadily as agricultural techniques advanced. Although there is still vociferous support in some quarters for the hypothesis that Sung agricultural estates were feudal in nature and that the majority of Sung peasants had the status of serfs,[b] we find it impossible to believe that this can have been true of any of the economically advanced regions; like Erkès (22) we are convinced that, if only for strictly practical reasons, serfdom was incompatible with sophisticated wet-rice cultivation. Lewin (1) also maintains that there is no evidence whatsoever for the existence of feudal institutions in the Sung. By far the largest category of landholders in the early Sung were independent smallholders, followed by the category known as *pan tzu keng nung*[1], 'farmers tilling land half their own', who leased in extra land to make ends meet: such farmers could clearly not have been serfs.[c]

If serfs did exist in Sung China, they would not have figured on the land registers. The status of workers on the large estates that still existed in North China[d] is uncertain, but whether they were treated as serfs, labourers or sharecroppers, it seems that their condition must have been less favoured than that of peasants in the southern provinces. The same was true generally of frontier regions and any other areas where population was sparse. There it was possible for rich men to lay claim to huge tracts of land, unhampered by official intervention; labour was in short supply, but as economic alternatives were non-existent, those unfortunate peasants who worked on such large estates found themselves with few privileges and a great many obligations.[e] The biography of Liu Shih-Tao[2] records that in 10th-century Szechwan and Shensi rich families took over almost all the land and obliged the original holders to act as their tenants, forcing

[a] Golas (1), p. 304, drawing principally on Miyazaki Ichisada (*1*), pp. 87–129 and Yanagida Setsuko (*1*).

[b] See Golas (1), pp. 306 ff., Grove & Esherick (1) for summaries of the crucial Japanese scholarship.

[c] On the five categories of landholders in the Sung, their status and relative size, see Lewin (1), pp. 71 ff., Golas (1), pp. 299 ff.

[d] Wang An-Shih's Land Survey and Equitable Tax (*fang thien chün shui*[3]) of +1072 aimed to control and reduce the abuses on such estates; James T. C. Liu (2), p. 5.

[e] Golas (1), p. 305; Kawahara Yoshio (*1*) p. 25; see also James C. Scott (1).

1 半自耕農 2 劉師道 3 方田均稅

them to perform all sorts of extra services; only through rebellion were the peasants eventually able to regain their rights.[a]

During the Sung all three forms of rental agreement known in later periods, namely sharecropping, fixed rent in kind and fixed rent in cash, are known to have existed. Significantly, sharecropping seems already to have been in decline: fixed rents paid in grain were common on the large dispersed estates of the Lower Yangtze, where supervision costs on sharecropping would have been disproportionately high.[b] Sharecropping agreements are most typical of subsistence economies, whereas fixed rents are generally believed to foster a spirit of economic enterprise among tenants.[c] During the Ming, in Fukien as in most of Central and Southern China, the rent was always paid as a fixed quantity and the landlord played no part at all, even supervisory, in the process of production:[d] 'The Fukienese landlord, unlike his north Chinese counterpart, did not generally provide either tools or seed to the cultivator. He did not participate in the farming process, and his only link with the land was the rent received from the tenant.' As landlords took less and less interest in the way their land was farmed, tenants acquired rights to greater security of tenure or even, eventually, to permanent tenancy. As early as the Sung measures were introduced to protect tenants against arbitrary eviction:[e] 'There are indications that the duration of Sung tenures was usually unspecified and that, in normal practice, a tenure tended to be semi-permanent or permanent. We have examples of tenant families who rented the same fields for generations, or even centuries.' So secure were the rights of tenants in Ming Fukien that on payment of a fee called 'manured soil money' (*fen thu yin*[1]), the tenant received transferable and negotiable cultivation rights over the topsoil, and could sublet or sell his right without the landlord's consent. This system of 'two owners of a single field' (*i thien liang chu*[2]) was common in many areas of South China right up until 1949.[f] The tenant had very strong customary rights; indeed, often the landlord could only raise rents if the tenant agreed, and it was quite common for tenants to withhold payment temporarily or even permanently. Absentee landlords had few forms of redress against defaulting tenants; as a last resort they would hire bands of strongmen to recuperate the rents due to them. Barrington Moore[g] sees this as evidence of the ferocity of landlord exploi-

[a] *Sung Shih* 304/11*a*, cited Lewin (1), p. 77. It is possible that where lineages were involved in opening up frontier lands, poor farmers belonging to the lineage were in a more privileged position; M. Freedman (3), (4). Whether the lineages opened up frontiers or the opening up of frontiers stimulated the organisation of lineage groups is a contentious issue. Freedman (3), (4) and Potter (2) hold the former view, B. Pasternak (1), (2) and James L. Watson (1) the latter.

[b] Golas (1), p. 308; J. P. McDermott (1), p. 208.

[c] For an account of the theories of the economic roles of sharecropping and other rental forms, see Bray & Robertson (1).

[d] E. S. Rawski (1), p. 18.

[e] Golas (1), p. 312.

[f] Fei (2); E. S. Rawski (1), pp. 18, 190 provides a useful bibliography.

[g] (1), p. 180.

[1] 糞土銀 [2] 一田兩主

tation; it also offers proof of their relative helplessness in the face of determined village solidarity.[a]

Despite the economic autonomy of his tenants, the landlord remained directly responsible for paying the land-tax (*tsu*[1]) to the government. Accurate land registers, recording the size, grade (and tax liability), and ownership of each plot, were compiled in the Sung and maintained through successive dynasties.[b] Anyone buying or selling land was required to report the transaction to the local authorities so that the land registers could be amended accordingly.[c] There is no doubt that many landlords did manage to wriggle out of paying at least a proportion of their taxes, especially during those periods when members of the landed gentry were made directly responsible for local tax collection. But that they did experience difficulty in evading their obligations is shown by the spread from Ming times of such practices as 'three owners of a single field' (*i thien san chu*[2]),[d] whereby the landlord transferred the tax payment obligation to a third party, who in exchange received part of the rent:[e] 'The wealthy households have much land and so their taxes are heavy. They transfer the land to other people; calculating the tax, they supply these culprits with just enough of the rent to cover the tax payments.'

By maintaining accurate land registers the government was able to keep track of the growth of large estates, and even to impose upper limits on the size of holdings. On several occasions the government simply confiscated land above a certain limit and redistributed it to landless peasants. Under the Public Fields Law (*kung thien fa*[3]) of +1260, for example, the state compulsorily purchased one-third of the land of all families owning over 100 *mu* in the six prefectures around Lake Thai in the Yangtze Delta;[f] an upper limit of 100 *mu* (about 15 acres), even in such a rich and productive region, hardly seems excessive. In any case, large estates rarely survived for long in economically advanced areas, where the risks and rewards of commercial activity ensured a rapid turnover. Under the system of fixed rents, skilled tenant farmers could confidently expect to save enough to buy some land of their own; Fei Hsiao-Thung reports that, even in a crowded 20th-century village near Shanghai, he met nobody who had been landless all his life.[g] A particular field would often be farmed by the same family for generations, but during this period its actual ownership might change several times. While the

[a] See Fei (2), p. 184 on the difficulties faced by landlords even in the Shanghai area in the 1930s; Ash (1), p. 43 documents the severe measures taken by landlords against rent defaulters in the same area, but the very fact that landlords were frequently obliged to resort to physical brutality to recover their dues points to the institutional weakness of their position.

[b] Golas (1), p. 313; James T. C. Liu (2), p. 4; Ray Huang (3), pp. 38–43.

[c] Golas (1), p. 299.

[d] E. S. Rawski (1), pp. 191–2.

[e] *Lung-Yen Hsien Chih*[4], 1558 ed., 1/46*a–b*; tr. E. S. Rawski (1), p. 192.

[f] Golas (1), p. 302; see also Ray Huang (3), p. 159.

[g] (2), pp. 177 ff.

[1] 租 [2] 一田三主 [3] 公田法 [4] 龍巖縣志

fixed rent system favoured tenants, it often meant that a landlord received very little from his land after paying the land tax. One did not make a fortune through being a landlord, one became a landlord through making a fortune. Land was safe, land was respectable, but profits on trade, moneylending, or commercial plantations of sugar, fruit or timber, were all much higher, and many wealthy families preferred to invest in these other sources of income.[a] Independent smallholders or tenants, on the other hand, could expect to make quite high profits from their land if they worked hard and played the market wisely, and these profits they might well invest in purchasing land. Although there were considerable fluctuations in the distribution of landholding, the majority of holdings in the economically advanced regions were either those of medium landowners, averaging 100 *mu* or so, or of smallholders or part-tenants. In the 11th century, the latter constituted over 50 % of the registered population and held a quarter of the land;[b] in the 17th-century Yangtze Delta, really large landowners were rare and probably three-quarters of the land was owned by medium landowners or smallholders.[c] Nor were these last excluded from the benefits of education or from access to political power: the Sung minister Wang An-Shih, for example, came from a smallholder family which had risen two or three generations previously to landlord status; the same was true of many of his colleagues, and perhaps the majority of bureaucrats from Sung times on were drawn from similar backgrounds.[d]

*Social stability of wet-rice systems*

It is easy to see that such improvements in tenant status as we have just described, together with a considerable degree of social mobility, would tend to reduce friction and antagonism between the landlord and tenant class, acting as a damper on class conflict precisely where changes in agricultural technology and economic development were most pronounced, and reinforcing rather than transforming the relations of production. This phenomenon is by no means peculiar to medieval China. Clifford Geertz (1) has described the social effects of 'agricultural involution' in Java, where tenure systems became more intricate, cooperative labour arrangements more complex, and mechanisms for the distribution of wealth (such little wealth as there was) were elaborated and extended; and although Javanese society was hardly egalitarian, there was no sign, even under such extreme conditions, of social polarisation into a class of large landlords and one of near-serfs. Of course, colonial Java as described by Geertz was not enjoying the type of economic expansion typical of medieval China, nor were there any significant technological advances such as those of the Sung 'green revolution' to

[a] See, for example, M. Cartier (1), p. 83.
[b] Golas (1), p. 303.
[c] Ray Huang (3), p. 158.
[d] Ho Ping-Ti (2); such career bureaucrats replaced the officials drawn from the hereditary landed aristocracy of North China, whose power derived ultimately from the size of their landholdings.

precipitate social change. But rice-growing societies in general seem remarkably resistant even to the pressures imposed by the introduction of modern scientific technology. Our own research in a rice-growing district of Malaysia, where a large irrigation scheme had recently been constructed and modern inputs introduced by the local agricultural development authority (and adopted with enthusiasm by the local farmers), revealed few signs of social polarisation resulting from the innovations: on the contrary, it seemed (for example) that tenancy contracts were being modified to the benefit of tenants rather than landlords.[a] The case of Tokugawa[1] Japan (1615–1868) is also striking in this respect.

Tokugawa Japan underwent an agricultural and economic expansion in many respects similar to that of Sung China, indeed, many agricultural innovations were probably based on Chinese precedent. In an effort to consolidate the authority of the state and reduce the power base of the feudal lords, the late 16th-century Tokugawa rulers organised detailed cadastral surveys to tighten their control over the land tax.[b] They opened up communications networks and encouraged the development of agriculture, industry and commerce, promulgating edicts exhorting the peasants to work harder, drink less, engage in handicrafts, and so on; by the late 17th century, a wholly new attitude towards change and innovation was apparent. Treatises on agricultural improvement were published, such as the *Nōgyō Zensho* of 1697 and the later *Nōgu Benri Ron*. These works were based on wide experience and intensive practical research, but their debt to the classical Chinese agricultural treatises is immediately apparent. There is evidence to show that these works were widely read by Japanese farmers, several of whom went on to write their own agricultural tracts. Whereas previously Japanese peasants had produced rice chiefly to pay their taxes and feudal dues, and themselves consumed only the produce of their dryland fields (barley and buckwheat), they were now beginning to extend the cultivation of irrigated crops and to produce rice for their own consumption, and also for the market. The number of crop varieties increased dramatically: one calculation gives 177 names for rice in the early 17th century and 2363 by the mid-19th century, while a 19th-century agricultural diary states that between 1808 and 1866 the breeding of improved rice varieties permitted an extension of the growing season by 17 days.[c] New irrigation works were built, both large- and small-scale, and new tools and commercial fertilisers became widely available. The increases in productivity were accompanied by marked social change.

The medieval Japanese family had included not only kinsmen but also hereditary and indentured servants, and a variety of dependants whose status varied between that of bondsman and tenant. As agricultural improvement progressed

[a] Bray & Robertson (1).
[b] The following account is chiefly based on T. C. Smith (1), Shimpo (1), Pauer (1) and Furushima (*1*).
[c] Shoji Kichinosuke (*1*), p. 40.

[1] 德川

under the Tokugawa and the economy expanded and became more commercialised, the social and legal status of the dependent rural classes improved rapidly. The servant classes decreased in numbers while the lower echelons of the landholding class grew, and tenants who had previously been obliged to provide their landlords with all sorts of labour services acquired a far greater measure of economic independence. The typical unit of production was now a nuclear family, individually responsible for the management of its landholding, supplementing its income by cottage industries such as the weaving of silk or cotton, wine-brewing, or the manufacture of beancurd or pickles. T. C. Smith sees this improvement in tenant status as the psychological well-spring of the impressive technological progress in agriculture during the Tokugawa period, and adds:[a] 'The incentive was not individual gain but the enrichment of the family, and *most technological innovations of this period tended to strengthen the solidarity of the nuclear family* [italics ours].' The phenomenon of improving tenant status was prolonged in Japan through the 19th and early 20th centuries, when landlords found to their dismay that their investments in irrigation schemes and other improvements, by increasing their tenants' financial security, brought them not more but less respect and influence in the village.[b]

As agriculture progressed in Tokugawa Japan, so too the economy expanded in other spheres. Here again the rapid expansion of textile and other commodity production was based largely on the increasing participation of peasant households in small-scale commodity manufacture. Some highly specialised traditional industries, such as the silk-brocade manufactures of Kyoto, found they now had serious rivals in provincial villages, where labour costs were rated on a very different scale. As in medieval China, there were fortunes to be made as merchants or middlemen,[c] but a very high proportion of manufacturing remained on the domestic level.

In Tokugawa Japan we can see plainly how advances in wet-rice technology contributed to the displacement of manorial estates by family-managed smallholdings. Little is known of pre-Sung tenurial relations in South China, but whether or not they were 'manorial' (in the sense that the predominant unit of production was a centrally-supervised manorial estate), it appears that one crucial result of the agricultural innovations of the Sung was firmly to establish the independently managed smallholding as the basic unit of agricultural production. Despite the high concentration of land-ownership in China's more productive rice-growing areas, large, centrally-run estates were almost unknown.

[a] (1), p. 92.

[b] Waswo (1). Rents actually fell during the course of the 19th century in Japan (Yamamura (1), p. 297). As Japan modernised its economy in the late 19th and early 20th century, however, landlord-tenant antagonisms grew increasingly bitter (Ilse Lenz (1)).

[c] The growth of the merchant class during this period, their wealth, luxurious lifestyle and patronage of such arts as the theatre and woodblock prints, are too well known to require detailed commentary.

*The role of petty commodity production*

The parallels between Sung China and Tokugawa Japan strongly suggest that technological advances and increased productivity in wet-rice agriculture are closely related to the development of petty commodity production. Indeed, similar links may be observed today in countries such as Malaysia and Indonesia[a] (though these cases are complicated by the rapid expansion of the non-agricultural sectors and the integration of the national into the world economy). We have observed that wet-rice agriculture is characterised by small, individually managed units of production, run essentially with family labour, at low capital costs. Technological advances do not take the form of capital-intensive improvements like large-scale machinery, which individual families would not be able to afford and which, with the exception of irrigation pumps, would in any case offer no long-term economic advantage where the optimal field size is well under half an acre. Most improvements are either relatively cheap (improved seeds, weeding tools and fertilisers) or require large investments not of capital but of labour (constructing new fields or canals, maintaining irrigation works).[b] Unlike dry-land agriculture, in wet-rice areas successful farming depends less on equipment than skill, less on capital investment and economies of scale than on labour shrewdly applied.

The labour requirements of wet-rice cultivation are high but largely concentrated in two peak periods (ploughing and transplanting, and harvesting). At these periods the whole family will be needed in the fields, indeed it may even be necessary to organise labour-exchanges or hire in extra labour, if available, to cope.[c] Interim tasks such as weeding and irrigation are necessary but less time-consuming, and there will be time to spare for other occupations. In near-subsistence economies there may be few means of absorbing the extra labour profitably. Farmers may take advantages of local differences in cropping-season to hire out their labour as harvesters; this sometimes involves travelling long distances, as does taking up seasonal employment in towns.[d] Or they may be fortunate enough to find some profitable work in or near their village, but such work in subsistence conditions is generally scarce.[e] In an expanding economy,

[a] Malaysian Government (1); I. Palmer (1), (2); J. S. Kahn (1).

[b] Tamaki Akira (*1*) makes the point that the huge investment in labour necessary to open up new rice-land of comparable productivity to that already in use will often push farmers to intensify cultivation techniques in their old fields rather than try to expand the arable area.

[c] On cooperative labour-exchange in rice-growing areas see Embree (1), A. H. Hill (1), Liefrinck (1). In Yunnan Chinese farmers would often hire aboriginals, whose hill farms left them some leisure and who asked very low wages, to cope with these temporary labour shortages; Fei & Chang (1).

[d] Farmers in Kelantan would travel right across the Malay Peninsula to harvest rice on the West Coast where the monsoon fell a month or two later. Nowadays they prefer to increase their off-season income by travelling several hundred miles south to the factories and building-sites of Singapore. In Java there is usually some rice to be harvested somewhere in every region, and many peasants, especially women, depend for their living on travelling from village to village with their reaping-knives; I. Palmer (2).

[e] In Laos many rice-farmers devote the off-season to logging, hunting or fishing; Taillard (1). In Kelantan villages part-time employments included brick-maker, builder, trishaw-driver, mat-weaver and *imam*, among others.

however, the production of handicrafts or commercial crops that before had merely provided a small supplement to the family income plays an increasingly important role. Such petty commodity production dovetails neatly with wet-rice farming: it requires very little capital to set up a family enterprise of the sort; it absorbs surplus labour without depriving the farm of workers at times of crisis;[a] it can be expanded, diversified or contracted to meet market demands, but the combination with the rice-farm guarantees the family's subsistence; the products can be conveniently conveyed to local or national markets by merchants, who pay the villagers for their labour and often provide raw materials as well as information on the state of the market. Under such conditions, then, agricultural intensification is more likely to absorb surplus labour within rural household industries than to create a landless proletariat.

The fact that petty commodity production is organised in individually-owned units naturally places a heavy restriction on capital investment, and thus on such improvements as mechanisation brings, but these decentralised, low-capital enterprises are not necessarily inefficient. In Sung and Ming China, as in Tokugawa and early Meiji Japan, there was almost no centrally-organised, large-scale, capital-intensive industry. Yet compared with medieval or early modern Europe consumer goods were plentiful, some of extremely high quality, but most designed for popular consumption. Merchants and middlemen amassed considerable fortunes through redistributing these goods. Petty commodity production in China and Japan had its roots in an agricultural technology that tended to buttress the relations of production.

We have perhaps given the impression that life in the countryside of Southern China and Japan was a rural idyll of egalitarian, family-centred prosperity. This is of course untrue. Where commerce thrives and entrepreneurs flourish, the timid and the foolhardy alike may find themselves ruined overnight, and their poverty will seem the more pitiful contrasted with their neighbours' affluence. Even in China's economic heyday many lived in desperate want, as foreign observers were quick to point out: 'Malgré l'industrie et la sobriété du peuple chinois', remarks Quesnay,[b] 'malgré la fertilité de ses terres et l'abondance qui y règne, il est peu de pays où il y a autant de pauvreté dans le menu peuple.' Naturally the problem was further aggravated as the limits to increase in agricultural production were reached and yet the population continued to grow. Ping-Ti Ho (4), Perkins (1) and Elvin (2) all agree that in China this point was reached in about 1800: all available arable land had by then been brought under cultivation, and no more significant increases in land productivity could be achieved with traditional methods of production. The 19th- and early 20th-century decay of the Chinese economy, hastened by Western military and commercial intervention,[c]

[a] The hours are naturally much more flexible than those worked in factories.
[b] (1), p. 579.
[c] Though the intrusion of Western manufactured goods on the Chinese market was less destructive than in colonial India, it nevertheless threatened the livelihood of many peasant producers, especially those in the textile industry; see Fei (2).

has perhaps given us an unjustifiably negative view of the traditional economy, its resilience and potential for expansion. Yet even when China's Southern-based economy was thriving and expanding, when the country was prosperous and entrepreneurial spirit abounded, the very nature of wet-rice agriculture was such that economic and technological innovation of the type we Westerners associate with 'progress' was virtually impossible. Explanations such as Elvin's 'high-level equilibrium trap' (2), which imputes technological stagnation to the diminishing returns typical of a system approaching its limits, are irrelevant to the Sung and Ming when the Chinese economy was still in a phase of rapid expansion. From the very outset, we believe, the fundamental constraints (both technical and social) of wet-rice cultivation significantly influenced China's peculiar path of development.

It is clear that the small size of units of production (and of individual fields) suited to efficient wet-rice production is a barrier to economies of scale and thus to technical inventiveness of the type to which we owe the European Agricultural and Industrial Revolutions, that is, a tendency towards mechanisation and improved returns to labour. The specialisation of labour, often regarded as a necessary precondition for mechanisation, is also limited where production units are small. This is borne out by the course of events in recent years. Even today, the mechanisation of wet-rice cultivation in Asia is proving a difficult and delicate task, with quite different requirements and constraints to those of dryland farming, and its technical advantages are not always immediately obvious.[a] The same is true of the manufacturing organisation associated with traditional wet-rice cultivation, namely petty commodity production. In traditional South China, the cost of labour in a family enterprise of this nature was negligible, and the social costs of organising production in sufficiently large units to make improved machinery an attractive investment would have been very high, even if it had occurred to anyone. It is not that the Chinese were unaware of the virtues of labour-saving: small labour-saving devices they had in plenty, in almost every household.[b] But gadgets are not machines: true mechanisation represents a quantum-leap, presupposing not only the concentration of labour but also its subdivision, and division of labour in petty commodity production is a contradiction in terms. The small-scale yet efficient machines that find their place in such systems of economic organisation today[c] are a very recent and sophisticated development, only made possible by enormous advances in engineering skills and by the use of such power sources as electricity and the internal combustion engine.

Here it is worth drawing attention to the timelag in Japan between industrial-

[a] In Northern Thailand, for example, local farmers were happy to hire tractors for opening up new land, but once the new rice-fields had been constructed they immediately reverted to the use of the buffalo-plough; Moerman (1).

[b] See Hommel (1), Elvin (3), and so on. A very telling example of the limited potential for mechanisation in the Chinese rural economy is detailed in D. Kuhn (4).

[c] Perhaps the best example is the small hand-tiller, now an indispensable item of equipment in so many rice-growing areas; several expensive, inefficient and unwieldy intermediate forms were produced before the development of this *n*-th generation descendant of the common tractor.

isation proper, which took off in the late 19th century, and rural mechanisation, both of agriculture and of household industries, which is very much a post-war development. Japan's 19th-century industrialisation campaign was principally directed towards heavy industry; it was inspired by Western example and was largely modelled upon it, though it did incorporate many elements of traditional Japanese technology.[a] Initially it was officially financed, mainly from funds obtained through heavy agricultural taxation; Japanese farmers eventually reaped the benefits in the form of increased markets and improved tools and fertilisers.[b] It is, however, worthy of note that the organisation of production, both agricultural and manufacturing, in the Japanese countryside is still dictated by the constraints of wet-rice agriculture even today.[c]

One might attempt to account for this by the fact that post-war land reforms in Japan, by distributing land to the tiller and limiting individual landholdings, imposed a smallholder regime that tended to preclude economies of scale. In the People's Republic of China considerable consolidation of landholdings has followed land reform: work teams or brigades have replaced individual holdings as the basic unit of production; for labour and finance they may draw upon the central resources of the commune, an administrative unit which generally corresponds in size to the old counties (*hsien*[1]). This is consolidation on a grand scale, and one might have expected radical changes in agricultural technology to result. It certainly has permitted the coordination of effort and resources for the improvement, for example, of irrigation facilities. Yet despite several centrally organised campaigns to mechanise agriculture, there have been few significant changes in rice-farming equipment.

In part this may be attributed to rural labour surpluses and cash shortages, which of course inhibit mechanisation. But the main problem is that it is very difficult to modify the sophisticated and labour-intensive techniques of traditional Chinese rice-farming without incurring, initially at least, significant falls in yields. Despite the high densities of population found in many South Chinese communes, labour shortages are still a serious problem at crucial periods such as transplanting or harvesting, yet it is difficult to develop machines that will transplant or weed as effectively as humans. Probably the most important technological improvements have been in irrigation, where canal networks have been extended and motorised pumps have largely replaced human labour. But despite land reform and the improvement of irrigation techniques, there has been little change in the South Chinese rice landscapes: experts still recommend one-sixth of an acre as the maximum size of wet-rice fields, and this limits the usefulness of large machinery like tractors and harvesters.[d] Increases in productivity of both land

[a] See, for example, T. Nakaoka (1). [b] T. C. Smith (1).
[c] Shimpo (1), Bray (6).
[d] Tractors and tillers on communes in South China are often used more for transport purposes than for tillage.

[1] 縣

and labour have been achieved, not so much by land consolidation and the introduction of machinery, as by improvements to the water supply, by the development of new crop varieties and pesticides, and by careful coordination of labour inputs. The labour requirements of rice cultivation have also affected the nature of recent rural industrialisation.[a] It seems that in South China even complete social revolution has been unable to erase the stamp of a long tradition of wet-rice cultivation.[b]

It is our contention, then, that while rural societies economically dependent on wet-rice cultivation are not incapable of mechanisation and industrialisation, they are unlikely to engender these processes spontaneously. Furthermore, in such societies refinements and innovations in agricultural technology have tended to reinforce the organisation of production in family units, acting as a further barrier to radical change.

## (3) Development or Change?

In the first part of our concluding section we demonstrated the enormous transformative potential of certain Chinese agricultural implements, and their role in stimulating a complete technical and social transformation in modern European agriculture. However, China's historical experience shows that the implements alone are not sufficient to produce real changes in the technological system: such transformations only occur where technical development and economic expansion arouse conflicts of interest sufficiently acute to destabilise the existing relations of production and stimulate further technical change.

Where agriculture is based on the extensive cultivation of dryland cereals, development is along very different lines from wet-rice societies. If as a result of technical innovations the economy enters a phase of expansion and land prices and productivity rise, landowners will try to accumulate centrally-managed, consolidated estates, and will generally find it more efficient to evict their tenants in preference for hired labour. There will eventually be an effort to bring down

[a] See, for example, T. G. Rawski (1), Stiefel & Wertheim (1) for elaboration of these points.

[b] The disintegration of the People's Communes and the official promulgation of the responsibility system (*jen tse chih*[1]) since 1981 marks a new phase of economic growth which has been particularly popular and effective, as one might expect, in the South and Southeast provinces (although it was initiated in Anhwei). There has been a marked shift from primary grain production organised on a communal basis, which achieved some of its most spectacular results under the Cultural Revolution, in the impoverished provinces of the Northwest. Today the primary units of management are once again the production team (which, as Wong (1) points out, like its institutional predecessor the Mutual Aid Team, is firmly rooted in the traditional cooperative groups of the South China villages), and the household. Generally speaking, individual households are responsible for basic agricultural production, while light industries, mostly devoted to the processing of such locally-produced crops as cotton or tobacco, are organised at team or brigade level. The situation is in many ways reminiscent of the agricultural improvements and concomitant growth of rural industry in the Sung. Overall production of economic crops and commodities has risen very rapidly, but regional disparities (between isolated regions and those near cities, and more especially between North and South) are becoming more and more marked (Kojima Reiitsu[2], pers. comm. October 1982).

[1] 任責制 [2] 小島麗兔

production costs by reducing the size of the labour force, and thus a tendency to invest in labour-saving equipment—here indeed is a stimulus to mechanical invention. Britain's Agricultural Revolution, the enclosures, the creation of a class of landless labourers, the concentration of labour on consolidated agricultural estates or in industrial towns and the consequent heavy capital investment in labour-saving equipment, that is to say industrial and agricultural machinery—these eventually resulted in a complete transformation not only of technology but of social and economic relations. Northern Chinese agrarian development in the early centuries of our era shows certain broad similarities to the British case: given the nature of class polarisation associated with such technological development, the North Chinese agricultural estates of that period might even be qualified as 'proto-capitalist'.[a] We do not suggest for a moment that medieval North China would, under more favourable conditions, have produced its own agricultural and industrial revolution, yet it is interesting to speculate on what might have happened if, for political or other reasons, the shift to the rice-producing regions of the south had not been possible.[b]

In wet-rice societies a contrary process is observed, for the natural course of development appears to be not towards capitalism but towards a social formation that may conveniently be called a petty commodity mode of production. Once this stage is reached, the relations of production in wet-rice cultivation have an internal dynamism that enables them to sustain not only significant increases in agricultural productivity, but also rapid economic diversification, without undergoing historical change. The Sung dynasty saw the emergence in the southern provinces of China of a petty commodity mode of production that was consolidated over nearly eight centuries: during this period economic expansion was sustained through a progressive development of the forces of production, as improved agricultural and manufacturing techniques were refined and the refinements spread from one region to another. Changes in the relations of production, however, were insignificant: the smallholding family remained the characteristic unit of management, and landlords and merchants had access to the means of production only through these family units. It seems, then, that while wet-rice agricultural systems can produce rapid, sustained technical and economic *development*, it is to the harsher dryland systems that we must look for real social transformation and historical *change*.

[a] Given his contribution to the breakdown of the North Chinese economy, perhaps we should now seriously consider Genghis Khan as a crisis of capitalism.

[b] Significantly, 'managerial farmers', who relied on hired labour to farm their lands, were the most common type of wealthy household in Northern Chinese villages in the first half of the 20th century; see Philip Huang (1), Hinton (1), Crook & Crook (1). This phenomenon was virtually unknown in Southern China; see Esherick (1).

# BIBLIOGRAPHIES

A CHINESE AND JAPANESE BOOKS BEFORE +1800

B CHINESE AND JAPANESE BOOKS AND JOURNAL ARTICLES SINCE +1800

C BOOKS AND JOURNAL ARTICLES IN WESTERN LANGUAGES

When obsolete or unusual romanisations of Chinese words occur in entries in Bibliography C, they are followed, wherever possible, by the romanisations adopted as standard in the present work. If inserted in the title, these are enclosed in square brackets; if they follow it, in round brackets. When Chinese words or phrases occur romanised according to the Wade-Giles system or related systems, they are assimilated to the system here adopted (cf. Vol. 1, p. 26) without indication of any change. Additional notes are added in round brackets. The reference numbers do not necessarily begin with (1), nor are they necessarily consecutive, because only those references required for this volume of the series are given.

Korean and Vietnamese books and papers are included in Bibliographies A and B. As explained in Vol. 1, pp. 221 ff., reference numbers in italics imply that the work is in one or other of the East Asian languages.

# CHINESE AND JAPANESE JOURNAL ABBREVIATIONS

*AHRA* — *Agricultural History Research Annual* (*Nung Shih Yen Chiu Chi Khan*; formerly *Nung Yeh I Chhan Yen Chiu Chi Khan*)

*AP/HJ* — *Historical Journal*, National Peiping Academy

*AS/BIHP* — *Bulletin of the Institute of History and Philology*, Academia Sinica (*Chungkuo Kho Hsüeh Yen Chiu Yuan, Li-Shih Yü-Yen Yen Chiu So Chi Khan*)

*BAFKU* — *Bulletin of the Agricultural Faculty of Kyushu University* (*Kyudai Nōgakubu Kiyō*)

*CHWSLT* — *Chung Hua Wen Shih Lun Tshung* (*Collected Essays on Chinese Literature and History*)

*CKKSH* — *Chungkuo Kho Hsüeh* (Academia Sinica)

*CKNP* — *Chungkuo Nung Pao* (*Chinese Agricultural Journal*)

*CKNYKH* — *Chungkuo Nung Yeh Kho Hsüeh* (*Chinese Agronomy*)

*CKSYC* — *Chungkuo Shih Yen Chiu* (*Research in Chinese History*)

*CLNK* — *Chin-Ling Nung Khan* (*Nanking Univ. Agricultural Journal*)

*FDKH* — *Fukuoka Daigaku Kenkyūjo Hō* (*Research Reports of Fukuoka University*)

*HCTHHP* — *Hang-Chou Ta Hsüeh Hsüeh Pao* (*Hangchou University Journal*)

*HNNYKH* — *Hua-Nan Nung Yeh Kho Hsüeh* (*South China University Journal of Agriculture*, Canton)

*HSYK* — *Hsüeh Shu Yüeh Khan* (*Technology Monthly*)

*ICHP* — *I Chhuan Hsüeh Pao* (*Hereditas*)

*ISTC* — *I Shih Tsa Chih* (*Chinese Journal of the History of Medicine*)

*JAAC* — *Journal of the Agricultural Association of China*

*KHNY* — *Kho Hsüeh Nung Yeh* (*Scientific Agriculture*)

*KHSC* — *Kho Hsüeh Shih Chi Khan* (*Chinese Journal of the History of Science*)

*KHSWC* — *Kho Hsüeh Shih Wen Chi* (*Papers in the History of Science*)

*KK* — *Khao Ku* (*Archaeology*)

*KKHP* — *Khao Ku Hsüeh Pao* (*Archaeological Bulletin*)

*KKHTC* — *Kōkogaku Zasshi* (*Journal of the Japanese Archaeological Society*)

*KKJRG* — *Kikan Jinruigaku* (*Anthropology Quarterly*)

*LHP* — *Ling-Nan Hsüeh Pao* (*Lingnan University Journal*)

*LSYC* — *Li Shih Yen Chiu* (Peking) (*Journal of Historical Research*)

*NACJ* — *Nanking Agricultural College Journal* (*Nan-Ching Nung Hsüeh Yuan Hsüeh Pao*)

*NSK* — *Nōgyo Sōgo Kenkyū* (*Collected Agricultural Studies*)

*NSYC* — *Nung Shih Yen Chiu* (*Research in Agricultural History*)

*NYCC* — *Nung Yeh Chi Chieh Hsüeh Pao* (*Acta Agromechanica Sinica*)

*NYKK* — *Nung Yeh Khao Ku* (*Agricultural Archaeology*)

*QBCB/C* — *Quarterly Bulletin of Chinese Bibliography* (Chinese edition) (*Thu Shu Chi Khan*)

*SB* — *Shizen to Bunka* (*Natura et Cultura*)

*SGKK* — *Shigaku Kenkyū* (*Review of Historical Studies*)

*SGZ* — *Shigaku Zasshi* (*Historical Journal of Japan*)

*SH* — *Shih Huo* (*Journal of the History of Economics*)

*SHYK* — *Shih Huo Yüeh Khan* (*Economics Monthly*, Taipei)

*SKS* — *Shakai Keizai Shigaku* (*Social and Economic History*)

*SSIP* — *Shanghai Science Institute Publications* (*Shang-Hai Tzu Jan Kho Hsüeh Yen Chiu Hui Pao*)

*TBK* — *Tōyō Bunka Kenkyūjo Kiyō* (*Memoirs of the Tōyō Bunka*, Univ. of Tokyo)

*TG* — *Tōhō Gakuhō* *TG/K*: *Kyoto Journal of Oriental Studies*; *TG/T*: *Tokyo Journal of Oriental Studies*

*TJTJKH* — *Tung-Pei Jen Min Ta Hsüeh Jen Wen Kho Hsüeh Hsüeh Pao* (*Journal of the Humanities Department of Shenyang University*)

*TK* — *Tōyōshi Kenkyū* (*Studies in Oriental History*)

*TLTC* — *Ta Lu Tsa Chih* (*Continental Magazine*, Taipei)

| | |
|---|---|
| *TSK* | *Thu Shu Kuan (Library)* |
| *TSWHP* | *Tso Wu Hsüeh Pao (Acta Agronomica Sinica)* |
| *WS* | *Wen Shih* (Peking) |
| *WSC* | *Wen Shih Che (Literature, History and Philosophy*, Shantung University) |
| *WUQSS* | *Wuhan University Quarterly Journal of Social Sciences* |
| *WW* | *Wen Wu*: see *WWTK* |
| *WWTK* | *Wen Wu Tshan Khao Tzu Liao (Reference Materials for History and Archaeology)*: later *WW* |

# TSHUNG SHU ABBREVIATIONS

| | |
|---|---|
| *HCCC* | *Huang Chhing Ching Chieh* |
| *MCHSAMUC* | *Mémoires Concernant l'Histoire, les Sciences, les Arts, les Moeurs et les Usages des Chinois, par les Missionnaires de Pékin.* |
| *SKCS* | *Ssu Khu Chhüan Shu* |
| *SSKTS* | *Shou Shan Ko Tshung Shu* |
| *TSCC/IW* | *(Ku Chin) Thu Shu Chi Chheng (I Wu)* |
| *TT* | *Tao Tsang* |
| *YCCC* | *Yün Chi Chhi Chhien* |

# WESTERN JOURNAL ABBREVIATIONS

| | |
|---|---|
| *AAAA* | *Archaeology* |
| *AAN* | *American Anthropologist* |
| *AGHR* | *Agricultural History Review* |
| *AGHST* | *Agricultural History* (Washington DC) |
| *AGT* | *Agronomie Tropicale* |
| *AHES/AESC* | *Annales; Economies, Sociétés, Civilisations* |
| *AHES/AHS* | *Annales d'Histoire Sociale* |
| *AHR* | *American Historical Review* |
| *ALTOF* | *Altorientalische Forschungen* (Berlin) |
| *AM* | *Asia Major* |
| *AMA* | *American Antiquity* |
| *AMBG* | *Annals of the Missouri Botanical Garden* |
| *AN* | *Anthropos* |
| *AQ* | *Antiquity* |
| *ARES* | *Annual Review of Ecology and Systematics* |
| *ARLC/DO* | *Annual Reports of the Librarian of Congress (Division of Orientalia)* |
| *ARS* | *Arabian Studies* |
| *ARSI* | *Annual Report of the Smithsonian Inst., Washington* |
| *ASP* | *Asian Perspectives* |
| *BEC* | *Bulletin de l'Ecole des Chartes* |
| *BEFEO* | *Bulletin de l'Ecole Française d'Extrême-Orient* |
| *BGSC* | *Bulletin of the Geological Survey of China* |
| *BHS* | *Bulletin of the Historical Society* |
| *BLSOAS* | *Bulletin of the London School of Oriental and African Studies* |
| *BSN* | *Behavior Science Notes* |
| *BT* | *Bulletin of Tibetology* |
| *CBOT* | *Chronica Botanica* |
| *GG* | *China Geographer* |
| *CJ* | *China Journal of Science and Arts* (Shanghai) |
| *CQ* | *Classical Quarterly* |
| *CQR* | *China Quarterly* |
| *CUA* | *Cuadernos Americanos* |
| *CURRA* | *Current Anthropology* |
| *DEC* | *Developing Economies* (Tokyo) |
| *DWAW/PH* | *Denkschriften d. Akad. d. Wissenschaften Wien* (Vienna): *Phil.Hist.Kl.* |
| *EC* | *Early China* |
| *ECB* | *Economic Botany* |
| *EH* | *Economic History* |
| *EHR* | *Economic History Review* |
| *ERDB* | *Erdball* |
| *ERU* | *Etudes Rurales* |
| *ESEA* | *Essays and Studies of the English Association* |
| *EUP* | *Euphytica* |
| *FEQ* | *Far Eastern Quarterly* |
| *FEL* | *Food and Flowers* (Dept. Agric. Hong Kong) |
| *GPO* | *Geographica Polonica* |
| *H* | *History* |
| *HEC* | *Human Ecology* |
| *HJAS* | *Harvard Journal of Asiatic Studies* |
| *HT* | *History of Technology* |
| *IEA* | *International Ethnographic Archives* |
| *IEJ* | *Israel Exploration Journal* |
| *IRRIR* | *International Rice Research Institute Review* |
| *IS* | *Islamic Studies* |
| *ISIS* | *Isis* |
| *JA* | *Journal Asiatique* |
| *JAS* | *Journal of Asian Studies* |
| *JASAG* | *Journal of the American Society of Agronomy* |
| *JATBA* | *Journal d'Agriculture Tropicale (Traditionelle) et de Botanique Appliquée* |
| *JCP* | *Journal of Chinese Philosophy* |
| *JESHO* | *Journal of the Economic and Social History of the Orient* |
| *JHKAS* | *Journal of the Hong Kong Archaeological Society* |

| | |
|---|---|
| *JOSHK* | *Journal of Oriental Studies* (Hong Kong University) |
| *JRAGS* | *Journal of the Royal Agricultural Society* |
| *JRAI* | *Journal of the Royal Anthropological Institute* |
| *JRAS/MB* | *Journal of the Royal Asiatic Society (Malay Branch)* |
| *JRCAS* | *Journal of the Royal Central Asiatic Society* (London) |
| *JSSI* | *Journal of the Shanghai Science Institute* |
| *LH* | *L'Homme* |
| *MAST* | *Modern Asian Studies* |
| *MIOF* | *Mitteilung des Instituts für Orientforschung* (Berlin) |
| *MJTG* | *Malay Journal of Tropical Geography* |
| *MODC* | *Modern China* |
| *MRDTB* | *Memoirs of the Research Dept. of Tōyō Bunka* (Tokyo) |
| *NCR* | *New China Review* |
| *NSN* | *New Statesman and Nation* (London) |
| *O* | *Observatory* |
| *OAZ* | *Ostasiatische Zeitschrift* |
| *OE* | *Oriens Extremus* (Hamburg) |
| *PHNB* | *Peking Natural History Bulletin* |
| *PP* | *Past and Present* |
| *PPHS* | *Proceedings of the Prehistoric Society* |
| *PSAS* | *Proceedings of the Society of Antiquarians* (Scotland) |
| *PTRSB* | *Philosophical Transactions of the Royal Society (Series B)* |
| *PV* | *Pacific Viewpoint* (New Zealand) |
| *RBS* | *Revue Bibliographique de Sinologie* |
| *S* | *Sinologica* (Basel) |
| *SAM* | *Scientific American* |
| *SC* | *Science* |
| *SCISA* | *Scientia Sinica* (Peking) |
| *SJA* | *Scottish Journal of Agriculture* |
| *SSCH* | *Social Sciences in China* |
| *SWAW/PH* | *Sitzungsber. d. preuss. Akad. d. Wissenschaften (Phil.Hist.Kl.)* |
| *TAPS* | *Transactions of the American Philosophical Society* |
| *TCULT* | *Technology & Culture* |
| *TH* | *Thien Hsia Monthly* (Shanghai) |
| *TP* | *T'oung Pao* |
| *TT* | *Tools & Tillage* |
| *VFIDMGNWT* | *Veröffentlichungen d. Forschungsinstituts der deutschen Museum für die Geschichte der Naturwissenschaften und der Technik* |
| *W* | *Weather* |
| *WARC* | *World Archaeology* |
| *ZAGA* | *Zeitschrift für Agrargeschichte und Agrarsoziologie* |
| *ZDGM* | *Zeitschrift der deutschen morgenländischen Gesellschaft* |
| *ZFE* | *Zeitschrift für Ethnologie* |

# A. CHINESE AND JAPANESE BOOKS BEFORE +1800

Each entry gives particulars in the following order:

(*a*) title, alphabetically arranged, with characters;
(*b*) alternative title, if any;
(*c*) translation of title;
(*d*) cross-reference to closely related book, if any;
(*e*) dynasty;
(*f*) date as accurate as possible;
(*g*) name of author or editor, with characters;
(*h*) title of other book, if the text of the work now exists only incorporated therein; or, in special cases, references to sinological studies of it;
(*i*) references to translations, if any, given by the name of the translator in Bibliography C;
(*j*) notice of any index or concordance to the book if such a work exists;
(*k*) reference to the number of the book in the *Tao Tsang* catalogue of Wieger (6), if applicable;
(*l*) reference to the number of the book in the *San Tsang* (Tripitaka) catalogues of Nanjio (1) and Takakusu & Watanabe, if applicable.

Words which assist in the translation of titles are added in round brackets.

Alternative titles or explanatory additions to the titles are added in square brackets.

Where there are any differences between the entries in these bibliographies and those in Vols. 1–5, the information here given is to be taken as more correct.

An interim list of references to the editions used in the present work, and to the *tshung-shu* collections in which books are available, has been given in Vol. 4, pt. 3, pp. 913 ff., and is available as a separate brochure.

ABBREVIATIONS

| | |
|---|---|
| C/Han | Former Han. |
| E/Wei | Eastern Wei. |
| H/Han | Later Han. |
| H/Shu | Later Shu (Wu Tai). |
| H/Thang | Later Thang (Wu Tai). |
| H/Chin | Later Chin (Wu Tai). |
| S/Han | Southern Han (Wu Tai). |
| S/Phing | Southern Phing (Wu Tai). |
| J/Chin | Jurchen Chin. |
| L/Sung | Liu Sung. |
| N/Chou | Northern Chou. |
| N/Chhi | Northern Chhi. |
| N/Sung | Northern Sung (before the removal of the capital to Hangchow). |
| N/Wei | Northern Wei. |
| S/Chhi | Southern Chhi. |
| S/Sung | Southern Sung (after the removal of the capital to Hangchow). |
| W/Wei | Western Wei. |


*Chao Hun* 招魂.
The Summons of the Soul [ode].
Chou (Chhu), *c.* −240.
Prob. Ching Chhai 景差.
Tr. Hawkes (1), p. 103.

*Chao Jen Pen Yeh* 兆人本業.
Fundamental Occupation of the Common People.
Thang, *c.* +686.
Official compilation.

*Chao Shih Shu* 趙氏書.
The [Agricultural] Treatise of Master Chao
C/Han, precise date unknown.
Attr. to Chao Kuo 趙過.
No longer extant except in quotations.

*Chen-Tsung Shou Shih Yao Lu* 眞宗授時要錄.
Recorded Works and Days of the Reign of Chen-Tsung.
Sung, *c.* +1000.
Official compilation.

*Cheng Lei Pen Tshao* 證類本草.
*Chhung Hsiu Cheng-Ho Ching Shih Cheng Lei Pei Yung Pen Tshao.*
Reorganised Pharmacopoeia.
N/Sung, +1108, enlarged +1116; re-edited in J/Chin, +1204, and definitively re-published in Yuan, +1249; re-printed many times afterwards, e.g. in Ming, +1468.
Original compiler: Thang Shen-Wei 唐愼微.
Cf. Hummel (13); Lung Po-Chien (*1*).

*Cheng Lun* 政論.
On Government.
H/Han, +155.
Tshui Shih 崔寔.
Ed. Yen Kho-Chün, +1815 嚴可均.
In *YHSF*, ch. 71, pp. 67*a* ff.

*Chha Ching* 茶經.
The Manual of Tea (*Camellia* (*Thea*) *sinensis*).
Thang, *c.* +770.
Lu Yü 陸羽.

*Chha Lu* 茶錄.
A Record of Tea (*Camellia* (*Thea*) *sinensis*).
Sung, *c.* +1060.
Tshai Hsiang 蔡襄.

*Chha Lun* 茶論.
On Tea.
Sung, +11th century.
Shen Kua 沈括.
See Hu Tao-Ching (*8*).

*Chhang Phing Fa* 常平法.
On the Administration of Ever-Normal Granaries.
Sung, 11th century.
Ssu-Ma Kuang 司馬光.
Repr. *SSTK* 55/4b ff.

*Chhen Fu Nung Shu* 陳旉農書.
See *Nung Shu.*

*Chhi Chhi Thu Shuo* 奇器圖說.
(*Yuan Hsi Chhi Chhi Thu Shuo Lu Tsui*)
Diagrams and Explanations of Wonderful Machines.
Ming, +1627.
Teng Yü-Han (Johann Schreck) 鄧玉函 & Wang Cheng 王徵.

*Chhi Min Ssu Shu* 齊民四術.
See Pao Shih-Chhen (*1*).

*Chhi Min Yao Shu* 齊民要術.
Essential Techniques for the Peasantry.
N. Wei, *c.* +535.
Chia Ssu-Hsieh 賈思勰.
Textual refs. are to the 1957 ed. of Shih Sheng-Han (*3*).

*Chhieh Yün* 切韻.
Dictionary of the Sounds of Characters [rhyming dictionary].
Sui, +601.
Lu Fa-Yen 陸法言.
See *Kuang Yün*.

*Chhien Han Shu* 前漢書.
History of the Former Han Dynasty [−206 to +24].
H/Han (begun about +65), *c.* +100.
Pan Ku 班固 and (after his death in +92) his sister Pan Chao 班昭.
Partial tr. Dubs (2), Pfizmaier (32–34, 37–51), Wylie (2, 3, 10), Swann (1). Yin-Te Index, no. 36.

*Chhien-Lung Huo-Shan Hsien Chih* 乾隆霍山縣志.
Gazetteer of Huo-Shan District (N. Anhwei) in the Chhien-Lung Period.
Chhing, 1776.

*Chhin Ting Hsü Thung Chih* 欽定續通志.
See *Hsü Thung Chih*.

*Chhin Ting Hsü Wen Hsien Thung Khao* 欽定續文獻通考.
Imperially Commissioned Continuation of the *Comprehensive Study of (the History of) Civilisation* (cf. *Wen Hsien Thung Khao* and *Hsü Wen Hsien Thung Khao*).
Chhing, ordered +1747, pr. +1772 (+1784).
Ed. Chhi Shao-Nan, Hsi Huang *et al.* 齊召南, 嵇璜.
This parallels, but does not replace, Wang Chhi's *Hsü Wen Hsien Thung Khao*.

*Chhin Ting Ssu Khu Chhüan Shu Tsung Mu Thi Yao* 欽定四庫全書總目提要.
Analytical Catalogue of the Books in the *Ssu Khu Chhüan Shu* Encyclopaedia, made by imperial order.
Chhing, +1782.
Ed. Chi Yün 紀昀.
Indexes by Yang Chia-Lo; Yü and Gillis. Yin-Te index, no. 7.

*Chhü Hsien Shen Yin Shu* 臞仙神隱書.
Book of Daily Occupations for Scholars in Rural Retirement, by the Emaciated Immortal.
Ming, *c.* 1440.
Compiler, Chu Chhüan (Ning Hsien Wang) 朱權(寧獻王).

*Chhu Tzhu* 楚辭.
Elegies of Chhu (State) [or, Songs of the South].
Chou, *c.* −300, (with Han additions).
Chhü Yuan 屈原 (& Chia I 賈誼, Yen Chi 嚴忌, Sung Yü 宋玉, Huainan Hsiao-Shan 淮南小山 *et al.*)
Partial tr. Waley (23); tr. Hawkes (1).

*Chhüan Thang Wên.*
See Tung Kao (*1*).

*Chhün Fang Phu* 羣芳譜.
The Assembly of Perfumes [thesaurus of botany].
Ming, *c.* +1621.
Ed. Wang Hsiang-Chin 王象晉.

*Chi Chu Thiao Chien* 積貯條件.
Desiderata for the Storage of Grains.
Ming, late +16th century.
Lü Khun 呂坤.
Repr. *SSTK* 57/5b–7*a*.

*Chi Chung Chou Shu* 汲冢周書.
The Books of (the) Chou Dynasty found in the Tomb at Chi.
= *I Chou Shu* (q.v.).

*Chi Chiu (Phien)* 急就(篇).
Handy Primer [orthographic word-lists intended for verbal exposition, connected with a continuous thread of text, and having some rhyme arrangements].
C/Han, between −48 and −33.
Shih Yu 史游 with +7th cent. commentary by Yen Shih-Ku 顏師古 and +13th cent. commentary by Wang Ying-Lin 王應麟.

*Chi Pei Shu* 吉貝疏.
Monograph on Cotton [also known as *Chung Mien Hua Fa*].
Ming, +17th century.
Hsü Kuang-Chhi 徐光啓.
The original monograph no longer survives, but was probably incorporated into the *Nung Cheng Chhüan Shu*.

*Chi Le Pien* 雞肋編.
The Chicken-Rib Registers.
S/Sung, *c.* 1130.
Chuang Chi-Yü 莊季裕.

*Chi Ni Tzu* 計倪子.
[*Fan Tzu Chi Jan* 范子計然].
The Book of Master Chi Ni
Chou (Yüeh), −4th century.
Attr. Fan Li 范蠡 recording the philosophy of his master Chi Jan 計然.

*Chi Yün* 集韻.
Complete Dictionary of the Sounds of Characters [cf. *Chhieh Yün* and *Kuang Yün*].
Sung, +1037.
Compiled by Ting Tu 丁度 *et al.*
Possibly completed in +1067 by Ssuma Kuang 司馬光.

*Chiao-Chou I Wu Chih* 交州異物志.
Strange Things (incl. plants and animals) in Chiao-chou (District), [mod. Annam].
H/Han, *c.* +90.
Yang Fu 楊孚.
Perhaps the later title of his *Nan I I Wu Chih*, q.v. Extant only in quotations.

*Chieh Tzu Yuan Hua Chuan* 芥子園畫傳.
The Mustard-Seed Garden Guide to Painting.
Chhing, +1679, with later continuations.
Li Li-Ong (preface) 李笠翁. Wang Kai (text and illustrations) 王概.

*Chih Fu Chhüan Shu* 致富全書.
The Complete Way to Wealth.

*Chih Fu Chhüan Shu* (*cont.*)
Late Ming or early Chhing, +17th century.
Attr. to Chhen Mei-Kung 陳眉公 with additions by the Recluse of Bell Mountain 鍾山逸叟, but in fact probably a bookseller's compilation.
*Chih Pen Thi Kang* 知本提綱.
Selected Guidelines to an Understanding of the Fundamental Occupation [Agriculture].
Chhing, +1747.
Yang Tshui 楊屾; comm. Cheng Shih-To 鄭世鐸. In Wang Yü-Hu (*2*).
*Chin Shu* 晉書.
History of the Chin Dynasty [+265 to +419].
Thang, +635.
Fang Hsüan-Ling 房玄齡.
A few chs. tr. Pfizmaier (54–57); the astronomical chs. tr. Ho Ping-Yü (1). For translations of passages see the index of Frankel (1).
*Ching Chhu Sui Shih Chi* 荊楚歲時記.
Annual Folk Customs of the States of Ching and Chhu [i.e. of the districts corresponding to those ancient states; Hupei, Hunan and Kiangsi].
Prob. Liang, *c.* +550, but perhaps partly Sui, *c.* +610.
Tsung Lin 宗懍.
Cf. des Rotours (1), p. cii.
*Chiu Chia Chi Chu Tu Shih* 九家集注杜詩.
Poems of Tu (Fu), Collected and Annotated by Nine Scholars.
Sung.
Kuo Chih-Ta (ed.) 郭知達.
*Chiu Huang Huo Min Shu* 救荒活民書.
The Rescue of the People; a Treatise on Famine Prevention and Relief.
S/Sung, late +12th century or later.
Ed. Tung Wei 董煟.
*Chiu Huang Pen Tshao* 救荒本草.
Treatise on Wild Food Plants for Use in Emergencies.
Ming, +1406, repr. 1525, 1555 etc.
Chu Hsiao 朱橚 (prince of the Ming) Chou Ting Wang 周定王.
Incl. as chs. 46 to 59 of *Nung Cheng Chhüan Shu* (q.v.).
*Chiu Ku Khao* 九穀考.
A Study of the Nine (Cereal) Grains.
Chhing, *c.*. +1790.
Chheng Yao-Thien 程瑤田.
(In *HCCC*, ch. 551.)
*Chiu Thang Shu* 舊唐書.
Old History of the Thang Dynasty [+618 to +906].
Wu Tai, +945.
Liu Hsü 劉昫.
Cf. des Rotours (2), p. 64.
For translations of passages see the index of Frankel (1).
*Chou Kuan* 周官.
See *Chou Li*.
*Chou Kuan I Su* 周官義疏.
Collected Commentaries and Text of the *Record of the Institutions (lit. Rites) of (the) Chou (Dynasty)* (imperially commissioned).
Chhing, +1748.
Ed. Fang Pao *et al.* 方苞.
*Chou Li* 周禮.
Record of the Institutions (lit. the Rites) of (the) Chou (Dynasty) [descriptions of all government official posts and their duties].
C/Han (perhaps containing some material from late Chou).
Compilers unknown.
Tr. E. Biot (1).
*Chü Chia Pi Yung Shih Lei Chhüan Chi* 居家必用事類全集.
Collection of Certain Sorts of Techniques Necessary for Households (encyclopaedia).
Yuan, +1301.
Prob. Hsiung Tsung-Li 熊宗立.
Ed. Ming, +1560, Thien Ju-Chheng 田汝成.
Partly repr. in Shinoda & Tanaka (*1*).
*SKCS/TMTY*, ch. 130, p. 75*a*.
*Chu Chhi Thu Shuo* 諸器圖說.
Diagrams and Explanations of a Number of Machines [mainly of his own invention or adaptation].
Ming, +1627.
Wang Cheng 王徵.
*Chu Tzu She Tshang Fa* 朱子社倉法.
Master Chu [Hsi] on Managing Communal Granaries.
Sung, +1182.
Chu Hsi 朱熹.
Repr. in *SSTK* 56/4*a*–6*b*.
*Chu Wen-Kung Wen Chi* 朱文公文集.
See *Hui-An Hsien-Sheng Chu Wen-Kung Wen Chi*.
*Chu Wei Shuo* 築圍說.
Remarks on the Construction of Dyked Fields.
Early Chhing.
Chhen Hu 陳瑚.
*Chu Yü Thu Shuo* 築圩圖說.
Illustrated Remarks on the Construction of Poldered Fields.
Chhing, probably 17th century.
Sun Chün 孫峻.
*Chün Tzu Thang Jih Hsün Shou Ching* 君子堂日詢手鏡.
A Hand-Mirror of Daily Deliberations in the Gentlemen's Hall.
Ming, +1522.
Wang Chi 王濟.
*Chung I Pi Yung* 種藝必用.
Everyman's Guide to Agriculture (lit. What one must Know and Do in the Art of Crop-Raising).
Sung, ca. +1250.
Wu I (or Tsuan)吳懌欑. With Supplement (*Pu I*) by Chang Fu 張福. (Yuan *c.* +1275).
In *Yung-Lo Ta Tien*, Ch. 13, 194. Ed. Hu Tao-Ching.
*Chung I Pi Yung Pu I* 種藝必用補遺.
An Expansion of *Everyman's Guide to Agriculture*.

*Chung I Pi Yung Pu I* (*cont.*)
Early Yuan, *c.* +1275.
Chang Fu 張福.
Ed. Hu Tao-Ching (*4*), Nung Yeh, 1963.

*Chung Mien Hua Fa* 種棉花法.
Cotton Growing.
Ming, +17th century.
Hsü Kuang-Chhi 徐光啓.
See *Chi Pei Shu*.

*Chung Shih Chih Shuo* 種蒔直說.
Plain Words on Agriculture.
Prob. Chin or early Yuan.
Author unknown.
No longer extant except in quotations.

*Chung Shu Shu* 種樹書.
Book of Tree Planting.
Ming.
Yü Tsung-Pen 俞宗本.

*Chung Yü Fa* 種芋法.
Treatise on Tuber Cultivation.
Ming, *c.* 1538.
Huang Hsing-Tseng 黃省曾.

*Erh Ya* 爾雅.
The Literary Expositor [dictionary].
Chou material, stabilised in Chhin or C/Han.
Compiler unknown.
Enlarged and commented on *c.* +300 by Kuo Pho 郭璞.
Yin-Te Index no. (suppl.) 18.

*Fan Sheng-Chih Shu* 氾勝之書.
[= *Chung Chih Shu*.]
The Book of Fan Sheng-Chih on Agriculture.
C/Han, late −1st century (*c.* −10).
Fan Sheng-Chih 氾勝之.
In *YHSF*, ch. 69, pp. 50 *a* ff.
Tr. Shih Sheng-Han (2).

*Fan Tzu Chi Jan* 范子計然.
See *Chi Ni Tzu*.

*Fang Yen* 方言.
Dictionary of Local Expressions.
C/Han, *c.* −15 (but much interpolated later).
Yang Hsiung 揚雄.

*Fu Huang Khao* 捕蝗考.
A Treatise on Catching Locusts.
Chhing, +18th century.
Chhen Fang-Sheng 陳芳生.

*Han Shih Chih Shuo* 韓氏直說.
Master Han's Plain Words [on agriculture].
Probably Chin or early Yuan.
Author unknown.
No longer extant except in quotations.

*Heng Chhan So Yen* 恆產瑣言.
Remarks on Real-Estate.
Chhing, *c.* 1697.
Chang Ying 張英.
Tr. H. Beattie (1).

*Ho Phu* 禾譜.
Monograph on Cereals.
Sung, late 11th century.
Tsheng An-Chih 曾安止.

*Hou Chi Shu* 反稷書.
The Book of Lord Millet.
Late Chou.
Author unknown.
No longer extant except for certain sections incorporated into later works such as the *Lü Shih Chhun Chhiu*.

*Hou Han Shu* 後漢書.
History of the Later Han Dynasty [+25 to +220].
L/Sung, +450.
Fan Yeh 范曄.
The monograph chapters by Ssuma Piao 司馬彪 (d. +305), with commentary by Liu Chao 劉昭 (*c.* +510), who first incorporated them into the work.
A few ch. trs. Chavannes (6, 16); Pfizmaier (52, 53). Yin-Te Index, no. 41.

*Hsi Han Hui Yao* 西漢會要.
History of the Administrative Statutes of the Former (Western) Han Dynasty.
Sung, +1211.
Ed. Hsü Thien-Lin 徐天麟.
Cf. Teng & Biggerstaff (1), p. 158.

*Hsia Hsiao Cheng* 夏小正.
Lesser Annuary of the Hsia Dynasty.
Chou, between −7th and −4th century.
Writers unknown.
Incorporated in *Ta Tai Li Chi*, q.v.
Tr. Grynpas (1).
Tr. R. Wilhelm (6), Soothill (5).

*Hsia Hsiao Cheng Su I* 夏小正疏義.
Commentary on the *Lesser Annuary of the Hsia Dynasty*.
Chhing.
Hung Chen-Hsuan 洪震煊.

*Hsiang-Shan Wen Chi* 象山文集.
The Collected Works of Hsiang-Shan [Lu Chiu-Yuan].
S/Sung, *c.* +1190.
Lu Chiu-Yuan 陸九淵.

*Hsiang Shan Yeh Lu* 湘山野錄.
Rustic Notes from Hsiang-Shan.
Sung, *c.* +1060.
Wen-Jung 文瑩.

*Hsiao Tai Li Chi*. See *Li Chi*.

*Hsin-An Chih* 新安志.
History of Hui-Chou (徽州) Prefecture in Anhwei.
S. Sung, +1175.
Lo Yuan 羅願.

*Hsin Hsiu Pen Tshao* 新修本草.
The New (lit. Newly Improved) Pharmacopoeia.
Thang, +659.
(Ed.) Su Ching (= Su Kung) 蘇敬 (蘇恭) and a commission of 22 collaborators under the direction first of Li Chi 李勣 and Yü Chih-Ning 于志寧, then of Chhangsun Wu-Chi 長孫無忌. This work was afterwards commonly but incorrectly known as *Thang Pen Tshao*. It was lost in China, apart from MS

*Hsin Hsiu Pen Tshao* (*cont.*)
fragments at Tunhuang, but copied by a Japanese in +731 and preserved in Japan though incompletely.

*Hsin Thang Shu* 新唐書.
New History of the Thang Dynasty [+618 to +906].
Sung, +1061.
Ouyang Hsiu 歐陽修 & Sung Chhi 宋祁.
Cf. des Rotours (2), p. 56.
Partial trs. des Rotours (1, 2); Pfizmaier (66–74). For translations of passages see the index of Frankel (1).
Yin-Te Index, no. 16.

*Hsü Hsien Chuan* 續仙傳.
Further Biographies of the Immortals.
Wu Tai (H/Chou). Between +923 and +936.
Shen Fen 沈汾.
In *YCCC*, ch. 113.

*Hsü Po Wu Chih* 續博物志.
Supplement to the *Record of the Investigation of Things*. (cf. *Po Wu Chih*).
Sung, mid +12th century.
Li Shih 李石.

*Hsü Thung Chih* 續通志.
The *Historical Collections* Continued (see *Thung Chih*) [to the end of the Ming Dynasty].
Chhing, commissioned +1767, pr. *c.* +1770.
Ed. Hsi Huang 嵇璜 *et al.*

*Hsü Wen Hsien Thung Khao* 續文獻通考.
Continuation of the Comprehensive Study of (the History of) Civilisation (cf. *Wen Hsien Thung Khao* and *Chhin Ting Hsü Wen Hsien Thung Khao*).
Ming, +1586, pr. +1603.
Ed. Wang Chhi 王圻.

*Hua Yung Yüeh Ling* 花傭月令.
Monthly Ordinances for the Flowers' Slaves.
Early Chhing.
Hsü Shih-Chhi 徐石麒.

*Huai Nan Tzu* 淮南子.
The Book of (the Prince of) Huainan [compendium of natural philosophy].
C/Han, *c.* −120.
Written by the group of scholars gathered by Liu An 劉安, Prince of Huai-Nan.
Partial trs. Morgan (1); Erkes (1); Hughes (1); Chatley (1); Wieger (2).
Chung-Fa Index, no. 5. *TT*/1170.

*Huang Chhing Ching Chieh* 皇清經解.
Collection of (more than 180) Monographs on Classical Subjects writen during the Chhing Dynasty.
See Yen Chieh (*1*) (ed.).

*Hui-An Hsien-Sheng Chu Wen-Kung Wen Chi* 晦庵先生朱文公文集.
The Collected Works of Chu Wen-Kung [Chu Hsi], Master of Hui-An.
Sung, late +12th century.
Chu Hsi 朱熹.

*Hyakushō Denki* 百姓傳記.
Peasants' Chronicle.
+1684, Japan.
Author unknown.

*I Chhang Phing Tshang Ao Shen Wen* 議常平倉廒申文.
A Report to the Emperor on the Institution of Ever-Normal Granaries.
Ming, *c.* 1570.
Chang Chhao-Jui 張朝瑞.
Repr. *SSTK* 55/6*b*–10*a*.

*I Ching* 易經.
The Classic of Changes [Book of Changes].
Chou with C/Han additions.
Compilers unknown.
See Li Ching-Chhih (*1*, *2*); Wu Shih-Chhang (1).
Tr. R. Wilhelm (2), Legge (9), de Harlez (1).
Yin-Te Index, no. (suppl.) 10.

*I Chou Shu* 逸周書.
[= *Chi Chung Chou Shu*]
Lost Records of the Chou (Dynasty).
Chou, −245 and before, such parts as are genuine. Found in the tomb of An Li Wang (r. −276 to −245), a prince of the Wei state, in +281.
Writers unknown.

*I Wu Chih* 異物志.
See *Nan I I Wu Chih*.

*Kan Shu Su* 甘藷疏.
On the Sweet Potato.
Ming, +1608.
Hsü Kuang-Chhi 徐光啓.

*Keng Chih Thu* 耕織圖.
Pictures of Tilling and Weaving.
Sung, presented in MS, +1145, and perhaps first printed from wood blocks at that time; engraved on stone, +1210, and probably then printed from wood blocks.
Lou Shou 樓璹.
The illustrations published by Franke (11) are those of +1462 and +1739; Pelliot (24) published a set based on an edition of +1237. The original illustrations are lost, but cannot have differed much from these last, which include the poems of Lou Shou. The first Chhing edition was in +1696.

*Khai-Chi Chih* 會稽志.
Gazetteer of Khai-Chi [Chekiang].
S/Sung, +1201 ed.
Compiled Shih Su 施宿 *et al.*

*Khao Kung Chi* 考工記.
The Artificers' Record [a section of the *Chou Li*, q.v.].
Chou and Han, perhaps originally an official document of Chhi State, incorporated *c.* −140.
Compiler unknown.
Tr. E. Biot (1).
Cf. Kuo Mo-Jo (*1*); Yang Lien-Sheng (7).

*Khao Kung [Chi] Chhuang Wu Hsiao Chi* 考工創物小記.

*Khao Kung [Chi] Chhuang Wu Hsiao Chi (cont.)*
See Chheng Yao-Thien (*2*).
*Khao Kung Chi Chieh* 考工記解.
The *Artificers' Record* Explicated.
Sung, *c.* +1235.
Lin Hsi-I 林希逸.
*Khao Kung Chi Thu* 考工記圖.
Illustrations for the *Artificers' Record* (of the *Chou Li*) (with a critical archaeological analysis).
Chhing, +1746.
Tai Chen 戴震.
In *HCCC*, Chs 563, 564; reprinted Shanghai, 1955.
See Kondō (*1*).
*Ku Chin Thu Shu Chi Chheng* 古今圖書集成.
See *Thu Shu Chi Chheng*.
*Kuan Tzu* 管子.
The Book of Master Kuan.
Chou and C/Han. Perhaps mainly compiled in the Chi-Hsia Academy (late −4th century) in part from older materials.
Attrib. Kuan Chung 管仲.
Partial trs. Haloun (2, 5); Than Po-Fu *et al.* (1).
*Kuang Chhün Fang Phu* 廣羣芳譜.
The *Assembly of Perfumes* Enlarged [thesaurus of botany].
Chhing, +1708.
Wang Hao (ed.) 王灝.
*Kuang Chih* 廣志.
Extensive Records of Remarkable Things.
Chin, late +4th century.
Kuo I-Kung 郭義恭.
*YHSF*, ch. 74.
*Kuang Tung Hsin Yü* 廣東新語.
New Descriptions of Kwangtung Province.
Chhing, late +17th century.
Chhü Ta-Chün 屈大均.
*Kuang Ya* 廣雅.
Enlargement of the *Erh Ya*; *Literary Expositor* [dictionary].
San Kuo (Wei) +230.
Chang I 張揖.
*Kuang Yün* 廣韻.
Revision and Enlargement of the *Dictionary of Characters arranged according to Their Sounds when Split* [rhyming phonetic dictionary based on, and including, the *Chhieh Yün* and the *Thang Yün*, q.v.]
Sung, +1011.
Chhen Pheng-Nien 陳彭年, Chhiu Yung 丘雍. *et al.*
T & B, p. 203.
*Kuo Yü* 國語.
Discourses of the (ancient feudal) States.
Late Chou, Chhin and C/Han, containing much material from ancient written records.
Writers unknown.

*Lei Ssu Ching* 耒耜經.
The Classic of the Plough.
Thang, *c.* +880.
Lu Kuei-Meng 陸龜蒙.
*Li Chi* 禮記.
[*Hsiao Tai Li Chi.*]
Record of Rites [compiled by Tai the Younger]. (Cf. *Ta Tai Li Chi.*)
Ascr. C/Han, *c.* −70 to −50, but really H/Han, between +80 and +105, though the earliest pieces included may date from the time of the *Analects* (*c.* −465 to 450).
Attrib. ed. Tai Sheng 戴聖.
Actual ed. Tshao Pao 曹褒.
Trs. Legge (7); Couvreur (3); R. Wilhelm (6).
Yin-Te Index, no. 27.
*Li Chih Phu* 荔枝譜.
Monograph on the Lichi (*Nephelium litchi*).
Sung, +1059.
Tshai Hsiang 蔡襄.
*Lieh Hsien Chhüan Chuan* 列仙全傳.
Complete Collection of the Biographies of the Immortals.
Ming, *c.* +1580.
Wang Shih-Chen 王世貞.
Collated and corrected by Wang Yün-Pheng 汪雲鵬.
*Lieh Hsien Chuan* 列仙傳.
Collection of the Biographies of the Immortals.
Chin, +3rd or +4th century.
Attr. Liu Hsiang 劉向.
*Lin-Chhuan Hsien-Sheng Wen Chi* 臨川先生文集.
The Collected Works of Master [Wang] Lin-Chhuan.
Sung, +1140.
Wang An-Shih 王安石.
Ed. Chan Ta-Ho 詹大和.
*Ling Piao Lu I* 嶺表錄異.
Strange Southern Ways of Men and Things [on the special characteristics and natural history of Kwangtung].
Thang & Wu Tai, between *c.* +895 and +915.
Liu Hsün 劉恂.
*Ling Wai Tai Ta* 嶺外代答.
Information on What is Beyond the Passes (lit. a book in lieu of individual replies to questions from friends).
Sung, +1178.
Chou Chhü-Fei 周去非.
*Liu Chhing Jih Cha* 劉青日札.
Diary on Bamboo Tablets.
Ming, +1579.
Thien I-Heng 田藝衡.
*Liu Meng-Te Wen Chi* 劉夢得文集.
The Collected Works of Liu Meng-Te.
Thang.
Liu Yü-Hsi 劉禹錫.
*Liu Pu Chheng Yü Chu Chieh* 六部成語註解.
The Terminology of the Six Boards, with Explanatory Notes.
Chhing, text +1742, notes *c.* 1875.
Writer unknown.
Commentator unknown.
Tr. E-Tu Zen Sun (1).
*Lü Shih Chhun Chhiu* 呂氏春秋.
Master Lü's Spring and Autumn Annals

*Lü Shih Chhun Chhiu* (*cont.*)
[compendium of natural philosophy].
Chou (Chhin), −239.
Written by the group of scholars gathered by Lü Pu-Wei 呂不韋.
Tr. R. Wilhelm (3).
Chung-Fa Index, no. 2.

*Lun Heng* 論衡.
Discourses Weighed in the Balance.
H/Han, +82 or +83.
Wang Chhung 王充.
Tr. Forke (4); cf. Leslie (3).
Chung-Fa Index, no. 1.

*Lun Yü* 論語.
Conversations and Discourses (of Confucius) [perhaps Discussed Sayings, Normative Sayings, or Selected Sayings]; Analects.
Chou (Lu), *c.* −465 to −450.
Compiled by disciples of Confucius (chs. 16, 17, 18 and 20 are later interpolations).
Tr. Legge (2); Lyall (2); Waley (5); Ku Hung-Ming (1).
Yin-Te Index no. (suppl.) 16.

*Lung-Yen Hsien Chih* 龍巖縣志.
Gazetteer of Lung-Yen Country [Fukien].
Ming, +1558.
Thang Hsiang-Hsiu, 湯相修.
Ed. Mo Khang 莫亢 *et al.*

*Ma Shih Thung Khao* 馬氏通考.
See *Wen Hsien Thung Khao*.

*Ma Shou Nung Yen* 馬首農言.
Farming Precepts of Horse-Head District [Shensi].
Chhing, +1836.
Chhi Chün-Tsao 祁嶲藻.
Repr. Wang Yü-Hu (*2*).

*Man Shu* 蠻書.
Documents Relating to the Man Tribes.
Thang, *c.* +860.
Fan Chho 樊綽.

*Meng Chhi Pi Than* 夢溪筆談.
Dream Pool Essays.
Sung, +1086; last supplement dated +1091.
Shen Kua 沈括.
Ed. Hu Tao-Ching (*1*); cf. Holzman (1).

*Meng Chhi Wang Huai Lu* 夢溪忘懷錄.
Things Forgotten and Remembered by the Dream Pool.
Sung, *c.* +1090.
Shen Kua 沈括.
Cf. Hu Tao-Ching (*13*).

*Meng Tzu* 孟子.
The Book of Master Meng (Mencius).
Chou, *c.* −290.
Meng Kho 孟軻.
Tr. Legge (3); Lyall (1).
Yin-Te Index, no. (suppl.) 17.

*Mien Yeh Thu Shuo* 棉業圖說.
See Min. Ag. Ind. & Trade (*1*).

*Min Shu* 閩書.
History of Fukien.
Ming, +1629.
Ho Chhiao-Yuan 何喬遠.

*Ming Shih* 明史.
History of the Ming Dynasty [+1368 to +1643].
Chhing, begun +1646, completed +1736, first pr. +1739.
Chang Thing-Yü 張廷玉 *et al.*

*Nan Fang Tshao Mu Chuang* 南方草木狀.
A Prospect of the Plants and Trees of the Southern Regions.
Chin, +304.
Hsi Han 嵇含.
Trs. H.-L. Li (11).

*Nan Fang Tshao Wu Chuang* 南方草物狀.
A Prospect of the Plants and Products of the Southern Regions.
Chin +3rd or +4th.
Hsü Chung 徐衷.
Extant only in questions, esp. in *Chhi Min Yao Shu* and *TPYL*.

*Nan I I Wu Chih* 南裔異物志.
Strange Things from the Southern Borders.
H/Han, *c.* +90.
Yang Fu 楊孚.
Perhaps the earlier title of his *Chiao-Chou I Wu Chih*, q.v. Extant only in quotations.

*Nan Shih* 南史.
History of the Southern Dynasties [Nan Pei Chhao period, +420 to +589].
Thang, *c.* +670.
Li Yen-Shou 李延壽.
For translations of passages see the index of Frankel (1).

*Nihon Eitai Gura* 日本永代蔵.
The Eternal Repository of Japan.
Edo, +1688.
Ihara Saikaku 井原西鶴.

*Nōgu Benri Ron* 農具便利論.
Treatise on Useful Farm Tools.
Edo, +1822.
Ōkura Nagatsune 大藏永常.

*Nōgyō Zensho* 農業全書.
Collected Writings on Agriculture.
Edo, +1697.
Miyazaki Yasusada 宮崎安貞.

*Nung Cheng Chhüan Shu* 農政全書.
Complete Treatise on Agricultural Administration.
Ming, composed 1625–28, printed 1639.
Hsü Kuang-Chhi 徐光啓; ed. Chhen Tzu-Lung 陳子龍.
Textual refs. are to the 1843 repr. of the Palace ed.; a new annotated ed. was published in Shanghai in 1979 by Shih Sheng-Han (*8*).

*Nung Chü Chi* 農具記.
A Record of Agricultural Implements.
Chhing, *c.* +1670 or later.
Chhen Yü-Chi 陳玉璂.

*Nung I Tsa Su* 農遺雜疏.
Miscellaneous Remarks on our Agricultural

*Nung I Tsa Su* (*cont.*)
Heritage.
Late Ming, pre-1620.
Hsü Kuang-Chhi 徐光啓.

*Nung Phu Pien Lan* 農圃便覽.
Handy Survey of Agriculture & Horticulture.
Chhing, +1755.
Ting I-Tseng 丁宜曾.
Ed. Wang Yü-Hu (*3*), Chung Hua, 1957.

*Nung Sang Chi Yao* 農桑輯要.
Fundamentals of Agriculture & Sericulture.
Yuan, +1273.
Preface by Wang Phan 王磐. Imperially commissioned, & produced by the Agricultural Extension Bureau 司農司.
Probable editor, Meng Chhi 孟祺. Probable later editors, Chhang Shih-Wen (*c.* 1286) 暢師文, Miao Hao-Chhien (*c.* 1318) 苗好謙.
Textual refs. are to the 1847 repr. of the Palace ed.

*Nung Sang I Shih Tsho Yao* 農桑衣食撮要.
Selected Essentials of Agriculture, Sericulture, Clothing & Food.
Yuan, *c.* +1314.
Ed. Lu Ming-Shan (Uighur) 魯明善.

*Nung Shu* 農書.
Agricultural Treatise.
Sung, +1149.
Chhen Fu 陳旉.

*Nung Shu* 農書.
Agricultural Treatise
Yuan, +1313.
Wang Chen 王禎.
Textual refs. are to the 22 *chüan* Palace ed. of 1783, prefaced 1774.

*Nung Shu* 農書.
Agricultural Treatise.
Late Ming.
Master Shen 沈氏.
Tr. into modern Chinese in Chhen Heng-Li & Wang Ta-Tshan (*1*), 210–50.

*Nung Shuo* 農說.
Agriculture Explained.
Ming, +16th century
Ma I-Lung 馬一龍.
Ed. Wang Yün-Wu (*1*), Commercial Press, 1936.

*Nung Thien Yü Hua* 農田餘話.
On Farming for Abundance.
Ming, +14th century, probably with later additions.
Chhang-Ku Chen-I 長谷眞逸.

*Nung Tshan Ching* 農蠶經.
Classic of Agriculture and Sericulture.
Chhing, +1705.
Compiled by Phu Sung-Ling 蒲松齡.

*Nung Ya* 農雅.
Agricultural Dictionary.
Chhing, +1813.
Compiled by Ni Cho 倪倬.
Reprinted Peking 1956.

*Nung Yen Chu Shih* 農言著實.
Farming Sayings Set Forth.
Later Chhing.
Yang Hsiu-Yuan 楊秀元.

*Pei Meng So Yen* 北夢瑣言.
Fragmentary Notes Indited North of (Lake) Meng.
Wu Tai (S/Phing) *c.* +950.
Sun Kuang-Hsien 孫光憲.
See des Rotours (4), p. 38.

*Pen Tshao Ching Chi Chu* 本草經集注.
Collected Commentaries on the Classical Pharmacopoeia (of the Heavenly Husbandman).
S/Chhi, +492.
Thao Hung-Ching 陶弘景.
Now extant only in fragmentary form as a Tunhuang or Turfan MS, apart from the many quotations in the pharmaceutical natural histories, under Thao Hung-Ching's name.

*Pen Tshao Kang Mu* 本草綱目.
The Great Pharmacopoeia; or, The Pandects of Pharmaceutical Natural History.
Ming, +1596.
Li Shih-Chen 李時珍.
Paraphrased and abridged tr. Read & collaborators (1–7) and Read & Pak (1), with indexes.

*Pen Tshao Thu Ching* 本草圖經.
Illustrated Pharmacopoeia; or, Illustrated Treatise of Pharmaceutical Natural History.
Sung, +1061.
Su Sung 蘇頌 *et al.*
Now preserved only in numerous quotations in the later pandects of pharmaceutical natural history.

*Pen Tshao Yen I* 本草衍義.
Dilations upon Pharmaceutical Natural History.
Sung, pref. +1116, pr. +1119, repr. +1185, +1195.
Khou Tsung-Shih 寇宗奭.
See also *Thu Ching Yen I Pen Tshao* (*TT*/761).

*Pien Min Thu Tsuan* 便民圖纂.
Everyman's Handy Illustrated Compendium; or, the Farmstead Manual.
Ming, +1502, repr. 1552 & 1593.
Author unknown, but could be Kuang Fan 廣璠.

*Po Wu Chih* 博物志.
Record of the Investigation of Things. (Cf. *Hsü Po Wu Chih.*)
Chin, *c.* +290 (begun about +270).
Chang Hua 張華.

*Pu Nung Shu* 補農書.
Supplement to the Treatise on Agriculture by Mr Shen [*Nung Shu*].
Ming, *c.* +1620.
Chang Lü-Hsiang 張履祥.
Tr. into modern Chinese in Chhen Heng-Li & Wang Ta-Tshan (*1*), 251–80.

*San Fu Huang Thu* 三輔黃圖.
Illustrated Description of the Three Cities of the Metropolitan Area (Chhang-an (mod. Sian), Feng-i and Fu-Feng).
Chin, original text late +3rd century, or perhaps H/Han; present version stabilised, including much older material, between +757 and +907.
Attr. Miao Chhang-Yen 苗昌言.
Cf. des Rotours (1), p. lxxxvi.

*San Nung Chi* 三農記.
Records of the Three Departments of Agriculture.
Chhing, preface +1760.
Chang Tsung-Fa 張宗法.

*San Tshai Thu Hui* 三才圖會.
Universal Encyclopaedia.
Ming, +1609.
Wang Chhi 王圻.

*Shan Hai Ching* 山海經.
Classic of the Mountains and Rivers.
Chou and C/Han.
Writers unknown.
Partial tr. de Rosny (1).
Chung-Fa Index, no. 9.

*Shang Chün Shu* 商君書.
Book of the Lord Shang.
Chou, −4th or −3rd century.
Kungsun Yang (attrib.) 公孫鞅.
Tr. Duyvendak (3).

*Shang-Hai Hsien Chih* 上海縣志.
Gazetteer of Shanghai County.
Chhing, +1684.
Shih Tshai-Hsiu, 史彩修, Yen Ying-Liu 葉映榴 *et al.*

*Shao-Hsing Chiao Ting Ching Shih Cheng Lei Pei Chi Pen Tshao* 紹興校定經史證類備急本草.
The Corrected Classified and Consolidated Armamentarium; Pharmacopoeia of the Shao-Hsing Reign-Period.
S/Sung, pres. +1157, pr. +1159, often copied and repr. especially in Japan.
Thang Shen-Wei 唐愼微.
Ed. Wang Chi-Hsien 王繼先 *et al.*
Cf. Nakao Manzō (*1*, 1); Swingle (11).
Illustrations reported in facsimile by Wada (*1*); Karrow (2).
Facsimile edition of a MS in the Library of Ryokoku University, Kyoto 龍谷大學圖館.
Ed. with an analytical and historical introduction, including contents-table and indexes (別册) by Okanishi Tameto 岡西爲人 (Shunyōdō, Tokyo, 1971).

*Shao-Hsing Fu Chih* 紹興府志
Gazetteer of Shao-Hsing (Chekiang).
Chhing, +1792.
Li Heng-The 李亨特.
Ed. Phing Shu 平恕 *et al.*

*She Tshang Fa* 社倉法.
The Administration of Communal (Village) Granaries.
Sung, +1182.
Chu Hsi 朱熹.
Repr. *SSTK* 56/4*a*–6*b*.

*Shen Hsien Chuan* 神仙傳.
Lives of the Holy Immortals. (Cf. *Lieh Hsien Chuan, Lieh Hsien Chhüan Chuan, Hsü Shen Hsien Chuan* and *Shen Hsien Thung Chien.*)
Chin, +4th century.
Attrib. Ko Hung 葛洪.

*Shen Hsien Thung Chien* 神仙通鑑.
(cf. (*Li Tai*) *Shen Hsien* (*Thung*) *Chien*).
General Survey of the Lives of the Holy Immortals.
Ming, +1640.
Hsüeh Ta-Hsün 薛大訓.

*Shen Nung Pen Tshao Ching* 神農本草經.
Classical Pharmacopoeia of the Heavenly Husbandman.
C/Han, based on Chou and Chhin material, but not reaching final form before the +2nd century.
Writers unknown.
Lost as a separate work, but the basis of all subsequent compendia of pharmaceutical natural history, in which it is constantly quoted.
Reconstituted and annotated by many scholars; see Lung Po-chien (*1*), pp. 2 ff, 12 ff.
Best reconstructions by:
Mori Tateyuki 森立之 (1845).
Liu Fu 劉復 (1942).

*Shen Nung Shu* 神農書.
The Book of the Heavenly Husbandman.
Prob. late Chou.
Author unknown.
No longer extant, but quoted extensively in later agricultural works, particularly the *Lü Shih Chhun Chhiu*.
*YHSF*, 69/2*a* ff.

*Shen Shih Nung Shu* 沈氏農書.
See *Nung Shu*.

*Shih Chi* 史記.
Historical Record (down to −99).
C/Han, *c.* −90.
Ssuma Chhien 司馬遷, and his father Ssuma Than 司馬談.
Partial trs. Chavannes (1); Pfizmaier (13–36); Hirth (2); Wu Khang (1); Swann (1) etc.
Yin-Te Index, no. 40.

*Shih Ching* 詩經.
Book of Odes [ancient folksongs].
Chou, −11th to −7th centuries (Dobson's dating).
Writers and compilers unknown.
Tr. Legge (8); Waley (1); Kalgren (14).

*Shih Ming* 釋名.
Expositor of Names.
Early +2nd century.
Liu Hsi 劉熙.

*Shih Pen* 世本.
Book of Origins [imperial genealogies, family names, and legendary inventors].
C/Han (incorporating Chou material), −2nd

*Shih Pen (cont.)*
century.
Ed. Sung Chung 宋衷 (H/Han).
*Shika Nōgyō Dan* 私家農業談.
Remarks on Smallholding.
Edo, +1788.
Miyanaga Masatsura 宮長正行.
*Shou Shan Ko Tshung Shu* 守山閣叢書.
See Chhien Hsi-Tsu (*1*) (ed.)
*Shou Shih Thung Khao* 授時通考.
Compendium of Works & Days.
Chhing, +1742.
Compiled by imperial order under the direction of O-Erh-Thai 鄂爾泰.
Textual refs. are to the 1847 repr. of the original 1742 Palace ed.
*Shu Hsü Pu* 樹畜部.
The Division of Trees and Livestock.
Ming, +1504.
Sung Hsü 宋詡.
*Shui Ching* 水經.
The Waterways Classic [geographical account of rivers and canals].
Ascr. C/Han, prob. San Kuo.
Attrib. Sang Chhin 桑欽.
*Shui Ching Chu* 水經注.
Commentary on the *Waterways Classic* [geographical account greatly extended].
N/Wei, late +5th/early +6th century.
Li Tao-Yuan 酈道元.
*Shuo Wen Chieh Tzu* 說文解字.
Analytical Dictionary of Characters (lit. Explanations of Simple Characters and Analysis of Composite Ones).
H/Han, +121.
Hsü Shen 許慎.
*So Shan Nung Phu* 梭山農譜.
A Survey of the Agriculture of Shuttle Mountain.
Chhing, +1717.
Liu Ying-Thang 劉應棠.
Ed. Wang Yü-Hu (7), Nung Yeh, 1960.
*Ssu Khu Chhüan Shu* 四庫全書.
See *Chhin Ting Ssu Khu Chhüan Shu*, etc.
*Ssu Min Yüeh Ling* 四民月令.
Monthly Ordinances for the Four Sorts of People.
H./Han, *c.* +160.
Tshui Shih 崔寔.
Tr. Herzer (1).
*Ssu Shih Tsuan Yao* 四時纂要.
Important Rules for the Four Seasons.
Thang, *c.* +750.
Compiled Han O 韓鄂.
*Sui Shih Kuang Chi* 歲時廣記.
Expanded Records of the Annual Seasons.
Sung.
Chhen Yuan-Ching 陳元靚.
*Sui Shu* 隋書.
History of the Sui Dynasty [+581 to +617].
Thang, +636 (annals and biographies); +656 (monographs and bibliography).
Wei Cheng 魏徵 *et al.*
Partial trs. Pfizmaier (61–65); Balazs (7, 8); Ware (1).
For translations of passages see the index of Frankel (1).
*Sung-Chiang Fu Hsü Chih* 松江府續志.
The Gazetteer of Sung-Chiang Prefecture (Kiangsu) Continued.
Chhing, pr. +1884.
Po Jun-Hsiu 博潤修.
Ed. Yao Kuang-Fa 姚光發 *et al.*
*Sung Hui Yao Kao* 宋會要稿.
Drafts for the *History of the Administrative Statutes of the Sung Dynasty*.
Sung.
Collected by Hsü Sung 徐松 (1809) from the *Yung-Lo Ta Tien*.
*Sung Kao Shan Chi* 嵩高山記.
Descriptions of Lofty Mountains.
Prob. N. Dynasties.
Lu Hsiu 盧㑃.
*Sung Shih* 宋史.
History of the Sung Dynasty [+960 to +1279].
Yuan, *c.* +1345.
Tho-Tho (Toktaga) 脫脫 & Ouyang Hsüan 歐陽玄.
Yin-Te Index, no. 34.

*Ta Chhing Hui Tien* 大淸會典.
History of the Administrative.
Statutes of the Chhing Dynasty.
Chhing: 1st ed. +1690; 2nd +1733; 3rd. +1767; 4th +1818; 5th +1899.
Ed. Wang An-Kuo 王安國 and many others.
*Ta-Li Fu Chih* 大理府志.
Gazetteer of Ta-Li [W. Yünnan].
Ming, +1563 ed.
Li Yuan-Yang 李元陽.
*Ta Tai Li Chi* 大戴禮記.
Record of Rites [compiled by Tai the Elder.] (Cf. *Hsiao Tai Li Chi*; *Li Chi*).
C/Han, *c.* −70 to −50 but really H/Han, between +80 and +105.
Attrib. ed. Tai Te 戴德; in fact probably ed. Tshao Pao 曹褒.
See Legge (7).
Tr. Douglas (1); R. Wilhelm (6).
*Tao Phin* 稻品.
The Varieties of Rice.
Ming, *c.* +1550.
Huang Hsing-Tseng 黃省曾.
*Tao Tsang* 道藏.
The Taoist Patrology [containing 1464 Taoist works].
All periods, but first collected in the Thang about +730, then again in +870 and definitively in +1019. First printed in the Sung (+1111 to +1117). Also printed in J/Chin (+1168 to +1191), Yuan (+1244), and Ming (+1445, +1598 and +1607).
Writers numerous.

*Tao Tsang* (*cont.*)
Indexes by Wieger (6), on which see Pelliot's review (58), and Ong Tu-Chien (*1*).
Yin-Te Index, no. 25.

*Thai Hsi Shui Fa* 泰西水法.
Hydraulic Machinery of the West.
Ming, +1612.
Hsiung San-Pa (Sabatino de Ursis). 熊三拔.
& Hsü Kuang-Chhi 徐光啓.

*Thai-Phing Yü Lan* 太平御覽.
Thai-Phing reign-period Imperial Encyclopaedia (lit. the Emperor's Daily Readings).
Sung, +983.
Ed. Li Fang 李昉.
Some chs. tr. Pfizmaier (84–106).
Yin-Te Index, no. 23.

*Thang Hui Yao* 唐會要.
History of the Administrative Statutes of the Thang Dynasty.
Sung, +961.
Wang Phu 王溥.
Cf. des Rotours (2), p. 92.

*Thang Liu Tien* 唐六典.
Institutes of the Thang Dynasty (lit. Administrative Regulations of the Six Ministries of the Thang).
Thang, +738 or +739.
Ed. Li Lin-Fu 李林甫.
Cf. des Rotours (2), p. 99.

*Thang Lü Su I* 唐律疏議.
Commentary on the Penal Code of the Thang Dynasty [imperially ordered].
Thang, +653.
Ed. Chhangsun Wu-Chi 長孫無忌.
Cf. des Rotours (2), p. 98.

*Thang Shu.*
See *Chiu Thang Shu* and *Hsin Thang Shu.*

*Thang Shuang Phu* 糖霜譜.
Monograph on Sugar.
Sung, +1154.
Wang Cho 王灼.

*Thang Yüeh Ling* 唐月令.
Monthly Ordinances of the Thang.
Thang.
Li Lung-Chi 李隆基.

*Thang Yün* 唐韻.
Thang Dictionary of Characters arranged according to their Sounds [rhyming phonetic dictionary based on, and including, the *Chhieh Yün*, q.v.].
Thang, +677, revised and republished +751.
Chhangsun No-Yen 長孫訥言(+7th) & Sun Mien 孫愐(+8th).
Now extant only within the *Kuang Yün*, q.v.

*Thien Chia Wu Hsing* 田家五行.
The Farmer's Book of Five-Element (Natural Philosophy).
Sung.
Lou Yuan-Shan 婁元善.
Cf. Ho & Needham (1).

*Thien Kung Khai Wu* 天工開物.
The Exploitation of the Works of Nature.
Ming, +1637.
Sung Ying-Hsing 宋應星.
Tr. Sung (1).

*Thu Ching (Pen Tshao)* 圖經(本草).
Illustrated Treatise (of Pharmaceutical Natural History). See *Pen Tshao Thu Ching.*
The term *Thu Ching* applied originally to one of the two illustrated parts (the other being a *Yao Thu*) of the *Hsin Hsiu Pen Tshao* of +659 (q.v.); cf. *Hsin Thang Shu*, ch. 59, p. 21*a* or *TSCCIW*, p. 273. By the middle of the +11th century these had become lost, so Su Sung's *Pen Tshao Thu Ching* was prepared as a replacement. The name *Thu Ching Pen Tshao* was often afterwards applied to Su Sung's work, but (according to the evidence of the *Sung Shih* bibliographies, *SSIW*, pp. 179, 529) wrongly.

*Thu Ching Chi Chu Yen I Pen Tshao* 圖經集注衍義本草.
Illustrations and Collected Commentaries for the *Dilations upon Pharmaceutical Natural History*
*TT*/*761*. (Ong index, no. 767).
See also *Thu Ching Yen I Pen Tshao.*
The *Tao Tsang* contains two separately catalogued books, but the *Thu Ching Chi Chu Yen I Pen Tshao* is in fact the introductory 5 chapters, and the *Thu Ching Yen I Pen Tshao* the remaining 42 chapters of a single work.

*Thu Ching Yen I Pen Tshao* 圖經衍義本草.
Illustrations (and Commentary) for the *Dilations upon Pharmaceutical Natural History.* (An abridged conflation of the *Cheng-Ho .... Cheng Lei .... Pen Tshao* with the *Pen Tshao Yen I.*)
Sung, *c.* +1223.
Thang Shen-Wei 唐愼微, Khou Tsung-Shih 寇宗奭, ed. Hsü Hung 許洪.
*TT*/761; see also *Thu Ching Chi-Chu Yen I Pen Tshao.*
Cf. Chang Tsan-Chhen (*2*); Lung Po-Chien (*1*), nos. 38, 39.

*Thu Shu Chi Chheng* 圖書集成.
Imperial Encyclopaedia [or: Imperially Commissioned Compendium of Literature and Illustrations, Ancient and Modern].
Chhing, +1726.
Ed. Chhen Meng-Lei 陳夢雷.
Index by L. Giles (2).

*Thung Chih* 通志.
Historical Collections.
Sung, *c.* +1150.
Cheng Chhiao 鄭樵.
Cf. des Rotours (2), p. 85.

*To Neng Pi Shih* 多能鄙事.
Rustic Skills Surveyed.
Ming, earliest extant ed. +1540.
Liu Chi 劉基.

*Tsa Yin Yang (Shu)* 雜陰陽(書).
A Miscellany of Divinations.
Han.
Author unknown.

*Tsa Yin Yang (Shu) (cont.)*
No longer extant except in a few quotations, for example in the *Chhi Min Yao Shu*.

*Tsai Shih Shu* 宰氏書.
The Treatise [on agricultural divination] of Master Tsai.
Possibly the same work as the *Fan Tzu Chi Jan*, q.v.
Attr. to Tsai Chi-Jan 宰計然.
Cf. Wang Yü-Hu (*1*), p. 4.
*YHSF* 69/17*a* ff.

*Tshai Kuei Shu* 蔡癸書.
The [Agricultural] Treatise of Tshai Kuei.
C/Han.
Tshai Kuei 蔡癸.
*YHSF* 69/66*a* ff.

*Tshan Luan Lu* 驂鸞錄.
Guiding the Reins (a narrative of a three month journey from the capital to Kueilin).
Sung, +1172.
Fan Chheng-Ta 范成大.

*Tshang Ao I* 倉廒議.
A Treatise on Granaries.
Ming, *c.* +1570.
Chang Chhao-Jui 張朝瑞.
Repr. *SSTK* 57/2*a*–5*b*.

*Tshang Hsieh (Phien)* 倉頡(篇).
Book of Tshang Hsieh [legendary inventor of writing; an orthographic primer].
Chhin, *c.* −220.
Li Ssu 李斯. Edited by Chang I 張揖 (San Kuo, Wei) and Kuo Pho 郭璞 (Chin).
Reconstruction in *YHSF* ch. 59, pp. 18*a* ff.

*Tsuan Wen* 纂文.
Lexicography.
+5th century.
Ho Chheng-Thien 何承天.

*Tung An-Kuo Shu* 董安國書.
The [Agricultural] Treatise of Tung An-Kuo.
C/Han.
Tung An-Kuo 董安國.

*Wang Chen Nung Shu* 王禎農書.
See *Nung Shu*.

*Wang Lin-Chhuan Chi* 王臨川集.
See *Lin-Chhuan Hsien-Sheng Wen Chi*.

*Wang Shih Shu* 王氏書.
The [Agricultural] Treatise of Master Wang.
C/Han.
Master Wang 王氏.

*Wei Lüeh* 魏略.
Memorable Things of the Wei Kingdom (San Kuo).
San Kuo (Wei) or Chin, +3rd or +4th century.
Yü Huan 魚豢.

*Wen Chhang Tsa Lu* 文昌雜錄.
Things Seen and Heard by an Official at Court (during service in the Department of Ministries).
Sung, +1056.
Phang Yuan-Ying 龐元英.

*Wen Hsien Thung Khao* 文獻通考.
Comprehensive Study of (the History of) Civilisation (lit: Complete Study of the Documentary Evidence of Cultural Achievements (in Chinese Civilisation)).
Sung & Yuan, begun perhaps as early as +1270 and finished before +1317, printed +1322.
Ma Tuan-Lin 馬端臨.
Cf. des Rotours (2), p. 87.
A few chs. tr. Julien (2); d'Hervey St Denys (1).

*Wen Hsüan* 文選.
General Anthology of Prose and Verse.
Liang, +530.
Ed. Hsiao Thung 蕭統 (prince of the Liang).
Comm. Li Shan 李善 *c.* +670.
Tr. von Zach (6).

*Wu Ching Su* 蕪菁疏.
On The Rape Turnip.
Late Ming, *c.* +1600.
Hsü Kuang-Chhi 徐光啓.

*Wu-Chün Chih* 吳郡志.
Local Gazeteer of Wu-Chün [modern Suchou].
Sung; preface +1229, text somewhat earlier.
Fan Chheng-Ta 范成大.
*SSKTS* 48.

*Wu-Hsing Chang Ku Chi* 吳興掌故集.
The Collected Historical Records of Wu-Hsing (Chekiang).
Ming, +1560.
Hsü Hsien-Chung 徐獻忠.

*Yang Yü Yüeh Ling* 養餘月令.
Monthly Ordinances for Superabundance.
Ming, +1633.
Compiled by Tai Hsi 戴羲.

*Yeh Lao Shu* 野老書.
The [Agricultural] Treatise of the Old Countryman.
Chou (Chhi or Chhu).
Author unknown.
*YHSF* 69/9*a* ff.

*Yen Shih Chia Hsün* 顏氏家訓.
Mr. Yen's Advice to his family.
Sui, *c.* +590.
Yen Chih-Thui 顏之推.

*Yin Tu-Wei Shu* 尹都尉書.
The [Agricultural] Treatise of Marshal Yin.
C/Han, probably −2nd century.
Marshal Yin 尹都尉.
*YHSF* 69/43*a* ff.

*Ying-Chou Chih* 潁州志.
Gazetteer of Ying-Chou (N. Anhwei).
Ming, +1511 ed.
Li I-Chhun 李宜春.

*Ying Tsao Fa Shih* 營造法式.
Treatise on Architectural Methods.
Sung, +1097; printed +1103; reprinted +1145.
Li Chieh 李誡.

*Yung Chhuang Hsiao Phin* 湧幢小品.
Bagatelles from the Billowing Screens.
Ming, +1622.
Chu Kuo-Chen 朱國禎.

*Yung-Lo Ta Tien* 永樂大典.
Great Encyclopaedia of the Yung-Lo reign-period [only in manuscript].
Amounting to 22,877 chapters in 11,095 volumes, only about 370 still being extant.
Ming, +1407.
Ed. Hsieh Chin 解縉.
See Yuan Thung-Li (*1*).

*Yü Hai* 玉海.
Ocean of Jade [encyclopaedia].
Sung, +1267, not pr. till +1337/+1340 or +1351.
Wang Ying-Lin 王應麟.
Cf. des Rotours (2), p. 96.

*Yü Han Shan Fang Chi I Shu.*
See Ma Kuo-Han (*1*).

*Yü I-Chhi Chien* 俞益期牋.
The Memoranda of Yü I-Chhi.
E/Chin, +4th to +5th century.
Yü I-Chhi 俞益期.

*Yü Phien* 玉篇.
Jade Page Dictionary.
Liang, +543.
Ku Yeh-Wang 顧野王.
Extended and edited in the Thang (+674) by Sun Chhiang 孫强.

*Yuan Shih* 元史.
History of the Yuan (Mongol) Dynasty [+1206 to +1367].
Ming, *c.* +1370.
Sung Lien 宋濂.
Yin-Te Index, no. 35.

*Yüeh Chüeh Shu* 越絕書.
Lost Records of the State of Yüeh.
H/Han, *c.* +52.
Attrib. Yuan Khang 袁康.

*Yün Chi Chhi Chhien* 雲笈七籤.
The Seven Bamboo Tablets of the Cloudy Satchel [an important collection of Taoist material made by the editor of the first definitive form of the *Tao Tsang* (+1019), and including much material which is not in the Patrology as we now have it].
Sung, +1022.
Chang Chün-Fang 張君房.
*TT*/1020.

*Yün-Nan Thung Chih* 雲南通志.
Historical Records of Yunnan.
Ming, +1574 ed.
Li Yuan-Yang 李元陽.

ADDENDA

Liu Ho-Tung Chi 柳河東集.
Collected Works of Liu Tsung-Yuan.
Thang, early +9th century.
Liu Tsung-Yuan 柳宗元.

*Nung Hsüeh Tsuan Yao* 農學纂要.
Essentials of Agronomy.
Chhing, +1901.
Fu Tseng-Hsiang 傅增湘.

*Shu Ching* 書經.
Historical Classic [Book of Documents].
Chou, with later additions.
Writers unknown.
Tr. Medhurst (1); Legge (1, 10); Karlgren (12).

*Tso Chuan* 左傳.
Master Tsochhiu's Enlargement of the *Chhun Chhiu* (Spring and Autumn Annals), [dealing with the period −722/−468].
Chou, between −400 and −250, but with additions by Chhin and Han scholars.
Attrib. Tsochhiu Ming 左邱明.
See Karlgren (8); Maspero (1).
Tr. Couvreur (1); Legge (11).

*Yen Thieh Lun* 鹽鐵論.
Discourses on Salt and Iron [record of the debate of −81 on state control of commerce and industry].
C/Han, *c.* −80 to −60.
Huan Khuan 桓寬.
Partial tr. Gale (1).

*Yuan Shih* 元史.
History of the Yuan (Mongol) Dynasty [+1206 to +1367].
Ming, *c.* +1370.
Sung Lien 宋濂.
Yin-Te Index, no. 35.

*Yüeh Chüeh Shu* 越絕書.
Lost Records of the State of Yüeh.
H/Han, *c.* +52.
Attrib. Yuan Khang 袁康.

## B. CHINESE AND JAPANESE BOOKS AND JOURNAL ARTICLES SINCE +1800


Amano Motonosuke (*4*) 天野元之助.
*Chūgoku Nōgyōshi Kenkyū* 中国農業史研究.
Researches into Chinese Agricultural History.
Tokyo, 1962; 2nd, expanded ed. 1979.

Amano Motonosuke (*6*) 天野元之助.
*Tenkō Kaibutsu to Mindai no Nōgyō* 天工開物と明代の農業.
The *Thien Kung Khai Wu* and Ming Agriculture.
In Yabuuchi (*11*), 47.

Amano Motonosuke (*7*) 天野元之助.
*KoGi no Ka Shi-Kyō 'Seimin Yōjutsu' no Kenkyū.*
後魏の賈思勰"斉民要術"の研究.
Researches on the Later Wei Author Chia Ssu-Hsieh's *Chhi Min Yao Shu.*
In Yamada (*3*), 369–570.

Amano Motonosuke (*8*), reviser 天野元之助.
*Chūgoku Nogaku Shoroku* 中国農学書録.
Bibliography of Chinese Agriculture (revised ed. of Wang Yü-Hu (*1*)).
Ryukei Press, Tokyo, 1975.

Amano Motonosuke (*9*) 天野元之助.
*Chūgoku Konōsho Kō* 中国古農書考.
Researches on Ancient Chinese Agricultural Works.
Ryukei Press, Tokyo, 1975.

Amano Motonosuke (*10*) 天野元之助.
*Mindai Nōgyō no Tenkai* 明代農業の展開.
The Development of Ming Agriculture.
*SKS* (1958) 23, **5–6**, 19–40.

Amano Motonosuke (*11*) 天野元之助.
*Chūgoku Nōgyō no Chiiki-teki Tenkai* 中国農業の地域的展開.
The Regional Development of Chinese Agriculture.
Ryukei Press, Tokyo, 1979.

An Chih-Min (*3*) 安志敏.
*Chung-Kuo Ku-Tai ti Shih-Tao* 中國古代的石刀.
Stone Knives of Ancient China.
*KKHP*, 1955, **10**, 27–51.

An Chih-Min (*4*) 安志敏.
*Chung-Kuo ti Hsin-Shih-Chhi Shih-Tai* 中國的新石器時代.
The Chinese Neolithic.
*KK*, 1981, **3**, pp. 252–60.

Andō Kōtarō (*1*) 安藤廣太郎.
*Nihon Kodai Inasaku Shi Kenkyū* 日本古代稲作史研究.
A Study of the Early History of Rice Cultivation in Japan.
Tokyo, 1959.

Anon. (*18*).
*Chhüan Kuo Nung Chü Chan-Lan Hui Thui-Chien Chan Phin* 全國農具展覽會推薦展品.
Catalogue of Recommended Designs at the National Exhibition of Agricultural Machinery.
Peking, 1958.

Anon. (*28*).
*Yün-Nan Chin-Ning Shi-Chai-Shan Ku-Mu-Chhün Fa-Chüeh Pao-Kao* 雲南晉寧石寨山古墓羣發掘報告.
Report on the Excavation of a Group of Tombs at Shih-Chai-Shan, Chin-Ning, Yunnan.
Wen-Wu, Peking, 1959.

Anon. (*42*).
*Kuang-Chou Chhu-Thu Han-Tai Thao Wu* 廣州出土漢代陶屋.
Pottery Models of Dwellings excavated from Cantonese Tombs (including Granaries, Wellheads and Stoves).
Wen-Wu, Peking, 1958.

Anon. (*43*).
*Hsin Chung-Kuo-ti Khao-Ku Shou-Huo* 新中國的考古收獲.
Successes of Archaeology in New China.
Wen-Wu, Peking, 1961.
Names of the 22 writers in this collective work are given on p. 135.

Anon. (*109*).
*Chung-Kuo Kao-Teng Chih-Wu Thu-Chien* 中國高等植物圖鑑.
Iconographia Cormophytorum Sinicorum (Flora of Chinese Higher Plants).
2 vols., Kho-Hsüeh, Peking, 1972.

Anon. (*501*).
*Chhang-Chiang Hsia-Yu Hsin-Shih-Chhi Shih-Tai Wen-Hua Jo-Kan Wen-Thi ti Than-Hsin* 長江下游新石器時代文化若干問題的探析.
Analysis of Some Problems Concerning the Neolithic Cultures of the Lower Yangtze.
*WW*, 1978, **4**, 46.

Anon. (*502*).
*Nung-Chü Thu-Phu* 農具圖譜.
An Illustrated Survey of Agricultural Implements.
4 vols.
Thung-Su Tu-Wu, Peking, 1958.

Anon. (*503*).
*Ho-Mu-Tu Fa-Hsien Yuan-Shih She-Hui Chung-Yao I-Chih* 河姆渡發現原始社會重要遺址.
Reconnaissance of the Neolithic Site at Ho-Mu-Tu in Yü-Yao County, Chekiang Province.
*WW*, 1976, **8**, 6.

Anon. (*504*).
*Ho-Mu-Tu I-Chih Ti-I-Chhi Fa-Chüeh Pao-Kao* 河姆渡遺址第一期發掘報告.
Report on the First Season's Excavations at Ho-Mu-Tu.
*KKHP*, 1978, **1**, 39.

Anon. (505).
*Chhang-Chiang Hsia-Yu Hsin-Shih-Chhi Shih-Tai Wen-Hua Jo-Kan Wen-Thi ti Than-Hsi* 長江下游新石器時代文化若干問題的探析.
Analysis of Some Problems Concerning the Neolithic Cultures of the Lower Yangtze.
*WW* (1978), **4**, 46.

Anon. (506).
*Shan-Hsi-Sheng Fa-Hsien ti Han-Tai Thieh-Hua ho Pi-Thu* 陝西省發現的漢代鐵鏵和鏵土.
Iron Ploughshares and Mouldboards from the Han Dynasty Discovered in Shensi Province.
*WW* (1966), **1**, 19.

Anon. (507).
*Ho-Nan Mien-Chhih Chiao-Tsang Thieh-Chhi Chien-Yen Pao-Kao* 河南澠池窖藏鐵器檢驗報告.
Report on an Examination of the Hoard of Iron Tools Found at Mien-Chhih in Honan.
*WW* (1976), **8**, 52.

Anon. (508).
*Mien-Chhih-Hsien Fa-Hsien ti Ku-Tai Chiao-Tsang Thieh-Chhi* 澠池縣發現的古代窖藏鐵器.
An Ancient Hoard of Iron Tools Discovered at Mien-Chhih County.
*WW* (1976), **8**, 45.

Anon. (509).
*Shan-Tung-Sheng Lai-Wu-Hsien Hsi-Han Nung-Chü Thieh-Fan* 山東省萊蕪縣西漢農具鐵范.
Iron Moulds for Agricultural Implements from Lai-Wu County, Shantung Province.
*WW* (1977), **7**, 68.

Anon. (510).
*Shan-Hsi-Sheng Fa-Hsien ti Han-Tai Thieh Hua ho Pi-Thu* 陝西省發現的漢代鐵鏵和鏵土.
Iron Ploughshares and Mouldboards of the Han Dynasty Discovered in Shensi Province.
*WW* (1966), **1**, 19.

Anon. (511).
*Chiang-Hsi Hsiu-Shui Chhu-Thu Chan-Kuo Chhing-Thung Yüeh-Chhi ho Han-Tai Thieh-Chhi* 江西修水出土戰國青銅樂器和漢代鐵器.
Warring States Bronze Musical Instruments and Han Iron Tools Excavated in Hsiu-Shui, Kiangsi.
*KK* (1965), **6**.

Anon. (512).
*Han-Thang Pi-Hua* 漢唐壁畫.
Murals from the Han to the Thang Dynasty.
Foreign Languages Press, Peking, 1974.

Anon. (514).
*Shang-Hai Ma-Chhiao I-Chih Ti-I, Erh-Tzhu Fa-Chüeh* 上海馬橋遺址第一,二次發掘.
Exavations (First and Second Seasons) at the Ma-Chhiao Site in Shanghai.
*KKHP* (1978), **1**, 109–37.

Anon. (515).
*Kuang-Hsi Nan-Pu Ti-Chü ti Hsin-Shih-Chhi Shih-Tai Wan-Chhi Wen-Hua I-Tshun* 廣西南部地區的新石器時代晚期文化遺存.
Cultural Remains from Late Neolithic Sites in Southern Kwangsi.
*WW* (1978), **9**, 14–24.

Anon. (516).
*Kuang-Tung Chhü-Chiang Shih-Hsia Mo-Tsang Fa-Chüeh Chien-Pao* 廣東曲江石峽墓葬發掘簡報.
A Brief Report of the Excavation of Grave Sites in Shih-Hsia, Chhü-Chiang County, Kwangtung.
*WW* (1978), **7**, 1–15.

Anon. (517).
*Shen-Hsi Lin-Tung Fa-Hsien 'Wu-Wang Cheng Shang' Kuei* 陝西臨潼發現武王征商簋.
Bronze *kuei* with the Inscription 'Wu Wang Conquered Shang' Found at Lintung, Shensi Province.
*WW* (1977), **8**, 1–13.

Anon. (519).
*Nung-Yeh Chih-Shih Shou-Tshe* 農業知識手册.
Agronomic Handbook.
Jen Min, Kansu, 1972.

Anon. (520).
*Kuei-Chou Hsing-I Hsing-Jen Han-Mu* 貴州興義興仁漢墓.
Han Graves in Hsing-Jen, Hsing-I District, Kweichow.
*WW* (1979), **5**, 20–35.

Anon. (521).
*Yen-Hsia-Tu Ti-22-Hao I-Chih Fa-Chüeh Pao-Kao* 燕下都第22號遺址發掘報告.
Report on the Excavations of Site 22 at Yen-hsia-tu (Lower Capital of the Yen State).
*KK* (1965), **11**, 562–70.

Anon. (522).
*Fang-She-Hsing Than-Su Tshe-Ting Nien-Tai Pao-Kao* 放射性碳素測定年代報告.
Report on $C_{14}$ Dates.
*KK*, 1974, **5**, 333–8.

Anon. (523).
*Chi-Yuan Ssu-Chien-Kou San-Tso Han-Mu ti Fa-Chüeh* 濟源泗澗溝三座漢墓的發掘.
Excavation of Three Han Tombs at Ssu-Chien-Kou, Chi-Yuan [Honan].
*WW* (1973), **2**, pp. 46–54.

Anon. (524).
*Ssu-Chhuan Hsin-Tu-Hsien Fa-Hsien I-Phi Hua-Hsiang-Chuan* 四川新都縣發現一批畫象磚.
A Set of Stelae Discovered in Hsin-Tu District, Szechwan.
*WW* (1980), **2**, 56–7.

Anon. (525).
*Lo-Yang Sui Thang Han-Chia Tshang ti Fa-Chüeh* 洛陽隋唐含嘉倉的發掘.
The Excavation of the Sui and Thang Han-Chia Granaries at Loyang.
*WW* (1972), **3**, pp. 49 ff.

Anon. (*526*).
*Lo-Yang Shao-Kou Han Mu* 洛陽燒溝漢墓.
The Han Graves at Shao-Kou near Loyang.
Peking, 1959.

Anon. (*527*).
*Kuang-Hsi Wu-Chou-Shih Chin-Nien-Lai Chhu-Thu ti I-Phi Han-Tai Wen-Wu* 廣西梧州市進年來出土的一批漢代文物.
A Set of Han Objects Excavated Recently at Wu-Chou in Kwangsi.
*WW* (1977), **2**, 70–1.

Anon. (*529*).
*Hsü Kuang-Chhi Chi-Nien Lun-Wen Chi* 徐光啓紀念論文集.
Collected Essays for the Tercentenary of Hsü Kuang-Chhi.
Academia Sinica public., Chung-Hua Press, Peking, 1963.

Anon. (*530*).
*Liao-Ning-Sheng Nan-Pu I-Wan-Nien Lai Tzu-Jan Huan-Ching chih Yen-Pien* 遼寧省南部一萬年來自然環境之演變
Changes in Natural Conditions in Southern Liaoning Province over the Last Ten Thousand Years.
*CKKSH*, 1977, **6**, pp. 603–14.

Anon. (*531*).
*Chung-Kuo Ku-Tai Nung-Yeh Kho-Chi* 中國古代農業科技.
Ancient Chinese Agricultural Science & Technology.
Nung Yeh, Peking, 1980.

Anon. (*532*).
*Shui-Tao Chi-Chhu Chih-Shih* 水稻基礎知識.
Fundamental Facts about Rice.
Jen-Min, Shanghai, 1976.

Anon. (*533*).
*Tsen-Yang Chung Shui-Tao* 怎樣種水稻.
How to Grow Rice.
Jen Min, Shanghai, 1971.

Anon. (*534*).
*Yu-Liao Tso-Wu Tsai-Phei* 油料作物栽培.
The Cultivation of Oil Crops.
Nung Yeh, Peking, 1963.

Anon. (*535*).
*Nung-Chia Fei-Liao* 農家肥料.
Farmers' Fertilizers.
Jen Min, Loyang, 1979.

Anon. (*536*).
*Ma-Lei Tsai-Phei* 麻類栽培.
The Cultivation of Fibre Crops.
Nung Yeh, Peking, 1962.

Anon. (*537*).
*Mien-Hua* 棉花.
Cotton.
Kho Hsüeh, Peking, 1977.

Anon. (*538*).
*Kuang-Chou Shu-Tshai Phin-Chung Chih* 廣州蔬菜品種志.
The Vegetable Varieties of Canton.
Jen Min, Shanghai, 1974.

Anon. (*539*).
*Kan-Su Yung-Ching Ta-Ho-Chuang I-Chih Fa-Chüeh Pao-Kao* 甘肅永靖大何莊遺址發掘報告.
Excavation of the Remains of the Chhi-Chia Culture at Ta-Ho-Chuang in Yung-Ching County, Kansu Province.
*KKHP* (1974), **2**, 29–62.

Aoyama Sadao (*12*) 青山定雄.
*Tōdai no Tonden to Eiden* 唐代の屯田と営田.
Military and Civilian Agricultural Colonies in the Thang Dynasty.
*SGZ*, 1954, 63, **1**, 17–57.

Chang Chen-Hsin (*1*) 張振新.
*Han-Tai ti Niu-Keng* 漢代的牛耕.
Ploughing with Oxen in the Han Dynasty.
*WW* (1977), **8**, 57.

Chang Cheng-Lang (*2*) 張政烺.
*Pu-Tzhu 'Phou-Thien' Chi Chhi Hsiang-Kuan Chu Wen-Thi* 卜辭裒田及其相關諸問題.
The 'Phou-Thien' of the Oracle Inscriptions and Some Related Problems.
*KKHP*, 1973, **1**, p. 93.

Chang Chieh (*1*) 章楷.
*Lun 'Thien Kung Khai Wu' Chung Chi-Shu ti Shui-Tao Chi Chhi-Tha* 論天工開物中記述的水稻及其他.
On Wet Rice and Other Topics as Recorded in *Thien Kung Khai Wu.*
*KHSC*, 1966, 9, p. 56.

Chang Ping-Chhüan (*2*) 張秉權.
*Yin-Tai ti Nung-Yeh yü Chhi-Hsiang* 殷代的農業與氣象.
Agriculture and Climate in The Yin Dynasty.
*AS/BIHP* (Taipei), 1970, **42**, 267.

Chang Tsan-Chhen (*2*) 張贊臣.
*Wo Kuo Li-Tai Pen-Tshao Pien-Chi* 我國歷代本草編輯.
A History of the Chinese Pharmaceutical Natural Histories.
*ISTC*, 1955, **7**, (no. 1), 3.

Chang Wei-Ju (*1*) 張偉如.
*Chung-Kuo Chih-Wu-Yu chi chhi Chien-Yen Fang-Fa Shou-Tshe* 中國植物油及其檢驗方法手册.
A Handbook of Chinese Vegetable Oils & their Testing Methods.
Chung Hua, Shanghai, 1953.

Chao Ya-Shu (*1*) 趙雅書.
*Mien-Hua Chhuan-Ju Chung-Kuo chih Ching-Kuo* 棉花傳入中國之經過.
The Stages of the Introduction of Cotton into China.
In Shen Tsung-Hua & Chao Ya-Shu (*1*) pp. 222–44.

Chekiang Agricultural College (*1*).
*'Chhi Min Yao Shu' chi chhi Tso-Che Chia Ssu-Hsieh* 齊民要術及其作者賈思勰.
The *Chhi Min Yao Shu* & its author Chia Ssu-Hsieh.
Jen-Min Press, Peking, 1976.

Chekiang Agricultural College (*2*).

Chekiang Agricultural College (*2*) (*cont.*)
*Nung-Yeh Chih-Wu Ping-Li Hsüeh* 農業植物病理學.
Control of Crop Diseases.
Vol. 1, Kho-Hsüeh, Shanghai, 1978.

Chhen Chiu-Chin (*1*) 陳久金.
*Li-Fa ti Chhi-Yuan ho Hsien-Chhin ti Ssu-Fen Li* 曆法的起源和先秦的四分曆.
The Origins of Calendrical Science and the Four-Seasonal Calendar of pre-Chhin Times.
*KHSWC* (1978), **1**.

Chhen Heng-Li & Wang Ta-Tshan (*1*) 陳恒力, 王達參.
*'Pu Nung Shu' Yen-Chiu* 補農書研究.
Reseaches on the *Nung Shu* of Master Shen & the *Pu Nung Shu*.
Chung-Hua Press, Peking, 1958.

Chhen Hsi-Chhen (*1*) 陳錫臣.
*Chung-Kuo ti Ma-Lei Tso-Wu* 中國的麻類作物.
Chinese Fibre Crops.
Commercial Press, Shanghai, 1952.

Chhen Liang-Tso (*1*) 陳艮佐.
*Wo-Kuo Shui-Tao Tsai-Phei ti Chi-Hsiang Chi-Shu chih Fa-Chan chi chhi Chung-Yao-Hsing* 我國水稻栽培的幾項技術之發展及其重要性.
The Development of some Chinese Wet Rice Cultivation Techniques and Their Importance.
*SHYK* (1977), 7, **11**, 537–46.

Chhen Liang-Tso (*2*) 陳艮佐.
*Wo-Kuo Li-Tai Nung-Thien Shih-Yung chih Lü-Fei* 我國歷代農田施用之綠肥.
A Historical Account of the Green Manures Used in Chinese Agriculture.
*TLTC*, 1973, 46, **5**, pp. 1–25.

Chhen Meng-Chia (*4*) 陳夢家.
*Yin-Hsü Pu-Tzhu Tsung-Shu* 殷虛卜辭綜述.
A Study of the Characters on the Shang Oracle Bones.
Kho-Hsüeh, Peking 1956.

Chhen Teng-Yuan (*1*) 陳登原.
*Chung-Kuo Thien-Fu Shih* 中國田賦史.
A History of Land Taxes in China.
Commercial Press, n.d.

Chhen Tsu-Kuei (*1*) 陳祖槼.
*Mien* 棉.
Cotton.
Chinese Agricultural Heritage Series no. 5, Chung-Hua, Peking, 1957.

Chhen Tsu-Kuei (*2*) 陳祖槼.
*Tao* 稻.
Rice.
Chinese Agricultural Heritage Series no. 1, Chung-Hua, Peking, 1958.

Chheng Yao-Thien (*2*) 程瑤田.
*Khao Kung Chhuang Wu Hsiao Chi* 考工創物小記.
Brief Notes on the Specifications (for the Manufacture of Objects) in the *Artificers' Record* (of the *Chou Li*).
Peking, *c.* 1805.
In *HCCC*, ch. 536–9.

Chhien Hsi-Tsu (*1*) (ed.) 錢熙祚.
*Shou Shan Ko Tshung Shu* 守山閣叢書.
Mountain Guardhouse Collection.
1894; repr. Po Ku Chai, Shanghai, 1922.

Chhin Chung-Hsing (*1*) 秦中行.
*Chi Han-Chung Chhu-Thu ti Han-Tai Pei-Chhih Mu-Hsing* 記漢中出土的漢代陂池模型.
Note on a Han Model of a Tank Excavated in Han-Chung District [S. Shensi].
*WW* (1976), **3**, 77–8.

Chhü Chih-Sheng (*1*) 曲直生.
*Chung-Kuo Ku Nung-Shu Chien-Chieh* 中國古農書簡介.
A Brief Introduction to the Agricultural Works of Traditional China.
Economics Press, Taipei, 1960.

Chin Shan-Pao (*1*) 全善寶.
*Huai-Pei Phing-Yuan ti Hsin-Shih-Chhi Shih-Tai Hsiao-Mai* 淮北平原的新石器時代小麥.
Neolithic Wheat from the Plains North of the Huai River.
*TSWHP* (1962), 1, **1**, 67–72.

Chinese Institute of Agronomy (*1*).
*Chung-Kuo Kuo-Shu Chih* 中國果樹志.
The Fruit Trees of China.
Kho-Hsüeh, Shanghai, 1963.

Chinese Institute of Agronomy (*2*).
*Chung-Kuo Shu-Tshai Yu-Liang Phin-Chung* 中國蔬菜優良品種.
Improved Chinese Vegetable Varieties.
Nung Yeh, Peking, 1959.

Chou Pen-Hsiung (*1*) 周本雄.
*Ho-Pei Wu-An Tzhu-Shan I-Chih ti Tung-Wu Ku-Hai* 河北武安磁山遺址的動物骨骸.
The Animal Remains Discovered at Tzhu-Shan Village, Wu-An, Hopei Province.
*KKHP*, 1981, **3**, 339–48.

Chu Kho-Chen (*8*) 竺可楨.
*Chu Kho-Chen Wen-Chi* 竺可楨文集.
Collected Works of Chu Kho-Chen.
Kho-Hsüeh, Peking, 1979.

Chu Kho-Chen (*9*) 竺可楨.
*Lun-Lun Yüeh-Ling* 論論月令.
On Yüeh-Ling.
In Chu Kho-Chen (*8*).

Chu Kho-Chen (*10*) 竺可楨.
*Wu-Hou-Hsüeh yü Nung-Yeh Sheng-Chhan* 物候學與農業生產.
Phenology and Agricultural Production.
In Chu Kho-Chen (*8*).

Chu Kho-Chen (*11*) 竺可楨.
*Chung-Kuo Chin Wu-Chhien-Nien Lai Chhi-Hou Pien-Chhien ti Chhu-Pu Yen-Chiu* 中國近五千年來氣候變遷的初步研究.
A Preliminary Study of Climatic Changes in China over the Last Five Thousand Years.
*KKHP*, 1972, **1**, pp. 15–38.

Fang Cheng-San *et al.* (*1*) 方正三.
*Huang-Ho Chung-Yu Huang-Thu Kao-Yuan Thi-Thien ti Tiao-Chha Yen-Chiu* 黃河中游黃土高原梯田的調查研究.
An Investigation into Terracing in the Loess Plateaux of the Central Reaches of the Yellow River.
Kho-Hsüeh, Peking, 1958.
Feng Tse-Fang (*1*) 馮澤芳.
*Chung-Kuo ti Mien-Hua* 中國的棉花.
Chinese Cotton.
Tshai Cheng Ching Chi Press, Peking, 1956.
Fu I-Ling (*1*) 傅衣凌.
*Ming Chhing Nung-Tshun She-Hui Ching-Chi* 明清農村社會經濟.
San-Lien, Peking, 1961.
Furushima Toshio (*1*) 古島敏雄.
*Nihon Nōgyō Gijutsu Shi* 日本農業技術史.
A History of Agricultural Technology in Japan.
2 vols., Tokyo, 1959.
Furushima Toshio (*2*) 古島敏雄.
*Nihon Nōgyō Shi* 日本農業史.
A History of Japanese Agriculture.
Iwanami Zensho, Tokyo, 1st edn. 1956, 16th edn. 1973.

Han Kuo-Pan (*1*) 韓國磐.
*Pei-Chhao Ching-Chi Shih-Than* 北朝經濟試探.
An Exploration of the Economy of the Northern Dynasties.
People's Press, Shanghai, 1958.
Han Kuo-Pan (*2*) 韓國磐.
*Thang-Tai ti Chün-Thien-Chih yü Tsu-Yung-Tiao* 唐代的均田制與租傭調.
The Equal Allotments System and Rental Arrangements in the Thang Dynasty.
*LSYC*, 1955, **5**, 79–90.
Han-tan CPAM (*1*).
*Ho-Pei Tzhu-Shan Hsin-Shih-Chhi I-Chih Shih-Chüeh* 河北磁山新石器遺址試掘.
Trial Diggings at the Neolithic Site of Tzhu-Shan, Hopei.
*KK*, 1977, **6**, pp. 361–72.
Hara Motoko (*1*) 原宗子.
*Iwayuru Daidenho no Kisai o meguru Shokaishaku ni tsuite* いわゆる代田法の記載をめぐる諸解釋について.
On the Interpretation of Literary References to the Tai-Thien System.
*SGZ*, 1974, **85**, p. 11.
Hayashi Minao (*4*) (ed.) 林巳奈夫.
*Kandai no Bunbutsu* 漢代の文物.
Han Culture.
Kyoto University Press, 1977.
Heilungkiang Agricultural Institute (*1*).
*Ta-Tou Tsai-Phei Chi-Shu* 大豆栽培技術.
Soybean Cultivation Methods.
Nung Yeh, Peking, 1978.
Higashi Ichio (*1*) 東一夫.
*Ō Anseki Shinpō no Kenkyū* 王安石新法の研究.
A Study of Wang An-Shih's New Policies.
Kazama Shobō, Tokyo, 1970.
Higuchi Seishi (ed.) (*1*) 樋口清之.
*Yayoi to Yamatai-koku* 弥生と邪馬台国.
Yayoi and the State of Yamatai.
Rediscovery of Ancient Japan Series, no. 2, Gakushū Kenkyūsha, Tokyo, 1977.
Ho Chhang-Chhün (*1*) 賀昌羣.
*Han Thang Feng-Chien Thu-Ti So-Yu-Chih Hsing-Shih Yen-Chiu* 漢唐封建土地所有之形勢研究.
A Study of the Landed Feudal Aristocracy from Han to Thang.
Peking, 1964.
Ho Ping-Ti (*1*) 何炳棣.
*Huang-Thu yü Chung-Kuo Nung-Yeh ti Chhi-Yüan* 黃土與中國農業的起源.
The Loess Lands and the Origins of Chinese Agriculture.
Hongkong University Press, Hongkong, 1969.
Hopei Agricultural College (*1*).
*Kuo-Shu Tsai-Phei Hsüeh* 果樹栽培學.
The Cultivation of Fruit-Trees.
Jen Min, Peking, 1976.
Hopei CPAM (*1*).
*Ho-Pei Wu-An Tzhu-Shan I-Chih* 河北武安磁山遺址.
The Tzhu-Shan Site in Wu-An, Hopei Province.
*KKHP*, 1981, **3**, 303–38.
Hsi Chheng (*1*) 奚誠.
*Keng Hsin Nung Hua* 耕心農話.
Thoughts and Words on Ploughing and Farming.
1852.
Hsia Nai (*6*) 夏鼐.
*Than-14 Tshe-Ting Nien-Tai ho Chung-Kuo Shih-Chhien Khao-Ku-Hsüeh* 碳-14測定年代和中國史前考古學.
Carbon-14 Dating and The Study of Chinese Prehistory.
*KK* (1977), **4**, p. 217–32.
Hsia Wei-Ying (*2*) 夏緯瑛.
*'Kuan Tzu' Ti-Yuan Phien Chiao-Shih* 「管子」地員篇校釋.
The Chapter in the *Kuan Tzu* Book on the 'Variety of Earth's Products' Emended and Explained.
Chung-Hua, Peking & Shanghai, 1958.
Hsia Wei-Ying (*3*) 夏緯瑛.
*'Lü Shih Chhun-Chhiu' Shang-Nung-Teng Ssu-Phien Chiao-Shih* 呂氏春秋上農等四篇校釋.
The Four Chapters in the *Lü Shih Chhun-Chhiu* on Agriculture Emended and Explained.
Chung-Hua, Peking, 1956.
Hsia Wei-Ying (*4*) 夏緯瑛.
*'Hsia Hsiao Cheng' Ching-Wen Chiao-Shih* 夏小正經文校釋.
A Critical Translation into Modern Chinese of The Canonical Text *Hsia Hsiao Cheng.*
Nung Yeh, Peking, in press.
Hsia Wei-Ying (*5*) 夏緯瑛.
*'Shih Ching' Chung Yu-Kuan Nung-Shih Chang-Chü ti*

Hsia Wei-Ying (*5*) (*cont.*)
*Chih-Shih* 詩經中有關農事章句的解釋.
An Exegesis of the Agricultural Passages in the *Shih Ching*.
Nung Yeh, Peking, 1981.

Hsia Wei-Ying & Fan Chhu-Yü (*1*) 夏緯瑛, 范楚玉.
*'Hsia Hsiao Cheng' chi chhi tsai Nung-Yeh-Shih shang-ti I-I* 夏小正及其在農業史上的意義.
The *Hsia Hsiao Cheng* & its Significance in Agricultural History.
*CKSYC* (1979), **3**, 141–8.

Hsü Cho-Yün (*1*) 許倬雲
*Liang-Chou Nung-Tso Chi-Shu* 兩周農作技術.
Agricultural Techniques of The Western and Eastern Chou.
*AS/BIHP* (Taipei), 1971, **42**, 803.

Hsü Chung-Shu (*10*) 徐中舒.
*Lei-ssu khao* 耒耜考.
An Investigation of Types of Early Plough.
*AS/BIHP*, 1930, **2** (no. 1), 11.

Hsü Fu-Wei, Ho Kuan-Pao (*1*) 徐扶危, 賀官保.
*Lo-Yang Tung-Kuan Tung-Han Hsün-Jen-Mu* 洛陽東關東漢殉人墓.
An Eastern Han Tomb Containing Grave Figures from the Eastern Gate of Loyang.
*WW* (1973), **2**, pp. 55–62.

Hsü Heng-Pin (*1*) 徐恒彬.
*Chien-Than Kuang-Tung Lien-Hsien Chhu-Thu ti Hsi-Chin Li-Thien Pa-Thien Mo-Hsing* 簡談廣東連縣出土的西晉犁田耙田模型.
A Brief Discussion of the Western Chin Models of Ploughing and Harrowing Excavated in Lien County, Kwangtung Province.
*WW* (1976), **3**, 75.

Hu Hou-Hsüan (*3*) 胡厚宣.
*Pu-Tzhu-Chung So-Chien-Chih Yin-Tai Nung-Yeh* 卜辭中所見之殷代農業.
Yin Dynasty Agriculture as Seen in The Oracle Inscriptions, in:
*Chia-Ku Hsüeh Shang-Shih Lun-Tsung* 甲骨學商史論叢.
Collected Essays on Shang History Based on the Study of the Oracle Bones.
Chengtu and Hong Kong 1944.

Hu Hou-Hsüan (*4*) 胡厚宣.
*Yin-Tai Fen-Thien Shuo* 殷代焚田說.
An Explanation of The Yin Term 'Field-Firing' in:
*Chia-Ku Hsüeh Shang-Shih Lun-Tshung* 甲骨學商史論叢.
Collected Essays on Shang History Based on The Study of The Oracle Bones.
Chengtu and Hong Kong, 1944.

Hu Hou-Hsüan (*5*) 胡厚宣.
*Yin-Tai Nung-Tso Shih-Fei Shuo* 殷代農作施肥說.
On The Use of Fertilisers in Agriculture in The Yin Period.
*LSYC*, 1955, **1**, 97–106.

Hu Hsi-Wen (*2*) 胡錫文.
*Mai* 麥.
Wheat and Barley.
Chinese Agricultural Heritage Series no. 2, Chung-Hua, Peking, 1958.

Hu Hsi-Wen (*3*) 胡錫文.
*Liang-Shih Tso-Wu* 糧食作物.
Cereal Crops.
Chinese Agricultural Heritage Series no. 3, Nung-Yeh Press, Peking, 1959.

Hu Tao-Ching (*1*) 胡道靜.
*'Meng Chhi Pi Than' Chiao Cheng* 夢溪筆談校證.
Complete Annotated and Collated Edition of the *Dream Pool Essays* (of Shen Kua, +1086). 2 vols.
Shanghai Pub. Co., Shanghai, 1956.
Analyt. rev. Nguyen Tran-Huan, *RHS*, 1957, **10**, 182.

Hua Tao-Ching (*4*) (ed.) 胡道靜.
*Chung I Pi Yung* 種藝必用.
Everyman's Guide to Agriculture.
Agricultural Press, Peking 1963.

Hu Tao-Ching (*5*) 胡道靜.
*'Chung I Pi Yung' tsai Chung-Kuo Nung-Hsüeh Shih Shang-ti Ti-Wei* 種藝必用在中國農學史上的地位.
The Place of the *Chung I Pi Yung* in Chinese Agronomic History.
*WW*, 1962, **1**, 39–42.

Hu Tao-Ching (*6*) 胡道靜.
*Shan-Tung ti Nung-Hsüeh Chhuan-Thung* 山東的農學傳統.
The Shantung Agriculturalist Tradition.
*WSC* (1962), **2**, 48–9.

Hu Tao-Ching (*7*) 胡道靜.
*Shen Kua tsai Ku Nung-Hsüeh Shang-ti Chheng-Chiu ho Kung-Hsien* 沈括在古農學上的成就和貢獻.
Shen Kua's Achievements in and Contributions to Ancient Agronomy.
*HSYK* (1966), **2**, 48–53.

Hu Tao-Ching (*8*) 胡道靜.
*Shen Kua ti Nung-Hsüeh Chu-Tso 'Meng Chhi Wang Huai Lu'* 沈括的農學著作夢溪忘懷錄.
Shen Kua's Agricultural Work, the *Meng Chhi Wang Huai Lu*.
*WS* (1963), **3**, 221–5.

Hu Tao-Ching (*9*) 胡道靜.
*Hsü Kuang-Chhi Nung-Hsüeh-Chu Shu-Khao* 徐光啓農學著述考.
An Examination of the Agronomic Writings of Hsü Kuang-Chhi.
*TSK* (1962), **3**, 32–41.

Hu Tao-Ching (*10*) 胡道靜.
*Hsü Kuang-Chhi Yen-Chiu Nung-Hsüeh Li-Chheng ti Than-So* 徐光啓研究農學歷程的探索.
An Investigation of the Successive Stages of Hsü Kuang-Chhi's Agricultural Research.
*LSYC*, 1980, **6**, 117–34.

Hu Tao-Ching (*11*) 胡道靜.
*Hsi-Chien Ku Nung-Shu Pieh-Lu* 稀見古農書

Hu Tao-Ching (*11*) (*cont.*)
別錄.
A List of Rare Ancient Agricultural Works.
*TSC* (1962), **4**, 38–42.
Hu Tao-Ching (*12*) 胡道靜.
*Wo-Kuo Ku-Tai Nung-Hsüeh Fa-Chan Kai-Khuang ho Jo-Kan Ku-Nung-Hsüeh Tzu-Liao Kai-Shu* 我國古代農學發展概況和若干古農學資料概述.
The Development of Early Chinese Agronomy and a Brief Account of Some Data on Early Agronomy.
*HSYK* (1963), **4**, 22.
Hu Tao-Ching (*13*) 胡道靜.
*'Meng Chhi Wang Huai Lu' Kou Shen* "夢溪忘懷錄"鈎沉.
An Investigation of Shen Kua's *Meng Chhi Wang Huai Lu.*
*HCTHHP* (1981), 11, **1**, 1–16.
Hu Tao-Ching (*14*) 胡道靜.
*Shih Shu Phien* 釋菽篇.
On the Interpretation of the Term *Shu* [Legume].
*CHWSLT*, 1963, **3**, p. 111.
Hua Chhüan (*1*) 華泉.
*Tui Ho-Mu-Tu I-Chih Ku-Chih Keng-Chü ti Chi-Tien Khan-Fa* 對河姆渡遺址骨制耕具的幾點看法.
Different Interpretations of the Bone Tillage Implements Found at Ho-Mu-Tu.
*WW* (1977), **7**, 51–3.
Huanan Agricultural College, Agricultural Department (*1*) 華南農學院農學系.
*Wo-Kuo Yeh-Sheng Tao chi chhi Ti-Li Fen-Pu* 我國野生稻及其地理分布.
Chinese Wild Rices and their Geographical Distribution.
*ICHP* (1975), **2**, 1.
Huanan Agricultural College Historical Research Group (*1*) 華南農學院農業歷史遺產研究室.
*San Chung Hsi-Chien Ku-Nung-Shu Ho-Khan* 三種希見古農書合刋.
Three Rare Ancient Agricultural Works.
Huanan Agricultural College, 1978.
Huang Chan-Yüeh (*1*) 黃展岳.
*Chin-Tai Chhu-Thhu ti Chan-Kuo Liang-Han Thieh-Chhi* 近代出土的戰國兩漢鐵器.
Recently Excavated Iron Implements of the Warring States & Han Periods.
*KKHP* (1957), **3**, 93–108.

Iinuma Jirō, Horio Hisashi (*1*) 飯沼二郎, 堀尾尚志.
*Nōgu* 農具.
Agricultural Implements.
Hosei University Press, 1976.

Kansu Agricultural Institute (*1*) 甘肅省農業科學院.
*Kansu-Sheng Chu-Yao Nung-Tso-Wu Yu-Liang Phin-Chung* 甘肅省主要農作物優良品種.
Chief Improved Crop Varieties of Kansu Province.
Jen Min, Kansu, 1977.
Katō Shigemoto *et al.* (*1*) 加藤茂苞.
*Zasshu Shokubutsu no Ketsujitsudo ni mirareru Inashu no Ruien ni tsuite* 雑種植物の結実度にみられる稲種の類縁について.
On the Affinity of Rice Varieties as Shown by the Fertility of Hybird Plants.
*BAFKU* (1928), **3**, 132–47 [Eng. tr. *BAKFU* (1930) 2, **9**, 241–76].
Katō Shigeshi (*2*) 加藤繁.
*Keizai-shi jō yori mitaru Kita Shina to Minami Shina* 経済史上より観たる北支那と南支那.
North and South China from the Perspective of Economic History.
*SKS* (1943), 12, **11–12**, 1–13.
Katō Shigeshi (*3*) 加藤繁.
*Shina Keizai Shi Kōshō* 支那経済史考証.
Studies in Chinese Economic History.
Tōyōbunka Series A, no. 34, Tokyo, 1953.
Kawahara Yoshirō (*1*) 河原由郎.
*Sōdai Tochi Shoyū no Kihon Mondai* 宋代土地所有の基本問題.
The Basic Questions of Land Tenure in the Sung.
*FDKH* (1964), **5**, 23–47.
Khai-feng CPAM *et al.* (*1*).
*Phei-Li-Kang I-Chih I-Chiu-Chhi-Pa Nien Fa-Chüeh Chien-Pao* 裴李崗遺址一九七八年發掘簡報.
A Summary of the Discoveries Made in 1978 at Phei-li-kang.
*KK*, 1979, **3**, pp. 197–205.
Khai-feng CPAM (*2*).
*Ho-Nan Khai-Feng Ti-Chü Hsin-Shih-Chhi Shih-Tai I-Chih Tiao-Chha Chien-Pao* 河南開封地區新石器時代遺址調查簡報.
A Reconnaissance of Neolithic Sites in the District of Khai-Feng, Honan.
*KK*, 1979, **3**, pp. 206–8.
Khai-Feng CPAM (*3*).
*Ho-Nan Hsin-Cheng Phei-Li-Kang Hsin-Shih-Chhi Shih-Tai I-Chih* 河南新鄭裴李崗新石器時代遺址.
The Neolithic Site of Phei-Li-Kang at Hsin-Cheng in Honan.
*KK*, 1978, **2**, pp. 73–9.
Khang Chheng-I (*1*) 康成懿.
*'Nung Cheng Chhüan Shu' Chheng-Yin Wen-Hsien Than-Yuan* 農政全書徵引文獻探原.
An Enquiry into the Literary Sources Quoted in the *Nung Cheng Chhüan Shu.*
Agricultural Press, Shanghai, 1960.
Kitamura Shirō (*1*) 北村四郎.
*Chūgoku Saibai Sakumotsu no Kigen* 中国栽培作物の起源.
The Origins of China's Domesticated Plants.
*TG/K* (1950), **19**, 82.
Ko Chin (*1*) 葛今.
*Ching-Yang Kao-Chia-Pao Tsao Chou Mu-Tshan Fa-Chüeh Chi* 涇陽高家堡早周墓葬

Ko Chin (*1*) (*cont.*)
發掘記.
Account of the Excavation of an Early Chou Tomb at Kao-Chia-Pao.
*WW*, 1972, **7**, 5–7.
Kondō Mitsuo (*1*) 近藤光男.
*Tai Shin no Kōkōkizu ni tsuite* 戴震の考工記図につひて.
On Tai Chen and his *Khao Kung Chi Thu.*
*TG/T*, 1955, **11**, 1; abstr. *RBS*, 1955, **1**, no-452.
Kuo Mo-Jo (*1*) 郭沫若.
*Shih Phi Phan Shu* 十批判書.
Ten Critical Essays.
Chün-i, Chungking, 1945.
Kuo Mo-Jo(*12*) 郭沫若.
*Ku-Tai Wen-Hsüeh chih Pien-Cheng Fa-Chan* 古代文學之辯證發展.
A Dialectical Proof of Literacy in Ancient China.
*KK* (1972), **3**, 2.
Kuo Yün-Sheng (*1*) 郭雲陞.
*Chiu Huang Chien I Shu* 救荒簡易書.
A Treatise on Simple Methods of Alleviating Famine.
1896.
Kwangtung Provincial Museum (*1*) 廣東省博物館.
*Kuang-Tung Chhü-Chiang Shih-Hsia Mu-Sang Fa-Chüeh Chien-Pao* 發掘簡報.
An Outline Report of the Excavations of Graves at Shih-Hsia on the Chhü River in Kwangtung.
*WW*, 1978, **7**, pp. 1–15.
Kweichow CPAM (*1*).
*Kuei-Chou Hsing-I, Hsing-Jen Han Mu* 桂州興義興仁漢墓.
Han Graves from Hsing I & Hsing-Jen in Kweichow.
*WW*, 1979, **5**, 20–35.

Lai-Yang (Shantung) Agricultural College (*1*) 萊陽(山東)農業學校.
*Hsiao Mai* 小麥.
Wheat.
Kho Hsüeh, Peking, 1975.
Li Chhang-Nien (*2*) (ed.) 李長年.
*Ma-Lei Tso-Wu* 麻類作物.
Fibre Crops.
Chinese Agricultural Heritage Series no. 8, Nung-Yeh Press, Peking, 1961.
Li Chhang-Nien (*3*) 李長年.
*'Chhi Min Yao Shu' Yen-Chiu* 齊民要術研究.
A Study of *Chhi Min Yao Shu.*
Agricultural Press, Peking, 1959.
Li Chhang-Nien (*4*) 李長年.
*Tou Lei* 豆類.
Legumes.
Chinese Agricultural Heritage Series no. 4, Chung-Hua, Peking, 1958.
Li Chien-Nung (*3*) 李劍農.
*Wei Chin Nan-Pei-Chhao Sui Thang Ching-Chi Shih-Kao* 魏晉南北朝隋唐經濟史稿.
A Draft Economic History of China from the Wei-Chin to the Thang Dynasty.
Chung-Hua, Peking, 1958; 2nd ed. 1963.
Li Chien-Nung (*4*) 李劍農.
*Hsien-Chhin Liang-Han Ching-Chi Shih-Kao* 先秦兩漢經濟史稿.
A Draft Economic History of China from pre-Chhin Times to the End of the Han.
Chung-Hua, Peking, 1962.
Li Chien-Nung (*5*) 李劍農.
*Sung Yuan Ming Ching-Chi Shih-Kao* 宋元明經濟史稿.
A Draft Economic History of the Sung, Yuan & Ming Dynasties.
San-Lien, Peking, 1957.
Li Ching-Chhih (*1*) 李鏡池.
*'Chou I' Kua Ming Khao Shih* 周易卦名考釋.
A Study of the Names of the Sixty-four Hexagrams in the *Book of Changes.*
*LHP*, 1948, **9**, (no. 1), 197 and 303.
Li Ching-Chhih (*2*) 李鏡池.
*'Chou I' Shih Tzhu Hsü Khao* 周易筮辭續考.
A Further Study of the Explicative Texts in the *Book of Changes.*
*LHP*, 1947, **8** (no. 1) 1 and 169.
Li Fan *et al.* (*1*) 李璠.
*Sheng-Wu Shih* 生物史.
A Natural History.
Vol. 5, Kho-Hsüeh, Peking, 1979.
Li Hui-Lin (*1*).
*Tung-Nan-Ya Tsai-Phei Chih-Wu chih Chhi-Yüan* 東南亞栽培植物之起源.
The Origin of Cultivated Plants in Southeast Asia.
Inaugural address, Chinese University of Hong Kong, 1966.
Li Lai-Jung (*1*) 李來榮.
*Nan-Fang ti Kuo-Shu Shang-Shan* 南方的果樹上山.
Mountain Cultivation of the Fruits of the South.
Kho-Hsüeh, Peking, 1956.
Li Ming-Chhi (*1*) 李明啓.
*'Kuan-Tzu Ti Yuan Phien' chung-ti Chih-Wu Sheng-Li-Hsüeh Chih-Shih* 管子地員篇中的植物生理學知識.
Knowledge of Botanical Physiology in the *Ti Yuan* chapter of the *Kuan Tzu.*
*NSYC* (1980), **1**, 71–4.
Li Tso-Hsien (*1*) 李佐賢.
*Ku Chhüan Hui* 古泉滙.
Ancient Springs and Fountains.
20 vols. in 3, Li Chin, Shantung, 1864.
Li Yen-Chang (*1*) 李彥章.
*Chiang-Nan Tshui-Keng Kho-Tao Pien* 江南催耕課稻編.
A Record of Early-Rice Cultivation in Southern China.
1834, repr. in Chhen Tsu-Kuei (*1*), pp. 374–430.
Liang Chia-Mien (*1*) 梁家勉.
*Wo-Kuo Tung-Chih-Wu Chih ti Chhu-Hsien chi chhi*

Liang Chia-Mien (*1*) (*cont.*)
*Fa-Chan* 我國動植物志的出現及其發展.
The Earliest Chinese Monographs on Animals & Plants & their Development.
Paper presented at the 8th Scientific Conference of the South China Agricultural Institute, n.d.

Liang Chia-Mien (*2*) 梁家勉.
*Chung-Kuo Nung-Chih-Wu Shih-Cheng Shu-Li* 中國農植物史證敍例.
A History of Chinese Agricultural Plants.
*CLNK*, 1949, **1** (no. 1), 17.

Liang Chia-Mien (*3*) 梁家勉.
*'Nung Cheng Chhüan Shu' Chuan-Shu Kuo-Chheng chi Jo-Kan Yu-Kuan Wen-Thi ti Than-Thao* 農政全書撰述過程及若干有關問題的探討.
An Enquiry into the Stages of Compilation of the *Nung Cheng Chhüan Shu* & Some Related Problems.
In Anon. (*529*), 75–109.

Liang Chia-Mien (*4*) 梁家勉.
*'Chhi Min Yao Shu' ti Chuan-Che Chu-Che ho Chuan-Chhi.* 齊民要術的撰者注者和撰期.
The Author, Commentators, and date of Composition of the *Chhi Min Yao Shu.*
*HNNYKH* (1957), **3**, 92–8.

Liang Chia-Mien & Chhi Ching-Wen (*1*) 梁家勉, 戚經文.
*Fan Shu Yin Chung Khao* 番薯引種考.
A Review of the Literature on Sweet Potatoes.
*HNNYKH* (1980), 1, **3**, 74–8.

Liang Kuang-Shang (*1*) 梁光商.
*'Chhi Min Yao Shu' chung-ti Sheng-Wu-Hsüeh Chih-Shih* 齊民要術中的生物學知識.
Biological knowledge in the *Chhi Min Yao Shu.*
*NSYC* (1980), **1**, 81–8.

Liang Kuang-Shang, Chhi Ching-Wen & Wu Wan-Chhun (*1*) 梁光商, 戚經文, 吳萬春.
*Wo-Kuo Hsien Keng Tao ti Chhi-Yuan ho Fen-Lei ti Than-Thao* 我國秈粳的起源和分類的探討.
An Investigation of the Origins & Diversification of Chinese *keng* and *hsien* Rice.
In Anon. (*531*), pp. 249–53.

Liu Chih-Yuan (*3*) 劉志遠.
*Ssu-Chhuan Han-Tai Hua-Hsiang-Chuan Fan-Yang ti She-Hui Sheng-Huo* 四川漢代畫象磚反映的社會生活.
Social Life as Reflected in Szechwanese Decorated Bricks of the Han Period.
*WW* (1975), **4**, 45–55.

Liu Chih-Yuan (*4*) 劉志遠.
*Khao-Ku Tshai-Liao So-Chien Han-Tai ti Ssu-Chhuan Nung-Yeh* 考古材料所見漢代的四川農業.
Agriculture in Szechwan in the Han Dynasty as seen from Archaeological Data.
*WW*, 1979, **12**, 61–9.

Liu Hsien-Chou (*7*) 劉仙洲.
*Chung-Kuo Chi-Hsieh Kung-Chheng Fa-Ming Shih* 中國機械工程發明史.
A History of Chinese Engineering Inventions.
Kho-Hsüeh, Peking, 1962.

Liu Hsien-Chou (*8*) 劉仙洲.
*Chung-Kuo Ku-Tai Nung-Yeh Chi-Hsieh Fa-Ming Shih* 中國古代農業機械發明史.
A History of Chinese Inventions in Agricultural Engineering.
Kho-Hsüeh, Peking, 1963.
First pub. in *NYCC*, 1962, **5** (nos. 1 & 2), (abridged).

Liu Hsien-Chou (*8a*). 劉仙洲.
*Chung-Kuo Ku-Tai tsai Nung-Yeh Chi-Hsieh Fang-Mien-ti Fa-Ming* 中國古代在農業機械方面的發明.
Ancient Chinese Agricultural Machines and Implements and their Invention.
*NYCC*, 1962, **5** (no. 1), 1.

Liu Pao-Nan (*1*) 劉寶楠.
*Shih Ku* 釋穀.
On the Cereal Grains [historical and philological].
Peking, 1855.
In *HCCC* (*HP*), chs. 1075–8.

Liu Tun-Chen (*4*) 劉敦楨.
*Chung-Kuo Chu Tse Kai Shuo* 中國住宅概說.
A Short Study of Chinese Domestic Architecture.
Based on material collected by the Chinese Architectural Research Unit, formed jointly by the Architectural Research Institute and the Nanking College of Engineering.
Architectural and Engineering Press, Peking, 1957.
Abridged translation, without illustrations, by Liao Hung-Yung & R. T. F. Skinner, Collet, London, 1957.

Liu Tzu-Ming (*2*) 柳子明.
*Chung-Kuo Tshai-Phei Tao ti Chhi-Yuan chi Fa-Chan* 中國栽培稻的起源及發展.
The Origins and Development of Chinese Cultivated Rice.
*ICHP* (1975), 2, **1**.

Liu Yü-Chhüan (*1*) 劉毓瑔.
*'Nung Sang Chi Yao' ti Tso-Che, Pan-Pen ho Nei-Jung* 農桑輯要的作者版本和內容.
The Author, Editions & Contents of the *Nung Sang Chi Yao.*
*AHRA*, 1958, **1**, 215–26.

Lo Hsiang-Lin (*6*) 羅香林.
*Chung-Kuo Tsu-Phu Yen-Chiu* 中國族譜研究.
A Study of Chinese Genealogies.
Chung-Kuo Hsüeh-She, Hong Kong, 1971.

Loyang Museum (*1*).
*Lo-Yang Sui Thang Tung-Tu Huang-Chheng-Nei ti Tshang-Chiao I-Chih* 洛陽隋唐東都皇城內的倉窖遺址.
Excavations of Four Granaries in the Imperial City of Loyang, Eastern Capital of the Sui and Thang Dynasties.
*KK*, 1981, **4**, 309–18.

Lung Po-Chien (*1*) 龍伯堅.
*Hsien Tshun Pen Tshao Shu Lu* 現存本草書錄.

Lung Po-Chien (*1*) (*cont.*)
Bibliographical Study of Extant Pharmacopoeias (from all periods).
Jen-min Wei-sheng, Peking, 1957.

Ma Kuo-Han (*1*) (ed.) 馬國翰.
*Yü Han Shan Fang Chi I Shu* 玉函山房輯佚書.
The Jade-Box Mountain Studio Collection of (Reconstituted) Lost Books.
1853.

Makino Tatsumi (*1*) 牧野巽.
*Kinsei Chūgoku Sōzoku Kenkyū* 近世中国宗族研究.
A Study of Recent Chinese Clans.
Nikkō Publishers, Tokyo, 1949.

Mao Yung (*1*) 毛雝.
*Chung-Kuo Nung-Shu Mu-Lu Hui-Pien* 中國農書目錄彙編.
Bibliography of Chinese Literature on Agriculture.
University Library, Nanking, 1924.

Miao Chhi-Yü (*1*) 繆啓愉.
*Wu Yüeh Chhien Shih tsai Thai-Hu Ti-Chü ti Yü-Thien Chih-Tu ho Shui-Li Hsi-Thung* 吳越錢氏在太湖地區的圩田制度和水利系統.
The Poldered Fields Regulations of Master Chhien of Wu & Yüeh in The Thai-Hu Area & the Classification of Water Control.
*AHRA*, 1960, **2**, 139–58.

Min. of Agriculture, Industry & Trade (*1*).
*Mien Yeh Thu Shuo* 棉業圖說.
Illustrated Account of Textile Production.
8 vols., Peking, 1911.

Miyazaki Ichisada (*3*) 宮崎市定.
*Ajia Shi Kenkyū* アジア史研究.
Studies in Oriental History.
Vol. 4, Tōyōshi kenkyūkai, Kyoto.

Mou Yung-Khang & Sung Chao-Lin (*1*) 牟永抗，宋兆麟.
*Chiang-Che ti Shih-Li ho Pho-Thu-Chhi—Shih-Lun Wo-Kuo Li-Keng ti Chhi-Yuan* 江浙的石犁和破土器—試論我國犁耕的起源.
The stone ploughshares & soil-breaking implements of the Yangtze Delta—a hypothesis on the origins of Chinese ploughing.
*NYKK* (1981), **2**, 75–84.

Mou Yung-Khang & Wei Cheng-Chin (*1*) 牟永抗，魏正瑾.
*Ma-Chia-Pang Wen-Hua ho Liang-Chu Wen-Hua* 馬家浜文化和良渚文化.
The Ma-Chia-Pang Cultures and the Liang-Chu Cultures.
*WW*, 1978, **4**, pp. 67–73.

Nakao Manzō (*1*) 中尾万三.
*Shokuryō Honsō no Kōsatsu* 食療本草の考察.
A Study of the [Tunhuang MS of the] *Shih Liao Pen Tshao* (Nutritional Therapy; a Pharmaceutical Natural History), [by Meng Sheng, *c.* +670].
*SSIP*, 1930, *1* (no. 3), 1–222.

Nanking Museum (*1*).
*Chhang-Chiang Hsia-Yu Hsin-Shih-Chhi Shih-Tai Wen-Hua Jo-Kan Wen-Thi ti Than-Che* 長江下游新石器時代文化若干問題的探析.
A Consideration of some Problems Concerning the Neolithic Cultures of the Lower Yangtze.
*WW*, 1978, **4**, pp. 46–57.

Niida Noboru (*2*) 仁井田陞.
*Tō Sō Hōritsu Bunsho no Kenkyū* 唐宋法律文書の研究.
The Critical Study of Legal Documents of the Thang & Sung Eras.
Tōhō Bunkwa Gakuin, Tokyo, 1937.

Niida Noboru (*3*) 仁井田陞.
*Chūgoku no Nōson Kazoku* 中国の農村家族.
Chinese Rural Families.
Tokyo University Press, 1952.

Nishijima Sadao (*1*) 西嶋定生.
*Chūgoku Keizai Shi Kenkyū* 中国経済史研究.
Studies in Chinese Economic History.
Tokyo Daigaku Bungakubu, Tokyo, 1966.

Nishiyama Buichi (*1*) 西山武一.
*Chūgoku-ni okeru Suito Nōgyō no Hattatsu* 中国における水稲農業の発達.
Development of Wet-Rice Cultivation in China.
*NSK* (1959), 3, **1**, 135–9.

Nishiyama Buichi & Kumashiro Yukio (*1*) (trs. & comm.) 西山武一，熊代幸雄.
*Seimin Yōjutsu* 斉民要術.
*Chhi Min Yao Shu*, translated into Japanese with Revisory and Explanatory Notes.
2nd ed. Ajya Keizai Press, Tokyo, 1969 (1st ed. 1957).

Okazaki Satoshi (*1*) 岡崎敬.
*Kandai Deisho ni arawareta Suiden, Suichi ni tsuite.*
漢代泥象にあらわれた水田，水池について.
Irrigated fields and tanks as seen in Han clay model:
*KKHTC* (1958), 44, **2**.

Ong Tu-Chien (*1*) 翁獨健.
'*Tao Tsang' Tzu Mu Yin Te* 道藏子目引得.
An Index to the Taoist Patrology.
Harvard-Yenching, Peiping, 1935.

Pao Shih-Chhen (*1*) 包世臣.
*Chhi Min Ssu Shu* 齊民四術.
Four Arts for The Common People.
1849.

Peking Agricultural College (*1*).
*Kho-Hsüeh Chung-Thien Shou-Tshe* 科學種田手册.
A Handbook of Scientific Farming.
Jen-Min, Peking, 1975.

Pheng Shih-Chiang (*1*) 彭世獎.
'*Nan Fang Tshao Mu Chuang' Chuan-Che Chuan-Chhi ti Jo-Kan Wen-Thi* 南方草木狀撰者撰期的若干問題.
Some Problems of Dating & Attribution of the Compilation of the *Nan Fang Tshao Mu*

Pheng Shih-Chiang (*1*) (*cont.*)
*Chuang.*
*NSYC* (1980), **1**, 75–80.

Pheng Shu-Lin & Chou Shih-Pao (*1*) 彭書琳, 周石保.
*Kuang-Hsi Pin-Yang Fa-Hsien Shih-Wan Nien-Chhien ti Hua-Sheng Shih-Hua* 廣西賓陽發現十萬年前的花生石化.
Hundred Thousand Year Old Peanut Fossils From Pin-Yang Country, Kwangsi.
*NYKK* (1981), **1**, 17–20.

Sasaki Kōmei (*1*) 佐々木高明.
*Inasaku Izen* 稲作以前.
Before Rice Cultivation.
Nihon Hōsō Shuppan Kyōkai (NHK Books), Tokyo, 1971.

Sekino Takeshi (*1*) 關野雄.
*Shin Raishi Kō* 新耒耜考.
New Researches on the *Lei-ssu.*
*TBK* (1954), **19**.

Shansi Agricultural Institute (*1*) 山西省農業科學院.
*Ku-Tzu Tsai-Phei Chi-Shu* 谷子栽培技術.
Millet Cultivation Techniques.
Nung Yeh, Peking, 1977.

Shansi (Chin-Tung) Technical Institute (*1*) 山西省晉東地區科技局.
*Ku-Tzu* 谷子.
Millet.
Kho Hsüeh, Peking, 1976.

Shao Chhi-Chhüan, Li-Chhang-Shen Pa-Sang-Tzhu-Jen (*1*) 邵啓全, 李長森巴桑次仁.
*Tsai-Phei Ta-Mai ti Chhi-Yuan yü Chin-Hua—Wo-Kuo Hsi-Tsang ho Ssu-Chhuan ti Yeh-Sheng Ta-Mai* 栽培大麥的起源與進化—我國西藏和四川的野生大麥.
The Origins & Evolution of Domesticated Barley—Wild Barleys in Tibet and Szechwan.
*ICHP* (1975), **2**, 123–8.

Shen Tsung-Han & Chao Ya-Shu (ed.) (*1*) 沈宗瀚, 趙雅書.
*Chung-Hua Nung-Yeh Shih—Lun Chi* 中華農業史—論集.
Collected Essays on the History of Chinese Agriculture.
Com. Press, Taipei, 1979.

Shen Wen-Cho (*1*) 沈文倬.
*Fu yü Chi* 艮與耤.
Slaves and Agricultural Slaves [in the Oracle Bone Inscriptions].
*KK*, 1977, **5**, 335.

Shiba Yoshinobu (*1*) 斯波義信.
*Sōdai Shōgyō Shi Kenkyū* 宋代商業史研究.
Research on Sung Commercial History.
Kazama Shobō, Tokyo, 1968.

Shih Hsing-Pan et al (*1*) 石興邦.
*Hsi-An Pan-Pho* 西安半坡.
The Neolithic Site of Pan-Pho, Sian.
Peking, 1963.

Shih Sheng-Han (*2*) 石聲漢.
*Ssu-Min Yüeh-Ling Chiao-Chu* 四民月令校注.
An Annotated Commentary on the *Ssu Min Yüeh Ling.*
China Press, Peking, 1965.

Shih Sheng-Han (*3*) 石聲漢.
*'Chhi Min Yao Shu' Chin Shih* 齊民要術今釋.
A Modern Translation of *Chhi Min Yao Shu.*
4 vols, Science Press, Peking, 1957.

Shih Sheng-Han (*4*) 石聲漢.
*Tshung 'Chhi Min Yao Shu' Khan Chung-Kuo Ku-Tai ti Nung-Yeh Kho-Hsüeh Chih-Shih* 從齊民要術看中國古代的農業科學知識.
Medieval Chinese Agricultural Science as seen in the *Chhi Min Yao Shu.*
Science Press, Shanghai, 1957.

Shih Sheng-Han (*5*) 石聲漢.
*Liang Han Nung-Shu Hsüan-Tu* 兩漢農書選讀.
Selected Readings from Two Han Agricultural Works [*Fan Sheng-Chih Shu* and *Ssu Min Yüeh Ling*].
Agriculture Press, Peking, 1979.

Shih Sheng-Han (comp.) (*6*) 石聲漢.
*Chi Hsü Chung Nan Fang Tshao Wu Chuang* 輯徐衷南方草物狀.
The Collated Text of Hsü Chung's *Nan Fang Tshao Wu Chuang.*
Historical Unit of the North Western Agricultural College, Hsi-an, 1973.

Shih Sheng-Han (*7*) 石聲漢.
*Chung-Kuo Ku-Tai Nung-Shu Phing-Chieh* 中國古代農書評介.
A Critical Introduction to the Medieval Chinese Agricultural Treatises.
Nung-Yeh, Peking, 1980.

Shih Sheng-Han (ed.) (*8*) 石聲漢.
*'Nung Cheng Chhüan Shu' Chiao-Chu* 農政全書校注.
Annotated Edition of the *Nung Cheng Chhüan Shu.*
Ku-Chi, Shanghai, 1979, 3 vols.

Shinoda Osamu (*1*) 篠田統.
*Gokoku no Kigen* 五穀の起源.
Origin of the Five cereal crops, *wu-ku*, in the Far East.
*SB*, 1951, **2**, 37.

Shinoda Osamu (*5*) 篠田統.
*Chūgoku Shokkei Sōsho* 中国食経叢書.
A Collection of Chinese Dietary Classics.
2 vols, Tokyo, 1972.

Shinoda Osamu (*6*) 篠田統.
*Mindai no Shoku Seikatsu* 明代の食生活.
Gastronomic Life in the Ming Dynasty.
In Yabuuchi (*11*), pp. 74–92.

Shinoda Osamu (*7*) 篠田統.
*Chūgoku Shokumotsu Shi* 中国食物史.
A History of Diet in China.
Shibata Shoten, Tokyo, 1974.

Shinoda Osamu & Tanaka Seiichi (*1*) 篠田統, 田中静一.
*Chūgoku Shokkei Sōsho* 中国食経叢書.
A Collection of Chinese Dietary Classics.
Shoseki Bunbutsu Ryūtsūkai, Tokyo, 1973.

Shōji Kichinosuke (*1*) 庄司吉之助.
*Meiji Ishin no Keizai Kōzō* 明治維新の経済構造.
The Economic Basis of the Meiji Reforms.
Tokyo, 1940.

Sinkiang Agricultural Institute (*1*).
*Hsin-Chiang Nung-Yeh Chi-Shu Shou-Tshe* 新疆農業技術手册.
Manual of Sinkiang Agricultural Practice.
People's Press, Sinkiang, 1976.

Su Ping-Chhi (*1*) 蘇秉琦.
*Shih-Hsia Wen-Hua Chhu-Lun* 石峡文化初論.
A Preliminary Discussion of the Shih-Hsia Culture.
*WW*, 1978, **7**, pp. 16–22.

Sudō Yoshiyuki (*4*) 周藤吉之.
*Sōdai Keizai Shi Kenkyū* 宋代経済史研究.
Researches in Sung Economic History.
Tokyo University Press, Tokyo, 1962.

Sun Chhang-Hsü (*1*) 孫常敍.
*Lei-Ssu ti Chhi-Yuan chi chhi Fa-Chan* 耒耜的起源及其發展.
The Origins and Development of the *Lei-Ssu*.
Jen-Min, Shanghai, 1959.

Sun Yün-Yü (*1*) 孫云蔚.
*Hsi-Pei ti Kuo-Shu* 西北的果樹.
Fruit-trees of Northwest-China.
Kho Hsüeh, Peking, 1962.

Sung Chao-Lin (*1*) 宋兆麟.
*Hsi-Han Shih-Chhi Nung-Yeh Chi-Shu ti Fa-Chan* 西漢時期農業技術的發展.
The Development of Agricultural Techniques During the Western Han.
*KK* (1976), **1**, 3.

Sung Chao-Lin (*2*) 宋兆麟.
*Ho-Mu-Tu I-Chih Chhu-Thu Ku-Ssu-ti Yen-Chiu* 河姆渡遺址出土骨耜的研究.
A Study of the Bone Tillage Implements Excavated at Ho-Mu-Tu.
*KK* (1979), **2**, 155–60.

Szechwan Agricultural Institute (*1*).
*Chung-Kuo Yu-Tshai Tsai-Phei* 中國油菜栽培.
The Cultivation of Brassicas for Oil in China.
Nung Yeh, Peking, 1964.

Tachai Brigade Discussion Group *et al.* (*1*).
*'Chhi Min Yao Shu' Hsüan Shih* 齊民要術選釋.
Selected Translations into Modern Chinese from the *Chhi Min Yao Shu*.
Science Press, Peking, 1975.

Tamaki Akira (*1*) 玉城哲.
*Mizu no Shisō* 水の思想.
The Philosophy of Water.
Ronsō, Tokyo, 1979.

Thang Han-Liang (ed.) (*1*) 唐漢良.
*Nung-Li chi chhi Pien-Suan* 農曆及其編算.
The Agricultural Calendar and its Calculation.
Jen-Min Press, Kiangsu, 1977.

Thung Chu-Chhen (*1*) 佟柱臣.
*Tshung Erh-Li-Thou Lei-Hsing Wen-Hua Shih-Than Chung-Kuo ti Kuo-Chia Chhi-Yuan Wen-Thi* 從二里頭類型文化試談中國的國家起源問題.
An Approach to the Origins of the Chinese State, Based on *Erh-li-thou*-type Cultures.
*WW* (1975), **6**, 29.

Ting Ying (ed.) (*1*) 丁穎.
*Chung-Kuo Shui-Tao Tsai-Phei Hsüeh* 中國水稻栽培學.
The Cultivation of Wet Rice in China.
Agricultural Press, Peking, 1961.

Ting Ying (*2*) 丁穎.
*Chung-Kuo Shui-Tao Phin-Chung ti Sheng-Thai Lei-Hsing chi chhi yü Sheng-Chhan Fa-Chan ti Kuan-Hsi* 中國水稻品種的生態類型及其與生產發展的關係.
An Ecological Classification of Chinese Wet Rice Varieties & their Relation to the Development of Production.
*CKNYKH* (1964), **10**.

Ting Ying & Chhi Ching-Wen (*1*) 丁穎,戚經文.
*Chung-Kuo chih Kan-Shu* 中國之甘藷.
The Sweet Potato in China.
*JAAC*, 1948, **186**, pp. 23–33.

Tseng Hsiang-Hsü (*1*) 增湘叙.
*Nung Hsüeh Tsuan Yao* 農學纂要.
The Essentials of Agronomy.
Szechwan, 1902.

Tsou Shu-Wen (*1*) 鄒樹文.
*'Li Chi Yüeh Ling' Pien Wei* 禮記月令辨僞.
Inauthentic Passages in the *Li Chi Yüeh Ling*.
*AHRA*, 1958, **1**, 183–213.

Tung Kao (ed.) (*1*) 董誥 *et al.*
*Chhüan Thang Wen* 全唐文.
Collected Literature of the Thang Dynasty.
1814.
Cf. des Rotours (2), p. 97.

Tung Khai-Chhen (*1*) 董愷忱.
*Shih-Lun Yüeh-Ling Thi-Tshai ti Chung-Kuo Nung-Shu* 試論月令體裁的中國農書.
A Preliminary Discussion of Chinese Agricultural Treatises in the Style of 'Monthly Ordinances'.
Peking Agricultural College Library, May 1979.
Tr. Tung Khai-Chhen (1).

Tung Tso-Pin (*1*) 董作賓.
*Yin Li Phu* 殷曆譜.
On the Calendar of the Yin (Shang) Period.
Academia Sinica, Lichuang, 1945.
(Prelim. notes *AS/BIHP*, 1936, 7, 45; *SSE*, 1941. **2**, 1.)
See also Tung Tso-Pin (*6*).
Rev. K. Yabuuchi (*13*).

Tung Tso-Pin (*6*) 董作賓.
*Yin Li Phu Hou Chi* 殷曆譜後記.
Supplementary Notes to the *Yin Li Phu* (*1*).
*AS/BIHP*, 1948, **13**, 183.

Utsunomiya Kiyoyoshi (*1*) 宇都宮清吉.
*Kandai Shakai Keizai Shi Kenkyū* 漢代社会経済史研究.
A Study of Han Social & Economic History.

Utsunomiya Kiyoyoshi (*1*) (*cont.*)
Kobundo, Tokyo, 1967.

Wada Toshihiko (ed.) (*1*) 和田利彦.
*Shao-Hsing Chiao-Ting Ching-Shih Cheng-Lei Pei-Chi Pen-Tshao* 紹興校定經史証類備急本草.
Facsimile Edition of the *Corrected, Classified and Consolidated Armamentarium, Pharmacopoeia of the Shao-Hsing reign period*, from a +12th century Manuscript (perhaps +1159) preserved in the Omori Memorial Library of the Kyoto Botanic Gardens.
Shunyōdō, Tokyo, 1933.
Cf. Karow (2).

Wan Kuo-Ting (*1*) 萬國鼎.
*'Fan Sheng-Chih Shu' Chi Shih* 氾勝之書輯釋.
An Annotated Translation into Modern Chinese of *Fan Sheng-Chih Shu*.
Chung-Hua Press, Peking, 1957.

Wan Kuo-Ting (*3*) 萬國鼎.
*Ou Thien Fa ti Yen-Chiu* 區田法的研究.
A Study of Pit- or Basin-Cultivation.
*AHRA*, 1958, **1**, 5.

Wan Kuo-Ting (*4*) 萬國鼎.
*'Chhi Min Yao Shu' So-Chi Nung-Yeh Chi-Shu chi chhi tsai Chung-Kuo Nung-Yeh Chi-Shu-Shih Shang-ti Ti-Wei* 齊民要術所記農業技術及其在中國農業技術史上的地位.
Agricultural Techniques Recorded in the *Chhi Min Yao Shu* and their Historical Significance.
*NACJ*, 1956, **1**, 89.

Wan Kuo-Ting (*5*) 萬國鼎.
*Ou-Keng Khao* 耦耕考.
An Investigation of the Term *Ou-Keng*.
*AHRA* (1959), **1**, 75.

Wan Kuo-Ting (*6*) 萬國鼎.
*Chhen Fu 'Nung Shu' Chiao-Chu* 陳旉農書校注.
A Comparative, Annotated Edition of Chhen Fu's *Nung Shu*.
Agriculture Press, Peking, 1965.

Wan Kuo-Ting (*7*) 萬國鼎.
*Kuang Fan 'Pien Min Thu Tsuan'* 廣璠便民圖纂.
Kuang Fan's *Pien Min Thu Tsuan*.
*CKNP* (1962), **11**.

Wan Kuo-Ting (*8*) 萬國鼎.
*Han O 'Ssu Shih Tsuan Yao'* 韓鄂四時纂要.
Han O's *Ssu Shih Tsuan Yao*.
*CKNP* (1962), **11**.

Wan Kuo-Ting (*9*) 萬國鼎.
*Wu Ku Shih Hua* 五穀史話.
A Historical Discussion of Chinese Cereals.
Chung-Hua, Peking, 1961.

Wan Kuo-Ting (*10*) 萬國鼎.
*Chung-Kuo Thien-Chih Shih* 中國田制史.
A History of Chinese Land Tenure.
Cheng-Chung, Nanking, 1934.

Wang Chih-Jui (*1*) 王志瑞.
*Sung Yuan Ching-Chi Shih* 宋元經濟史.
An Economic History of the Sung & Yuan Dynasties.
Taipei, 1964.

Wang Chin-Ling (*1*) 王金陵.
*Ta-Tou* 大豆.
Soybeans.
Kho Hsüeh, Peking, 1966.

Wang Chung-Min (ed.) (*4*) 王重民.
*Hsü Kuang-Chhi Chi* 徐光啓集.
Collected Works of Hsü Kuang-Chhi.
Chung-Hua Press, Peking, 1963; 2 vols.

Wang Ning-Sheng (*1*) 汪寧生.
*Ou-Keng Hsin-Chieh* 耦耕新解.
A New Interpretation of the Term *Ou-Keng*.
*WW* (1977), **4**, 74.

Wang Ning-Sheng (*2*) 汪寧生.
*Han-Chin Hsi-Yüeh yü Tsu-Kuo Wen-Ming* 漢晉西域與祖國文明.
The Western Frontiers and Chinese Civilization during the Han & Chin Dynasties.
*KKHP* (1977), **1**, 23.

Wang Yü-Hu (*1*) 王毓瑚.
*Chung-Kuo Nung-Hsüeh Shu-Lu* 中國農學書錄.
Bibliography of Chinese Agriculture.
Agriculture Press, Peking, 1964; 2nd ed. 1979; Jap. ed. Amano (*8*)

Wang Yü-Hu (ed.) (*2*) 王毓瑚.
*Chhin Chin Nung Yen* 秦晉農言.
Agricultural Precepts of Chhin and Chin.
Incl. 知本提綱.
農言著實.
馬首農言.
Chung-Hua, Peking, 1957.

Wang Yü-Hu (ed.) (*3*) 王毓瑚.
*Nung Phu Pien Lan* 農圃便覽.
The *Nung Phu Pien Lan* [by Ting I-Tseng, Chhing].
Chung-Hua Press, Peking, 1957.

Wang Yü-Hu (ed.) (*4*) 王毓瑚.
*Nung Sang I-Shih Tsho Yao* 農桑衣食撮要.
Selected Essentials of Agriculture, Sericulture, Clothing & Food.
Agriculture Press, Peking; 1st ed. 1962, repr. 1979.

Wang Yü-Hu (ed.) (*5*) 王毓瑚.
*Ou-Thien Shih Chung* 區田十種.
Ten Methods of Pit Cultivation.
Tshai-Cheng Ching-Chi Press, Peking, 1955.

Wang Yü-Hu (*6*) 王毓瑚.
*Wo-Kuo Tzu-Ku I-Lai ti Chung-Yao Nung-Tso-Wu* 我國自古以來的重要農作物.
The History of China's Main Crop Plants.
Drafted 1975, published by Peking Agricultural College 1980.

Wang Yü-Hu (ed.) (*7*) 王毓瑚.
*So Shan Nung Phu* 梭山農譜.
A Survey of the Agriculture of Shuttle Mountain.
Nung Yeh, Peking, 1960.

Wang Yün-Wu (ed.) (*1*) 王雲五.
*Nung Shuo, Shen Shih Nung Shu, Lei Ssu Ching* 農說，沈氏農書，耒耜經.

Wang Yün-Wu (ed.) (*1*) (*cont.*)
Three Agricultural Classics.
Commercial Press, Shanghai, 1936.

Watanabe Tadayo *et al.* (*1*) 渡部忠世.
*Inasaku no Kigen to sono Tenkai o megutte* 稲作の起源とその展開をめぐつて.
On the Origins and Development of rice Cultivation.
*KKJRG* (1976), 7, **2**.

Wei Ching-Chhao (*1*) 魏景超.
*Shui-Tao Ping-Yuan Shou-Tshe* 水稻病原手册.
A Manual of Rice Diseases.
Kho Hsüeh, Peking, 1975 (1st ed. 1957).

Wu Chheng-Lo (*2*) 吳承洛.
*Chung-Kuo Tu-Liang-Heng Shih* 中國度量衡史.
A History of Chinese Weights and Measures.
2nd ed., Shanghai, 1957.

Wu Chhi-Chhang (*4*) 吳其昌.
*Chhin-Yi-Chhien Chung-Kuo Thien-Chih Shih* 秦以前中國田制史.
On the History of the Chinese Land System before the Chhin Dynasty.
*WUQJSS*, 1935, **5**, 543 & 833.
Abridged tr. Sun & de Francis (1), p. 55.

Wu Chhi-Chün (*1*) 吳其濬.
*Chih Wu Ming Shih Thu Khao* 植物名實圖考
Illustrated Investigation of the Names and Natures of Plants.
Peking, 1848.
Repr. Comm. Press, Shanghai, 1919 (with index).

Wu Chhi-Chün (*2*) 吳其濬.
*Chih Wu Ming Shih Thu Khao Chhang Phien* 植物名實圖考長編.
Comprehensive Treatise on the Names and Natures of Plants.
Peking, 1848.
Repr. Comm. Press, Shanghai, 1919 (with index).

Wu Shan-Chhing (*1*) 吳山菁.
*Lüeh Lun Chhing-Lien-Kang Wen-Hua* 略論青蓮崗文化.
A Brief Discussion of the Chhing-Lien-Kang Culture.
*WW* (1973), **6**, 45.

Wu Shih-Chhang (*1*) 吳世昌.
*Mi Tsung Su Hsiang Shuo Lüeh* 密宗塑像說略.
A Brief Discussion of Tantric (Buddhist) Images.
*AP/HJ*, 1935, **1**.

Yabuuchi Kiyoshi, ed. (*11*) 薮内清.
*Tenkō Kaibutsu no Kenkyū* 天工開物の研究.
Studies on the *Thien Kung Khai Wu*.
Tokyo, 1955.

Yamada Keiji (ed.) (*3*) 山田慶兒.
*Chūgoku no Kagaku to Kagakusha* 中国の科学と科学者.
Chinese Science and Scientists.
Humanities Institute of Kyōtō Univ., Kyōtō, 1979.

Yanagida Setsuko (*1*) 柳田節子.
*Sōdai Tochi Shoyū Sei ni mirareru futatsu no Kata* 宋代土地所有性に見られる二つの型.
Two forms of Sung landholding.
*TBK*, 1963, **29**, pp. 95–130.

Yang Chia-Lo (ed.) (*1*) 楊家駱.
*'Ssu Khu Chhüan Shu' Hsüeh Tien* 四庫全書學典.
Bibliographical Index of the *Ssu Khu Chhüan Shu* Encyclopaedia.
World Book Co., Shanghai, 1946.

Yang Chien-Fang (*1*) 楊建芳.
*An-Hui Tiao-Yü-Thai Chhu-Thu Hsiao-Mai Nien-Tai Shang-Chüeh* 安徽釣魚台出土小麥年代商榷.
Discussion of the Dating of the Wheat Excavated at Tiao-Yü-Thai in Anhwei.
*KK*, 1963, **11**, 630–1.

Yang Chih-Min *et al.* (*1*) 楊直民.
*Chung-Kuo Nung-Shu chi chhi Fen-Lei Hsi-Thung* 中國農書及其分類系統.
Classificatory Systems of Chinese Agricultural Works.
Peking Agricultural College Library, May 1979.

Yang Chih-Min & Tung Khai-Chhen (*1*) 楊直民, 董愷忱.
*Wo-Kuo Ku-Tai Tshai-Phei Chih-Wu Chhi-Yuan Fang-Mien ti Kung-Hsien* 我國古代栽培植物起源方面的貢獻.
A Contribution on the Origins of Ancient Chinese Domesticated Plants.
In Anon. (*531*) pp. 254–83.

Yang Khuan (*11*) 楊寬.
*Ku-Shih Hsin-Than* 古史新探.
New Enquiries into Ancient History.
Chung Hua, Peking, 1965.

Yang Lien-Sheng (*1*) 楊聯陞.
*Tshung 'Ssu Min Yüeh Ling' So Chien-Tao ti Han-Tai Chia-Tsu ti Sheng-Chhan* 從四民月令所見到的漢代家族的生產.
The Growth of Clans in the Han Dynasty, as seen in the *Ssu Min Yüeh Ling*.
*SH*, 1935, 1, **6**, pp. 8 ff.

Yang Min (*1*) 楊旻.
*Ku-Chin Shih-Wu Kho-Hsüeh Tsa-Than* 古今事物科學雜談.
Miscellaneous Talks on Scientific Matters, Old & New.
Com. Press, Peking, 1959 repr. 1960.

Yang Shih-Thing (*1*) 楊式挺.
*Than-Than Shih-Hsia Fa-Hsien ti Tshai-Phei Tao I-Chih* 談談石峽發現的栽培稻遺址.
On the Remains of Domesticated Rice Discovered at Shih-Hsia.
*WW*, 1978, **7**, pp. 23–8.

Yeh Ching-Yuan (*2*) 葉靜淵.
*Kan-Chü* 柑橘.
Citrus Fruits.
Chinese Agricultural Heritage Series no. 14, Chung-Hua, Peking, 1958.

Yen Chieh (ed.) (*1*) 嚴杰.

Yen Chieh (ed.) (*1*) (*cont.*)
*Huang Chhing Ching Chieh* 皇清經解.
Collection of [more than 180] Monographs on Classical Subjects written during the Chhing Dynasty.
1829; 2nd ed. Keng Shen Pu Khan, 1860.

Yen Ju-Yü (*2*) 嚴如熤.
*Miao-Fang Pei-Lan* 苗方備覽.
A complete Survey of Miao Country.
1820.

Yen Tun-Chieh (*18*) 嚴敦傑.
*Ku-Kung so Tshang Chhing-Tai Chi-Suan I-Chhi* 故宮所藏清代計算儀器.
Computing Apparatus of the Chhing Period preserved in the Palace Museum Collections [including two calculating machines of Pascal].
*WWTK*, 1962, **3**, 19.

Yoneda Kenjirō (*1*) 米田賢次郎.
*Seimin Yōjutsu to Ninen Sanmōsaku* 斉民要術と二年三毛作.
The *Chhi Min Yao Shu* and the System of Three Crops in Two Years.
*TK* (1959), 17, **4**, 407–30.

Yoshioka Yoshinobu (*1*) 吉岡義信.
*Sōdai no Kannōshi ni tsuite* 宋代の勧農史に付いて.
Agricultural Encouragement in the Sung.
*SGKK*, 1955, **60**, pp. 43–9.

Yu Hsiu-Ling (*1*) 游修齡.
*Tshung Ho-Mu-Tu I-Chih Chhu-Thu Tao-Ku Shih-Lun Wo-Kuo Tsai-Phei Tao ti Chhi-Yuan, Fen-Hua yü Chuan-Po* 從河姆渡遺址出土稻谷試論我國栽培稻的起源,分化與傳播.
A Preliminary Discussion of the Origins, Diversification and Dissemination of Rice Cultivation in China, Based on the Archaeological Rice Remains from Ho-Mu-Tu.
*TSWHP* (1979), 5, **3**, 1–10.

Yu Yü (*1*) 友于.
*'Kuan Tzu Tu Ti Phien' Than-Wei* 管子度地篇探微.
An Investigation of the *Tu Ti* chapter of the *Kuan Tzu*.
*AHRA*, 1959, **1**, 1–15.

Yu Yü (*2*) 友于.
*'Kuan Tzu Ti Yuan Phien' Yen-Chiu* 管子地員篇研究.
Research on the *Ti Yuan* chapter of the *Kuan Tzu*.
*AHRA*, 1959, **1**, 17–36.

Yu Yü (*3*) 友于.
*Yu Hsi-Chou Tao Chhien-Han ti Keng-Tso Chih-Tu Yen-Ko* 由西周到前漢的耕作制度沿革.
Successive Changes in Tillage Regulations from Western Chou to Former Han Times.
*AHRA* (1960), **2**, 1.

Yü Ching-Jang (*1*) 于景讓.
*Chung-Kuo Tsai-Phei Chan-Chheng Tao ti Yen-Ko* 中國栽培占城稻的沿革.
The Stages of Champa Rice Cultivation in China.
*KHNY* (1956), **4**, 12.

Yü Hsing-Wu (*2*) 于省吾.
*Tshung Chia-Ku-Wen Khan Shang-Tai ti Nung-Thien Khen-Chih* 從甲骨文看商代的農田墾殖.
Shang Land Clearance and Cultivation as Seen in the Oracle Bone Inscriptions.
*KK* (1972), **4**, 39.

Yü Hsing-Wu (*3*) 于省吾.
*Shang-Tai ti Ku-Lei Tso-Wu* 商代的穀類作物.
Cereal Crops of the Shang Dynasty.
*TJTJKH* (1957), **1**, 100.

Yuan Thung-Li (*1*) 袁同禮.
*'Yung-Lo Ta Tien' Hsien Tshun Chüan Mu Piao* 永樂大典現存卷目表.
Census of the Locations and Contents of the still existing Volumes of the *Yung-Lo Ta Tien* Encyclopaedia.
*QBCB/C*, 1939 (NS), **1**, 246.

ADDENDUM

*Nagahiro Toshio* (*1*) (ed.) 長廣敏雄.
*Kandai Gazō no Kenkyū* 漢代畵象の研究.
The Representational Art of the Han Dynasty.
Report of the Humanistic Studies Research Institute, Kyoto; Chuo-Koron Bijutsu Shuppan, Tokyo, 1965.

# C. BOOKS AND JOURNAL ARTICLES IN WESTERN LANGUAGES

ABEL, WILHELM (1). *Geschichte der deutschen Landwirtschaft vom frühen Mittelatter bis zum 19. Jahrhundert.* Deutsche Agrargeschichte II, Eugen Ulmer, Stuttgart, 1962.

ADAIR, C. R. *et al.* (1). 'Rice breeding and testing methods in the United States.' In *Rice in the United States: Varieties and Production*, US Dept. Agric., Agric. Handbook no. 289, 1966.

AIGNER, J. (1). 'Pleistocene remains from South China.' *ASP* (1974), **16**, 16–38.

ALEX, W. (1). *Japanese Architecture.* Prentice-Hall, London; Braziller, New York; 1963.

ALLAN, WILLIAM (1). *The African Husbandman.* Oliver and Boyd, Edinburgh and London, 1967.

ALLAN, WILLIAM (2). 'Ecology, Techniques and Settlement Patterns.' In Ucko, Tringham & Dimbleby (1), p. 211.

ALLCHIN, F. R. (1). 'Early cultivated plants in India and Pakistan.' In Ucko and Dimbleby (1) 323.

ALLEN, JIM (1). 'The Hunting Neolithic: Adaptations to the Food Quest in Prehistoric Papua New Guinea.' In J. V. S. Megaw (1), pp. 167–88.

ALLEN, J., J. GOLSON & R. JONES, ed. (1). *Sunda and Sahel: Prehistoric Studies in Southeast Asia, Melanesia and Australia.* Academic Press, London & New York, 1977.

ALLEY, REWI and C. C. BOJESEN (1). 'Agricultural Implements used in Southern Kiangsu.' *CJ* (Feb. 1937), 26, **2**, 87.

AMANO, MOTONOSUKE (1). 'Dry Farming and *Chhi Min Yao Shu*'. Silver Jubilee Volume of The Zinbun-Kagaku-Kenkyusyo, Kyoto University, 1954, pp. 451–65.

AMES, OAKES (1). *Economic Annuals and Human Cultures.* Botanical Museum of Harvard University, Cambridge, Mass., 1939; repr. 1953.

AMIOT, J. J. M. (10). 'Serres chinoises.' *MCHSAMUC*, 1778, **3**, 423–37.

AMMERMAN, A. J. & L. L. CAVALLI-SFORZA (1). 'The wave of advance model for the spread of agriculture in Europe.' In Renfew & Cooke (1), pp. 275–94.

ANDERSON, E. N. Jr & M. J. (1). 'Modern China, South.' In K. C. Chang (3), pp. 317–82.

ANDERSON, RUSSELL H. (2). 'Grain drills through thirty-nine centuries.' *AGHST* (1936), 10, 4, 157–205.

ANDERSSON, V. G. (1). *Children of the Yellow Earth: Studies in Prehistoric China.* Kegan Paul, Trench, Trübner & Co, London, 1934; repr. MIT Press, Cambridge Mass., 1973. Tr. from the Swedish by E. Classen.

ANDERSSON, J. G. (3). 'An early Chinese culture.' *BGSC*, 1923, 5, 1, pp 1–68.

ANDERSSON, J. G. (4). 'Preliminary report on archaeological research in Kansu.' *BGSC* (1925), Memoirs, Ser. A, 5.

ANGLADETTE, A. (1). *Le riz.* G-P. Maisonneuve et Larose, Paris, 1966.

ANON. (160). *Historical Relics Unearthed in New China.* Foreign Languages Press, Peking, 1972.

ANON. (161). *Curiosities of Nature and Art in Husbandry and Gardening.* London, 1707.

ARCHER, M. (1). *Natural History Drawings in the India Office Library.* H.M.S.O., London, 1962.

ARNON, I. (1). *Crop Production in Dry Regions.* 2 vols, Leonard Hill, London, 1972.

ARNOTT, MARGARET L., ed. (1). *Gastronomy: the Anthropology of Food and Food Habits.* Mouton Publishers, The Hague & Paris, 1975.

ASCHMANN, H. (1). 'Evaluations of Dry Land Environments by Societies at Various Levels of Technical Competence.' In R. B. Woodbury (ed.), *Civilisations in Desert Lands*, University of Utah Press, Utah, 1962, p. 1.

ASH, ROBERT (1). *Land Tenure in Pre-Revolutionary China: Kiangsu Province in the 1920s and 1930s.* Res. Notes and Studies no, 1. Contemporary China Institute, School of Oriental and African Studies, London, 1976.

ASSOCIATION OF JAPANESE AGRICULTURAL SCIENTIFIC SOCIETIES, ed. (1). *Rice in Asia.* University of Tokyo Press, Tokyo, 1975.

AUBERT, C., MAUREL, F. & PAIRAULT , T. (1). *Comptabilité rurale et Répartition du Revenu.* Centre de Recherche et de Documentation sur la Chine Contemporaire, Ecole des Hautes Etudes en Sciences Sociales, Paris, 1975.

BAKER, A. R. H. & R. A. BUTLIN, eds. (1). *Studies of Field Systems in the British Isles.* Cambridge University Press, Cambridge, 1973.

BAKER, A. R. H. & R. A. BUTLIN (2). 'Conclusion: problems and perspectives.' In Baker & Butlin (1), pp. 619–56.

BALASSA, I. (1). 'The Earliest Ploughshares in Central Europe.' *TT* (1975), 2, **4**, 242.

BALASZ, E. (=S.) (7) (tr.). 'Le traité économique du *Souei-Chou* [*Sui-Shu*] (Etudes sur la société et l'économie de la Chine médiévale). *TP* (1953), **42**, 113. Also sep. issued, Brill, Leiden, 1953.

BALASZ, E. (=S) (8) (tr.). 'Le traité juridique du *Souei-Chou* [*Sui Shu*].' *TP* (1954). Sep. pub. as *Etudes sur la société et l'économie de la Chine médiévale*, no. 2. Brill, Leiden, 1954 (Bibliotheque de l'Inst. des Hautes Etudes Chinoises, no. 9).

BARCHAEUS, A. G. (1). *Utdrag utur A.G.B.'s anteckningar 1–2.* Uppsala, 1828–9.

BARNARD, NOEL, ed. (2). *Early Chinese Art and its Possible Influence in the Pacific Basin.* Proceedings of a Symposium Arranged by the Department of Art History & Archaeology, Columbia University, New York City, August 21–5, 1967; 3 vols; Intercultural Arts Press, New York, 1972.

BARRAU, JACQUES (1). 'La région indo-pacifique comme centre de mise en culture et de domestication des végétaux.' *JATBA* (1970), **17**, 487–504.

BARRAU, J. (1a). 'The Indo-Pacific area as a centre of origin of plant cultivation and domestication.' Paper given at a Symposium on Ethnobotany, Peabody Museum of Natural History, Yale, 1966.

BARRAU, JACQUES, ed. (2). *Plants and the Migrations of Pacific Peoples.* 10th Pacific Science Congress; Bishop Museum Press, Honolulu, 1961.

BAYARD, D. T. (1). 'On Chang's interpretation of Chinese radiocarbon dates.' *CURRA*, 1975, 16, **1**, pp. 167–9.

BEADLE, GEORGE W. (1). 'The ancestry of corn.' *SAM*, 1980, **1**, 96–103.

BENDER, BARBARA (1). *Farming in Prehistory: from Hunter-Gatherer to Food-Producer.* John Baker, London, 1975.

BEATTIE, HILARY J. (1). *Land and Lineage in China: a Study of Thung-Chheng Country, Anhwei, in the Ming & Chhing Dynasties.* Cambridge University Press, 1979.

BENEDICT, PAUL K. (1). 'Austro-Thai.' *BSN* (1966) **1**, 227–61.

BENEDICT, PAUL K. (2). 'Austro-Thai Studies, 3: Austro-Thai and Chinese.' *BSN* (1967), **2**, 275–336.

BERCH, A. (1). *Anmerkungen über die Schwedischen Pflüge.* Proceedings of the Royal Swedish Academy of Sciences vol. 21 (1759), Stockholm.

BERG, GÖSTA (2). 'Den Svenska Sadesharpan och dem Kinesiska' (on the coming of the rotary winnowing-fan from China to Europe). Art. in *Nosdiskt Folkminne; Studien tillagnade C. W. von Sydow*, Stockholm 1928.

BERG, GÖSTA (3). 'The Introduction of the Winnowing-Machine in Europe in the 18th Century.' *TT* (1976), 3, **1**, 25–46.

BEUTLER, C. (1). 'Un chapitre de la sensibilité collective: la littérature agricole en Europe continentale au XVI$^{e}$ siècle.' *AHES/AESC*, 1973, **5**, 1280–1301.

BINFORD, LEWIS R. (1) 'Post-Pleistocene adaptation.' In R. S. Binford & L. R. Binford, ed., *New Perspectives in Archaeology*, Aldine, Chicago, 1968; pp. 313–41.

BIOT, E. (1) (tr.). *Le Tcheou-Li on Rites des Tcheou.* 3 vols., Imp. Nat., Paris, 1851 (photographically reproduced, Wentienko, Peiping, 1930).

BISHOP, C. W. (4). 'The Neolithic Age in Ancient China.' *AQ*, 1938, **7**, p. 369.

BISHOP, CARL W. (9). 'The Ritual Bullfight.' *ARSI* (1926), 447–55; repr. from *CJ* (1925), **3**, 630–7.

BISHOP, C. W. (15). 'The Origin and Early Diffusion of the Traction Plough.' *AQ*, 1936, **10**, 261.

BISHOP, T. A. M. (1). 'Assarting and the growth of the open fields.' *EHR* (1935–6), **6**, pp. 13–29.

BLITH, WALTER (1). *English Improver.* London, 1649.

BLITH, WALTER (2). *English Improver Improved.* London, 1653.

BLOCH, JULES (2). 'La Charrue Védique.' *BLSOAS* (1936), **8**, 411–18.

BLOCH, MARC (7). *Les caractères originaux de l'histoire rurale française.* 2 vols, A. Colin, Paris, 1952–6.

BODDE, D. (12) (tr.). *Annual Customs and Festivals in Peking, as Recorded in the Yen-Ching Sui-Shih-Chi by Tun Li-Chhen.* Henri Vetch, Peiping, 1936; 2nd ed. Hong Kong Univ. Press, Hong Kong, 1965.

BODDE, D. (24). 'Henry A. Wallace and the Ever-Normal Granary.' *FEQ* (1946), **5**, 411–26

BODDE, D. (25). *Festivals in Classical China: New Year and Other Annual Observances During the Han Dynasty 206 BC–AD 220.* Princeton University Press/Chinese University of Hongkong, Princeton, 1975.

BOLENS, L. (2). *Les méthodes culturales au Moyen Âge d'après les traités d'agronomie andalous; traditions et techniques.* Inaug. Diss. Geneva (Médecine et Hygiène Publique), 1974.

BONEBAKKER, S. A. (1). *The Kitab Naqd al-Shi'r of Qudāma b. Ja'far.* Brill, Leiden, 1956.

BOSERUP, E. (1). *The Conditions of Agricultural Growth: The Economics of Agrarian Change under Population Pressure.* Allen & Unwin, London, 1965.

BOWEN, H. C. (1). *Ancient Fields: a Tentative Analysis of Vanishing Earthworks & Landscapes.* British Assocn. for the Advancement of Science, London, 1962.

BOURDE, ANDRÉ J. (1). *Agronomie et agronomes en France au XVIII$^{e}$ siècle.* Série Les Hommes et la Terre, S.E.V.P.E.N., Paris, 1967, 3 vols.

BRAIDWOOD, ROBERT J. (1). *Prehistoric Men.* Chicago Natural History Museum Popular Series, Anthropology, no. 37, Chicago, 1961 (5th edn).

BRANDENBURG, DAVID J. (1). 'Agriculture in the *Encyclopédie*: an essay in French intellectual history.' *AGH* (1950), 24, **2**, 96–108.

BRATANIĆ, B. (1). 'Some Similarities between Ards of the Balkans, Scandinavia, and Anterior Asia, and their Methodological Significance.' In A. F. C. Wallace (ed.), *Selected Papers of the Fifth International Congress of Anthropological and Ethnological Sciences, Philadelphia, September 1–9, 1956: Men and Cultures*, Philadelphia, 1960, pp. 221–8.

BRAUDEL, F. (1). *La Méditerranée et le monde méditerranéen à l'époque de Philippe II.* Colin, Paris, 1949.

BRAY, F. (1). 'Swords into Ploughshares: a Study of Agricultural Technology and Society in Early China.' *TCULT*, (1978), 19, **1**, 1–31.

BRAY, F. (2). 'Recent Changes in Padi Farming in Kelantan, Malaysia.' Unpublished report for the British Academy & Royal Society, 1977.

BRAY, F. (3). 'Agricultural Development and Agrarian Change in Han China.' *EC* (1980) 5, 1–13.
BRAY, F. (4). 'The Green Revolution: a new perspective.' *MAST* (1979), 13, **4**, 681–8.
BRAY, F. (5). 'Essential Techniques for the Peasantry: an annotated translation of the 6th century Chinese agricultural treatise *Chhi Min Yao Shu*.' In preparation.
BRAY, F. (6). 'A slight technical hitch: universal theory versus specific applications in technological development, as seen in Asian agriculture.' Paper presented at the United Nations University Symposium on Universality & Specificity, Tokyo, November 1981.
BRAY, F. (7). 'The Chinese contribution to Europe's Agricultural Revolution: a technology transfomed.' In Li *et al.* (1), 597–637.
BRAY, F. (8). 'The evolution of the mouldboard plough in China.' *TT* (1979), 3, **4**, 227–40.
BRAY, F. and A. F. ROBERTSON (1). 'Sharecropping in Kelantan.' In G. Dalton (ed.), *Research in Economic Anthropology*, vol. 3, J.A.I. Inc., Greenwich, Conn., 1980, 209–44.
BRAY, WARWICK (1). 'From foraging to farming in early Mexico.' In J. V. S. Megaw, ed. (1), pp. 225–50.
BREEZE, D. J. (1). 'Plough Marks at Carrawburgh on Hadrian's Wall.' *TT* (1974), 2, **3**, 188.
BRETSCHNEIDER, E. (1). *Botanicum Sinicum: Notes on Chinese Botany from Native and Western Sources.* 3 vols, Trübner & Co. London, 1882.
BRETSCHNEIDER, E. (6). *On the Study & Value of Chinese Botanical Works, with Notes on the History of Plants and Geographical Botany from Chinese Sources.* Foochow, 1870.
BRONSON, BENNET (1). 'The earliest farming: demography as cause and consequence.' In Charles A. Reed (2), pp. 23–48.
BROOK, TIMOTHY (1). 'The social limits to technological transfer: the spread of rice cultivation into the Hopei region in the Ming and Chhing dynasties.' In Li *et al.* (1), 659–79.
BUCHANAN, FRANCIS (1). *A Journey from Madras Through the Countries of Mysore, Canara, and Malabar.* London, 1807.
BUCHANAN, K. (1). *The Chinese People and the Chinese Earth.* G. Bell & Sons, London, 1966.
BUCHANAN, K. (2). *The Transformation of the Chinese Earth: Perspectives on Modern China.* G. Bell & Sons, London, 1970.
BUCK, JOHN LOSSING (1). *Chinese Farm Economy: A Study of 2866 Farms in Seventeen Localities and Seven Provinces in China.* Publ. for the Univ. of Nanking & the China Council of the Inst. of Pacific Relations by the Univ. of Chicago Press, Chicago, 1930.
BUCK, JOHN LOSSING (2). *Land Utilisation in China.* Commercial Press, Shanghai, 1937.
BUDDENHAGEN, I. W. & G. J. PERSLEY (ed.) (1). *Rice in Africa.* Academic Press, London, 1978.
BURKILL, I. H. (1). *A Dictionary of the Economic Products of the Malay Peninsula.* 2 vols, published for the Malay Govt. by Crown Agents, London, 1935.
BYERS, D. S. (1) ed. *The Prehistory of the Tehuacan Valley, vol. 1: Environment and Subsistence.* University of Texas Press, Austin, 1967.

CALLEN, E. O. (1). 'The first New World cereal.' *AMA*, 1967, 32, **4**, pp. 535–8.
CAMERON, J. W. & R. K. SOOST (1). 'Citrus.' In Simmonds (1), pp. 261–5.
DE CANDOLLE, ALPHONSE (1). *The Origin of Cultivated Plants.* Kegan Paul, London, 1884 (International Scientific Series, no. 49). Translated from the French edition, Geneva, 1883. Engl. 2nd ed. London, 1886 reproduced photolithographically, Hafner, New York, 1959.
CAREFOOT, G. L. and E. R. SPROTT (1). *Famine on the Wind: Plant Diseases and Human History.* New York, 1967.
CARO BAROJA, J. (1). *Los Pueblos del Norte de la Península Ibérica: Análisis Histórico-Cultural.* Editorial Txertoa, San Sebastian, 1973.
CARTER, GEORGE F. (2). 'Movement of people and ideas across the Pacific.' In J. Barrau (2), pp. 7–22.
CARTER, GEORGE F. (10). 'A Hypothesis suggesting a Single Origin of Agriculture.' In Charles A. Reed (2), pp. 89–134.
CARTIER, M. (1). *Une réforme locale en Chine au XVI^e siècle: Hai Jui à Chun'an 1558–1562.* Mouton, Paris & The Hague, 1973.
CATO, MARCUS PORCIUS (1). *De Agri Cultura.* Tr. W. D. Hooper, revised H. B. Ash, Loeb Classical Library, Heinemann, London, 1954.
CATON-THOMPSON, G. & E. W. GARDNER (1). *The Desert Fayum.* 2 vols, Royal Anthropological Institute, London, 1937.
CHAGNON, N. A. (1). *Yqnamamö, the Fierce People.* Holt, Rinehart and Winston, New York & London, 1968.
CHAMBERS, J. D. (1). 'Enclosures and labour supply to the Industrial Revolution.' In E. L. Jones, ed. (1), pp. 94–127.
CHANG, CHUNG-LI (1). *The Income of the Chinese Gentry.* University of Washington Press, Seattle, 1962.
CHANG, K. C. (1). *The Archaeology of Ancient China.* Yale University Press, 1st ed. 1963; 3rd revised ed. 1977.
CHANG, K. C. (2). *Fengpitou, Tapenkeng, and The Prehistory of Taiwan.* Yale University Publications in Anthropology no. 73, New Haven, 1969.
CHANG, K. C., ed. (3). *Food in Chinese Culture: Anthropological and Historical Perspectives.* Yale University Press, New Haven & London, 1977.
CHANG, K. C. (4). 'Ancient China.' In K. C. Chang, ed. (3), pp. 25–52.

CHANG, K. C. (5). *Shang Civilisation.* Yale University Press, New Haven & London, 1980.

CHANG, K. C. (6). 'The beginnings of agriculture in the Far East.' *AQ*, 1970, **64**, pp. 175–85.

CHANG, TE-TZU (1). 'The Rice Cultures.' In Hutchinson, Clarke, Jope & Riley (1), 143–57.

CHANG, TE-TZU (2). 'The origin, evolution, cultivation, dissemination, and diversification of Asian and African rices.' *EUP* (1976), **25**, 425–41.

CHATLEY, H. (1). MS. translation of the astronomical chapter (Ch. 3, Thien Wen) of *Huai Nan Tzu*. Unpublished. (Cf. note in *O*, 1952, **72**, 84.)

CHAVANNES, E. (1). *Les Mémoires Historiques de Se-Ma Ts'ien* [*Ssuma Chhien*]. 5 vols. Leroux, Paris, 1895–1905. (Photographically reproduced, in China, without imprint and undated.)

1895 vol. 1 tr. *Shih Chi*, chs. 1, 2, 3, 4.
1897 vol. 2 tr. *Shih Chi*, chs. 5, 6, 7, 8, 9, 10, 11, 12.
1898 vol. 3 (i) tr. *Shih Chi*, chs. 13, 14, 15, 16, 17, 18, 19, 20, 21, 22.
vol. 3 (ii) tr. *Shih Chi*, chs. 23, 24, 25, 26, 27, 28, 29, 30.
1901 vol. 4 tr. *Shih Chi*, chs. 31, 32, 33, 34, 35, 36, 37, 38, 39, 40, 41, 42.
1905 vol. 5 tr. *Shih Chi*, chs. 43, 44, 45, 46, 47.

CHAVANNES, E. (6) (tr.). 'Les Pays d'Occident d'après le Heou Han Chou.' *TP*, 1907, **8**, 149. (Ch. 118, on the Western Countries, from *Hou Han Shu*.)

CHAVANNES, E. (16). 'Trois Généraux Chinois de la Dynastie des Han Orientaux.' *TP*, 1906, **7**, 210. (Tr. ch. 77 of the *Hou Han Shu* on Pan Chhao, Pan Yung and Liang Chhin.)

CHEN TSU-LUNG (1). 'Note on Wang Fu's *Chha Chiu Lun*.' *S* (1953), 6, **4**, 271.

CHENG, SIOK-HWA (1). *The Rice Industry of Burma, 1852–1940.* University of Malaya Press, Kuala Lumpur and Singapore, 1968.

CHENG, TE-KHUN (17). 'Metallurgy in Shang China.' *TP* (1974), 60, **4–5**, 109.

CHESNEAUX, JEAN, ed. (1). *Popular Movements and Secret Societies in China, 1840–1950.* Stanford University Press, Stanford, 1972.

CHEVALIER, H. (3). 'Les Anciennes Charrues de l'Europe.' *IEA* (1912), **1**, 41.

CHHÜ THUNG-TSU (1). *Han Social Structure.* Washington University Press, Seattle, 1972.

CHI CHHAO-TING (1). *Key Economic Areas in Chinese History.* Allen and Unwin, London, 1936.

CHILDE, V. GORDON (2). *Man Makes Himself.* Watts, London, 1941.

CHILDE, V. GORDON (4). *What Happened in History.* Penguin, Harmondsworth, 1st ed. 1942; revised ed. 1954.

CHU KHO-CHEN (9). 'A preliminary study of climatic fluctuations during the last five thousand years in China.' *SCISA*, 1973, 14, **2**.

CIPOLLA, CARLO M., ed. (4). *The Fontana Economic History of Europe: The Industrial Revolution.* Fontana/Collins, London, 1973.

CLARK, GRAHAME (3). *World Prehistory: an Outline.* Cambridge University Press, Cambridge, 1961.

CLARKE, CUTHBERT (1). *True Theory and Practice of Husbandry.* London, 1777.

CLARKE, D. (1). *Analytical Archaeology.* Methuen, London, 1968.

CLARKE, D. V. (1). 'A Plough Pebble from Colstoun, Scotland.' *TT* (1972), 2, **1**, 50.

CLÉMENT-MULLET, J.-J., tr. (1). *Le livre d'agriculture d'Ibn al-'Awwām.* Paris, 1864–7.

COBBETT, WILLIAM (1). *Rural Rides.* 1st ed. 1830; revised Penguin, Harmondsworth, 1967.

COEDÈS, G. (5). *Les états hindouisés d'Indochine et d'Indonésie.* Boccard, Paris, 1948. (*Histoire du monde*, ed. E. Cavaignac, vol. 8, pt. 2.)

COHEN, M. N. (1). *The Food Crisis in Prehistory: Overpopulation and the Origins of Agriculture.* Yale University Press, New Haven & London, 1977.

COHEN, M. N. (2). 'Population pressure and the origins of agriculture: an archaeological example from the coast of Peru.' In Charles A. Reed (2), pp. 135–8.

COLANI, MADELEINE (7). *Emploi de la pierre en des temps reculés. Annam—Indonésie—Assam.* Publication des Amis du Vieux Hué, Hanoi, 1940.

COLER, JOHANN (1). *Oeconomia ruralis et domestica, oder Hausbuch.* 6 vols, Silesia, 1591–1607.

COLLINS, E. J. T. (1). 'The diffusion of the threshing machine in Britain, 1790–1880.' *TT* (1972), 2, **1**, 16–23.

COLUMELLA, LUCIUS JUNIUS MODERATUS (1) (Trs. H. B. Ash). *Res Rustica* (On Agriculture). Loeb Classical Library, Heinemann, London, 1948.

CONDOMINAS, G. (1). *Nous avons mangé la forêt.* Paris, 1957.

COPELAND, E. B. (1) *Rice.* MacMillan, London, 1924.

COPLAND, SAMUEL ['The Old Norfolk Farmer'] (1). *Agriculture Ancient and Modern: a Historical Account of its Principles and Practice exemplified in their Rise Progress and Development.* James S. Virtue, London, 1866.

COURSEY, D. G. (1). *Yams.* Longmans, Green, London, 1967.

COURSEY, D. G. (2). 'The origins and domestication of yams in Africa.' In Harlan, de Wet & Stemler (1), pp. 383–408.

COURSEY, D. G. (3). 'Yams.' In Simmonds (1), pp. 70–4.

COUVREUR, F. S. (3) (tr.). *'Li Ki' [Li Chi], ou Mémoires sur les Bienséances et les Cérémonies.* 2 vols. Hochienfu, 1913.

CRESCENZI, PIETRO DE (1). *Opus ruralium commodorum.* Presented to Charles II, King of Sicily, in 1304; first printed ed. in 1471 (Augsburg).
DE CRESPIGNY, R. (1). *Official Titles of the Former Han Dynasty.* Australian National University, Canberra, 1967.
CROOK, ISABEL & DAVID CROOK (1). *Revolution in a Chinese Village: Ten Mile Inn.* Routledge & Kegan Paul, London, 1959.
CTARIKOV, V. S. (1). 'K Istorii Zemledel' cheskix Orudii Hanizev na Cevero-vostokye Kitaya' [Towards a history of Han agricultural implements in Northeast China]. In *Iz Istorii Nauki i Texniki v Stranax Vostoka*, Moscow, 1960, pp. 81–126.
CUMMINS, J. S. (1). 'Fray Domingo Navarrete: a source for Quesnay.' *BHS* (1959), **36**, 37–50.
CURWEN, E. C. (1). 'The plough and the origin of strip lynchets.' *AQ*, 1939, **13**, p. 45.
CURWEN, E. C. & G. HATT (1): *Plough and Pasture: the Early History of Farming*, Henry Schuman, New York, 1953.

DAHLMAN, CARL J. (1). *The Open Field System & Beyond: a Property Rights Analysis of an Economic Institution.* Cambridge University Press, 1980.
DALTON, G. (1). 'Economic Surplus, Once Again.' *AAN* (1963), 65, **2**, 389.
DAVIDSON, JEREMY (1). 'Recent Archaeological Activity in Vietnam.' *JHKAS* (1975), **6**, 80–100.
DAVIES, D. ROY (1). 'Peas.' In Simmonds (1), pp. 172–4.
DAVIES, NIGEL (1). *The Aztecs: a History.* Abacus (Sphere), London, 1977.
VON DEWALL, M. (2). 'Decorative Concepts and Stylistic Principles in the Bronze Art of Tien.' In N. Barnard (2), 329–72.
VON DEWALL, M. (3). 'The Tien culture of South-West China.' *AQ*, 1967, **40**, pp. 8–21.
DHARAMPAL (1). *Indian Science and Technology in the Eighteenth Century.* Impex India, Delhi, 1971.
DOBBY, E. H. G. (1). 'Paddy landscapes of Malaya: Kelantan.' *MJTG* (1957), **10**, i–42.
DOBSON, W. A. C. H. (1). 'Linguistic Evidence and the Dating of the Book of Songs.' *TP* (1964), LT, 322–34.
DOGGETT, H. (1). 'Sorghum.' In N. W. Simmonds (1), pp. 112–16.
DONKIN, R. A. (1). *Agricultural Terracing in the Aboriginal New World.* Viking Fund Publications in Anthropology, University of Arizona Press, Tucson, 1979.
DOUGLAS, MARY (1). *Purity and Danger: An Analysis of Concepts of Pollution and Taboo.* Routledge and Kegan Paul, London, 1966.
DOUGLAS, R. K. (1). *Orientalia Antiqua.* 1882.
DUHAMEL DU MONCEAU, H.-L. (1). *Expériences et réflexions relatives au traité de la culture des terres.* Paris, 1751.
DUBY, G. (1). *Guerriers et paysans: VII^e–XII^e siècle, premier essor de l'économie européenne.* *NRF*, Paris, 1973.
DUBY, G. (2). *L'économie rurale et la vie des campagnes dans l'Occident médiéval.* 2 vols., Aubier, Paris, 1962.
DUBY, G. & A. WALLON (eds.) (1). *Histoire de la France rurale.* 4 vols., Seuil, Paris, 1975–6.
DUMAN, L. I. (1). 'On the Social and Economic System of China in the Western Han Period.' Paper presented at the 24th International Congress of Orientalists, Moscow, 1957.
DUYVENDAK, J. L. L. (3) (tr.). *The Book of the Lord Shang; a Classic of the Chinese School of Law.* Probsthain, London, 1928.

EBERHARD, W. (2) *The Local Cultures of South and East China.* Brill, Leiden, 1968.
EBERHARD, WOLFRAM (26). *Das Toba-Reich Nordchinas.* Brill, Leiden, 1949.
EBERHARD, WOLFRAM (28). *Social Mobility in Traditional China.* E. J. Brill, Leiden, 1962.
EBREY, P. B. (1). *The Aristocratic Families of Early Imperial China: a Case Study of the Po-Ling Tshui Family.* Cambridge Studies in Chinese History, Literature and Institutions, Cambridge University Press, Cambridge, 1978.
EBREY, P. B. (2). 'Estate & family management in the Later Han as seen in the *Monthly Instructions for the Four Classes of People.*' *JESHO*, 1974, 17, **2**, 173 ff.
ELVIN, MARK (2). *The Pattern of the Chinese Past.* Methuen, London, 1973.
ELVIN, MARK (3). 'Skills and resources in late traditional China.' In Dwight H. Perkins, ed. (2), pp. 85–113.
EMBREE, J. F. (1). *A Japanese Village: Suye Mura.* Kegan Paul, Trench, Trubner & Co., London, 1946.
ERKES, E. (1) (tr.). 'Das Weltbild d. Huai-nan-tze' (tr. of ch. 4). *OAZ*, 1918, **5**, 27.
ERKES, E. (22). *Die Entwicklung der chinesischen Gesellschaft von der Urzeit bis zur Gegenwart.* Berlin, 1953.
ERNLE, LORD (K. E. PROTHERO) (1). *English Farming Past & Present.* 1st ed. London 1917; reprinted Benjamin Blom, Inc., New York, 1972.
ERNLE, LORD (K. E. PROTHERO) (2). *The Pioneers and Progress of English Farming.* Longmans, Green & Co, London, 1888.
ESHERICK, JOSEPH W. (1). 'Number games: a note on land distribution in prerevolutionary China.' *MODC* (1981), 7, **4**, 387–412.
ESTIENNE, CHARLES (1). *Praedium rusticum.* Paris, 1554.
ESTIENNE, CHARLES & LIEBAULT, JEAN (1). *Maison Rustique, or the Countrie Farme*, tr. Richard Surfleet, London, 1600; ed. by Gervase Markham, London, 1616 (1st French ed. Paris, 1567).
EVANS, ALICE M. (1). 'Beans.' In Simmonds (1), pp. 168–72.

EVANS, E. ESTYN (1). 'Introduction.' In Gailey & Fenton (1), p. 1.

EVANS, GEORGE EWART (1). *Ask The Fellows who Cut the Hay*. Faber & Faber, London, 1956.

EVANS, L. T. and W. J. PEACOCK eds. (1). *Wheat Science—Today and Tomorrow*. Cambridge University Press, Cambridge, 1981.

FEI, HSIAO-THUNG (2). *Peasant Life in China: a Field Study of Country Life in the Yangtze Valley*. George Routledge and Sons, London, 1939.

FEI, HSIAO-THUNG and CHANG, CHIH-I (1). *Earthbound China: a Study of Rural Economy in Yünnan*. Routledge and Kegan Paul, London, 1948; rep. Chicago University Press, Chicago, 1975.

FELDMAN, MOSHE (1). 'Wheats.' In N. W. Simmonds (1), pp. 120–8.

FELDMAN, MOSHE & E. R. SEARS (1). 'The wild gene resources of wheat.' *SAM*, 1981, **1**, 98–109

FENTON, A. (1). 'Early and Traditional Cultivating Implements in Scotland.' *PSAS*, 1962–3, **96**, 312.

FENTON, A. (2). 'The Cas-Chrom, a Review of the Scottish Evidence.' *TT* (1974), 2, **3**, 131.

FENTON, A. (3). 'The Plough-Song: a Scottish Source for Medieval Plough History.' *TT* (1970), 1, **3**, 175.

FENTON, A. (4). Review of *Stone Shares of Ploughing Implements from the Bronze Age of Syria* by Axel Steensberg (4). *TCULT* (1978), 19, **3**, 514.

FINSTERBUSCH, K. (1). *Verzeichnis und Motivindex der Han-Darstellung*. 2 vols, Otto Harrassowitz, Wiesbaden, 1971.

FIRTH, RAYMOND (1). 'Faith & Scepticism in Kelantan Village Magic.' In W. R. Roff (1), pp. 192 ff.

FITZHERBERT, JOHN (1). *The Boke of Husbandrye Verye Profytable and Necessarye for Al Maner of Persons Newlye Corrected and Amended by the Auctor Fitzherbard*. Richard Kele, Lumbard St., London, 1523.

FLANNERY, KENT V. (1). 'Origins and Ecological Effects of Early Domestication in Iran and the Near East.' In Ucko & Dimbleby (1), pp. 73–100.

FLANNERY, KENT V. (2). 'Archaeological systems theory and early meso-America.' In Betty J. Meggers, ed., *Anthropological Archaeology in the Americas*, Anthrop. Soc. of Washington, Washington D.C., 1968; pp. 67–87.

FLANNERY, KENT V. (3). 'The cultural evolution of civilisation.' *ARES* (1972), **3**, 399–426.

FOGG, WAYNE H. (1). 'The domestication of *Setaria italica* (L.) Beauv.; a study of the process and origin of cereal agriculture in China.' In Keightley (4), pp. 95–115.

FOOD & FERTILISER TECHNOLOGY CENTRE FOR THE ASIAN & PACIFIC REGION (1). *Multiple Cropping Systems in Taiwan*. Taipei, 1974.

FORKE, A. (4) (tr.). *'Lun-Hêng', Philosophical Essays of Wang Chhung*. Vol. 1, 1907. Kelly & Walsh, Shanghai; Luzac, London: Harrassowitz, Leipzig. Vol. 2, 1911 (with the addition of Reimer, Berlin). Photolitho re-issue, Paragon, New York, 1962. (*MSOS*, Beibände, **10** and **14**.) Crit. P. Pelliot, *JA*, 1912 (10ᵉ sér.), **20**, 156.

FORTUNE, ROBERT (4). *A Residence among the Chinese: Inland, on the Coast, and at Sea, Being a Narrative of Scenes & Adventures During a Third Visit to China, from 1853 to 1856*. John Murray, London, 1857.

FRANKE, O. (11). *Kêng Tschi T'u: Ackerbau und Seidengewinnung in China*. L. Friederichsen & Co, Hamburg, 1913.

FRANKEL, FRANCINE (1). *India's Green Revolution: Economic Gains and Political Costs*. Princeton University Press, Princeton, 1971.

FRANKEL, H. H. (1). *Catalogue of Translations from the Chinese Dynastic Histories for the Period +220 to +960*. Univ. Calif. Press, Berkeley and Los Angeles, 1957. (Inst. Internat. Studies, Univ. of California, East Asia Studies, Chinese Dynastic Histories Translations, Suppl. no. 1.)

FRAZER, J. G. (1). *The Golden Bough: A Study in Magic and Religion*. Abridged edition, Macmillan & Co. Ltd, London, 1923.

FREAM, W. (1). *Elements of Agriculture*. 15th ed., edited by D. H. Robinson, revised & metricated by Neil F. McCann; John Murray, London, 1977 (1st ed. 1892).

FREEDMAN, M. (3). *Chinese Lineage and Society: Fukien and Kwangtung*. University of London, Athlone Press, London, 1966.

FREEDMAN, M. (4). *Lineage Organisation in Southeastern China*. University of London, Athlone Press, London; 1st ed. 1958, repr. 1970.

FREEDMAN, M., ed. (5). *Family and Kinship in Chinese Society*. Stanford University Press, Stanford, 1970.

FREEMAN, DEREK (1). *Report on the Iban*. 2nd ed., University of London, Athlone Press, London, 1970.

FREEMAN, MICHAEL (2). 'Sung.' In K. C. Chang (3), pp. 141–92.

FUCHS, LEONHARD (1). *De Historia Stirpium* ... Isingrin, Basel, 1542. Repr. 1545. German ed. *New Kreüterbůch*, Isingrin, Basel, 1543.

FUKUSHIMA, Y. (1). Review of E. Pauer (1). *TCULT* (1975), 16, **4**, 628–30.

FUSSELL, G. E. (1). *Farming Technique from Prehistoric to Modern Times*. The Commonwealth and International Library of Science, Technology, Engineering and Liberal Studies, Pergamon Press, Oxford, 1965.

FUSSELL, G. E. (2). *The Farmer's Tools: 1500–1900*. Andrew Melrose, London, 1952.

FUSSELL, G. E. (3). *Crop Nutrition: Science and Practice Before Liebig*. Coronado Press, Kansas, 1971.

FUSSELL, G. E. (4). 'The Agricultural Revolution, 1600–1850.' From *Technology in Western Civilisation*, vol. 1, ed. Melvin Kranzberg & Carroll W. Pursell Jr., University of Wisconsin Press, Madison, 1967.

FUSSELL, G. E. (5). *The Classical Tradition in West European Farming*. David & Charles, Newton Abbot, 1972.

FÜZES, E. (1). 'Die Getreidespeicher im Südlichen Teil des Karpatenbeckens.' *Akadémiai Kiadó*, Budapest (1972) pp. 583–619.

FÜZES, E. (2). 'Die traditionelle Getreideaufbewahrung im Karpatenbecken.' In Gast & Sigaut (1), vol. 2, 66–83.

GABEL, C. (1). *Analysis of Prehistoric Economic Patterns*. Holt, Rinehart and Winston, New York & London, 1967.

GAILEY, A. (1). 'Spade Tillage in South-West Ulster and North Connacht.' *TT* (1971), 1, **4**, 225.

GAILEY, A. (2). 'The Typology of the Irish Spade.' In Gailey & Fenton (1), p. 45.

GAILEY, A. & A. FENTON (1) (eds.). *The Spade in Northern and Atlantic Europe*. Ulster Folk Museum Institute of Irish Studies, Queens University, Belfast, 1970.

GAIR, R., J. E. E. JENKINS & E. LESTER (1). *Cereal Pests and Diseases*. Farming Press, Ipswich, 1976.

GALE, E. M. (1) tr. *Discourses on Salt and Iron*. Brill, Leyden, 1931.

GALLO, AGOSTINO (1). *La dieci giornata dell vera agricoltura*. Italy, 1556.

GAMBLE, SIDNEY T. (1). *Ting Hsien: A North China Rural Community*. Stanford University Press, Stanford; 1st ed. 1954, reissued 1968.

GARINE, I. DE (1). 'Greniers à mil dans l'arrondissement de Thienaba, région de Thiès (Sénégal).' In Gast & Sigaut (1), vol. 2, 85–97.

GAST, M. & F. SIGAUT (1), eds. *Les techniques de conservation des grains à long terme*. CNRS, Paris; vol. 1, 1979; vol. 2, 1981.

GEDDES, W. R. (1). *Migrants of the Mountains: the Cultural Ecology of the Blue Miao (Hmong Njua) of Thailand*. Clarendon Press, Oxford, 1976.

GEERTZ, C. (1). *Agricultural Involution: The Processes of Ecological Change in Indonesia*. University of California Press, Berkeley and Los Angeles, 1963.

GERARD, JOHN (1). *The Herbal or General History of Plants*. Complete 1633 ed. as revised and enlarged by Thomas Johnson, Dover Publications, New York, 1975.

GIBSON, MCGUIRE (1). 'Population shift and the rise of Mesopotamian civilisation.' In Colin Renfew (2) pp. 447–63.

GILES, L. (2). *An Alphabetical Index to the Chinese Encyclopaedia (Chin Ting Ku Chin Thu Shu Chi Chheng)*. British Museum, London, 1911.

GILLE, B. (15). 'Recherches sur les Instruments du Labour au Moyen Âge.' *BEC*, 1962, **120**, 5.

GILLE, B., ed. (16). *Histoire des techniques: technique et civilisations, technique et sciences*. Encyclopédie de la Pléiade, N. R. F., Paris, 1978.

GILLE, B. (17). 'Les systèmes bloqués.' In B. Gille (16), 441–507.

GLOVER, I. C. (1). 'The Hoabinhian: hunter-gatherers or early agriculturalists in Southeast Asia?' In Megaw (1) pp. 145–66.

GOLAS, PETER J. (1). 'Rural China in the Song.' *JAS*, 1980, 39, **2**, 291–325.

GOLSON, JACK (1). 'No room at the top: agricultural intensification in the New Guinea Highlands.' In Allen, Golson & Jones (1), pp. 601–38.

GOMEZ-TABANERA, J. M. (1). 'El hórreo hispanico y las técnicas de conservación de grano en el N. W. de la peninsula iberica.' In Gast & Sigaut (1), vol. 2, 97–117.

GONZALES DE MENDOZA, PADRE JUAN (1). *The History of the Great and Mighty Kingdom of China and the Situation Thereof*. Tr. R. Parke, ed. Sir George Staunton, Bart., Hakluyt Society, London, 1853 (1st published in Rome in 1585).

GOODY, J. R. & S. J. TAMBIAH (1). *Bridewealth and Dowry*. Cambridge University Press, Cambridge, 1973.

GOODY, J. R., J. THIRSK & E. P. THOMPSON, eds. (1). *Family & Inheritance: Rural Society in Western Europe 1200–1800*. Cambridge University Press, 1976.

GORMAN, CHESTER (1). '*A priori* models and Thai prehistory: beginnings of agriculture.' In C. A. Reed (2), pp. 321–55.

GORMAN, CHESTER (2). 'The Hoabinhian and after: subsistence patterns in Southeast Asia during the Late Pleistocene and early Recent periods.' *WARC* (1971), 2, **3**, 300–20.

GRAHAM, D. C. (1). *Statistical Report on the Principality of Kolhapoor*. Selections from the Records of the Bombay Government no. VIII, new series, Bombay, 1854.

GREGORY, W. C. & M. P. GREGORY (1). 'Groundnut.' In Simmonds (1), pp. 151–4.

GRIEVE, M. (1). *A Modern Herbal*. Edited & introduced by C. F. Leyel, 2nd ed., Jonathan Cape, 1974 (1st ed. 1931).

GRIGG, D. B. (1). *The Agricultural Systems of the World; an Evolutionary Approach*. Cambridge University Press, 1974.

GRIST, D. H. (1). *Rice*. 5th ed., Longman, London, 1975.

GROSSER, MARTIN (1). *Kurze und gar einfeltige Anleitung zu der Landwirtschaft*. 1st ed. 1590; new ed. Fischer Verlag, Stuttgart, 1965.

GROVE, LINDA & ESHERICK, JOSEPH W. (1). 'From feudalism to capitalism: Japanese scholarship on the transformation of Chinese rural society.' *MODC* (1980), 6, **4**, 397–438.

GRYNPAS, B. (1). *Les Écrits de Tai l'Ancien et le Petit Calendrier des Hia: Textes Confucéens Taoisants*. Librairie d'Amérique et d'Orient, Adrien Maisonneuve, Paris, 1972.

GUTHRIE, CHESTER L. (1). 'A Seventeenth Century 'Ever-Normal Granary', the Alhóndiga of Colonial Mexico City.' *AGHST* (1941), 15, 37–43.

HAGERTY, M. J. (17). 'Comments on writings concerning Chinese sorghums.' *HJAS*, Jan. 1941, 5, **3–4**, pp. 234–60.

HAHN, E. (1). *Von der Hacke zum Pflug*. Quelle & Mayer, Leipzig, 1914.

HALCOTT, THOMAS (1). 'On the Drill Husbandry of Southern India.' 1797; reproduced in Dharampal (1), 210–14.

DU HALDE, J. B. ed. (2). *Lettres édifiantes et curieuses écrites des missions étrangères par quelques missionnaires de la Compagnie de Jésus*. 18 vols, Paris, 1711–43.

HALOUN, G. (2). Translations of *Kuan Tzu* and other ancient texts made with Joseph Needham, unpub. MSS.

HALOUN, G. (5). 'Legalist Fragments, I; *Kuan Tzu* ch. 55, and related texts.' *AM*, 1951 (n.s.), **2**, 85.

HAMMOND, J. L. & B. HAMMOND (1). *The Village Labourer*. 1st ed. 1911; new ed. introduced & edited by G. E. Mingay, Longman, London, 1978.

HANSEN, H.-O. (1). 'Experimental Ploughing with a Døstrup Ard Replica.' *TT* (1969), 1, **2**, 67.

HARLAN, JACK R. (1). 'The origins of cereal agriculture in the Old World.' In Charles A. Reed (2), 357–83.

HARLAN, JACK R. (2). 'Barley.' In N. W. Simmonds (1), 93–8.

HARLAN, JACK R. (3). 'The early history of wheat: earliest traces to the sack of Rome.' In Evans & Peacock (1), pp. 1–19.

HARLAN, JACK R. (4). 'Plant breeding and genetics.' In L. A. Orleans (1), pp. 295–312.

HARLAN, JACK R. (5). *Crops and Man*. American Society of Agronomy, Crop Science Society of America, Madison, Wisconsin, 1975.

HARLAN, JACK R. (6). 'Agricultural origins: centres and noncentres.' *SC*, 1971, **174**, pp. 468–74.

HARLAN, JACK R. & ANN STEMLER (1). 'The races of Sorghum in Africa.' In Harlan, de Wet & Stemler (1); pp. 465–78.

HARLAN, JACK R., J. M. J. DE WET & ANN B. L. STEMLER, eds. (1). *Origins of African Plant Domestication*. Mouton Publishers, The Hague & Paris, 1976.

HARLAN, J. R. & ZOHARY, D. (1). 'Distribution of wild wheats and barley.' *SC* (1966), **153**, 1074–80.

HÅRLEMAN, C. (1). *Dagbok öfver en Resa 1749*. Stockholm, 1751.

DE HARLEZ, C. (1). *Le Yih-King, Texte primitif rétabli, traduit et commenté*. Hayez, Bruxelles, 1889.

HARRIS, D. R. (1). 'Swidden systems and settlement.' In Ucko, Tringham & Dimbleby (1), p. 245.

HARRIS, D. R. (2). 'The prehistory of tropical agriculture: an ethnoecological model.' In Colin Renfrew (2), pp. 391–417.

HARRIS, D. R. (3). 'Alternative pathways to agriculture.' In Charles A. Reed (2), pp. 179–243.

HART, KEITH (1). 'The development of commercial agriculture in West Africa.' Discussion paper prepared for the United States Agency for International Development, 1979.

HARTLIB, S. (1). *His Legacie or an Enlargement of the Discourse of Husbandry used in Brabant and Flanders*. London, 1657.

HARTMANN, F. (1). *L'Agriculture dans l'Ancienne Égypte*. Librairies-Imprimeries Réunies, Paris, 1923.

HAUDRICOURT, A. (13). 'Nature et Culture dans la Civilisation de l'Igname.' *LH* (1964), 4, **1**, 93.

HAUDRICOURT, A. (14). 'Domestication des animaux, cultures et plantes, et civilisation d'autrui.' *LH* (1962), **2**, 40–50.

HAUDRICOURT, A. & M. J-B. DELAMARRE (1). *L'Homme et la Charrue à travers le Monde*. 2nd. ed., Gallimard, Paris, 1955.

HAWKES, DAVID (1) (tr.). *Chhu Tzhu; the Songs of the South—an Ancient Chinese Anthology*. Oxford, 1959 (rev. J. Needham, *NSN*, 18 Jul. 1959).

HAWKES, JACQUETTA & SIR LEONARD WOOLLEY (1). *Prehistory and the Beginnings of Civilisations. History of Mankind*, vol. I, George Allen & Unwin, London, 1963.

HELBAEK, H. (1). 'The plant husbandry of Hacilar.' In Mellaart (1), pp. 189–244.

HERESBACH, CONRAD (1). *Rei Rusticae libri quatuor*. Cologne, 1570.

HERKLOTS, G. A. C. (3). *Vegetables in South-East Asia*. George Allen & Unwin, London, 1972.

HERKLOTS, G. A. C. (4). 'Rice Cultivation in Hong Kong.' *FFL* (1948), 2, 1–20.

D'HERVEY ST DENYS, M. J. L. (1) (tr.) *Ethnographie des Peuples Étrangers à la Chine; ouvrage composé au 13e siècle de notre ère par Ma Touan-Lin . . . . avec un commentaire perpétuel*. Georg & Mueller, Geneva, 1876–1883. 4 vols. [Translation of chs. 324–48 of the *Wen Hsien Thung Khao* of Ma Tuan-Lin.] Vol. 1. Eastern Peoples; Korea, Japan, Kamchatka, Thaiwan, Pacific Islands (chs. 324–27). Vol. 2 Southern Peoples; Hainan, Tongking, Siam, Cambodia, Burma, Sumatra, Borneo, Philippines, Moluccas, New Guinea (chs. 328–32). Vol. 3. Western Peoples (chs. 333–9) Vol. 4. Northern Peoples (chs. 340–8).

HERZER, C. (1). 'Das *Ssu-Min Yüeh-Ling* des Tshui Shi: ein Bauern-Kalendar aus der Späteren Han-Zeit.' Ph.D. thesis, Hamburg, 1963.

HERZER, C. (2). 'Chia Ssu-Hsieh, der Verfasser des *Chhi Min Yao Shu*.' *OE* (1972), 19, **1–2**, 27–30.

HEYERDAHL, T. (7). 'Plant evidence for contacts with America before Columbus.' *AQ*, 1964, **38**, pp. 120–33.

HICKEY, G. C. (1). *Village in Vietnam*. Yale University Press, New Haven & London, 1964.

HIGGS, E. S. (1). ed. *Papers in Economic Prehistory*. Cambridge University Press, Cambridge, 1972.
HIGGS, E. S. & M. R. JARMAN (1). 'The origins of animal & plant husbandry.' In Higgs (1), pp. 3–13.
HIGHAM, C. F. W. (1). 'Initial model formation *in terra incognita*.' In D. L. Clarke (ed.), *Models in Archaeology*, Methuen, London, 1972; pp. 453–76.
HIGHAM, C. F. W. (2). 'Economic change in prehistoric Thailand.' In Charles A. Reed (2), pp. 385–412.
HILL, A. H. (1). 'Kelantan Padi Planting.' *JRAS/MB* (1957), 24, **1**, pp. 56–76.
HILL, R. D. (1). 'On the origins of domesticated rice.' *JOSHK*, 1976, 14, **1**, 35–44.
HILL, R. D. (2). *Rice in Malaya: a Study in Historical Geography*. Oxford University Press, Kuala Lumpur & Oxford, 1977.
HILL, R. D. (3). 'Peasant rice cultivation systems with some Malaysian examples.' *GPO* (1970), 19, 91–8.
HILL, R. D. (4). 'Note sur la culture du riz sec dans les états malaisiens de Kelantan et de Trengganu.' *AGT* (1964), 19, 499–504.
HINTON, W. (1). *Fanshen*. Penguin, Harmondsworth, 1972.
HIRSCHBERG, W. & A. JANATA (1). *Technologie und Ergologie in der Völkerkunde*. Bibliographisches Institut, Mannheim, 1966.
HIRTH, F. (2) (tr.). 'The Story of Chang Chhien, China's Pioneer in West Asia.' *JAOS*, 1917, **37**, 89. (Translation of ch. 123 of the *Shih Chi*, containing Chang Chhien's Report; from §18–52 inclusive and 101 to 103. §98 runs on to §104, 99 and 100 being a separate interpolation. Also tr. of ch. ? containing the biogr. of Chang Chhien.)
HO, PING-TI (1). 'The introduction of American food plants into China.' *AAN*, 1955, 57, **2**, pp. 191–201.
HO, PING-TI (2). *The Ladder of Success in Imperial China: Aspects of Social Mobility, 1368–1911*. Science Editions, John Wiley & Sons, New York, 1964.
HO, PING-TI (4). *Studies on the Population of China, 1368–1953*. Harvard University Press, Cambridge Mass., 1959.
HO, PING-TI (5). *The Cradle of the East; an Inquiry into the Indigenous Origins of Techniques and Ideas of Neolithic and Early Historic China, 5000–1000 BC*. Chinese University Publications Office, Hong Kong, 1975.
HO, PING-TI (6). 'The Loess and the Origin of Chinese Agriculture.' *AHR*, 1969, **75**, 1.
HO, PING-TI (7). 'Early ripening rice in Chinese history.' *EHR*, 2nd series, 1956, 9, **2**, 200–18.
HO, PING-YÜ (1). 'The Astronomical Chapters of the *Chin Shu*, with Amendments, Full Translation and Annotations.' Inaug. Diss., Singapore, 1957. Univ. Malaya Press, Kuala Lumpur, 1966; Mouton, Paris and the Hague, 1966. (Ecole Pratique des Hautes Etudes, VI^e Section, Sciences Economiques et Sociales, 'Le Monde d'Outre-Mer Passé et Présent', 2^e série, Documents, no. 9).
HO, PING-YÜ & JOSEPH NEEDHAM (1). 'Ancient Chinese Observations of Solar Haloes and Parhelia.' *W*, 1959, **14**, 124.
HOLE, F., K. FLANNERY & J. A. NEELY (1). *Prehistory and Human Ecology of the Deh Luran Plain*. Mem. Mus. Anth., University of Michigan Press, Ann Arbor, 1969.
HOLZMAN, D. (1). 'Shen Kua and his *Meng Chhi Pi Than*.' *TP*, 1958, **46**, 260.
HOMANS, G. C. (1). *English Villagers of the Thirteenth Century*. New York, 1941.
HOMMEL, R. P. (1). *China at Work: an Illustrated Record of the Primitive Industries of China's Masses, whose Life is Toil, and thus an Account of Chinese Civilization*. Bucks Country Historical Society, Doylestown Pa./John Day, New York, 1937. Repr. MIT Press, Cambridge Mass., 1969.
HOPFEN, H. J. (1). *Farm Implements for Arid and Tropical Regions*. F. A. O., Rome, 1963.
HOPFEN, H. J. (1a). *L'outillage agricole pour les régions arides et tropicales*. Revised ed., F. A. O., Rome, 1970.
HORIO, HISASHI (1). 'Farm tools in the "Nōgu-Benri-Ron"; intensive hoe-farming during the Edo Period in Japan.' *TT* (1974), 2, **3**, 167–85.
HOSHI AYAO (1) (trsl. Mark Elvin). *The Ming Tribute Grain System*. Michigan Abstracts of Chinese & Japanese Works on Chinese History, no. 1, Ann Arbor, 1969.
HOWELL, CICELY (1). 'Peasant inheritance customs in the Midlands, 1280–1700.' In Goody, Thirsk & Thompson (1), 112–56.
HSIAO, K. C. (6). *Rural China: Imperial Control in the Nineteenth Century*. University of Washington Press, Seattle, 1967.
HSÜ, CHO-YÜN (1), ed. JACK L. DULL. *Han Agriculture: the Formation of Early Chinese Agrarian Economy (206 B.C.–A.D. 220)*. University of Washington Press, Seattle & London, 1980.
HSÜ, CHO-YÜN (2). 'Agricultural intensification and marketing agrarianism in the Han dynasty.' In Roy & Tsien, ed. (1).
HUANG, PHILIP (1). 'Managerial farming and leasing landlordism in North China, 1890s to 1940s.' Paper presented at the Annual Meeting of the Association for Asian Studies, Los Angeles, 1979.
HUANG, PHILIP (2). 'Analysing the Twentieth-century Chinese countryside: revolutionaries versus Western scholarship.' *MODC* (1975), 1, **2**, 132–60.
HUANG, RAY (3). *Taxation and Government Finance in Sixteenth Century Ming China*. Cambridge University Press, 1974.
HUARD P. & M. DURAND (1). *Connaissance du Viêt-nam*. Ecole Française d'Extrême-Orient, Imprimerie Nationale, Hanoi & Paris, 1954.
HUARD, P. & M. WONG (5). 'Les enquêtes françaises sur la science et la technologie chinoises au 18^e siècle.' *BEFEO*, 1966, 13, **1**, pp. 137–226.

HUBER, LOUISA G. FITZGERALD (1). 'The Relationship of the Painted Pottery and Lung-Shan Cultures.' In Keightley (4), pp. 177–216.

HUGHES, E. R. (1) (tr.). *Chinese Philosophy in Classical Times*. Dent, London, 1942 (Everyman Library, no. 973.).

HUMMEL, A. W. (13). 'The printed herbal of +1249.' *ISIS*, 1941, **33**, 439; *ARLC/DO*, 1940, 155.

HUSĀM QAWĀM EL-SĀMARRĀIE (1). *Agriculture in Iraq during the 3rd Century A. H.* Librairie du Liban, Beirut, 1972.

HUTCHINSON, J. (1). 'India: local and introduced crops.' In Hutchinson, Clarke, Jope & Riley (1), 129–42.

HUTCHINSON, J. (ed.) (2). *Evolutionary Studies on World Crops: Diversity and Change in the Indian Subcontinent*. Cambridge University Press, 1974.

HUTCHINSON, J., GRAHAME CLARKE, E. M. JOPE & R. RILEY (eds.) (1). *The Early History of Agriculture, a Joint Symposium of the Royal Society & The British Academy*. Oxford University Press, Oxford, 1977.

HUTCHINSON, J., GRAHAME CLARKE, E. M. JOPE & R. RILEY (eds.) (1a). 'The Early History of Agriculture.' *PTRSB*, 1976, **275**, 1.

HYMOWITZ, T. (1). 'On the domestication of the soybean.' *ECB*, 1970, **24**, pp. 408–21.

HYMOWITZ, T. (2). 'Soybeans'. In Simmonds (1) pp. 159–162.

IINUMA, JIRŌ (1). 'The Nenohi-karasuki of Shōsōin.' *TT* (1969), 1, **2**, 105.

IMAMUDDIN, S. M. (1). 'Al-Filāḥah in Muslim Spain.' *IS* (1962), 1, **4**, 57–89.

INTERNATIONAL RICE RESEARCH INSTITUTE (1). 'Methods of Weed Control.' *IRRIR* (1966), 2, **2**, pp. 1–4.

JACK, H. W. (1). *Rice in Malaya*. Dep. Agriculture, Malaya Bulletin no. 35, Kuala Lumpur, 1923.

JACOBS, P. & B. J. STERN (1). *Outline of Anthropology*. Barnes & Noble, New York, 1947.

JÄGER, F. (4). 'Der angebliche Steindruck des Kêng-Tschi-T'u [*Keng Chih Thu*] von Jahre +1210.' *OAZ*, 1933, **9** (*19*), 1.

JEFFREYS, M. D. W. (1). 'Pre-Columbian maize in the Old World.' In M. L. Arnott (1), pp. 23–66.

JEKYLL, GERTRUDE (1). *Old West Surrey*. London, 1904.

JONES, E. L. (ed.) (1). *Agriculture and Economic Growth in England 1650–1815*. Methuen, London, 1967.

JONES, L. J. (1). 'The Early History of Mechanical Harvesting.' *HT* (1979), 4, 101–48.

JOYCE, C. R. B & S. H. CURRY, (ed.) (1). *The Botany and Chemistry of Cannabis*. London, 1970.

JULIEN, STANISLAS (2). *Mélanges de Geographie Asiatique et de Philologie Sinico-Indienne*. Paris, 1864.

KAHN, JOEL S. (1). *Minangkabau Social Formations: Indonesian Peasants in the World Economy*. Cambridge University Press, 1980.

KANO TADAO (1). 'Cereals cultivated in Indonesia' (in Japanese). *Ethnological & Prehistoric Studies of Southeast Asia*, 1946, **1**, pp. 278–95.

KARLGREN, B. (1). *Grammata Serica Recensa*. Museum of Far Eastern Antiquities, Bull. no. 29, Stockholm, 1957.

KARLGREN, B. (14). *The Book of Odes*. Museum of Far Eastern Antiquities, Stockholm, 1950.

KARROW, OTTO (2). *Die Illustrationen des Arzneibuches der Periode Shao-hsing (Shao-hsing pen-tshao hua-thu) vom Jahre 1159*. Leverkusen, 1956.

KEESING, FELIX M. (1). *The Ethnohistory of Northern Luzon*. Stanford University Press, Stanford, 1962.

KEIGHTLEY, D. (1). 'The Late Shang State: When, Where and What.' Paper presented at the Conference on the Origins of Chinese Civilisation, University of California, Berkeley, 26–30 June 1978.

KEIGHTLEY, D. (2). *Sources of Shang History: the Oracle-Bone Inscriptions of Bronze Age China*. University of California Press, Berkeley, Los Angeles and London, 1978.

KEIGHTLEY, D. (3). 'Public work in ancient China: a study of forced labour in the Shang and Northern Chou.' PhD thesis, Columbia, 1969.

KEIGHTLEY, D. (ed.) (4). *The Origins of Chinese Civilisation*. University of California Press, Berkeley, 1983.

KENYON, K. M. (1). *Digging up Jericho*. Ernest Benn, London, 1957.

KING, F. H. (1). *Farmers of Forty Centuries, or Permanent Agriculture in China, Korea and Japan*. Jonathan Cape, London, 1926, 1st ed. 1911; repr. Rodale Press, Pennsylvania, 1972.

KING, L. J. (1). *Weeds of the World: Biology and Control*. Plant Science Monographs (ed. N. Polunin), Leonard Hill, London, and Interscience, New York, 1966.

KIRBY, R. H. (1). *Vegetable Fibres*. London, 1963.

KOLENDO, J. (1). 'Pourquoi la moissonneuse antique était-elle utilisée seulement en Gaule?' In H. J. Diener, R. Günther & G. Schrot (eds.), Deutsche Historiker-Gesellschaft, *Sozialökonomische Verhältnisse im Alten Orient und im Klassischen Altertum*, Akademie-Verlag, Berlin, 1961.

KOUL, A. K. (1). 'Job's tears.' In Hutchinson (2), pp. 64 ff.

KRAYBILL, NANCY (1). 'Pre-agricultural Tools for the Preparation of Foods in the Old World.' In Charles A. Reed (2), pp. 485–521.

KU HUNG-MING (1) (tr.). *The Discourses and Sayings of Confucius*. Kelly and Walsh, Shanghai, 1898.

KUHN, DIETER (1). *Chinese Baskets and Mats*. Publikationen der Abteilung Asien Kunsthistorisches Institut der Universität Köln, no. 4, Franz Steiner Verlag, Wiesbaden, 1980.

KUHN, DIETER (2). 'Die Darstellung des *Keng Chih Thu* und ihre Wiedergabe in populär-enzyklopädischen Werken der Ming-Zeit.' *ZDMG*, 1976, 126, **2**, 336–67.

KUHN, DIETER (3). 'Silk technology in the Sung period (960–1278 AD).' *TP*, 1981, 67, **1–2**, pp. 48–90.

KUHN, DIETER (4). 'Harvesting ramie three times a year: an approach towards an understanding of the water-powered multiple spinning-frame.' Paper presented at the First International Colloquium on the History of Chinese Science, Katholieke Universiteit, Louvain, Belgium, August 1982.

KULA, WITOLD (1). *Théorie économique du système féodal: pour un modèle de l'économie polonaise 16ᵉ–18ᵉ siècles.* Mouton, Paris & The Hague, 1970 (1st Polish ed. 1962).

KUMASHIRO, YUKIO (1). 'Recent Developments in Scholarship on the *Chhi Min Yao Shu* in Japan and China.' *DEC* (1971), 9, **4**, 422–48.

KUMASHIRO, YUKIO (2). 'Empirical Principles of Dry-Land Farming in the *Chhi Min Yao Shu*, Compared with the Modern Experimental Principles.' In Nishiyama & Kumashiro (*1*), pp. ii–xvi.

KUNZ, L. (1). 'Subterranean grain stores in pre-war Czechoslovakia.' Paper presented (with film) at the 2nd CNRS Seminar on *Les Techniques de Conservation des Grains à Long Terme*, Arudy, France, June 1978.

LACH, DONALD F. (1). *The Preface to Leibniz' Novissima Sinica.* University of Hawaii Press, Honolulu, 1957.

LACH, DONALD F. (5). *Asia in the Making of Europe.* 3 vols, University of Chicago Press, Chicago, 1965.

LAMB, CHARLES (1). *Essays of Elia.* 1st ed. London, 1823.

LAMB, H. H. (1). *Climate: Past, Present and Future.* 2 vols, Methuen, London & Barnes & Noble, New York, 1977.

LAMBTON, A. K. S. (1). *Landlord and Peasant in Persia.* Oxford University Press, Oxford, 1953.

LANDÍVAR, RAFAEL (1). *Rusticatio Mexicana* [1781]. English prose translation by Graydon W. Regents, Middle American Research Institute, Tulane University, New Orleans, publicn. no. 11, 1948.

LASTEYRIE, GRAF (1). *Sammlung von Maschinen, Gerätschaften, Gebäuden, Apparaten usw. für landwirtschaftliche, häusliche und industrielle Ökonomie.* Stuttgart and Tübingen, 1821–3.

LATHRAP, D. W. (1). *The Upper Amazon.* Thames & Hudson, London, 1970.

LATHRAP, D. W. (2). 'Our father the cayman, our mother the gourd.' In Charles A. Reed (2), pp. 713–51.

LATTIMORE, OWEN (9). *Studies in Frontier History. Collected Papers, 1928–1958.* Oxford University Press, London, 1962.

LATTIMORE, OWEN (10). 'The Chinese as a dominant race.' *JRCAS*, 1928, 15, **3**; repr. Lattimore (9), pp. 200–20.

LATTIMORE, OWEN (11). 'On the wickedness of being nomads.' *TH* 1935, 1, **2**; repr. Lattimore (9), pp. 415–26.

LATTIMORE, OWEN (12). 'Herdsmen, farmers, urban culture.' In Equipe écologie et anthropologie des sociétés pastorales: *Pastoral Production and Society*, Maison des Sciences de l'Homme & Cambridge University Press, Paris & Cambridge, 1979; pp. 479 ff.

LAUFER, B. (1). *Sino-Iranica: Chinese Contributions to the History of Civilisation in Iran, with Special Reference to the History of Cultivated Plants and Products.* Anthropological Series vol. XV, **3**, Field Museum of Natural History, Chicago, 1919.

LAUFER, B. (3). *Chinese Pottery of the Han Dynasty.* E. J. Brill, Leiden, 1909.

LAUFER, B. (36). *The Introduction of Maize into Eastern Asia.* Proc. XVth Internat. Congr. Americanists, Quebec, 1906 (1907), vol. 2, p. 223.

LEACH, E. R. (2). *Pul Eliya: A Village in Ceylon: A Study of Land Tenure and Kinship.* Cambridge University Press, Cambridge, 1961; repr. 1968, 1971.

LEACH, E. R. (3). *Political Systems of Highland Burma: a Study of Kachin Social Structure.* London School of Economics Monographs on Social Anthropology no. 44, Athlone Press, London, 1954.

LEE, JOSEPH (1). *Εὐτάξια τοῦ ἄγρου: or a Vindication of a Regulated Enclosure.* London, 1656.

LEE, RICHARD B. (1). 'What hunters do for a living, or how to make out on scarce resources.' In Richard B. Lee & Irven DeVore, ed. (1), pp. 30–48.

LEE, RICHARD B. (2). 'Population growth and the beginnings of sedentary life among the !Kung bushmen.' In B. Spooner (1), pp. 329–42.

LEE, RICHARD B. (3). 'The intensification of social life among the !Kung bushmen.' In B. Spooner (1), pp. 343–50.

LEE, RICHARD B. & IRVEN DEVORE (eds.) (1). *Man the Hunter.* Aldine, Chicago, 1968.

LEEMING, F. (1). 'Official landscapes in traditional China.' *JESHO*, 1980, **23**, 153–204.

LEEMING, F. (2). 'New farmland terracing in contemporary China.' *CG* (1978), **10**, 29–41.

LEFEBVRE, L. (1). 'Les surprises d'Hérodote, ou: les acquisitions de l'agriculture méditerranée.' *AHES/AHS*, 1940, **2**, 29.

LEGGE, A. J. (1). 'Prehistoric Exploitation of the Gazelle in Palestine.' In E. S. Higgs (1), pp. 119–24.

LEGGE, A. J. (2). 'The origins of agriculture in the Near East.' In Megaw (1), pp. 51–67.

LEGGE, J. (2) (tr.). *The Chinese Classics, etc.*: Vol. 1. *Confucian Analects, The Great Learning, and the Doctrine of the Mean.* Legge, Hongkong, Trübner, London, 1861. Photolitho re-issue, Hongkong Univ. Press, Hongkong 1960 with supplementary volume of concordance tables, etc.

LEGGE, J. (3) (tr.). *The Chinese Classics, etc.*: Vol. 2. *The Works of Mencius.* Legge, Hong Kong, Trübner, London, 1861. Photo-litho re-issue, Hong Kong Univ. Press, Hong Kong, 1960, with suppl. vol. of concordance tables, & notes by A. Waley.

LEGGE, JAMES (7) (tr.). *The Texts of Confucianism*, Pt. III. *The Li Ki*. 2 vols. Oxford, 1585; reprint 1926. (SBE 27 and 28.)

LEGGE, JAMES (8) (tr.). *The Chinese Classics, etc.*: vol. 4, pts 1 & 2: *The She-King, with a Translation, Critical and Exegetical Notes, Prologomena, and Copious Indexes*. Lane Crawford, Hong Kong, Trübner, London, 1871. Repr., without notes, Com. Press, Shanghai, n.d.. Photolitho re-issue, Hong Kong Univ. Press, Hong Kong, 1960, with suppl. vol. of concordance tables, etc.

LEGGE, J. (9) (tr.). *The Texts of Confucianism*. Pt. II. *The 'Yi King'* [*I Ching*]. Oxford, 1882, 1899. (*SBE*, no. 16.)

LENZ, ILSE (1). 'The deformation of agricultural subsistence production during the initial phases of Japanese industrialisation (1880–1930).' Paper presented at the Seminar on Underdevelopment and Subsistence Production in Southeast Asia, Bielefeld University, W. Germany, 1978.

LÉON, PIERRE ed. (1). *Structures économiques et problèmes sociaux du monde rural dans la France du Sud-Est (fin du XVII^e^ siècle—1835)*. Bibliothèque de la Faculté des Lettres et Sciences Humaines de Lyon, Société d'Édition 'Les Belles Lettres', Paris, 1966.

LEONARD, W. H. & J. H. MARTIN (1). *Cereal Crops*. Macmillan, London, 1963.

LERCHE, G. (1). 'The Ploughs of Mediaeval Denmark.' *TT* (1970), 1, **3**, 131.

LERCHE, G. (2). 'Pebbles from Wheelploughs.' *TT* (1970), 1, **3**, 150.

LERCHE, G. (3). 'Observations on Harvesting with Sickles in Iran.' *TT* (1968), 1, **1**, 33–49.

LERCHE, G. & A. STEENSBERG (1). 'Observations on Spade Cultivation in the New Guinea Highlands.' *TT* (1973), 2, **2**, 87.

LEROI-GOURHAN, A. (1). *Evolution et Techniques; vol. I: L'Homme et la Matière; vol II: Milieu et Techniques*. Albin Michel, Paris, 1943, 1945.

LE ROY LADURIE, E. (1). *Montaillou, village occitan de 1294 à 1324*. NRF, Gallimard, Paris 1975; English eds. Scolar Press, London, 1975; Penguin, 1980.

LE ROY LADURIE, E. (2). *Histoire du climat depuis l'an mil*. Flammarion, Paris, 1967; trs. into English as *Times of Feast, Times of Famine*, Doubleday, New York, 1971.

LESER, P. (1). *Entstehung und Verbreitung des Pfluges*. Aschendorff, Münster, 1931. (Anthropos Bibliothek, no. 3.)

LESER, P. (2). 'Westöstlichen Landwirtschaft; Kulturbeziehungen zwischen Europa, dem vord. Orient n.d. Fern-Osten, aufgezeigt an Landwirtschaftlichen Geräten und Arbeitsvorgängen.' In *P. W. Schmidt Festschrift*, ed. W. Koppers, pp. 416 ff. Vienna, 1928.

LESER, P. (3). 'Vom Mittelmeer nach Südostasien.' *ERDB*, 1931, **5**, (no. 6), 1.

LESER, P. (5). 'Plough Complex, Culture Change and Cultural Stability.' Art. in *Selected Papers of the Vth International Congress of Anthropological and Ethnological Sciences, 1956*. Philadelphia, 1960, p. 292.

LESLIE, D. (3). 'Contribution to a new translation of the *Lun Heng*.' *TP* (1956), **44**, 100.

LEWIN, GÜNTER (1). *Die Erste Fünfzig Jahre der Song-Dynastie in China*. Akademie-Verlag, Berlin, 1973.

LEWIS, B., C. H. PELLAT & J. SCHACHT (eds.) (1). *The Encyclopaedia of Islam*. 2nd ed., Brill, Leiden, 1965.

LEWIS, HENRY T. (1). *Ilocano Rice Farmers: a Comparative Study of Two Philippine Barrios*. University of Hawaii Press, Honolulu, 1971.

LI, GUOHAO, ZHANG MENGWEN & CAO TIANQIN, ed. (1). *Explorations in the History of Science and Technology in China*. Special ed. of *CHWSLT* in honour of Joseph Needham's 80th birthday. Shanghai Classics Publishing House, Shanghai, 1982.

LI, HSIEN-WEN, C. H. LI & W. K. PAO (1). 'Cytological and genetical studies of the interspecific cross of the cultivated foxtail millet, *Setaria italica* (L.) Beauv., & the green foxtail, *Setaria viridis* (L.) Beauv.'. *JASAG* (1945), 37, **1**, 32–54.

LI, HSIEN-WEN, C. J. MENG & T. N. LIU (1). 'Problems in the breeding of millet (*Setaria italica* (L.) Beauv.).' *JASAG* (1935), 27, **12**, 963–70.

LI, HUI-LIN (1). 'A case for pre-Columbian transatlantic travel by Arab ships.' *HJAS*, 1961, **23**, pp. 114–26.

LI, HUI-LIN (6). 'An archaeological and historical account of Cannabis in China.' *ECB*, 1974, 28, **4**, 437–48.

LI, HUI-LIN (11). *Nan Fang Tshao Mu Chuang: a Fourth Century Flora of Southeast Asia*. The Chinese University Press, Hong Kong, 1979.

LI, HUI-LIN (14). 'The vegetables of ancient China.' *ECB*, 1969, 23, **3**, 253–60.

LI, HUI-LIN (15). 'The origin of cultivated plants in Southeast Asia.' *ECB*, 1970, 24, **1**, pp. 3–19.

LIEBIG, JUSTUS (1). *Organic Chemistry in its Application to Agriculture and Physiology*. Playfair, London, 1840 (1st ed.)

LIEFRINCK, F. A. (1). 'Rice cultivation in Northern Bali.' In J. van Baal (ed.), *Bali, Further Studies in Life, Thought and Ritual, Selected Studies on Indonesia* vol. 8, W. van Hoeve, The Hague, 1969, pp. 1–74 (1st published 1886–7).

LIU, HUI-CHEN WANG (1). *The Traditional Chinese Clan Rules*. Monographs of the Association for Asian Studies VII, J. J. Augustin Inc., New York, 1959.

LIU, JAMES T. C. (2). *Reform in Sung China: Wang An-Shih (1021–1086) and his New Policies*. Harvard University Press, Cambridge Mass., 1959.

LIU, WU-CHI & IRVING LO (ed.). (1). *Sunflower Splendor: Three Thousand Years of Chinese Poetry*. Indiana University Press, Bloomington, 1975.

LOOFS, H. H. E. (1). 'A new winnowing method from Northern West-Malaysia.' *TT* (1976), 3, **1**, 20–4.

DE LOUREIRO, JUAN (1). *Flora Cochinchinensis, sistens Plantas in Regno Cochinchina nascentes; Quibus accedunt aliae Observatae in Sinensi Imperio, Africa Orientali, Indiaeque Locis Variis; Omnes dispositae secundum Systema Sexuale Linneanum.* Acad. Sci. Lisbon, 1790. See Merrill (2).

LU GWEI-DJEN (3). Abstract of Hu Hou-Hsüan (5). *RBS*, 1957, **1**, p. 41.

LYALL, L. A. (1) (tr.). *Mencius.* Longmans Green, London, 1932.

LYALL, L. A. (2) (tr.). *The Sayings of Confucius* [*Lun Yü*]. Longmans Green, London, 1935 (this edition superseded earlier editions).

MCDERMOTT, J. P. (1). 'Land tenure and rural control in the Liangche region during the Southern Sung.' Unpublished PhD Thesis, Cambridge, 1978.

MACDONALD, J. (1). *General View of the Agriculture of the Hebrides.* Edinburgh, 1811.

MACFARLANE, ALAN (1). *The Origins of English Individualism.* Blackwell, Oxford, 1978.

MACFARLANE, A., S. HARRISON & C. JARDINE (1). *Reconstructing Historical Communities.* Cambridge University Press, 1977.

MCLENNAN, GREGOR (1). *Marxism and the Methodologies of History.* Verso Editions and NLB, London, 1981.

MACNEISH, R. S. (1). 'First Annual Report of the Ayacucho Archaeological-Botanical Project.' Robert S. Peabody Foundation for Archaeology, Andover, Mass., 1969.

MACNEISH, R. S. (2). 'Second Annual Report of the Ayacucho Archaeological-Botanical Project.' Robert S. Peabody Foundation for Archaeology, Andover, Mass., 1970.

MACNEISH, R. S. (ed.) (3). *The Prehistory of the Tehuacán Valley, vol. 5: Excavations and Reconnaissance.* University of Texas Press, Austin, 1975.

MAJOR, JOHN S. (3). 'Myth, Cosmology, and the Origins of Chinese Science.' *JCP* (1978), **5**, 1.

MAJOR, J. S. & D. C. MAJOR (1). 'Mark Elvin; *The Pattern of The Chinese Past*' [review]. *TCULT*, 1974, 15, **3**, pp. 511–74.

MALAYSIAN GOVERNMENT (1). *Third Malaysia Plan 1976–1980.* Government Press, Kuala Lumpur, 1976.

MALDEN, W. J. (1). *Tillage and Implements.* Bell's Agricultural Series, George Bell & Sons, London, 1891.

MANGELSDORF, P. C. (4). 'Genetic Potentials for Increasing Yields of Food Crops and Animals.' In *Prospects of the World Food Supply—A Symposium*, National Academy of Science, 1966, Washington D. C., pp. 66–71.

MANGELSDORF, P. C. (5). *Corn: its Origin, Evolution and Improvement.* The Belknap Press of Harvard University Press, Cambridge Mass, 1974.

MARÉCHAL, R., J.-M. MASCHERRA, & F. STAINIER (1). *Etude taxonomique d'un groupe complexe d'espèces des genres Phaseolus et Vigna (Papilionaceae) sur la base des données morphologiques et polliniques, traitées par l'analyse informatique.* Boissiera vol. 28, Mémoires des Conservatoire et Jardin Botanique de la Ville de Genève, 1978.

MARINOV, V. (1). 'On the terminology and classification of Bulgarian plough irons.' *TT* (1973), 2, **2**, 119.

MARKHAM, GERVASE (1). *Markhams Farewell to Husbandry or, the Enriching of All Sorts of Barren and Sterile Grounds in our Kingdome.* London, 1631.

MARKHAM, GERVASE (2). *The English Husbandman.* 2nd ed., London, 1635.

MARKHAM, GERVASE (3). *A Way to Get Wealth.* 5th ed., London, 1631.

MARKHAM, GERVASE (4). *The Inrichment of the Weald of Kent: or, a Direction to the Husbandman, for the true ordering, manuring, & inriching of all the Grounds within the Wealds of Kent and Sussex.* Roger Jackson, London, 1625; facsimile ed. Theatrum Orbis Terrarum, Amsterdam & Da Capo Press Inc., New York, 1973.

MARKHAM, GERVASE (5). *Markhams Maister-Peece.* London, 1610.

MARX, KARL (1). *Capital.* 1st published as *Das Kapital: Kritik der politischen Oekonomie*, Hamburg, 1867; 1st English ed., ed. Friedrich Engels, 1887; Penguin ed., Harmondsworth, 1976.

MASPÉRO, H. & E. BALASZ (1). *La Chine ancienne.* Paris, 1967.

MATSUO, TAKANE (1). 'Rice culture in China.' In *Assoc. of Japanese Ag. Sci. Socs.* (1), 157–69.

MATTHEWS, J. M. (1). *A Checklist of Hoabinhian Sites Excavated in Malaya 1860–1939.* Singapore, 1961.

MATTHEWS, J. M. (2). 'A review of the Hoabinhian in Indo-China.' *ASP* (1966), **9**, 86–95.

MAVERICK, L. A. (1). 'Hsü Kuang-Ch'i, a Chinese Authority on Agriculture.' *AGHST* (1940), **14**, 143.

MAVERICK, L. A. (2). 'Chinese influences upon the Physiocrats.' *EH* (1938), 54–67.

MAXEY, EDWARD (1). *A New Instruction of Plowing and Setting of Corne.* Felix Kyngston, London, 1601.

MAXWELL, NEVILLE (ed.) (1). *China's Road to Development.* Pergamon Press, Oxford; 1st ed. 1976; 2nd enlarged ed. 1979.

MAYER, L. T. (1). *Een Blik in het Javansche Volksleven.* E. G. Brill, Leiden, 1897.

MAYERSON, P. (1). *The Ancient Agricultural Regime of Nessana and the Central Negeb.* British School of Archaeology in Jerusalem, Jerusalem, 1960.

MEACHAM, WILLIAM (2). 'Continuity and local evolution in the Neolithic of South China: a non-nuclear approach.' *CURRA* (1977), 15, **3**, pp. 419–40.

MEDLEY, MARGARET (3). *The Chinese Potter: a Practical History of Chinese Ceramics.* Phaidon, Oxford, 1976.

MEGAW, J. V. S. (ed.) (1). *Hunters, Gatherers and First Farmers Beyond Europe: an Archaeological Survey.* Leicester University Press, Leicester, 1977.

MELLAART, J. (ed.) (1). *Excavations at Hacilar.* Edinburgh University Press, Edinburgh, 1970.

MEMON, A. A. (1). *Indigenous Agricultural Implements in Bombay State*. Government Press, Baroda, 1955.

MERLE, LOUIS (1). *La métairie et l'évolution agraire de la Gâtine poitevine de la fin du Moyen Age à la Révolution*. Série Les Hommes et la Terre, S. E. V. P. E. N., Paris, 1958.

MERRILL, E. D. (2). 'A commentary on Loureiro's *Flora Cochinchinensis*.' *TAPS* (1935), **24**, 1. Repr. in abridged form as. 'On Loureiro's *Flora Cochinchinensis*' in Merrill (5), p. 243.

MERRILL, E. D. (5). 'Merrilleana; a selection from the general writings of Elmer Drew Merrill ...' *CBOT* (1946), 10, **3–4**, 127–394.

MERTENS, J. (1). 'Eine Antike Mähmaschine.' *ZAGA* (1959), 7, 1–3.

MESKILL, J. (tr. & ed.) (1). *Ch'oe Pu's Diary: A Record of Drifting Across the Sea*. University of Arizona Press, Tucson, 1965.

MESKILL, JOHN, (ed.) (2). *Wang An-Shih, Practical Reformer?* D. C. Heath & Co. Boston, 1963.

MEUVRET, J. (1). 'Agronomie et jardinage au XVI^e^ et au XVII^e^ siècle,' *Cahiers des Annales* 32, Paris, 1971, 153–161.

MINGAY, G. E. (1). *The Agricultural Revolution: Changes in Agriculture 1650–1880*. Documents in Economic History, A. & C. Black, London, 1977.

MINGAY, G. E. (2). *Rural Life in Victorian England*. Futura Publications, London, 1979.

MOERMAN, M. (1). *Agricultural Change and Peasant Choice in a Thai Village*. University of California Press, Berkeley & Los Angeles, 1968.

MOISE, EDWIN E. (1). 'The Moral Economy dispute.' *Bull. Concerned Asian Scholars* 14, **1** (Jan.–Mar. 1982), pp. 72–7.

MONTANDON, G. (1). *Traité d'Ethnologie Culturelle*. Payot, Paris, 1934.

MOORE, BARRINGTON (1). *Social Origins of Dictatorship & Democracy*. Penguin, Harmondsworth, 1967.

MOORE, JOHN (1). *The Crying Sin of England of not Caring for the Poor, Wherein Inclosure is Arraigned, Convicted and Condemned by the Word of God*. London, 1653.

MORE, SIR THOMAS (1). *Utopia*. London, 1516.

MORGAN, E. (tr.) (1). *Tao the Great Luminant; Essays from 'Huai Nan Tzu', with introductory articles, notes and analyses*. Kelly & Walsh, Shanghai, n.d. (1933?).

MORITZ, L. A. (2). '*ΆΛΦΙΤΑ*—a note' *CQ* (1949), 43, **3–4**, pp. 113–17.

MORITZ, L. A. (3). 'Husked and "Naked" Grain.' *CQ* (1955) 5, **3–4**, 129–34.

MORITZ, L. A. (4). 'Corn.' *CQ* (1955) 5, **3–4**, 135–41.

MORTIMER, JOHN (1). *The Whole Art of Husbandry*. London, 1707.

MOTE, FREDERICK W. (4). 'Yüan and Ming.' In K. C. Chang (3), pp. 193–257.

MOULE, A. C. (5). 'The Wonder of the Capital' [the Sung books *Tu Chheng Chi Sheng* and *Meng Liang Lu* about Hangchou]. *NCR* (1921), **3**, 12, 356.

M[OULE], [A.] C. (11). 'The Fire-Proof Warehouses of Lin-An [+13th Century Hangchou].' *NCR* (1920) **2**, 207–10.

MOULE, A. C. (15). *Quinsai, with other Notes on Marco Polo*. Cambridge, 1957. An extension of a number of previous papers, notably 'Marco Polo's Description of Quinsai', *TP* (1937), **33**, 105.

MUHLY, J. D. (1). 'The origin of agriculture and technology—West or East Asia? Summary of a conference on *The Origin of Agriculture and Technology*: Aarhus, Denmark, November 21–25, 1978.' *TCULT* (1981), 22, **1**, pp. 125–45.

MULVANEY, D. J. & J. GOLSON (ed.) (1). *Aboriginal Man and Environment in Australia*. Australian National University Press, Canberra, 1971.

MYERS, RAMON H. (1). *The Chinese Peasant Economy: Agricultural Development in Hopei and Shangtung, 1890–1949*. Harvard University Press, Cambridge Mass., 1970.

MYERS, RAMON H. (2). 'Wheat in China—past, present and future.' *CQR*, (1978), **74**, 297–333.

NAKAMURA, HISASHI (1). 'Village community and paddy agriculture in South India.' *DEC* (1972), 10, **2**, 141–65.

NAKAO MANZŌ (1). 'Notes on the *Shao-Hsing Chiao-Ting Ting-Shih Cheng-Lei Pen Tshao* [The Classified and Consolidated Armamentarium; Pharmacopoeia of the Shao-Hsing Reign-Period]—the ancient Chinese Materia Medica revised in the Sung Dynasty (+1131 to +1162).' *JSSI* (1933), Sect. III, **1**, 1. (English version of the introduction to Nakao (2).)

NAKAOKA, TETSURŌ (1). 'Science and technology in the history of modern Japan—imitation or endogenous creativity?' Paper presented at the First International Seminar on Science and Technology in the Transformation of the World, United Nations University, Belgrade, October 1979.

NAYAR, N. M. (1). 'Sesame.' In Simmonds (1), 231–3.

NAYAR, N. M. & K. L. MEHRA (1). 'Sesame: its uses, botany, cytogenetics and origin.' *ECB* (1970), **24**, 20–31.

NEEDHAM, JOSEPH (32). *The Development of Iron and Steel Technology in China*. 2nd Biennial Dickinson Memorial Lecture to the Newcomen Society, 1956. Newcomen Society, London, 1958.

NIELSEN, S. (1). 'The first reaping machines in Denmark.' *TT* (1970), 1, **3**, 166–74.

NIELSEN, VIGGO (1). 'Iron Age Plough-Marks in Store Vildmose, Jutland.' *TT* (1970), 1, **8**, 151.

NOPSCA, F. (1). 'Zur Genese der primitiven Pflugtypen.' *ZFE (1919)*, **51**, 234.

NOY, T., A. J. LEGGE & E. S. HIGGS (1). 'Recent excavations at Nahal Oren, Israel.' *PPHS* (1973), **39**, 75–99.

OATES, D. & J. (1). 'Early irrigation agriculture in Mesopotamia'. In G. de G. Sieveking, I. H. Longworth & K. W. Wilson (eds.), *Problems in Social and Economic Archaeology*, Duckworth, London, 1976.

OATES, J. (1). 'Prehistoric Settlement Patterns in Mesopotamia.' In Ucko, Tringham & Dimbleby (1), p. 299.

OJEA, HERNÁNDO (1). *Libro tercero de la historia religiosa de la Provincia de México de la Orden de Santo Domingo* [c. +1610]. Mexico, 1897.

OKA, HIKO-ICHI (1). 'The origin of cultivated rice and its adaptive evolution.' In Assoc. of Japanese Ag. Sci. Socs (1), 21–34.

OLSON, LOIS (1). 'Pietro de Crescenzi: the founder of modern agronomy.' *AGHST* (1944), **18**, 35–40.

ORLEANS, LEO A. (ed.) (1). *Science in Contemporary China*. Stanford University Press, Stanford, 1980.

ORME, BRYONY (1). 'The advantages of agriculture.' In J. V. S. Megaw (1), pp. 41–9.

ORWIN, C. S. & C. S. (1). *The Open Fields*. Oxford, 1938; repr. 1967.

OSBECK, PETER (1). *A Voyage to China and the East Indies*, by Peter Osbeck, Rector of Hasloef and Woxtorp, Member of the Academy of Stockholm, and of the Society of Upsal; together with *A Voyage to Seratte*, by Olof Toreen, Chaplain of the Gothic Lion East-Indiaman; and *An Account of the Chinese Husbandry*, by Captain Charles Gustavus Eckeberg [sic]; translated from the German by John Reinhold Foster, F. A. S., to which are added, *A Faunula and Flora Sinensis*. 2 vols., London, 1771.

OSCHINSKY, D. (1). *Walter of Henley and Other Treatises on Estate Management and Accounting*. Clarendon Press, Oxford, 1971.

PALERIN, ANGEL (1). *Obras Hidráulicas Prehispánicas en el Sistema Lacustre del Valle de México*. Instituto Nacional de Antropología e Historia México, Sep Inah, Mexico City, 1973.

PALLADIUS, RUTILIUS TAURUS AEMILIANUS (tr. T. Owen) (1). *The Fourteen Books of Palladius Rutilius Taurus Aemilianus, on Agriculture* [De Re Rustica]. London, 1807.

PALMER, INGRID (1). *The Indonesian Economy Since 1965: A Case Study of Political Economy*. Frank Cass, London, 1978.

PALMER, INGRID (2). *The New Rice in Indonesia*. United Nations Research Institute for Social Development, Geneva, 1977.

PARANAVITANA, S. (5). 'Ploughing as a Ritual of Royal Consecration in Ceylon.' *R. C. Majumdar Felicitation Volume*, Mukhopadhyay, Calcutta, 1970.

PARTRIDGE, M. (1). *Farm Tools Through the Ages*. Osprey, London, 1973; repr. 1976.

PASTERNAK, B. (1). *Kinship and Community in Two Chinese Villages*. Stanford University Press, Stanford, 1972.

PASTERNAK, B. (2). 'The role of the frontier in Chinese lineage development.' *JAS* (1969), **28**, pp. 551–61.

PASTERNAK, B. (3). 'The sociology of irrigation: two Taiwanese villages.' In W. E. Willmott (ed.), *Economic Organisation in Chinese Society*, Stanford University Press, Stanford, 1972, pp. 193–213.

PAUER, E. (1). *Technik, Wirtschaft, Gesellschaft: der Einfluss wirtschaftlicher und gesellschaftlicher Veränderung auf die Entwicklung der landwirtschaftlichen Geräte in der vorindustriellen Epoche Japans ab dem 17. Jahrhundert*. Beiträge zur Japanologie vol. 10, University of Vienna Press, Vienna, 1973. Review: Fukushima (1).

PAYNE, F. G. (1). 'The British Plough: Some Stages in its Development.' *AGHR* (1957), 5, **2**, 74.

PELLIOT, P. (24). *A propos du 'Keng Tche T'ou'*. (from: *Mémoires concernant l'Asie orientale: Ind, Asie Centrale, Extrême-Orient*; publiées par l'Academie des Inscriptions et Belles-Lettres, Tome I). Leroux, Paris 1913.

PERCIVAL, JOHN (1). *The Roman Villa: an Historical Introduction*. Batsford, London, 1976.

PEREIRA, H. C. (1). *Land Use and Water Resources in Temperate and Tropical Climates*. Cambridge University Press, Cambridge, 1973.

PERKINS, DWIGHT H. (1). *Agricultural Development in China 1368–1968*. Edinburgh University Press, Edinburgh, 1969.

PERKINS, DWIGHT H. (ed.) (2) *China's Modern Economy in Historical Perspective*. Stanford University Press, Stanford, 1975.

PERNÈS, J., J. BELLIARD & G. MÉTAILIÉ (1). 'Mission agronomique en Chine "ressources génétiques".' Roneographed report, Paris, 1979.

PETERS, MATTHEW (1). *Agricultura or the Good Husbandman*. London, 1776.

PFIZMAIER, A. (13) (tr.). 'Die Geschichte des Reiches U' (Wu). *DWAW/PH*, 1857, **8**, 123. (Tr. ch. 31 *Shih Chi*; cf. Chavannes (1), vol. 4.)

PFIZMAIER, A. (14) (tr.). 'Die Geschichte des Hauses Thai Kung' (of Chhi). *SWAW/PH*, 1862, **40**, 645. (Tr. ch. 32, *Shih Chi*; cf. Chavannes (1), vol. 4.)

PFIZMAIER, A. (15) (tr.). 'Die Geschichte des Hauses Tscheu Kung' (Chou Kung). *SWAW/PH*, 1863, **41**, 90. (Tr. ch. 33, *Shih Chi*; Cf. Chavannes (1), vol. 4.)

PFIZMAIER, A. (16) (tr.). 'Die Geschichte des Hauses Schao-Kung u. Khang-Scho' (of Yen and Wei). *SWAW/PH*, 1863, **41**, 435, 454. (Tr. chs. 34, 37 *Shih Chi*; of Chavannes (1), vol. 4.)

PFIZMAIER, A. (17) (tr.). 'Die Geschichte des Fürstenlandes Tsin' (Chin). *SWAW/PH*, 1863, **43**, 74. (Tr. ch. 39 *Shih Chi*; cf. Chavannes (1), vol. 4.)

PFIZMAIER, A. (18) (tr.). 'Die Geschichte des Fürstenlandes Tsu' (Chhu). *SWAW/PH*, 1863, **44**, 68. (Tr. ch. 40, *Shih Chi*; cf. Chavannes (1), vol. 4.)

PFIZMAIER, A. (19) (tr.). 'Keu-Tsien, König von Yue, und dessen Haus (Kou Chien of Yueh and Fan Li). *SWAW/PH*, 1863, **44**, 197. (Tr. ch. 41. *Shih Chi*; cf. Chavannes (1), vol. 4.)

PFIZMAIER, A. (20) (tr.). 'Geschichte d. Hauses Tschao' (Chao). *DWAW/PH*, 1859, **9**, 45. (Tr. ch. 43, *Shih Chi*; cf. Chavannes (1), vol. 5.)

PFIZMAIER, A. (21) (tr.). 'Das Leben des Feldherrn U-Khi' (Wu Chi). *SWAW/PH*, 1859, **30**, 267. (Tr. ch. 65. *Shih Chi*; not in Chavannes (1).)

PFIZMAIER, A. (22) (tr.). 'Der Landesherr von Schang' (Shang Yang). *SWAW/PH*, 1858, **29**, 98. (Tr. ch. 68. *Shih Chi*; not in Chavannes (1). cf. Duyvendak (3).)

PFIZMAIER, A. (23) (tr.). 'Das Rednergeschlecht Su' (Su Chhin). *SWAW/PH*, 1860, **32**, 642. (Tr. ch. 69. *Shih Chi*; not in Chavannes (1).)

PFIZMAIER, A. (24) (tr.). 'Der Redner Tschang I und einige seiner Zeitgenossen' (Chang I and Chhu Li Tzu). *SWAW/PH*, 1860, **33**, 525, 566. (Tr. chs. 70, 71, *Shih Chi*; not in Chavannes (1).)

PFIZMAIER, A. (25) (tr.). 'Wei-Jen, Fürst von Jang.' *SWAW/PH*, 1859, **30**, 155. (Tr. ch. 72. *Shih Chi*; not in Chavannes (1).)

PFIZMAIER, A. (26) (tr.). 'Zur Geschichte von Entsatzes von Han Tan' *SWAW/PH*, 1859, **31**, 65, 87, 104, 120. (Tr. chs. 75, 76, 78, 83, *Shih Chi*; includes life of the Prince of Phing Yuan; not in Chavannes (1).)

PFIZMAIER, A. (27) (tr.). 'Das Leben des Prinzen Wu Ki [Wu Chi] von Wei.' *SWAW/PH*, 1858, **28**, 171. (Tr. ch. 77. *Shih Chi*; not in Chavannes (1).)

PFIZMAIER, A. (28) (tr.). 'Das Leben des Redners Fan Hoei' (Fan Hui). *SWAW/PH*, 1859, **30**, 227. (Tr. ch. 80 (in part), *Shih Chi*; not in Chavannes (1).)

PFIZMAIER, A. (29) (tr.). 'Die Feldherren des Reiches Tschao.' *SWAW/PH*, 1858, **28**, 55, 65, 69. (Tr. chs. 80 (in part), 81, 82, *Shih Chi*; not in Chavannes (1).)

PFIZMAIER, A. (30) (tr.). 'Li Sse, Der Minister des ersten Kaisers' (Li Ssu). *SWAW/PH*, 1859, **31**, 120, 311. (Tr. chs. 83, 87, *Shih Chi*; not in Chavannes (1).)

PFIZMAIER, A. (31) (tr.). 'Das Ende Mung Tien's' (Meng Thien). *SWAW/PH*, 1860, **32**, 134. (Tr. ch. 88, *Shih Chi*; not in Chavannes (1).)

PFIZMAIER, A. (32) (tr.). 'Die Genossen des Königs Tschin Sching' (Chang Erh and Chhen Yü). *SWAW/PH*, 1860, **32**, 333. (Tr. ch. 89, *Shih Chi*, ch. 32, *Chhien Han Shu*; not in Chavannes (1).)

PFIZMAIER, A. (33) (tr.). 'Die Nachkommen der Könige von Wei, Tsi [Chhi] und Han.' *SWAW/PH*; 1860, **32**, 529, 533, 542, 551, 562, 567. (Tr. chs. 90, 93, 94, 97, *Shi Chi*, ch. 33, *Chhien Han Shu*; not in Chavannes (1).)

PFIZMAIER, A. (34) (tr.). 'Die Feldherren Han Sin, Peng Yue, und King Pu' (Han Hsin, Pheng Yüeh & Ching Pu). *SWAW/PH* (1860), **34**, 371, 411, 418. (Tr. chs. 90 (in part), 91, 92, *Shih Chi*; ch. 34, *Chhien Han Shu*; not in Chavannes (1).)

PFIZMAIER, A. (35) (tr.). 'Der Abfall des Königs Pi von U' (Wu). *SWAW/PH*, 1861, **36**, 17. (Tr. ch. 106, *Shih Chi*; not in Chavannes (1).)

PFIZMAIER, A. (36) (tr.). 'Sse-ma Ki-Tschü, der Wahrsager von Tschang-ngan' (Ssuma Chi-Chu in the Chapter on diviners, *Jih Che Lieh Chuan*). *SWAW/PH* (1861), **37**, 408. (Tr. ch. 127, *Shih Chi*; not in Chavannes (1).)

PFIZMAIER, A. (37) (tr.). 'Die Gewaltherrschaft Hiang Yü's' (Hsiang Yü). *SWAW/PH* (1860), **32**, 7. (Tr. ch. 31, *Chhien Han Shu.*)

PFIZMAIER, A. (38) (tr.). 'Die Anfänge des Aufstandes gegen das Herrscherhaus Thsin' (Chhin). *SWAW/PH*, 1860, **32**, 273. (Tr. ch. 33, *Chhien Han Shu.*)

PFIZMAIER, A. (39) (tr.). 'Die Könige von Hoai Nan aus dem Hause Han.' (Huai Nan Tzu) *SWAW/PH*, 1862, **39**, 575. (Tr. ch. 44, *Chhien Han Shu.*)

PFIZMAIER, A. (40) (tr.). 'Das Erreigniss des Wurmfrasses der Beschwörer.' *SWAW/PH*, 1862, **39**, 50, 55, 58, 65, 76, 89. (Tr. chs. 45, 63, 66, 74, *Chhien Han Shu.*)

PFIZMAIER, A. (41) (tr.). 'Worte des Tadels in dem Reiche der Han.' *SWAW/PH*, 1861, **35**, 206. (Tr. ch. 51, *Chhien Han Shu.*)

PFIZMAIER, A. (42) (tr.). 'Die Heerführer Li Kuang und Li Ling.' *SWAW/PH* (1863), **44**, 511. (Tr. ch. 54, *Chhien Han Shu.*)

PFIZMAIER, A. (43) (tr.). 'Die Geschichte einer Gesandtschaft bei den Hiung-Nu's' (Su Wu). *SWAW/PH*, 1863, **44**, 581. (Tr. ch. 54 (second part), *Chhien Han Shu.*)

PFIZMAIER, A. (44) (tr.). 'Die Heerführer Wei Tsing und Ho Khiu-Ping' (Wei Chhing and Ho Chhü-Ping). *SWAW/PH* (1864), **45**, 139. (Tr. ch. 55, *Chhien Han Shu.*)

PFIZMAIER, A. (45) (tr.). 'Die Antworten Tung Tschung-Schü's [Tung Chung Shu] auf die Umfragen des Himmelssohnes.' *SWAW/PH*, 1862, **39**, 345. (Tr. ch. 56, *Chhien Han Shu.*)

PFIZMAIER, A. (46) (tr.). 'Zwei Statthalter der Landschaft Kuei Ki.' *SWAW/PH*, 1861, **37**, 304. (Tr. ch. 64 A (in part), *Chhien Han Shu.*)

PFIZMAIER, A. (47) (tr.). 'Die Bevorzugten des Allhalters Hiao Wu' [emperor Hsien Wu Ti]. *SWAW/PH*, 1861, **38**, 213, 234. (Tr. ch. 64 A (second part), 64 B, *Chhien Han Shu.*)

PFIZMAIER, A. (48) (tr.). 'Die Würdenträger Tsiuen Pu-I, Su Kuang, Yü Ting-Kue, und deren Gesinnungsgenossen' (Chien Pu-I, Su Kuang, Yü Ting-Kuo). *SWAW/PH*, 1862, **40**, 131. (Tr. ch. 71, *Chhien Han Shu.*)

PFIZMAIER, A. (49) (tr.). 'Tschin Thang, Fürst-Zertrümmerer von Hu' (Chhen Thang). *SWAW/PH*, 1862, **40**, 396. (Tr. ch. 70 (in part), *Chhien Han Shu.*)

PFIZMAIER, A. (50) (tr.) 'Die Menschenabtheilung der wandernden Schirmgewaltigen' (*yü hsia*; Knights errant; soldiers of fortune). *SWAW/PH*, 1861, **37**, 103. (Tr. ch. 92, *Chhien Han Shu.*)

PFIZMAIER, A. (51) (tr.). 'Die Eroberung der beiden Yue [Yüeh] und des Landes Tschao Sien [Chao-Hsien, Korea] durch Han.' *SWAW/PH*, 1864, **46**, 481. (Tr. ch. 95, *Chhien Han Shu.*)

PFIZMAIER, A. (52) (tr.). 'Zur Geschichte d. Zwischenreiches von Han.' *SWAW/PH*, 1869, **61**, 275, 309. (Tr. chs. 41, 42, *Hou Han Shu.*)

PFIZMAIER, A. (53) (tr.). 'Die Aufstände Wei-Ngao's und Kungsun Scho's' (Wei Ao and Kungsun Shu). *SWAW/PH*, 1869, **62**, 159. (Tr. ch. 43, *Hou Han Shu.*)

PFIZMAIER, A. (54) (tr.). 'Aus der Geschichte d. Zeitraumes Yuen-Khang von Tsin' (+292/+299). *SWAW/PH*, 1876, **82**, 179, 205, 212, 223, 230, 232. (Tr. chs. 31 (in part), 40, 58, 60, 100, *Chin Shu.*)

PFIZMAIER, A. (55) (tr.). 'Aus der Geschichte des Hofes von Tsin.' *SWAW/PH*, 1876, **81**, 545, 561, 568. (Tr. chs. 31 (in part), 53, 59, *Chin Shu.*)

PFIZMAIER, A. (56) (tr.). 'Über einige Wundermänner Chinas' (magicians and technicians such as Chhen Hsün, Tai Yang, Wang Chia, Shunyu Chih, etc.). *SWAW/PH* (1877), **85**, 37. (Tr. ch. 95, *Chin Shu.*)

PFIZMAIER, A. (57) (tr.). 'Die Machthaber Hoan Wen und Hoan Hiuen' (Huan Wen and Huan Hsüan). *SWAW/PH*, 1877, **85**, 603, 632. (Tr. chs. 98, 99, *Chin Shu.*)

PFIZMAIER, A. (61) (tr.). 'Darlegungen a. d. Gesch. d. Hauses Sui.' *SWAW/PH*, 1881, **97**, 627, 649, 653, 658, 686, 702. (Tr. chs. 36, 40, 45, 48, 79, *Sui Shu.*)

PFIZMAIER, A. (62) (tr.). 'Lebensbeschreibungen von Heerführern und Würdentragern des Hauses Sui.' *DWAW/PH*, 1882, **32**, 281, 301, 320, 351, 369. (Tr. chs. 37, 38, 39, 40 (in part), 41 (in part), *Sui Shu.*)

PFIZMAIER, A. (63) (tr.). 'Fortsetzungen a. d. Gesch. d. Hauses Sui.' *SWAW/PH*, 1882, **101**, 187, 201, 207, 230, 249. (Tr. chs. 41 (in part), 70, 73 (in part), 74, 85, *Sui Shu.*)

PFIZMAIER, A. (64) (tr.). 'Die fremdländischen Reiche zu den Zeiten d. Sui.' *SWAW/PH*, 1881, **97**, 411, 418, 422, 429, 444, 477, 483. (Tr. chs. 64, 81, 82, 83, 84, *Sui Shu.*)

PFIZMAIER, A. (65) (tr.). 'Die Classe der Wahrhaftigen in China.' *SWAW/PH*, 1881, **98**, 983, 1001, 1036. (Tr. chs. 71, 73, 77, *Sui Shu.*)

PFIZMAIER, A. (66) (tr.). 'Zur Geschichte d. Aufstände gegen das Haus Sui.' *SWAW/PH*, 1878, **88**, 729, 743, 766, 799. (Tr. chs. 1, 84, 85, 86, *Hsin Thang Shu.*)

PFIZMAIER, A. (67) (tr.). 'Seltsamkeiten aus den Zeiten d. Thang' I & II. I, *SWAW/PH*, 1879, **94**, 7, 11, 19. II, *SWAW/PH*, 1881, **96**, 293. (Tr. chs. 34–6, (*Wu Hsing Chih*) 88, 89, *Hsin Thang Shu.*)

PFIZMAIER, A. (68) (tr.). 'Darlegung der chinesischen Ämter.' *DWAW/PH* (1879), **29**, 141, 170, 213; (1880), **30**, 305, 341. (Tr. chs. 46, 47, 48, 49A, *Hsin Thang Shu*; cf. des Rotours (1).)

PFIZMAIER, A. (69) (tr.). 'Die Sammelhäuser der Lehenkönige Chinas.' *SWAW/PH* (1880), **95**, 919. (Tr. ch. 49B, *Hsin Thang Shu*; cf. des Rotours (1).)

PFIZMAIER, A. (70) (tr.). 'Über einige chinesische Schriftwerke des siebenten und achten Jahrhunderts n. Chr.' *SWAW/PH* (1879), **93**, 127, 159. (Tr. chs. 57, 59 (in part: *I Wen Chih*, agric., astron., maths., war, *wu hsing*), *Hsin Thang Shu.*)

PFIZMAIER, A. (71) (tr.). 'Die philosophischen Werke China's in dem Zeitalter der Thang.' *SWAW/PH*, 1878, **89**, 237. (Tr. ch. 59 (in part: *I Wen Chih*, philosophical sect., incl. Buddh.), *Hsin Thang Shu.*)

PFIZMAIER, A. (72) (tr.). 'Der Stand der chinesische Geschichtsschreibung in dem Zeitalter der Thang' (original has Sung as misprint). *DWAW/PH*, 1877, **27**, 309, 383. (Tr. chs. 57 (in part), 58 (*I Wen Chih*, history and classics Section), *Hsin Thang Shu.*)

PFIZMAIER, A. (73) (tr.). Zur Geschichte d. Gründung d. Hauses Thang.' *SWAW/PH* 1878, **91**, 21, 46, 71. (Tr. chs. 86 (in part), 87, 88, (in part), *Hsin Thang Shu.*)

PFIZMAIER, A. (74) (tr.). 'Nachrichten von Gelehrten China's.' (Scholars such as Khung Ying-Ta, Ouyang Hsün, etc.) *SWAW/PH*, 1878, **91**, 694, 734, 758. (Tr. chs. 198, 199, 200, *Hsin Thang Shu.*)

PFIZMAIER, A. (84) (tr.). 'Aus dem Traumleben der Chinesenen.' *SWAW/PH* (1870), **64**, 69, 711, 722, 733. (Tr. chs. 397, 398, 399, 400, *Thai-Phing Yü Lan.*)

PFIZMAIER, A. (85) (tr.). 'Geschichtliches ü. einige Seelenzustände u. Leidenschaften.' *SWAW/PH* (1868), **59**, 248, 258, 271, 274, 289, 302, 315. (Tr. *Thai-Ping Yü Lan* chs. 469 (Furcht), 483 (Zorn), 489 (Vergesslichkeit u. lrrtum), 491 (Beschämung), 493 (Verschwendung), 498 (Hochmut), 499 (Dummheit).)

PFIZMAIER, A. (86) (tr.). 'Reichtum und Armut in dem alten China.' *SWAW/PH* (1868), **58**, 61, 69, 84, 104, 110. (Tr. chs. 471, 472, 484, 485, 486, *Thai-Phing Yü Lan.*)

PFIZMAIER, A. (87) (tr.). 'Die Taolehre v. den wahren Menschen u.d. Unsterblichen.' *SWAW/PH*, 1869, **63**, 217, 235, 252, 268. (Tr. chs. 660, 661, 662, 663, *Thai-Phing Yü Lan.*)

PFIZMAIER, A. (88) (tr.). 'Die Lösung d. Leichnam und Schwerter, ein Beitrag zur Kenntnis d. Taoglaubens.' *SWAW/PH*, 1870, **64**, 26, 45, 60, 79. (Tr. chs. 664, 665, 666, 667, *Thai-Phing Yü Lan.*)

PFIZMAIER, A. (89) (tr.). 'Die Lebensverlängerungen d. Männer des Weges' (*Tao Shih*). *SWAW/PH*, 1870, **65**, 311, 334, 346, 359. (Tr. chs. 668, 669, 670, 671, *Thai-Phing Yü Lan.*)

PFIZMAIER, A. (90) (tr.). 'Über einige Kleidertrachten d. chinesischen Altertums.' *SWAW/PH* (1872) **71**, 567,

578, 588, 593, 605, 609, 616, 627, 636, 640. (Tr. chs. 690, 691, 692, 693, 694, 695, 696, 715, 716, *Thai-Phing Yü Lan*.)

**Pfizmaier**, A. (91) (tr.). 'Denkwürdigkeiten v. chinesischen Werkzeugen und Geräthen.' *SWAW/PH*, 1872, **72**, 247, 265, 272, 275, 295, 308, 313, 315. (Tr. chs. 701 (screens), 702 (fans), 703 (whisks, sceptres, censers, etc.), 707 (pillows), 711 (boxes and baskets), 713 (chests), 714 (combs and brushes), 717 (mirrors), *Thai-Phing Yü Lan*.)

**Pfizmaier**, A. (92) (tr.). 'Kunstfertigkeiten u. Künste d. alten Chinesen.' *SWAW/PH*, 1871, **69**, 147, 164, 178, 202, 208. (Tr. chs. 736, 737 (magic), 750, 751 (painting) and 752 (inventions and automata), *Thai-Phing Yü Lan*.)

**Pfizmaier**, A. (93) (tr.). 'Zur Geschichte d. Erfindung u.d. Gebrauches d. chinesischen Schriftgattungen.' *SWAW/PH*, 1872, **70** 10, 28, 46. (Tr. chs. 747, 748, 749, *Thai Phing Yü Lan*.)

**Pfizmaier**, A. (94) (tr.). 'Beiträge z. Geschichte d. Perlen.' *SWAW/PH*, 1867, **57**, 617, 629. (Tr. *Thai-Phing Yü Lan*, chs. 802 (in part), 803.)

**Pfizmaier**, A. (95) (tr.). 'Beiträge z. Geschichte d. Edelsteine u.des Goldes.' *SWAW/PH*, 1867, **58**, 181, 194, 211, 217, 218, 223, 237. (Tr. chs. 807 (coral), 808 (amber), 809 (gems), 810, 811 (gold), 813 (in part), *Thai-Phing Yü Lan*.)

**Pfizmaier**, A. (96) (tr.). 'Zur Geschichte der alten Metalle.' *SWAW/PH* (1868), **60**, 7, 26, 44, 50, 67. (Tr. chs. 802 (in part) (gems), 804, 805 (gems), 812 (Hg, Ag, yellow Ag, Pb, Sn), 813 (in part) (Cu, Fe, brass), *Thai-Phing Yü Lan*.)

**Pfizmaier**, A. (97) (tr.). 'Alte Nachrichten u. Denkwürdigkeiten von einigen Lebensmitteln Chinas.' *SWAW/PH* (1871), **67**, 413, 418, 432, 441, 453, 459. (Tr. *Thai-Phing Yü Lan*, chs. 857 (honey), 860 (cakes), 861 (soups), 863 (meat dishes), 865 (salt), 867 (tea).)

**Pfizmaier**, A. (98) (tr.). 'Die Anwendung u.d. Zufälligkeiten des Feuers in d. alten China.' *SWAW/PH* (1870), **65**, 767, 777, 786, 799. (Tr. chs. 868, 869 (fire and fire-wells), 870 (lamps, candles and torches), 871 (coal), of *Thai-Phing Yü Lan*.)

**Pfizmaier**, A. (99) (tr.). 'Der Geisterglaube in dem alten China.' *SWAW/PH*, 1871, **68**, 641, 652, 665, 679, 695. (Tr. *Thai-Phing Yü Lan*, chs. 881, 882, 883, 884, 887 (in part).)

**Pfizmaier**, A. (100) (tr.). 'Zur Geschichte d. Wunder in dem alten China.' *SWAW/PH* (1871), **68**, 809, 828, 844, 848. (Tr. *Thai-Phing Yü Lan*, chs. 885, 886, 887 (in part), 888 (quasi-biological metamorphoses).)

**Pfizmaier**, A. (101) (tr.). 'Denkwürdigkeiten aus dem Tier-reich Chinas.' *SWAW/PH* (1875), **80**, 6, 8, 17, 22, 35, 41, 51, 57, 68, 73, 79. (Tr. *Thai-Phing Yü Lan*, chs. 901, 902, 903, 904, 905, 909, 912 (mammals), 918, 919 (birds), 950 (in part), 951 (insects).)

**Pfizmaier**, A. (102) (tr.). 'Über einige Gegenstände des Taoglaubens.' *SWAW/PH* (1875), **79**, 5, 16, 29, 42, 50, 59, 61, 68, 73, 78. (Tr. *Thai-Phing Yü Lan*, chs. 929, 930 (dragons), 931, 932 (tortoises), 933, 934 (snakes), 984, 985, 986, 990 (miscellaneous stones).)

**Pfizmaier**, A. (103) (tr.). 'Denkwürdigkeiten von dem Insekten Chinas.' *SWAW/PH* (1874), **78**, 345, 356, 368, 378, 387, 397, 410. (Tr. *Thai-Phing Yü Lan*, chs. 944, 945, 946, 947, 948, 949, 950 (in part).)

**Pfizmaier**, A. (104) (tr.). 'Denkwürdigkeiten von den Bäumen Chinas.' *SWAW/PH* (1875), **80**, 191, 198, 205, 213, 220, 234, 240, 251, 264. (Tr. *Thai-Phing Yü Lan*, chs. 952, 953, 954, 955, 956, 957, 958, 959, 960 (in part).)

**Pfizmaier**, A. (105) (tr.). 'Ergänzungen zu d. Abhandlung von den Bäumen Chinas.' *SWAW/PH* (1875), **81**, 143, 160, 167, 177, 188, 189, 192, 196. (Tr. *Thai-Phing Yü Lan*, chs. 960 (in part), 961, 962, 963, 969 (in part), 972 (in part), 973 (in part), 974 (in part).)

**Pfizmaier**, A. (106) (tr.). 'Denkwürdigkeiten v. den Früchten Chinas.' *SWAW/PH* (1874), **78**, 195, 202, 214, 222, 230, 238, 244, 249, 260, 267, 274, 280. (Tr. *Thai-Phing Yü Lan*, chs. 964, 965, 966, 967, 968, 969 (in part), 970, 971, 972 (in part), 973 (in part), 974 (in part), 975.)

**Phillips**, L. L. (1). 'Cotton.' In Simmonds (1), 197–200.

**Pickersgill**, B. & C. B. **Heiser** (1). 'Origins and distribution of plants domesticated in the New World tropics.' In Charles A. Reed (2), pp. 803–36.

**Pinto, Fernaõ Mendes** (1). *Peregrinaçam de Fernam Mendez Pinto em que da conta de muytas e muyto estranhas cousas que vio e ouvio no reyno da China, no da Tartaria* ... Crasbeec, Lisbon, 1614. Abridged Eng. tr. by H. Cogan: '*The Voyages and Adventures of Ferdinand Mendez Pinto, a Portugal, During his Travels for the space of one and twenty years in the kingdoms of Ethiopia, China, Tartaria, etc.* Herringman, London 1653, 1663, repr. 1692. Still further abridged edition, Unwin, London, 1891. Full French tr. by B. Figuier: *Les Voyages Advantureux de Fernand Mendez Pinto* ... Cotinet and Roger, Paris 1628, repr. 1645. Cf. M. Collis (1): *The Grand Peregrination* (paraphrase and interpretation), Faber and Faber, London, 1949.

**Plat, Hugh** (1). *The Newe and Admirable Art of Setting of Corne*. London, 1601.

**Pliny** (trs. H. Rackham) (1). *Natural History*. Loeb Classical Library, Heinemann, London, 1950.

**Plucknett**, D. L. (1). 'Edible aroids.' In Simmonds (1), pp. 10–12.

**Postan**, M. M. (2). *Essays on Medieval Agriculture & General Problems of The Medieval Economy*. Cambridge University Press, 1973.

**Postan**, M. M., E. E. **Rich** & E. **Miller** (eds.) (1). *The Cambridge Economic History of Europe*, vol. II. Cambridge University Press.

POTTER, JACK M. (1). *Thai Peasant Social Structure.* University of Chicago Press, Chicago & London, 1976.
POTTER, JACK M. (2). 'Land and lineage in traditional China.' In M. Freedman (5).
POYNTER, F. N. L. (1). 'Gervase Markham.' *ESEA* (1962), **15**, 27–39.
PRESCOTT, W. (1). *History of the Conquest of Peru.* Everyman, London, 1968 (1st ed. 1847).
PUHVEL, J. (1). 'The Indo-European and Indo-Aryan Plough: A Linguistic Study of Technological Diffusion.' *TCULT* (1964), 5, **2**, pp. 176–90.
PULESTON, D. E. & O. S. (1). 'An Ecological Approach to the Origins of Maya Civilisation.' *AAAA* (1974), 24, **4**, 330.
PURCAL, J. T. (1). *Rice Economy: Employment and Income in Malaysia.* East-West Centre, University Press of Hawaii, Honolulu, 1972.
PURCELL, V. (1). *The Chinese in South-East Asia.* Oxford University Press (for the Royal Institute of International Affairs and the Institute of Pacific Relations), Oxford 1951; 2nd ed. 1965.
PURSEGLOVE, J. W. (1). *Tropical Crops: Dicotyledons.* Longman, London, 1968.
PURSEGLOVE, J. W. (2). *Tropical Crops: Monocotyledons.* Longman, London, 1972.
PUSEY, PHILIP (1). 'Report on The Royal Agricultural Society's Show.' *JRAGS* (1857), **12**, 642–4; cited Mingay (1), pp. 191–4.

QUESNAY, FRANÇOIS (1). 'Despotisme de la Chine.' In Quesnay (2), pp. 563–660.
QUESNAY, FRANÇOIS (2). *Oeuvres économiques et philosophiques.* Frankfurt, 1888: repr. Scientia Verlag, Aalen, 1965.

RAFFLES, T. S. (1). *The History of Java.* 2 vols.; Black, Parbury & Allen, London, 1817. Facsimile ed., Oxford in Asia Historical Reprints, Oxford University Press, Oxford, 1978.
RAFTIS, J. A. (1). *Tenure & Mobility: Studies in The Social History of the Medieval English Village.* Toronto, 1964.
RAIKES, R. (1). *Water, Weather and Prehistory.* London, 1967.
RANDALL, JOHN (1). *Observations on The Structure and Use of The Spiky Roller in Museum Rusticum.* 1766.
RANSOME, J. ALLEN (1). *The Implements of Agriculture.* London, 1843.
RASMUSSEN, H. (1). 'Grain harvest and threshing in Calabria.' *TT* (1969), 1, **2**, 93–104.
RAU, K. H. (1). *Geschichte des Pfluges.* Winter, Heidelberg, 1845; reprinted by Gerstenberg, Hildesheim, 1972.
RAVENSTONE, PIERCY (1). *Thoughts on The Funding System and its Effects.* London, 1824.
RAWSKI, EVELYN SAKAKIDA (1). *Agricultural Change and the Peasant Economy of South China.* Harvard University Press, Cambridge Mass., 1972.
RAWSKI, EVELYN SAKAKIDA (2). 'Chinese and American Frontier Agriculture in The Eighteenth Century.' Seminar paper, n.d.
RAWSKI, T. G. (1). *Economic Growth and Employment in China.* World Bank Research Publication, Oxford University Press, 1979.
READ, BERNARD E. (with LIU JU-CHHIANG) (1). *Chinese Medicinal Plants from the 'Pen Ts'ao Kang Mu' A.D. 1596 ... a Botanical, Chemical and Pharmacological Reference List.* (Publication of the Peking Nat. Hist. Bull.). French Bookstore, Peking, 1936 (chs. 12–37 of *Pen Tshao Kang Mu*) (rev. W. T. Swingle, *ARLC/DO*, 1937, 191).
READ, BERNARD E. (2) (with LI YÜ-THIEN). *Chinese Materia Medica; Animal Drugs.*

| | | Serial nos. | Corresp. with chaps. of *Pen Tshao Kang Mu* |
|---|---|---|---|
| Pt. I | Domestic Animals | 322–349 | 50 |
| II | Wild Animals | 350–387 | 51 *A* and *B* |
| III | Rodentia | 388–399 | 51 *B* |
| IV | Monkeys and Supernatural Beings | 400–407 | 51 *B* |
| V | Man as a Medicine | 408–444 | 52 |

*PNHB*, 1931, **5** (no. 4), 37–80; **6** (no. 1), 1–102. (Sep. issued, French Bookstore, Peking, 1931.)
READ, BERNARD E. (3) (with LI YÜ-THIEN). *Chinese Materia Medica; Avian Drugs.*

| | | | |
|---|---|---|---|
| Pt. VI | Birds | 245–321 | 47, 48, 49 |

*PNHB*, 1932, **6** (no. 4), 1–101. (Sep. issued, French Bookstore, Peking, 1932.)
READ, BERNARD E. (4) (with LI YÜ-THIEN). *Chinese Materia Medica; Dragon and Snake Drugs.*

| | | | |
|---|---|---|---|
| Pt. VII | Reptiles | 102–127 | 43 |

*PNHB*, 1934, **8** (no. 4), 297–357. (Sep. issued, French Bookstore, Peking, 1934.)
READ, BERNARD E. (5) (with YU CHING-MEI). *Chinese Materia Medica; Turtle and Shellfish Drugs.*

| | | | |
|---|---|---|---|
| Pt. VIII | Reptiles and Invertebrates | 199–244 | 45, 46 |

*PNHB* (Suppl.), 1939, 1–136. (Sep. issued, French Bookstore, Peking, 1937.)
READ, BERNARD E. (6) (with YU CHING-MEI). *Chinese Materia Medica; Fish Drugs.*

| | | | |
|---|---|---|---|
| Pt. IX | Fishes (incl. some amphibia, octopoda and crustacea) | 128–198 | 44 |

*PNHB* (Suppl.), 1939. (Sep. issued, French Bookstore, Peking, n.d. prob. 1939.)
READ, BERNARD E. (7) (with YU CHING-MEI). *Chinese Materia Medica; Insect Drugs.*

| | | Serial nos. | Corresp. with chaps. of *Pen Tshao Kang Mu* |
|---|---|---|---|
| Pt. X | Insects (incl. arachnidae etc.) | 1–101 | 39, 40, 41, 42 |

*PNHB* (Suppl.), 1941. (Sep. issued, Lynn, Peking, 1941.)

READ, BERNARD E. (8). *Famine Foods listed in the 'Chiu Huang Pen Tshao'.* Lester Institute, Shanghai, 1946.

READ, BERNARD E. & PAK KYEBYŎNG (1). *A Compendium of Minerals and Stones used in Chinese Medicine, from the 'Pen Tshao Kang Mu'. PNHB*, 1928, **3** (no. 2), i–vii, 1–120. (Revised and enlarged, issued separately, French Bookstore, Peking, 1936 (2nd ed.).) Serial nos. 1–135, corresp. with chs. of *Pen Tshao Kang Mu*, 8, 9, 10, 11.

REED, CHARLES A. (1). 'The Pattern of Animal Domestication in the Prehistoric Near East.' In Ucko & Dimbleby (1), p. 361.

REED, CHARLES A. (ed.) (2). *Origins of Agriculture.* Mouton, the Hague, 1977.

REED, CHARLES A. (3). 'A Model for the Origin of Agriculture in the Near East.' In Charles A. Reed (2), pp. 543–67.

REED, CHARLES A. (4). 'Origins of agriculture: discussion and some conclusions.' In Charles A. Reed, ed. (2), 879–956.

RENFREW, C., ed. (2). *The Explanation of Culture Change: Models in Prehistory.* Duckworth, London, 1973.

RENFREW, C. (3). 'Monuments, Mobilisation and Social Organisation in Neolithic Wessex.' In Renfrew (2), p. 539.

RENFREW, C. (4). *Before Civilisation: the Radiocarbon Revolution and Prehistoric Europe.* Jonathan Cape, London, 1973.

RENFREW, C. & K. L. COOKE (eds.) (1). *Transformations: Mathematical Approaches to Culture Change.* Academic Press, New York, 1979.

REYNOLDS, P. (1). 'A general report of underground grain storage experiments at the Butser ancient farm research project.' In Gast & Sigaut (1), vol. 1, 70–80.

RICHARDSON, H. G. (1). 'The Medieval Plough Team.' *H* (1942), **26**, 287.

RICKMAN, G. (1). *Roman Granaries and Store Buildings.* Cambridge University Press, 1971.

W. R. ROFF (ed.) (1). *Kelantan: Religion, Society and Politics in a Malay State.* Oxford University Press, Kuala Lumpur, 1974.

DE ROSNY, L. (1) (tr.). *Chan-Hai-King; Antique Géographie Chinoise.* Maisonneuve, Paris, 1891.

ROSSITER, M. W. (1). *The Emergence of Agricultural Science: Justus Liebig and the Americans, 1840–1880.* Yale University Press, New Haven & London, 1975.

DES ROTOURS, R. (1). *Traité des Fonctionnaires et Traité de l'Armée, traduits de la Nouvelle Histoire des T'ang* (ch. 46–50). 2 vols. Brill, Leiden, 1948 (Bibl. de l'Inst. des Hautes Etudes Chinoises, vol. 6). (Rev. P. Demiéville, *JA*, 1950, **238**, 395.)

DES ROTOURS, R. (2) (tr.). *Traité des Examens* (translation of chs. 44 and 45 of the *Hsin Thang Shu*). Leroux, Paris, 1932. (Bibl. de l'Inst. des Hautes Etudes Chinoises, no. 2.)

DES ROTOURS, R. (4) (tr.). *Courtisanes chinoises à la fin des Thang, entre c. +789 et le 8 janvier, +881; Pei Li Tche* [*Chih*] *(Anecdotes du Quartier du Nord) par Souen K'i* [*Sun Chhi*] ... Presses Univ. de France, Paris, 1968. (Bibl. de l'Inst. des Hautes Etudes Chinoises, no. 22.)

ROY, D. T. & T. H. TSIEN (eds.) (1). *Ancient China: Studies in Early Civilisation.* Chinese University Press, Hong Kong, 1978.

RUSSELL, E. J. (1). *A Student's Book on Soils and Manures.* Cambridge University Press, 1915, repr. 1951.

ŠACH, F. (1). 'Proposal for the Classification of Pre-Industrial Tilling Implements.' *TT* (1968), 1, **1**, 1.

SAHLINS, MARSHALL (1). *Stone Age Economics.* Aldine Atherton, Chicago, 1972.

SALONEN, ARMAS (1). *Agricultura Mesopotamica: nach sumerisch-akkadischen Quellen, eine lexikalische und kulturgeschichtliche Untersuchung.* Annales Academiae Scientiarum Fennicae, B. vol. 149, Helsinki, 1968.

SAUER, CARL O. (1). *Seeds, Spades, Hearths and Herds: the Domestication of Animals and Foodstuffs.* American Geographical Society, 1952; 2nd ed. MIT Press, Cambridge Mass., 1969.

SAUSSURE, [N.] TH. DE (1). *Recherches chimiques sur la végétation.* V. Nyon, Paris, 1804.

SCHAFER, E. H. (13). *The Golden Peaches of Samarkand: a Study of Thang Exotics.* University of California Press, Berkeley & Los Angeles, 1963.

SCHAFER, E. H. (16). *The Vermilion Bird: Thang Images of the South.* University of California Press, Berkeley & Los Angeles, 1967.

SCHAFER, E. H. (25). 'Thang.' In K. C. Chang (3).

SCHMAUDERER, E. (1). 'Kenntnisse über Fette und Öle bei den alten Kulturvolken; II, Die Entwicklung technologischer Verfahren zur Herstellung von Butter und pflanzlichen Ölen im Altertum.' *VFIDMGNWT* (1968), **40**, 1.

SCHULTES, R. E. (2). 'Random thoughts and queries on the botany of *Cannabis*.' In Joyce & Curry (1), 11–38.

SCOTT, J. C. (1). *Health & Agriculture in China.* Faber & Faber, London, 1952.

SCOTT, JAMES C. (1). *The Moral Economy of the Peasant: Rebellion and Subsistence in Southeast-Asia.* Yale University Press, New Haven & London, 1976.

SCOTT, JOHN (1). *The Complete Test-Book of Farm Engineering.* Crosby, Lockwood & Co, London, 1885.
SEKINO, TAKESHI (1). 'New Researches on the *Lei-Ssu.*' *MRDTB* (1967), **25**, 59.
SERJEANT, R. B. (1). 'The Cultivation of cereals in mediaeval Yemen.' *ARS* (1974), **1**, 24–74.
SERRES, OLIVER DE, SIEUR DE PRADEL (1). *Le Théâtre d'Agriculture et Mesnage des Champs.* Paris, 1600.
SHA, J. (1). *Certaine Plaine and Easie Demonstrations of Divers Easie Wayes and Meanes for the Improving of Any Manner of Barren Land.* London, 1657.
SHAW, THURSTAN (1). 'Hunters, Gatherers and First Farmers in West Africa.' In J. V. S. Megaw (1), pp. 69–126.
SHEN, T. H. (1). *Agricultural Resources of China.* Cornell University Press, Ithaca N. Y., 1951.
SHIBA, YOSHINOBU (1) (trs. Mark Elvin). *Commerce and Society in Sung China.* Michigan Abstracts of Chinese & Japanese Works on Chinese History no. 2, Ann Arbor, 1970.
SHIH SHENG-HAN (1). *A Preliminary Survey of the book 'Chhi Min Yao Shu', an Agricultural Encyclopaedia of the +6th Century.* Science Press, Peking, 1958.
SHIH SHENG-HAN (2). *On the 'Fan Sheng-Chih Shu', an Agricultural Book written by Fan Sheng-Chih in −1st Century China.* Science Press, Peking, 1959.
SHIMPO, MITSURO (1). *Three Decades in Shiwa: Economic Development and Social Change in a Japanese Farming Community.* U. British Columbia Press, Vancouver, 1976.
SHURTLEFF, W. & A. AOYAGI (1). *The Book of Miso: Food for Mankind.* Ballantine Books, New York, 1976.
SICKMAN, L. & A. SOPER (1). *The Art and Architecture of China.* The Pelican History of Art, Penguin, Harmondsworth, 1956.
SIGAUT, F. (1). 'Identification des Techniques de Récolte des Graines Alimentaines.' *JATBA* (1978), 25, **3**, 147–61.
SIGAUT, F. (2). 'Identification des techniques de conservation et de stockage des grains.' In Gast & Sigaut (1), vol. 2, 156–81.
SIGAUT, F. (3). *L'agriculture et le feu: rôle et place du feu dans les techniques de préparation du champ de l'ancienne agriculture européenne.* Mouton & Co, Paris, The Hague, 1975.
SIGAUT, F. (4) 'Possibilités et limites de la recherche, de l'interprétation et de la représentation des instruments agricoles dans les musées d'agriculture.' Paper given at the 5th International Congress of Agricultural Museums, n.d.
SIMMONDS, N. W. (ed.) (1). *Evolution of Crop Plants.* Longman, London & New York, 1976.
SIMMONDS, N. W. (2). 'Hemp.' In Simmonds (1), pp. 203–4.
SINGER, C., E. J. HOLMYARD, A. R. HALL & T. I. WILLIAMS (eds.) (1). *A History of Technology.* 7 vols., Clarendon Press, Oxford, 1954–78.
SIVIN, N. (17). 'Imperial China: has its present past a future?' [review article of M. Elvin (2)]. *HJAS*, 1978, 38, **2**, pp. 449–80.
SKOCPOL, THEDA (1). *States and Social Revolutions: a Comparative Analysis of France, Russia and China.* Cambridge University Press, 1979.
SLICHER VAN BATH, B. H. (1). *The Agrarian History of Western Europe, A.D. 500–1850.* Edward Arnold, London, 1963.
SLICHER VAN BATH, B. H. (2). 'Agriculture in the Low Countries (*c.* 1600–1800).' In *Relazione del X Congresso Internazionale di Scienze Storiche,* IV (1955), pp. 169–203; Florence, 1955.
SMALL, JAMES (1). *Treatise of Ploughs and Wheel Carriages.* Edinburgh, 1784.
SMITH, ADAM (1). *An Inquiry into the Nature and Causes of the Wealth of Nations.* (1st ed. 1776); Nelson & Sons, London, 1901 (there have been many subsequent edns.).
SMITH, P. E. L. & J. C. YOUNG (1). 'The evolution of early agriculture and culture in Greater Mesopotamia: a trial model.' In B. Spooner (1), pp. 1–59.
SMITH, R. B. & W. WATSON (eds.) (1). *Early Southeast Asia.* Oxford University Press, Oxford, New York & Kuala Lumpur, 1979.
SMITH, THOMAS C. (1). *The Agrarian Origins of Modern Japan.* Stanford University Press, Stanford, 1959.
SOLHEIM, WILHELM G. II (1). 'Northern Thailand, Southeast Asia, and world prehistory' *ASP* (1970), **13**, 145–62.
SOLHEIM, WILHELM G. II (2). 'Southeast Asia and the West.' *SC* (1967), **157**, pp. 896–902.
SOLHEIM, WILHELM G. II (3). 'Remarks on the neolithic in South China and Southeast Asia.' *JHKAS* (1973), **4**, 25–9.
SOOTHILL, W. E. (5) (posthumous). *The Hall of Light; a Study of Early Chinese Kingship.* Lutterworth, London, 1951. (On the Ming Thang; also contains discussion of the *Pu Thien Ko* and transl. of *Hsia Hsiao Cheng.*)
SOULET, J.-F. (1). *La vie quotidienne dans les Pyrénées sous l'Ancien Régime du XVI^e^ au XVIII^e^ siècle.* Hachette, Paris. 1974.
SPENCE, JONATHAN D. (1) *The Death of Woman Wang.* Viking Press, New York, 1978; Penguin, 1979.
SPENCER, A. J. & J. B. PASSMORE (1). *Handbook of the Collections Illustrating Agricultural Implements & Machinery: A Brief Survey of the Machines & Implements Which are Available to the Farmer with Notes on their Development.* Science Museum, H.M.S.O., London, 1930.
SPENCER, J. E. (4). *Shifting Cultivation in Southeastern Asia.* University of California Publicns. in Geography no. 19, 1966.

**Spencer**, J. E. (5). 'The development and spread of agricultural terracing in China.' In S. G. Davis (ed.), *Symposium on Land Use and Mineral Deposits in Hong Kong, Southern China and South East Asia*, Hong Kong, 1964, pp. 105–10.

**Spencer**, J. E. & G. A. **Hale** (1). 'The origin, nature and distribution of agricultural terracing.' *PV* (1961), **2**, 1–40.

**Spooner**, B. (ed.) (1). *Population Growth: Anthropological Implications*. MIT Press, Cambridge, Mass., 1972.

**Šramko**, B. A. (1). 'Der Hakenpflug der Bronzezeit in der Ukraine.' *TT* (1971), 1, **4**, 223.

**Steensberg**, A. (2). 'A 6,000 Year Old Ploughing Implement from Satrup Moor.' *TT* (1973), 2, **2**, 105.

**Steensberg**, A. (3). 'The Husbandry of Food Production.' In Hutchinson, Clarke, Jope & Riley (1), p. 43.

**Steensberg**, A. (4). *Stone Shares of Ploughing Implements from the Bronze Age of Syria: A Contribution to the Early History of the Ard-Plough*. Royal Danish Academy of Sciences and Letters, Copenhagen, 1977. Reviewed A. Fenton (4).

**Steensberg**, A. (5) (trs. W. E. Calvert). *Ancient Harvesting Implements*. Copenhagen, 1943.

**Steensberg**, A. (6). 'Drill-Sowing and Threshing in Southern India Compared with Sowing Practices in Other Parts of Asia.' *TT* (1971), 1, **4**, 241–56.

**Steensberg**, A. (7). *New Guinea Gardens: A Study of Husbandry with Parallels in Prehistoric Europe*. Academic Press, London & New York, 1980.

**Stein**, R. A. (6). *Tibetan Civilisation*. Faber & Faber, London, 1972.

**Stephens, Charles** & **Iohn Liebault, Doctors of Physicke** (1). *Maison Rustique, or, the Country Farme*. Translated into English by Richard Surfleet; reviewed, corrected and augmented by Gervase Markham; London, 1616.

**Stern, Harold** P. (1). *Birds, Beasts, Blossoms, and Bugs: the Nature of Japan*. Harry N. Abrams, Inc., New York, 1976; in assoc. with the UCLA Art Council & the Frederick S. Wight Gallery, Los Angeles.

**Stiefel**, M. & W. F. **Wertheim** (1). *Production, Equality and Participation in Rural China*. United Nations Research Institute for Social Development (Popular Participation Programme), Geneva, 1982.

**Stonor**, C. R. & E. **Anderson** (1). 'Maize among the hill peoples of Assam.' *AMBG* (1949), 36, 355–404.

**Stuart**, G. A. (1). *Chinese Materia Medica, Vegetable Kingdom*. American Presbyterian Mission Press, Shanghai, 1911.

**Sun**, E-Tu **Zen** (1). *Chhing Administrative Terms*. Harvard University Press, 1961.

**Sung**, **Ying-Hsing** (1), tr. E.-T. Z. **Sun**. & S.-C. **Sun**. *Thien-Kung Khai-Wu: Chinese Technology in the Seventeenth Century*. Pennsylvania State University Press, University Park and London, 1966.

**Swann**, N. (1) (trs). *Food and Money in Ancient China*. Princeton University Press, Princeton N.J., 1950.

**Swingle**, W. T. (11). 'Chinese and other East Asiatic books added to the Library of Congress, 1926–27.' *ARLC/DO* (1926–7), 245. [On editions of the *Pen Tshao Kang Mu*, the *Pen Tshao Kang Mu Shi I*, the *Shen Nung Pen Tshao Ching Su*, and the *Shao-Hsing Pen Tshao*.]

**Tanabe, Shigeharu** (1). 'Land reclamation in the Chao Phya Delta.' In Y. Ishii, ed., *Thailand: A Rice Growing Society*. Monog. of the Center of Southeast Asian Studies, Kyoto; Univ. of Hawaii Press, Honolulu, 1978.

**Taillard**, C. (1). 'Les berges de la Nam Ngum et du Mekong: systèmes économiques villageois et organisation de l'espace dans la plaine de Vientiane (Laos).' *ERU* (1974), 119–68.

**Teng Ssu-Yü** & K. **Biggerstaff** (1). *An Annotated Bibliography of Selected Chinese Reference Works*. Harvard-Yenching Inst., Peiping, 1936. (Yenching Journ. Chin. Studies, monograph no. 12.)

**Than Po-Fu** *et al.* (1). *The Kuan-Tzu: Economic Dialogues in Ancient China*. Far Eastern Publications, Yale University, New Haven, 1954.

**Thapa**, J. K. (1). 'Primitive maize with the Lepchas.' *BT* (1966), **3**, 29–31.

**Thilo**, T. (1). 'Die Kapitel 1 und 4 (Ackerbau und Weiterarbeitung der Ackerbauprodukte) des *Tiangong kaiwu* von Song Yingxing: Übersetzung and Kommentar.' Unpublished doctoral thesis, Humboldt-Universität, Berlin, 1964.

**Thilo, Thomas** (2). 'Eine problematische Darstellung einer chinesischen Windfege.' *MIOF* (1966), 12, **3**.

**Thilo, Thomas** (3). 'Die Schrift vom Pflug (*Leisijing*) und das Verhältnis ihres Verfassers Lu Guimeng zur Landwirtschaft.' *ALTOF* (1980), **7**.

**Thirsk, Joan** (1). 'The common fields.' *PP*, (1964), **29**, p. 9.

**Torr, Cecil** (1). *Small Talk at Wreyland*. 1st published 1918; reissued by Oxford University Press, Oxford, 1979.

**Tregear**, T. R. (2). *China: A Geographical Survey*. Hodder & Stoughton, London, 1980.

**Tringham**, R. (1). *Hunters, Fishers and Farmers of Eastern Europe, 6000–3000 BC*. Hutchinson University Library, London, 1971.

**Trippner**, J. (1). 'Das "Röstmehl" bei den Ackerbauern in Tsinghai, China.' *AN* (1957), **52**, 603–16.

**Truong Van Binh** (1). 'Vietnamese Feasts and Holidays (Customs of Viet Nam V).' *Times of Vietnam Magazine* (1963), 5, **39**, pp. 11–12, 16.

**Tull, Jethro** (1). *Horse Hoeing Husbandry*. 1st edition, London, 1733. (Page refs. are to Cobbett ed. of 1812; repr. in extract in Mingay (1) 117–21.)

**Tun Li-Chhen**. See D. Bodde (12).

**Tung Khai-Chhen** (1) (tr. F. A. Bray). 'A preliminary discussion of Chinese agricultural treatises in the style of "monthly ordinances" *yüeh ling*.' *JATBA* (1981), 28, **3–4**.

TUSSER, THOMAS (1). *A Hundred Good Pointes of Husbandrie.* London, 1557.
TUSSER, THOMAS (2). *Five Hundred Good Pointes of Husbandrie.* London, 1573.
TWITCHETT, D.C. (4). *Financial Administration under the Thang Dynasty.* Cambridge University Press, 1st ed. 1963, 2nd ed. 1970.
TWITCHETT, D. C. (6). 'Some remarks on irrigation under the Thang.' *TP* (1961), 48, **1–3**, 175–94.
TWITCHETT, D. C. (9). 'Documents on Clan Administration, I: The Rules of Administration of the Charitable Estate of the Fan Clan; Annotated Translation of the I-Chuang Kuei-Chü 義莊規矩' *AM* (1960), 8, **1**, pp. 1–35.
TWITCHETT, D. C. (10). 'Lands under state cultivation during the Thang dynasty.' *JESHO* (1959), 2, **2**, 162–203; 2, **3**, 335–6.

UCKO, P. J., & G. W. DIMBLEBY (eds.) (1). *The Domestication and Exploitation of Plants and Animals.* Duckworth, London, 1969.
UCKO, P. J., R. TRINGHAM, & G. W. DIMBLEBY (eds.) (1). *Man, Settlement and Urbanism.* Duckworth, London, 1972.

VALLICROSA, MILLAS & 'AZIMAN (ed. & trs.) (1). *Ibn Baṣṣal: 'Kitāb al Filāḥa'.* Tetúan, 1955.
VAMPLEW, WRAY (1). 'The progress of agricultural mechanics: the cost of best practice in the mid nineteenth century.' *TT* (1979), 3, **4**, 204–14.
VAN ZEIST, W. (1). 'Reflections on prehistoric environments in the Near East.' In Ucko & Dimbleby (1), pp. 35–46.
VARRO, MARCUS TERRENTIUS (1). *De Re Rustica.* Tr. W. D. Hooper, revised H. B. Ash, Loeb Classical Library, Heinemann, London, 1954.
VAVILOV, N. I. (1). 'The problem of the origin of the world's agriculture in the light of the latest investigations.' Address to the Second International Congress on History of Science & Technology, London, July 1931; KNIGA (England) Ltd, London.
VAVILOV, N. I. (2). *The Origin, Variation, Immunity and Breeding of Cultivated Plants.* Tr. K. Starr Chester, Chronica Botanica vol. 13, Waltham, Mass., 1949–50.
VAVILOV, N. I. (4). *Studies on the Origins of Cultivated Plants.* Institut de Botanique Appliquée et d'Amélioration des Plantes, Leningrad, 1926.
VERMUYDEN, SIR CORNELIUS (1). *Discourse Touching the Draining of the Great Fennes.* London, 1642.
VINOGRADOFF, P. (2). *Villainage in England.* Oxford, 1892.
VIRGIL (trs. T. F. Royds) (1). *The Eclogues and Georgics of Virgil.* Everyman's Library, Dent, London, n.d.
VISHNU-MITTRE (1). 'The beginnings of agriculture: palaeographic evidence from India.' In J. Hutchinson (2), 3–33.
VISHNU-MITTRE (2). 'Changing economy in ancient India.' In Charles A. Reed (2), 569–88.
VITA-FINZI, C. & E. S. HIGGS (1). 'Prehistoric Economy in the Mount Carmel Area of Palestine: Site Catchment Analysis.' *PPHS* (1971), 36, 1–37.

WAGNER, W. (1). *Die Chinesische Landwirtschaft.* Paul Parey, Berlin, 1926.
WAKEMAN, FREDERIC, JR. & CAROLYN GRANT (eds.) (1). *Conflict and Control in Late Imperial China.* University of California Press, Berkeley, 1975.
WALEY, ARTHUR (1). *The Book of Songs.* George Allen & Unwin, London, 1937.
WALEY, A. (5) (tr.). *The Analects of Confucius.* Allen & Unwin, London, 1938.
WALEY, A. (23). *The Nine Songs; a Study of Shamanism in Ancient China* [the *Chiu Ko* attributed traditionally to Chhü Yuan]. Allen & Unwin, London, 1955.
WALKER, ALEXANDER (1). 'Indian Agriculture.' *c.* 1820, reproduced in Dharampal (1), 179–209.
WARE, J.-R. (1). 'The *Wei Shu* and the *Sui Shu* on Taoism.' *JAOS*, 1933, **53**, 215. Corrections and emendations in *JAOS*, 1934, **54**, 290. Emendations by H. Maspero, *JA*, 1935, **226**, 313.
WASWO, A. (1). *Japanese Landlords: the Decline of a Rural Elite.* University of California Press, Berkeley & Los Angeles, 1977.
WATABE, TADAO (1). *Glutinous Rice in Northern Thailand.* The Centre for Southeast Asian Studies, Kyoto University, Kyoto, 1967.
WATSON, J. A. S. (1). 'Farm Implements in Scotland; Historical Notes, 2.' *SJA* (1926), **9**.
WATSON, JAMES A. S. & JAMES. A. MORE (1). *Agriculture: the Science and Practice of British Farming.* 6th ed., Oliver and Boyd, Edinburgh, 1942.
WATSON, JAMES L. (1). *Emigration and the Chinese Lineage: the Mans in Hong Kong and London.* University of California Press, Berkeley, Los Angeles & London, 1975.
WATSON, WILLIAM (6). *Cultural Frontiers in Ancient East Asia.* Edinburgh University Press, Edinburgh, 1971.
WEBER, MAX (4). *The Agrarian Sociology of Ancient Civilisations.* Tr. R. I. Frank, NLB, London, 1976; 1st ed. entitled 'Agrarverhältnisse im Altertum', in *Handwörterbuch der Staatwissenschaften*, 1909.
WEST, ROBERT C. & PEDRO ARMILLAS (1). 'Las Chinampas de Mexico: poesia y realidad de los 'Jardines Flotantes'.' *CUA* (1950), 165–82.

WESTERMANN, W. L. (1). 'Egyptian agricultural labour under Ptolemy Philadelphus.' *AGHST* (1927), **1** (no. 2), 34.

WESTON, SIR RICHARD (1). *A Discours of Husbandrie used in Brabant and Flanders.* Manuscript bequeathed to his sons in 1645; piratically published by Hartlib (1) in 1651.

DE WET, J. M. J., J. R. HARLAN & E. G. PRICE (1). 'Variability in *Sorghum bicolor.*' In Harlan, de Wet & Stemler (1), pp. 453–64.

WEULERSSE, J. (3). *Paysans de Syrie et du Proche-Orient.* NRF, Gallimard, Paris, 1946.

WHEATLEY, PAUL (2). *The Pivot of the Four Quarters: a Preliminary Enquiry into the Origins and Character of the Ancient Chinese City.* Edinburgh University Press, Edinburgh, 1971.

WHEATLEY, PAUL (4). 'Agricultural terracing: discursive scholia on recent papers on agricultural terracing and on related matters pertaining to Northern Indo-China and neighbouring areas.' *PV* (1965), **6**, 123–44.

WHITE, K. D. (1). *Agricultural Implements of the Roman World.* Cambridge University Press, 1967.

WHITE, K. D. (2). *Roman Farming.* Thames & Hudson, London, 1970.

WHITE, K. D. (3). *Farming Equipment of the Roman World.* Cambridge University Press, 1975.

WHITE, LYNN (7). *Medieval Technology and Social Change.* Clarendon Press, Oxford, 1962.

WHYTE, R. O. (1). *Rural Nutrition in China.* Oxford University Press, Hong Kong, 1972.

WHYTE, R. O. (2). *Rural Nutrition in Monsoon Asia.* Oxford University Press, Kuala Lumpur & London, 1974.

WHYTE, R. O. (3). 'The botanical Neolithic Revolution.' *HEC* (1977), 5, **3**, 209–22.

WHYTE, R. O., G. NILSSON-LEISSNER & H. C. TRUMBLE (1). *Legumes in Agriculture.* FAO Agricultural Studies no. 21, Rome, 1953.

WIEGER, L. (2). *Textes Philosophiques.* (Ch. and Fr.). Mission Press, Hsienhsien, 1930.

WIEGER, L. (6). *Taoisme.* Vol. I. *Bibliographie générale*: (1) Le Canon (Patrologie); (2) Les Index officiels et privés. Mission Press, Hsienhsien, 1911. (Crit. by P. Pelliot, JA (1912) **20**, 141.)

WILHELM, R. (2) (tr.). *I Ging [I Ching]: das Buch der Wandlungen.* 2 vols. (3 books, pagination of 1 and 2 continuous in first volume). Diederichs, Jena, 1924. Eng. tr. C. F. Baynes (2 vols.) Bollingen Pantheon, New York, 1950.

WILHELM, RICHARD (3) (tr.). *Frühling u. Herbst d. Lü Bu-We* (the *Lü Shih Chhun Chhiu*). Diederichs, Jena, 1928.

WILHELM, RICHARD (6) (tr.). *Li Gi, , das Buch der Sitte des älteren und jungeren Dai* [i.e. both *Li Chi* and *Ta Tai Li Chi*]. Diederichs, Jena, 1930.

WILL, P.-E. (1). *Bureaucratie et famine en Chine au 18e siècle.* Mouton/Ecole des Hautes Etudes en Sciences Sociales, Paris & The Hague, 1980.

WILLIAMS, S. WELLS (1). *The Middle Kingdom: a Survey of the Geography, Government, Literature, Social Life, Arts and History of The Chinese Empire and its Inhabitants.* 2 vols., Wiley, New York, 1848; later eds. 1861, 1900; London, 1883.

WILLIAMSON, H. R. (1). *Wang An-Shih, Chinese Stateman and Educationalist of the Sung Dynasty.* 2 vols., Arthur Probsthain, London, 1935.

WINCH, DONALD (1). 'The emergence of economics as a science, 1750–1870.' In Cipolla (4), 507–73.

WISSLER, CLARK (1). *The Cereals and Civilisation.* Science Guide no. 129, American Museum of Natural History, New York, n.d.

WITTFOGEL, K. A. (9). *Oriental Despotism: a Comparative Study of Total Power.* Yale Univ. Press, New Haven; Oxford Univ. Press, London, 1957.

WOLF, ERIC R. (1). *Peasants.* Foundation of Modern Anthropology Series, Prentice-Hall, New Jersey, 1966.

WONG, JOHN (1). *Land Reform in the People's Republic of China: Institutional Transformation in Agriculture.* Praeger, New York, 1973.

WU KHANG (1). *Les Trois Politiques du Tchounn Tsieou interprétées par Tong Tchong-Chou d'après les principes de l'école de Kong-Yang.* Leroux, Paris, 1932.

WULFF, HANS E. (1). *The Traditional Crafts of Persia.* MIT Press, Cambridge Mass., 1966.

WYLIE A. (2). 'History of the Hsiung-Nu' (tr. of the chapter on the Huns in the *Chhien Han Shu*, ch. 94). *JRAI*, 1874, **3**, 401; 1875, **5**, 41.

WYLIE, A. (3). 'The History of the South-western Barbarians and Chao Sëen' [Chao-Hsien, Korea]. (tr. of ch. 95 of the *Chhien Han Shu.*) *JRAI*, 1880, **9**, 53.

WYLIE, A. (10) (tr.). 'Notes on the Western Regions, translated from the "Ts'een Han Shoo" [*Chhien Han Shu*] Bk. 96.' *JRAI*, 1881, **10**, 20; 1882, **11**, 83. (Ch. 96 A and B, as also the biography of Chang Chhien in ch. 61, pp. 1–6, and the biography of Chhen Thang in ch. 70.)

YAMAMURA, KOZO (1). 'Pre-industrial landholding patterns in Japan and England.' In: Albert M. Craig (ed.), *Japan: a Comparative View*, Princeton University Press, 1979.

YANG, LIEN-SHENG (7). 'Notes on N. L. Swann's "Food and Money in Ancient China".' *HJAS*, 1950, **13**, 524. Repr. in Yang Lien-Sheng (9), p. 85 with additions and corrections.

YANG, LIEN-SHENG (9). *Studies in Chinese Institutional History.* Harvard-Yenching Institute Series no. 20, Harvard University Press, Harvard, 1961.

YARRANTON, ANDREW (1). *The Great Improvements of Lands by Clover.* London, 1663.

YEH HSIEN-EN (1). 'The tenant-servant system in Huizhou Prefecture, Anhui.' *SSCH* (1981), **1**, 90–119.

YEN, D. E. (1). *The Sweet Potato and Oceania: an Essay in Ethnobotany*. Bernice P. Bishop Museum Bulletin, 236, Honolulu, 1974.

YEN, D. E. (2). 'Sweet Potato.' In Simmonds (1), 42–5.

YOUNG, ARTHUR (1). *Rural Oeconomy: or Essays on the Practical Parts of Husbandry*. 1st published 1770; reproduced in part in G. E. Mingay (1) 123–7.

YOUNG, ARTHUR (2). *Observations on The Present State of Waste Lands of Great Britain*. London, 1773.

YÜ PING-YAO & I. V. GILLIS (1). *Title Index to the Ssu Khu Chhüan Shu*. French Bookstore, Peiping, 1934.

YÜ, YING-SHIH (1). 'Han.' In K. C. Chang (3), 23–52.

VON ZACH, E. (6). *Die Chinesische Anthologie; Übersetzungen aus dem 'Wen Hsüan'*. 2 vols. Ed. I. M. Fang. Harvard Univ. Press, Cambridge, Mass., 1958. (Harvard-Yenching Studies, no. 18.)

ZOHARY, D. (1). 'Notes on Ancient agriculture in the Central Negev.' *IEJ* (1954), **4**, 17–25.

ZOHARY, D. (2). 'The progenitors of wheat and barley in relation to domestication and agricultural dispersal in the Old World.' In Ucko & Dimbleby (1), 47.

ZOHARY, D. (3). 'Centres of diversity and centres of origin.' In O. H. Frankel & E. Bennet, ed., *Genetic Resources in Plants—their Exploration and Conservation*, Blackwell, Oxford, 1970; pp. 33–42.

ZOHARY, D. (4). 'Lentil.' In Simmonds (1), pp. 163–4.

ZOHARY, D. & M. HOPF (1). 'Domestication of pulses in the Old World.' *SC*, 1973, **182**, pp. 887–94.

ADDENDUM

CRESSEY, G. B. (1). *China's Geographic Foundations; A Survey of the Land and its People*. McGraw-Hill, New York, 1934.

# GENERAL INDEX

# TABLE OF CHINESE DYNASTIES

| | | | | |
|---|---|---|---|---|
| | 夏 | Hsia kingdom (legendary?) | | c. −2000 to c. −1520 |
| | 商 | Shang (Yin) kingdom | | c. −1520 to c. −1030 |
| | 周 | Chou dynasty (Feudal Age) | Early Chou period | c. −1030 to −722 |
| | | | Chhun Chhiu period 春秋 | −722 to −480 |
| | | | Warring States (Chan Kuo) period 戰國 | −480 to −221 |
| First Unification | 秦 | Chhin dynasty | | −221 to −207 |
| | 漢 | Han dynasty | Chhien Han (Earlier or Western) | −202 to +9 |
| | | | Hsin interregnum | +9 to +23 |
| | | | Hou Han (Later or Eastern) | +25 to +220 |
| | 三國 | San Kuo (Three Kingdoms period) | | +221 to +265 |
| First Partition | 蜀 | Shu (Han) | +221 to +264 | |
| | 魏 | Wei | +220 to +265 | |
| | 吳 | Wu | +222 to +280 | |
| Second Unification | 晉 | Chin dynasty: Western | | +265 to +317 |
| | | Eastern | | +317 to +420 |
| | 劉宋 | (Liu) Sung dynasty | | +420 to +479 |
| Second Partition | | Northern and Southern Dynasties (Nan Pei chhao) | | |
| | 齊 | Chhi dynasty | | +479 to +502 |
| | 梁 | Liang dynasty | | +502 to +557 |
| | 陳 | Chhen dynasty | | +557 to +589 |
| | 魏 | Northern (Thopa) Wei dynasty | | +386 to +535 |
| | | Western (Thopa) Wei dynasty | | +535 to +556 |
| | | Eastern (Thopa) Wei dynasty | | +534 to +550 |
| | 北齊 | Northern Chhi dynasty | | +550 to +577 |
| | 北周 | Northern Chou (Hsienpi) dynasty | | +557 to +581 |
| Third Unification | 隋 | Sui dynasty | | +581 to +618 |
| | 唐 | Thang dynasty | | +618 to +906 |
| Third Partition | 五代 | Wu Tai (Five Dynasty period) (Later Liang, Later Thang (Turkic), Later Chin (Turkic), Later Han (Turkic) and Later Chou) | | +907 to +960 |
| | 遼 | Liao (Chhitan Tartar) dynasty | | +907 to +1124 |
| | | West Liao dynasty (Qarā-Khiṭāi) | | +1124 to +1211 |
| | 西夏 | Hsi Hsia (Tangut Tibetan) state | | +986 to +1227 |
| Fourth Unification | 宋 | Northern Sung dynasty | | +960 to +1126 |
| | 宋 | Southern Sung dynasty | | +1127 to +1279 |
| | 金 | Chin (Jurchen Tartar) dynasty | | +1115 to +1234 |
| | 元 | Yuan (Mongol) dynasty | | +1260 to +1368 |
| | 明 | Ming dynasty | | +1368 to +1644 |
| | 清 | Chhing (Manchu) dynasty | | +1644 to +1911 |
| | 民國 | Republic | | +1912 |

N.B. When no modifying term in brackets is given, the dynasty was purely Chinese. Where the overlapping of dynasties and independent states becomes particularly confused, the tables of Wieger (1) will be found useful. For such periods, especially the Second and Third Partitions, the best guide is Eberhard (9). During the Eastern Chin period there were no less than eighteen independent States (Hunnish, Tibetan, Hsienpi, Turkic, etc.) in the north. The term 'Liu chhao' (Six Dynasties) is often used by historians of literature. It refers to the south and covers the period from the beginning of the +3rd to the end of the +6th centuries, including (San Kuo) Wu, Chin, (Liu) Sung, Chhi, Liang and Chhen. For all details of reigns and rulers see Moule & Yetts (1).

# ROMANISATION CONVERSION TABLES

BY ROBIN BRILLIANT

## PINYIN/MODIFIED WADE–GILES

| Pinyin | Modified Wade–Giles | Pinyin | Modified Wade–Giles |
|---|---|---|---|
| a | a | chou | chhou |
| ai | ai | chu | chhu |
| an | an | chuai | chhuai |
| ang | ang | chuan | chhuan |
| ao | ao | chuang | chhuang |
| ba | pa | chui | chhui |
| bai | pai | chun | chhun |
| ban | pan | chuo | chho |
| bang | pang | ci | tzhu |
| bao | pao | cong | tshung |
| bei | pei | cou | tshou |
| ben | pên | cu | tshu |
| beng | pêng | cuan | tshuan |
| bi | pi | cui | tshui |
| bian | pien | cun | tshun |
| biao | piao | cuo | tsho |
| bie | pieh | da | ta |
| bin | pin | dai | tai |
| bing | ping | dan | tan |
| bo | po | dang | tang |
| bu | pu | dao | tao |
| ca | tsha | de | tê |
| cai | tshai | dei | tei |
| can | tshan | den | tên |
| cang | tshang | deng | têng |
| cao | tshao | di | ti |
| ce | tshê | dian | tien |
| cen | tshên | diao | tiao |
| ceng | tshêng | die | dieh |
| cha | chha | ding | ting |
| chai | chhai | diu | tiu |
| chan | chhan | dong | tung |
| chang | chhang | dou | tou |
| chao | chhao | du | tu |
| che | chhê | duan | tuan |
| chen | chhên | dui | tui |
| cheng | chhêng | dun | tun |
| chi | chhih | duo | to |
| chong | chhung | e | ê, o |

| Pinyin | Modified Wade–Giles | Pinyin | Modified Wade–Giles |
|---|---|---|---|
| en | ên | jia | chia |
| eng | êng | jian | chien |
| er | êrh | jiang | chiang |
| fa | fa | jiao | chiao |
| fan | fan | jie | chieh |
| fang | fang | jin | chin |
| fei | fei | jing | ching |
| fen | fên | jiong | chiung |
| feng | fêng | jiu | chiu |
| fo | fo | ju | chü |
| fou | fou | juan | chüan |
| fu | fu | jue | chüeh, chio |
| ga | ka | jun | chün |
| gai | kai | ka | kha |
| gan | kan | kai | khai |
| gang | kang | kan | khan |
| gao | kao | kang | khang |
| ge | ko | kao | khao |
| gei | kei | ke | kho |
| gen | kên | kei | khei |
| geng | kêng | ken | khên |
| gong | kung | keng | khêng |
| gou | kou | kong | khung |
| gu | ku | kou | khou |
| gua | kua | ku | khu |
| guai | kuai | kua | khua |
| guan | kuan | kuai | khuai |
| guang | kuang | kuan | khuan |
| gui | kuei | kuang | khuang |
| gun | kun | kui | khuei |
| guo | kuo | kun | khun |
| ha | ha | kuo | khuo |
| hai | hai | la | la |
| han | han | lai | lai |
| hang | hang | lan | lan |
| hao | hao | lang | lang |
| he | ho | lao | lao |
| hei | hei | le | lê |
| hen | hên | lei | lei |
| heng | hêng | leng | lêng |
| hong | hung | li | li |
| hou | hou | lia | lia |
| hu | hu | lian | lien |
| hua | hua | liang | liang |
| huai | huai | liao | liao |
| huan | huan | lie | lieh |
| huang | huang | lin | lin |
| hui | hui | ling | ling |
| hun | hun | liu | liu |
| huo | huo | lo | lo |
| ji | chi | long | lung |

| Pinyin | Modified Wade–Giles | Pinyin | Modified Wade–Giles |
|---|---|---|---|
| lou | lou | pa | pha |
| lu | lu | pai | phai |
| lü | lü | pan | phan |
| luan | luan | pang | phang |
| lüe | lüeh | pao | phao |
| lun | lun | pei | phei |
| luo | lo | pen | phên |
| ma | ma | peng | phêng |
| mai | mai | pi | phi |
| man | man | pian | phien |
| mang | mang | piao | phiao |
| mao | mao | pie | phieh |
| mei | mei | pin | phin |
| men | mên | ping | phing |
| meng | mêng | po | pho |
| mi | mi | pou | phou |
| mian | mien | pu | phu |
| miao | miao | qi | chhi |
| mie | mieh | qia | chhia |
| min | min | qian | chhien |
| ming | ming | qiang | chhiang |
| miu | miu | qiao | chhiao |
| mo | mo | qie | chhieh |
| mou | mou | qin | chhin |
| mu | mu | qing | chhing |
| na | na | qiong | chhiung |
| nai | nai | qiu | chhiu |
| nan | nan | qu | chhü |
| nang | nang | quan | chhüan |
| nao | nao | que | chhüeh, chhio |
| nei | nei | qun | chhün |
| nen | nên | ran | jan |
| neng | nêng | rang | jang |
| ng | ng | rao | jao |
| ni | ni | re | jê |
| nian | nien | ren | jên |
| niang | niang | reng | jêng |
| niao | niao | ri | jih |
| nie | nieh | rong | jung |
| nin | nin | rou | jou |
| ning | ning | ru | ju |
| niu | niu | rua | jua |
| nong | nung | ruan | juan |
| nou | nou | rui | jui |
| nu | nu | run | jun |
| nü | nü | ruo | jo |
| nuan | nuan | sa | sa |
| nüe | nio | sai | sai |
| nuo | no | san | san |
| o | o, ê | sang | sang |
| ou | ou | sao | sao |

| Pinyin | Modified Wade–Giles | Pinyin | Modified Wade–Giles |
|---|---|---|---|
| se | sê | wan | wan |
| sen | sên | wang | wang |
| seng | sêng | wei | wei |
| sha | sha | wen | wên |
| shai | shai | weng | ong |
| shan | shan | wo | wo |
| shang | shang | wu | wu |
| shao | shao | xi | hsi |
| she | shê | xia | hsia |
| shei | shei | xian | hsien |
| shen | shen | xiang | hsiang |
| sheng | shêng, sêng | xiao | hsiao |
| shi | shih | xie | hsieh |
| shou | shou | xin | hsin |
| shu | shu | xing | hsing |
| shua | shua | xiong | hsiung |
| shuai | shuai | xiu | hsiu |
| shuan | shuan | xu | hsü |
| shuang | shuang | xuan | hsüan |
| shui | shui | xue | hsüeh, hsio |
| shun | shun | xun | hsün |
| shuo | shuo | ya | ya |
| si | ssu | yan | yen |
| song | sung | yang | yang |
| sou | sou | yao | yao |
| su | su | ye | yeh |
| suan | suan | yi | i |
| sui | sui | yin | yin |
| sun | sun | ying | ying |
| suo | so | yo | yo |
| ta | tha | yong | yung |
| tai | thai | you | yu |
| tan | than | yu | yü |
| tang | thang | yuan | yüan |
| tao | thao | yue | yüeh, yo |
| te | thê | yun | yün |
| teng | thêng | za | tsa |
| ti | thi | zai | tsai |
| tian | thien | zan | tsan |
| tiao | thiao | zang | tsang |
| tie | thieh | zao | tsao |
| ting | thing | ze | tsê |
| tong | thung | zei | tsei |
| tou | thou | zen | tsên |
| tu | thu | zeng | tsêng |
| tuan | thuan | zha | cha |
| tui | thui | zhai | chai |
| tun | thun | zhan | chan |
| tuo | tho | zhang | chang |
| wa | wa | zhao | chao |
| wai | wai | zhe | chê |

| Pinyin | Modified Wade–Giles | Pinyin | Modified Wade–Giles |
|---|---|---|---|
| zhei | chei | zhui | chui |
| zhen | chên | zhun | chun |
| zheng | chêng | zhuo | cho |
| zhi | chih | zi | tzu |
| zhong | chung | zong | tsung |
| zhou | chou | zou | tsou |
| zhu | chu | zu | tsu |
| zhua | chua | zuan | tsuan |
| zhuai | chuai | zui | tsui |
| zhuan | chuan | zun | tsun |
| zhuang | chuang | zuo | tso |

## MODIFIED WADE–GILES/PINYIN

| Modified Wade–Giles | Pinyin | Modified Wade–Giles | Pinyin |
|---|---|---|---|
| a | a | chhio | que |
| ai | ai | chhiu | qiu |
| an | an | chhiung | qiong |
| ang | ang | chho | chuo |
| ao | ao | chhou | chou |
| cha | zha | chhu | chu |
| chai | chai | chhuai | chuai |
| chan | zhan | chhuan | chuan |
| chang | zhang | chhuang | chuang |
| chao | zhao | chhui | chui |
| chê | zhe | chhun | chun |
| chei | zhei | chhung | chong |
| chên | zhen | chhü | qu |
| chêng | zheng | chhüan | quan |
| chha | cha | chhüeh | que |
| chhai | chai | chhün | qun |
| chhan | chan | chi | ji |
| chhang | chang | chia | jia |
| chhao | chao | chiang | jiang |
| chhê | che | chiao | jiao |
| chhên | chen | chieh | jie |
| chhêng | cheng | chien | jian |
| chhi | qi | chih | zhi |
| chhia | qia | chin | jin |
| chhiang | qiang | ching | jing |
| chhiao | qiao | chio | jue |
| chhieh | qie | chiu | jiu |
| chhien | qian | chiung | jiong |
| chhih | chi | cho | zhuo |
| chhin | qin | chou | zhou |
| chhing | qing | chu | zhu |

| Modified Wade–Giles | Pinyin | Modified Wade–Giles | Pinyin |
|---|---|---|---|
| chua | zhua | huan | huan |
| chuai | zhuai | huang | huang |
| chuan | zhuan | hui | hui |
| chuang | zhuang | hun | hun |
| chui | zhui | hung | hong |
| chun | zhun | huo | huo |
| chung | zhong | i | yi |
| chü | ju | jan | ran |
| chüan | juan | jang | rang |
| chüeh | jue | jao | rao |
| chün | jun | jê | re |
| ê | e, o | jên | ren |
| ên | en | jêng | reng |
| êng | eng | jih | ri |
| êrh | er | jo | ruo |
| fa | fa | jou | rou |
| fan | fan | ju | ru |
| fang | fang | jua | rua |
| fei | fei | juan | ruan |
| fên | fen | jui | rui |
| fêng | feng | jun | run |
| fo | fo | jung | rong |
| fou | fou | ka | ga |
| fu | fu | kai | gai |
| ha | ha | kan | gan |
| hai | hai | kang | gang |
| han | han | kao | gao |
| hang | hang | kei | gei |
| hao | hao | kên | gen |
| hên | hen | kêng | geng |
| hêng | heng | kha | ka |
| ho | he | khai | kai |
| hou | hou | khan | kan |
| hsi | xi | khang | kang |
| hsia | xia | khao | kao |
| hsiang | xiang | khei | kei |
| hsiao | xiao | khên | ken |
| hsieh | xie | khêng | keng |
| hsien | xian | kho | ke |
| hsin | xin | khou | kou |
| hsing | xing | khu | ku |
| hsio | xue | khua | kua |
| hsiu | xiu | khuai | kuai |
| hsiung | xiong | khuan | kuan |
| hsü | xu | khuang | kuang |
| hsüan | xuan | khuei | kui |
| hsüeh | xue | khun | kun |
| hsün | xun | khung | kong |
| hu | hu | khuo | kuo |
| hua | hua | ko | ge |
| huai | huai | kou | gou |

| Modified Wade–Giles | Pinyin | Modified Wade–Giles | Pinyin |
|---|---|---|---|
| ku | gu | mu | mu |
| kua | gua | na | na |
| kuai | guai | nai | nai |
| kuan | guan | nan | nan |
| kuang | guang | nang | nang |
| kuei | gui | nao | nao |
| kun | gun | nei | nei |
| kung | gong | nên | nen |
| kuo | guo | nêng | neng |
| la | la | ni | ni |
| lai | lai | niang | niang |
| lan | lan | niao | niao |
| lang | lang | nieh | nie |
| lao | lao | nien | nian |
| lê | le | nin | nin |
| lei | lei | ning | ning |
| lêng | leng | niu | nüe |
| li | li | niu | niu |
| lia | lia | no | nuo |
| liang | liang | nou | nou |
| liao | liao | nu | nu |
| lieh | lie | nuan | nuan |
| lien | lian | nung | nong |
| lin | lin | nü | nü |
| ling | ling | o | e, o |
| liu | liu | ong | weng |
| lo | luo, lo | ou | ou |
| lou | lou | pa | ba |
| lu | lu | pai | bai |
| luan | luan | pan | ban |
| lun | lun | pang | bang |
| lung | long | pao | bao |
| lü | lü | pei | bei |
| lüeh | lüe | pên | ben |
| ma | ma | pêng | beng |
| mai | mai | pha | pa |
| man | man | phai | pai |
| mang | mang | phan | pan |
| mao | mao | phang | pang |
| mei | mei | phao | pao |
| mên | men | phei | pei |
| mêng | meng | phên | pen |
| mi | mi | phêng | peng |
| miao | miao | phi | pi |
| mieh | mie | phiao | piao |
| mien | mian | phieh | pie |
| min | min | phien | pian |
| ming | ming | phin | pin |
| miu | miu | phing | ping |
| mo | mo | pho | po |
| mou | mou | phou | pou |

| Modified Wade–Giles | Pinyin | Modified Wade–Giles | Pinyin |
|---|---|---|---|
| phu | pu | tên | den |
| pi | bi | têng | deng |
| piao | biao | tha | ta |
| pieh | bie | thai | tai |
| pien | bian | than | tan |
| pin | bin | thang | tang |
| ping | bing | thao | tao |
| po | bo | thê | te |
| pu | bu | thêng | teng |
| sa | sa | thi | ti |
| sai | sai | thiao | tiao |
| san | san | thieh | tie |
| sang | sang | thien | tian |
| sao | sao | thing | ting |
| sê | se | tho | tuo |
| sên | sen | thou | tou |
| sêng | seng, sheng | thu | tu |
| sha | sha | thuan | tuan |
| shai | shai | thui | tui |
| shan | shan | thun | tun |
| shang | shang | thung | tong |
| shao | shao | ti | di |
| shê | she | tiao | diao |
| shei | shei | tieh | die |
| shên | shen | tien | dian |
| shêng | sheng | ting | ding |
| shih | shi | tiu | diu |
| shou | shou | to | duo |
| shu | shu | tou | dou |
| shua | shua | tsa | za |
| shuai | shuai | tsai | zai |
| shuan | shuan | tsan | zan |
| shuang | shuang | tsang | zang |
| shui | shui | tsao | zao |
| shun | shun | tsê | ze |
| shuo | shuo | tsei | zei |
| so | suo | tsên | zen |
| sou | sou | tsêng | zeng |
| ssu | si | tsha | ca |
| su | su | tshai | cai |
| suan | suan | tshan | can |
| sui | sui | tshang | cang |
| sun | sun | tshao | cao |
| sung | song | tshê | ce |
| ta | da | tshên | cen |
| tai | dai | tshêng | ceng |
| tan | dan | tsho | cuo |
| tang | dang | tshou | cou |
| tao | dao | tshu | cu |
| tê | de | tshuan | cuan |
| tei | dei | tshui | cui |

| Modified Wade–Giles | Pinyin | Modified Wade–Giles | Pinyin |
|---|---|---|---|
| tshun | cun | wang | wang |
| tshung | cong | wei | wei |
| tso | zuo | wên | wen |
| tsou | zou | wo | wo |
| tsu | zu | wu | wu |
| tsuan | zuan | ya | ya |
| tsui | zui | yang | yang |
| tsun | zun | yao | yao |
| tsung | zong | yeh | ye |
| tu | du | yen | yan |
| tuan | duan | yin | yin |
| tui | dui | ying | ying |
| tun | dun | yo | yue, yo |
| tung | dong | yu | you |
| tzhu | ci | yung | yong |
| tzu | zi | yü | yu |
| wa | wa | yüan | yuan |
| wai | wai | yüeh | yue |
| wan | wan | yün | yun |